T0181544

# Solar System Dynamics

The solar system is a complex and fascinating dynamical system. This is the first textbook to describe comprehensively the dynamical features of the Solar System and to provide students with all the necessary mathematical tools and physical models they need to understand how it works.

Clearly written and well illustrated, *Solar System Dynamics* provides students with a complete introduction to understanding the intricate and often beautiful resonant structure of the solar system. Step by step, it shows how a basic knowledge of the two- and three-body problems and perturbation theory can be combined to understand features as diverse as the tidal heating of Jupiter's moon Io, the unusual rotation of Saturn's moon Hyperion, the origin of the Kirkwood gaps in the asteroid belt, the radial structure of Saturn's A ring, and the long-term stability of the solar system. Problems at the end of each chapter and a free Internet Mathematica® software package (that includes animations and computational tools) are provided to help students to test and develop their understanding.

*Solar System Dynamics* provides students with a clear, comprehensive, and authoritative textbook for courses on solar system dynamics, planetary dynamics, and celestial mechanics. It also provides the necessary mathematical tools for them to tackle more general courses on dynamics, dynamical systems, applications of chaos theory, and non-linear dynamics. This is a benchmark publication in the field of planetary dynamics and destined to become a classic.

CARL MURRAY is Professor of Mathematics and Astronomy at Queen Mary and Westfield College, London. In addition to scientific papers, he enjoys writing popular articles on the solar system. He is an associate editor of the journals *Icarus* and *Celestial Mechanics and Dynamical Astronomy* as well as being a member of the Imaging Team of the *Cassini* mission to Saturn. Asteroid 5598 was officially named Carlmurray in recognition of his contribution to planetary science.

STAN DERMOTT is Professor of Astronomy and chairman of the Department of Astronomy at the University of Florida, Gainesville. He is currently also a University of Florida Research Foundation Professor and a coinvestigator on the Cosmic Dust Experiment on the *Galileo* spacecraft. Asteroid 3647 was officially named Dermott in his honour.

# SOLAR SYSTEM DYNAMICS

## CARL D. MURRAY

Queen Mary and Westfield College,
University of London

## STANLEY F. DERMOTT

University of Florida, Gainesville

CAMBRIDGE
UNIVERSITY PRESS

CAMBRIDGE UNIVERSITY PRESS
Cambridge, New York, Melbourne, Madrid, Cape Town, Singapore,
São Paulo, Delhi, Dubai, Tokyo

Cambridge University Press
32 Avenue of the Americas, New York, NY 10013-2473, USA

www.cambridge.org
Information on this title: www.cambridge.org/9780521575973

© Carl D. Murray and Stanley F. Dermott 1999

This publication is in copyright. Subject to statutory exception
and to the provisions of relevant collective licensing agreements,
no reproduction of any part may take place without the written
permission of Cambridge University Press.

First published 1999
Reprinted 2001, 2004, 2005, 2006, 2008

A catalog record for this publication is available from the British Library

Library of Congress Cataloging in Publication data
Murray, Carl D.
Solar system dynamics / Carl D. Murray, Stanley F. Dermott.
p.   cm.
ISBN 0-521-57295-9 (hc.). – ISBN 0-521-57597-4 (pbk.)
1. Solar system.   2. Celestial mechanics.   I. Dermot, S. F.
II. Title.
QB500.5.M87            1999
523.2 – dc21           99-19679
                        CIP

ISBN 978-0-521-57295-8 Hardback
ISBN 978-0-521-57597-3 Paperback

Transferred to digital printing 2009

Cambridge University Press has no responsibility for the persistence or
accuracy of URLs for external or third-party Internet websites referred to in
this publication, and does not guarantee that any content on such websites is,
or will remain, accurate or appropriate. Information regarding prices, travel
timetables and other factual information given in this work are correct at
the time of first printing but Cambridge University Press does not guarantee
the accuracy of such information thereafter.

Acht chena is álaind cech nderg,
is gel cach núa,
is caín cech ard, is serb cech gnáth.
Caíd cech n-écmais, is faill cech n-aichnid
co festar cech n-éolas.

All that is red is beautiful,
and all that is new is bright,
all that is high is lovely, all that is familiar is bitter.
The unknown is honoured, the known is neglected,
until all knowledge is known.

Anonymous, Irish, ninth century, *The Sick-Bed of Cú Chulainn*

**In Memory of**

**Frank Murray**

He was a man, take him for all in all,
I shall not look upon his like again.
William Shakespeare, *Hamlet, I, ii*

and

**Geraldine Murphy**

At the end we preferred to travel all night,
Sleeping in snatches
With the voices singing in our ears, saying
That this was all folly.
T. S. Eliot, *Journey of the Magi*

# Contents

# Preface

What is a Man,
If his chief good and market of his time
Be but to sleep and feed? A beast, no more.
Sure, he that made us with such large discourse,
Looking before and after, gave us not
That capability and god-like reason
To fust in us unused.

William Shakespeare, *Hamlet, IV, iv*

We are living in a new age of discovery. The major voyages of exploration in the fifteenth and sixteenth centuries have modern parallels in the interplanetary spacecraft missions that have "discovered" our solar system. The data from these spacecraft combined with ground-based observations have revealed a solar system that is more than a collection of planets, satellites, asteroids, comets, and dust distributed in some arbitrary fashion: It has an intricate dynamical structure, which can be largely understood by the application of a simple inverse square law of force to its constituent bodies. To understand the dynamical structure and evolution of the solar system we must therefore understand the qualitative and quantitative effects of the universal law of gravitation.

We consider solar system dynamics to be the application of the techniques of celestial mechanics to solve real problems in planetary science. There are several classical texts on celestial mechanics and many are still in use today. These include the books by Plummer (1918), Brown & Shook (1933), Brouwer & Clemence, (1961) and, more recently Danby (1988). The books by Hagihara (1970, 1972a,b, 1974a,b, 1975a,b, 1976a,b) are authoritative works of reference but make little attempt to convey understanding. In many ways our own effort is an extension of the book by Roy (1988) and should, perhaps, be read in conjunction with it, even though we have attempted to make our book as self-contained as possible. Although most of the subjects covered in this book are discussed in the scientific literature, they are not conveniently located in one comprehensive (and comprehensible) source. Therefore, our overriding aim was to write a book that we would have liked to have read when we were starting out as researchers.

By writing this book we have attempted to give a comprehensive outline of the basic techniques of solar system dynamics together with their application to actual problems. This field, like the solar system itself, is continually evolving and so this book has to be regarded as our personal perception of the important principles and areas of research in the subject at this time; in our choice of subject matter and examples we have been biased toward areas that we have worked on. As a consequence topics such as lunar theory, geophysics, and Cassini states are not covered in this edition. Nevertheless, we believe that our selection is a representative survey of the field.

We have included exercise questions at the end of every chapter. It should be clear that access to a computer and some programming ability will be required to answer some of these questions. This is deliberate and reflects the fact that many of the recent breakthroughs in solar system dynamics have resulted from the use of computers.

In the course of producing this book we have developed a variety of programs written in *Mathematica*®. In order to enhance the educational value of the book the source code of these programs together with several animations that help to illustrate dynamical phenomena in the solar system are available at *http://ssdbook.maths.qmul.ac.uk*. The site is also being used to document known errors in the text. Readers of the book are encouraged to consult the site regularly.

The ultimate goal of solar system dynamics is to understand the dynamical origin, evolution, and stability of the bodies that make up our local environment in space. Even without new observations many problems remain to be solved, but solar system dynamicists believe that they now understand the basic mechanisms that have determined the structure of our own and other planetary systems. No doubt the next generation of planetary spacecraft missions will discover new phenomena requiring new explanations from the next generation of dynamicists. Our only hope is that this book will still serve as a useful source of information and methods when that time comes.

## Acknowledgments

This book has been prepared from the lecture notes of courses we have given at our respective institutions and it is intended for postgraduate students or researchers new to this field. We are grateful to the many students and colleagues who have pointed out errors and suggested improvements to the various drafts of the text that have appeared over the years. In particular we wish to thank Apostolos Christou, Keren Ellis, Mitch Gordon, Sean Greaves, Tom Kehoe, Helena Morais, and Othon Winter for their assistance and important contributions to the text. Doug Hamilton and Carolyn Porco read and commented on early drafts, and Phil Nicholson supplied some material for Chapters 4 and 6 as well as allowed us to use exercise questions from his graduate course. Sumita Jayaraman and Jer-Chyi Liou provided data and calculations from their own work. Fathi Namouni provided constructive comments on all aspects of the book, and we thank him for suggesting several major improvements, particularly to Chapters 8 and 10. We are grateful to Faber & Faber Limited and Harcourt, Inc. for giving us permission to use an extract from T. S. Eliot's *Journey of the Magi* and to Donal O'Ceallaigh for helping with the translation of the Ninth Century quotation in Irish. Finally, we thank Kim and Margaret for their patience and understanding throughout the many years it took to complete this book.

# 1

# Structure of the Solar System

There's not the smallest orb which thou behold'st
But in his motion like an angel sings,
Still quiring to the young-eyed cherubins;
Such harmony is in immortal souls;

William Shakespeare, *Merchant of Venice, V, i*

## 1.1 Introduction

It is a laudable human pursuit to try to perceive order out of the apparent randomness of nature; science is, after all, an attempt to make sense of the world around us. Moving against the background of the "fixed" stars, the regularity of the Moon and planets demanded a dynamical explanation.

The history of astronomy is the history of a growing awareness of our position (or lack of it) in the universe. Observing, exploring, and ultimately understanding our solar system is the first step towards understanding the rest of the universe. The key discovery in this process was Newton's formulation of the universal law of gravitation; this made sense of the orbits of planets, satellites, and comets, and their future motion could be predicted: The Newtonian universe was a deterministic system. The *Voyager* missions increased our knowledge of the outer solar system by several orders of magnitude, and yet they would not have been possible without knowledge of Newton's laws and their consequences. However, advances in mathematics and computer technology have now revealed that, even though our system is deterministic, it is not necessarily predictable. The study of nonlinear dynamics has revealed a solar system even more intricately structured than Newton could have imagined.

In this chapter we review some of the observations that have motivated the quest for an understanding of the dynamical structure of the solar system.

## 1.2 The Belief in Number

The desire to perceive order in the distribution of objects in the solar system can be traced to early Greece, although it may have had its roots in Babylonian astronomy. Anaximander of Miletus (611–547 B.C.) claimed that the relative distances of the stars, Moon, and Sun from the Earth were in the ratio 1:2:3 (Bernal 1969). The importance of whole numbers to members of the Pythagorean school led them to believe that the distances of the heavenly bodies from the Earth corresponded to a sequence of musical notes and this gave rise to the concept of the "harmony of the spheres". This, in turn, influenced Plato (427–347 B.C.) whose work was to have a great effect on Johannes Kepler nearly two thousand years later (Field 1988).

Kepler was obsessed with the belief that numbers and geometry could be used to explain the spacing of the planetary orbits. He firmly believed in the Copernican rather than the Ptolemaic system, but his views on planetary orbits had foundations in numerology and astrology (Field 1988) rather than scientific method. In the first edition of his book *Mysterium Cosmographicum*, Kepler (1596) described his model of the solar system, which consisted of the six known planets (Mercury, Venus, Earth, Mars, Jupiter, and Saturn) moving within spherical shells whose inner and outer surfaces had precise separations determined by the circumspheres and inspheres of the five regular polyhedra (cube, tetrahedron, dodecahedron, icosahedron, and octahedron). Kepler believed that the widths of these shells were related to the orbital eccentricities. This is illustrated in Fig. 1.1 for the outer solar system. He also developed a similar theory to explain the relative spacings of the newly discovered moons of Jupiter (Kepler 1610). Between the first and second editions of *Mysterium Cosmographicum*, Kepler had empirically deduced the first two of his laws of planetary motion (Kepler 1609), and the notes accompanying the second edition (Kepler 1621) make it clear that his belief in astrology was waning (Field 1988). Although it is unlikely that he had a literal belief in musical notes emanating from the planets, Kepler persisted in his search for harmonic relationships between orbits.

Despite its metaphysical origins, Kepler's geometrical model was a surprisingly good fit to the available data (Field 1988). Although Kepler looked unsuccessfully for simple numerical relationships between the orbital distances of the planets, it was his fascination with numbers that ultimately led to the discovery of his third law of motion, which relates the orbital period of a planet to its average distance from the Sun. On 15 May 1618 he became convinced that *"it is most certain and most exact that the proportion between the periods of any two planets is precisely three halves the proportion of the mean distances"*. Kepler's most important legacy was not his intricate geometrical model of the spacing of the planets, but his empirical derivation of his laws of planetary motion.

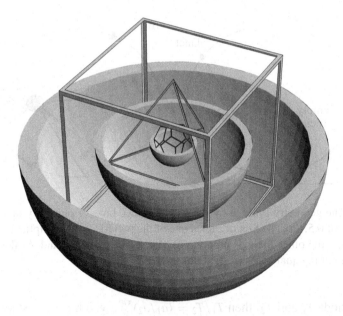

Fig. 1.1.  Kepler's geometrical model of the relative distances of the planets.  Each planet moved within a spherical shell with inner and outer radii defined by the limiting spheres of the regular polyhedra.  For the outer solar system the orbits of Saturn, Jupiter, and Mars are on spheres separated by a cube, a tetrahedron, and a dodecahedron.

## 1.3  Kepler's Laws of Planetary Motion

Kepler (1609, 1619) derived his three laws of planetary motion using an empirical approach.  From observations, including those made by Tycho Brahe, Kepler deduced that:

1) The planets move in ellipses with the Sun at one focus.
2) A radius vector from the Sun to a planet sweeps out equal areas in equal times.
3) The square of the orbital period of a planet is proportional to the cube of its semi-major axis.

The geometry implied by the first two laws is illustrated in Fig. 1.2. An ellipse has two foci and according to the first law the Sun occupies one focus while the other one is empty (Fig. 1.2a). In Fig. 1.2b each shaded region represents the area swept out by a line from the Sun to an orbiting planet in equal time intervals, and the second law states that these areas are equal. The geometry of the ellipse will be considered further in Chapter 2.

Half the length of the long axis of the ellipse is called the semi-major axis, $a$ (see Fig. 1.2a). Kepler's third law relates $a$ to the period $T$ of the planet's orbit. He deduced that $T^2 \propto a^3$, so that if two planets have semi-major axes $a_1$ and

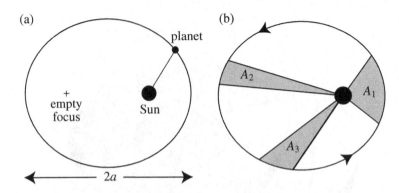

Fig. 1.2. The geometry implied by Kepler's first two laws of planetary motion for an eccentricity of 0.5. (a) The Sun occupies one of the two foci of the elliptical path traced by the planet; the other focus is empty. (b) The regions $A_1$, $A_2$, and $A_3$ denote equal areas swept out in equal times by the radius vector.

$a_2$ and periods $T_1$ and $T_2$, then $T_1/T_2 = (a_1/a_2)^{3/2}$, which is consistent with his original formulation of the law.

It is important to remember that Kepler's laws were purely empirical: He had no physical understanding of why the planets obeyed these laws, although he did propose a "magnetic vortex" to explain planetary orbits.

## 1.4 Newton's Universal Law of Gravitation

In the seventeenth century Isaac Newton (1687) proved that a simple, inverse square law of force gives rise to all motion in the solar system. There is good evidence that Robert Hooke, a contemporary and rival of Newton, had proposed the inverse square law of force before Newton (Westfall 1980) but Newton's great achievement was to show that Kepler's laws of motion are a natural consequence of this force and that the resulting motion is described by a conic section.

In scalar form, Newton proposed that the magnitude of the force $F$ between any two masses in the universe, $m_1$ and $m_2$, separated by a distance $d$ is given by

$$F = \mathcal{G}\frac{m_1 m_2}{d^2},\tag{1.1}$$

where $\mathcal{G}$ is the universal constant of gravitation.

In his *Principia* Newton (1687) also propounded his three laws of motion:

1) Bodies remain in a state of rest or uniform motion in a straight line unless acted upon by a force.
2) The force experienced by a body is equal to the rate of change of its momentum.
3) To every action there is an equal and opposite reaction.

The combination of these laws with the universal law of gravitation was to have a profound effect on our understanding of the universe. Although Newton *"stood on the shoulders of giants"* such as Copernicus, Kepler, and Galileo, his discoveries revolutionised science in general and dynamical astronomy in particular. By extending Newtonian gravitation to more than two bodies it was shown that the mutual planetary interactions result in ellipses that are no longer fixed. Instead the orbits of the planets slowly rotate or *precess* in space over timescales of $\sim 10^5$ y. For example, calculations based on the Newtonian model have shown that the orbit of Mercury should currently be precessing at a rate of $531''$ century$^{-1}$.

However, observations show that Mercury's orbit is precessing at a rate that is $43''$ century$^{-1}$ greater than that predicted by the Newtonian model. We now know that Newton's universal law of gravitation is only an approximation, albeit a very good one, and that a better model of gravity is given by Einstein's general theory of relativity. Applied to the precession of Mercury's perihelion this predicts an additional contribution of $43''$ century$^{-1}$, and the combination of the relativistic contribution to the Newtonian model gives an agreement that is within the current observational limitations (Roseveare 1982).

## 1.5 The Titius–Bode "Law"

The regularity in the spacing of the planetary orbits led to the formulation of a simple mnemonic by Johann Titius in 1766 (Nieto 1972). Titius pointed out that the mean distance $d$ in astronomical units (AU) from the Sun to each of the six known planets was well approximated by the equation

$$d = 0.4 + 0.3\,(2^i), \quad \text{where } i = -\infty, 0, 1, 2, 4, 5. \tag{1.2}$$

The "law" was soon popularised by Johann Bode and is now commonly referred to as Bode's law. Although the "law" had no physical foundation, Bode claimed that an undiscovered planet orbited at the $i = 3$ location. The subsequent discovery of the planet Uranus in 1781 at 19.18 AU ($i = 6$) and the first asteroid (1) Ceres in 1801 at 2.77 AU ($i = 3$) were considered triumphs of the "law" (see Table 1.1 and Nieto 1972).

Such was the success of the Titius–Bode "law" that both John Couch Adams (1847) and Urbain Le Verrier (1847) used it as a basis for their calculations on the predicted orbit of the eighth planet (Grosser 1979). Using $i = 7$ in Eq. (1.2) the "law" predicts a semi-major axis of 38.8 AU; the planet Neptune was discovered in 1846, but it has a semi-major axis of 30.1 AU. The breakdown of the "law" was complete with the discovery of Pluto in 1929 at 39.4 AU compared with a predicted distance of 77.2 AU ($i = 8$). Of course, it could be argued that Pluto is too small to be considered a planet and should therefore be excluded from the

Table 1.1. A comparison of the semi-major axes of the planets, including the minor planet Ceres, with the values predicted by the Titius–Bode law.

| Planet | $i$ | Semi-major Axis (AU) | Titius–Bode Law (AU) |
|--------|-----|-----------------------|------------------------|
| Mercury | $-\infty$ | 0.39 | 0.4 |
| Venus | 0 | 0.72 | 0.7 |
| Earth | 1 | 1.00 | 1.0 |
| Mars | 2 | 1.52 | 1.6 |
| *Ceres* | 3 | 2.77 | 2.8 |
| Jupiter | 4 | 5.20 | 5.2 |
| Saturn | 5 | 9.54 | 10.0 |
| Uranus | 6 | 19.18 | 19.6 |
| Neptune | 7 | 30.06 | 38.8 |
| Pluto | 8 | 39.44 | 77.2 |

calculation. However, if every value of $i$ is filled we should expect an infinite number of planets between Mercury and Venus!

Some of the regular satellite systems of the outer planets appear to have nonrandom distributions because there are a number of simple numerical relationships between their periods (see Sects. 1.6 and 1.7). In an attempt to incorporate orbital resonances into a Titius–Bode "law", Dermott (1972, 1973) discussed the significance of a modified form of the law using a two-parameter, geometric progression of orbital periods rather than orbital distances. Writing

$$T_i = T_0 A^i,  \tag{1.3}$$

where the satellites are numbered in order of increasing period, $T_i$ is the predicted orbital period of the $i$th satellite, and $T_0$ and $A$ are arbitrary constants, Dermott pointed out that if the possibility of "empty orbitals" is excluded from the system, then such a relationship can only hold for the regular satellites of Uranus. Taking the logarithm of each side of Eq. (1.3), we have

$$\log T_i = \log T_0 + i \log A.  \tag{1.4}$$

The observed data can then be fitted to this model using a standard linear regression technique where one measure of the goodness of fit is the root mean square (rms) value, $\chi$, of the residuals. This is given by

$$\chi^2 = \frac{1}{n} \sum_{i=1}^{n} (\log T_i - \log T_0 - i \log A)^2,  \tag{1.5}$$

where $T_i$ is now understood to be the observed orbital period. The comparison of the predictions with the observed data shown in Table 1.2 and Fig. 1.3 seem remarkably favourable. Using the fitted parameters of $T_0 = 0.7919$, $A = 1.777$

Table 1.2. A comparison of the observed orbital period (in days) of the large satellites of Uranus with values calculated using a geometric progression of the form of Eq. (1.3).

| Satellite | $i$ | $T_i$ (obs.) | $T_i$ (calc.) |
|-----------|-----|--------------|---------------|
| Miranda | 1 | 1.413 | 1.407 |
| Ariel | 2 | 2.520 | 2.500 |
| Umbriel | 3 | 4.144 | 4.442 |
| Titania | 4 | 8.706 | 7.893 |
| Oberon | 5 | 13.46 | 14.02 |

we obtain $\chi = 0.0247$. However, it is not enough to be impressed by the seemingly remarkable fits. We need to subject the data to a statistical test and to calculate whether or not the value of $\chi$ is small enough to be statistically significant. This can be addressed using a Monte Carlo approach.

Using a technique similar to that of Dermott (1972, 1973) we have generated a series of sets of five satellites subject to certain restrictions on the distribution of their orbital periods. The innermost satellite was always chosen to have the observed orbital period of Miranda. The remaining periods were then generated using the relationship

$$\frac{T_{i+1}}{T_i} = L + x_i(U - L) \qquad (i = 1, 2, 3, 4), \tag{1.6}$$

where $x_i$ is a random number in the range $0 \le x_i \le 1$, and $L$ and $U$ ($> L$) represent the fixed lower and upper limits on the ratio of successive orbital periods in the system. For each such system of five satellites the parameters $\log T_0$ and $\log A$ and the rms deviation $\chi$ are determined.

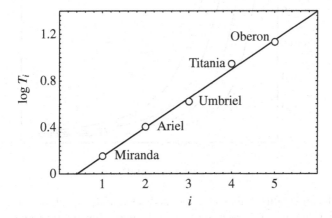

Fig. 1.3. A linear, least squares fit for the orbital periods, $T_i$, of the five major uranian satellites.

The choice of $L$ and $U$ is to some extent arbitrary. For the five uranian satellites the observed values are $L = 1.546$ and $U = 2.101$. By generating $10^5$ sets of five satellites with periods given by the formula in Eq. (1.6), we can determine the number that have a value of $\chi < 0.0247$ for these values of $L$ and $U$. Hence we estimate that the probability that the current configuration has arisen by chance is 0.79. Thus, despite the seemingly impressive fit displayed in Table 1.2 and Fig. 1.3, almost any distribution of periods, subject to the same constraints on $L$ and $U$, would fit into a Titius–Bode "law" equally well.

We have investigated the sensitivity of this estimate to the values of $L$ and $U$ by repeating the above procedure for $L = 1.0$, 1.2, 1.4, and 1.6 using values of $U$ in the range $1.2 < U < 5$. For every value of $L$ and $U$ we have calculated the value of $\chi$ for each of the set of $10^5$ systems of five satellites. In each case we calculated the number of systems, $N$, for which $\chi < 0.0247$ and hence we estimated the probability $P(\chi < 0.0247) = 10^{-5}N$ for those values of $L$ and $U$. The results are shown in Fig. 1.4. It is clear that $P \to 1$ as $U \to L$ when the ratios of successive periods are nearly equal. However, $P$ only becomes small ($P < 0.01$) for large values of $U$, corresponding to widely spaced satellites.

These results suggest that the apparent regular spacing of the orbital periods shown in Table 1.2 is not significant. There is no compelling evidence that the uranian satellite system is obeying any relation similar to the Titius–Bode "law", beyond what would be expected by chance. This leads us to suggest that the "law" as applied to other systems, including the planets themselves, is also without significance. However, even though there are no grounds for belief in a Titius–

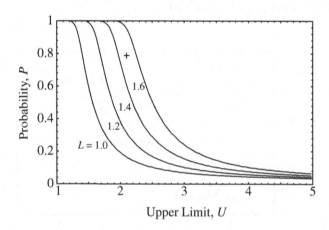

Fig. 1.4. The probability $P$ that the calculated value of the rms deviation $\chi$ is less than the observed value of 0.0247 as a function of the upper limit $U$ for values of the lower limit, $L = 1.0$, 1.2, 1.4, and 1.6. The cross marks the value appropriate for the actual uranian system.

Bode "law", the various bodies of the solar system do exhibit some remarkable numerical relationships and these can be shown to be dynamically significant.

## 1.6 Resonance in the Solar System

Our knowledge of the solar system has increased dramatically in recent years. Although no new planets have been discovered since 1930, there have been a number of advances in the study of the minor members of the solar system. By the start of the twentieth century 22 planetary satellites had been discovered and now there are known to be more than 60 satellites (see Appendix A) with indirect evidence for the existence of others. In addition there are currently more than 10,000 catalogued asteroid orbits and more than 500 reliable orbits for comets. Numerous bodies have been discovered with orbits beyond that of Pluto in the Edgeworth–Kuiper belt. Some estimates suggest that there may be as many as $2 \times 10^8$ objects with radii $\sim 10$ km in this region (Cochran et al. 1995). Observations by the Infra-Red Astronomical Satellite (IRAS) have revealed the presence of dust bands in the asteroid belt and dust trails associated with comets. The study of planetary rings has also undergone radical changes; prior to 1977 it was believed that Saturn was the only ringed planet, whereas now we know that all the giant planets possess ring systems, each with unique characteristics.

The avalanche of planetary data in recent years has provided striking confirmation that our solar system *is* a highly structured assembly of orbiting bodies, but the structure is not as simple as Kepler's geometrical model nor as crude as that implied by the Titius–Bode "law". It is Newton's laws that are at work and the subtle gravitational effect that determines the dynamical structure of our solar system is the phenomenon of *resonance*.

In basic terms a resonance can arise when there is a simple numerical relationship between frequencies or periods. The periods involved could be the rotational and orbital periods of a single body, as in the case of *spin–orbit coupling*, or perhaps the orbital periods of two or more bodies, as in the case of *orbit–orbit coupling*. Other, more complicated resonant relationships are also possible. We now know that dissipative forces are driving evolutionary processes in the solar system and that these are connected with the origins of some of these resonances.

The most obvious example of a spin–orbit resonance is the Moon, which has an orbital period that is equal to its rotational period, resulting in the Moon keeping one face towards the Earth. Most of the major, natural satellites in the solar system are in a 1:1 or *synchronous* spin–orbit resonance. However, other spin–orbit states are also possible and radar observations of Mercury by Pettengill & Dyce (1965) showed that the planet Mercury is in a 3:2 spin–orbit resonance.

In the following sections we discuss each of the subsystems of the solar system.

### *1.6.1 The Planetary System*

The orbital elements of Jupiter and Saturn are modified on a $\sim 900$ y timescale by a 5:2 near-resonance between their orbital periods, which French astronomers called *la grande inégalité* (the great inequality). Although the two planets are not actually in a 5:2 resonance they are sufficiently close to it for significant perturbations to be experienced by both bodies.

The planets Neptune and Pluto are in a peculiar 3:2 orbit–orbit resonance that maximises their separation at conjunction with the result that they avoid a close approach. The complexity of the Neptune–Pluto resonance was discovered and has been studied, not by a prolonged series of observations, but by direct numerical integration of the appropriate equations of motion. This is the only possible technique because observations of Pluto span less than a third of its orbital period.

As well as the resonances involving their orbital periods, some of the planets are also involved in long-term or *secular* resonances associated with the precession of the planetary orbits in space.

### *1.6.2 The Jupiter System*

Perhaps the most striking example of orbit–orbit resonance occurs amongst three of the Galilean satellites of Jupiter (see Fig. 1.5). Io is in a 2:1 resonance with Europa, which is itself in a 2:1 resonance with Ganymede, resulting in all three satellites being involved in a configuration known as a *Laplace resonance*. The average orbital angular velocity or *mean motion n* in $^\circ\mathrm{d}^{-1}$ is defined by $n = 360/T$, where $T$ is the orbital period of the body in days. The mean motions of Io, Europa, and Ganymede are

$$n_\mathrm{I} = 203.488992435^\circ\mathrm{d}^{-1}, \qquad (1.7)$$

Fig. 1.5.   A montage of images of the Galilean satellites of Jupiter shown to the same relative sizes. The satellites are (from left to right) Io, Europa, Ganymede, and Callisto. Ganymede has a mean radius of 2,634 km and is the largest moon in the solar system. The images were taken by the *Galileo* spacecraft. *(Image courtesy of NASA/JPL.)*

$$n_E = 101.374761672°d^{-1}, \tag{1.8}$$

and

$$n_G = 50.317646290°d^{-1} \tag{1.9}$$

respectively. Hence

$$\frac{n_I}{n_E} = 2.007294411 \tag{1.10}$$

and

$$\frac{n_E}{n_G} = 2.014696018 \tag{1.11}$$

or

$$\frac{n_G}{n_E} = \frac{1}{2}(1 - 0.007294411). \tag{1.12}$$

From this we deduce that, to within observational error ($10^{-9}\,°d^{-1}$),

$$n_I - 3n_E + 2n_G = 0. \tag{1.13}$$

This relation, which is known as the Laplace relation, prevents the occurrence of triple conjunctions of the three satellites. The geometry of the resonance ensures that when a conjunction takes place between any pair of satellites, the third satellite is always at least 60° away. The 2:1 Io–Europa resonance is directly responsible for the active vulcanism on Io that was first observed by the *Voyager 1* spacecraft in 1979, three weeks after the publication of the paper by Peale et al. (1979) in which the phenomenon was predicted.

The planet Jupiter has a thin ring, the structure of which is thought to have been created by resonances; in this case the resonances involve numerical relationships between frequencies associated with the motions of dust particles in Jupiter's gravitational field and the rotation of the magnetic field of the planet (Burns et al. 1984).

### 1.6.3 The Saturn System

The saturnian system (Fig. 1.6) has perhaps the widest variety of resonant phenomena. For example, the satellites Mimas and Tethys are in a 4:2 orbit–orbit resonance and the ratio of their mean motions, $n_M$ and $n_{Te}$, is

$$\frac{n_M}{n_{Te}} = 2.003139. \tag{1.14}$$

Enceladus and Dione are in a 2:1 orbit–orbit resonance and the ratio of their mean motions, $n_E$ and $n_D$, is given by

$$\frac{n_E}{n_D} = 1.997431. \tag{1.15}$$

Fig. 1.6.   A *Voyager 1* montage of Saturn, its ring system, and six of its satellites. The satellites are Dione, Titan, Mimas, Tethys, Rhea, and Enceladus. *(Image courtesy of NASA/JPL.)*

Similarly, Titan and Hyperion are in a 4:3 orbit–orbit resonance and the ratio of their mean motions, $n_{\text{Ti}}$ and $n_{\text{H}}$, is given by

$$\frac{n_{\text{Ti}}}{n_{\text{H}}} = 1.334342. \tag{1.16}$$

As well as being involved in resonances with other major satellites, Dione and Tethys each maintain smaller objects in their orbits by means of a 1:1 orbit–orbit resonance. The satellites Janus and Epimetheus move on *horseshoe* orbits and periodically change their radial positions (either closer or further from Saturn) owing to their 1:1 orbital resonance.

The 2:1 resonant perturbations of Mimas produce a gap, the Cassini division, between Saturn's A and B rings, while most of the structure visible in the A ring of Saturn can be explained by the resonant effects of the small satellites Pandora, Prometheus, and Janus in orbit just beyond the main ring system. The radial extent of the Encke gap in the A ring can also be understood in terms of the gravitational effect of a small satellite that acts to clear the gap. A dramatic confirmation of this mechanism came with the discovery by Showalter (1991) of Pan, Saturn's eighteenth satellite, based on predictions made by Cuzzi and Scargle (1985).

### 1.6.4 The Uranus System

The ring system of Uranus has a number of examples of resonant phenomena, although none involve the five major satellites. However, if we consider the small satellites we find that Rosalind and Cordelia are close to a 5:3 resonance (Murray & Thompson 1990) and Cordelia and Ophelia bound the narrow $\epsilon$ ring by means of a 24:25 and a 14:13 resonance with its inner and outer edges (Porco & Goldreich 1987). Cordelia is also involved in resonances with other ring edges. There is also good evidence for the existence of other small satellites in the main ring system based on the observed resonant structure of the rings (Murray & Thompson 1990).

In marked contrast to the jovian and saturnian systems, there are, at present, no resonances between the major satellites of the uranian system. However, some characteristics of the system, in particular the anomalously high orbital inclination of Miranda, provide strong evidence that resonances may have existed in the past and may have been responsible for resurfacing events on a number of moons of Uranus (Dermott 1984; Dermott et al. 1988; Tittemore & Wisdom 1988, 1989, 1990).

### 1.6.5 The Neptune System

Although the peculiar ring system of Neptune is not yet fully understood, it is likely that an explanation for the "arcs" of optically thicker material in the outermost ring, the Adams ring, will involve the resonant effects of small satellites. In particular, the satellite Galatea may be involved in shepherding and providing azimuthal confinement for material in the Adams ring (Porco 1991).

### 1.6.6 The Pluto System

Pluto and Charon are each in the synchronous spin state, and the system is said to be *totally tidally despun*. The result of the process is that the planet and its moon keep the same face towards each other. Consequently, viewed from Pluto, Charon would keep almost the same position, directly over a fixed point on the surface.

### 1.6.7 The Asteroid Belt

The asteroid belt also exhibits resonant structure. Explaining the gaps in the radial distribution of the asteroid orbits remains an important problem in solar system dynamics. Using a sample of < 100 asteroids, Kirkwood (1867) was the first to notice gaps in the asteroid belt corresponding to important resonances with Jupiter. The distribution of asteroids that have been discovered since Kirkwood's time show a number of cleared regions, most notably at the 4:1, 3:1, 5:2, and 2:1 jovian resonances, but there are also concentrations of asteroids at the 3:2 and 1:1 resonances (Fig. 1.7).

The Kirkwood gaps are not entirely empty and a small number of asteroids are known to be in resonance with Jupiter at these locations. In addition, numerical

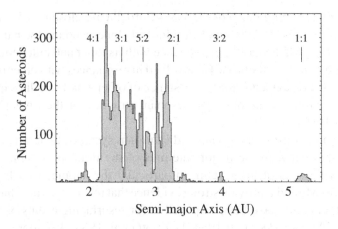

Fig. 1.7.   A histogram showing the number of catalogued asteroids (per 0.02 AU) as a function of semi-major axis and the location of the major jovian resonances.

investigations have revealed several asteroids that have been or will be trapped in jovian resonances.

### 1.6.8 Comets, Meteors, and Dust

Numerical integrations of cometary and meteor orbits have shown that resonant phenomena can exist in such systems, although these are difficult to handle analytically because of the large eccentricities of the orbits involved.

The existence of a belt of comets in the outer solar system was postulated independently by Edgeworth (1943) and Kuiper (1951) on cosmogonical grounds. Recent numerical modelling by Duncan, Quinn, and Tremaine (1987) suggests that the *Edgeworth–Kuiper belt* is a more likely source of the short-period comets associated with Jupiter than the more distant *Oort cloud*. Objects in the Edgeworth–Kuiper belt were first detected in 1992 and more than sixty such objects are currently known to exist. Curiously enough, knowledge of their orbits suggests that almost a third of these could be involved in a 3:2 orbit–orbit resonance with Neptune and this could be evidence of evolution of planetary orbits in the early solar system (Malhotra 1995).

There is also some marginal evidence for gaps in the distribution of the semi-major axes of meteor streams (Murray 1996), those particles of dust generally associated with cometary orbits.

The discovery of a number of dust bands in the solar system (Low et al. 1984) and the subsequent identification of their close association with groupings of asteroids (Dermott et al. 1984) has highlighted the effects of secular perturbations on the orbits of interplanetary dust particles in the solar system. It has now been shown that the Earth is embedded in a dust ring composed of asteroidal particles temporarily trapped in a series of strong orbit–orbit resonances (Dermott et al. 1994).

## 1.7 The Preference for Commensurability

In Sect. 1.6 we saw that the solar system contains a large number of resonant phenomena involving precise numerical relationships between various periods. Is the number of such relationships any more or less than we would expect to occur by chance in a random distribution of objects? This particularly fruitful question was first examined by Roy and Ovenden (1954) and subsequently by Goldreich (1965). These authors considered *commensurabilities* of the form

$$\frac{n_1}{n_2} \approx \frac{i_1}{i_2}, \tag{1.17}$$

where $n_1$ and $n_2$ are the mean motions (or average angular velocities) of the two objects ($n_1 < n_2$), using integers $i_1, i_2 \in \{1, 2, \ldots, i_{max}\}$ with $i_1 < i_2$ and $i_{max} = 7$ but excluding the case $i_1 = i_2 = 1$, the 1:1 commensurability. However, the observed near-commensurabilities in the solar system are all of the form

$$\frac{n_1}{n_2} \approx \frac{p}{p+1}, \tag{1.18}$$

where $p$ is an integer. The more particular question of the preference of ratios of mean motions in the solar system to be close to fractions of the form $p/(p+1)$ was considered by Dermott (1973) and we reproduce those results here.

If $p/(p+1)$ and $p'/(p'+1)$, where $p$ and $p'$ are integers, are the two fractions that bound $n_1/n_2$ most closely from above and below, then we define

$$a = \frac{n_1/n_2 - p'/(p'+1)}{p/(p+1) - p'/(p'+1)}, \tag{1.19}$$

$$b = \begin{cases} 0 & \text{if } a \leq 1/2, & (1.20) \\ 1 & \text{if } a > 1/2, & (1.21) \end{cases}$$

$$c = 2\pi(a - b). \tag{1.22}$$

Thus $-\pi \leq c \leq +\pi$ and the preference for near-commensurability increases as $c$ tends to zero. If the mean motions were randomly distributed then the distribution of $c$ would be rectangular. We also allow that if $1/3 < n_1/n_2 < 1/2$, then for $p'/(p'+1)$ in Eq. (1.19) we substitute the value $1/3$.

If we define the solar system to consist of the nine major planets and all the regular (Goldreich 1965) satellites of the jovian, saturnian, and uranian systems with diameters > 150 km (these are the large, regular satellites known prior to the exploration of the outer solar system by the *Voyager* spacecraft) then there are nineteen independent ratios of mean motions.

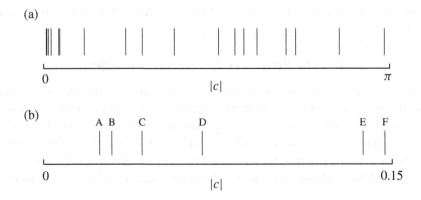

Fig. 1.8.   (a) The distribution of $|c|$ for the nineteen pairs of commensurabilities. (b) The distribution of the six smallest values of $|c|$. The labels correspond to the (A) 2:1 Enceladus–Dione, (B) 2:1 Mimas–Tethys, (C) 4:3 Titan–Hyperion, (D) 2:1 Io–Europa, (E) 2:1 Europa–Ganymede, and (F) 3:2 Neptune–Pluto commensurabilities.

The distribution of $|c|$ is shown in Fig. 1.8a. If this is compared to the distribution of points on a circle, then by treating each value of $c$ as a unit vector $\hat{\mathbf{c}}_i = (\cos c_i, \sin c_i)$ and comparing the magnitude of the sum of the unit vectors, $|\sum \hat{\mathbf{c}}_i|$, with $\sqrt{N}$, where $N$ is the number of vectors or ratios, we find that $|\sum \hat{\mathbf{c}}_i|/\sqrt{N} = 1.01$, which is not large enough to be interesting (to be of significance at the "$3\sigma$ level" we require $|\sum \hat{\mathbf{c}}_i|/\sqrt{N} > 3$).

However, we note that six values of $|c|$ in Fig. 1.8a are particularly close to zero. The expanded view of this section is shown in Fig. 1.8b. We have six ratios with $|c| < 0.15$, and the probability of this occurring by chance in a sample of nineteen ratios is 0.00017. This result may have some statistical significance.

However, statistical arguments alone are not always compelling. There are many curious near-resonant relations between mean motions in the solar system. For example, the mean motions of the uranian satellites Miranda, Ariel, and Umbriel are

$$n_{\text{M}} = 254.6906654°\text{d}^{-1}, \tag{1.23}$$
$$n_{\text{A}} = 142.8356540°\text{d}^{-1}, \tag{1.24}$$

and

$$n_{\text{U}} = \phantom{0}86.8688800°\text{d}^{-1}, \tag{1.25}$$

and we have

$$n_{\text{M}} - 3n_{\text{A}} + 2n_{\text{U}} = -0.0785°\text{d}^{-1} \tag{1.26}$$

Table 1.3. Examples of near-resonances among triads of mean motions in the solar system.

| Triad (1,2,3) | $\|(n_3 - n_2)/(n_1 - n_2)\|$ | $p/(p+1)$ |
|---|---|---|
| Miranda, Ariel, Umbriel | 0.50035 | 1/2 |
| Venus, Earth, Mars | 0.74865 | 3/4 |
| Mimas, Enceladus, Titan | 0.49661 | 1/2 |

and thus the satellites are extremely close to a Laplace-type resonance. Equation (1.26) can be written as the ratio of relative mean motions, that is,

$$\left| \frac{n_U - n_A}{n_M - n_A} \right| = 0.50035 \approx \frac{1}{2}. \tag{1.27}$$

Other examples of near-resonance among triads of mean motions in the solar system are shown in Table 1.3.

Dermott (1973) has shown that there is a preference for commensurability among triads of mean motions (or pairs of relative mean motions) in the solar system and that the probability that this has occurred by chance is as low as 0.006. However, there is no evidence that these particular near-commensurabilities have any dynamical significance. This is in marked contrast to the near-commensurabilities observed among pairs of mean motions shown in Fig. 1.8b.

The mean motions listed in that figure are not merely near-commensurate but are part of other relations that are exact and have real dynamical significance. For example, the mean motions of Titan and Hyperion, $n_{Ti}$ and $n_H$, are involved in a resonance of the form

$$3n_{Ti} - 4n_H + \dot{\varpi}_H = 0, \tag{1.28}$$

where $\dot{\varpi}_H$ is the mean rate of precession of the line of apses of Hyperion's orbit. It is this relationship that has dynamical significance; the fact that $3n_{Ti} \approx 4n_H$ is merely a consequence of the fact that precession rates are small and $\dot{\varpi}_H \ll n_H$.

## 1.8 Recent Developments

The development of solar system dynamics has been driven by observations and the constant search for anomalies. Observed deviations of the planets, particularly Mars, from their predicted motions ultimately led to the discovery of Kepler's laws, universal gravitation, new planets, and finally relativity.

There are a number of intriguing questions concerning the dynamics of solar system bodies. For example, why are there pronounced gaps at most of the major jovian resonances in the asteroid belt but a clustering of asteroids at the 3:2 resonance? Where do short-period comets come from? Why do the orbital elements of some groups of asteroids have common values? Why are there numerous resonances among the satellites in the jovian and saturnian systems but none in the uranian system? Why are the eccentricities and inclinations of some satellite orbits too large to fit in with current understanding of tidal evolution? What produced the Cassini division in Saturn's rings? How are narrow rings maintained despite the spreading effects of collisions and drag forces? Are planetary rings transient phenomena or can they survive for billions of years? Satisfactory explanations for most of these anomalies have been obtained in recent years, but each new generation of instruments and spacecraft provides more puzzling phenomena for each new generation of dynamicists to explain. Solar system dynamics is a subject driven by the need to explain anomalies.

In the opinion of many astronomers, celestial mechanics reached its zenith with the successful prediction by Adams and Le Verrier of the discovery of the planet Neptune. There has been a widely held belief that nothing new remains to be discovered in celestial mechanics and that all the important problems have been solved. That view has had to be revised in recent years owing to the advent of the digital computer and the application of nonlinear dynamics. For several centuries progress in understanding the dynamics of the solar system was hampered, not by a lack of knowledge of the basic equations describing the motion of the bodies, but by the inability to solve those equations for anything but the shortest of time intervals. With the widespread availability of digital computers in the 1970s it became possible to carry out numerical investigations of the dynamical evolution of solar system bodies in realistic timescales. Most studies have concentrated on the evolution of the outer solar system. One of the longest full integrations to date has been carried out by Sussman & Wisdom (1988) who used a specially built computer, the Digital Orrery, to integrate the orbits of the five outer planets for 845 My, or 20% of the age of the solar system.

Analytic perturbation theory requires extensive algebraic manipulation. For example, the Fourier series expansion to fourth order in the eccentricity and inclination of the standard perturbing potential experienced by one orbiting body due to another involves 79 separate cosine arguments and 144 terms (see Appendix B). Consequently the availability of computer algebra systems has greatly enhanced the speed and reliability of algebraic calculations. One of the first successes of such packages was the discovery of minor errors in Delaunay's lunar theory by Deprit et al. (1971).

The most recent revolution in solar system dynamics occurred in the 1980s with the realisation that chaos has played an important rôle in the dynamical evolution of the solar system (see, e.g., Wisdom 1987a,b). The fact that deterministic systems of nonlinear equations can give rise to unpredictable, chaotic solutions was known to Poincaré (1892, 1893, 1899), but the consequences of his work were not fully appreciated until the second half of this century, and it is only in the past decade that the "new methods of celestial mechanics" have been adopted by solar system dynamicists. The major successes of this new approach include preliminary explanations for such diverse phenomena as the

Kirkwood gaps in the asteroid belt and the unusual rotation of the saturnian moon Hyperion. Recent work has provided evidence that even the orbital motion of Earth may be chaotic. A consequence of this renewed research activity has been the development of a variety of new numerical techniques to study long-term motion in the solar system. In particular, dynamicists make frequent use of algebraic mappings, which offer significant improvements in speed over standard numerical methods.

It is interesting to note that in the nineteenth century Poincaré, considered by many to be the founder of dynamical systems theory, set out to solve the important, practical problem of the gravitational interactions of a system of three orbiting bodies (the three-body problem). Partly because of the new techniques he developed, dynamical systems theory has become a branch of mathematics in its own right, with a body of work that has diverged significantly from its ancestor, celestial mechanics. Therefore it seems only appropriate that the "prodigal child" has made a triumphant return by the successful application of the results of nonlinear dynamics to the solar system.

Over the years our understanding of solar system dynamics has evolved from the regular, deterministic model of the planets envisaged by Newton and Laplace to the chaotic model revealed by modern numerical and analytical studies. However, this progress has made us realise that the solar system itself is evolving on a variety of timescales. For example, the current orbits of the planets and satellites may be significantly different from those 4.5 billion years ago, soon after the solar system was formed. We now know that the effect of mutual perturbations and dissipative forces can lead to extensive orbital evolution and that this evolution continues today. Although our knowledge of these processes is far from complete, at least we now appreciate that a better understanding of the orbital evolution of those asteroids, comets, and dust particles that may impact the Earth could have important consequences for the evolution and possible extinction of life on our planet (Sagan 1994).

## Exercise Questions

**1.1** In a simplified form of Kepler's geometrical model of the planetary distances (see Fig. 1.1 and Sect. 1.2), the insphere and circumsphere of the five platonic solids determine the separations of the planets with the five spaces between the orbits of Mercury, Venus, Earth, Mars, Jupiter, and Saturn being separated by an octahedron, icosahedron, dodecahedron, tetrahedron, and cube, respectively. The ratio of the radii of the insphere to the circumsphere for an icosahedron and dodecahedron is $\sqrt{15 - 6\sqrt{5}}$, for an octahedron and cube it is $\sqrt{3}$, and for a tetrahedron it is 3. (a) Taking Kepler's values for the observed semi-major axes $a$ of the planets to be 0.38806, 0.72414, 1, 1.52350, 5.2, and 9.51 AU, start with the semi-major axis of the Earth and calculate the semi-major axes of Mercury, Venus, Mars, Jupiter, and Saturn using Kepler's model. Hence show that the root mean squared (rms) deviation of the observed–theoretical values is 1.350 AU. (b) Calculate the rms deviation for each of the remaining twenty-nine possible orderings of the five ratios (only three of which are unique) and hence find the ordering of the solids that gives the largest and smallest rms deviation. Do the results for parts (a) and (b) change if the semi-major axes listed in Table A.2 for the J2000 epoch are used instead of Kepler's values?

**1.2**     The actual geometrical model used by Kepler also involved the eccentricities of the planets. (a) Taking Kepler's values for the observed semi-major axes given in Question 1.1 and the corresponding eccentricities $e$ as 0.21001, 0.00692, 0.01800, 0.09265, 0.04822, and 0.05700, calculate the observed aphelion distance, $a(1+e)$, of Mercury, Venus, and Earth and the perihelion distance, $a(1-e)$, of Mars, Jupiter, and Saturn. (b) Calculate the same set of six quantities using Kepler's actual model starting with the observed aphelion distance of Earth. For Mars, Jupiter, and Saturn multiply the relevant aphelion distance by the appropriate sphere ratio to calculate the theoretical perihelion distance of the next planet and use Kepler's value of the eccentricity to calculate that planet's aphelion distance. For Venus and Mercury divide the relevant perihelion distance by the appropriate ratio to calculate the aphelion distance of the next planet and use Kepler's value of the eccentricity to calculate the relevant perihelion distance. Hence show that the rms deviation of the observed–theoretical values is 0.151 AU. (c) Calculate the rms deviation for each of the remaining twenty-nine possible orderings and hence show that, subject to an interchange of solids with the same insphere–circumsphere ratio, Kepler's choice of ordering the solids gives the smallest rms value (i.e., the best fit to observations). Find the ordering with the largest rms value. Do the results for parts (b) and (c) change if the semi-major axes and eccentricities listed in Table A.2 for the J2000 epoch are used instead of Kepler's values?

**1.3**     Use Eq. (1.6) with $T_1 = 1.413$ d, $L = 1.546$, $U = 2.101$, and $i = 1, 2$ to generate a large number ($\sim 10^5$) of random sets of orbital periods for a model satellite system, similar to the three inner satellites of Uranus. For each set of three periods calculate the mean motion, $n_i = 360°/T_i$, for each satellite, and the quantity $\delta n = |n_1 - 3n_2 + 2n_3|$. Hence show that the probability that $\delta n \leq 0.1°\mathrm{d}^{-1}$ is $\sim 0.3\%$.

**1.4**     Taking the orbital periods listed in Table A.9 but excluding the values for Epimetheus, Telesto, Calypso, and Helene, use the criteria given in Sect. 1.7 to show that there are twenty-eight ratios of mean motions to consider in the Saturn system. Use Eqs. (1.19)–(1.22) to calculate the closest ratio $p/p+1$ and the value of $|c|$ for each pair of mean motions and hence show that as well as the Enceladus–Dione, Mimas–Tethys, and Titan–Hyperion commensurabilities (see Fig. 1.8b) there are two additional pairs with $|c| < 0.15$.

**1.5**     The periods (in years) of six planets orbiting a star are deduced to be $T_1 = 0.984027$, $T_2 = 1.83248$, $T_3 = 5.80493$, $T_4 = 6.76471$, $T_5 = 13.9359$, and $T_6 = 19.6679$. By considering the fifteen possible ratios $T_i/T_j$ ($i < j$) of orbital periods and the ten first-order commensurabilities of the form $p/(p+1)$, where $1 \leq p \leq 10$ is an integer, identify the pair of planets with orbital periods such that $|(T_i/T_j) - p/(p+1)| < 0.001$ for a particular value of $p$. How would you estimate the probability of this occurring by chance if the six orbital periods were randomly distributed in the range $0 \leq T \leq 20$ years?

**1.6**     Consider the study of the preference for commensurability in the solar system carried out by Roy & Ovenden (see Sect. 1.7). Pairs of integers, $i_1$ and $i_2$ with $i_1 < i_2 \leq i_{max}$, can be used to generate $N_r$ rationals of the form $i_1/i_2$. (a) What is the relationship between $N_r$ and $i_{max}$? (b) If $\epsilon_{max}$ is defined to be half the separation of the two closest of the $N_r$ rationals, write down an expression for $\epsilon_{max}$ in terms of $i_{max}$. (c) Any ratio of orbital periods, $T_1/T_2$ (with $T_1 < T_2$), that lies in the permitted range $r_{min} - \epsilon_{max} \leq T_1/T_2 \leq r_{max} + \epsilon_{max}$ (where $r_{min}$ and $r_{max}$ are the smallest and largest of the $N_r$ rationals) differs by a quantity $\epsilon = |T_1/T_2 - j/k|$ from the nearest rational, $j/k$, belonging to the set. Show that the probability $p$ that a given ratio $T_1/T_2$ is commensurable (i.e., has $\epsilon < \epsilon_{max}$) is

$$p = \frac{2\epsilon_{max} N_r}{(i_{max} - 2)/i_{max} + 2\epsilon_{max}}.$$

(d) If $N_p$ pairs of orbital periods in a system are found to lie within the permitted range and $N_{obs}$ of these are observed to be commensurable within a tolerance $\epsilon_{max}$, show that the probability $P$ of this occurring by chance is

$$P = \frac{N_p!}{(N_p - N_{obs})! \, N_{obs}!} p^{N_{obs}} (1 - p)^{N_p - N_{obs}}.$$

(e) Use the data in Appendix A to find the ratio of all possible pairs of periods among the planets and the prograde satellites of Mars, Jupiter, Saturn, and Uranus with mean radii $> 100$ km and orbital eccentricities $< 0.1$. Taking $i_{max} = 7$ show that thirty pairs of objects have ratios of orbital periods within $\epsilon_{max}$ of a permitted commensurability. According to the theory given above, what is the probability that this number of pairs is due to chance? What would be the likely effect of increasing the sample to include the small satellites?

# 2

# The Two-Body Problem

So we grew together,
Like to a double cherry, seeming parted,
But yet an union in partition –
Two lovely berries moulded on one stem;
So, with two seeming bodies, but one heart.

William Shakespeare, *A Midsummer Night's Dream, II, ii*

## 2.1 Introduction

The two-body problem is perhaps the simplest, integrable problem in solar system dynamics. It concerns the interaction of two point masses moving under a mutual gravitational attraction described by Newton's universal law of gravitation, Eq. (1.1). The wide variety of masses in the solar system permits the orbits of most planets and satellites to be approximated by two-body motion, consisting of a smaller body moving around a much larger central body. The effects of other bodies can usually be treated as perturbations to the two-body system. For example, the path of Jupiter (mass $m_J = 1.9 \times 10^{27}$ kg) around the Sun (mass $m_{Su} = 2.0 \times 10^{30}$ kg $= 1000\,m_J$) is basically an ellipse with the principal perturbations coming from the other planets, notably Saturn (mass $m_{Sa} = 5.7 \times 10^{26}$ kg).

In Volume I of his *Principia*, Isaac Newton (1687) showed that only two types of central force could give rise to the observed elliptical motion of the planets. The first was a linear force directed towards the centre of the ellipse, and the second was an inverse square force directed towards one focus of the ellipse. However, only the second type of force can give rise to Kepler's empirical laws of planetary motion (Sect. 1.3). In this chapter we derive the basic equations of planetary motion and solve the two-body problem, showing how Kepler's laws arise.

## 2.2  Equations of Motion

Consider the motion of two masses $m_1$ and $m_2$ with position vectors $\mathbf{r}_1$ and $\mathbf{r}_2$ referred to some origin $O$ fixed in inertial space (see Fig. 2.1).

The vector $\mathbf{r} = \mathbf{r}_2 - \mathbf{r}_1$ denotes the relative position of the mass $m_2$ with respect to $m_1$. The gravitational forces and the consequent accelerations experienced by the two masses are

$$\mathbf{F}_1 = +\mathcal{G}\frac{m_1 m_2}{r^3}\mathbf{r} = m_1\ddot{\mathbf{r}}_1 \qquad \text{and} \qquad \mathbf{F}_2 = -\mathcal{G}\frac{m_1 m_2}{r^3}\mathbf{r} = m_2\ddot{\mathbf{r}}_2 \qquad (2.1)$$

respectively, where $\mathcal{G} = 6.67260 \times 10^{-11}\,\mathrm{Nm^2kg^{-2}}$ is the *universal gravitational constant*. Thus

$$m_1\ddot{\mathbf{r}}_1 + m_2\ddot{\mathbf{r}}_2 = 0, \qquad (2.2)$$

which can be integrated directly twice to give

$$m_1\dot{\mathbf{r}}_1 + m_2\dot{\mathbf{r}}_2 = \mathbf{a} \qquad \text{and} \qquad m_1\mathbf{r}_1 + m_2\mathbf{r}_2 = \mathbf{a}t + \mathbf{b}, \qquad (2.3)$$

where $\mathbf{a}$ and $\mathbf{b}$ are constant vectors. If $\mathbf{R} = (m_1\mathbf{r}_1 + m_2\mathbf{r}_2)/(m_1 + m_2)$ denotes the position vector of the centre of mass, then Eqs. (2.3) can be written

$$\dot{\mathbf{R}} = \frac{\mathbf{a}}{m_1 + m_2} \qquad \text{and} \qquad \mathbf{R} = \frac{\mathbf{a}t + \mathbf{b}}{m_1 + m_2}. \qquad (2.4)$$

This implies that either the centre of mass is stationary (if $\mathbf{a} = 0$) or it is moving with a constant velocity in a straight line with respect to the origin $O$. Note that this result is not specific to the inverse square law of force.

Now consider the motion of $m_2$ with respect to $m_1$. This allows us to simplify the problem without losing any of its essential features. In Sect. 2.7 we shall revert to considering motion in the centre of mass frame. Writing $\ddot{\mathbf{r}} = \ddot{\mathbf{r}}_2 - \ddot{\mathbf{r}}_1$, and using Eq. (2.1), we obtain

$$\frac{d^2\mathbf{r}}{dt^2} + \mu\frac{\mathbf{r}}{r^3} = 0, \qquad (2.5)$$

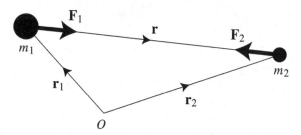

Fig. 2.1.  A vector diagram for the forces acting on two masses, $m_1$ and $m_2$, with position vectors $\mathbf{r}_1$ and $\mathbf{r}_2$.

where $\mu = \mathcal{G}(m_1 + m_2)$. This is the *equation of relative motion*. In order to solve it and find the path of $m_2$ relative to $m_1$ we must first derive several constants of the motion.

Taking the vector product of $\mathbf{r}$ with Eq. (2.5) we have $\mathbf{r} \times \ddot{\mathbf{r}} = 0$, which can be integrated directly to give

$$\mathbf{r} \times \dot{\mathbf{r}} = \mathbf{h}, \tag{2.6}$$

where $\mathbf{h}$ is a constant vector perpendicular to both $\mathbf{r}$ and $\dot{\mathbf{r}}$. Hence the motion of $m_2$ about $m_1$ lies in a plane perpendicular to the direction defined by $\mathbf{h}$. This also implies that the position and velocity vectors always lie in the same plane (see Fig. 2.2). Equation (2.6) is commonly referred to as the *angular momentum integral*. However, although $h = |\mathbf{h}|$ for systems in which $m_2 \ll m_1$ is approximately equal to the orbital angular momentum per unit mass of the body $m_2$, it is not the actual angular momentum in the inertial system since this is calculated using the position and velocity vectors referred to the centre of mass. We consider this in more detail in Sect. 2.7.

Since $\mathbf{r}$ and $\dot{\mathbf{r}}$ always lie in the same plane (the *orbit plane*) it is natural that we now restrict ourselves to considering motion in that plane; the motion referred to an arbitrary reference frame is considered in Sect. 2.8. We now transform to a polar coordinate system $(r, \theta)$ referred to an origin centred on the mass $m_1$ and an arbitrary reference line corresponding to $\theta = 0$. Note that even though the centre of mass of $m_1$ and $m_2$ could be moving in inertial space, the direction of the reference line remains fixed. If we let $\hat{\mathbf{r}}$ and $\hat{\boldsymbol{\theta}}$ denote unit vectors along and perpendicular to the radius vector respectively, then the position, velocity, and acceleration vectors can be written in polar coordinates as

$$\mathbf{r} = r\hat{\mathbf{r}}, \qquad \dot{\mathbf{r}} = \dot{r}\hat{\mathbf{r}} + r\dot{\theta}\hat{\boldsymbol{\theta}}, \qquad \ddot{\mathbf{r}} = (\ddot{r} - r\dot{\theta}^2)\hat{\mathbf{r}} + \left[\frac{1}{r}\frac{\mathrm{d}}{\mathrm{d}t}\left(r^2\dot{\theta}\right)\right]\hat{\boldsymbol{\theta}}. \tag{2.7}$$

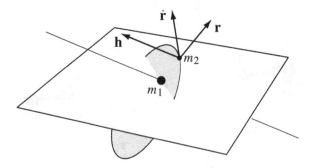

Fig. 2.2.   The motion of $m_2$ with respect to $m_1$ defines an orbital plane (shaded region), because $\mathbf{r} \times \dot{\mathbf{r}}$ is a constant vector, $\mathbf{h}$, the angular momentum vector, and this is always perpendicular to the orbit plane.

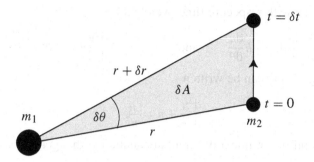

Fig. 2.3. The area $\delta A$ swept out in a time $\delta t$ as a position vector moves through an angle $\delta\theta$.

Substituting the expression for $\dot{\mathbf{r}}$ into Eq. (2.6) gives $\mathbf{h} = r^2\dot{\theta}\hat{\mathbf{z}}$, where $\hat{\mathbf{z}}$ is a unit vector perpendicular to the plane of the orbit forming a right-handed triad with $\hat{\mathbf{r}}$ and $\hat{\boldsymbol{\theta}}$. Hence

$$h = r^2\dot{\theta}. \tag{2.8}$$

Consider the motion of the mass $m_2$ during a time interval $\delta t$ (see Fig. 2.3). At time $t = 0$ it has polar coordinates $(r, \theta)$, while at time $t + \delta t$ its polar coordinates have changed to $(r + \delta r, \theta + \delta\theta)$. The area swept out by the radius vector in time $\delta t$ is

$$\delta A \approx \frac{1}{2}r\,(r + \delta r)\sin\delta\theta \approx \frac{1}{2}r^2\delta\theta, \tag{2.9}$$

where we have neglected second- and higher-order terms in the small quantities. Hence, by dividing each side by $\delta t$ and taking the limit as $\delta t \to 0$ we have

$$\frac{dA}{dt} = \frac{1}{2}r^2\frac{d\theta}{dt} = \frac{1}{2}h. \tag{2.10}$$

Since $h$ is a constant this implies that equal areas are swept out in equal times and hence Eq. (2.10) is the mathematical form of Kepler's second law of planetary motion (see Sect. 1.3). Note that this does not require an inverse square law of force, but only that the force is directed along the line joining the two masses.

## 2.3 Orbital Position and Velocity

We obtain a scalar equation for the relative motion by substituting the expression for $\ddot{\mathbf{r}}$ from Eq. (2.7) into Eq. (2.5); comparing the $\hat{\mathbf{r}}$ components gives

$$\ddot{r} - r\dot{\theta}^2 = -\frac{\mu}{r^2}. \tag{2.11}$$

To solve this equation and find $r$ as a function of $\theta$ we need to make the substitution $u = 1/r$ and to eliminate the time by making use of the constant $h = r^2\dot{\theta}$. By differentiating $r$ with respect to time, we obtain

$$\dot{r} = -\frac{1}{u^2}\frac{\mathrm{d}u}{\mathrm{d}\theta}\dot{\theta} = -h\frac{\mathrm{d}u}{\mathrm{d}\theta} \qquad \text{and} \qquad \ddot{r} = -h\frac{\mathrm{d}^2u}{\mathrm{d}\theta^2}\dot{\theta} = -h^2u^2\frac{\mathrm{d}^2u}{\mathrm{d}\theta^2} \qquad (2.12)$$

and hence Eq. (2.11) can be written

$$\frac{\mathrm{d}^2u}{\mathrm{d}\theta^2} + u = \frac{\mu}{h^2} . \qquad (2.13)$$

This is a second-order, linear differential equation with a general solution

$$u = \frac{\mu}{h^2}\left[1 + e\cos(\theta - \varpi)\right], \qquad (2.14)$$

where $e$ (an amplitude) and $\varpi$ (a phase) are two constants of integration. Substituting back for $r$ we have

$$r = \frac{p}{1 + e\cos(\theta - \varpi)}, \qquad (2.15)$$

which is the general equation of a conic in polar coordinates where $e$ is the *eccentricity* and $p$ is the *semilatus rectum* given by

$$p = h^2/\mu . \qquad (2.16)$$

The four possible conics are:

$$\begin{array}{lll}
\text{circle:} & e = 0, & p = a; \\
\text{ellipse:} & 0 < e < 1, & p = a(1 - e^2); \\
\text{parabola:} & e = 1, & p = 2q; \\
\text{hyperbola:} & e > 1, & p = a(e^2 - 1),
\end{array} \qquad (2.17)$$

where the constant $a$ is the *semi-major axis* of the conic. In the special case of the parabola $p$ is defined in terms of $q$, the distance to the central mass at closest approach. The conic section curves derive their name from the curves formed by the intersection of various planes with the surface of a cone (see Fig. 2.4).

The type of conic is determined by the angle the plane makes with the horizontal. If the plane is horizontal, that is, perpendicular to the axis of symmetry of the cone, then the resulting curve is a circle. If the angle is less than the slope angle of the cone then an ellipse results, whereas if the plane is parallel to the slope of the cone a parabola results. A hyperbola results if the angle the plane makes with the horizontal is anywhere between the slope angle of the cone and the vertical.

In the context of the two-body problem the path of a planet about the Sun is elliptical and closed in inertial space (see Fig. 2.5), and hence Kepler's first law of planetary motion (see Sect. 1.3) is a consequence of the inverse square law of force. Note that the mass $m_1$ fills one focus of the ellipse while the other focus is empty.

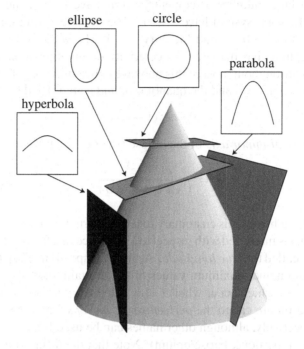

Fig. 2.4. The intersections of planes at different angles with the surface of a cone form the family of curves known as the conic sections.

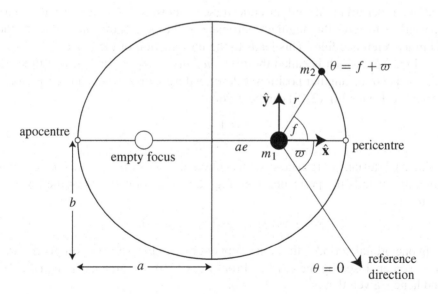

Fig. 2.5. The geometry of the ellipse of semi-major axis $a$, semi-minor axis $b$, eccentricity $e$, and longitude of pericentre $\varpi$.

Although a large number of cometary orbits have $e \approx 1$, most permanent members of the solar system have $e \ll 1$. The notable exceptions among the planets are Pluto ($e = 0.25$) and Mercury ($e = 0.21$), while Nereid ($e = 0.75$), a moon of Neptune, has the largest eccentricity of any known natural satellite. Consequently, throughout most of this book we concentrate on elliptical motion. In this case $p = a(1 - e^2)$ and the quantities $a$ and $e$ are related by

$$b^2 = a^2(1 - e^2), \tag{2.18}$$

where $b$ is the *semi-minor axis* of the ellipse (see Fig. 2.5); we also have

$$r = \frac{a(1 - e^2)}{1 + e \cos(\theta - \varpi)}. \tag{2.19}$$

In celestial mechanics it is customary to use the term longitude when referring to any angle that is measured with respect to a reference line fixed in inertial space. The angle $\theta$ is called the *true longitude*. A simple inspection of Eq. (2.19) shows that the minimum and maximum values of the orbital radius are $r_p = a(1 - e)$ and $r_a = a(1 + e)$, which occur when $\theta = \varpi$ and $\theta = \varpi + \pi$ respectively. These points in the orbit are called the *pericentre* (or *periapse*) and the *apocentre* (or *apoapse*) respectively, although other names can be used for particular systems (e.g., perihelion, perijove, periselenium). Note that the distance of either focus from the centre of the ellipse is $ae$.

The angle $\varpi$ (pronounced "curly pi") is called the *longitude of pericentre*. Although this is a constant for the two-body problem, it can vary with time when additional perturbations are introduced (see Chaps. 3, 6–8). It is usually more convenient to refer the angular coordinate to the pericentre rather than to the arbitrary reference line. This leads to the introduction of the angle $f = \theta - \varpi$ (see Fig. 2.5), which is called the *true anomaly*. Since $\varpi$ is constant the path is closed and the angular position is described by $f$ or $\theta$, which are $2\pi$-periodic variables. Hence Eq. (2.19) can be written

$$r = \frac{a(1 - e^2)}{1 + e \cos f}. \tag{2.20}$$

Using a Cartesian coordinate system centred on the central mass with the $x$ axis pointing towards the pericentre (see Fig. 2.5), the components of the position vector are

$$x = r \cos f \qquad \text{and} \qquad y = r \sin f. \tag{2.21}$$

In one orbital period $T$ the area swept out by a radius vector is simply the area $A = \pi ab$ enclosed by the ellipse. From Eq. (2.10) this area has to equal $hT/2$ and hence, given that $h^2 = \mu a(1 - e^2)$,

$$T^2 = \frac{4\pi^2}{\mu}a^3, \tag{2.22}$$

which corresponds to Kepler's third law of planetary motion (see Sect. 1.3). Note that the period of the orbit is independent of $e$ and is a function of $\mu$ and $a$ only.

Consider the case of two objects of mass $m$ and $m'$, orbiting a central object of mass $m_c$. Let the orbiting objects have semi-major axes $a$ and $a'$ and orbital periods $T$ and $T'$. Equation (2.22) gives

$$\frac{m_c + m}{m_c + m'} = \left(\frac{a}{a'}\right)^3 \left(\frac{T'}{T}\right)^2. \tag{2.23}$$

In the case of planets orbiting the Sun we have $m, m' \ll m_c$ and hence $(a/a')^3 \approx (T/T')^2$. Therefore, if $a$ and $T$ denote the values of the semi-major axis and the period of the Earth's orbit and the unit of length is taken to be the *astronomical unit AU* (1 AU is the approximate semi-major axis of the Earth's orbit) and the unit of time is taken to be the year (the approximate period of the Earth's orbit), we have $T' \approx a'^{3/2}$.

If any solar system object (e.g., an asteroid or a comet) has a small natural or artificial satellite, then observations of the distance and period of the satellite can be used with Kepler's third law to derive an estimate of the mass of the object. Consider Eq. (2.22) applied to the Sun–object and object–satellite systems. Let $m_c$, $m$, and $m'$ now denote the masses of the Sun, object, and satellite respectively with similar definitions for the semi-major axes and orbital periods. This gives

$$\frac{m + m'}{m_c + m} \approx \frac{m}{m_c} = \left(\frac{a'}{a}\right)^3 \left(\frac{T}{T'}\right)^2, \tag{2.24}$$

where we have taken $m' \ll m$ and $m \ll m_c$. This means that the mass of the object can be estimated from the orbital properties of its satellite.

Figure 2.6 shows an image of the asteroid (243) Ida and its moon Dactyl taken by the *Galileo* spacecraft on its way to Jupiter. Direct estimates of the masses of asteroids are notoriously difficult because of their small size (the largest asteroid, (1) Ceres, has a diameter of 913 km) and hence their small perturbations on other objects. Although analysis of the *Viking* data (Standish & Hellings 1989) has permitted mass determinations of the larger objects due to their direct perturbations on the orbit of Mars, similar calculations for the smaller asteroids are almost impossible. However, observations of Dactyl's motion using images obtained by the *Galileo* spacecraft have resulted in an estimated density of $2.6 \pm 0.5$ g cm$^{-3}$ (Belton et al. 1995).

Since the angle $\theta$ covers $2\pi$ radians in one orbital period we can define the "average" angular velocity, or the *mean motion, n* as

$$n = \frac{2\pi}{T} \tag{2.25}$$

Fig. 2.6.   An image of the asteroid (243) Ida and its moon Dactyl taken by the *Galileo* spacecraft on 28th August 1993. Ida is approximately $56 \times 24 \times 21$ km in size and its moon Dactyl is about 1.4 km across. *(Image courtesy of NASA/JPL.)*

and we can write

$$\mu = n^2 a^3 \quad \text{and} \quad h = na^2\sqrt{1 - e^2} = \sqrt{\mu a(1 - e^2)}. \tag{2.26}$$

Although the mean motion is constant in the two-body problem, the actual angular velocity $\dot{f}$ of the orbiting body *is* a function of the longitude.

We can derive another constant of the motion by taking the scalar product of $\dot{\mathbf{r}}$ with Eq. (2.5) and using the expressions for $\mathbf{r}$ and $\dot{\mathbf{r}}$ from Eq. (2.7). This gives the scalar equation

$$\dot{\mathbf{r}} \cdot \ddot{\mathbf{r}} + \mu\frac{\dot{r}}{r^2} = 0, \tag{2.27}$$

which can be integrated to give

$$\frac{1}{2}v^2 - \frac{\mu}{r} = C, \tag{2.28}$$

where $v^2 = \dot{\mathbf{r}} \cdot \dot{\mathbf{r}}$ is the square of the velocity and $C$ is a constant of the motion. Equation (2.28), often called the *vis viva integral*, shows that the orbital energy per unit mass is conserved. Thus the two-body problem has four constants of the motion: the energy integral $C$ and the three components of the angular momentum integral, $\mathbf{h}$. Note that it is also possible to express these constants in different forms such as the orbital elements, or quantities such as the eccentricity vector (see Question 2.4).

By finding another expression for $v^2$ we can derive an expression for $C$. Since $\varpi$ is fixed we have $\dot{\theta} = \mathrm{d}(f + \varpi)/\mathrm{d}t = \dot{f}$ and using the definition of $\dot{\mathbf{r}}$ from Eq. (2.7) gives

$$v^2 = \dot{\mathbf{r}} \cdot \dot{\mathbf{r}} = \dot{r}^2 + r^2\dot{f}^2. \tag{2.29}$$

Differentiating Eq. (2.20) we have

$$\dot{r} = \frac{r \dot{f} e \sin f}{1 + e \cos f} . \qquad (2.30)$$

Using $r^2 \dot{f} = h = na^2 \sqrt{1 - e^2}$, we can write

$$\dot{r} = \frac{na}{\sqrt{1 - e^2}} e \sin f \qquad (2.31)$$

and

$$r\dot{f} = \frac{na}{\sqrt{1 - e^2}} (1 + e \cos f), \qquad (2.32)$$

so that Eq. (2.29) can be written

$$v^2 = \frac{n^2 a^2}{1 - e^2} (1 + 2e \cos f + e^2) = \frac{n^2 a^2}{1 - e^2} \left( \frac{2a(1 - e^2)}{r} - (1 - e^2) \right) . \qquad (2.33)$$

Hence

$$v^2 = \mu \left( \frac{2}{r} - \frac{1}{a} \right) . \qquad (2.34)$$

Consequently, the velocity of the orbiting body is a maximum at pericentre ($f = 0$) and a minimum at apocentre ($f = \pi$). The respective values are

$$v_p = na \sqrt{\frac{1 + e}{1 - e}} \quad \text{and} \quad v_a = na \sqrt{\frac{1 - e}{1 + e}} . \qquad (2.35)$$

We can also find the $x$ and $y$ components of the velocity vector by taking the time derivatives of the expressions for $x$ and $y$ in Eq. (2.21) and substituting the expressions for $\dot{r}$ and $r\dot{f}$ given in Eqs. (2.31) and (2.32). This gives

$$\dot{x} = -\frac{na}{\sqrt{1 - e^2}} \sin f,$$

$$\dot{y} = +\frac{na}{\sqrt{1 - e^2}} (e + \cos f) . \qquad (2.36)$$

By comparing Eq. (2.34) with Eq. (2.28) we see that the energy constant can be written as

$$C = -\frac{\mu}{2a}, \qquad (2.37)$$

and hence the energy of an elliptical orbit is a function of its semi-major axis alone and is independent of the eccentricity. Similar quantities can be defined for parabolic and hyperbolic orbits. It can be shown that

$$C_{\text{para}} = 0, \quad \text{and} \quad C_{\text{hyper}} = \frac{\mu}{2a} . \qquad (2.38)$$

## 2.4 The Mean and Eccentric Anomalies

In the previous section we showed that, given the value of the true anomaly $f$, we can calculate the orbital radius and velocity of a body provided we know the eccentricity and semi-major axis of its orbit. However, in practice we usually want to calculate the location of a body at a given time and our solution to the two-body problem (Eq. (2.20)) does not contain the time explicitly. Although $f$ and $r$ are functions of $t$, we have not shown the nature of this dependence, although it is obviously nonlinear for $e \neq 0$.

Ideally we would like to make use of an angle that is not only $2\pi$-periodic but also a linear function of the time. This will be particularly useful later on when we have to calculate time averages of various quantities. Using our definition of the mean motion $n$ in Eq. (2.25) we can define the *mean anomaly* $M$ by

$$M = n(t - \tau), \tag{2.39}$$

where the constant $\tau$ is the *time of pericentre passage*. Although $M$ has the dimensions of an angle, and it increases linearly with time at a constant rate equal to the mean motion, it has no simple geometrical interpretation. However, from our definition of $M$ and Eq. (2.20) it is clear that when $t = \tau$ (pericentre passage), $M = f = 0$, and when $t = \tau + T/2$ (apocentre passage), $M = f = \pi$; similar relationships will hold for additive multiples of the orbital period $T$.

Although $M$ has no simple geometrical interpretation, it can be related to an angle that does. Consider a circumscribed circle, radius $a$, that is concentric with an orbital ellipse of semi-major axis $a$ and eccentricity $e$ (see Fig. 2.7). A line perpendicular to the major axis of the ellipse is extended through the point on the orbit and intersects the circle. We can define $E$, the *eccentric anomaly*, to be the angle between the major axis of the ellipse and the radius from the centre

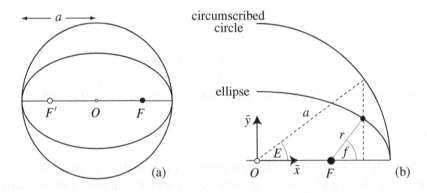

Fig. 2.7. (a) The circumscribed, concentric circle has a radius $a$ equal to the semi-major axis of the ellipse. (b) The relationship between the true anomaly $f$ and the eccentric anomaly $E$.

to the intersection point on the circumscribed circle. Hence, $E = 0$ corresponds to $f = 0$ and $E = \pi$ corresponds to $f = \pi$.

The equation of a centred ellipse in rectangular coordinates is

$$\left(\frac{\bar{x}}{a}\right)^2 + \left(\frac{\bar{y}}{b}\right)^2 = 1. \tag{2.40}$$

But from Fig. 2.7 we have $\bar{x} = a \cos E$ and therefore $\bar{y}^2 = b^2 \sin^2 E$ and hence from Eq. (2.18), $\bar{y} = a\sqrt{1 - e^2} \sin E$. Thus the projections of $r$ in the horizontal and vertical directions are

$$x = a(\cos E - e) \qquad \text{and} \qquad y = a\sqrt{1 - e^2} \sin E \tag{2.41}$$

(cf. Eq. (2.21)). By adding the squares of these expressions and then taking the square root we have

$$r = a(1 - e \cos E) \tag{2.42}$$

and

$$\cos f = \frac{\cos E - e}{1 - e \cos E}. \tag{2.43}$$

We can derive a simpler relationship between $E$ and $f$ by writing

$$1 - \cos f = \frac{(1 + e)(1 - \cos E)}{1 - e \cos E}, \qquad 1 + \cos f = \frac{(1 - e)(1 + \cos E)}{1 - e \cos E}. \tag{2.44}$$

Using the standard double angle formulae, these equations can be written as

$$2 \sin^2 \frac{f}{2} = \frac{1 + e}{1 - e \cos E} 2 \sin^2 \frac{E}{2}, \qquad 2 \cos^2 \frac{f}{2} = \frac{1 - e}{1 - e \cos E} 2 \cos^2 \frac{E}{2} \tag{2.45}$$

and hence

$$\tan \frac{f}{2} = \sqrt{\frac{1 + e}{1 - e}} \tan \frac{E}{2}. \tag{2.46}$$

Thus, knowing $E$, we can determine $r$ and $f$ uniquely from Eqs. (2.42) and (2.43), since $E$ and $f$ will always lie in the same half of the ellipse. However, to locate a body in its orbit at some time $t$, we need to derive a relationship between $M$ and $E$.

Using $v^2 = \dot{r}^2 + (r\dot{f})^2$ and Eqs. (2.32) and (2.34), we have

$$\dot{r}^2 = n^2 a^3 \left(\frac{2}{r} - \frac{1}{a}\right) - \frac{n^2 a^4 (1 - e^2)}{r^2}. \tag{2.47}$$

Hence

$$\frac{dr}{dt} = \frac{na}{r} \sqrt{a^2 e^2 - (r - a)^2}. \tag{2.48}$$

This can then be integrated by making the substitution

$$r - a = -ae \cos E \tag{2.49}$$

from Eq. (2.42). Hence Eq. (2.48) can be written as

$$\frac{dE}{dt} = \frac{n}{1 - e \cos E}.\tag{2.50}$$

This equation can also be derived by differentiating Eq. (2.41) to find $\dot{x}$ and $\dot{y}$, finding $\mathbf{r} \times \dot{\mathbf{r}}$ and equating the magnitude, $h$, with $na^2\sqrt{1 - e^2}$. The resulting equation, Eq. (2.50), can be easily integrated to give

$$n(t - \tau) = E - e \sin E\tag{2.51}$$

where we have taken $\tau$ to be the constant of integration and used the boundary condition $E = 0$ when $t = \tau$. Hence, from Eq. (2.39) we have

$$M = E - e \sin E.\tag{2.52}$$

This is *Kepler's equation* and its solution is fundamental to the problem of finding the orbital position at a given time. At a particular time $t$ we can (i) find $M$ from Eq. (2.39), (ii) solve Eq. (2.52) for $E$, (iii) use Eq. (2.41), or Eqs. (2.43) and (2.20), to find $r$ and $f$.

So far we have defined the true longitude ($\theta$), the true anomaly ($f$), the mean anomaly ($M$), the eccentric anomaly ($E$), and the longitude of pericentre ($\varpi$). To complete this set we define the *mean longitude* $\lambda$ by

$$\lambda = M + \varpi.\tag{2.53}$$

Therefore $\lambda$ is a linear function of time and, since it is derived from $M$, it has no geometrical interpretation, except in the special case of a circular orbit. It is important to note that all longitudes ($\theta$, $\varpi$, $\lambda$) are defined with respect to a common, arbitrary reference direction (see Fig. 2.5).

Colwell (1993) points out that papers have been published about the solution of Kepler's equation in virtually every decade since 1650 and that many eminent scientists have attempted solutions. Kepler's equation cannot be solved directly because it is transcendental in $E$ and therefore, apart from the trivial solutions $E = j\pi$ when $M = j\pi$ for integer $j$, it is impossible to express $E$ as a simple function of $M$. We briefly consider two iterative techniques: one producing a series solution and the other a numerical solution. In each case we assume that $M$ and $E$ are expressed in radians.

We can derive a series solution by using an iterative method of the form

$$E_{i+1} = M + e \sin E_i, \qquad i = 0, 1, \ldots,\tag{2.54}$$

where we take $E_0 = M$ as our first approximation. Using the formula $\sin(A + B) = \sin A \cos B + \cos A \sin B$ and the fact that $\sin x \approx x - \frac{1}{6}x^3 + \mathcal{O}(x^5)$ and

$\cos x \approx 1 - \frac{1}{2}x^2 + \mathcal{O}(x^4)$ for small $x$, we obtain

$$E_1 = M + e \sin M,$$

$$E_2 = M + e \sin(M + e \sin M) \approx M + e \sin M + \frac{1}{2}e^2 \sin 2M,$$

$$E_3 = M + e \sin(M + e \sin M + \frac{1}{2}e^2 \sin 2M)$$
(2.55)

$$\approx M + \left(e - \frac{1}{8}e^3\right) \sin M + \frac{1}{2}e^2 \sin 2M + \frac{3}{8}e^3 \sin 3M$$

for the first three steps, where we introduced only one additional term in $e$ at each step. It is clear from this approach that the final series for $E - M$ will have the form

$$E - M = \sum_{s=1}^{\infty} b_s(e) \sin sM, \qquad (2.56)$$

where the lowest order term in $b_s(e)$ is $\mathcal{O}(e^s)$. The form of Eq. (2.56) suggests that we have expressed $E - M$ as a Fourier sine series in $M$. We study such useful expansions for elliptical motion in more detail in Sect. 2.5, where we derive expressions for the $b_s(e)$ terms.

It is important to note that the series solution of Kepler's equation diverges for values of $e > 0.6627434$ (see Hagihara 1970, for a detailed explanation). Not only is this property a fundamental limitation to deriving a useful series solution to Kepler's equation, it also has important implications for other series, such as the planetary disturbing function (see Chap. 6), that make use of this solution. However, numerical solutions of Kepler's equation are not affected by this limitation.

Danby (1988) gives a variety of numerical methods for solving Kepler's equation. By writing Eq. (2.52) as

$$f(E) = E - e \sin E - M \qquad (2.57)$$

we can use the Newton–Raphson method to find the root of the nonlinear equation $f(E) = 0$. The iteration scheme is

$$E_{i+1} = E_i - \frac{f(E_i)}{f'(E_i)}, \qquad i = 0, 1, 2, \ldots, \qquad (2.58)$$

where $f'(E_i) = \mathrm{d}f(E_i)/\mathrm{d}E_i = 1 - e \cos E_i$. Danby (1988) points out that the convergence of the Newton–Raphson scheme is quadratic but that quartic convergence is also possible with a modified scheme. Using Danby's notation and a Taylor series expansion we can write

$$0 = f(E_i + \epsilon_i) = f(E_i) + \epsilon_i f'(E_i) + \frac{1}{2}\epsilon_i^2 f''(E_i) + \frac{1}{6}\epsilon_i^3 f'''(E_i) + \mathcal{O}(\epsilon_i^4). \quad (2.59)$$

Neglecting the higher order terms in $\epsilon_i$ we can write

$$0 = f_i + \delta_i f_i' + \frac{1}{2}\delta_i^2 f_i'' + \frac{1}{6}\delta_i^3 f_i''', \tag{2.60}$$

where $f_i = f(E_i)$, $f_i' = f'(E_i)$, etc. Hence

$$\delta_i = -\frac{f_i}{f_i' + \frac{1}{2}\delta_i f_i'' + \frac{1}{6}\delta_i^2 f_i'''}. \tag{2.61}$$

This can be solved for $\delta_i$ by defining

$$\delta_{i1} = -\frac{f_i}{f_i'}, \quad \delta_{i2} = -\frac{f_i}{f_i' + \frac{1}{2}\delta_{i1}f_i''}, \quad \delta_{i3} = -\frac{f_i}{f_i' + \frac{1}{2}\delta_{i2}f_i'' + \frac{1}{6}\delta_{i2}^2 f_i'''} \tag{2.62}$$

and then using the iteration scheme

$$E_{i+1} = E_i + \delta_{i3}. \tag{2.63}$$

Although this method has more arithmetic operations per iteration than the standard Newton–Raphson scheme given above, it is more efficient since (a) it can be programmed to make use of quantities that have already been calculated at each iteration and (b) it will converge faster.

An important consideration in either of these numerical schemes is a suitable starting value, $E_0$. Obviously for small $e$ we have $E \approx M$ and so $E_0 = M$ seems appropriate. However, this guess is only correct in the cases where $e = 0$ or $M$ is a multiple of $\pi$. Danby (1988) points out that, by first reducing $M$ to the range $0 \leq M \leq 2\pi$, the initial guess

$$E_0 = M + \text{sign}(\sin M)\,ke, \qquad 0 \leq k \leq 1 \tag{2.64}$$

has a better chance of being correct and improves the convergence; the recommended value is $k = 0.85$. Further details of this and other methods are discussed in Danby & Burkardt (1983), Burkardt & Danby (1983), and Danby (1987).

The solution of Kepler's equation to find $E$ for a given value of $M$ allows the calculation of the position and velocity at any time $t$ for an object in an elliptical orbit. If the object has a position vector $\mathbf{r}_0 = \mathbf{r}(t_0)$ and a velocity vector $\mathbf{v}_0 = \mathbf{v}(t_0)$ at time $t_0$ then this process can be simplified by the introduction of two special functions and their time derivatives. Provided that the initial vectors $\mathbf{r}_0$ and $\mathbf{v}_0$ are not parallel, $\mathbf{r}(t)$ can be written as

$$\mathbf{r}(t) = f(t, t_0)\mathbf{r}_0 + g(t, t_0)\mathbf{v}_0, \tag{2.65}$$

where $f(t, t_0)$ and $g(t, t_0)$ are referred to as the *f and g functions*.

Separating the $x$ and $y$ components we have

$$x = f(t, t_0)x_0 + g(t, t_0)\dot{x}_0 \qquad \text{and} \qquad y = f(t, t_0)y_0 + g(t, t_0)\dot{y}_0, \tag{2.66}$$

where we have taken $\mathbf{r}_0 = (x_0, y_0)$ and $\mathbf{v}_0 = (\dot{x}_0, \dot{y}_0)$. This gives two simultaneous linear equations, which we can solve for $f$ and $g$. The solution is

$$f(t, t_0) = \frac{x\dot{y}_0 - y\dot{x}_0}{x_0\dot{y}_0 - y_0\dot{x}_0} \quad \text{and} \quad g(t, t_0) = \frac{yx_0 - xy_0}{x_0\dot{y}_0 - y_0\dot{x}_0}. \tag{2.67}$$

Since $\cos f = x/r$ and $\sin f = y/r$ we can write Eq. (2.36) in terms of the eccentric anomaly instead of the true anomaly. This gives

$$\dot{x} = -\frac{na^2}{r} \sin E \quad \text{and} \quad \dot{y} = \frac{na^2\sqrt{1 - e^2}}{r} \cos E. \tag{2.68}$$

Substituting Eqs. (2.41) and (2.68), with appropriate expressions for $x_0$, $y_0$, $\dot{x}_0$, and $\dot{y}_0$, and making use of Eqs. (2.42) and (2.51) gives

$$f(t, t_0) = \frac{a}{r_0}\{\cos(E - E_0) - 1\} + 1,$$

$$g(t, t_0) = (t - t_0) + \frac{1}{n}\{\sin(E - E_0) - (E - E_0)\}. \tag{2.69}$$

The velocity at time $t$ can be written as

$$\mathbf{v}(t) = \dot{f}\mathbf{r}_0 + \dot{g}\mathbf{v}_0, \tag{2.70}$$

where $\dot{f}$ and $\dot{g}$, the partial derivatives of $f$ and $g$ with respect to time, can be derived from Eq. (2.67) by making use of the expression for $\dot{E}$ given in Eq. (2.50). We then have

$$\dot{f}(t, t_0) = -\frac{a^2}{rr_0}n \sin(E - E_0),$$

$$\dot{g}(t, t_0) = \frac{a}{r}\{\cos(E - E_0) - 1\} + 1. \tag{2.71}$$

The use of the $f$ and $g$ functions means that once $E$ is known from the solution of Kepler's equation, we can readily find $\mathbf{r}$ and $\mathbf{v}$. Although we have formulated the expressions for the scalar quantities $f$, $g$, $\dot{f}$, and $\dot{g}$ by considering motion in the plane of the orbit, the formulae are equally applicable in other reference systems. In particular the $f$ and $g$ functions obviate the need to transform to and from a coordinate system in the orbital plane to one in a more general three-dimensional reference frame (see Sect. 2.8). This introduces considerable computational savings in numerical work.

## 2.5 Elliptic Expansions

Since there are so few integrable problems in solar system dynamics, frequently we have to resort to approximations in order to achieve a practical solution to a particular problem. The small quantities inherent in most branches of solar system dynamics are the eccentricity and inclination (the angle the orbit plane makes with a reference plane) of an orbit. For example, in Sect. 2.4 we have shown how Kepler's equation can be solved by means of a series in powers of the eccentricity. In Chapter 6 we deal with an expansion of the perturbing potential experienced by one planet or satellite due to another. In that case the expansion is in terms of the eccentricity and inclination of the bodies involved. Throughout this book we make use of a number of expansions. Typically we make an expansion, neglect higher order terms in some quantity, and then apply the resulting series to a problem of interest. In this section we derive a number of fundamental expansions that will be needed later on.

In the previous section we saw how it was possible to derive a simple series solution for $E$ in terms of $M$. We can now formalise the result we obtained. If we write Eq. (2.52) as $E - M = e \sin E$ then, since $E - M$ is an odd periodic function, it can be expanded as a Fourier sine series,

$$e \sin E = \sum_{s=1}^{\infty} b_s(e) \sin sM, \tag{2.72}$$

where the coefficients $b_s(e)$ are given by

$$b_s(e) = \frac{2}{\pi} \int_0^\pi e \sin E \, \sin sM \, dM$$
$$= \left[ -\frac{2}{s\pi} e \sin E \cos sM \right]_0^\pi + \frac{2}{s\pi} \int_0^\pi \cos sM \, d(e \sin E). \tag{2.73}$$

The first part of this equation evaluates to zero and by using Kepler's equation to write $d(e \sin E) = d(E - M)$ we have

$$b_s(e) = -\frac{2}{s\pi} \int_0^\pi \cos sM \, dM + \frac{2}{s\pi} \int_0^\pi \cos sM \, dE. \tag{2.74}$$

The first integral evaluates to zero and we can use Kepler's equation again to obtain

$$b_s(e) = \frac{2}{s\pi} \int_0^\pi \cos(sE - se \sin E) \, dE. \tag{2.75}$$

This integral can be written in terms of a standard function called the *Bessel function* of the first kind (see, e.g., Bowman 1958). We can write

$$b_s(e) = \frac{2}{s} J_s(se), \tag{2.76}$$

where

$$J_s(se) = \frac{1}{\pi} \int_0^\pi \cos(sE - se \sin E) \, dE \tag{2.77}$$

is the Bessel function. For positive values of $s$, we can write

$$J_s(x) = \frac{1}{s!} \left(\frac{x}{2}\right)^s \sum_{\beta=0}^{\infty} (-1)^\beta \frac{(x/2)^{2\beta}}{\beta!(s+1)(s+2)\ldots(s+\beta)}. \tag{2.78}$$

This series is absolutely convergent for all values of $x$. The series for $J_s(x)$ for $s = 1, \ldots, 5$ including terms up to $\mathcal{O}(x^5)$ are given below:

$$J_1(x) = \frac{1}{2}x - \frac{1}{16}x^3 + \frac{1}{384}x^5 + \mathcal{O}(x^7),$$

$$J_2(x) = \frac{1}{8}x^2 - \frac{1}{96}x^4 + \mathcal{O}(x^6),$$

$$J_3(x) = \frac{1}{48}x^3 - \frac{1}{768}x^5 + \mathcal{O}(x^7), \tag{2.79}$$

$$J_4(x) = \frac{1}{384}x^4 + \mathcal{O}(x^6),$$

$$J_5(x) = \frac{1}{3840}x^5 + \mathcal{O}(x^7).$$

We can now write the solution of Kepler's equation as

$$E = M + 2\sum_{s=1}^{\infty} \frac{1}{s} J_s(se) \sin sM$$

$$= M + e \sin M + e^2 \left( \frac{1}{2} \sin 2M \right) + e^3 \left( \frac{3}{8} \sin 3M - \frac{1}{8} \sin M \right)$$

$$+ e^4 \left( \frac{1}{3} \sin 4M - \frac{1}{6} \sin 2M \right) + \mathcal{O}(e^5), \tag{2.80}$$

which is consistent with our result in Sect. 2.4. It is important to repeat the warning, mentioned in Sect. 2.4, that although this series solution rapidly converges for small values of $e$, the series is divergent for values of $e > 0.6627434$. This implies that all the series in this section that make use of this series are also divergent for sufficiently large values of $e$.

In addition to the series solution of Kepler's equation we will also need a number of other series expansions, all of which can be expressed in terms of Bessel functions. In particular we need series for $r/a$, $\cos E$, $(a/r)^3$, $\sin f$, $\cos f$, and $f - M$. The derivation of the results that follow are given in Brouwer & Clemence (1961).

The series for $r/a$ is given by

$$\frac{r}{a} = 1 + \frac{1}{2}e^2 - 2e \sum_{s=1}^{\infty} \frac{1}{s^2} \frac{\mathrm{d}}{\mathrm{d}e} J_s(se) \cos sM$$

$$= 1 - e \cos M + \frac{e^2}{2} \left( 1 - \cos 2M \right) + \frac{3e^3}{8} \left( \cos M - \cos 3M \right)$$

$$+ \frac{e^4}{3} \left( \cos 2M - \cos 4M \right) + \mathcal{O}(e^5). \tag{2.81}$$

This series is used in Sect. 2.6 in the guiding centre approximation and in Sect. 6.3 and 6.5 in our expansion of the planetary disturbing function.

We can use the fact that $\cos E = (1 - r/a)/e$ and the series for $(r/a)$ to derive the series for $\cos E$. It is given by

$$\cos E = -\frac{1}{2}e + 2\sum_{s=1}^{\infty}\frac{1}{s^2}\frac{d}{de}J_s(se)\cos sM$$

$$= \cos M + \frac{e}{2}(\cos 2M - 1) + \frac{3e^2}{8}(\cos 3M - \cos M)$$

$$+ e^3\left(\frac{1}{3}\cos 4M - \frac{1}{3}\cos 2M\right)$$

$$+ e^4\left(\frac{5}{192}\cos M - \frac{45}{128}\cos 3M + \frac{125}{384}\cos 5M\right) + \mathcal{O}(e^5). \quad (2.82)$$

We can also use the series for $r/a$ to derive the series for $(a/r)^3$. It is given by

$$\left(\frac{a}{r}\right)^3 = 1 + 3e\cos M + e^2\left(\frac{3}{2} + \frac{9}{2}\cos 2M\right)$$

$$+ e^3\left(\frac{27}{8}\cos M + \frac{53}{8}\cos 3M\right)$$

$$+ e^4\left(\frac{15}{8} + \frac{7}{2}\cos 2M + \frac{77}{8}\cos 4M\right) + \mathcal{O}(e^5). \quad (2.83)$$

This series is used in Sect. 6.3 in our expansion of the planetary disturbing function.

The series for $\sin f$ and $\cos f$ are given by

$$\sin f = 2\sqrt{1 - e^2}\sum_{s=1}^{\infty}\frac{1}{s}\frac{d}{de}J_s(se)\sin sM$$

$$= \sin M + e\sin 2M + e^2\left(\frac{9}{8}\sin 3M - \frac{7}{8}\sin M\right)$$

$$+ e^3\left(\frac{4}{3}\sin 4M - \frac{7}{6}\sin 2M\right)$$

$$+ e^4\left(\frac{17}{192}\sin M - \frac{207}{128}\sin 3M + \frac{625}{384}\sin 5M\right) + \mathcal{O}(e^5) \quad (2.84)$$

and

$$\cos f = -e + \frac{2(1 - e^2)}{e}\sum_{s=1}^{\infty}J_s(se)\cos sM$$

$$= \cos M + e(\cos 2M - 1) + \frac{9e^2}{8}(\cos 3M - \cos M)$$

$$+ \frac{4e^3}{3}(\cos 4M - \cos 2M)$$

$$+ e^4\left(\frac{25}{192}\cos M - \frac{225}{128}\cos 3M + \frac{625}{384}\cos 5M\right) + \mathcal{O}(e^5). \quad (2.85)$$

These series are used in Sect. 5.4 in our study of spin–orbit resonance and in Sect. 6.5 as part of the expansion of the planetary disturbing function.

We can derive a series for $f - M$, also called the *equation of the centre*. From Eqs. (2.20) and (2.32) we obtain

$$r^2 \dot{f} = na^2(1 - e^2)^{1/2}. \tag{2.86}$$

Using $dM = n\,dt$, $r = a(1 - e \cos E)$, and Kepler's equation, we obtain

$$df = \frac{\sqrt{1 - e^2}}{(1 - e \cos E)^2}dM = \sqrt{1 - e^2}\left(\frac{dE}{dM}\right)^2 dM. \tag{2.87}$$

This can be integrated term by term using the series solution for Kepler's equation to give

$$f - M = 2e \sin M + \frac{5}{4}e^2 \sin 2M + e^3 \left(\frac{13}{12}\sin 3M - \frac{1}{4}\sin M\right)$$
$$+ e^4 \left(\frac{103}{96}\sin 4M - \frac{11}{24}\sin 2M\right) + \mathcal{O}(e^5). \tag{2.88}$$

This series is used in Sect. 2.6 in an analysis of the guiding centre approximation and in Sect. 5.2 where we investigate tidal de-spinning.

Lagrange developed a useful method for inverting series expansions, which has applications to this work. He showed that if a variable $z$ can be expressed as a function of $\zeta$ of the form

$$\zeta = z + e\phi(\zeta) \qquad (e < 1) \tag{2.89}$$

then $\zeta$ can also be expressed as a function of $z$ using the relationship

$$\zeta = z + \sum_{j=1}^{\infty} \frac{e^j}{j!} \frac{d^{j-1}}{dz^{j-1}}[\phi(z)]^j. \tag{2.90}$$

For example, to obtain the expression for $f$ in terms of $M$, given in Eq. (2.88), we start from Kepler's second law, Eq. (2.8), and Eq. (2.26) giving,

$$h = r^2 \dot{f} = na^2(1 - e^2)^{1/2}. \tag{2.91}$$

Integrating this equation and substituting Eq. (2.20) for $r$ gives

$$M = (1 - e^2)^{3/2} \int_0^f \frac{df}{(1 + e \cos f)^2}. \tag{2.92}$$

Expanding binomially and integrating term by term, we obtain

$$M = f - 2e \sin f + \frac{3}{4}e^2 \sin 2f + \mathcal{O}(e^3). \tag{2.93}$$

This can be written as

$$f = M + e\left(2 \sin f - \frac{3}{4}e \sin 2f + \cdots\right). \tag{2.94}$$

Lagrange's inverse theorem then gives

$$f = M + \sum_{j=1}^{\infty} \frac{e^j}{j!} \frac{d^{j-1}}{dM^{j-1}} \left[ 2\sin M - \frac{3}{4} e \sin 2M + \cdots \right]^j, \qquad (2.95)$$

which agrees with Eq. (2.88) after expansion. We make frequent use of Lagrange's method in Sect. 3.6 where we derive the locations of the collinear equilibrium points in the circular restricted problem.

## 2.6 The Guiding Centre Approximation

In many applications in solar system dynamics the eccentricity is very small and approximations that are accurate to order $e$ are useful, particularly in some systems that are best viewed in a rotating reference frame. This approach is also appropriate when considering systems such as perturbed motion in the vicinity of equilibrium points (Sect. 3.8), the effects of planetary oblateness on near-circular, near-equatorial orbits (Sect. 6.11), and its applications to planetary rings (Chapter 10). In all these cases it is useful to characterise the extent of the departure from circular motion.

In the *guiding centre approximation,* the motion of a particle $P$ moving in an elliptical orbit about a focus $F$ (see Fig. 2.8) is viewed in a reference frame that is centred on a point $G$, the guiding centre, that rotates about the focus in a circle of radius $a$ equal to the particle's semi-major axis, with angular speed equal to the particle's mean motion $n$.

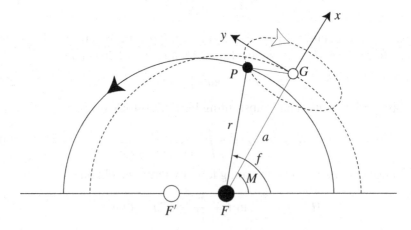

Fig. 2.8. The relationship between the true anomaly $f$ and the mean anomaly $M$ in the guiding centre approximation. $G$ denotes the guiding centre, $P$ the particle, $F$ the focus, and $F'$ the empty focus. The guiding centre moves on a circle of radius $a$ centred on $F$.

If we transform to a rectangular coordinate system centred on $G$, then the coordinates of $P$ are

$$x = r\cos(f - M) - a \qquad \text{and} \qquad y = r\sin(f - M). \qquad (2.96)$$

To order $e$, the expansion of $f - M$ from Eq. (2.88) is

$$f - M \approx 2e\sin M. \qquad (2.97)$$

Hence

$$x \approx -ae\cos M \qquad \text{and} \qquad y \approx 2ae\sin M \qquad (2.98)$$

and

$$\frac{x^2}{(ae)^2} + \frac{y^2}{(2ae)^2} \approx 1. \qquad (2.99)$$

It follows that while $G$ moves about $F$ in a circle of radius $a$ with mean motion $n$ and period $2\pi/n$, $P$ moves about $G$ in the opposite sense on a 2:1 ellipse of semi-major axis $2ae$, semi-minor axis $ae$, and period $2\pi/n$. The motion of $P$ with respect to $F$ is a Lissajou figure obtained by the superposition of two harmonic motions with a common frequency $n$, a phase difference of $\pi/2$, and a 2:1 amplitude ratio.

At this level of approximation there are two other features of the motion that are worth noting. The distance $R$ of $P$ from the centre of the ellipse (see Fig. 2.9) can be obtained from

$$R^2 = r^2 + (ae)^2 + 2aer\cos f. \qquad (2.100)$$

Hence

$$R \approx a\left(1 - \frac{1}{2}e^2\sin^2 f\right) \approx a\left(1 - \frac{1}{2}e^2\sin^2 M\right) \qquad (2.101)$$

since, from Eq. (2.97), $f = M + \mathcal{O}(e)$. Thus, to order $e$, the path of $P$ is a circle with centre at $O$. Therefore the path and the circumscribed circle (see Fig. 2.9) coincide and thus the angle $P\hat{O}F$ is the eccentric anomaly $E$. In fact, as we show below, in this approximation the angle $P\hat{F'}F$, where $F'$ is the empty focus, is the mean anomaly $M$.

Consider the true elliptical path of $P$ and denote the angle $F\hat{F'}P$ by $g$. Applying the cosine rule to the triangle $FF'P$ we obtain

$$r^2 = (2a - r)^2 + 4(ae)^2 - 4ae(2a - r)\cos g. \qquad (2.102)$$

Hence

$$\cos g = \frac{(1 - r/a) + e^2}{e(1 - r/a) + e}. \qquad (2.103)$$

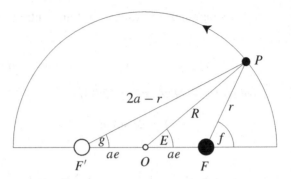

Fig. 2.9.   The relationships among the true, mean, and eccentric anomalies in the guiding centre approximation. Note that the diagram exaggerates the actual case: In reality the eccentricity is small and $F$ and $F'$ are close to $O$.

Here we have used the fact that since $F$ and $F'$ are the foci of the ellipse, $FP + F'P = 2a$. From the expansion of $r/a$ in Eq. (2.81) we obtain

$$1 - \frac{r}{a} \approx e \cos M - \frac{1}{2}e^2(1 - \cos 2M) - \frac{3}{8}e^3(\cos M - \cos 3M) \qquad (2.104)$$

and hence

$$\cos g \approx \cos M - \frac{1}{8}e^2(\cos M - \cos 3M) + \mathcal{O}(e^3), \qquad (2.105)$$

so that to $\mathcal{O}(e)$ we have $g = M$. Therefore the line joining the orbiting mass to the empty focus must rotate at the same rate as the mean motion of the orbiting mass.

This result has an interesting consequence if we apply it to the motion of a satellite that has a spin period equal to its orbital period (a *synchronously rotating* satellite). Since the line drawn from the satellite to the empty focus rotates with frequency $n$, equal to the mean motion, it follows that a synchronously rotating satellite rotates with one face pointing towards the empty focus of its orbit. This proves to be useful in understanding the origin of the librational tide on a synchronously rotating satellite like the Moon or Io. It also follows that the line drawn from the guiding centre to the central mass is parallel to the line joining the orbiting mass to the empty focus (Fig. 2.10).

It is interesting to note that Ptolemy's scheme for the motion of the Sun about the Earth had the Sun moving in a circle with uniform angular motion about an equant with the Earth displaced from the centre of the circle. If we place the Earth at the focus $F$ and identify the equant with the point $F'$, then we see that Ptolemy's scheme was accurate to order $e$. The triumph of Kepler was to produce a theory that was accurate to order $e^2$ (Hoyle 1974).

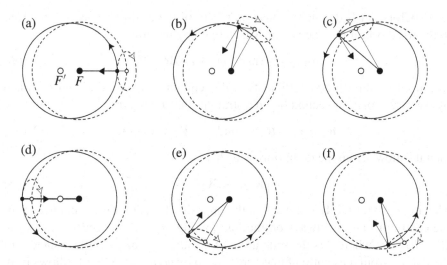

Fig. 2.10. An illustration of the guiding centre approximation for an ellipse of eccentricity $e = 0.2$. The position of the orbiting mass (small filled circle) with respect to the central mass (large filled circle) and the empty focus (large white circle) is shown at equal intervals of mean anomaly $M$. (a) $M = 0$, (b) $M = \pi/3$, (c) $M = 2\pi/3$, (d) $M = \pi$, (e) $M = 4\pi/3$, and (f) $M = 5\pi/3$. The solid curve denotes the keplerian ellipse while the dashed circle with a radius equal to the semi-major axis of the ellipse denotes the path of the guiding centre; the circle is centred on the primary focus of the ellipse.

## 2.7 Barycentric Orbits

We have shown that, with suitable starting conditions, the motion of the mass $m_2$ with respect to the mass $m_1$ describes a conic section in space. We now return to the formulation of the two-body problem using a centre of mass coordinate system with origin at the point $O'$ (see Fig. 2.11).

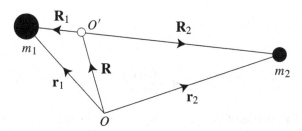

Fig. 2.11. The position vectors of the two masses with respect to the origin, $O$, and with respect to the centre of mass, $O'$.

As before, let $\mathbf{R}$ denote the position vector of the centre of mass $O'$, referred to the fixed origin $O$. The vector $\mathbf{R}$ is defined by the equation

$$m_1 \mathbf{r}_1 + m_2 \mathbf{r}_2 - (m_1 + m_2)\mathbf{R} = 0. \tag{2.106}$$

Because the sum of the coefficients in this equation is zero, the points defined by the three position vectors lie on a straight line. If we write

$$\mathbf{R}_1 = \mathbf{r}_1 - \mathbf{R} \qquad \text{and} \qquad \mathbf{R}_2 = \mathbf{r}_2 - \mathbf{R} \tag{2.107}$$

then it follows from the definition of $\mathbf{R}$ that

$$m_1 \mathbf{R}_1 + m_2 \mathbf{R}_2 = 0. \tag{2.108}$$

This implies that (i) $\mathbf{R}_1$ is always in the opposite direction to $\mathbf{R}_2$, and hence (ii) the centre of mass is always on the line joining $m_1$ and $m_2$, and we can write $R_1 + R_2 = r$, where $r$ is the separation of $m_1$ and $m_2$, and (iii) the distances of the masses from the centre of mass are related by $m_1 R_1 = m_2 R_2$. It follows from this that

$$R_1 = \frac{m_2}{m_1 + m_2} r \qquad \text{and} \qquad R_2 = \frac{m_1}{m_1 + m_2} r. \tag{2.109}$$

Therefore, whichever conic section describes the relative motion of the two masses, each mass will also orbit the centre of mass of the system in a path described by the same conic section reduced in scale by $m_1/(m_1+m_2)$ or $m_2/(m_1+m_2)$ (see Fig. 2.12).

A coordinate system referred to the centre of mass is called a *barycentric* system. In Sect. 2.2 we showed that for the path of the relative motion of $m_2$

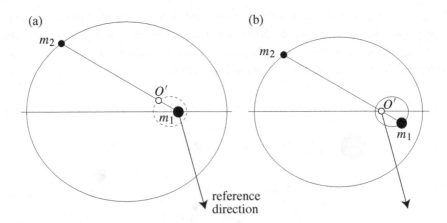

Fig. 2.12. (a) The motion of the mass $m_2$ with respect to the mass $m_1$ in the two-body problem; the dashed curve denotes the elliptical path of the centre of mass, $O'$. (b) The motion of the masses $m_1$ and $m_2$ with respect to the centre of mass, $O'$. In each case a mass ratio of $m_2/m_1 = 0.2$ and an eccentricity of 0.5 were used.

with respect to $m_1$, there exists a constant $h = r^2\dot\theta$. Since $R_1$ and $R_2$ are both proportional to $r$ we have

$$R_1^2\dot\theta = \text{constant} = h_1 = \left(\frac{m_2}{m_1 + m_2}\right)^2 h,$$

$$R_2^2\dot\theta = \text{constant} = h_2 = \left(\frac{m_1}{m_1 + m_2}\right)^2 h,$$

(2.110)

and the total orbital angular momentum of the system is given by

$$L^* = m_1 h_1 + m_2 h_2 = \frac{m_1 m_2}{m_1 + m_2}h.$$

(2.111)

Thus

$$h = \left(\frac{1}{m_1} + \frac{1}{m_2}\right)L^*$$

(2.112)

and hence, when $m_2 < m_1$, $h \approx h_2$ and $h$ is approximately equal to the angular momentum of the system per unit mass of the body $m_2$.

It is clear from the above and Fig. 2.12 that the period of the orbit of each mass about the centre of mass is equal to $T$, the period of the mass $m_2$ about the mass $m_1$. Therefore the mean motions are also equal, $n_1 = n_2 = n$, although the semi-major axes are not. The relationship between the semi-major axes is suggested by considering Eq. (2.109) in the case of circular orbits; it is

$$a_1 = \frac{m_2}{m_1 + m_2}a \quad \text{and} \quad a_2 = \frac{m_1}{m_1 + m_2}a.$$

(2.113)

Consequently, since $h = na^2\sqrt{1 - e^2}$, we have

$$h_1 = na_1^2\sqrt{1 - e^2} \quad \text{and} \quad h_2 = na_2^2\sqrt{1 - e^2},$$

(2.114)

which shows that, although $m_1 a_1 = m_2 a_2$, the eccentricities of the ellipses are equal and therefore all the ellipses are similar. Although each mass moves on its own elliptical orbit with respect to the common centre of mass, the pericentres of their orbits differ by $\pi$ (see Fig. 2.12b).

We can now consider the total energy of the system, $E^*$, which is the sum of the kinetic energy (referred to the inertial, barycentric coordinate system) and the potential energy. We have

$$E^* = \frac{1}{2}m_1 v_1^2 + \frac{1}{2}m_2 v_2^2 - \mathcal{G}\frac{m_1 m_2}{r}$$

$$= \frac{1}{2}m_1\left[\dot{R}_1^2 + (R_1\dot{f})^2\right] + \frac{1}{2}m_2\left[\dot{R}_2^2 + (R_2\dot{f})^2\right] - \mathcal{G}\frac{m_1 m_2}{r}. \quad (2.115)$$

We can simplify this using Eq. (2.109) to get

$$E^* = \frac{m_1 m_2}{m_1 + m_2}C = -\mathcal{G}\frac{m_1 m_2}{2a},$$

(2.116)

where $C$ is the energy constant from Eq. (2.37). Consequently, the total energy of the system is also purely a function of the semi-major axis of the orbit of $m_2$ with respect to $m_1$. Note that

$$C = \left( \frac{1}{m_1} + \frac{1}{m_2} \right) E^* \tag{2.117}$$

and hence, when $m_2 \ll m_1$, $C \approx E^*/m_2$, that is, the total energy per unit mass of the body $m_2$.

## 2.8 The Orbit in Space

In Sect. 2.2 we showed that the position and velocity vectors of the mass $m_2$ with respect to the mass $m_1$ always lie in a plane perpendicular to the angular momentum vector. The values of $\mathbf{r} = (x, y)$ and $\dot{\mathbf{r}} = (\dot{x}, \dot{y})$ (or alternatively, $r$, $\theta$, $\dot{r}$, and $\dot{\theta}$) of the mass $m_2$ with respect to $m_1$ at any time define a unique orbit and a location on that orbit by means of the three constants $a$, $e$, and $\varpi$ and the variable $f$. Our subsequent analysis was concerned with understanding the motion in the orbital plane. However, motion in the solar system is not confined to a single plane and we now consider the three-dimensional representation of an orbit in space (see Fig. 2.13).

Although we have shown that motion is confined to a fixed orbital plane, we consider a three-dimensional Cartesian coordinate system with respect to which an arbitrary point has a position vector $\mathbf{r} = (x, y, z) = x\hat{\mathbf{x}} + y\hat{\mathbf{y}} + z\hat{\mathbf{z}}$. The $x$ axis is taken to lie along the major axis of the ellipse in the direction of pericentre, the $y$ axis is perpendicular to the $x$ axis and lies in the orbital plane, while the

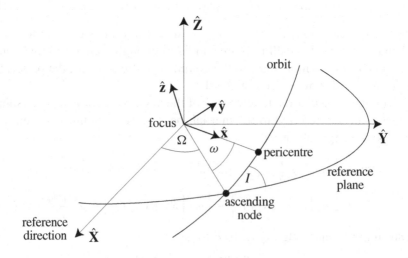

Fig. 2.13.   Orbital motion with respect to the reference plane in three-dimensional space.

$z$ axis is mutually perpendicular to the $x$ and $y$ axes such that all three form a right-handed triad.

We now wish to refer this orbital plane to a standard reference plane. The direction of the reference line in the reference plane forms the $X$ axis of our standard coordinate system. The $Y$ axis is in the reference plane at right angles to the $X$ axis, while the $Z$ axis is perpendicular to both the $X$ and $Y$ axes forming a right-handed triad. For example, when considering the motion of planets around the Sun, it is customary to use a Sun-centred, or *heliocentric coordinate system* where the reference plane is the plane of the Earth's orbit (the *ecliptic*) and the reference line is in the direction of the *vernal equinox*, along the line of intersection of the plane of the Earth's equator and the ecliptic. It should be pointed out that this particular reference frame varies with time because of perturbations by other bodies (see, for example, Standish et al. (1992) for a thorough discussion of coordinate systems and reference frames or Montenbruck (1989) for a set of coordinate transformations).

In general the orbital plane will be inclined to the reference plane at an angle $I$ called the *inclination* of the orbit. The line of intersection between the orbital plane and the standard reference frame is called the *line of nodes*. The point in both planes where the orbit crosses the reference plane moving from below to above the plane is called the *ascending node* while the angle between the reference line and the radius vector to the ascending node is called the *longitude of ascending node*, $\Omega$. The angle between this same radius vector and the pericentre of the orbit is called the *argument of pericentre*, $\omega$.

The inclination is always in the range $0 \le I \le 180°$. If $I < 90°$ the motion is said to be *prograde* whereas if $I \ge 90°$ the motion is *retrograde*. In the limit as $I \to 0$ the orbital plane coincides with the reference plane and we have

$$\varpi = \Omega + \omega, \tag{2.118}$$

where $\varpi$ is the longitude of pericentre, which was introduced in Sect. 2.3. However, the definition of $\varpi$ in (2.118) is also used in the inclined case, despite the fact that the angles $\Omega$ and $\omega$ lie in different planes. In general, therefore, $\varpi$ is a "dogleg" angle.

Figure 2.14 shows the relationship between the orbital plane coordinate system and the reference plane system. It is clear that coordinates in one system can be expressed in terms of the other by means of a series of three rotations about various axes.

To transform from the $(x, y, z)$ orbital plane system to the general $(X, Y, Z)$ reference system we have to carry out (i) a rotation about the $z$ axis through an angle $\omega$ so that the $x$ axis coincides with the line of nodes, (ii) a rotation about the $x$ axis through an angle $I$ so that the two planes are coincident, and finally (iii) a rotation about the $z$ axis through an angle $\Omega$ (see Fig. 2.13 and Fig. 2.14).

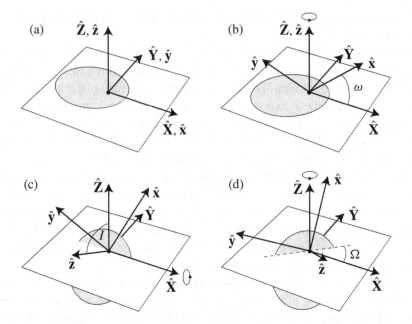

Fig. 2.14.   The relationship between the unit vectors $\hat{\mathbf{x}}$, $\hat{\mathbf{y}}$, $\hat{\mathbf{z}}$, $\hat{\mathbf{X}}$, $\hat{\mathbf{Y}}$, $\hat{\mathbf{Z}}$ and the angles $\omega$, $I$, and $\Omega$. (a) The transformation can be achieved through a series of three rotations applied to originally coincident axes. (b) The first rotation is through a positive angle $\omega$ about the $\hat{\mathbf{Z}}$ axis. (c) The second is through a positive angle $I$ about the $\hat{\mathbf{X}}$ axis. (d) The final rotation is through a positive angle $\Omega$ about the $\hat{\mathbf{Z}}$ axis.

We can represent these transformations by three $3 \times 3$ rotation matrices, denoted by $\mathbf{P}_1$, $\mathbf{P}_2$, and $\mathbf{P}_3$ respectively with elements

$$\mathbf{P}_1 = \begin{pmatrix} \cos\omega & -\sin\omega & 0 \\ \sin\omega & \cos\omega & 0 \\ 0 & 0 & 1 \end{pmatrix}, \qquad \mathbf{P}_2 = \begin{pmatrix} 1 & 0 & 0 \\ 0 & \cos I & -\sin I \\ 0 & \sin I & \cos I \end{pmatrix}, \qquad (2.119)$$

and

$$\mathbf{P}_3 = \begin{pmatrix} \cos\Omega & -\sin\Omega & 0 \\ \sin\Omega & \cos\Omega & 0 \\ 0 & 0 & 1 \end{pmatrix}. \qquad (2.120)$$

Consequently,

$$\begin{pmatrix} X \\ Y \\ Z \end{pmatrix} = \mathbf{P}_3\mathbf{P}_2\mathbf{P}_1 \begin{pmatrix} x \\ y \\ z \end{pmatrix} \quad \text{and} \quad \begin{pmatrix} x \\ y \\ z \end{pmatrix} = \mathbf{P}_1^{-1}\mathbf{P}_2^{-1}\mathbf{P}_3^{-1} \begin{pmatrix} X \\ Y \\ Z \end{pmatrix}, \qquad (2.121)$$

where $\mathbf{P}_1^{-1}$ is the inverse of the matrix $\mathbf{P}_1$ etc. Because all rotation matrices are orthogonal, the inverse of each matrix is simply equal to its transpose.

If we now restrict ourselves to coordinates that lie in the orbital plane, we have

$$\begin{pmatrix} X \\ Y \\ Z \end{pmatrix} = \mathbf{P}_3\mathbf{P}_2\mathbf{P}_1 \begin{pmatrix} r\cos f \\ r\sin f \\ 0 \end{pmatrix}$$

$$= r \begin{pmatrix} \cos\Omega\cos(\omega+f) - \sin\Omega\sin(\omega+f)\cos I \\ \sin\Omega\cos(\omega+f) + \cos\Omega\sin(\omega+f)\cos I \\ \sin(\omega+f)\sin I \end{pmatrix}. \qquad (2.122)$$

Note that the values of $a$ and $e$ are unchanged by considering the ellipse in this new coordinate system, since rotational transformations preserve lengths.

As an example of the use of these formulae we consider the problem of finding the positions of the planets at a given time, say September 25, 1993 at 5.32 PM British Summer Time. Appendix A gives formulae for the calculation of the orbital elements of the planets at any time referred to the mean ecliptic and equinox of the epoch of noon on 1st January 2000; this is called the *J2000 epoch*. The formulae give the corrections as a function of $T$, the interval in centuries between the given date, and the J2000 epoch date. In these calculations it is convenient to express each date as a *Julian date*; this is the number of days measured from noon on 1st January 4713 B.C. A Julian century is defined to have 36,525 days. The Julian date of the J2000 epoch is 2451545.0 and that of our given date is 2449256.189, so in our case $T = -0.06266423$.

If we consider the orbital elements for one planet, say Jupiter, the formulae in Appendix A give $a_j = 5.20332$ AU, $e_j = 0.0484007$, $I_j = 1°30537$, $\Omega_j = 100°535$, $\varpi_j = 14°7392$, and $\lambda_j = 204°234$, where the subscript j refers to the values for Jupiter. Hence $M_j = \lambda_j - \varpi_j = 189°495$. The numerical solution of Kepler's equation (see Sect. 2.4) gives $E_j = 189°059$ and hence, from Eq. (2.41), we have

$$x_j = -5.39027 \text{AU} \qquad \text{and} \qquad y_j = -0.818277 \text{AU}. \qquad (2.123)$$

Substituting the values of $I_j$, $\Omega_j$, and $\varpi_j$ in Eqs. (2.119) and (2.120) we obtain

$$\mathbf{P}_j = \mathbf{P}_3\mathbf{P}_2\mathbf{P}_1 = \begin{pmatrix} 0.966839 & -0.254401 & 0.0223971 \\ 0.254373 & 0.967097 & 0.00416519 \\ -0.0227198 & 0.00167014 & 0.99974 \end{pmatrix} \qquad (2.124)$$

for the transformation matrix, and hence the coordinates of Jupiter in the J2000 reference frame are

$$X_j = -5.00336, \qquad Y_j = -2.16249, \qquad Z_j = 0.121099. \qquad (2.125)$$

The procedure can be applied to find the positions of the other planets. The results are illustrated in Fig. 2.15.

We can now summarise an algorithm for transforming the position $(X, Y, Z)$ and velocity $(\dot{X}, \dot{Y}, \dot{Z})$ of an object in an elliptical orbit in the standard reference

(a)                                      (b)

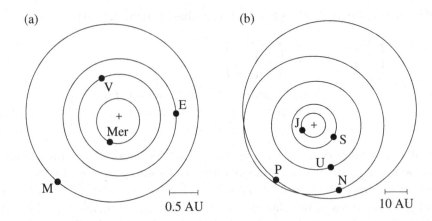

Fig. 2.15.   The positions and orbits of the planets on September 25, 1993 at 5.32 PM BST, projected onto the ecliptic of J2000. (a) The inner solar system showing the positions of the planets Mercury (Mer), Venus (V), Earth (E), and Mars (M). (b) The outer solar system showing the positions of the planets Jupiter (J), Saturn (S), Uranus (U), Neptune (N), and Pluto (P). The filled circles denote the positions of the planets and are not to scale. The Sun is denoted by the cross. The lines show the scale of each plot.

plane at a time $t$ to a set of six orbital elements, $a$, $e$, $I$, $\Omega$, $\omega$, and $f$, and a time of pericentre passage, $\tau$. We assume that the masses of the central and orbiting objects are $m_1$ and $m_2$ respectively. We have

$$R^2 = X^2 + Y^2 + Z^2, \tag{2.126}$$
$$V^2 = \dot{X}^2 + \dot{Y}^2 + \dot{Z}^2, \tag{2.127}$$
$$\mathbf{R} \cdot \dot{\mathbf{R}} = X\dot{X} + Y\dot{Y} + Z\dot{Z}, \tag{2.128}$$
$$\mathbf{h} = (Y\dot{Z} - Z\dot{Y},\ Z\dot{X} - X\dot{Z},\ X\dot{Y} - Y\dot{X}), \tag{2.129}$$
$$\dot{R} = \pm\sqrt{V^2 - \frac{h^2}{R^2}}, \tag{2.130}$$

where $R = r$ now denotes the length of the radius vector and $\dot{R}$ is its rate of change. The sign of $\dot{R}$ is taken to be the sign of $\mathbf{R} \cdot \dot{\mathbf{R}}$, since $R$ is always positive. Taking the projections of $\mathbf{h} = (h_X, h_Y, h_Z)$ onto the three planes we have

$$h \cos I = h_Z, \tag{2.131}$$
$$h \sin I \sin \Omega = \pm h_X, \tag{2.132}$$
$$h \sin I \cos \Omega = \mp h_Y, \tag{2.133}$$

where the upper sign in Eqs. (2.132) and (2.133) is taken if $h_Z > 0$ and the lower sign is taken if $h_Z < 0$.

The procedure is as follows:

1) Calculate $a$ using Eqs. (2.34), (2.126), and (2.127):

$$a = \left( \frac{2}{R} - \frac{V^2}{\mathcal{G}(m_1 + m_2)} \right)^{-1}. \tag{2.134}$$

2) Calculate $e$ using Eqs. (2.26) and (2.134):

$$e = \sqrt{1 - \frac{h^2}{\mathcal{G}(m_1 + m_2)a}}. \tag{2.135}$$

3) Calculate $I$ using Eq. (2.131):

$$I = \cos^{-1} \left( \frac{h_Z}{h} \right). \tag{2.136}$$

4) Calculate $\Omega$ using the expressions for $\sin \Omega$ and $\cos \Omega$ given in Eqs. (2.132) and (2.133):

$$\sin \Omega = \frac{\pm h_X}{h \sin I} \quad \text{and} \quad \cos \Omega = \frac{\mp h_Y}{h \sin I}. \tag{2.137}$$

The choice of sign is determined by the sign of $h_Z$ (see above).

5) Calculate $\omega + f$ from the expressions for $Z/R$ and $X/R$ in Eq. (2.122), recalling that $r = R$:

$$\sin(\omega + f) = \frac{Z}{R \sin I},$$

$$\cos(\omega + f) = \sec \Omega \left( \frac{X}{R} + \sin \Omega \sin(\omega + f) \cos I \right). \tag{2.138}$$

6) Calculate $f$ and hence $\omega$ from the expressions for $\sin f$ and $\cos f$ derived from Eqs. (2.20) and (2.31), recalling that $\dot{r} = \dot{R}$:

$$\sin f = \frac{a(1 - e^2)}{he} \dot{R} \quad \text{and} \quad \cos f = \frac{1}{e} \left( \frac{a(1 - e^2)}{R} - 1 \right). \tag{2.139}$$

7) Calculate $\tau$ by first calculating $E$ from Eq. (2.42) and then using Eqs. (2.26) and (2.51):

$$\tau = t - \frac{E - e \sin E}{\sqrt{\mathcal{G}(m_1 + m_2)a^{-3}}}. \tag{2.140}$$

Although these equations define a correct procedure for deriving orbital elements, for numerical purposes it is desirable to eliminate $\mathcal{G}(m_1 + m_2)$ from the equations and choose a more practical system of units rather than the SI or any other standard system. This can be achieved by scaling the independent variable, $t$, by a factor $\sqrt{\mu} = \sqrt{\mathcal{G}(m_1 + m_2)}$ and using a new time variable, $\bar{t}$, such that

$$\sqrt{\mu} \, dt = d\bar{t}. \tag{2.141}$$

From Eq. (2.5) it can be seen that this has the same effect as setting $\mu = 1$. If, in addition, the unit of length is chosen so that $a = 1$ then we are dealing with a two-body system in which the orbit has unit mean motion and an orbital period of $2\pi$ time units. This is a common system of units to adopt when dealing with the circular restricted three-body problem (see Chapter 3).

## 2.9 Perturbed Orbits

We have seen in Sect. 2.8 that in the two-body problem the orbital elements $a$, $e$, $I$, $\omega$, $\Omega$, and $\tau$ are constants that are uniquely determined from the position and velocity of the orbiting mass. Even if there is a perturbing force acting on the system, any instantaneous set of position and velocity vectors always defines a set of six orbital elements that would give the shape and orientation of the orbit that the mass would follow if the perturbing force were to disappear at that instant. These are called *osculating elements*, from the Latin verb *osculare*, to kiss. Although we shall deal with perturbations to orbits in future sections, at this stage it is useful to examine some of the basic effects that perturbing forces exert on orbits.

Burns (1976) showed how the equations for the time derivatives of $a$, $e$, $I$, $\omega$, $\Omega$, and $\tau$ could be derived in a straightforward manner using elementary dynamics. Following his approach we consider a small disturbing force

$$\mathbf{dF} = \bar{R}\hat{\mathbf{r}} + \bar{T}\hat{\boldsymbol{\theta}} + \bar{N}\hat{\mathbf{z}}, \tag{2.142}$$

where $\bar{R}$, $\bar{T}$, and $\bar{N}$ are the magnitudes of the radial, tangential, and normal components of the force respectively and $\hat{\mathbf{r}}, \hat{\boldsymbol{\theta}}$, and $\hat{\mathbf{z}}$ are the standard unit vectors introduced in Sect. 2.2. In the remainder of this section we derive expressions for $\dot{a}$, $\dot{e}$, $\dot{I}$, $\dot{\omega}$, $\dot{\Omega}$, and $\dot{\tau}$ as functions of these components and show which parts of the force give rise to changes in particular orbital elements.

We can equate the time derivative of the energy constant, $C$, with the work done on the orbiting body per unit mass per unit time. Hence

$$\dot{C} = \dot{\mathbf{r}} \cdot \mathbf{dF} = \dot{r}\bar{R} + r\dot{\theta}\bar{T} \tag{2.143}$$

and, from Eq. (2.37),

$$\dot{C} = \frac{\mu}{2a^2}\dot{a}. \tag{2.144}$$

Hence, from the definitions of $\dot{r}$ and $r\dot{\theta}(= r\dot{f})$ in Eqs. (2.31) and (2.32),

$$\frac{da}{dt} = 2\frac{a^{3/2}}{\sqrt{\mu(1 - e^2)}} \left[ \bar{R}e \sin f + \bar{T}(1 + e \cos f) \right]. \tag{2.145}$$

This implies that only forces in the plane of the orbit can change its semi-major axis.

Using Eqs. (2.135) and (2.37) we can write

$$e = \sqrt{1 + 2h^2 C \mu^{-2}} \tag{2.146}$$

and hence

$$\frac{de}{dt} = \frac{e^2 - 1}{2e}(2\dot{h}/h + \dot{C}/C). \tag{2.147}$$

Since the rate of change of angular momentum is equal to the applied moment, we have

$$\frac{d\mathbf{h}}{dt} = \mathbf{r} \times d\mathbf{F} = r\bar{T}\hat{\mathbf{z}} - r\bar{N}\hat{\boldsymbol{\theta}}. \tag{2.148}$$

This implies that

$$\frac{dh}{dt} = r\bar{T} \tag{2.149}$$

since the $-r\bar{N}\hat{\boldsymbol{\theta}}$ component changes the direction of $\mathbf{h}$ but does not affect its magnitude. From Eq. (2.42) and the formulae for $C, h, \dot{C}, \dot{a}$, and $\dot{h}$ in Eqs. (2.37), (2.26), (2.144), (2.145), and (2.149) we have

$$\frac{de}{dt} = \sqrt{a\mu^{-1}(1-e^2)} \left[ \bar{R}\sin f + \bar{T}(\cos f + \cos E) \right]. \tag{2.150}$$

This implies that the eccentricity can only be changed by the application of forces in the orbital plane.

Differentiating Eq. (2.131) gives

$$\frac{dI}{dt} = \frac{\dot{h}/h - \dot{h}_Z/h_Z}{\sqrt{(h/h_Z)^2 - 1}}. \tag{2.151}$$

We can express the $X, Y$, and $Z$ components of $\dot{\mathbf{h}}$ using

$$\begin{pmatrix} \dot{h}_X \\ \dot{h}_Y \\ \dot{h}_Z \end{pmatrix} = \mathbf{P}_3\mathbf{P}_2 \begin{pmatrix} \cos(\omega + f) & -\sin(\omega + f) & 0 \\ \sin(\omega + f) & \cos(\omega + f) & 0 \\ 0 & 0 & 1 \end{pmatrix} \begin{pmatrix} 0 \\ -r\bar{N} \\ r\bar{T} \end{pmatrix}, \tag{2.152}$$

where the matrices $\mathbf{P}_2$ and $\mathbf{P}_3$ are given in Eqs. (2.119) and (2.120). This gives

$$\begin{aligned} \dot{h}_X = r(\bar{T}\sin I \sin \Omega &+ \bar{N}\sin(\omega + f)\cos \Omega \\ &+ \bar{N}\cos(\omega + f)\cos I \sin \Omega), \end{aligned} \tag{2.153}$$

$$\begin{aligned} \dot{h}_Y = r(-\bar{T}\sin I \cos \Omega &+ \bar{N}\sin(\omega + f)\sin \Omega \\ &- \bar{N}\cos(\omega + f)\cos I \cos \Omega), \end{aligned} \tag{2.154}$$

$$\dot{h}_Z = r(\bar{T}\cos I - \bar{N}\cos(\omega + f)\sin I). \tag{2.155}$$

Substituting Eqs. (2.26), (2.131), (2.149), and (2.155) in Eq. (2.151) gives

$$\frac{dI}{dt} = \frac{\sqrt{a\mu^{-1}(1-e^2)}\bar{N}\cos(\omega + f)}{1 + e\cos f}, \tag{2.156}$$

which can also be written as

$$\frac{dI}{dt} = \frac{r\bar{N}\cos(\omega + f)}{h}. \tag{2.157}$$

Since $dI/dt$ only depends on $\bar{N}$, only forces normal to the orbital plane can change the inclination. Note that $r\bar{N}\cos(\omega + f)$ is the component of the torque that rotates the angular momentum vector about the line of nodes.

Dividing Eq. (2.132) by Eq. (2.133) gives

$$\tan\Omega = -h_X/h_Y, \tag{2.158}$$

which can then be differentiated with respect to time to give

$$\frac{d\Omega}{dt} = \frac{h_X\dot{h}_Y - h_Y\dot{h}_X}{h^2 - h_Z^2} \tag{2.159}$$

or

$$\frac{d\Omega}{dt} = \frac{\sin\Omega\dot{h}_Y + \cos\Omega\dot{h}_X}{h\sin I}. \tag{2.160}$$

Substituting Eqs. (2.26), (2.20), (2.153), and (2.154) into Eq. (2.160) gives

$$\frac{d\Omega}{dt} = \sqrt{a\mu^{-1}(1 - e^2)}\frac{\bar{N}\sin(\omega + f)}{\sin I(1 + e\cos f)} \tag{2.161}$$

or

$$\frac{d\Omega}{dt} = \frac{r\bar{N}\sin(\omega + f)}{h\sin I}. \tag{2.162}$$

In this equation $r\bar{N}\sin(\omega + f)$ is the moment acting to precess the plane of the orbit and $h\sin I$ is the component of the angular momentum vector that lies normal to the line of nodes in the $X$–$Y$ plane.

In order to derive an expression for $\dot{\omega}$ it is necessary to return to the equation of the ellipse, Eq. (2.20), and make use of our expressions for $e$ and $h$ given in Eqs. (2.146) and (2.26). This gives

$$h^2 = \mu r\left[1 + \sqrt{1 + 2Ch^2\mu^{-2}}\cos(\theta - \omega)\right], \tag{2.163}$$

where $\theta = \omega + f$ and we are choosing $\theta$ to be the position angle measured from the line of nodes. If we are interested in the change in the orbital elements due to the instantaneous application of a perturbing force $d\mathbf{F}$ then $C$, $\mathbf{h}$, and $\omega$ change but $r$ remains fixed. Differentiation of Eq. (2.163) gives

$$\frac{d\omega}{dt} = 2h\dot{h}\frac{r^{-1} + C(e\mu)^{-1}\cos(\theta - \omega)}{e\mu\sin(\theta - \omega)}$$

$$+ \dot{\theta} - \frac{h^2}{e^2\mu^2}\dot{C}\cot(\theta - \omega). \tag{2.164}$$

Substituting Eqs. (2.144) and (2.149) in Eq. (2.164) gives

$$\frac{d\omega}{dt} = e^{-1}\sqrt{a\mu^{-1}(1-e^2)}\left[-\bar{R}\cos f + \bar{T}\sin f\frac{2+e\cos f}{1+e\cos f}\right]$$
$$-\dot{\Omega}\cos I. \tag{2.165}$$

The last term arises from the $\dot{\theta}$ term in Eq. (2.164) using the fact that the instantaneous change in $\theta$ is due to the change in the longitude of ascending node, since $\theta$ is referred to the nodal position (Burns 1976).

The equation for $\dot{\tau}$ is derived from the differentiation of Kepler's equation, Eq. (2.51). Setting $\chi = n\tau$ we have

$$\frac{d\chi}{dt} = \frac{\dot{C}}{C}\left(-\frac{3}{2}nt + \frac{(1-e^2)^{3/2}(2e - \cos f - e\cos^2 f)}{2e^2\sin f(1+e\cos f)}\right)$$
$$-\frac{\dot{h}}{h}\frac{(1-e^2)^{3/2}}{e^2}\cot f. \tag{2.166}$$

Writing $\dot{\chi} = -n\dot{\tau} - \dot{n}\tau$ we have

$$\frac{d\tau}{dt} = \left[3(\tau - t)\frac{\sqrt{a}}{\sqrt{\mu(1-e^2)}}e\sin f\right.$$
$$\left.+a^2\mu^{-1}(1-e^2)\left(\frac{-\cos f}{e} + \frac{2}{1+e\cos f}\right)\right]\bar{R}$$
$$+\left[3(\tau - t)\frac{\sqrt{a}}{\sqrt{\mu(1-e^2)}}(1+e\cos f)\right.$$
$$\left.+a^2\mu^{-1}(1-e^2)\left(\frac{\sin f(2+e\cos f)}{e(1+e\cos f)}\right)\right]\bar{T}. \tag{2.167}$$

Again only forces in the orbital plane can change the time of pericentre passage.

Note that the time $t$ occurs on the right-hand side of Eqs. (2.166) and (2.167). This causes practical difficulties in the use of these derivatives since they grow with time. This problem also occurs with other forms of the perturbation equations and it will be discussed further in Sect. 6.8.

## 2.10 Hamiltonian Formulation

The approach taken in formulating and solving the equations of motion of the two-body problem and its perturbed motion is not unique and alternative formulations are not only possible but in some circumstances preferable. For most of the applications discussed in this book the classical approach is adequate, but there are some topics, notably the derivation of Lagrange's equations (Sect. 6.7), the discussion of the dynamics of resonance (Sect. 8.8), resonance encounters (Sect. 8.12), and algebraic mappings (Sect. 9.5), where a different mathematical approach is required. This is why we demonstrate the Hamiltonian formulation of the two-body problem in this section.

In Sect. 2.2 we formulated the equations of motion of the two-body problem (two objects of mass $m_1$ and $m_2$ moving under their mutual gravitational attraction) in terms of the Cartesian position $(x, y)$ and velocity $(\dot{x}, \dot{y})$ of $m_2$ with respect to $m_1$, and we derived the differential equation

$$\frac{d^2\mathbf{r}}{dt^2} + \mu\frac{\mathbf{r}}{r^3} = 0, \tag{2.168}$$

the solution of which is a conic section. Early on in our analysis we chose the reference plane to be the orbit plane. However, this vector equation is equally applicable to motion in three dimensions, with $\mathbf{r} = (x, y, z)$ and $\dot{\mathbf{r}} = (\dot{x}, \dot{y}, \dot{z})$. In what follows we use the variables $\mathbf{r}$ and $\mathbf{p}$ where, using a slightly different notation,

$$\mathbf{r} = r_x\mathbf{i} + r_y\mathbf{j} + r_z\mathbf{k} \qquad \text{and} \qquad \mathbf{p} = p_x\mathbf{i} + p_y\mathbf{j} + p_z\mathbf{k}. \tag{2.169}$$

Here $\mathbf{r}$ is the relative position vector, $\mathbf{p} = [m_1m_2/(m_1 + m_2)]\mathbf{v}$ is the linear momentum of the system, and, as usual, $\mathbf{v} = \dot{\mathbf{r}}$ is the velocity.

Now we can write the vector equations of motion in the form

$$\dot{\mathbf{r}} = +\nabla_\mathbf{p}\mathcal{H}_{\text{Kepler}} \qquad \text{and} \qquad \dot{\mathbf{p}} = -\nabla_\mathbf{r}\mathcal{H}_{\text{Kepler}} \tag{2.170}$$

where $\nabla_\mathbf{p}$ and $\nabla_\mathbf{r}$ are the vector differential operators given by

$$\nabla_\mathbf{p} = \mathbf{i}\frac{\partial}{\partial p_x} + \mathbf{j}\frac{\partial}{\partial p_y} + \mathbf{k}\frac{\partial}{\partial p_z} \quad \text{and} \quad \nabla_\mathbf{r} = \mathbf{i}\frac{\partial}{\partial r_x} + \mathbf{j}\frac{\partial}{\partial r_y} + \mathbf{k}\frac{\partial}{\partial r_z} \tag{2.171}$$

and

$$\mathcal{H}_{\text{Kepler}} = \frac{p^2}{2\mu^*} - \frac{\mu\mu^*}{r}. \tag{2.172}$$

Here, as previously, $\mu = \mathcal{G}(m_1 + m_2)$. Also, $p = |\mathbf{p}|$ and $r = |\mathbf{r}|$. The new quantity

$$\mu^* = \frac{m_1m_2}{m_1 + m_2} \tag{2.173}$$

is called the *reduced mass* of the system. The quantity $\mathcal{H}_{\text{Kepler}}$ is referred to as the *Hamiltonian* of the Kepler (i.e., two-body) problem. We have now replaced the three, coupled, second-order differential equations by an analogous system of six, coupled, first-order differential equations given by

$$\dot{\mathbf{r}} = \frac{\mathbf{p}}{\mu^*} \qquad \text{and} \qquad \dot{\mathbf{p}} = -\frac{\mu\mu^*}{r^3}\mathbf{r}. \tag{2.174}$$

If we compare Eq. (2.172) with Eq. (2.28) it is clear that $\mathcal{H}_{\text{Kepler}} = \mu^*C$. In our new formulation $\mathcal{H}_{\text{Kepler}}$ is the sum of the kinetic and potential energy of the system and so equals the total energy, a constant of the system.

In general, any system of equations that can be written in the form

$$\frac{dq_i}{dt} = +\frac{\partial \mathcal{H}}{\partial p_i}, \qquad \frac{dp_i}{dt} = -\frac{\partial \mathcal{H}}{\partial q_i} \qquad (i = 1, 2, \ldots, n), \tag{2.175}$$

where $\mathcal{H} = \mathcal{H}(q_i, p_i, t)$, is said to be a *Hamiltonian* system of *order 2n*, or, equivalently, of *n degrees of freedom*. The function $\mathcal{H}$ is called the *Hamiltonian* of the system. The variables $q_i$ $(i = 1, 2, \ldots, n)$ are called the *coordinates* while the $p_i$ $(i = 1, 2, \ldots, n)$ are called the *momenta*, which are said to be *conjugate* to the $q_i$. Although in our case above the $q_i$ are the coordinates of the body and the $p_i$ are the components of the momentum, there is no requirement that the description of the variables has to conform to their role. Note that $\mathcal{H}$ is determined only up to an arbitrary additive constant, since $\mathcal{H} + k$, where $k$ is a constant, also satisfies Eq. (2.175).

At first glance it may seem perverse to express the equations of motion in this form. However, the properties of such systems make coordinate transformations easier to carry out and this will be particularly useful when we investigate resonant phenomena. In addition, such transformations may simplify the problem and lead to a solution. It is beyond the scope of this book to provide a rigorous introduction to the theory of Hamiltonian systems; indeed only a few results are required for the subjects we cover. Elementary treatments are given by Brouwer & Clemence (1961) and Roy (1988).

We have already demonstrated that the variables $\mathbf{r}$ and $\mathbf{p}$ form a conjugate set. Unfortunately, it is easy to show that the same is not true of the orbital elements $a$, $e$, $I$, $\Omega$, $\omega$, and $f$ that can be derived from them. However, certain functions of the orbital elements can form conjugate sets. Here we mention two such sets: the *Delaunay variables* and the *Poincaré variables*.

The Delaunay variables are defined by

$$l = M, \qquad g = \omega, \qquad h = \Omega,$$

$$L = \mu^* \sqrt{\mu a}, \quad G = \mu^* \sqrt{\mu a (1 - e^2)}, \quad H = \mu^* \sqrt{\mu a (1 - e^2)} \cos I, \tag{2.176}$$

where $l$, $g$, and $h$ are the coordinates and $L$, $G$, and $H$ are the corresponding conjugate momenta. The Hamiltonian for the two-body problem expressed in terms of the Delaunay variables is

$$\mathcal{H} = -\frac{\mu^2 \mu^{*3}}{2L^2}. \tag{2.177}$$

Since $\mathcal{H}$ is only a function of $L$ it implies that $g$ and $h$ are constants (clearly the case from Sect. 2.8) and $L$, $G$, and $H$ are also constants (also clear since $L = L(a)$, $G$ is the angular momentum, and $H$ is the vertical component of the angular momentum vector).

The variation of $l$ is given by

$$\frac{dl}{dt} = +\frac{\partial \mathcal{H}}{\partial L} = \frac{\mu^2 \mu^{*3}}{L^3} = \sqrt{\frac{\mu}{a^3}}, \tag{2.178}$$

which is to be expected since $\mathrm{d}l/\mathrm{d}t = \mathrm{d}n(t - \tau)/\mathrm{d}t = n$.

The Poincaré variables are defined by

$$\lambda = M + \omega + \Omega, \qquad \Lambda = \mu^*\sqrt{\mu a},$$

$$\gamma = -\omega - \Omega, \qquad \Gamma = \mu^*\sqrt{\mu a}(1 - \sqrt{1 - e^2}),$$

$$z = -\Omega, \qquad Z = \mu^*\sqrt{\mu a(1 - e^2)}(1 - \cos I), \qquad (2.179)$$

where $\lambda$, $\gamma$, and $z$ are the coordinates and $\Lambda$, $\Gamma$, and $Z$ are the corresponding conjugate momenta. The angle $\lambda = M + \varpi$ is the mean longitude. The Poincaré variables can be derived from the Delaunay variables by the transformation

$$\Lambda = L, \qquad \Gamma = L - G, \qquad Z = G - H,$$

$$\lambda = l + g + h, \qquad \gamma = -g - h, \qquad z = -h. \qquad (2.180)$$

Note that the following relationship between the two sets of variables holds:

$$\Lambda\lambda + \Gamma\gamma + Zz = Ll + Gg + Hh. \qquad (2.181)$$

This is an example of what is called a *contact transformation* (see, for example, Brouwer & Clemence 1961) and such a transformation always preserves the canonical nature of the equations without change in the Hamiltonian. The Hamiltonian for the two-body problem expressed in terms of the Poincaré variables is

$$\mathcal{H} = -\frac{\mu^2 \mu^{*3}}{2\Lambda^2}. \qquad (2.182)$$

There is no requirement for a new set of variables to consist of coordinates and momenta: Mixed systems are also possible. Consider the variables

$$\xi = \sqrt{2\Gamma}\cos\gamma, \quad \eta = \sqrt{2\Gamma}\sin\gamma, \quad p = \sqrt{2Z}\cos z, \quad q = \sqrt{2Z}\sin z. \quad (2.183)$$

This is another example of a contact transformation. In this case the variables $\xi$ and $\eta$ are called the *eccentric variables* whereas $p$ and $q$ are called the *oblique variables*. When $e$ and $I$ are small these are the variables $h$, $k$, $p$, and $q$ used in the discussion of secular perturbations in Chapter 7.

### Exercise Questions

**2.1**    Consider a two-body problem in which the force of attraction is proportional to the radial separation of the two bodies (rather than the square of its inverse). Show that the motion of one mass with respect to the other is a centred ellipse.

**2.2**    Use the semi-major axes listed in Table A.2 to determine the average interval between times of orbital conjunction between the Earth and Mars. Show that for fixed orbits the *minimum* distance between Earth and Mars varies by almost a factor of two. What is the approximate interval between successive "very close" oppositions? Use the values of $a_0$, $e_0$, $\varpi_0$, and $\lambda_0$ given in Table A.2 (but not their variation given in Table A.3) and numerical solutions of Kepler's equation to determine the orbital motions of Earth and Mars over the period 1985–2002. Neglecting the relative orbital inclinations of the Earth and Mars, show that the closest opposition during this period occurred in September 1988 and the farthest in February 1995, and determine the minimum distances at these times.

**2.3** A test particle approaches a planet of mass $M$ and radius $R$ from infinity with speed $v_\infty$ and an impact parameter $p$. Use the particle's energy and angular momentum with respect to the planet to derive expressions for the semi-major axis and eccentricity of the hyperbolic orbit followed by the test particle about $M$, and for the pericentre distance $r_0$. Show that the eccentricity may be written $e = 1 + 2v_\infty^2/v_0^2$, where $v_0$ is the escape velocity at $r_0$. Use the expression for the true anomaly corresponding to the asymptote of the hyperbola ($r \to \infty$) to show that the overall deflection of the test particle's orbit after it leaves the vicinity of the planet, $\psi$, is given by $\sin(\psi/2) = e^{-1}$. Given that $r_0$ must be greater than $R$ to avoid a physical collision, calculate the maximum deflection angles for (i) a spacecraft skimming Jupiter, with $v_\infty = 10$ km s$^{-1}$, and (ii) the *Cassini* orbiter skimming Saturn's large moon Titan, at $v_\infty = 5$ km s$^{-1}$.

**2.4** Consider the relative motion form of the two-body problem. Explain why the vector $\mathbf{e} = -(\mathbf{h} \times \mathbf{v})/(\mathcal{G}(m_1 + m_2)) - \hat{\mathbf{r}}$ lies in the orbital plane, where $\mathbf{h}$ is the angular momentum per unit mass, $\mathbf{v}$ is the velocity, and $\hat{\mathbf{r}}$ is the unit radial vector. Show that $\mathbf{e}$ is a vector constant of the motion. By expressing the dot product $\mathbf{e} \cdot \mathbf{r}$ in two different forms, where $\mathbf{r}$ is the position vector, show that the orbit is an ellipse of eccentricity $e = |\mathbf{e}|$ and longitude of pericentre $\varpi$ defined as the angle that $\mathbf{e}$ makes with the reference $\hat{\mathbf{x}}$ direction.

**2.5** In July 1994 the *Galileo* spacecraft, on its way to Jupiter, observed the fragments of Comet Shoemaker-Levy 9 impacting the planet. The first impact occurred on 16 July 1994 when the spacecraft was moving along a near-zero inclination heliocentric orbit with orbital elements $a = 3.137$ AU, $e = 0.690$, and $\varpi = 82.2°$, where $\varpi$ is the longitude of perihelion. The mean anomaly of the spacecraft was 45.7° at an epoch 322 days prior to the first impact. Find the spacecraft's eccentric anomaly, true anomaly, radial distance from the Sun, and true longitude (in the standard reference system) on 16 July 1994. At the time of the first impact Jupiter had a true longitude of 225.4 degrees and a heliocentric distance of 5.417 AU. Assuming that Jupiter's orbit has zero inclination, show that in terms of longitude coverage on the surface of the planet, approximately 28% of the unilluminated side was observable by the *Galileo* spacecraft.

**2.6**      There have been suggestions that Comet P/Swift-Tuttle could have
a close approach to the Earth in AD 2126.  The Earth was at the perihelion
of its orbit on 4 January 1993 at 3h UT. Assuming that the inclination of the
Earth's orbit is zero and taking its semi-major axis, eccentricity, and longitude
of perihelion to have the fixed values 1 AU, 0.0167, and 102.996° respectively,
calculate its position vector 48,799.375 days later at noon on 14 August 2126.
Observations suggest that Comet P/Swift-Tuttle was at the perihelion of its orbit
on 12 December 1992 at 21h 23m UT with the following orbital elements:
$a = 26.35441$ AU, $e = 0.96362$, $I = 113.408°$, $\omega = 152.979°$, $\Omega = 139.430°$,
where $a$ is the semi-major axis, $e$ is the eccentricity, $I$ is the inclination, $\omega$ is the
argument of perihelion, and $\Omega$ is the longitude of ascending node. Calculate the
position vector of the comet 48,821.609 days later at noon on 14 August 2126.
Use your answer from the first part of the question to calculate the separation of
the comet and the Earth (in AU) at this date in the future. What are the major
sources of error in determining this separation?

# 3

# The Restricted Three-Body Problem

Two's company, three's a crowd.

Proverb

## 3.1 Introduction

In Chapter 2 we showed how the problem of the motion of two masses moving under their mutual gravitational attraction can be solved analytically and that the resulting motion is always confined to fixed geometrical paths that are closed in inertial space. We will now extend our analysis to consider the gravitational interaction of three bodies, paying particular attention to the problem in which the third body has negligible mass compared with the other two.

The simplicity and elusiveness of the three-body problem in its various forms have attracted the attention of mathematicians for centuries. Among the giants of mathematics who have tackled the problem and made important contributions are Euler, Lagrange, Laplace, Jacobi, Le Verrier, Hamilton, Poincaré, and Birkhoff. The books by Szebehely (1967) and Marchal (1990) provide authoritative coverage of the literature on the subject as well as derivations of the important results. Today the three-body problem is as enigmatic as ever and although much has been discovered already, the recent developments in nonlinear dynamics and the spur of new observations in the solar system have meant a resurgence of interest in the problem and the derivation of new results.

If two of the bodies in the problem move in circular, coplanar orbits about their common centre of mass and the mass of the third body is too small to affect the motion of the other two bodies, the problem of the motion of the third body is called the *circular, restricted, three-body problem*. At first glance this problem may seem to have little application to motion in the solar system. After all, the observed orbits of solar system objects are noncoplanar and noncircular. However, the hierarchy of orbits and masses in the solar system (e.g., Sun, planet, satellite, ring particle) means that the restricted three-body problem provides a good approximation for certain systems and that the qualitative behaviour of the motion can be understood by relatively simple analysis. In this context we will study systems that have applications to typical problems in the solar system, thereby excluding the more exotic variations of the three-body problem, such as the Copenhagen problem (where the two large masses are equal), the Pythagorean problem (where the three bodies have initial masses in the ratio 3:4:5 and initial positions on a 3,4,5 right-angled triangle), and triple collisions (where the three bodies can "pass through" one another).

In this chapter we describe the equations of motion of the three-body problem and discuss the location and stability of equilibrium points with particular reference to a constant of the motion, the Jacobi integral, in the circular restricted case. We demonstrate the relationship between curves defined by the Jacobi integral and the orbital path of the particle. We derive Hill's equations to study the motion of the particle in the vicinity of one of the masses and show the general properties of such a system. Finally, we discuss the effects of drag forces in the three-body problem.

## 3.2 Equations of Motion

We consider the motion of a small particle of negligible mass moving under the gravitational influence of two masses $m_1$ and $m_2$. We assume that the two masses have circular orbits about their common centre of mass and that they exert a force on the particle although the particle cannot affect the two masses.

Consider a set of axes $\xi$, $\eta$, $\zeta$ in the inertial frame referred to the centre of mass of the system (see Fig. 3.1). Let the $\xi$ axis lie along the line from $m_1$ to $m_2$ at time $t = 0$ with the $\eta$ axis perpendicular to it and in the orbital plane of the two masses and the $\zeta$ axis perpendicular to the $\xi$–$\eta$ plane, along the angular momentum vector. Let the coordinates of the two masses in this reference frame be $(\xi_1, \eta_1, \zeta_1)$ and $(\xi_2, \eta_2, \zeta_2)$. The two masses have a constant separation and the same angular velocity about each other and their common centre of mass. Let the unit of mass be chosen such that $\mu = \mathcal{G}(m_1 + m_2) = 1$. If we now assume that $m_1 > m_2$ and define

$$\bar{\mu} = \frac{m_2}{m_1 + m_2} \tag{3.1}$$

then in this system of units the two masses are

$$\mu_1 = \mathcal{G}m_1 = 1 - \bar{\mu} \quad \text{and} \quad \mu_2 = \mathcal{G}m_2 = \bar{\mu}, \tag{3.2}$$

where $\bar{\mu} < 1/2$. The unit of length is chosen such that the constant separation of the two masses is unity. It then follows that the common mean motion, $n$, of the two masses is also unity.

Let the coordinates of the particle in the *inertial, or sidereal, system*, be $(\xi, \eta, \zeta)$. By applying the vector form of the inverse square law, the equations of

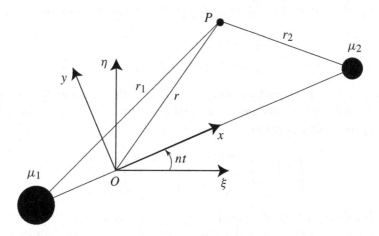

Fig. 3.1. A planar view of the relationship between the sidereal coordinates $(\xi, \eta, \zeta)$ and the synodic coordinates $(x, y, z)$ of the particle at the point $P$. The origin $O$ is located at the centre of mass of the two bodies. The $\zeta$ and $z$ axes coincide with the axis of rotation and the arrow indicates the direction of positive rotation.

motion of the particle are

$$\ddot{\xi} = \mu_1 \frac{\xi_1 - \xi}{r_1^3} + \mu_2 \frac{\xi_2 - \xi}{r_2^3}, \tag{3.3}$$

$$\ddot{\eta} = \mu_1 \frac{\eta_1 - \eta}{r_1^3} + \mu_2 \frac{\eta_2 - \eta}{r_2^3}, \tag{3.4}$$

$$\ddot{\zeta} = \mu_1 \frac{\zeta_1 - \zeta}{r_1^3} + \mu_2 \frac{\zeta_2 - \zeta}{r_2^3}, \tag{3.5}$$

where, from Fig. 3.1,

$$r_1^2 = (\xi_1 - \xi)^2 + (\eta_1 - \eta)^2 + (\zeta_1 - \zeta)^2, \tag{3.6}$$
$$r_2^2 = (\xi_2 - \xi)^2 + (\eta_2 - \eta)^2 + (\zeta_1 - \zeta)^2. \tag{3.7}$$

Note that these equations are also valid in the general three-body problem since they do not require any assumptions about the paths of the two masses.

If the two masses are moving in circular orbits then the distance between them is fixed and they move about their common centre of mass at a fixed angular velocity, the mean motion $n$. In these circumstances it is natural to consider the motion of the particle in a rotating reference frame in which the locations of the two masses are also fixed. Consider a new, rotating coordinate system that has the same origin as the $\xi$, $\eta$ system but which is rotating at a uniform rate $n$ in the positive direction (see Fig. 3.1). The direction of the $x$ axis is chosen such that the two masses always lie along it with coordinates $(x_1, y_1, z_1) = (-\mu_2, 0, 0)$ and

$(x_2, y_2, z_2) = (\mu_1, 0, 0)$. Hence, from Eq. (3.2) and Fig. 3.1 we have

$$r_1^2 = (x + \mu_2)^2 + y^2 + z^2, \tag{3.8}$$
$$r_2^2 = (x - \mu_1)^2 + y^2 + z^2, \tag{3.9}$$

where $(x, y, z)$ are the coordinates of the particle with respect to the rotating, or *synodic, system*. These coordinates are related to the coordinates in the sidereal system by the simple rotation

$$\begin{pmatrix} \xi \\ \eta \\ \zeta \end{pmatrix} = \begin{pmatrix} \cos nt & -\sin nt & 0 \\ \sin nt & \cos nt & 0 \\ 0 & 0 & 1 \end{pmatrix} \begin{pmatrix} x \\ y \\ z \end{pmatrix}. \tag{3.10}$$

Although in our system of units $n = 1$, we will retain $n$ in the equations to emphasise that all the terms in the equations of motion are accelerations.

If we now differentiate each component in Eq. (3.10) twice we obtain

$$\begin{pmatrix} \dot{\xi} \\ \dot{\eta} \\ \dot{\zeta} \end{pmatrix} = \begin{pmatrix} \cos nt & -\sin nt & 0 \\ \sin nt & \cos nt & 0 \\ 0 & 0 & 1 \end{pmatrix} \begin{pmatrix} \dot{x} - ny \\ \dot{y} + nx \\ \dot{z} \end{pmatrix} \tag{3.11}$$

and

$$\begin{pmatrix} \ddot{\xi} \\ \ddot{\eta} \\ \ddot{\zeta} \end{pmatrix} = \begin{pmatrix} \cos nt & -\sin nt & 0 \\ \sin nt & \cos nt & 0 \\ 0 & 0 & 1 \end{pmatrix} \begin{pmatrix} \ddot{x} - 2n\dot{y} - n^2x \\ \ddot{y} + 2n\dot{x} - n^2y \\ \ddot{z} \end{pmatrix}. \tag{3.12}$$

Note that the switch to a rotating reference frame has introduced terms in $n\dot{x}$ and $n\dot{y}$ (the *Corioli's acceleration*) and $n^2x$ and $n^2y$ (the *centrifugal acceleration*) into the equations of motion. Using these substitutions for $\xi$, $\eta$, $\zeta$, $\ddot{\xi}$, $\ddot{\eta}$, and $\ddot{\zeta}$, Eqs. (3.3), (3.4), and (3.5) become

$$(\ddot{x} - 2n\dot{y} - n^2x)\cos nt - (\ddot{y} + 2n\dot{x} - n^2y)\sin nt =$$
$$\left[ \mu_1 \frac{x_1 - x}{r_1^3} + \mu_2 \frac{x_2 - x}{r_2^3} \right] \cos nt + \left[ \frac{\mu_1}{r_1^3} + \frac{\mu_2}{r_2^3} \right] y \sin nt, \tag{3.13}$$

$$(\ddot{x} - 2n\dot{y} - n^2x)\sin nt + (\ddot{y} + 2n\dot{x} - n^2y)\cos nt =$$
$$\left[ \mu_1 \frac{x_1 - x}{r_1^3} + \mu_2 \frac{x_2 - x}{r_2^3} \right] \sin nt - \left[ \frac{\mu_1}{r_1^3} + \frac{\mu_2}{r_2^3} \right] y \cos nt, \tag{3.14}$$

$$\ddot{z} = -\left[ \frac{\mu_1}{r_1^3} + \frac{\mu_2}{r_2^3} \right] z. \tag{3.15}$$

If we multiply Eq. (3.13) by $\cos nt$, and Eq. (3.14) by $\sin nt$, and add the results, and then multiply Eq. (3.13) by $-\sin nt$, and Eq. (3.14) by $\cos nt$, and add these together, the equations of motion in the synodic system become

$$\ddot{x} - 2n\dot{y} - n^2 x = -\left[\mu_1 \frac{x + \mu_2}{r_1^3} + \mu_2 \frac{x - \mu_1}{r_2^3}\right], \tag{3.16}$$

$$\ddot{y} + 2n\dot{x} - n^2 y = -\left[\frac{\mu_1}{r_1^3} + \frac{\mu_2}{r_2^3}\right] y, \tag{3.17}$$

$$\ddot{z} = -\left[\frac{\mu_1}{r_1^3} + \frac{\mu_2}{r_2^3}\right] z. \tag{3.18}$$

These accelerations can also be written as the gradient of a scalar function $U$:

$$\ddot{x} - 2n\dot{y} = \frac{\partial U}{\partial x}, \tag{3.19}$$

$$\ddot{y} + 2n\dot{x} = \frac{\partial U}{\partial y}, \tag{3.20}$$

$$\ddot{z} = \frac{\partial U}{\partial z}, \tag{3.21}$$

where $U = U(x, y, z)$ is given by

$$U = \frac{n^2}{2}(x^2 + y^2) + \frac{\mu_1}{r_1} + \frac{\mu_2}{r_2}. \tag{3.22}$$

In this equation the term in $x^2 + y^2$ is the centrifugal potential and the term in $1/r_2$ and $1/r_2$ is the gravitational potential, the partial derivatives of which give rise to the centrifugal and gravitational forces, respectively.

The $-2n\dot{y}$ and $+2n\dot{x}$ terms in Eqs. (3.19) and (3.20) are the Corioli's terms, which depend on the velocity of the particle in the rotating reference frame. The resulting Corioli's force is at right angles to the velocity and therefore does no work.

Note that in our definition $U$ is positive. However, this is opposite to the practice in physics and is purely a convention in celestial mechanics. We could equally well have taken it to be negative, say $U^* = -U$, and the equations of motion would become

$$\ddot{x} - 2n\dot{y} = -\frac{\partial U^*}{\partial x}, \tag{3.23}$$

$$\ddot{y} + 2n\dot{x} = -\frac{\partial U^*}{\partial y}, \tag{3.24}$$

$$\ddot{z} = -\frac{\partial U^*}{\partial z}. \tag{3.25}$$

Note also that $U$ is not a true potential and it is best referred to as a scalar function from which some (but not all) of the accelerations experienced by the particle in the rotating frame can be derived. $U$ is called a "pseudo-potential."

### 3.3 The Jacobi Integral

Multiplying Eq. (3.19) by $\dot{x}$, and Eq. (3.20) by $\dot{y}$, and Eq. (3.21) by $\dot{z}$ and adding we have

$$\dot{x}\ddot{x} + \dot{y}\ddot{y} + \dot{z}\ddot{z} = \frac{\partial U}{\partial x}\dot{x} + \frac{\partial U}{\partial y}\dot{y} + \frac{\partial U}{\partial z}\dot{z} = \frac{dU}{dt}. \tag{3.26}$$

This can be integrated to give

$$\dot{x}^2 + \dot{y}^2 + \dot{z}^2 = 2U - C_J, \tag{3.27}$$

where $C_J$ is a constant of integration. Since $\dot{x}^2 + \dot{y}^2 + \dot{z}^2 = v^2$, the square of the velocity of the particle in the rotating frame, we have

$$v^2 = 2U - C_J \tag{3.28}$$

or, using Eq. (3.22),

$$C_J = n^2(x^2 + y^2) + 2\left(\frac{\mu_1}{r_1} + \frac{\mu_2}{r_2}\right) - \dot{x}^2 - \dot{y}^2 - \dot{z}^2. \tag{3.29}$$

This demonstrates that the quantity $2U - v^2 = C_J$ is a constant of the motion. This is the *Jacobi integral*, or Jacobi constant, sometimes called the *integral of relative energy*. It is important to note that this is not an energy integral because in the restricted problem neither energy nor angular momentum is conserved. The Jacobi integral is the only integral of the circular restricted three-body problem and this means that the problem cannot be solved in closed form for general cases.

The expression for $C_J$ can also be written in terms of the position and velocity of the particle in the nonrotating, sidereal frame. For the position vectors we can use Eq. (3.10) to obtain

$$\begin{pmatrix} x \\ y \\ z \end{pmatrix} = \begin{pmatrix} \cos nt & \sin nt & 0 \\ -\sin nt & \cos nt & 0 \\ 0 & 0 & 1 \end{pmatrix} \begin{pmatrix} \xi \\ \eta \\ \zeta \end{pmatrix}. \tag{3.30}$$

For the velocity vectors we can use Eq. (3.11) to obtain

$$\begin{pmatrix} \dot{x} - ny \\ \dot{y} + nx \\ \dot{z} \end{pmatrix} = \begin{pmatrix} \cos nt & \sin nt & 0 \\ -\sin nt & \cos nt & 0 \\ 0 & 0 & 1 \end{pmatrix} \begin{pmatrix} \dot{\xi} \\ \dot{\eta} \\ \dot{\zeta} \end{pmatrix}. \tag{3.31}$$

However,

$$\begin{pmatrix} \dot{x} - ny \\ \dot{y} + nx \\ \dot{z} \end{pmatrix} = \begin{pmatrix} \dot{x} \\ \dot{y} \\ \dot{z} \end{pmatrix} + n \begin{pmatrix} \sin nt & -\cos nt & 0 \\ \cos nt & \sin nt & 0 \\ 0 & 0 & 0 \end{pmatrix} \begin{pmatrix} \xi \\ \eta \\ \zeta \end{pmatrix} \tag{3.32}$$

and hence

$$
\begin{pmatrix} \dot{x} \\ \dot{y} \\ \dot{z} \end{pmatrix} = \begin{pmatrix} \cos nt & \sin nt & 0 \\ -\sin nt & \cos nt & 0 \\ 0 & 0 & 1 \end{pmatrix} \begin{pmatrix} \dot{\xi} \\ \dot{\eta} \\ \dot{\zeta} \end{pmatrix} - n \begin{pmatrix} \sin nt & -\cos nt & 0 \\ \cos nt & \sin nt & 0 \\ 0 & 0 & 0 \end{pmatrix} \begin{pmatrix} \xi \\ \eta \\ \zeta \end{pmatrix}.
$$

$$ (3.33) $$

If we let

$$
\mathbf{A} = \begin{pmatrix} \cos nt & \sin nt & 0 \\ -\sin nt & \cos nt & 0 \\ 0 & 0 & 1 \end{pmatrix} \quad \text{and} \quad \mathbf{B} = \begin{pmatrix} \sin nt & -\cos nt & 0 \\ \cos nt & \sin nt & 0 \\ 0 & 0 & 0 \end{pmatrix} \quad (3.34)
$$

then, from Eq. (3.33),

$$
\dot{x}^2 + \dot{y}^2 + \dot{z}^2 = (\dot{x} \quad \dot{y} \quad \dot{z}) \begin{pmatrix} \dot{x} \\ \dot{y} \\ \dot{z} \end{pmatrix}
$$

$$
= (\dot{\xi} \quad \dot{\eta} \quad \dot{\zeta}) \mathbf{A}^{\mathrm{T}} \mathbf{A} \begin{pmatrix} \dot{\xi} \\ \dot{\eta} \\ \dot{\zeta} \end{pmatrix} - n (\dot{\xi} \quad \dot{\eta} \quad \dot{\zeta}) \mathbf{A}^{\mathrm{T}} \mathbf{B} \begin{pmatrix} \xi \\ \eta \\ \zeta \end{pmatrix}
$$

$$
- n (\xi \quad \eta \quad \zeta) \mathbf{B}^{\mathrm{T}} \mathbf{A} \begin{pmatrix} \dot{\xi} \\ \dot{\eta} \\ \dot{\zeta} \end{pmatrix} + n^2 (\xi \quad \eta \quad \zeta) \mathbf{B}^{\mathrm{T}} \mathbf{B} \begin{pmatrix} \xi \\ \eta \\ \zeta \end{pmatrix}
$$

$$
= \dot{\xi}^2 + \dot{\eta}^2 + \dot{\zeta}^2 + n^2(\xi^2 + \eta^2) + 2n(\dot{\xi}\eta - \dot{\eta}\xi), \qquad (3.35)
$$

where $\mathbf{A}^{\mathrm{T}}$ and $\mathbf{B}^{\mathrm{T}}$ denote the transposes of the matrices $\mathbf{A}$ and $\mathbf{B}$. Since $\mathbf{A}$ and $\mathbf{B}$ are both orthogonal matrices, their inverses are simply their transposes. Because distances are always unchanged by rotations (or, equivalently, since the determinants of orthogonal matrices are equal to unity) we also have $x^2+y^2+z^2 = \xi^2 + \eta^2 + \zeta^2$; this can also be obtained from Eq. (3.10). We obtain

$$
C_J = 2 \left( \frac{\mu_1}{r_1} + \frac{\mu_2}{r_2} \right) + 2n(\xi\dot{\eta} - \eta\dot{\xi}) - \dot{\xi}^2 - \dot{\eta}^2 - \dot{\zeta}^2 \qquad (3.36)
$$

for the expression for the Jacobi constant in terms of the sidereal coordinates. We can rewrite this as

$$
\frac{1}{2} \left( \dot{\xi}^2 + \dot{\eta}^2 + \dot{\zeta}^2 \right) - \left( \frac{\mu_1}{r_1} + \frac{\mu_2}{r_2} \right) = \mathbf{h} \cdot \mathbf{n} - \frac{1}{2} C_J, \qquad (3.37)
$$

where $\mathbf{n} = (0, 0, n)$ and the left-hand side of the equation is the total, or mechanical, energy per unit mass of the particle. Because $\mathbf{h} \cdot \mathbf{n}$ is not a constant, this explains why energy is not conserved in the restricted three-body problem.

The measurement of the particle's position and velocity in either frame determines the value of the Jacobi constant associated with the motion of the particle. The existence of the angular momentum and energy integrals in the two-body problem enabled us to solve for the motion of one mass with respect to the other (see Sect. 2.3). The Jacobi constant is the only integral of the motion in the

restricted three-body problem. We cannot make use of it to provide an exact solution for the orbital motion, but we can use it to determine regions from which
the particle is excluded.

The usefulness of the Jacobi constant can be appreciated by considering the
locations where the velocity of the particle is zero. In this case we have

$$2U = C_J \tag{3.38}$$

or

$$n^2(x^2 + y^2) + 2\left(\frac{\mu_1}{r_1} + \frac{\mu_2}{r_2}\right) = C_J. \tag{3.39}$$

Equation (3.39) defines a set of surfaces for particular values of $C_J$. These
surfaces, known as the *zero-velocity surfaces*, play an important role in placing
bounds on the motion of the particle. For simplicity we restrict ourselves to the
$x$–$y$ plane. In this case the intersection of the zero-velocity surfaces with the
$x$–$y$ plane produces a set of *zero-velocity curves*. Figure 3.2 shows examples
of these curves for the case $\mu_2 = 0.2$; the mean motion $n$ is taken to be unity.
From our expression for the Jacobi constant, Eq. (3.27), it is clear that we must
always have $2U \geq C_J$ since otherwise the velocity $v$ would be complex. Thus
Eq. (3.39) defines the boundary curves of regions where particle motion is not
possible, in other words excluded regions. Hence, although the restricted three-
body problem is not integrable (we cannot solve for the motion of the particle
for arbitrary starting conditions), the existence of the Jacobi integral does permit
us to find regions of $x$–$y$ space where the particle cannot be. The result is easily
extended to three dimensions.

Since the shaded areas in Fig. 3.2 denote regions where motion is impossible,
we can see from Fig. 3.2a, for example, that if a particle with that value of $C_J$ is in
orbit in the unshaded region around the mass $\mu_1$ then it can never orbit the mass
$\mu_2$ or escape from the system since it would have to cross the excluded region to

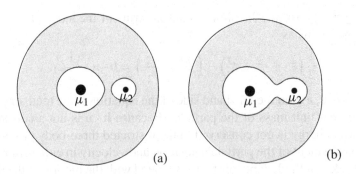

Fig. 3.2. Zero-velocity curves for two values of the Jacobi constant for the case when
$\mu_2 = 0.2$. The values of $C_J$ are (a) $C_J = 3.9$, (b) $C_J = 3.7$. The shaded areas denote the
excluded regions. (See also Fig. 9.11 for the case $\mu_2 = 10^{-3}$.)

do so. Similarly, in Fig. 3.2b, if the particle is originally orbiting the mass $\mu_1$ then it is possible that it could eventually orbit the mass $\mu_2$, but it could never escape from the system. This is the concept of *Hill's stability*. However, it is useful to recall that such statements are only valid under the assumptions inherent in the restricted three-body problem (i.e., the two masses move in circular orbits about their common centre of mass, and the third mass has no gravitational influence on the other two).

## 3.4 The Tisserand Relation

Consider a comet with initial semi-major axis $a$, eccentricity $e$, and inclination $I$. Following a close approach to Jupiter the comet's orbital elements become $a'$, $e'$, and $I'$. We can use the Jacobi integral and some simple approximations to relate these two sets of elements. The Jacobi integral, given by $C_J = 2U - v^2$, remains constant throughout the encounter. In three-dimensional, inertial space the comet has a position vector $\mathbf{r} = (\xi, \eta, \zeta)$ and velocity vector $\dot{\mathbf{r}} = (\dot{\xi}, \dot{\eta}, \dot{\zeta})$. In this system we can use Eq. (3.37) to write the Jacobi constant as

$$C_J = 2\left(\frac{\mu_1}{r_1} + \frac{\mu_2}{r_2}\right) + 2n(\xi\dot{\eta} - \eta\dot{\xi}) - \dot{\xi}^2 - \dot{\eta}^2 - \dot{\zeta}^2, \tag{3.40}$$

where $r_1$ and $r_2$ are the distances of the comet from the Sun and Jupiter respectively. We will also choose units such that the semi-major axis and mean motion of Jupiter's orbit are unity and, since the mass of Jupiter and the comet are much smaller than the Sun's mass,

$$\mathcal{G}\left(m_{\text{Sun}} + m_{\text{comet}}\right) \approx \mathcal{G}\left(m_{\text{Sun}} + m_{\text{Jupiter}}\right) = 1, \tag{3.41}$$

where $m_{\text{Sun}}$, $m_{\text{Jupiter}}$, and $m_{\text{comet}}$ are the masses of the Sun, Jupiter, and the comet respectively. From the energy integral of the Sun–comet two-body problem (see Eq. (2.34)) we have

$$\dot{\xi}^2 + \dot{\eta}^2 + \dot{\zeta}^2 = \frac{2}{r} - \frac{1}{a}, \tag{3.42}$$

where we have taken $\mu = 1$ in accordance with our system of units and where we are assuming $r_1 \approx r$ since the mass of the comet and Jupiter are effectively negligible compared with the mass of the Sun. The angular momentum per unit mass of the comet's orbit is given by

$$\mathbf{h} = \mathbf{r} \times \dot{\mathbf{r}}. \tag{3.43}$$

If $I$ is the inclination of the comet's orbit relative to the plane of Jupiter's orbit, then the $\zeta$ component of the angular momentum vector is given by

$$\xi\dot{\eta} - \eta\dot{\xi} = h \cos I, \tag{3.44}$$

where $h^2 = a(1 - e^2)$ in our system of units. Therefore the form of the Jacobi integral expressed in Eq. (3.40) implies

$$\frac{2}{r} - \frac{1}{a} - 2\sqrt{a(1 - e^2)}\cos I = \frac{2}{r} - 2\mu_2\left(\frac{1}{r} - \frac{1}{r_2}\right) - C_J. \tag{3.45}$$

If we assume that the comet is not close to Jupiter so that $1/r_2$ is always a small quantity and neglect the $\mu_2$ terms, we have

$$\frac{1}{2a} + \sqrt{a(1 - e^2)}\cos I \approx \text{constant}. \tag{3.46}$$

Therefore the approximate relationship between the orbital elements of the comet before and after the encounter with Jupiter is given by

$$\frac{1}{2a} + \sqrt{a(1 - e^2)}\cos I = \frac{1}{2a'} + \sqrt{a'(1 - e'^2)}\cos I'. \tag{3.47}$$

This is known as the *Tisserand relation* (Tisserand 1896) and it can be used to determine whether or not a newly discovered comet is a previously known object that has had its orbital elements changed by a close approach to a planet.

An example of such an encounter for a hypothetical comet is shown in Fig. 3.3. A close approach to Jupiter alters the orbital elements of the comet with the semi-major axis increasing by almost 8 AU. The initial orbital elements of the comet

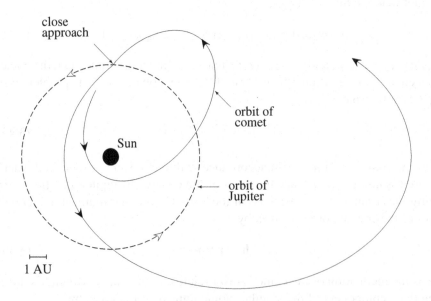

Fig. 3.3. The changing orbit of a hypothetical comet that has a close approach to Jupiter. The encounter produces large changes in the orbital elements of the comet.

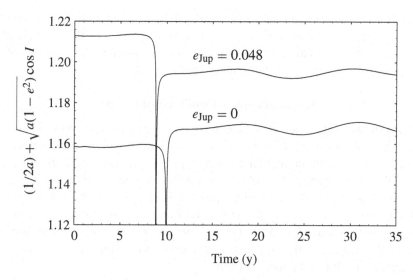

Fig. 3.4. The variation of the Tisserand constant throughout the 35 y integration of the two cometary orbits for the cases where $e_{Jup} = 0$ and $e_{Jup} = 0.048$.

are $a = 4.81$ AU ($= 0.924$ in our units), $e = 0.763$, and $I = 7°\!.47$ while the final elements are $a' = 10.8$ AU, $e' = 0.731$, and $I' = 21°\!.4$.

Although the Tisserand relation is only an approximation to the Jacobi constant and is derived by assuming that Jupiter is in a circular orbit, the quantity $(1/2a) + \sqrt{a(1 - e^2)} \cos I$ is still an approximate constant of the motion in the case where the eccentricity of Jupiter is nonzero. This is illustrated in Fig. 3.4 where the variation in this quantity is plotted as a function of time for two separate numerical integrations. In the first case (the lower curve) Jupiter is taken to move in a circular orbit, while in the second case the eccentricity of Jupiter is taken to have its current value of 0.048. Apart from a short time interval near closest approach where the approximation is invalid, the value of the Tisserand constant changes by less than 1% in the former case and less than 2% in the latter. The regular variations away from closest approach have a period equal to Jupiter's period of 12 y.

The difference in the orbit of the comet before and after its close approach to Jupiter shown in Fig. 3.3 illustrates the dramatic effects of planetary encounters. The same technique is used to obtain gravitational assists for spacecraft such as *Voyager*, *Galileo*, and *Cassini* journeying to the outer planets of the solar system. The above analysis is based on the circular restricted problem where the mass of the third body (the comet or spacecraft) is negligible and hence there is no conservation of energy. However, in the actual situation there is energy conservation and a close approach that increases the semi-major axis of the third body will result in a decay of the semi-major axis of the perturbing body. Because the spacecraft to planet mass ratio is so small, the effect on the planet is undetectable. However, in the early solar system close encounters between the major planets and planetesimals resulted in the formation of the Oort cloud of comets and the cumulative effects of these close encounters on the orbits of the planets were probably large (Fernandez 1997).

### 3.5 Lagrangian Equilibrium Points

We have shown that in the case where the two masses $m_1$ and $m_2$ move in circular orbits about their common centre of mass, $O$, their positions are stationary in a frame rotating with an angular velocity equal to the mean motion $n$ of either mass. We will approach the subject of *equilibrium points* by considering the problem of finding the location of the points where a particle $P$ could be placed, with the appropriate velocity in the inertial frame, where it remains stationary in the rotating frame. It is important to remember that at such an equilibrium position the particle is still subject to a number of forces and that it is still moving in a keplerian orbit in the inertial frame.

Let $\mathbf{a}$, $\mathbf{b}$, and $\mathbf{c}$ denote the location of the mass $m_1$, the centre of mass, $O$, and the mass $m_2$ with respect to the point $P$ (see Fig. 3.5). Let $\mathbf{F}_1$ and $\mathbf{F}_2$ denote the forces per unit mass on the particle directed towards the masses $m_1$ and $m_2$ respectively. For $P$ to be at a fixed location in the rotating frame, it must be at a fixed distance $b$ from $O$, which is the only fixed point in the inertial frame. Therefore $P$ is subject to a centrifugal acceleration in the $-\mathbf{b}$ direction and this is balanced by the vector sum

$$\mathbf{F} = \mathbf{F}_1 + \mathbf{F}_2, \tag{3.48}$$

which lies in the direction of $\mathbf{b}$ and passes through the centre of mass. Note that here we do not need to take the Coriolis force into account because the particle is stationary in the rotating frame.

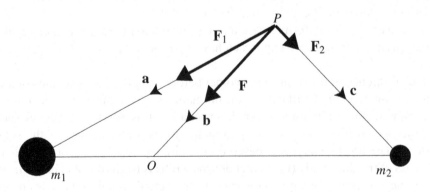

Fig. 3.5. The forces experienced by a test particle $P$ due to the gravitational attraction of the two masses $m_1$ and $m_2$. The point $O$ denotes the location of the centre of mass of $m_1$ and $m_2$.

The position of $O$ is given by

$$\mathbf{b} = \frac{m_1\mathbf{a} + m_2\mathbf{c}}{m_1 + m_2} \tag{3.49}$$

or, rearranging,

$$m_1(\mathbf{a} - \mathbf{b}) = m_2(\mathbf{b} - \mathbf{c}). \tag{3.50}$$

Taking the vector product of $\mathbf{F}_1 + \mathbf{F}_2$ with Eq. (3.50) gives

$$m_2(\mathbf{F}_1 \times \mathbf{c}) + m_1(\mathbf{F}_2 \times \mathbf{a}) = \mathbf{0}. \tag{3.51}$$

Since the angle between $\mathbf{F}_1$ and $\mathbf{c}$ is minus the angle between $\mathbf{F}_2$ and $\mathbf{a}$, we can write the scalar form of Eq. (3.51) as

$$m_2 F_1 c = m_1 F_2 a. \tag{3.52}$$

In the case of gravitational forces, $F_1 = \mathcal{G}m_1/a^2$ and $F_2 = \mathcal{G}m_2/c^2$ and hence, from Eq. (3.52), $a = c$. Therefore the triangle formed by joining the particle to the two masses must be an isosceles triangle. This implies that the locus of all points $P$ for which $\mathbf{F}$ passes through the centre of mass is the perpendicular bisector of the line joining $m_1$ and $m_2$ (the dashed line in Fig. 3.6).

To balance the centrifugal acceleration of $P$ with the force per unit mass directed towards the centre of mass, we must have

$$n^2 b = F_1 \cos \beta + F_2 \cos \gamma, \tag{3.53}$$

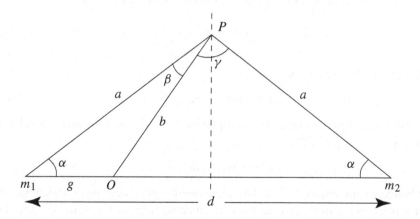

Fig. 3.6.   The geometry of the balance of forces where $P$ denotes the location of a test particle at an equilibrium position. The dashed line denotes the perpendicular bisector of the line joining the two masses, $m_1$ and $m_2$; this is the locus of equilibrium positions in the case of gravitational forces.

where $\beta$ is the angle between $\mathbf{F}_1$ and $\mathbf{b}$, and $\gamma$ is the angle between $\mathbf{F}_2$ and $\mathbf{b}$ (see Fig. 3.6). Therefore

$$n^2 = \frac{\mathcal{G}}{a^2 b^2}(m_1 b \cos\beta + m_2 b \cos\gamma). \tag{3.54}$$

However, by inspection of the triangles formed by $O$, $P$, and the two masses, we have

$$b\cos\beta = a - g\cos\alpha,$$
$$b\cos\gamma = a - (d - g)\cos\alpha, \tag{3.55}$$

where $d$ is the distance between $m_1$ and $m_2$ and $g$ is the distance between $m_1$ and $O$, and

$$\cos\alpha = \frac{d}{2a}. \tag{3.56}$$

Furthermore, from the definition of the centre of mass, we have

$$g = \frac{m_2}{m_1 + m_2}d,$$
$$d - g = \frac{m_1}{m_1 + m_2}d. \tag{3.57}$$

Thus Eq. (3.54) becomes

$$n^2 = \frac{\mathcal{G}(m_1 + m_2)}{a^3 b^2}\left(a^2 - \frac{m_1 m_2}{(m_1 + m_2)^2}d^2\right). \tag{3.58}$$

From the cosine rule,

$$b^2 = a^2 + g^2 - 2ag\cos\alpha = a^2 + g^2 - gd. \tag{3.59}$$

Substituting in this equation the expression for $g$ from Eq. (3.57) and rearranging gives

$$b^2 = a^2 - \frac{m_1 m_2}{(m_1 + m_2)^2}d^2, \tag{3.60}$$

and so Eq. (3.58) becomes

$$n^2 = \mathcal{G}(m_1 + m_2)/a^3. \tag{3.61}$$

However, the reference frame is rotating in inertial space with an angular velocity $n$, where, from Eq. (2.22),

$$n^2 = \mathcal{G}(m_1 + m_2)/d^3, \tag{3.62}$$

and hence we must have $a = d$. The same result can be derived by balancing the vector force in an arbitrary direction. If this direction is chosen to be normal to line joining $m_1$ and $m_2$ then the result follows immediately.

Therefore, in the case of the gravitational force exerted by $m_1$ and $m_2$, the system has an equilibrium point at the apex of an equilateral triangle with a base formed by the line joining the two masses. This result implies the existence of another equilibrium point located below the same line, also lying at the apex of an equilateral triangle. These are the *Lagrangian equilibrium points* $L_4$ and $L_5$ respectively. In the classical problem there are three more equilibrium points, $L_1$, $L_2$, and $L_3$, which lie along the line joining the two masses.

Although we have only considered gravitational forces, it is important to realise that a similar analysis can be carried out for any time-independent force. For example, in the presence of drag forces the location of the shifted equilibrium points can be located using the fact that the resultant of all the forces must lie along a line pointing towards the centre of mass of the system. This is just another way of stating that the sum of all accelerations must be equal and opposite to the centrifugal acceleration in the rotating frame.

## 3.6 Location of Equilibrium Points

Although the circular restricted three-body problem is not integrable we can find a number of special solutions. This can be achieved by searching for points where the particle has zero velocity and zero acceleration in the rotating frame. Such points are called equilibrium points of the system (cf. Sect. 3.5). From now on we assume that all motion is confined to the $x$–$y$ plane. We also choose the unit of distance to be the constant separation of the two masses. It then follows that $n = 1$. None of these assumptions changes the essential dynamics of the system.

To facilitate the calculation of the locations of the equilibrium points we follow the example of Brouwer & Clemence (1961) and rewrite $U$ in a different form. From the definitions of $r_1$ and $r_2$ in Eqs. (3.8) and (3.9), and using the fact that $\mu_1 + \mu_2 = 1$, we have

$$\mu_1 r_1^2 + \mu_2 r_2^2 = x^2 + y^2 + \mu_1 \mu_2 \tag{3.63}$$

and hence

$$U = \mu_1 \left( \frac{1}{r_1} + \frac{r_1^2}{2} \right) + \mu_2 \left( \frac{1}{r_2} + \frac{r_2^2}{2} \right) - \frac{1}{2} \mu_1 \mu_2. \tag{3.64}$$

The advantage of this expression for $U$ is that the explicit dependence on $x$ and $y$ is removed, and so the partial derivatives become simpler. Note that $r_1$ and $r_2$, unlike $x$ and $y$, are always positive quantities.

Now consider the equations of motion, Eqs. (3.19) and (3.20), with $\ddot{x} = \ddot{y} = \dot{x} = \dot{y} = 0$. To find the locations of the equilibrium points we must solve the simultaneous nonlinear equations

$$\frac{\partial U}{\partial x} = \frac{\partial U}{\partial r_1} \frac{\partial r_1}{\partial x} + \frac{\partial U}{\partial r_2} \frac{\partial r_2}{\partial x} = 0, \tag{3.65}$$

$$\frac{\partial U}{\partial y} = \frac{\partial U}{\partial r_1} \frac{\partial r_1}{\partial y} + \frac{\partial U}{\partial r_2} \frac{\partial r_2}{\partial y} = 0 \tag{3.66}$$

using the form of $U = U(r_1, r_2)$ given in Eq. (3.64). Evaluating the partial derivatives, we can write the equations for the location of the equilibrium points as

$$\mu_1 \left( -\frac{1}{r_1^2} + r_1 \right) \frac{x + \mu_2}{r_1} + \mu_2 \left( -\frac{1}{r_2^2} + r_2 \right) \frac{x - \mu_1}{r_2} = 0, \tag{3.67}$$

$$\mu_1 \left( -\frac{1}{r_1^2} + r_1 \right) \frac{y}{r_1} + \mu_2 \left( -\frac{1}{r_2^2} + r_2 \right) \frac{y}{r_2} = 0. \tag{3.68}$$

Inspection of Eqs. (3.65) and (3.66) shows the existence of the trivial solution

$$\frac{\partial U}{\partial r_1} = \mu_1 \left( -\frac{1}{r_1^2} + r_1 \right) = 0, \qquad \frac{\partial U}{\partial r_2} = \mu_2 \left( -\frac{1}{r_2^2} + r_2 \right) = 0, \tag{3.69}$$

which gives $r_1 = r_2 = 1$ in our system of units. Using Eqs. (3.8) and (3.9) this implies

$$(x + \mu_2)^2 + y^2 = 1, \qquad (x - \mu_1)^2 + y^2 = 1 \tag{3.70}$$

with the two solutions

$$x = \frac{1}{2} - \mu_2, \qquad y = \pm \frac{\sqrt{3}}{2}. \tag{3.71}$$

Since $r_1 = r_2 = 1$, each of the two points defined by these equations forms an equilateral triangle with the masses $\mu_1$ and $\mu_2$. These are the *triangular Lagrangian equilibrium points* referred to in Sect. 3.5 as $L_4$ and $L_5$. By convention the leading triangular point is taken to be $L_4$ and the trailing point $L_5$. The analysis given in Sect. 3.5 rules out any additional off-axis equilibrium points.

It is clear from Eq. (3.68) that $y = 0$ is a simple solution of Eq. (3.66), implying that the remaining equilibrium points lie along the $x$ axis and satisfy Eq. (3.65). There are, in fact, three such solutions corresponding to the *collinear Lagrangian equilibrium points* denoted by $L_1$, $L_2$, and $L_3$. The $L_1$ point lies between the masses $\mu_1$ and $\mu_2$, the $L_2$ point lies outside the mass $\mu_2$, and the $L_3$ point lies on the negative $x$ axis. We now derive an approximate location for each of the collinear points.

At the $L_1$ point we have

$$r_1 + r_2 = 1, \quad r_1 = x + \mu_2, \quad r_2 = -x + \mu_1, \quad \frac{\partial r_1}{\partial x} = -\frac{\partial r_2}{\partial x} = 1. \tag{3.72}$$

Hence, substituting in Eq. (3.67) gives us

$$\mu_1 \left( -\frac{1}{(1 - r_2)^2} + 1 - r_2 \right) - \mu_2 \left( -\frac{1}{r_2^2} + r_2 \right) = 0 \tag{3.73}$$

or

$$\frac{\mu_2}{\mu_1} = 3r_2^3 \frac{(1 - r_2 + r_2^2/3)}{(1 + r_2 + r_2^2)(1 - r_2)^3}. \tag{3.74}$$

If we define

$$\alpha = \left(\frac{\mu_2}{3\mu_1}\right)^{1/3} \tag{3.75}$$

then for small $r_2$ it is clear that there is a solution close to $r_2 = \alpha$. We will see later on (see Sect. 3.13) that $\alpha$ is a parameter that occurs naturally in Hill's equations, an approximation to the equations of motion in the vicinity of $\mu_2$. From Eq. (3.74) we have

$$\alpha = r_2 + \frac{1}{3}r_2^2 + \frac{1}{3}r_2^3 + \frac{53}{81}r_2^4 + \mathcal{O}(r_2^5). \tag{3.76}$$

We can use Lagrange's inversion method (see Sect. 2.5) to invert this series and express $r_2$ as a function of $\alpha$. Comparing Eq. (3.76) with Eq. (2.89) we can write

$$r_2 = \alpha + (-1/3)\,\phi(r_2), \tag{3.77}$$

where the function $\phi$ is defined by

$$\phi(r_2) = r_2^2 + r_2^3 + \frac{53}{27}r_2^4 + \mathcal{O}(r_2^5), \tag{3.78}$$

and hence

$$[\phi(\alpha)]^2 = \alpha^4 + 2\alpha^5 + \mathcal{O}(\alpha^6), \tag{3.79}$$

$$\frac{d}{d\alpha}[\phi(\alpha)]^2 = 4\alpha^3 + 10\alpha^4 + \mathcal{O}(\alpha^5), \tag{3.80}$$

$$[\phi(\alpha)]^3 = \alpha^6 + \mathcal{O}(\alpha^7), \tag{3.81}$$

$$\frac{d^2}{d\alpha^2}[\phi(\alpha)]^3 = 30\alpha^4 + \mathcal{O}(\alpha^5). \tag{3.82}$$

Therefore, using these expressions and Eq. (2.90) we have

$$r_2 = \alpha + \sum_{j=1}^{\infty} \frac{(-1/3)^j}{j!}\frac{d^{j-1}}{d\alpha^{j-1}}[\phi(\alpha)]^j$$

$$= \alpha - \frac{1}{3}\alpha^2 - \frac{1}{9}\alpha^3 - \frac{23}{81}\alpha^4 + \mathcal{O}(\alpha^5). \tag{3.83}$$

At the $L_2$ point we have

$$r_1 - r_2 = 1, \quad r_1 = x + \mu_2, \quad r_2 = x - \mu_1, \quad \frac{\partial r_1}{\partial x} = \frac{\partial r_2}{\partial x} = 1. \tag{3.84}$$

Hence substituting for $r_1$ in Eq. (3.67) we have

$$\mu_1\left(-\frac{1}{(1+r_2)^2} + 1 + r_2\right) + \mu_2\left(-\frac{1}{r_2^2} + r_2\right) = 0 \tag{3.85}$$

or

$$\frac{\mu_2}{\mu_1} = 3r_2^3\,\frac{(1 + r_2 + r_2^2/3)}{(1+r_2)^2(1 - r_2^3)}. \tag{3.86}$$

Using the definition of $\alpha$ in Eq. (3.75) gives us

$$\alpha = r_2 - \frac{1}{3}r_2^2 + \frac{1}{3}r_2^3 + \frac{1}{81}r_2^4 + \mathcal{O}(r_2^5), \tag{3.87}$$

$$r_2 = \alpha + \frac{1}{3}\alpha^2 - \frac{1}{9}\alpha^3 - \frac{31}{81}\alpha^4 + \mathcal{O}(\alpha^5). \tag{3.88}$$

At the $L_3$ point we have

$$r_2 - r_1 = 1, \quad r_1 = -x - \mu_2, \quad r_2 = -x + \mu_1, \quad \frac{\partial r_1}{\partial x} = \frac{\partial r_2}{\partial x} = -1. \tag{3.89}$$

Hence substituting for $r_2$ in Eq. (3.67) gives us

$$\mu_1\left(-\frac{1}{r_1^2} + r_1\right) + \mu_2\left(-\frac{1}{(1+r_1)^2} + 1 + r_1\right) = 0 \tag{3.90}$$

or

$$\frac{\mu_2}{\mu_1} = \frac{(1 - r_1^3)(1 + r_1)^2}{r_1^3(r_1^2 + 3r_1 + 3)}. \tag{3.91}$$

If we let $r_1 = 1 + \beta$ (which implies $r_2 = 2 + \beta$) we then have

$$\frac{\mu_2}{\mu_1} = -\frac{12}{7}\beta + \frac{144}{49}\beta^2 - \frac{1567}{343}\beta^3 + \mathcal{O}(\beta^4), \tag{3.92}$$

$$\beta = -\frac{7}{12}\left(\frac{\mu_2}{\mu_1}\right) + \frac{7}{12}\left(\frac{\mu_2}{\mu_1}\right)^2 - \frac{13223}{20736}\left(\frac{\mu_2}{\mu_1}\right)^3 + \mathcal{O}\left(\frac{\mu_2}{\mu_1}\right)^4. \tag{3.93}$$

    The locations of all the Lagrangian points and zero-velocity curves for three critical values of the Jacobi constant for a mass ratio $\mu_2 = 0.2$ are shown in Fig. 3.7. In Fig. 3.8 the surface defined by the zero-velocity curves is shown for the same mass ratio.

    Referring to Figs. 3.7 and 3.8 we see that the $L_1$ point has the largest value of the Jacobi constant ($C_J = 3.805$ for $\mu_2 = 0.2$) and lies at a critical point of the innermost curve. There is another branch of the same curve (approximately circular in shape) that lies beyond the $L_2$ point. It is clear from a comparison with Fig. 3.2 that the $L_1$ point separates those trajectories confined to orbits either around the primary or the secondary from those that can orbit both masses (cf. Fig. 3.2). The $L_2$ point ($C_J = 3.552$ for $\mu_2 = 0.2$) is another saddle point on the zero-velocity curve. A particle with $C_J < C_{L_2}$ could orbit in the interior or exterior regions of the plane. Two branches of the zero-velocity curves also meet at the $L_3$ point ($C_J = 3.197$ for $\mu_2 = 0.2$). The triangular equilibrium points $L_4$ and $L_5$ have the lowest value of the Jacobi constant ($C_J = 2.84$ for $\mu_2 = 0.2$); if

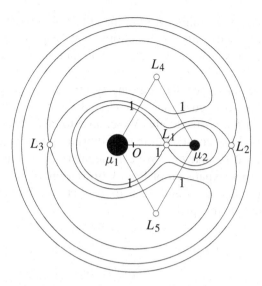

Fig. 3.7. The location of the Lagrangian equilibrium points (denoted by the small, open circles) and associated zero-velocity curves for $\mu_2 = 0.2$. The plot shows the zero-velocity curves for the three critical values of the Jacobi constant (3.805, 3.552, 3.197) that pass through the points $L_1$, $L_2$, and $L_3$ for this value of $\mu_2$. The point $O$ denotes the centre of mass of the system.

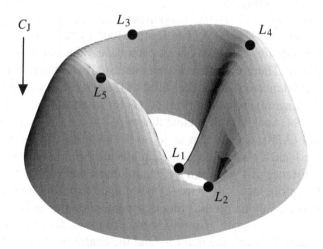

Fig. 3.8. The three-dimensional surface defined by $C_J = 2U$ and the location of the Lagrangian equilibrium points for $\mu_2 = 0.2$, the same as that used in Fig. 3.7. The points $L_1$, $L_2$, and $L_3$ all lie at saddle points on the surface. The surface is drawn such that the vertical height represents the negative value of the Jacobi constant $C_J$.

a particle has a Jacobi constant $C_J < C_{L_{4,5}}$ then there are no regions of the plane from which it is excluded. This does not imply that the particle will travel to all accessible regions, only that the Jacobi constant cannot provide bounds to the motion.

There is no universally accepted ordering system for the Lagrangian equilibrium points, although it is customary for $L_4$ and $L_5$ to denote the leading and trailing triangular points respectively. We choose to label the points according to the value of the Jacobi constant at the equilibrium point, from the largest (at $L_1$) to the smallest (at $L_4$ and $L_5$). We can calculate these values of the Jacobi constant using a series expansion of Eq. (3.39) (taking $n = 1$) with the appropriate values for $x$ and $y$, or $r_1$ and $r_2$, as derived above. Including terms up to $\mathcal{O}(\mu_2)$ in the expansion, this gives

$$C_{L_1} \approx 3 + 3^{4/3}\mu_2^{2/3} - 10\mu_2/3, \tag{3.94}$$

$$C_{L_2} \approx 3 + 3^{4/3}\mu_2^{2/3} - 14\mu_2/3, \tag{3.95}$$

$$C_{L_3} \approx 3 + \mu_2, \tag{3.96}$$

$$C_{L_4} \approx 3 - \mu_2, \tag{3.97}$$

$$C_{L_5} \approx 3 - \mu_2. \tag{3.98}$$

Note that Eqs. (3.8), (3.9), and (3.39) are identical if $y$ is replaced with $-y$ and hence the zero-velocity curves must be symmetric about the $x$ axis (this is already clear from Figs. 3.2 and 3.7), and thus the values of the Jacobi constant at the $L_4$ and $L_5$ points are the same.

The largest value of $\mu_2 = m_2/(m_1+m_2)$ in the solar system occurs in the Pluto–Charon system where $\mu_2 \approx 10^{-1}$. The Earth–Moon system has $\mu_2 \approx 10^{-2}$ but all other planet–satellite and Sun–planet pairs have values at least an order of magnitude smaller. Therefore, since we are interested in applications to the solar system, it is important to consider the shape of the zero-velocity curves and the positions of the Lagrangian equilibrium points for small values of $\mu_2$.

From Eqs. (3.94) and (3.95) it is clear that as $\mu_2 \to 0$, $C_{L_1} \to C_{L_2}$. Also, from Eqs. (3.83) and (3.88), we see that as $\mu_2 \to 0$ the terms of $\mathcal{O}(\alpha^2)$ and higher can be neglected and $L_1$ and $L_2$ become equidistant from the mass $\mu_2$. The $L_3$ point lies at a distance of $1 + \beta$ from the central mass, where $\beta$ ($< 0$) is given by Eq. (3.93). Therefore, as $\mu_2 \to 0$ the $L_3$ point approaches the unit circle. The triangular points already lie on the unit circle centred on the mass $\mu_1$ and are at a distance $r = (1 - \mu_2 + \mu_2^2)^{1/2}$ from the centre of mass. Since the quantity $\mu_2$ is also the distance between the mass $\mu_1$ and the centre of mass of the system, as $\mu_2 \to 0$ the unit circle centred on $\mu_1$ approaches the unit circle centred on the centre of mass.

Figure 3.9 shows a selection of zero velocity curves and the locations of the Lagrangian equilibrium points for the case where $\mu_2 = 0.01$, a value comparable

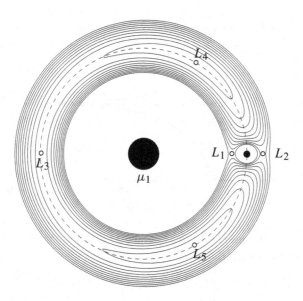

Fig. 3.9. The location of the Lagrangian equilibrium points (open circles) and associated zero-velocity curves for a mass $\mu_2 = 0.01$. The dashed line denotes the circle of unit radius centred on the mass $\mu_1$.

to that in the Earth–Moon system. Note that the $L_1$ and $L_2$ points are almost equidistant from the mass $\mu_2$ and that the $L_3$ point lies close to the unit circle. Although the dashed line in Fig. 3.9 denotes the unit circle centred on the mass $\mu_1$, a shift of 1% of its radius to the right would produce a unit circle centred on the centre of mass.

To illustrate the degree of symmetry about the unit circle when $\mu_2 \to 0$ we have calculated the radial distances of the individual equilibrium points from the centre of mass of the system for values of $\mu_2$ between $10^{-1}$ and $10^{-10}$ (see Fig. 3.10). The results show the approach of first $L_3$, $L_4$, and $L_5$, and then $L_1$ and $L_2$, to the unit radius as $\mu_2 \to 0$. This property of the zero-velocity curves and the equilibrium points allows us to make useful approximations when discussing the motion of the test particle in the case of a system with a small mass ratio.

## 3.7 Stability of Equilibrium Points

It is not enough to know that a number of equilibrium points exist for a dynamical system; we also need to determine the stability of such points. When a system is subjected to a force derived from a potential, the sum of the kinetic energy and potential energy remains constant. In such a system "stable" motion occurs at equilibrium positions that are potential minima. Objects placed at these positions will remain in the vicinity despite small displacements. We can demonstrate this by considering a test object given a small displacement from an equilibrium position at a potential minimum. Since the potential energy must increase away from

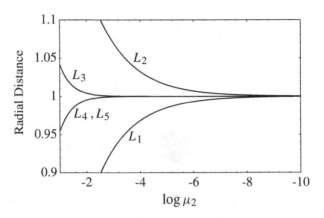

Fig. 3.10.   The radial distance of the five Lagrangian equilibrium points from the centre of mass as a function of $\log \mu_2$.

the minimum, the kinetic energy must decrease until it reaches zero. However, the sum of the two forms of energy has to be a constant and so at this point the motion has to reverse, leading to an increase in kinetic energy and a decrease in potential energy. In the presence of a dissipative force that opposes the motion the particle moves towards the equilibrium position. Note, however, that in the case of two-body motion such a drag force actually increases the speed of the orbiting particle because loss of energy implies a decrease in orbital radius and hence an increase in orbital speed.

Consider the zero-velocity curves in the vicinity of a triangular equilibrium point. We have already shown that $C_J = 2U - v^2$ is a minimum at $L_4$ and $L_5$. If we adopted the convention of making the quantity $U$ negative by transforming to a function $U^* = -U$ (see Eqs. (3.23) and (3.24)), then we would make $C_J$ a maximum at the triangular points. However, all that really matters is the direction of motion of the particle and this is not determined by convention. Figure 3.11 shows two zero-velocity curves for the case $\mu_2 = 0.1$. At various points around each curve we have used Eqs. (3.16) and (3.17) to calculate values of the acceleration, $(\ddot{x}, \ddot{y})$, of the particle in the rotating reference frame; although $\dot{x} = \dot{y} = 0$ (since the curves are zero-velocity curves) the vector $(\ddot{x}, \ddot{y})$ is nonzero and is related to the initial direction of the particle's velocity. Notice that this vector is always perpendicular to the zero-velocity curve and that its magnitude, on a given zero-velocity curve, is largest in the vicinity of the triangular point.

Since all the directions of the initial velocity vectors in Fig. 3.11 are pointing away from the $L_4$ point and towards those regions associated with higher values of $C_J$, it would be natural to assume that the point is an unstable equilibrium point. However, appearances can be deceptive and in order to carry out a proper study of the stability we need to examine the behaviour of a particle that is given

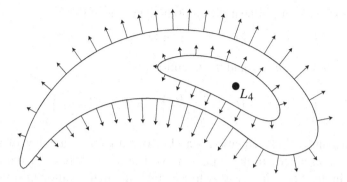

Fig. 3.11. The relative magnitudes and directions of initial motion around two zero-velocity curves ($C_J = 2.95$ and $C_J = 3.09$) in the vicinity of the $L_4$ point for the case $\mu_2 = 0.1$.

a small displacement from an equilibrium position. We can do this by linearising the equations of motion and carrying out a *linear stability analysis*.

Let the location of an equilibrium point in the circular restricted problem be denoted by $(x_0, y_0)$ and consider a small displacement $(X, Y)$ from the point such that $x = x_0 + X$ and $y = y_0 + Y$. Then, by substituting in Eqs. (3.19) and (3.20), and expanding in a Taylor series, we have

$$\ddot{X} - 2n\dot{Y} \approx \left(\frac{\partial U}{\partial x}\right)_0 + X \left(\frac{\partial}{\partial x}\left(\frac{\partial U}{\partial x}\right)\right)_0 + Y \left(\frac{\partial}{\partial y}\left(\frac{\partial U}{\partial x}\right)\right)_0$$
$$= X \left(\frac{\partial^2 U}{\partial x^2}\right)_0 + Y \left(\frac{\partial^2 U}{\partial x \partial y}\right)_0 \tag{3.99}$$

and

$$\ddot{Y} + 2n\dot{X} \approx \left(\frac{\partial U}{\partial y}\right)_0 + X \left(\frac{\partial}{\partial x}\left(\frac{\partial U}{\partial y}\right)\right)_0 + Y \left(\frac{\partial}{\partial y}\left(\frac{\partial U}{\partial y}\right)\right)_0$$
$$= X \left(\frac{\partial^2 U}{\partial x \partial y}\right)_0 + Y \left(\frac{\partial^2 U}{\partial y^2}\right)_0 , \tag{3.100}$$

where the subscript 0 denotes evaluation of the partial derivatives at the equilibrium point and we have used the fact that $(\partial U/\partial x)_0 = (\partial U/\partial y)_0 = 0$ (recall that the equilibrium points are derived by solving the equations $\partial U/\partial x = \partial U/\partial y = 0$). We are only interested in the motion in the immediate vicinity of the equilibrium point, and so it is legitimate to neglect the higher order terms since these will always be small quantities, providing we only consider a small initial displacement from the point $(x_0, y_0)$; obviously the assumption no longer holds if the displacement is too large. From our initial set of nonlinear differential equations, we have carried out a standard procedure to *linearise* the equations of motion.

The end result is a set of linear differential equations of the form

$$\ddot{X} - 2\dot{Y} = XU_{xx} + YU_{xy}, \qquad \ddot{Y} + 2\dot{X} = XU_{xy} + YU_{yy}, \qquad (3.101)$$

where we have taken $n = 1$ and the quantities

$$U_{xx} = \left(\frac{\partial^2 U}{\partial x^2}\right)_0, \quad U_{xy} = \left(\frac{\partial^2 U}{\partial x \partial y}\right)_0, \quad U_{yy} = \left(\frac{\partial^2 U}{\partial y^2}\right)_0 \qquad (3.102)$$

are all constants. These equations can be solved using a number of standard techniques, and since similar equations will arise in different circumstances elsewhere in this book, it is worthwhile examining their solution in some detail.

We can write the equations in matrix form as

$$\begin{pmatrix} \dot{X} \\ \dot{Y} \\ \ddot{X} \\ \ddot{Y} \end{pmatrix} = \begin{pmatrix} 0 & 0 & 1 & 0 \\ 0 & 0 & 0 & 1 \\ U_{xx} & U_{xy} & 0 & 2 \\ U_{xy} & U_{yy} & -2 & 0 \end{pmatrix} \begin{pmatrix} X \\ Y \\ \dot{X} \\ \dot{Y} \end{pmatrix}, \qquad (3.103)$$

thereby changing the problem to the solution of four, simultaneous, first-order differential equations instead of two, simultaneous, second-order differential equations. The equations now have the form

$$\dot{\mathbf{X}} = \mathbf{A}\mathbf{X}, \qquad (3.104)$$

where

$$\mathbf{X} = \begin{pmatrix} X \\ Y \\ \dot{X} \\ \dot{Y} \end{pmatrix} \quad \text{and} \quad \mathbf{A} = \begin{pmatrix} 0 & 0 & 1 & 0 \\ 0 & 0 & 0 & 1 \\ U_{xx} & U_{xy} & 0 & 2 \\ U_{xy} & U_{yy} & -2 & 0 \end{pmatrix}. \qquad (3.105)$$

Before proceeding let us consider a general matrix equation of the form given in Eq. (3.104) where $\mathbf{A}$ is an $n \times n$ matrix of constants and $\mathbf{X}$ is an $n$-dimensional vector with elements $X_i$ $(i = 1, 2, \ldots n)$.

If any vector $\mathbf{x}$ satisfies the equation

$$\mathbf{A}\mathbf{x} = \lambda\mathbf{x}, \qquad (3.106)$$

where $\lambda$ is a scalar constant, then $\mathbf{x}$ is said to be an *eigenvector* of the matrix $\mathbf{A}$ and $\lambda$ is its corresponding *eigenvalue*. If $\mathbf{A}$ is thought of as a transformation matrix, then the result of applying $\mathbf{A}$ to the particular vector $\mathbf{x}$ satisfying Eq. (3.106) is to produce a vector in the same direction as $\mathbf{x}$ but of a different magnitude.

The first step in the solution of Eq. (3.104) is to find the eigenvalues of $\mathbf{A}$. We can rewrite Eq. (3.106) as

$$(\mathbf{A} - \lambda\mathbf{I})\mathbf{x} = 0, \qquad (3.107)$$

where $\mathbf{I}$ is the $n \times n$ unit matrix. This set of $n$ simultaneous, linear equations in $n$ unknowns (the $n$ elements of $\mathbf{x}$) will have nontrivial solutions provided the determinant of the matrix $\mathbf{A} - \lambda\mathbf{I}$ is zero. Hence, we must have

$$\det(\mathbf{A} - \lambda\mathbf{I}) = 0. \tag{3.108}$$

This produces the *characteristic equation*, which is a polynomial equation of degree $n$ in $\lambda$ with $n$ possible complex roots. The $n$ corresponding eigenvectors can be found by substituting each $\lambda$ into Eq. (3.106) and solving for the components of each $\mathbf{x}$.

Now let us return to the differential equations given in Eq. (3.104). The equations are *coupled*, in the sense that the equation for the time derivative of a general element $X_i$ depends, in general, on various elements of the vector. For example, in our own problem, the equation for the time variation of $\dot{X}$ is a function of $X$, $Y$, and $\dot{Y}$. The equations would be much easier to solve if the coupling could be removed by a suitable transformation to a new set of $n$ variables, $Y_i$, $(i = 1, 2, \ldots n)$ say, where $\dot{Y}_i$ is a function of $Y_i$ alone. This can be achieved by a standard technique known as a *similarity transformation*. Let the transformation from $\mathbf{X}$ to $\mathbf{Y}$ be represented by

$$\mathbf{Y} = \mathbf{BX}, \tag{3.109}$$

where the constant matrix $\mathbf{B}$ has yet to be determined. We have

$$\mathbf{X} = \mathbf{B}^{-1}\mathbf{Y} \quad \text{and} \quad \dot{\mathbf{X}} = \mathbf{B}^{-1}\dot{\mathbf{Y}}, \tag{3.110}$$

where $\mathbf{B}^{-1}$ is the inverse of the matrix $\mathbf{B}$. Hence Eq. (3.104) can be written

$$\mathbf{B}^{-1}\dot{\mathbf{Y}} = \mathbf{AB}^{-1}\mathbf{Y} \tag{3.111}$$

or, premultiplying each side by $\mathbf{B}$,

$$\dot{\mathbf{Y}} = \mathbf{BB}^{-1}\dot{\mathbf{Y}} = \mathbf{BAB}^{-1}\mathbf{Y}. \tag{3.112}$$

We want the new differential equations given in Eq. (3.112) to be uncoupled. This can be achieved if the matrix $\mathbf{B}$ is chosen such that

$$\mathbf{BAB}^{-1} = \mathbf{\Lambda}, \tag{3.113}$$

where $\mathbf{\Lambda}$ denotes a diagonal matrix (i.e., a matrix with all elements not on the leading diagonal equal to zero). It can easily be shown that if the columns of the matrix $\mathbf{B}$ are constructed from the $n$ (column) eigenvectors of the matrix $\mathbf{A}$, then the resulting matrix $\mathbf{\Lambda}$ is a diagonal matrix composed of the eigenvalues of $\mathbf{A}$ with

$$\mathbf{\Lambda} = \begin{pmatrix} \lambda_1 & 0 & \cdots & 0 \\ 0 & \lambda_2 & \cdots & 0 \\ \vdots & \vdots & \ddots & \vdots \\ 0 & 0 & \cdots & \lambda_n \end{pmatrix}. \tag{3.114}$$

The transformed system of equations is now

$$\dot{\mathbf{Y}} = \mathbf{\Lambda Y} \tag{3.115}$$

or, alternatively, in component form,

$$\dot{Y}_i = \lambda_i Y_i \quad (i = 1, 2, \ldots, n). \tag{3.116}$$

The solutions of Eq. (3.116) are easily found; they are

$$Y_i = c_i e^{\lambda_i t} \quad (i = 1, 2, \ldots, n), \tag{3.117}$$

where the $c_i$ are $n$ constants of integration. However, although the solution of the equations in the $Y_i$ is relatively simple, we must now transform back to obtain our solutions in terms of our original variables, $X_i$. From Eq. (3.110) we have

$$\mathbf{X} = \mathbf{B}^{-1}\mathbf{Y} = \mathbf{B}^{-1}\begin{pmatrix} c_1 e^{\lambda_1 t} \\ c_2 e^{\lambda_2 t} \\ \vdots \\ c_n e^{\lambda_n t} \end{pmatrix}. \tag{3.118}$$

The $n$ constants of integration, $c_i$, can now be determined from the starting values by solving the $n$ simultaneous, linear equations given in Eq. (3.118).

We can now apply the method given above to our problem. In our case $n = 4$ and the characteristic equation is given by

$$\det(\mathbf{A} - \lambda\mathbf{I}) = \begin{vmatrix} -\lambda & 0 & 1 & 0 \\ 0 & -\lambda & 0 & 1 \\ U_{xx} & U_{xy} & -\lambda & 2 \\ U_{xy} & U_{yy} & -2 & -\lambda \end{vmatrix} = 0, \tag{3.119}$$

which reduces to the polynomial equation

$$\lambda^4 + (4 - U_{xx} - U_{yy})\lambda^2 + U_{xx}U_{yy} - U_{xy}^2 = 0. \tag{3.120}$$

This is a quartic in the variable $\lambda$ but it is also a *biquadratic*, or a quadratic equation in $\lambda^2$. This makes it relatively easy to find the four roots. They are

$$\lambda_{1,2} = \pm\left[\frac{1}{2}(U_{xx} + U_{yy} - 4)\right.$$
$$\left. -\frac{1}{2}\left[(4 - U_{xx} - U_{yy})^2 - 4(U_{xx}U_{yy} - U_{xy}^2)\right]^{1/2}\right]^{1/2} \tag{3.121}$$

and

$$\lambda_{3,4} = \pm\left[\frac{1}{2}(U_{xx} + U_{yy} - 4)\right.$$
$$\left. +\frac{1}{2}\left[(4 - U_{xx} - U_{yy})^2 - 4(U_{xx}U_{yy} - U_{xy}^2)\right]^{1/2}\right]^{1/2}. \tag{3.122}$$

Although our general solution given in Eq. (3.118) appears to involve knowing the four eigenvectors of $\mathbf{A}$, in practice this is not required. It is clear from

Eq. (3.118) that we can write the solution for $X$ and $\dot{X}$ as

$$X = \sum_{j=1}^{4} \bar{\alpha}_j e^{\lambda_j t}, \qquad \dot{X} = \sum_{j=1}^{4} \bar{\alpha}_j \lambda_j e^{\lambda_j t}, \qquad (3.123)$$

where the $\bar{\alpha}_j$ are constants. Similarly we can derive expressions for $Y$ and $\dot{Y}$ in terms of constants $\bar{\beta}_j$. We have

$$Y = \sum_{j=1}^{4} \bar{\beta}_j e^{\lambda_j t}, \qquad \dot{Y} = \sum_{j=1}^{4} \bar{\beta}_j \lambda_j e^{\lambda_j t}, \qquad (3.124)$$

where the constants $\bar{\beta}_j$ are functions of the $\bar{\alpha}_j$ since there can only be four constants in the solution. The relationship between the $\bar{\alpha}_j$ and the $\bar{\beta}_j$ can be derived from either of the equations in Eq. (3.101). Substituting the expressions for $X$, $Y$, and $\dot{Y}$ into Eq. (3.101) we have

$$\sum_{j=1}^{4} \left( \bar{\alpha}_j \lambda_j^2 - 2\bar{\beta}_j \lambda_j - U_{xx}\bar{\alpha}_j - U_{xy}\bar{\beta}_j \right) e^{\lambda_j t} = 0. \qquad (3.125)$$

The trivial solution of this equation gives the relationship between the $\bar{\alpha}_j$ and the $\bar{\beta}_j$:

$$\bar{\beta}_j = \frac{\lambda_j^2 - U_{xx}}{2\lambda_j + U_{xy}} \bar{\alpha}_j. \qquad (3.126)$$

If, at time $t = 0$, we have the boundary conditions $X = X_0$, $Y = Y_0$, $\dot{X} = \dot{X}_0$, and $\dot{Y} = \dot{Y}_0$, then the four $\bar{\alpha}_j$ (and hence the related $\bar{\beta}_j$) are determined from the solution of the four simultaneous linear equations

$$\sum_{j=1}^{4} \bar{\alpha}_j = X_0, \qquad \sum_{j=1}^{4} \lambda_j \bar{\alpha}_j = \dot{X}_0, \qquad \sum_{j=1}^{4} \bar{\beta}_j = Y_0, \qquad \sum_{j=1}^{4} \lambda_j \bar{\beta}_j = \dot{Y}_0. \quad (3.127)$$

Although the complete solution is described by Eqs. (3.123) and (3.124) with the constants $\bar{\alpha}_j$ and $\bar{\beta}_j$ given by the solutions to Eqs. (3.127), the stability of the equilibrium points can be determined from an inspection of the eigenvalues alone. To investigate the nature of the eigenvalues we define the quantities

$$\bar{A} = \frac{\mu_1}{(r_1^3)_0} + \frac{\mu_2}{(r_2^3)_0}, \qquad (3.128)$$

$$\bar{B} = 3 \left[ \frac{\mu_1}{(r_1^5)_0} + \frac{\mu_2}{(r_2^5)_0} \right] y_0^2, \qquad (3.129)$$

$$\bar{C} = 3 \left[ \mu_1 \frac{(x_0 + \mu_2)}{(r_1^5)_0} + \mu_2 \frac{(x_0 - \mu_1)}{(r_2^5)_0} \right] y_0, \qquad (3.130)$$

$$\bar{D} = 3 \left[ \mu_1 \frac{(x_0 + \mu_2)^2}{(r_1^5)_0} + \mu_2 \frac{(x_0 - \mu_1)^2}{(r_2^5)_0} \right]. \qquad (3.131)$$

In terms of these constants we can write

$$U_{xx} = 1 - \bar{A} + \bar{D}, \tag{3.132}$$
$$U_{yy} = 1 - \bar{A} + \bar{B}, \tag{3.133}$$
$$U_{xy} = \bar{C}. \tag{3.134}$$

Note that all of these quantities are real numbers and that $X$, $Y$, $\dot{X}$, and $\dot{Y}$ must also be real, even though the eigenvalues $\lambda_j$ and the constants $\bar{\alpha}_j$ and $\bar{\beta}_j$ can be complex.

The general form of the eigenvalues given by Eqs. (3.121) and (3.122) is

$$\lambda_{1,2} = \pm(j_1 + ik_1), \qquad \lambda_{3,4} = \pm(j_2 + ik_2), \tag{3.135}$$

where $j_1$, $k_1$, $j_2$, and $k_2$ are real quantities and $i = \sqrt{-1}$. Given that the form of the general solution for the components of the position and velocity vectors relative to the equilibrium point (see Eqs. (3.123) and (3.124)) involves a linear combination of terms of the form $e^{\lambda_j t}$, this implies that each of the two $e^{+(j+ik)t}$ terms will be matched by an $e^{-(j+ik)t}$ term. If $j = 0$ then an oscillatory solution will result because terms with $e^{+ikt}$ and $e^{-ikt}$ will reduce to sines and cosines (since $e^{\pm i\theta} = \cos\theta \pm i\sin\theta$). However, if $j$ is positive then there will always be exponential growth in at least one mode and so the perturbed solution is unstable. Therefore the equilibrium point is stable if all the eigenvalues are purely imaginary. We will see how this works in practice by considering the general and specific cases.

### 3.7.1 The Collinear Points

Consider the collinear solutions corresponding to the Lagrangian points $L_1$, $L_2$, and $L_3$. For these points we have $y_0 = 0$, $(r_1^2)_0 = (x_0 + \mu_2)^2$, and $(r_2^2)_0 = (x_0 - \mu_1)^2$, and hence

$$U_{xx} = 1 + 2\bar{A}, \qquad U_{yy} = 1 - \bar{A}, \qquad U_{xy} = 0. \tag{3.136}$$

The characteristic equation becomes

$$\lambda^4 + (2 - \bar{A})\lambda^2 + (1 + \bar{A} - 2\bar{A}^2) = 0. \tag{3.137}$$

As a specific example consider the stability of the $L_1$ point for the case when $\mu_2 = 0.01$ using the theory developed in Sect. 3.6. From Sect. 3.6 we have $x_0 = 0.848$ and $y_0 = 0$; we will assume an initial displacement of $X_0 = Y_0 = 10^{-5}$ and take $\dot{X}_0 = \dot{Y}_0 = 0$. The resulting eigenvalues are $\pm 2.90$ and $\pm 2.32i$, indicating that the point is linearly unstable. Solving for the $\bar{\alpha}_j$ and $\bar{\beta}_j$ we derive the solution

$$X(t) = 6.99 \times 10^{-6} e^{-2.90t} + 4.96 \times 10^{-6} e^{+2.90t}$$
$$+ 1.96 \times 10^{-6} \cos 2.32t + 2.54 \times 10^{-6} \sin 2.32t,$$
$$Y(t) = 3.25 \times 10^{-6} e^{-2.90t} - 2.31 \times 10^{-6} e^{+2.90t}$$
$$+ 9.06 \times 10^{-6} \cos 2.32t + 6.96 \times 10^{-6} \sin 2.32t. \tag{3.138}$$

The second term in each of these equations will eventually dominate and lead to exponential growth in the $X$ and $Y$. The e-folding timescale for growth in this case is $1/2.90$ orbital periods of the mass $\mu_2$.

The nonzero values of the real and imaginary parts of the solutions to the quartic for the case of the $L_1$ point are shown in Fig. 3.12 for values of $\mu_2$ between 0.1 and 0.001. Notice, as in the specific case shown above, that the roots are always of the form $\pm j$ and $\pm ik$, where $j$ and $k$ are real quantities. A similar plot can be obtained for $L_2$ and $L_3$ but with different numerical values.

From the properties of polynomials we know that the product of the roots must equal the constant term in the polynomial. Hence the characteristic equation must satisfy

$$(\lambda_1\lambda_2)(\lambda_3\lambda_4) = 1 + \bar{A} - 2\bar{A}^2, \tag{3.139}$$

where, from Eqs. (3.121) and (3.122), $\lambda_1 = -\lambda_2$ and $\lambda_3 = -\lambda_4$. In order for all the roots to be purely imaginary (the condition for stability) we must have $\lambda_1^2 = \lambda_2^2 < 0$ and $\lambda_3^2 = \lambda_4^2 < 0$; this implies that $1 + \bar{A} - 2\bar{A}^2 = (1 - \bar{A})(1 + 2\bar{A}) > 0$. Therefore the collinear points are stable provided $-1/2 < \bar{A} < 1$. However, substituting the values of $r_1$ and $r_2$ for the collinear points into Eq. (3.128) shows that $\bar{A} > 1$ (given that $\mu_2 < 1/2$) and so the collinear points are unstable for all values of $\mu_2$. In fact the solution will always have the form given in Eq. (3.138) but with decreasing values of the eigenvalues as the mass decreases (see Fig. 3.12). Physically we can understand the existence of the unstable collinear points by considering the changes in gravitational acceleration and centrifugal force at different locations along the $x$ axis. Moving away from both masses gravity drops off and the centrifugal acceleration increases. Thus

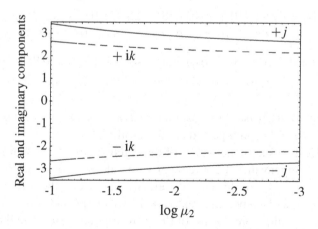

Fig. 3.12. Numerical values of the real (solid lines) and imaginary (dashed lines) components of the roots of the characteristic equation for the $L_1$ Lagrangian point as a function of the value of $\mu_2$.

there are exactly two equilibrium points with $m_1$ and $m_2$ between them. Similar reasoning shows that there is just one equilibrium point between the two masses. The force on a particle anywhere along the $x$ axis is directed away from these equilibrium points and hence they are all unstable.

However, with special starting conditions it is possible to find stable, periodic orbits in the vicinity of the equilibrium points (see, for example, Szebehely 1967). In fact, the *SOHO* spacecraft (Domingo et al. 1995) was placed in such an orbit near the $L_1$ point in the Earth–Sun system in order to observe the Sun continuously.

### 3.7.2 The Triangular Points

Now consider the motion in the vicinity of $L_4$ and $L_5$. In this case, $r_1 = r_2 = 1$, $x = 1/2 - \mu_2$, $y = \pm\sqrt{3}/2$, and we have

$$U_{xx} = 3/4, \qquad U_{yy} = 9/4, \qquad U_{xy} = \pm 3\sqrt{3}(1 - 2\mu_2)/4. \tag{3.140}$$

The characteristic equation becomes

$$\lambda^4 + \lambda^2 + \frac{27}{4}\mu_2(1 - \mu_2) = 0. \tag{3.141}$$

As above we now consider a specific example. At the $L_4$ point with $\mu_2 = 0.01$ we have $x_0 = 0.49$ and $y_0 = \sqrt{3}/2$ with the same initial conditions as before. The resulting eigenvalues are $\pm 0.963i$ and $\pm 0.268i$, indicating that the point is linearly stable. The solution for the perturbed motion is

$$X(t) = 3.45 \times 10^{-5} \cos 0.268t - 2.45 \times 10^{-5} \cos 0.963t$$
$$+ 3.07 \times 10^{-4} \sin 0.268t - 8.55 \times 10^{-5} \sin 0.963t,$$

$$Y(t) = 5.20 \times 10^{-5} \cos 0.268t - 4.20 \times 10^{-5} \cos 0.963t$$
$$- 1.76 \times 10^{-4} \sin 0.268t + 4.90 \times 10^{-5} \sin 0.963t. \tag{3.142}$$

Therefore the solution is of the oscillatory type with fundamental periods of $1/0.268$ and $1/0.963$ orbital periods of the orbiting body.

Figure 3.13 shows how the nature of the eigenvalues varies with the mass. It is clear that the eigenvalues have real parts for values of $\mu_2$ larger than a critical value ($\log \mu_2 \approx -1.4$). In this case the eigenvalues are of the form $\pm j \pm ik$ and so there will always be a positive real part. However, provided the mass is small enough the eigenvalues are always of the form $\pm ik_1$ and $\pm ik_2$, and the perturbed motion is stable.

We can understand the behaviour shown in Fig. 3.13 by considering the analytical solutions to the characteristic equation. Substituting the expressions for

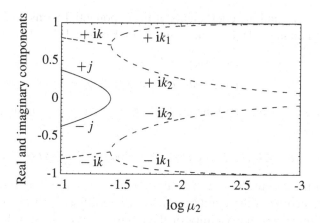

Fig. 3.13. Numerical values of the real (solid lines) and imaginary (dashed lines) components of the roots of the characteristic equation for the $L_4$ and $L_5$ Lagrangian points as a function of the value of $\mu_2$. Note the change in the nature of the eigenvalues at $\log \mu_2 \approx -1.4$.

$U_{xx}$, $U_{yy}$, and $U_{xy}$ into Eqs. (3.121) and (3.122) gives

$$\lambda_{1,2} = \pm \frac{\sqrt{-1 - \sqrt{1 - 27(1 - \mu_2)\mu_2}}}{\sqrt{2}} \tag{3.143}$$

and

$$\lambda_{3,4} = \pm \frac{\sqrt{-1 + \sqrt{1 - 27(1 - \mu_2)\mu_2}}}{\sqrt{2}}. \tag{3.144}$$

Therefore all the eigenvalues will be purely imaginary if and only if the condition

$$1 - 27(1 - \mu_2)\mu_2 \geq 0$$

is satisfied. This implies that the condition for linear stability reduces to

$$\mu_2 \leq \frac{27 - \sqrt{621}}{54} \approx 0.0385. \tag{3.145}$$

When the eigenvalues are purely imaginary they will occur in pairs of the form $\lambda_{1,2} = \pm ik_1$ and $\lambda_{3,4} = \pm ik_2$, where $k_1$ and $k_2$ are real numbers. If we write $\bar{\alpha}_j = \bar{a}_j + i\bar{b}_j$, where $\bar{a}_j$ and $\bar{b}_j$ are real, then, from (3.123), the solution for $X(t)$ is of the form

$$X(t) = (\bar{a}_1 + i\bar{b}_1)e^{+ik_1t} + (\bar{a}_2 + i\bar{b}_2)e^{-ik_1t}$$
$$+ (\bar{a}_3 + i\bar{b}_3)e^{+ik_2t} + (\bar{a}_4 + i\bar{b}_4)e^{-ik_2t}, \tag{3.146}$$

with a similar equation for $Y(t)$. We can use Eq. (3.127) and the fact that $X$, $Y$, $\dot{X}$, and $\dot{Y}$ must all be real to show that $\bar{a}_1 = \bar{a}_2 = a_1$, $\bar{a}_3 = \bar{a}_4 = a_2$, $\bar{b}_1 = -\bar{b}_2 = b_1$, and $\bar{b}_3 = -\bar{b}_4 = b_2$ and the coefficients of the exponential terms consist of pairs of complex conjugates. Hence our solution for $X(t)$ can be written

$$X(t) = (a_1 + ib_1)e^{+ik_1t} + (a_1 - ib_1)e^{-ik_1t}$$
$$+ (a_2 + ib_2)e^{+ik_2t} + (a_2 - ib_2)e^{-ik_2t}. \qquad (3.147)$$

Since $e^{i\theta} = \cos\theta + i\sin\theta$ and the coefficients are complex conjugates this can be rewritten as

$$X(t) = 2a_1 \cos k_1 t + 2a_2 \cos k_2 t - 2b_1 \sin k_1 t - 2b_2 \sin k_2 t. \qquad (3.148)$$

Therefore, if the eigenvalues are purely imaginary the resulting motion of the particle displaced from the equilibrium point is oscillatory in form; hence the particle will remain in the vicinity of the equilibrium point and the motion is stable. This confirms the numerical result given above.

If the condition in Eq. (3.145) is not satisfied then the eigenvalues have real parts and have the form $\pm(j \pm ik)$ (see the left part of Fig. 3.13). In this case we can write

$$X(t) = (a_1 + ib_1)e^{(j+ik)t} + (a_1 - ib_1)e^{(j-ik)t}$$
$$+ (a_2 + ib_2)e^{(-j+ik)t} + (a_2 - ib_2)e^{(-j-ik)t}, \qquad (3.149)$$

which reduces to

$$X(t) = 2\left(a_1 e^{jt} + a_2 e^{-jt}\right)\cos kt - 2\left(b_1 e^{jt} + b_2 e^{-jt}\right)\sin kt. \qquad (3.150)$$

Therefore there will always be exponential growth in the variable $X$ arising from at least two of the terms. The same is true for the expressions for $Y$, $\dot{X}$, and $\dot{Y}$. This results in a trajectory that spirals away from the equilibrium point and hence the point is said to be linearly unstable.

Now let us return to our analysis of the stable case ($\mu_2 < 0.0385$). From Eqs. (3.143) and (3.144) we see that for small values of $\mu_2$ the eigenvalues can be written

$$\lambda_{1,2} \approx \pm\sqrt{-1 + \frac{27}{4}\mu_2}, \qquad \lambda_{3,4} \approx \pm\sqrt{-\frac{27}{4}\mu_2}. \qquad (3.151)$$

These are both purely imaginary numbers for small mass ratios and we have seen numerically and analytically that the moduli of these values of $\lambda$ are the two frequencies of the motion of the particle in its oscillations about the triangular equilibrium point. The existence of these two frequencies in the solution to the perturbed motion near the $L_4$ and $L_5$ points gives rise to a further problem, which reduces the generality of the condition shown in Eq. (3.145). The possible existence of *resonances* (i.e., simple numerical relationships) between these frequencies means that for a finite number of specific mass ratios the points are unstable, even though the relationship in Eq. (3.145) is satisfied (Deprit & Deprit-Bartholomé 1967).

## 3.8 Motion near $L_4$ and $L_5$

We have chosen a system of units such that the mean motion of the mass $\mu_2$ is unity and its orbital period is $2\pi$. The corresponding periods of the particle motion are $2\pi/|\lambda_{1,2}|$ and $2\pi/|\lambda_{3,4}|$. Therefore the resulting motion of the particle is composed of two different motions:

- a short-period motion with a period $2\pi/|\lambda_{1,2}| \approx 2\pi$ (i.e., a period that is close to the orbital period of the mass $\mu_2$) and
- a superimposed longer period motion of period $2\pi/|\lambda_{3,4}|$, known as a *libration*, about the equilibrium point.

The amplitudes of these motions are determined by the constants $\bar{\alpha}_j$ and $\bar{\beta}_j$, which are in turn determined by the starting conditions. In this context the motion can be thought of as a long-period motion of an epicentre around the equilibrium point with the particle simultaneously executing a short-period motion around the epicentre (see Fig. 3.14). The resulting motion of the particle for the numerical solution given by Eq. (3.142) is shown in Fig. 3.15.

The peculiar, looping nature of the particle's path shown in Fig. 3.15 results from the two different types of motion that contribute to the perturbed orbit about the equilibrium point. The nature of this motion can be further simplified by rotating the coordinate system by $30°$ ($\pi/6$ radians) clockwise about the equilibrium point such that the transformed $X$ axis is nearly tangential to the unit circle. The new coordinates, $(X', Y')$, are given by

$$\begin{pmatrix} X'(t) \\ Y'(t) \end{pmatrix} = \begin{pmatrix} \cos 30° & -\sin 30° \\ \sin 30° & \cos 30° \end{pmatrix} \begin{pmatrix} X(t) \\ Y(t) \end{pmatrix}, \tag{3.152}$$

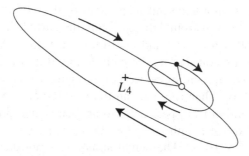

Fig. 3.14. The epicyclic motion (small ellipse) and the motion of the epicentre (large ellipse) for the solution given in Eq. (3.142). The epicentre is denoted by the small empty circle. The cross denotes the location of the $L_4$ equilibrium point. The particle's path around the equilibrium is a combination of the epicyclic motion and the motion of the epicentre.

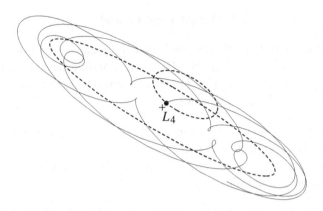

Fig. 3.15.    The trajectory in the rotating frame of a particle librating about the $L_4$ Lagrangian equilibrium point. The trajectory is derived from the analytical solution to the stability problem given in Eq. (3.142) and the motion is shown for $25\pi$ time units (12.5 orbital periods of the mass $\mu_2$). The dashed lines denote the epicyclic motion and the path of the epicentre shown in Fig. 3.14.

where $X(t)$ and $Y(t)$ are given by Eq. (3.142). Hence

$$X'(t) \approx 3.54 \times 10^{-4} \sin 0.268t - 9.85 \times 10^{-5} \sin 0.963t,$$
$$Y'(t) \approx 6.23 \times 10^{-5} \cos 0.268t - 4.86 \times 10^{-5} \cos 0.963t.$$

(3.153)

Although the values of the constants (apart from the frequencies) in these equations depend on the specific boundary conditions, the rotation serves to highlight several additional properties of the orbital solution. By separating the two types of motion we see that each is of the form given in Eq. (2.40) and so each describes a centred ellipse. In the case of the motion of the epicentre the path is an elongated ellipse. In Sect. 3.10 we will show that the dimensions of this ellipse are determined by the dimensions of the *associated zero-velocity curve* and that the ratio of the semi-minor and semi-major axes is given by $b/a = (3\mu_2)^{1/2}$.

For the epicyclic motion around the epicentre the particle traces a path that is a centred ellipse with semi-major and semi-minor axes in the approximate ratio 2:1. Given that we are considering the path in a rotating frame, this motion is identical to the epicycle or guiding centre approximation to the two-body problem described in Sect. 2.6. Therefore we can think of the epicyclic motion as the regular keplerian motion from pericentre to apocentre that produces a centred ellipse with semi-axes of length $2e$ and $e$. In our particular case this suggests that we have $e \approx 5 \times 10^{-5}$. The actual situation is not quite so simple and we have to consider the osculating eccentricity of the particle as a combination of a "forced" eccentricity from the mass $\mu_2$ (reflected in part by the changing shape of the elongated ellipse of the epicentre's path) and a "free" eccentricity giving the 2:1 epicyclic motion. We will return to this concept in Chapter 7 when we consider secular perturbations. We note in passing that by a judicious choice of starting conditions it is possible to find orbital solutions where the epicyclic motion is entirely suppressed and the particle's path is just the path of the epicentre.

## 3.9 Tadpole and Horseshoe Orbits

It is important to remember that the types of motion described by the solution of the equations for perturbed motion around $L_4$ and $L_5$ are only valid in the vicinity of these equilibrium points and, by implication, only for small amplitudes of libration. This means that we cannot deduce anything about the solution for large displacements from the equilibrium points. However, it is always possible to resort to numerical integration of the full equations of motion of the particle in the rotating frame; these were derived in Sect. 3.2 (see Eqs. (3.16) and (3.17)).

Figure 3.16 shows two separate trajectories obtained by integrating the full equations of motion for starting conditions in the vicinity of the $L_4$ point for $\mu_2 = 0.001$, a value similar to the Sun–Jupiter mass ratio. The trajectory shown in Fig. 3.16a has been started slightly closer to $L_4$ than the one shown in Fig. 3.16b. Note that in each case the path is more elongated ahead of the $L_4$ point. Furthermore, in Fig. 3.16a the orbit extends over 86°; the orbit that was started further from $L_4$ shown in Fig. 3.16b extends over 115°. Recalling that the two masses and $L_4$ form an equilateral triangle, the side of which is the unit of length, it is clear from Fig. 3.16 that both orbits move close to the unit circle centred on the mass $\mu_1$ (cf. Fig. 3.9). The orbits shown in Fig. 3.16 are referred to as *tadpole orbits* because of the elongated shapes of the orbits (and the zero-velocity curves). The similarity is even more enhanced when orbits with zero free eccentricity are considered. Note that tadpole orbits also describe particle paths around the $L_5$ point.

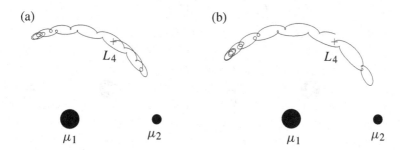

Fig. 3.16. Two examples of tadpole orbits librating about the $L_4$ equilibrium point (denoted by a cross and located at $x_0 = 1/2 - \mu_2$, $y_0 = \sqrt{3}/2$) for $\mu_2 = 0.001$. The masses $\mu_1$ and $\mu_2$ are denoted by the filled circles. (a) The starting conditions are $x = x_0 + 0.0065$, $y = y_0 + 0.0065$ with $\dot{x} = \dot{y} = 0$ and the orbit is followed for 15 orbital periods of $\mu_2$. (b) The starting conditions are $x = x_0 + 0.008$, $y = y_0 + 0.008$ with $\dot{x} = \dot{y} = 0$ and the orbit is followed for 15.5 orbital periods of $\mu_2$.

The question now arises as to what kind of orbit would we expect if we increase the initial radial separation from $L_4$ or $L_5$ even more? The answer is that provided the initial distance is not too large, the resulting orbit will encompass both $L_4$ and $L_5$. These are referred to as *horseshoe orbits* and two examples are shown in Fig. 3.17 using starting conditions taken from the paper by Taylor (1981). Note that the orbit shown in Fig. 3.17b has starting conditions that almost suppress the epicyclic motion, giving a smooth appearance to the path.

Comparison of the paths of the epicentres deduced from Figs. 3.16 and 3.17 with the form of the zero-velocity curves shown in Fig. 3.9 reveals striking similarities. However, it is important to note that the zero-velocity curves do not define the orbital paths, although, as we shall see in Sect. 3.10, there is a close connection between the two when $\mu_2$ is small. Note that those orbits shown in Figs. 3.16 and 3.17 have zero initial velocities. Hence there is a zero-velocity curve associated with each of the orbits, although these have not been drawn. Such curves only guarantee that the orbits do not get too close to the $L_4$ and $L_5$ points and give no indication of the long-term stability of the orbits. However, remember that if the Jacobi constant of an orbit is less than that for $L_4$ and $L_5$, then a zero-velocity curve cannot be drawn and there are no excluded regions.

We can illustrate the properties of horseshoe and tadpole orbits for small mass ratios with a number of examples obtained from direct numerical integrations of the equations of motion. Figure 3.18 shows three trajectories (one horseshoe and two tadpole) for the case where $\mu_2 = 10^{-3}$. Here $\theta$ is the angle around the orbit, with $\theta = 0$ corresponding to the line from the primary mass to the secondary mass. All the orbits are started with eccentricity $e \approx 0$. Note that there is considerable variation of the semi-major axis around the orbit with the

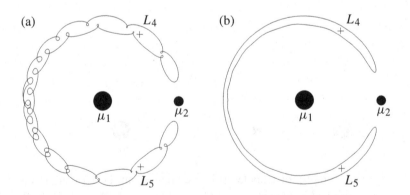

Fig. 3.17. Two examples of near-periodic horseshoe orbits librating about the $L_4$ equilibrium point for $\mu_2 = 0.000953875$, taken from data given by Taylor (1981). (a) The starting conditions are $x = -0.97668$, $y = \dot{x} = 0$, $\dot{y} = -0.06118$. (b) The starting conditions are $x = -1.02745$, $y = \dot{x} = 0$, $\dot{y} = 0.04032$.

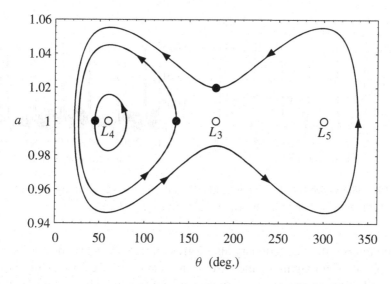

Fig. 3.18. The variation of semi-major axis $a$ with the angle $\theta$ around the orbit for three trajectories with $\mu_2 = 10^{-3}$. The starting positions are indicated by filled circles and the enclosed equilibrium points with open circles. All orbits were followed for 100 orbital periods of the secondary mass and arrows indicate the direction of the motion.

maximum deviation from the unit circle occurring at the $L_4$ and $L_5$ equilibrium points. At these points the horseshoe and tadpoles have their maximum width. Note too that there is considerable asymmetry in the path. This is demonstrated by the fact that the $L_3$ point does not lie midway between the outer and inner branches of the horseshoe at $\theta = 180°$. Although we are plotting the variation of $a$ and not $r$ around the path, there would be little difference because we have chosen $e \approx 0$. Therefore the loops apparent in Figs. 3.16 and 3.17 do not appear in this example.

The changes in $a$ and $e$ as a function of time for the horseshoe orbit (with $\mu_2 = 10^{-3}$) are shown in Fig. 3.19. The relatively sudden changes in $a$ denote the effect of encounters with the secondary mass when the orbit switches from $a > 1$ to $a < 1$, or vice versa (see Fig. 3.19a). If we write $a = 1 + \Delta a$ and measure $\Delta a$ every time $\theta = 180°$ then an interesting phenomenon emerges. We chose an initial value of $\Delta a = 0.020$. After one encounter with the secondary $\Delta a = -0.0143$, but after two encounters measurements show that $\Delta a = 0.0198$, with the pattern repeating itself at subsequent encounters. This demonstrates that the asymmetry in $|\Delta a|$ is almost cancelled out after one complete cycle of the horseshoe. As we shall see in Sect. 3.15, this property can be useful in maintaining horseshoe orbits against the orbital decay produced by dissipation (Dermott et al. 1980, Dermott & Murray 1981a). It is clear from Fig. 3.19b that

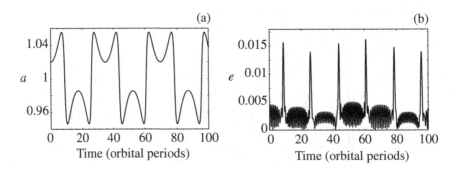

Fig. 3.19.  The variation of (a) semi-major axis $a$, and (b) eccentricity $e$ as a function of time for the horseshoe orbit shown in Fig. 3.18 for $\mu_2 = 10^{-3}$.

the encounters with the secondary also correspond to times of sudden changes in $e$, with an order of magnitude increase on approach and decrease on retreat. The small oscillations in eccentricity between encounters (which, in accordance with Tisserand's relation also have a maximum amplitude at the $L_4$ and $L_5$ points) imply that phase effects could be important. This helps to explain why there are variations in the magnitude of the eccentricity "impulse" at encounters, rather than the symmetry (after two encounters) shown in the variation in $a$.

The effect of a decrease in the mass ratio is shown in Figs. 3.20 and 3.21, where we show two of the equivalent plots for the case $\mu_2 = 10^{-6}$. Because of the decrease in $\mu_2$ we have chosen the initial semi-major axis to be $a = 1.002$, (i.e., a value of $\Delta a$ that is an order of magnitude smaller). The plot of the variation of $a$ with $\theta$ shows, as before, that the paths have maximum width in $a$ at $\theta = 60°$ and $\theta = 300°$, the locations of the $L_4$ and $L_5$ points. However, there are two subtle differences. Firstly, the degree of symmetry is more pronounced with $L_3$ now lying approximately midway between the values of $a$ at $\theta = 180°$. Secondly, although we have started both tadpoles with the same values of $\theta$ as before ($135°$ and $45°$) they have a smaller relative extent in $a$ than their counterparts for $\mu_2 = 10^{-3}$. This implies that the radial extent of the tadpole zone with respect to the horseshoe zone has decreased with decreasing $\mu_2$. For the tadpole orbit we have added its zero-velocity curve for comparison. Note that its radial width is always half that of the tadpole orbit, in keeping with our analytical result.

The extent of symmetry in the small-$\mu_2$ case is clearly shown in Fig. 3.21. The integration shows that that while initially $\Delta a = 0.00200$, after one encounter $\Delta a = -0.00199$, and after two encounters $\Delta a = 0.00200$. If the long-term stability of a horseshoe orbit can be deduced from its proximity to a symmetric periodic orbit then this suggests that horseshoe orbits in the small-$\mu_2$ case are highly stable. Examination of the changes in $e$ shows that order of magnitude

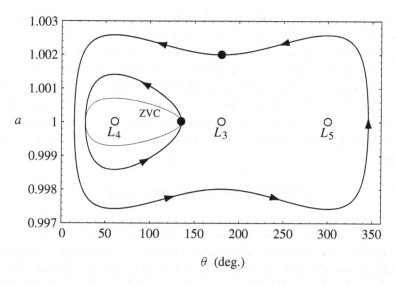

Fig. 3.20. The variation of semi-major axis $a$ with the angle $\theta$ around the orbit for horseshoe and tadpole trajectories with $\mu_2 = 10^{-6}$. The starting positions are indicated by filled circles and the enclosed equilibrium points with open circles. The zero-velocity curve for the tadpole trajectory is the thin curve labelled ZVC. All orbits were followed for 1,000 orbital periods of the secondary mass and arrows indicate the direction of the motion.

changes still occur at encounters but the symmetry is much more pronounced than in the large $\mu_2$ case.

If a starting condition is chosen that corresponds to a near-circular orbit either interior or exterior to the unit circle, then it is clear that the nature of the resulting orbit depends on the radial separation of the orbit from the unit circle. We have shown analytically and numerically how particles started close

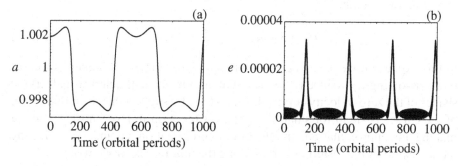

Fig. 3.21. The variation of (a) semi-major axis $a$, and (b) eccentricity $e$ as a function of time for the horseshoe orbit shown in Fig. 3.20 for $\mu_2 = 10^{-6}$.

to $L_4$ and $L_5$ execute a small-amplitude libration about the equilibrium point. By increasing the initial separation the resulting orbit becomes more elongated in the direction of the $L_3$ point (tadpole orbits). Eventually particles started sufficiently far from the unit circle will librate about $L_4$, $L_3$, and $L_5$ (horseshoe orbits). However, particles moving sufficiently far from the unit circle will not give rise to orbits with a change in the direction of motion of the epicentre and the orbits are said to *circulate* interior or exterior to the unit circle. Depending on the value of the Jacobi constant it may be possible for the particle to encounter the mass $\mu_2$.

## 3.10 Orbits and Zero-Velocity Curves

The zero-velocity curve that defines the limits of the forbidden region in Fig. 3.15 has an elongated shape and is tilted at an angle of $\sim 30°$ to the horizontal. We can study the behaviour of such curves in the vicinity of the triangular equilibrium points by means of a translation of the origin, a rotation of $30°$, and an expansion about the new origin. At the $L_4$ point, for example, we have $x = 1/2 - \mu_2$ and $y = \sqrt{3}/2$. The translation of the origin is achieved with the substitutions $x \to (1/2 - \mu_2) + x$ and $y \to \sqrt{3}/2 + y$ in Eqs. (3.8) and (3.9). A rotation of the coordinate system by $30°$ about the new origin is achieved by a substitution $x \to \sqrt{3}x'/2 + y'/2$ and $y \to -x'/2 + \sqrt{3}y'/2$, where $x'$ and $y'$ are the new values of $x$ and $y$. This gives

$$r_1^2 = 1 + 2y' + x'^2 + y'^2, \tag{3.154}$$
$$r_2^2 = 1 - \sqrt{3}x' + y' + x'^2 + y'^2. \tag{3.155}$$

From Eq. (3.22) the transformed equation defining the zero-velocity curves, $C_J = 2U$, is given by

$$C_J = 1 - \mu_2 - \sqrt{3}\mu_2 x' + (2 - \mu_2)y' + x'^2 + y'^2 + 2\left(\frac{(1-\mu_2)}{r_1} + \frac{\mu_2}{r_2}\right), \tag{3.156}$$

where terms of $\mathcal{O}(\mu_2^2)$ have been neglected. Expanding Eq. (3.156) about the new origin we obtain

$$C_J \approx 3 - \mu_2 + \frac{9}{4}\mu_2 x'^2 + 3y'^2. \tag{3.157}$$

In this expansion we have neglected terms of order three and higher as well as terms involving $\mu_2 y'$ since (i) $\mu_2$ is assumed to be a small quantity and (ii) the shapes of the zero-velocity curves shown in Fig. 3.9 suggest that the radial extent of the curves is small for small $\mu_2$; note that the term in $\mu_2 x$ vanishes without making any approximation. In effect our approximation amounts to neglecting the curvature of the resulting curves along the unit radius. If we write

$$C_J = 3 + \gamma\mu_2, \tag{3.158}$$

where $\gamma$ is a small quantity equal to $-1$ at the $L_4$ and $L_5$ points (see Eqs. (3.97) and (3.98)), then Eq. (3.157) can be written

$$\frac{x'^2}{(4/9)(1+\gamma)} + \frac{y'^2}{(\mu_2/3)(1+\gamma)} = 1. \tag{3.159}$$

If we compare this with Eq. (2.40) we see that the resulting zero-velocity curves are ellipses centred on the $L_4$ point with $a' = (2/3)\sqrt{1+\gamma}$ and $b' = \sqrt{\mu_2/3}\sqrt{1+\gamma}$ as the values of the semi-major and semi-minor axes respectively. Since $b'/a' = (1/2)\sqrt{3\mu_2}$ it is clear that these ellipses become highly elongated as $\mu_2 \to 0$. Note also that the expression for $b'/a'$ is twice the value we gave in Sect. 3.8 for the motion of the guiding centre, suggesting that it is the shape of the zero-velocity curve that determines the particle path. We now investigate the relationship between the zero-velocity curves and the actual orbits in the case where $\mu_2 \ll \mu_2^{1/2} \ll \mu_2^{1/3} \ll 1$. We also assume that the eccentricity of the particle's orbit is near zero and thus, in effect, that the epicyclic motion of the particle about its guiding centre is negligible.

We start by noting from Eqs. (3.94)–(3.98) that the zero-velocity curves that give rise to the tadpole-shaped curves can be characterised by the quantity $\gamma$ defined in Eq. (3.158), where

$$-1 \le \gamma \le +1. \tag{3.160}$$

The lower limit of $\gamma$ corresponds to the value of $C_J$ at the $L_4$ and $L_5$ points whereas the upper limit corresponds to the value at the $L_3$ point. In the case of the horseshoe-shaped curves we can write

$$C_J = 3 + \zeta\mu_2^{2/3} + \mathcal{O}(\mu_2), \tag{3.161}$$

where

$$0 \le \zeta \le 3^{4/3}. \tag{3.162}$$

In this case the lower limit of $\zeta$ corresponds to the value of $C_J$ (to $\mathcal{O}(\mu_2)$) at the $L_3$ point and the upper limit to the value at the $L_1$ and $L_2$ points. Therefore the value of either $\gamma$ or $\zeta$ serves to parameterise the tadpole or horseshoe nature of the zero-velocity curve.

Using the particular form of the function $U$ given in Eq. (3.64) and taking $n = 1$ we have

$$2U = 2\frac{\mu_1}{r_1} + \mu_1 r_1^2 + 2\frac{\mu_2}{r_2} + \mu_2 r_2^2 - \mu_1\mu_2 \tag{3.163}$$

with, from Eq. (3.28),

$$v^2 = \dot{r}^2 + (r\dot{\theta})^2 = 2U - C_J. \tag{3.164}$$

Because we are considering $\mu_2 \ll 1$ we make no distinction between $r_1$ and $r$.

If we now restrict our attention to those orbits that are nearly circular and lie close to the unit circle we can write

$$r = 1 + \delta r, \tag{3.165}$$

where $\delta r \ll 1$. In these circumstances $|\dot{r}| \ll |r\dot{\theta}|$ except near the turning points where we have $\dot{\theta} = 0$ but $\dot{r} \neq 0$. Hence, for most of the time the motion is near-keplerian and

$$v \approx r\dot{\theta} = -\frac{3}{2}\delta r. \tag{3.166}$$

Using the same type of approximations we can simplify the expression for $2U$ given in Eq. (3.163). Including terms up to $\mathcal{O}(\mu_2)$ gives

$$2U = 3 + 3\delta r^2 + \mu_2 H, \tag{3.167}$$

where

$$H = \frac{2}{r_2} + r_2^2 - 4. \tag{3.168}$$

Hence, using $v^2 = 2U - C_J$ and Eq. (3.164), we have

$$\frac{9}{4}\delta r^2 = 3 + 3\delta r^2 + \mu_2 H - C_J. \tag{3.169}$$

If we consider the equation for the zero-velocity curves then $v^2 = 0$ and the curves are defined by the equation

$$0 = 3 + 3\delta r_{zv}^2 + \mu_2 H_{zv} - C_J, \tag{3.170}$$

where the subscript "zv" denotes that the quantities described refer to the zero-velocity curves, not the orbits. Given that $C_J$ is a constant, Eqs. (3.169) and (3.170) give

$$\delta r^2 = \left(2\delta r_{zv}\right)^2 + \frac{4}{3}\mu_2 \left(H_{zv} - H\right). \tag{3.171}$$

For motion in both tadpole and horseshoe orbits, $H_{zv} - H \sim \delta r$ and $\mu_2 \delta r \ll \delta r^2$. Hence, for the motion of the guiding centre

$$\delta r = 2\delta r_{zv}, \tag{3.172}$$

and the radial displacement of the guiding centre from the unit circle is always twice that of the associated zero-velocity curve (Dermott & Murray 1981a).

In the case of horseshoe orbits Eqs. (3.161) and (3.162) apply and thus

$$3 + \frac{3}{4}\delta r^2 + \mu_2 H = 3 + \varsigma\mu_2^{2/3} + \mathcal{O}(\mu_2),$$
$$3 + 3\delta r_{zv}^2 + \mu_2 H_{zv} = 3 + \varsigma\mu_2^{2/3} + \mathcal{O}(\mu_2). \tag{3.173}$$

Hence, since $\mu_2^{2/3} \gg \mu_2$ for small values of $\mu_2$, we have

$$\delta r = 2\delta r_{zv} = 2(\varsigma/3)^{1/2}\mu_2^{1/3}. \tag{3.174}$$

For tadpole orbits Eqs. (3.158) and (3.160) apply and so

$$3 + \frac{3}{4}\delta r^2 + \mu_2 H = 3 + \gamma\mu_2, \qquad 3 + 3\delta r_{zv}^2 + \mu_2 H_{zv} = 3 + \gamma\mu_2. \qquad (3.175)$$

Hence

$$\delta r \approx 2\delta r_{zv} = 2[(\gamma - H)/3]^{1/2}\mu_2^{1/2}, \qquad (3.176)$$

and again the radial width of the orbit is twice the width of its associated zero-velocity curve.

We have already seen in Sect. 3.8 that for small oscillations about $L_4$ or $L_5$, the semi-minor axis of the elliptical path of the guiding centre is twice that of the semi-minor axis of the ellipse describing the zero-velocity curve associated with the guiding centre. The argument given above suggests, firstly, that the motion of the particle described numerically in Eq. (3.153) is given by

$$X'(t) = a \sin\lambda_3(t - t_3) - 2e \sin\lambda_1(t - t_1),$$
$$Y'(t) = (3\mu_2)^{1/2}a \cos\lambda_3(t - t_3) - e \cos\lambda_1(t - t_1), \qquad (3.177)$$

where $\lambda_1$ and $\lambda_3$ are the two fundamental eigenfrequencies and the four arbitrary constants, $a$, $e$, $t_1$, and $t_3$, are determined by the initial conditions, $X'(0)$, $Y'(0)$, $\dot{X}'(0)$, and $\dot{Y}'(0)$. Secondly, it suggests that for small $e$, the motion of the guiding centre is independent of $e$. Thirdly, and most importantly, it suggests that the relationship between the path of the guiding centre and its associated zero-velocity curve that was first encountered when we considered small oscillations about $L_4$ and $L_5$ is true in general and can be applied to motions of any amplitude including motions in horseshoe orbits. We now use this result to derive further properties of the motion.

Once we consider motion far from $L_4$ or $L_5$, the orbits become more tadpolelike with "tails" extending towards $L_3$. We can calculate the minimum and maximum angular extent of these orbits by making use of the Jacobi constant. Equations (3.173) and (3.175) can be rewritten as

$$\frac{3}{4}\delta r^2 + \mu_2 H(\theta) = \text{constant}, \qquad (3.178)$$

where $\theta$ is the angular separation of the particle from the secondary mass; using the geometry of the isosceles triangle $r_2 = 2\sin\theta/2$ and hence

$$H(\theta) = \left(\sin\frac{\theta}{2}\right)^{-1} - 2\cos\theta - 2, \qquad (3.179)$$

where $H$ is the same function as defined in Eq. (3.168). The constant in Eq. (3.178) is $\gamma\mu_2$ for tadpole orbits and $\zeta\mu_2^{2/3}$ for horseshoe orbits.

Consider any two points $(r_i, \theta_i)$ and $(r_j, \theta_j)$ on the particle's path. These polar coordinates must satisfy the relation

$$\delta r_i^2 - \delta r_j^2 = -\frac{4}{3}\mu_2\left[H(\theta_i) - H(\theta_j)\right] \qquad (3.180)$$

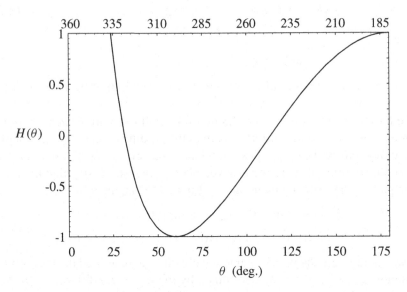

Fig. 3.22.   A plot of $H(\theta)$, defined in Eq. (3.179), as a function of $\theta$ for tadpole orbits around $L_4$ (lower scale) and $L_5$ (upper scale).

regardless of whether the particle is in a tadpole or horseshoe orbit. Figure 3.22 shows a plot of $H(\theta)$ as a function of $\theta$. In the case of tadpole orbits we know that at the extremes of the motion $\delta r_i = \delta r_j = 0$ and hence $H(\theta_i) = H(\theta_j)$. The two solutions of this equation, $\theta_{\min}$ and $\theta_{\max}$, give the minimum and maximum angular separations of the particle from the secondary mass, and their difference, $D = \theta_{\max} - \theta_{\min}$, gives the amplitude of the libration. The shape of the $H(\theta)$ curve in Fig. 3.22 helps to explain the observation from our numerical integrations that the tadpole orbits become more elongated at larger amplitudes (see Figs. 3.18 and 3.20).

Consider the case when $\theta_i = 180°$ and $\delta r_i = 0$. This is the orbit of the critical tadpole. With $H(180°) = 1$, Eq. (3.180) gives

$$\delta r_j^2 = \frac{4}{3}\mu_2 \left[1 - H(\theta_j)\right],    \tag{3.181}$$

and because $H(\theta) \geq -1$, any tadpole orbit must satisfy the equation

$$\delta r \leq \delta r_{\text{crit}} = \left(\frac{8}{3}\right)^{1/2} \mu_2^{1/2}.    \tag{3.182}$$

Therefore, provided $\mu_2$ is known and $e$ is small, a single observation of a $\delta r$ (i.e., the radial separation of the particle from the secondary) is sufficient to determine whether or not the particle moves in a tadpole orbit.

Consider the case when $\theta_i = 180°$ and $\delta r_i = \delta r_{180} \neq 0$. Provided $\zeta \leq 3^{4/3}$ (cf. Eq. (3.162)) the particle is moving on a horseshoe orbit and the turning point of its trajectory occurs when $\delta r_j = 0$ and $\theta_j = \theta_{min}$. These are related by

$$\delta r_{180}^2 = \frac{4}{3}\mu_2 \left[ H(\theta_{min}) - 1 \right],$$ (3.183)

and this can be used to determine the closest approach distance between the particle and the secondary and in the case of horseshoe motion an observation of $\theta_{min}$ could be used to determine the mass $\mu_2$. In the case of tadpole motion $\delta r_{180} = 0$ and $\theta_{min}$ is given by $H(\theta_{min}) = 1$ from which we deduce that $\theta_{min} = 23.5°$ for all $\mu_2$.

### 3.11 Trojan Asteroids and Satellites

The condition Eq. (3.145) is satisfied for every Sun–planet pair in the solar system. The same condition is also satisfied for every planet–satellite pair, with the exception of the Pluto–Charon system where $\mu_2 \sim 0.1$. However, although the behaviour of objects librating about the stable equilibrium points was known to Lagrange, such objects were not discovered until this century. Table 3.1 gives a list of some of the asteroids that are known to move in tadpole orbits about the triangular equilibrium points in the Sun–Jupiter system. These are referred to as the *Trojan asteroids* and the first, (588) Achilles, was discovered in 1906 librating around Jupiter's $L_4$ point.

Up to the end of 1998 more than 450 Trojan asteroids had been discovered librating about Jupiter's $L_4$ and $L_5$ points. The leading ($L_4$) group are commonly referred to as "Greeks" and the trailing ($L_5$) group as "Trojans" with appropriate names taken from the respective sides that are mentioned in Homer's *Iliad* with the added complication of an enemy spy in each camp.

The amplitudes of libration (the quantity $D$ in Table 3.1) can exceed 30° but the mean value of the amplitude is 14°. In such cases the motion of the guiding centre begins to resemble the more elongated zero-velocity curves shown in Fig. 3.9. The physical and dynamical properties of the Trojan asteroids are summarised by Shoemaker et al. (1989).

Figure 3.23 shows the distribution of the Sun–Jupiter Trojan asteroids in December 1997. The projected positions onto the plane of the ecliptic are shown in Fig. 3.23a, together with the orbit and position of Jupiter with respect to the Sun. Although there are two distinct clusterings about the triangular points, the large libration amplitudes are apparent. Figure 3.23b shows the view along the Sun–Jupiter line, illustrating the vertical extent of the Trojan groups.

The presence of Trojan asteroids is not unique to the Sun–Jupiter system: The first Sun–Mars Trojan, (5261) Eureka, was discovered in 1990; Eureka has been shown to be librating about the $L_5$ point of Mars (Mikkola et al. 1994). Asteroid

Table 3.1. Orbital properties of the Trojan asteroids with numbers less than 2,000 that are associated with the Sun–Jupiter system. $D$ is the amplitude of libration; $e_{pr}$ and $I_{pr}$ are the proper eccentricity and inclination respectively as determined by Milani (1993). The final entry indicates whether libration takes place about the $L_4$ or $L_5$ point.

| Asteroid | $D\,(°)$ | $e_{pr}$ | $\sin I_{pr}$ | $L_4$ | $L_5$ |
|---|---|---|---|---|---|
| (588) Achilles | 6.45 | 0.1032 | 0.1967 | • | |
| (617) Patroclus | 5.02 | 0.1005 | 0.3662 | | • |
| (624) Hektor | 18.99 | 0.0543 | 0.3259 | • | |
| (659) Nestor | 10.03 | 0.1297 | 0.0870 | • | |
| (884) Priamus | 10.82 | 0.0883 | 0.1739 | | • |
| (911) Agamemnon | 16.95 | 0.0207 | 0.3857 | • | |
| (1143) Odysseus | 9.84 | 0.0521 | 0.0689 | • | |
| (1172) Aeneas | 10.15 | 0.0602 | 0.3056 | | • |
| (1173) Anchises | 23.99 | 0.0914 | 0.1404 | | • |
| (1208) Troilus | 10.63 | 0.0354 | 0.5446 | | • |
| (1404) Ajax | 19.98 | 0.0761 | 0.3270 | • | |
| (1437) Diomedes | 28.73 | 0.0179 | 0.3653 | • | |
| (1583) Antilochus | 24.36 | 0.0183 | 0.4858 | • | |
| (1647) Menelaus | 7.93 | 0.0587 | 0.1168 | • | |
| (1749) Telamon | 13.61 | 0.0686 | 0.1185 | • | |
| (1867) Deiphobus | 17.50 | 0.0294 | 0.4738 | | • |
| (1868) Thersites | 22.88 | 0.0979 | 0.2906 | • | |
| (1869) Philoctetes | 21.04 | 0.0576 | 0.0596 | • | |
| (1870) Glaukos | 9.49 | 0.0169 | 0.1114 | | • |
| (1871) Astyanax | 27.76 | 0.0142 | 0.1299 | | • |
| (1872) Helenos | 23.55 | 0.0148 | 0.2538 | | • |
| (1873) Agenor | 12.08 | 0.1168 | 0.3791 | | • |

(3753) Cruithne, originally designated 1986TO, appears to be involved in an unusual horseshoe libration in the Sun–Earth system (Wiegert et al. 1997). Work by Namouni (1999), Christou (1999) and Namouni et al. (1999) shows that the true nature of Cruithne's behaviour, and that of dynamically similar asteroids can be understood in the context of a comprehensive theory of coorbital motion. For example, both Cruithne and (3362) Khufu can become temporary retrograde satellites of Earth whereas 1989VA is involved in a variety of coorbital modes with Venus.

The Trojan asteroids belonging to Jupiter and Mars are examples of librational motion about the $L_4$ and $L_5$ points of Sun–planet systems. However, the dynamics are identical if we also consider motion in the vicinity of the triangular points of a planet–satellite system. These are usually referred to as *coorbital satellites*,

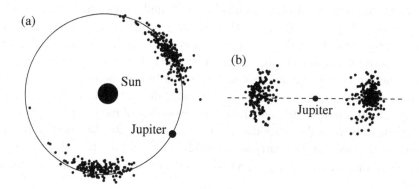

Fig. 3.23. (a) The distribution of asteroids in the vicinity of the orbit of Jupiter on December 18, 1997 at $0^h$ UT (Julian Date 2450800.5). The plot denotes the positions of the asteroids projected onto the plane of the ecliptic. (b) The vertical distribution of the same asteroids viewed along the Jupiter–Sun line. The dashed line denotes the plane of Jupiter's orbit.

although they have also been called *Trojan satellites*. The first coorbital satellites were discovered in 1980 using ground-based CCD observations of the Saturn system. The three objects lie in the orbits of Tethys and Dione (see Table 3.2) and all were directly imaged by the *Voyager* spacecraft during their Saturn flybys in 1980 and 1981.

Although the term *coorbital* was first used to describe the configuration of Janus and Epimetheus (see Sect. 3.12 below), it is now used in connection with any material that shares the same orbit with a larger perturber. Despite the fact that coorbital satellites are common in the saturnian system, it is interesting to note that none exist in the jovian system. A possible explanation for these differences may lie in the relative widths of the tadpole and horseshoe regions.

We have already had some indication of this phenomenon in Sect. 3.9 where we undertook numerical investigations of horseshoe and tadpole orbits for two different mass ratios. Comparing Eqs. (3.174) and (3.176) we see that the width of the horseshoe region is $\sim \mu_2^{1/3}$ whereas that of the tadpole region is $\sim \mu_2^{1/2}$.

Table 3.2. A list of known coorbital satellites, the objects involved, and the extent of the libration. The libration entry indicates the extent of motion with respect to the $L_4$ or $L_5$ point.

| Satellite | Primary | Secondary | $\mu_2$ | $L_4$ | $L_5$ | Libration (°) |
|-----------|---------|-----------|---------|-------|-------|---------------|
| Telesto | Saturn | Tethys | $1.34 \times 10^{-6}$ | • | | $-2°$ to $+2°$ |
| Calypso | Saturn | Tethys | $1.34 \times 10^{-6}$ | | • | $-4°$ to $+4°$ |
| Helene | Saturn | Dione | $1.85 \times 10^{-6}$ | • | | $-13°$ to $+17°$ |

Therefore the ratio, $R$, of these widths is $\sim \mu_2^{1/6}$ and so $R$ decreases (slowly) as $\mu_2$ decreases. For $\mu_2 \approx 10^{-3}$, $R \approx 0.3$, but for $\mu_2 \approx 10^{-9}$, $R \approx 0.03$. This may explain why horseshoe orbits are more likely for small mass ratios. The lack of any coorbital satellites in the jovian system may be attributable to the larger mass ratios involved, the suggestion being that horseshoe orbits have shorter lifetimes than their tadpole counterparts because they have closer approaches to the perturbing satellite. By considering the evolution of the particle's $|\Delta a|$ after two encounters with the secondary to be a random walk, Dermott & Murray (1981a) estimated that the particles would be lost from the horseshoe configuration on a timescale $\Gamma = T/\mu_2^{5/3}$, where $T$ is the orbital period of the secondary mass.

### 3.12 Janus and Epimetheus

All Trojan asteroids and the above mentioned saturnian satellites are known to be moving in tadpole orbits. However, as shown above, another type of motion is also possible, where libration takes place about $L_4$, $L_5$, and $L_3$ with a guiding centre trajectory similar in shape to some of the zero-velocity curves shown in Fig. 3.9. These are the horseshoe orbits and although they were known to exist in theory and had been studied in numerical experiments, no such orbits had been found in the solar system until relatively recently.

In 1980 two satellites of Saturn, now named Janus and Epimetheus, were imaged by the *Voyager 1* spacecraft. At the time Janus had a semi-major axis $a_J = 151,472$ km while Epimetheus, the smaller of the two satellites, had $a_E = 151,422$ km (i.e. an orbital separation of 50 km). They have approximate mean diameters of 175 km and 105 km respectively. Since they were $\sim 180°$ apart when they were discovered in February 1980, a naïve analysis would have suggested a collision some time in 1982. However, it was quickly realised that the orbits are performing a variation on the horseshoe solution of the circular restricted problem. Taking $\mu_2 = 5 \times 10^{-9}$ and $\delta r = 3 \times 10^{-4}$, values appropriate for the Janus–Epimetheus system, application of Eq. (3.182) shows that $\delta r > \delta r_{crit} \approx 17$ km, whereas Eq. (3.174) gives $\zeta = 0.02 < 3^{4/3}$. Therefore we would expect Epimetheus to be moving in a horseshoe orbit.

However, resolved images of the two satellites showed that the mass of Epimetheus could not be considered as negligible compared to that of Janus if their densities are comparable (we now know that their mass ratio is $\sim 0.25$), and therefore mutual perturbations will be important when they approach one another. In fact, as we shall see, approaches between the two satellites lead to a simple modification of the horseshoe configuration. In a reference frame centred on Saturn and rotating with the average mean motion of either satellite, Janus and Epimetheus each librate on their own horseshoe path about longitudes 180° apart. If $W_J$ and $W_E$ represent the average widths of the librational arcs of Janus and Epimetheus respectively, then by assuming circular orbits for each satellite

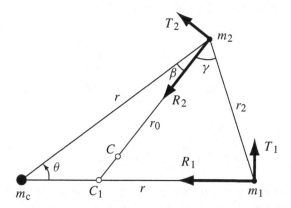

Fig. 3.24. The geometry of the modified horseshoe orbit configuration for two masses $m_1$ and $m_2$ orbiting a central mass $m_c$. $C_1$ is the centre of mass of $m_c$ and $m_1$ while $C$ is the centre of mass of the system. (cf. Fig. 3.6.)

and conservation of the total orbital angular momentum it is easy to show that

$$m_J W_J = m_E W_E. \tag{3.184}$$

We can study the dynamics of this system by taking an approach similar to that in Sect. 3.5. Consider two masses $m_1$ and $m_2$, with $m_2 < m_1$, moving in near-circular orbits with a small radial separation about a central mass $m_c$ (see Fig. 3.24). Let $C_1$ denote the centre of mass of $m_c$ and $m_1$, and $C$ the centre of mass of the system. Note that $C_1$, $C$, and $m_2$ lie on a straight line. The mass $m_2$ experiences an acceleration due to the gravitational attraction of $m_c$ and $m_1$. This can be resolved into a radial component of magnitude $R_2$ directed towards $C$ and a tangential component of magnitude $T_2$ perpendicular to it. We have already seen in Sect. 2.9 (see Eq. (2.149)) that only the tangential component of the force changes the angular momentum of the orbit. Using Eq. (2.145) and assuming $e_2 = 0$ we have, to a good approximation,

$$\dot{a}_2 = 2T_2/n_2, \tag{3.185}$$

where $a_2$ and $n_2$ denote the semi-major axis and the mean motion respectively of $m_2$. Again, we are only concerned with the motion of the guiding centre of the epicycle and we assume that if the free eccentricity is small, then the eccentricity does not have a significant effect on the motion of the guiding centre. The results of numerical experiments support these assumptions.

The tangential acceleration experienced by the mass $m_2$ has contributions from the masses $m_c$ and $m_1$. From Fig. 3.24 this gives

$$T_2 = \left(-\mathcal{G}m_1/r_2^2\right)\sin\gamma + \left(\mathcal{G}m_c/r^2\right)\sin\beta, \tag{3.186}$$

where, from the definitions of $C_1$ and the assumption that $m_2$ is close to the unit circle,

$$\sin \gamma = \frac{m_c}{m_c + m_1} \frac{r}{r_0} \cos \frac{\theta}{2}, \tag{3.187}$$

$$\sin \beta = \frac{m_1}{m_c + m_1} \frac{r}{r_0} \sin \theta, \tag{3.188}$$

and

$$\frac{r}{r_0} = \frac{m_c + m_1}{m_c} \left[ 1 + 4 \frac{m_1}{m_c} \left( \frac{m_c + m_1}{m_c} \right) \sin^2 \frac{\theta}{2} \right]^{-1/2} \approx \frac{m_c + m_1}{m_c}. \tag{3.189}$$

Obviously the latter approximation is only good when $\theta$ is small or $m_1 \ll m_c$, but the error does not affect our results. Therefore

$$T_2 = - \left( \mathcal{G} m_1 / r^2 \right) \bar{H}(\theta), \tag{3.190}$$

where

$$\bar{H}(\theta) = \frac{\cos(\theta/2)}{4 \sin^2(\theta/2)} - \sin \theta, \tag{3.191}$$

and we note that $\bar{H}(\theta) = -(1/2) \mathrm{d}H(\theta)/\mathrm{d}\theta$. We can go through a similar procedure to find the tangential acceleration, $T_1$, experienced by the mass $m_1$. This gives

$$T_1 = + \left( \mathcal{G} m_2 / r^2 \right) \bar{H}(\theta). \tag{3.192}$$

We are now in a position to calculate how the difference in the semi-major axes of the orbiting masses changes with angular separation, $\theta$. Let the difference be

$$s = a_2 - a_1. \tag{3.193}$$

Hence, from Eq. (3.185),

$$\dot{s} = \dot{\theta} \mathrm{d}s/\mathrm{d}\theta = -2 \left( T_1 - T_2 \right) / n, \tag{3.194}$$

where $n$ is the average mean motion of either mass. From Kepler's third law $\dot{\theta}/n = -(3/2)s/a$, where $a$ is the average semi-major axis of either mass. Solving the resulting equation for $\mathrm{d}s/\mathrm{d}\theta$ with boundary values $(s_i, \theta_i)$ and $(s_j, \theta_j)$ we obtain

$$\left( \frac{s_i}{a} \right)^2 - \left( \frac{s_j}{a} \right)^2 = \frac{8}{3} \mathcal{G} \frac{m_1 + m_2}{n^2 a^3} \int_{\theta_i}^{\theta_j} \bar{H}(\theta) \, \mathrm{d}\theta = -\frac{4}{3} \mathcal{G} \frac{m_1 + m_2}{n^2 a^3} \left[ H(\theta_i) - H(\theta_j) \right]. \tag{3.195}$$

Note from Eq. (3.191) that $\bar{H}(\theta)$, and hence $T_1$ and $T_2$, are zero at $\theta = \pm 60°$ and therefore the equilibrium configuration is an equilateral triangle of side $r$, with all bodies stationary in a reference frame rotating with mean motion $n$ where

$$n^2 r^3 = \mathcal{G} \left( m_c + m_1 + m_2 \right). \tag{3.196}$$

However, we have assumed $r \approx a$ and so in order to be consistent we also assume $m_c + m_1 + m_2 \approx m_c$. Therefore we can write Eq. (3.195) as

$$\left(\frac{s_i}{a}\right)^2 - \left(\frac{s_j}{a}\right)^2 = -\frac{4}{3}\frac{m_1 + m_2}{m_c}\left[H(\theta_i) - H(\theta_j)\right].\tag{3.197}$$

Note that this reduces to Eq. (3.180) in the case when $m_2 \ll m_1$. The equivalence of the two equations shows the generality of the relationship between the orbit and the zero-velocity curve. Similarly, the generalised versions of Eqs. (3.182) and (3.183) are

$$\left(\frac{s}{a}\right)_{\text{crit}} = \left(\frac{8}{3}\right)^{1/2}\left(\frac{m_1 + m_2}{m_c}\right)^{1/2}\tag{3.198}$$

and

$$\left(\frac{s_{180}}{a}\right)^2 = \frac{4}{3}\left(\frac{m_1 + m_2}{m_c}\right)\left[H(\theta_{\min}) - 1\right].\tag{3.199}$$

Now we are in a position to apply these analytical results to the Janus and Epimetheus system. When the satellites were discovered they had a radial separation of $s_i = \Delta a_0 = 50$ km and an angular separation $\theta_i = 180°$. Substituting these values in Eq. (3.197) with $m_2/m_1 = 0.25$ we can find the variation of $s$ with $\theta$. Writing the semi-major axes of each satellite as $a_J = a + \Delta a_J$ and $a_E = a + \Delta a_E$ we can make use of Eq. (3.184) and the fact that the ratio of the length of each orbital arc is also the ratio of the masses to show the variation of $\Delta a_J$ and $\Delta a_E$ around the orbits in the rotating frame (see Fig. 3.25). Note that the smaller arc or horseshoe is associated with Janus, the larger satellite. As noted above the existence of the triangular equilibrium configuration carries over into the nonrestricted problem. Just as the orbits in the restricted problem had maximum radial width when the test particle had a separation of $\pm 60°$ from the secondary mass (see Figs. 3.18 and 3.20), so in the more general case we see that the radial width of each orbital path is a maximum when the separation of the satellites is $\pm 60°$.

Figure 3.26 shows the paths of the satellites in a frame rotating with the mean motion of either satellite. Here we have exaggerated the radial variation – in reality the half-widths of the Janus and Epimetheus arcs are 10 km and 40 km respectively and each orbit has a mean semi-major axis of 150,432 km. The paths show that the satellites never pass one another. Instead, at four-year intervals their mutual gravitational perturbations cause an exchange of angular momentum: The satellite on the outer path moves to an inner one and vice versa.

Using the theory described above, Dermott & Murray (1981b) showed how observations of the motions of Janus and Epimetheus could be used to determine the sum of the masses as well as their ratio. This can be achieved by making use of Eq. (3.199). Figure 3.27 shows a plot of the minimum angular separation of the two guiding centres as a function of the combined mass of

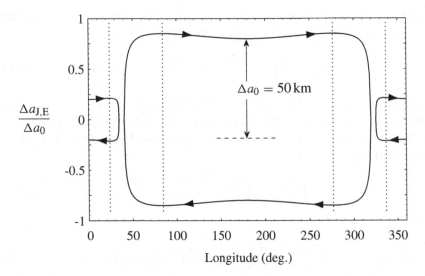

Fig. 3.25.   The variation of $\Delta a_J / \Delta_0$ and $\Delta a_E / \Delta a_0$ as a function of longitude in the rotating frame for the Janus–Epimetheus system, where $\Delta_0 = 50$ km is the separation in semi-major axis of the two orbits when the satellites are 180° apart. The dotted lines denote the longitudes of the maximum radial width. Note that these occur when the angular separation of Janus and Epimetheus is ±60°.

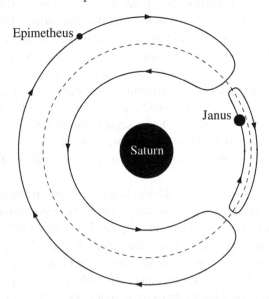

Fig. 3.26.   A schematic diagram of the librational behaviour of the Janus and Epimetheus coorbital system in a frame rotating with the average mean motion of either satellite. The radial extent of the librational arcs are exaggerated; the ratio of the radial widths of the arcs is equal to the Janus–Epimetheus mass ratio (∼ 0.25).

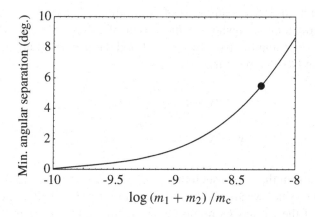

Fig. 3.27. The variation of the minimum angular separation of the two masses as a function of their combined mass. The point denotes the values that have been fitted to the Janus–Epimetheus system.

the satellites, taking $s_{180}/a = 3.32 \times 10^{-4}$. Although this is only an approximation, Dermott & Murray (1981b) showed by numerical integration that the results are accurate to within 0.001°. Nicholson et al. (1992) combined their own ground-based astrometric observations with existing data and derived a new solution for the orbits. This showed that Janus and Epimetheus can approach one another to within 5.64° and that the resulting masses yield densities of $0.65 \pm 0.08$ g cm$^{-3}$ and $0.63 \pm 0.11$ g cm$^{-3}$ respectively. Given that all the evidence from *Voyager* images suggests that these are icy bodies, such low values for the densities points towards the possibility that the small satellites of Saturn may be made of porous ice. This illustrates how knowledge of the dynamics of satellite orbits can lead directly to constraints on their internal properties.

## 3.13 Hill's Equations

For the small particle moving around the central, primary mass in the circular restricted problem the major orbital perturbations will only occur when it encounters the secondary mass. We have already seen examples of such behaviour in the integrations shown in Sect. 3.9; most of the time the particle moves on an unperturbed keplerian orbit. Therefore, instead of dealing with the full equations of the circular restricted three-body problem, it makes more sense to work with a system of equations that describe the motion of the particle in the vicinity of the secondary mass. Such a system was originally derived by Hill (1878) and we will make use of it in the derivation of an encounter map in Sect. 9.5.3 as well as in our study of the shepherding of narrow rings in Sect. 10.5.2.

It is possible to derive such a set of approximate equations by making various assumptions and transferring the origin of the coordinate system to the second mass. For small mass ratios $\mu_1 \approx 1$ and the planar equations of motion, Eqs. (3.16) and (3.17), become

$$\ddot{x} - 2\dot{y} - x = -\frac{x}{r_1^3} - \mu_2\frac{x-1}{r_2^3}, \tag{3.200}$$

$$\ddot{y} + 2\dot{x} - y = -\frac{y}{r_1^3} - \mu_2\frac{y}{r_2^3}. \tag{3.201}$$

We now transform the $x$ axis such that $x \to 1 + x$ leaving the $y$ axis unchanged, and let $\Delta = r_2$. Since we are now considering motion close to the satellite (i.e., in the vicinity of the $L_1$ and $L_2$ points) we can assume that $x$, $y$, and $\Delta$ are small quantities of $\mathcal{O}(\mu_2^{1/3})$. Neglecting higher powers of $\mu_2$ we have $r_1 \approx (1 + 2x)^{1/2}$ and Eqs. (3.200) and (3.201) can be written

$$\ddot{x} - 2\dot{y} = \left(3 - \frac{\mu_2}{\Delta^3}\right)x = \frac{\partial U_H}{\partial x}, \tag{3.202}$$

$$\ddot{y} + 2\dot{x} = -\frac{\mu_2}{\Delta^3}y = \frac{\partial U_H}{\partial y}, \tag{3.203}$$

where

$$U_H = \frac{3}{2}x^2 + \frac{\mu_2}{\Delta} \quad \text{and} \quad \Delta^2 = x^2 + y^2, \tag{3.204}$$

and the modified Jacobi constant, $C_H$, is given by

$$C_H = 3x^2 + 2\frac{\mu_2}{\Delta} - \dot{x}^2 - \dot{y}^2. \tag{3.205}$$

We can compare this with the Jacobi constant in the full problem by setting $n = 1$ and ignoring the $z$ motion in Eq. (3.29). This gives

$$C_J = x^2 + y^2 + 2\left(\frac{\mu_1}{r_1} + \frac{\mu_2}{r_2}\right) - \dot{x}^2 - \dot{y}^2. \tag{3.206}$$

Equations (3.202) and (3.203) are called *Hill's equations* and were first derived in connection with Hill's work on lunar theory.

Inspection of Eq. (3.202) reveals that the radial force vanishes when $3\Delta^3 = \mu_2$, expressing the equilibrium between the tidal force and the mutual attraction (cf. Sect. 4.6). This leads to the definition of the *Hill's sphere* as the sphere of radius

$$\Delta_H = \left(\frac{\mu_2}{3}\right)^{\frac{1}{3}} \tag{3.207}$$

surrounding the secondary mass (cf. Eq. (3.75)).

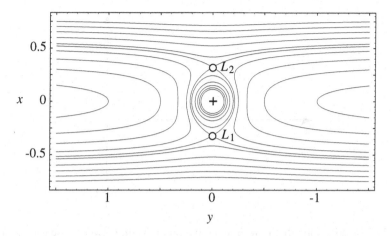

Fig. 3.28. The zero-velocity curves defined by the equation $C_H = 2U_H$ in the vicinity of the Lagrangian points $L_1$ and $L_2$ for a mass $\mu_2 = 0.1$. Note that in the Hill's approximation the equilibrium points are now equidistant from the mass $\mu_2$ (denoted by the cross at the origin).

By setting $\dot{x} = \dot{y} = \ddot{x} = \ddot{y} = 0$ with $(x \neq 0)$ in Eqs. (3.202) and (3.203) we can find the location of the Lagrangian points $L_1$ and $L_2$. Equation (3.202) gives $\Delta_{L_{1,2}} = (\mu_2/3)^{1/3}$ and Eq. (3.205) gives the corresponding value of the modified Jacobi constant as $C_H = 3^{4/3} \mu_2^{2/3}$, in agreement with the values derived for the limiting mass case in Sect. 3.6. Note that the $L_1$ and $L_2$ points lie on the Hill's sphere as we defined it in Eq. (3.207) above. If we write

$$C_H = \zeta \mu_2^{2/3} \tag{3.208}$$

then horseshoe motion is possible in the region where $\zeta < 3^{4/3}$. The shapes of the resulting zero-velocity curves in the vicinity of $L_1$ and $L_2$ are shown in Fig. 3.28.

We can use the above definitions to study the relationship between horseshoe orbits and their associated zero-velocity curves in the case where the orbits are near circular. For motion in a circle we have $\dot{x} = \ddot{x} = \ddot{y} = 0$ and $x$ and $\dot{y}$ are constants related by

$$\dot{y}^2 = 3x^2 - \zeta \mu_2^{2/3}, \tag{3.209}$$

where we have used Eqs. (3.205) and (3.208) and assumed that $\Delta$ is large. If $x_{zv}$ denotes the $x$ value (i.e., the half-width) of the zero-velocity curve associated with this horseshoe orbit, then by setting $\dot{y} = 0$ we have

$$x_{zv}^2 = \frac{1}{3} \zeta \mu_2^{2/3}. \tag{3.210}$$

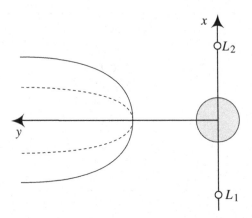

Fig. 3.29. A schematic diagram showing the relationship between a near-circular particle orbit (solid line) and its associated zero-velocity curve (dashed line) using the approximations of Hill's equations.

But $n^2a^3 = 1$ and hence $\dot{y} = -(3/2)x$, giving

$$x^2 = \frac{4}{3}\zeta\mu_2^{2/3}. \tag{3.211}$$

Therefore $x = 2x_{zv}$ and the width of the particle orbit is twice the width of its associated zero-velocity curve, as we already have seen in the case of the full equations (see Sect. 3.10). This is illustrated schematically in Fig. 3.29.

Although these results were derived assuming a near-circular orbit, they are equally applicable in the eccentric case if the orbit is taken to represent the motion of the guiding centre, rather than the particle path. Further details can be found in the work of Dermott & Murray (1981a,b).

We can use Tisserand's criterion to derive a relationship between the orbital elements before and after a satellite encounter using the approximate equations of motion derived above. Let the initial orbital elements be $a_1 = 1 + \Delta a_1, e = \Delta e_1$ and the final elements be $a_2 = 1 + \Delta a_2, e = \Delta e_2$, where $\Delta a_1, \Delta a_2, \Delta e_1$, and $\Delta e_2$ are all small quantities. From Sect. 3.4 Tisserand's criterion gives

$$\frac{1}{1 + \Delta a} + 2(1 + \Delta a)^{1/2}(1 - \Delta e^2)^{1/2} \approx \text{constant}, \tag{3.212}$$

which can be expanded binomially to give

$$\frac{3}{4}\Delta a^2 - \Delta e^2 \approx \text{constant} \tag{3.213}$$

or

$$\frac{3}{4}\Delta a_1^2 - \Delta e_1^2 \approx \frac{3}{4}\Delta a_2^2 - \Delta e_2^2. \tag{3.214}$$

Thus, if the orbit is symmetric about the unit semi-major axis (i.e., we have $\Delta a_1 \approx -\Delta a_2$) then $\Delta e_1 \approx \Delta e_2$.

We can also derive the same relationship by considering Hill's equations. If $\Delta$ is large in Eqs. (3.202) and (3.203) we have

$$\ddot{x} - 2\dot{y} = 3x, \tag{3.215}$$

$$\ddot{y} + 2\dot{x} = 0, \tag{3.216}$$

where we can use Eqs. (3.205) and (3.208) to write

$$\dot{x}^2 + \dot{y}^2 = 3x^2 - \zeta \mu_2^{2/3}. \tag{3.217}$$

If we now use the guiding centre approximation, then the radial excursions of the particle are harmonic with frequency $n = 1$ and amplitude equal to the epicyclic eccentricity and we can write $x = \Delta a + e \sin t$, $\dot{x} = e \cos t$, and $\ddot{x} = -e \sin t$. Hence, from Eqs. (3.215) and (3.216) we have $\ddot{y} = -2e \cos t$ and

$$\dot{y} = -\frac{3}{2}\Delta a - 2e \sin t. \tag{3.218}$$

The first term on the right-hand side of Eq. (3.218) represents the (constant) velocity of the guiding centre while the second term represents the varying velocity on the particle's elliptical path about the guiding centre. Using these expressions for $\dot{x}$, $\dot{y}$, and $x$ in Eq. (3.217) we have

$$e^2 \cos^2 t + \left(-\frac{3}{2}\Delta a - 2e \sin t\right)^2 = 3(\Delta a + e \sin t)^2 - \zeta \mu_2^{2/3}, \tag{3.219}$$

from which we can easily derive

$$\frac{3}{4}\Delta a^2 - \Delta e^2 = \zeta \mu_2^{2/3}, \tag{3.220}$$

where the right-hand side is a constant.

The symmetry of the particle's trajectory (or, if the orbit is eccentric, that of its guiding centre) about the $y$ axis allows us to find an expression for its minimum separation, $\Delta_{\min}$, from the secondary mass; this occurs when the particle crosses the $y$ axis (see Fig. 3.29). Consider a particle initially moving on a circular orbit with semi-major axis $1 + \Delta a_0$. In Hill's system the particle's initial position is $(x_0, y_0)$ and Eq. (3.218) gives

$$\dot{y}_0 = -\frac{3}{2}x_0 = -\frac{3}{2}\Delta a_0. \tag{3.221}$$

Hence, from Eq. (3.205),

$$|\Delta a_0| = 2(\zeta/3)^{1/2}\mu_2^{1/3}. \tag{3.222}$$

If we assume that the particle's path is symmetrical about the $y$ axis then at closest approach to the secondary $x = 0$, $\dot{y} = 0$, $\dot{x} = \dot{x}_{\min}$, and $y = \Delta_{\min}$, where

$\dot{x}_{\min}$ is the value of $\dot{x}$ at the crossing point. Equation (3.205) gives

$$\dot{x}_{\min}^2 = 2\frac{\mu_2}{\Delta_{\min}} - \zeta\mu_2^{2/3}. \tag{3.223}$$

In practice $\dot{x}_{\min}^2 \ll \dot{y}_0^2$ (see Dermott & Murray 1981a) and hence

$$y_{\min} \approx (2/\zeta)^{1/2}\mu_2^{1/3} \tag{3.224}$$

or, in terms of $\Delta a_0$,

$$y_{\min} \approx \frac{8}{3}\Delta a_0^2\mu_2. \tag{3.225}$$

Here we have made use of the fact that the guiding centre of the epicyclic path grazes its associated zero-velocity curve when it crosses the $y$ axis. Remember that we have to distinguish between the zero-velocity curve associated with the motion of the guiding centre, which has a fundamental role in determining the particle's trajectory, and the zero-velocity curve associated with the total motion, which, for these purposes, is largely irrelevant.

Another interesting property of Hill's equations is that the system equations scale as $\mu_2^{1/3}$. If we carry out the transformations $x \to x'(\mu_2/3)^{1/3}$, $y \to y'(\mu_2/3)^{1/3}$, $\Delta \to \Delta'(\mu_2/3)^{1/3}$ in Eqs. (3.202) and (3.203) we obtain the equations

$$\ddot{x}' - 2\dot{y}' = 3x'\left(1 - \frac{1}{\Delta'^3}\right), \tag{3.226}$$

$$\ddot{y}' + 2\dot{x}' = -3\frac{y'}{\Delta'^3}. \tag{3.227}$$

These equations contain no parameters, which implies that provided $\mu_2^{1/3} \gg \mu_2$, the particle paths will scale appropriately. In this system of scaled units the $L_1$ and $L_2$ points are at unit distance from the mass $\mu_2$. A selection of librating and circulating trajectories for the scaled equations is shown in Fig. 3.30. All the trajectories in Fig. 3.30 were started on circular orbits at large positive and negative values of $y'$. Note that the particles that were started close to the $y'$ axis ($|x'| < 1.7$) are all "reflected" and move in horseshoe orbits, although those that get close enough to the perturber can achieve a significant eccentricity as a result of the encounter. However, as the initial value of $|x'|$ is increased, there is a zone in which significant perturbations can occur with the result that particles obtain large eccentricities; most pass the perturber but some can still be "reflected" into horseshoe orbits. For the larger values of $|x'|$ the particles pass the perturber and move in circulating orbits. The eccentricity acquired during the encounter becomes negligible as the initial value of $|x'|$ increases. We will consider this type of behaviour in the context of narrow planetary rings in Sect. 10.5.

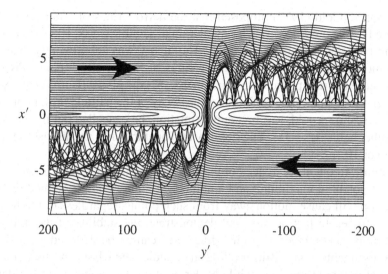

Fig. 3.30. Particle trajectories obtained by solving the scaled form of Hill's equations. The perturbing mass is located at the origin and the $L_1$ and $L_2$ points are at $y' = 0$, $x' = \pm 1$. The particles were all started with $\dot{x}' = 0$ (i.e., in circular orbits) at $y' = \pm 200$. The arrows indicate their direction of motion before encountering the perturber.

## 3.14 The Effects of Drag

The discovery of narrow, sharp-edged rings around Uranus (Elliot et al. 1977) posed a number of severe dynamical problems (see Sect. 10.2.3 and Sect. 10.5.1). One of these concerned the fact that narrow rings should spiral in towards the planet under the effects of Poynting–Robertson (PR) light drag (see Burns et al. (1979) for a comprehensive review of the various radiation forces that can be experienced by small particles in the solar system). PR drag is caused by the nonuniform reemission of the sunlight that a particle absorbs. The particle experiences a drag force, which causes its orbit to decay at a rate dependent on its size, the force being most effective for particles of a size comparable to the wavelength of the incident radiation (i.e., $\sim 10^{-6}$m). In the case where the central mass is the source of the radiation, the PR drag force has the form

$$\mathbf{F} = -\frac{\beta \mathcal{G} m_{\mathrm{c}}}{a^2 r^2} \left( \dot{x} - y + \frac{x}{r^2}(x\dot{x} + y\dot{y}), \ \dot{y} + x + \frac{y}{r^2}(x\dot{x} + y\dot{y}) \right), \tag{3.228}$$

where $\beta$ is the ratio of the force due to radiation pressure to the gravitational force, $m_{\mathrm{c}}$ is the central mass, and $a$ is the semi-major axis of the orbit. Note that the value of $\beta$ depends on such quantities as the luminosity of the source and the radiation pressure cross section. The term involving $\mathbf{r} \cdot \dot{\mathbf{r}}$ in each component of the force in Eq. (3.228) represents the Doppler shift of the incident radiation and the second term is the Poynting–Robertson drag (see Schuerman 1980).

Dermott & Gold (1977) proposed that the narrow rings of Uranus could be explained by the presence in each ring of a small satellite ($\mu_2 < 10^{-10}$) maintaining ring particles in horseshoe orbits. The theory was subsequently extended by Dermott et al. (1980) to the narrow rings of Saturn and Jupiter. PR drag should cause the ring particles to spiral in towards the planet (Burns et al. 1979). However, in the horseshoe orbit model the decay of the particle orbit in one half of the horseshoe would, to a first approximation, be cancelled out by the decay in the other half (Fig. 3.31). The stability of planetary rings are discussed further in Chapter 10.

There are frequent references in the literature to the fact that $L_4$ and $L_5$ are points of potential maximum, and a number of authors have assumed that any dissipation will cause motion away from such points, regardless of the form of the drag force. In fact, as we shall demonstrate, the stability of the triangular equilibrium points under a specific drag force cannot be deduced from simple energy arguments alone. Blitzer (1982) has studied the whole question of stable motion at potential maxima, while Yoder et al. (1983) pointed out the danger of using such arguments in the case of the libration amplitude of the Janus–Epimetheus system.

The particle in the restricted three-body problem can rarely be considered in isolation. In reality, any small particle is subjected to a number of external forces of varying magnitudes in addition to gravitational perturbations. For example, a number of studies have been carried out on the effects of radiation pressure in the three-body problem (see, e.g., Colombo et al. 1966, Schuerman 1980, and

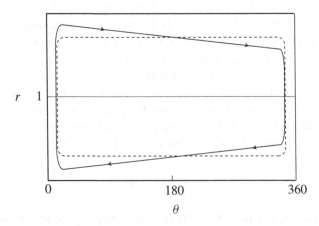

Fig. 3.31. A schematic diagram of the path in the rotating frame (radius $r$ as a function of angle $\theta$) of a particle in a horseshoe orbit subjected to a drag force (solid line) compared with its original path when no drag forces are operating (dashed line). The particle is assumed to have an eccentricity $e \approx 0$. The particle encounters the perturbing mass at $r = 1; \theta \approx 0°, 360°$.

Simmons et al. 1985). Below we consider the dynamical effects of drag forces on the particle using two different approaches: (i) a study of the behaviour of zero-velocity curves and (ii) a study of the location and stability of the Lagrangian equilibrium points. These are the approaches taken by Murray (1994b) in his study of the effects of drag in the circular restricted three-body problem.

In what follows the particle is considered to move under the gravitational attraction of $\mu_1$ and $\mu_2$ and an arbitrary external force, $\mathbf{F} = (F_x, F_y)$, which is a function of the particle's position and velocity. It is assumed that $|\mathbf{F}| = \mathcal{O}(k)$, where $k$ is a small quantity.

### 3.14.1 Analysis of the Jacobi Constant

From Sect. 3.2, the equations of motion of the particle in the rotating frame are

$$\ddot{x} - 2\dot{y} = \frac{\partial U}{\partial x} + F_x, \tag{3.229}$$

$$\ddot{y} + 2\dot{x} = \frac{\partial U}{\partial y} + F_y. \tag{3.230}$$

If we multiply Eq. (3.229) by $\dot{x}$ and Eq. (3.230) by $\dot{y}$ and, then add we get

$$\dot{x}\ddot{x} + \dot{y}\ddot{y} - \left(\dot{x}\frac{\partial U}{\partial x} + \dot{y}\frac{\partial U}{\partial y}\right) = \dot{x}F_x + \dot{y}F_y. \tag{3.231}$$

Because the Jacobi constant in the standard problem can be written as $C_J = 2U - \dot{x}^2 - \dot{y}^2$ we have

$$\frac{dC_J}{dt} = -2\left(\dot{x}F_x + \dot{y}F_y\right). \tag{3.232}$$

Note that $\dot{x}F_x + \dot{y}F_y$ is the work done by $\mathbf{F}$ per unit time. If we assume that $k < 0$ for a force that opposes the motion, the sign of $\dot{C}_J$ is opposite to the sign of $\dot{x}F_x + \dot{y}F_y$. From our study of zero-velocity curves in Sect. 3.3 we know that increasing values of $C_J$ imply increasing areas of exclusion and motion away from the $L_4$ and $L_5$ points. We will consider two illustrative examples.

If the drag force is proportional to the velocity of the particle in the rotating, synodic frame then $\mathbf{F} = k\mathbf{v} = k(\dot{x}, \dot{y})$ and

$$\dot{x}F_x + \dot{y}F_y = k(\dot{x}^2 + \dot{y}^2) < 0. \tag{3.233}$$

Hence this drag force, sometimes called *nebular drag*, causes motion away from the $L_4$ and $L_5$ points, independent of the path of the particle. This result was known to Jeffreys (1929).

In the case of the PR drag component of the radiation force, we have

$$\mathbf{F} = k(\dot{x} - y, \dot{y} + x)/r^2 \tag{3.234}$$

and hence

$$\dot{x}F_x + \dot{y}F_y = k(\dot{x}^2 + \dot{y}^2) - k(\dot{x}y - \dot{y}x). \tag{3.235}$$

However, the sign of this expression cannot be determined by simple inspection – we must know the path of the particle and this cannot be determined for arbitrary starting conditions. Therefore, there are certain drag forces for which this analysis of the Jacobi constant cannot be used to determine stability.

### 3.14.2  *Linear Stability of the $L_4$ and $L_5$ Points*

We now consider the alternative approach of using a linear stability analysis of the Lagrangian points to establish their local stability behaviour. Here we are concerned with the triangular points since these are the ones that are linearly stable in the zero-drag case. Our approach is based on that by Schuerman (1980), but we follow Murray (1994b) and generalise it to deal with arbitrary drag forces. Murray (1994b) derived expressions for the shift in the five Lagrangian equilibrium points as a function of the drag constant and then showed that, to $\mathcal{O}(\mu_2)$, the characteristic equation for perturbed motion around $L_4$ and $L_5$ can be written as

$$\lambda^4 + a_3\lambda^3 + (1 + a_2)\lambda^2 + a_1\lambda + \left(\frac{27}{4}\mu_2 + a_0\right) = 0, \tag{3.236}$$

where the constants $a_i$ $(i = 0, 1, 2, 3)$ are all $\mathcal{O}(k)$ and are given by

$$a_0 = \frac{9}{4}k_{x,x} + \frac{3}{4}k_{y,y} \mp \frac{3\sqrt{3}}{4}(k_{x,y} + k_{y,x}), \tag{3.237}$$

$$a_1 = \frac{9}{4}k_{x,\dot{x}} + \frac{3}{4}k_{y,\dot{y}} + 2(k_{x,y} - k_{y,x}) \mp \frac{3\sqrt{3}}{4}(k_{x,\dot{y}} + k_{y,\dot{x}}), \tag{3.238}$$

$$a_2 = -k_{x,x} - k_{y,y} + 2(k_{x,\dot{y}} - k_{y,\dot{x}}), \tag{3.239}$$

$$a_3 = -k_{x,\dot{x}} - k_{y,\dot{y}}, \tag{3.240}$$

and where the upper and lower signs in Eqs. (3.237) and (3.238) refer to the $L_4$ and $L_5$ points respectively. These definitions depend on the following partial derivatives:

$$k_{x,x} = \left[\frac{\partial F_x}{\partial x}\right]_0, \quad k_{y,x} = \left[\frac{\partial F_y}{\partial x}\right]_0, \quad k_{x,\dot{x}} = \left[\frac{\partial F_x}{\partial \dot{x}}\right]_0, \quad k_{y,\dot{x}} = \left[\frac{\partial F_y}{\partial \dot{x}}\right]_0,$$

$$k_{x,y} = \left[\frac{\partial F_x}{\partial y}\right]_0, \quad k_{y,y} = \left[\frac{\partial F_y}{\partial y}\right]_0, \quad k_{x,\dot{y}} = \left[\frac{\partial F_x}{\partial \dot{y}}\right]_0, \quad k_{y,\dot{y}} = \left[\frac{\partial F_y}{\partial \dot{y}}\right]_0, \tag{3.241}$$

where the subscript 0 denotes evaluation of the partial derivatives at the classical equilibrium positions, $x_0 = 1/2 - \mu_2$, $y_0 = \sqrt{3}/2$, rather than their displaced counterparts. This is because all these quantities are constants of $\mathcal{O}(k)$ and because the displacements are of a similar magnitude and assumed small compared with $x_0$ and $y_0$.

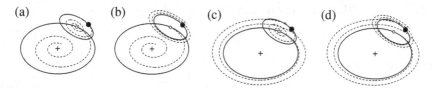

Fig. 3.32. A schematic diagram illustrating the different types of oscillation of the two modes that are possible in the perturbed solution around $L_4$ and $L_5$. The solid lines represent the paths in the absence of drag. The dashed lines show the spiral paths followed in the presence of drag. (a) All real parts of the eigenvalues are negative (asymptotic stability). (b) The real parts of the eigenvalues associated with the motion of the epicentre are negative, but those associated with the epicyclic motion are positive (unstable). (c) The real parts of the eigenvalues associated with the motion of the epicentre are positive, but those associated with the epicyclic motion are negative (unstable). (d) All real parts of the eigenvalues are positive (unstable).

Comparison of Eq. (3.236) with Eq. (3.237) shows that they are equivalent in the case of no drag, because all the $a_i$ are zero and to $\mathcal{O}(\mu_2)$ Eq. (3.236) reduces to the standard bi-quadratic for the classical problem. Without drag the solutions of the characteristic equation are purely imaginary provided that the usual condition on the mass ratio, Eq. (3.145), is satisfied. The situation is more complicated in the presence of drag. Four possibilities arise and each is illustrated in Fig. 3.32.

1) All real parts of the eigenvalues are negative. This gives asymptotic stability.
2) The real parts of the eigenvalues associated with the long-period motion of the epicentre are negative, but those associated with the short-period motion are positive. This is linearly unstable.
3) The real parts of the eigenvalues associated with the long-period motion of the epicentre are positive, but those associated with the short-period motion are negative. This is linearly unstable.
4) All real parts of the eigenvalues are positive. This is linearly unstable.

Note that although one of the eigenmodes may be damped (as in cases 2 and 3 of this list) this does not produce stability since the other mode is exponentially increasing. Murray (1994b) showed that the conditions for each of the real parts of $\lambda$ to be negative, and hence for the points to be asymptotically stable in the limit as $\mu_2 \to 0$, reduce to

$$0 < a_1 < a_3, \tag{3.242}$$

where $a_1$ and $a_3$ are defined in Eqs. (3.238) and (3.240). Note that, with these approximations, the stability is independent of the values of $a_0$ and $a_2$ and thus the linear stability of $L_4$ and $L_5$ does not depend on the values of $k_{x,x}$ and $k_{y,y}$. We can now consider the two drag forces studied above from the point of view of the linear stability of the $L_4$ and $L_5$ points.

If $\mathbf{F} = k(\dot{x}, \dot{y})$ then the only nonzero values of the constants in Eq. (3.241) are

$$k_{x,\dot{x}} = k, \qquad k_{y,\dot{y}} = k, \tag{3.243}$$

and hence

$$a_1 = 3k, \qquad a_3 = -2k. \tag{3.244}$$

Since $k < 0$ we have $a_1 < 0$, the condition in Eq. (3.242) is not satisfied, and the triangular equilibrium points are unstable under this drag force, as we deduced from our study of the Jacobi constant.

In the case of PR drag $\mathbf{F} = k(\dot{x} - y, \dot{y} + x)/r^2$ and the only relevant nonzero values of the constants in Eq. (3.241) are

$$k_{x,\dot{x}} = k_{y,x} = k_{y,\dot{y}} = k, \quad k_{x,y} = -k, \tag{3.245}$$

and hence

$$a_1 = -5k, \qquad a_3 = -2k. \tag{3.246}$$

Therefore the points are linearly unstable to PR drag.

However, it is interesting to note that if we consider a force $\mathbf{F} = k(\dot{x} - y, \dot{y} + x)$ then the only relevant nonzero values of the constants in Eq. (3.241) are

$$k_{x,\dot{x}} = k_{y,x} = k_{y,\dot{y}} = k, \quad k_{x,y} = -k, \tag{3.247}$$

and hence

$$a_1 = -k, \qquad a_3 = -2k. \tag{3.248}$$

In this case the condition in Eq. (3.242) is satisfied and the triangular equilibrium points are asymptotically stable under this drag force. Of course, this gives us no information on the stability of large-amplitude tadpole or horseshoe orbits about these points since our analysis always assumes a small displacement from the equilibrium point.

### 3.14.3 Inertial Drag Forces

Murray (1994b) considered a general drag force per unit mass of the form

$$\mathbf{F_i} = k\mathbf{V}g(x, y, \dot{x}, \dot{y}), \tag{3.249}$$

where $k < 0$, $\mathbf{V} = (\dot{x} - y, \dot{y} + x)$ is the particle's velocity in the inertial frame, and $g(x, y, \dot{x}, \dot{y})$ is a scalar function of its position and velocity; Murray (1994b) referred to this as an *inertial drag*. Eliminating the drag constant, $k$, from the equations of motion gives

$$r^2 r_1^3 r_2^3 + \mu_2 r_1^3(\mu_1 x - r^2) - \mu_1 r_2^3(\mu_2 x + r^2) = 0 \tag{3.250}$$

for the paths along which the five equilibrium points must move. Figure 3.33a shows these paths for the case $\mu_2 = 0.2$.

In the limit as $\mu_2 \rightarrow 0$ Murray showed that $L_3$, $L_4$, and $L_5$ move along the unit circle whereas $L_1$ and $L_2$ move along circles close to the secondary. In cases where $g^* = g(x, y, 0, 0) = g^*(r)$, a function of the radius only, the angular position of $L_4$ and $L_5$ for small $\mu_2$ can be derived from a plot of

$$\frac{\bar{k}}{\mu_2} = \sin\theta \left( \frac{1}{(2 - 2\cos\theta)^{3/2}} - 1 \right), \qquad (3.251)$$

where $\bar{k} = kg^*(1)$. This function is shown in Fig. 3.33b. It has a minimum at $\theta = 108.4°$ for $\bar{k}/\mu_2 = -0.7265$ and a maximum at $\theta = 251.6°$ for $\bar{k}/\mu_2 = +0.7265$. Therefore the $L_3$ and $L_4$ points can shift from position angles of $180°$ and $60°$ respectively to a position angle of $108.4°$ for $\bar{k}/\mu_2 = -0.7265$ before disappearing for all values of $\bar{k}/\mu_2 < -0.7265$. This behaviour is independent of the form of $g_r$. In all systems where $g^*$ is a function of $r$ these results show that there are regimes where $L_5$ can survive even though $L_4$ has ceased to exist.

The stability of the shifted $L_4$ and $L_5$ points under inertial drag can be considered using techniques derived in Sect. 3.15.2. For example, Murray (1994b) considered an inertial drag force of the form

$$\mathbf{F}_i = k \mathbf{V} V^i r^j, \qquad (3.252)$$

where $i$ and $j$ are real numbers. He showed that the triangular equilibrium points will be asymptotically stable to this drag force provided $k < 0$ and the conditions

$$0 < 1 - i + 2j < 2 + i \qquad (3.253)$$

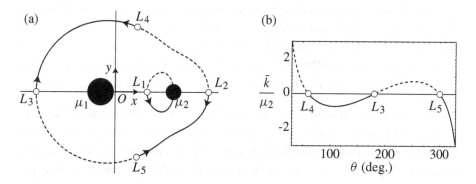

Fig. 3.33. (a) The paths of the shifted Lagrangian equilibrium points in the presence of a general inertial drag force of the form given in Eq. (3.249) for $\mu_2 = 0.2$. (b) The relationship between $\bar{k}/\mu_2$ (where $\bar{k}$ is the modified drag constant defined in the text) and the shifted position angle, $\theta$ (between $30°$ and $330°$) of the $L_3$, $L_4$, and $L_5$ equilibrium points in the case of small mass ratio for an inertial drag. In each diagram the solid lines denote the case when $k < 0$ and the dashed lines the case when $k > 0$. (Adapted from Murray 1994b.)

are satisfied. If we consider the specific case $i = 0$, $j = n$ then this drag force will produce asymptotic stability in $L_4$ and $L_5$ provided $-1/2 < n < 1/2$. This highlights the need to consider each drag force separately when discussing the stability of the triangular equilibrium points. Murray (1994b) also showed that in cases where the drag force is large it is possible for the $L_4$ and $L_5$ points to both be unstable but with different $e$-folding timescales. He suggested that this could explain possible asymmetries in the distribution of the Trojan asteroids. The existence of the maximum shifted longitude of $L_4$ at $\theta = 108.4°$ had been observed in numerical experiments by Peale (1993b).

### Exercise Questions

**3.1**     Using the standard system of units for the planar, circular, restricted three-body problem, the equation defining the zero-velocity curves is $C_J = x^2 + y^2 + 2(\mu_1/r_1 + \mu_2/r_2)$, where $C_J$ is the value of the Jacobi constant and $r_1 = \sqrt{(x + \mu_2)^2 + y^2}$ and $r_2 = \sqrt{(x - \mu_1)^2 + y^2}$ are the distances to the masses $\mu_1$ and $\mu_2$, respectively. The critical zero-velocity curve that passes through the $L_3$ equilibrium point (where $C_J \approx 3 + \mu_2$) has two branches. Use polar coordinates to show that for small mass ratios each of the curves crosses the unit semi-major axis at points with an angular separation of $23.5°$ from the secondary mass.

**3.2**     In the circular restricted three-body problem the condition for the three collinear Lagrangian equilibrium points to be linearly unstable is $\bar{A} > 1$, where $\bar{A} = \mu_1/r_1^3 + \mu_2/r_2^3$ and $r_1$ and $r_2$ (both positive quantities) are the distances to the masses $\mu_1$ and $\mu_2$. Taking $\alpha = ((1/3)\mu_2/\mu_1)^{1/3}$ and $\beta = -(7/12)\mu_2/\mu_1$, where $\mu_2 \ll \mu_1$, derive expressions for $\bar{A}$ including terms up to $\mathcal{O}(\alpha, \beta)$ and show that the condition $\bar{A} > 1$ is satisfied for (i) $L_1$, where $r_1 \approx 1 - \alpha$, $r_2 \approx \alpha$, (ii) $L_2$, where $r_1 \approx 1 + \alpha$, $r_2 \approx \alpha$, and (iii) $L_3$, where $r_1 \approx 1 + \beta$, $r_2 \approx 2 + \beta$. In the case of the $L_3$ point show that the real roots of the characteristic equation, and hence the eigenvalues that give rise to the instability, are approximately $\pm\sqrt{(21/8)\mu_2}$.

**3.3**     The centre of mass of a spacecraft is moved to within 1 m of the inner ($L_1$) Lagrangian equilibrium point of Io, one of the moons of Jupiter. Assuming that an analytical solution based on a linear stability analysis is always valid, use the data in Appendix A to calculate how long it would be before the spacecraft hits the surface of Io.

**3.4**     Two planetary satellites of masses $m_1$ and $m_2$, semi-major axes $a_1$ and $a_2$, and eccentricities $e_1$ and $e_2$ orbit a planet of mass $m_p$ (where $m_p \gg m_1, m_2$) such that at the same time interval before and after their closest approach their semi-major axes are $a_1 = a_0 \pm \Delta a_1$ and $a_2 = a_0 \mp \Delta a_2$, where $a_0$ is the mean semi-major axis of both satellites; their eccentricities are unchanged by the encounter.

Derive expressions for the total orbital angular momentum of the system before and after closest approach by carrying out expansions to second order in $\Delta a_1/a_0$, $\Delta a_2/a_0$, $e_1$, and $e_2$. If the total orbital angular momentum is preserved, use your expressions to show that $\Delta a_1/\Delta a_2 \approx m_2/m_1$.

**3.5** The parameterless form of Hill's equations for the motion of a test particle in the vicinity of the secondary mass are $\ddot{x} - 2\dot{y} = 3x\left(1 - 1/\Delta^3\right)$, $\ddot{y} + 2\dot{x} = -3y/\Delta^3$, where $\Delta = \sqrt{x^2 + y^2}$ is the distance from the secondary. In this system of units the $L_1$ and $L_2$ points are at $y = 0$, $x = \mp 1$. Write a computer program to solve the equations of motion numerically, taking initial values of $x_0$ at intervals of 0.01 between 0.01 and 5 with $y_0 = 200$, $\dot{x}_0 = 0$, and $\dot{y}_0 = -(3/2)x_0$ in all cases. Use your results to find (i) the smallest value of $x_0$ below which the test particle is always "reflected" by the secondary mass, (ii) the smallest value of $x_0$ above which the particle always passes the secondary mass, and (iii) the value of $x_0$ that gives the largest value of $|x|$. How would you characterise the trajectories for values of $x_0$ given in your answers to parts (i) and (ii).

**3.6** The test particle in the circular restricted three-body problem experiences a drag force $\mathbf{F} = kvv^i r^j$, where $k < 0$ is a constant, $i$ and $j$ are real numbers, and $\mathbf{r} = (x, y)$, and $\mathbf{v} = (\dot{x} - y, \dot{y} + x)$ are the position and velocity vectors (both in the rotating frame) with magnitudes $r$ and $v$. Use the method given in Sect. 3.15.2 to show that for small values of $k$ this drag force results in asymptotic, linear stability of the $L_4$ and $L_5$ equilibrium points provided $0 < 1 - i + 2j < 2 + i$.

# 4

# Tides, Rotation, and Shape

What fates impose, that men must needs abide;
It boots not to resist both wind and tide.

William Shakespeare, *Henry VI (3), IV, iii*

## 4.1 Introduction

So far we have considered all objects as being point masses with no physical dimensions. Since this is evidently not the case for real bodies, we must now consider the effects of the application of universal gravitation to the matter that forms the bodies of the solar system. A *tide* is raised on one body by another because of the effect of the gravitational gradient or the variation of the gravitational force across the body. For example, if we consider the tide raised on a planet by an orbiting satellite, the force experienced by the side of the planet facing the satellite is stronger than that experienced by the far side of the planet. Since none of the bodies that make up the solar system is perfectly rigid, there will be a distortion that gives rise to a *tidal bulge*.

The magnitude of the tidal bulge on a body is determined in part by its internal density distribution and thus, in principle, a measurement of the tidal amplitude could lead to a determination of the internal structure. Such measurements are not possible for any of the planets in the solar system other than the Earth. However, the deforming potential associated with planetary rotation acts in a similar way as that due to tides and a measurement of the rotational deformation of a planet can be used to determine its internal density distribution; this knowledge can then be used to estimate the response of the planet to a tidal potential. Satellites in the solar system that are in synchronous rotation are deformed by both rotational and tidal forces and measurements of their triaxial ellipsoidal figures have been used to determine their internal structures.

The response of the satellite to the tide it has raised can result in dynamical evolution of the system. Since friction is always present to some degree, tides are a dissipative phenomenon and the tide raised by a satellite on a planet can lead to orbital evolution of the satellite and a change in the spin rate of the planet. Just as the satellite raises a tide on the planet, so the planet also raises a tide on the satellite. This can be especially important when the satellite's orbit is eccentric. In some cases the effect of tidal dissipation in a satellite can lead to dramatic consequences, such as the runaway tidal heating of the jovian satellite Io.

## 4.2 The Tidal Bulge

Consider the case of the tides raised on a planet of mass $m_p$ by a satellite of mass $m_s$. If we represent the objects as point masses, then Newton's law of gravitation gives the mean magnitude of the mutual force, $\langle F \rangle$, as

$$\langle F \rangle = \mathcal{G} \frac{m_p m_s}{r^2}, \tag{4.1}$$

where $r$ is the separation of the centres. If we assume that the bodies move in circular orbits about their common centre of mass (Fig. 4.1), then we have already seen in Sect. 2.7 that the semi-major axes of the orbits are related to the masses by

$$a_s/a_p = m_p/m_s, \tag{4.2}$$

where the bodies have a constant separation $a = a_p + a_s$. The motion of the particle $P_1$ at the centre of the planet with respect to the centre of mass, $C_1$, is a circle

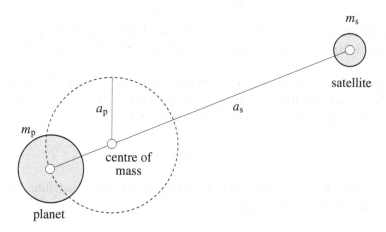

**Fig. 4.1.** A planet and satellite moving about their common centre of mass in circular orbits. Their semi-major axes with respect to the centre of mass are $a_p$ and $a_s$, while the semi-major axis of the satellite with respect to the planet is $a = a_p + a_s$.

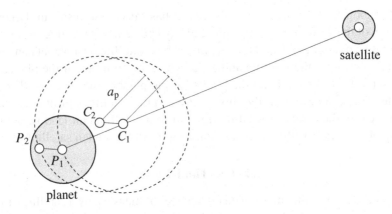

Fig. 4.2.   All the particles in the planet move in similar circles of identical radii $a_p$, but with different centres. The particles $P_1$ and $P_2$ are on circles with centres $C_1$ and $C_2$, respectively.

of radius $a_p$. If we neglect rotation it follows that the motion of any other point, $P_2$, in the planet is a circle of the same radius but with a centre, $C_2$, displaced from $C_1$ to the same extent that $P_2$ is displaced from $P_1$ (see Fig. 4.2). It follows that all particles that make up the planet are acted on by equal (in magnitude and direction) centrifugal forces but not by equal gravitational forces, $\mathbf{F}$. The common centrifugal force is equal to the mean gravitational force, $\langle \mathbf{F} \rangle$, that is,

$$\langle \mathbf{F} \rangle = \text{centrifugal force} \neq \mathbf{F}. \tag{4.3}$$

The tide-generating force, $\mathbf{F}_{\text{tidal}}$, that deforms the planet is defined by

$$\mathbf{F}_{\text{tidal}} = \mathbf{F} - \langle \mathbf{F} \rangle. \tag{4.4}$$

Rotational forces can also deform a body (see Sect. 4.7), but if the tidal and rotational deformations are small, they can be calculated separately and added linearly.

We approach the problem of determining the shape of the tidal bulge on the planet by considering the potential $V$ at some point $P$ on the surface of the planet due to the satellite (treated as a point mass). We have

$$V = -\mathcal{G}\frac{m_s}{\Delta}, \tag{4.5}$$

where $\Delta$ is the distance of the point $P$ from the centre of the satellite. From the cosine rule (see Fig. 4.3) we have

$$\Delta = a\left[1 - 2\left(\frac{R_p}{a}\right)\cos\psi + \left(\frac{R_p}{a}\right)^2\right]^{\frac{1}{2}}, \tag{4.6}$$

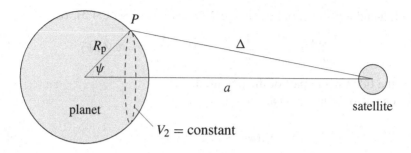

Fig. 4.3. The relationships among the radius of the planet, $R_p$, the semi-major axis of the satellite, $a$, and the distance $\Delta$ from a point $P$ on the surface of the planet. The dashed line denotes the plane defined by the equipotential surface $V_2 = -\mathcal{G}(m_s/a^2)R_p \cos \psi =$ constant.

where $\psi$ is measured from the line joining the centres of the two bodies. In most cases of interest, $R_p/a \ll 1$. For example, the equatorial radius of the Earth is 6,378 km and the semi-major axis of the Moon's orbit is 384,400 km (see Tables A.4 and A.5). Consequently we can expand Eq. (4.6) binomially to obtain

$$V = -\mathcal{G}\frac{m_s}{a}\left[1 + \left(\frac{R_p}{a}\right)\cos\psi + \left(\frac{R_p}{a}\right)^2\frac{1}{2}\left(3\cos^2\psi - 1\right) + \cdots\right] \tag{4.7}$$
$$\approx V_1 + V_2 + V_3,$$

where we have neglected the higher order terms in the expansion.

The first term in Eq. (4.7), $V_1 = -\mathcal{G}m_s/a$, is a constant and since $\mathbf{F}/m_p = -\nabla V$ this term produces no force on the planet. The second term in Eq. (4.7), $V_2 = -\mathcal{G}(m_s/a^2)R_p \cos\psi$, gives rise to the force on the particle at the point $P$ needed for motion in a circle. The force arising at any point from a potential is in the direction of the potential gradient and perpendicular to the equipotential surface that passes through that point. In this case, this is the plane perpendicular to the line connecting the centres of the gravitating bodies and the potential gradient is given by

$$-\frac{\partial V_2}{\partial(R_p\cos\psi)} = \mathcal{G}\frac{m_s}{a^2} = \frac{\langle F\rangle}{m_p}. \tag{4.8}$$

The potential at the surface of the planet due to the third term in Eq. (4.7) can be written as

$$V_3(\psi) = -\mathcal{G}\frac{m_s}{a^3}R_p^2\mathcal{P}_2(\cos\psi), \tag{4.9}$$

where

$$\mathcal{P}_2(\cos\psi) = \frac{1}{2}\left(3\cos^2\psi - 1\right) \tag{4.10}$$

is the Legendre polynomial of degree 2 in $\cos \psi$ (see Sect. 4.3). Because

$$\frac{\mathbf{F}_{tidal}}{m_p} = \frac{\mathbf{F}}{m_p} - \frac{\langle \mathbf{F} \rangle}{m_p} = -\nabla V - \frac{\langle \mathbf{F} \rangle}{m_p} \approx -\nabla V_3(\psi), \tag{4.11}$$

this is the tide raising part of the potential.

We can also write $V_3(\psi)$ as

$$V_3(\psi) = -\zeta g P_2(\cos \psi), \tag{4.12}$$

where

$$\zeta = \frac{m_s}{m_p} \left( \frac{R_p}{a} \right)^3 R_p \tag{4.13}$$

and

$$g = \frac{\mathcal{G} m_p}{R_p^2} \tag{4.14}$$

is the gravitational acceleration, or surface gravity, of the planet. In this case $\zeta P_2(\cos \psi)$ is said to be the amplitude of the *equilibrium tide* for any value of $\psi$ on the planet's surface. Note that $P_2(\cos \psi)$ is a maximum at $\theta = 0$ or $\pi$ and a minimum at $\theta = \pi/2$ or $3\pi/2$. Given that the Earth rotates on its axis with respect to the stars once every twenty-four hours, this explains why the Moon produces two high tides and two low tides on the Earth approximately every day.

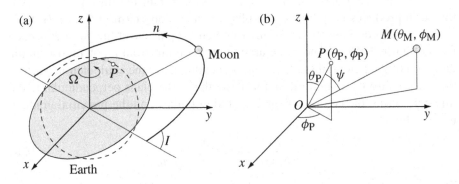

Fig. 4.4. Schematic diagram of the shape of the tidal distortion (solid line) arising from the $V_3(\psi)$ term in the gravitational potential compared with the circular, zero-distortion shape (dashed line). (a) The axis of symmetry of the tidal distortion passes through the centres of Earth and Moon, while the Earth rotates about the $z$ axis with angular speed $\Omega$; the Moon has a mean motion $n$ and an orbital inclination $I$ with respect to the equatorial plane of the Earth. (b) The colatitudes and longitudes of the Moon ($\theta_M$, $\phi_M$) and a point $P$ ($\theta_P$, $\phi_P$) on the surface of the Earth. Note that the longitude $\phi_P$ (and $\phi_M$) is measured from a fixed direction in space and not from a direction that rotates with the Earth.

The tidal deformation of the Earth is made more complicated by the facts that (i) both the Sun and the Moon raise significant tides and (ii) in a geocentric frame both of these bodies orbit the Earth in paths that are eccentric and inclined with respect to the Earth's equator. If we neglect the orbital eccentricities, then the Sun and Moon both raise three principal tides on the Earth. In Fig. 4.4a the axis of symmetry of the tidal distortion passes through the centres of the Earth and Moon, the Earth rotates about the $z$ axis with angular speed $\Omega$, and we allow that the Moon has a mean motion $n$ and an inclination $I$ with respect to the Earth's equatorial plane. In Fig. 4.4b we show the colatitudes and longitudes of the Moon ($\theta_M$, $\phi_M$) and of a point P ($\theta_P$, $\phi_P$) on the surface of the Earth. Note that the longitudes $\phi_P$ and $\phi_M$ are measured from a fixed direction in space and not from a direction that rotates with the Earth.

Using the cosine rule we can show that the angle $\psi$ between the position vectors $OP$ and $OM$ is given by

$$\cos \psi = \cos \theta_P \cos \theta_M + \sin \theta_P \sin \theta_M \cos(\phi_P - \phi_M). \tag{4.15}$$

Hence

$$\frac{1}{2}\left(3\cos^2\psi - 1\right) = \frac{1}{2}\left(3\cos^2\theta_P - 1\right)\frac{1}{2}\left(3\cos^2\theta_M - 1\right)$$
$$+ \frac{3}{4}\sin^2\theta_P \sin^2\theta_M \cos 2(\phi_P - \phi_M)$$
$$+ \frac{3}{4}\sin 2\theta_P \sin 2\theta_M \cos(\phi_P - \phi_M). \tag{4.16}$$

Given that the colatitude $\theta_P$ of a fixed point on the Earth is constant, the tidal amplitude at $P$ varies with $\phi_P$, $\theta_M$, and $\phi_M$. The variation of the first term in Eq. (4.16) is determined by the variation with time of $\cos^2\theta_M = (1/2)(1 + \cos 2\theta_M)$. Hence, this term varies with frequency $2n$ and gives rise to a *fortnightly tide*. The second term varies with frequency $2(\Omega - n) \approx 2\Omega$ and gives rise to a *semidiurnal tide*, while the third terms varies with frequency $(\Omega - n) \approx \Omega$ and gives rise to a *diurnal tide*. Because the latter term contains the factor $\sin 2\theta_M$, the diurnal tide has a strong fortnightly modulation. Other tidal terms are associated with the Moon's orbital eccentricity.

The corresponding solar *semiannual*, *semidiurnal*, and *diurnal tides* have frequencies $2n_{Sun}$, $2(\Omega - n_{Sun}) \approx 2\Omega$, and $(\Omega - n_{Sun}) \approx \Omega$, where $n_{Sun}$ is the solar (or the Earth's) mean motion. It follows from Eq. (4.13) that the ratio of the amplitudes of the corresponding solar and lunar tides is given, in each case, by

$$\frac{m_{Sun}}{m_{Moon}}\left(\frac{a_{Moon}}{a_{Sun}}\right)^3 \approx 0.46. \tag{4.17}$$

For the tide raised by the Moon on the Earth, $\zeta = 0.36$ m, while for the solar tide $\zeta = 0.16$ m.

## 4.3 Potential Theory

Before proceeding to the calculation of tidal and rotational deformation it is useful to summarise some results derived from potential theory. The gravitational potential of a homogeneous, spherical body of density $\gamma$ and radius $C$ can be found by considering the internal and external potentials of a thin spherical shell of radius $r$, thickness $\delta r$, and mass $\delta m$ (Ramsey 1940). The potential exterior to the shell at some point distant $r'$ from its centre is given by

$$V_{\text{ext}}(r') = -\frac{\mathcal{G}\delta m}{r'} \tag{4.18}$$

and is the same as if all the mass were concentrated at the centre of the shell. Thus the external potential at the surface of a uniform sphere is given by

$$V_{\text{ext}}(C) = -\frac{\mathcal{G}\sum \delta m}{C} = -\frac{4}{3}\pi\gamma\mathcal{G}C^2. \tag{4.19}$$

Because the gravitational force is described by an inverse square law, it follows that the force on a particle interior to the shell is zero. Hence, the gravitational potential interior to the shell must be a constant and can be determined by calculating the potential at any point in the interior. By calculating the potential at the centre of the shell, we obtain

$$V_{\text{int}}(r) = -\frac{\mathcal{G}\sum \delta m}{r} = -4\pi\gamma\mathcal{G}r\,\delta r. \tag{4.20}$$

It follows that the internal potential of a uniform shell of external radius $C$ and internal radius $r$ is given by

$$V_{\text{int}}(C, r) = -4\pi\gamma\mathcal{G}\int_r^C r\,dr = -2\pi\gamma\mathcal{G}\left(C^2 - r^2\right). \tag{4.21}$$

Hence, the interior $(r < C)$ and exterior $(r > C)$ potentials of a homogeneous spherical body at some point distant $r$ from its centre are given by

$$V_{\text{int}}(r) = -\frac{4}{3}\pi\gamma\mathcal{G}r^2 - 2\pi\gamma\mathcal{G}\left(C^2 - r^2\right) = -\frac{2}{3}\pi\gamma\mathcal{G}\left(3C^2 - r^2\right), \tag{4.22}$$

$$V_{\text{ext}}(r) = -\frac{4}{3}\pi\gamma\mathcal{G}\frac{C^3}{r}. \tag{4.23}$$

The internal and external potentials of a deformed body can be expressed in terms of spherical harmonic functions. The following brief discussion of the use of these functions in potential theory is based on the fuller accounts given by Ramsey (1940), MacRobert (1967), Bullen (1975), and Blakely (1995).

The gravitational potential $V$ in free space satisfies Laplace's equation:

$$\nabla^2 V = 0. \tag{4.24}$$

A function $V$ is said to be *homogeneous of degree n* if it satisfies *Euler's equation*:

$$x\frac{\partial V}{\partial x} + y\frac{\partial V}{\partial y} + z\frac{\partial V}{\partial z} = nV. \tag{4.25}$$

Homogeneous functions that also satisfy Laplace's equations are called *spherical solid harmonics*. They have the important property that when transformed into spherical coordinates, they can be factored into three functions, each of which depends on only one of the three variables $r$, $\theta$, or $\phi$. (For a good account of spherical harmonic analysis, see Blakely (1995).)

Using spherical polar coordinates $(r, \theta, \phi)$, where $r$ is the radial distance from the centre of mass, $\theta$ is the colatitude measured from the polar axis, and $\phi$ is a longitude measured from some arbitrary fixed direction, we can write Laplace's equation as

$$\frac{\partial}{\partial r}\left(r^2\frac{\partial V}{\partial r}\right) + \frac{\partial}{\partial \mu}\left((1-\mu^2)\frac{\partial V}{\partial \mu}\right) + \frac{1}{(1-\mu^2)}\frac{\partial^2 V}{\partial \phi^2} = 0, \qquad (4.26)$$

where $\mu = \cos\theta$. Laplace's equation can be solved by substituting the trial solution $V = r^n S_n(\mu, \phi)$, where $S_n(\mu, \phi)$ is independent of $r$, into Eq. (4.26). This gives

$$\frac{\partial}{\partial r}\left(r^2\frac{\partial V}{\partial r}\right) = n(n+1)r^n S_n, \qquad (4.27)$$

and Eq. (4.26) reduces to

$$\frac{\partial}{\partial \mu}\left((1-\mu^2)\frac{\partial S_n}{\partial \mu}\right) + \frac{1}{1-\mu^2}\frac{\partial^2 S_n}{\partial \phi^2} + n(n+1)S_n = 0. \qquad (4.28)$$

This is called *Legendre's equation* and any function $S_n$ that satisfies this equation is called a *spherical surface harmonic* of degree $n$. Because $n(n+1)$ remains unchanged when we write $-(n+1)$ for $n$, the general solution of Laplace's equation can be written as

$$V = \sum_{n=0}^{\infty}\left(A_n r^n + B_n r^{-(n+1)}\right) S_n(\mu, \phi). \qquad (4.29)$$

Each of the terms in this equation is called a *solid harmonic* of degree $n$ and $-(n+1)$, respectively (Ramsey 1940).

In the applications discussed in this chapter, where the deformations, either tidal or rotational, have axial symmetry, the solution of Legendre's equation reduces to

$$(1-\mu^2)\frac{\partial^2 \mathcal{P}_n(\mu)}{\partial \mu^2} - 2\mu\frac{\partial \mathcal{P}_n(\mu)}{\partial \mu} + n(n+1)\mathcal{P}_n(\mu) = 0, \qquad (4.30)$$

where Legendre's polynomial, $\mathcal{P}_n(\mu)$, is given by the terminating series

$$\mathcal{P}_n(\mu) = \frac{1\cdot 3\cdot 5\ldots(2n-1)}{n!}\left[\mu^n - \frac{n(n-1)}{2(2n-1)}\mu^{n-2}\right.$$

$$\left. + \frac{n(n-1)(n-2)(n-3)}{2\cdot 4\cdot(2n-1)(2n-3)}\mu^{n-4} + \cdots\right] \qquad (4.31)$$

or by Rodrigues's formula

$$P_n(\mu) = \frac{1}{2^n n!} \frac{d^n (\mu^2 - 1)^n}{d\mu^n} \tag{4.32}$$

and the colatitude $\theta$ is measured from the axis of symmetry of the deformation. Legendre polynomials are *zonal harmonics*; the first five are given by

$$P_0(\mu) = 1, \tag{4.33}$$

$$P_1(\mu) = \mu = \cos \theta, \tag{4.34}$$

$$P_2(\mu) = \frac{1}{2}(3\mu^2 - 1) = \frac{1}{4}(3 \cos 2\theta + 1), \tag{4.35}$$

$$P_3(\mu) = \frac{1}{2}(5\mu^3 - 3\mu) = \frac{1}{8}(5 \cos 3\theta + 3 \cos \theta), \tag{4.36}$$

$$P_4(\mu) = \frac{1}{8}(35\mu^4 - 30\mu^2 + 3) = \frac{1}{64}(35 \cos 4\theta + 20 \cos 2\theta + 9). \tag{4.37}$$

Surface harmonics are orthogonal functions and double integrals of their products over the surface of the sphere have the following useful properties. The element of area on a unit sphere ($r = 1$) is given by $\sin \theta \, d\theta \, d\phi = -d\mu \, d\phi$. If $Y_m(\mu, \phi)$ and $S_n(\mu, \phi)$ are two surface harmonics of degrees $m$ and $n$, respectively, and $m \neq n$, then

$$\int_0^{2\pi} \int_{-1}^{+1} Y_m(\mu, \phi) \, S_n(\mu, \phi) \, d\mu \, d\phi = 0. \tag{4.38}$$

If the two surface harmonics have the same degree, $n$, and one is a zonal harmonic, $P_n(\mu)$, then

$$\int_0^{2\pi} \int_{-1}^{+1} S_n(\mu, \phi) \, P_n(\mu) \, d\mu \, d\phi = \frac{4\pi}{2n+1} S_n(1), \tag{4.39}$$

where $S_n(1)$ is the value of $S_n(\mu, \phi)$ at the pole of $P_n(\mu)$.

Consider two points on a unit sphere: a variable point $(\theta', \phi')$ and a fixed point $(\theta, \phi)$. Let $\psi$ be the angle subtended by these points at the centre of the sphere, and consider the integral

$$\int_0^{2\pi} \int_{-1}^{+1} S_n(\theta', \phi') \, P_n(\cos \psi) \, d\mu' \, d\phi'. \tag{4.40}$$

Let the axes of the coordinates of the variable point be changed such that the new axis defining the variable colatitude passes through the fixed point $(\theta, \phi)$, and let the new angular coordinates of the variable point be $(\Theta', \Phi')$, such that $\Theta' = \psi$. Also let $S_n(\theta', \phi')$ become $Y_n(\Theta', \Phi')$. Then, from Eq. (4.39), we obtain

$$\int_0^{2\pi} \int_{-1}^{+1} Y_n(\Theta', \Phi') \, P_n(\cos \Theta') \, d(\cos \Theta') \, d\Phi' = \frac{4\pi}{2n+1} Y_n(1). \tag{4.41}$$

However,

$$Y_n(1) = S_n(\theta, \phi). \tag{4.42}$$

It follows from this that

$$\int_0^{2\pi} \int_{-1}^{+1} S_n(\mu', \phi') \, \mathcal{P}_n(\cos\psi) \, d\mu' \, d\phi' = \frac{4\pi}{2n+1} S_n(\mu, \phi), \tag{4.43}$$

where $S_n(\mu, \phi)$ is the same function of $(\mu, \phi)$ that $S_n(\mu', \phi')$ is of $(\mu', \phi')$ (Mac-Robert 1967).

Now consider the potential at some fixed point $P$ due to a homogeneous, nearly spherical body whose surface is defined by

$$R(\theta') = C \left[ 1 + \epsilon_2 \mathcal{P}_2(\cos\theta') \right], \tag{4.44}$$

where $\epsilon_2 (\ll 1)$ is a constant and $C$ is now the mean radius. The point $P$ can be either internal ($r < C$) or external ($r > C$) and has spherical coordinates ($r, \mu, \phi$), where $\mu = \cos\theta$ and the colatitude $\theta$ is measured from the axis of symmetry of the tidal bulge (Fig. 4.5). The total gravitational potential at $P$ is the sum of two parts. The part due to the spherical body is given by Eqs. (4.22) and (4.23), while the other, noncentral, part of the potential is due to the thin distribution of matter between the surface of the deformed body and the mean sphere. At some point $P'(r', \mu', \phi')$, the radial thickness of this thin layer of matter is $\epsilon_2 C \mathcal{P}_2(\mu')$ and the element of volume at that point is $\epsilon_2 C^3 \mathcal{P}_2(\mu') \, d\mu' \, d\phi'$. The potential at $P$ due to the element of mass is determined by the distance $PP' = \Delta$, where

$$\frac{1}{\Delta} = \left( C^2 + r^2 - 2Cr\cos\psi \right)^{-1/2}. \tag{4.45}$$

If $r < C$, then

$$\frac{1}{\Delta} = \frac{1}{C} \left[ 1 + \left( \frac{r}{C} \right)^2 - 2\mu \left( \frac{r}{C} \right) \right]^{-1/2}, \tag{4.46}$$

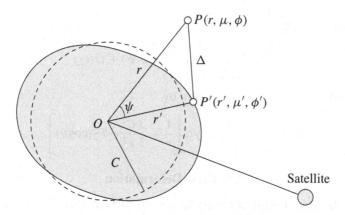

Fig. 4.5. The potential at the point $P$ due to a deformed central planet arises from the spherical body of mean radius $C$ and the distribution of matter associated with the tidal bulge.

and on expanding binomially and ordering the terms in powers of $r/C$ we obtain

$$
\frac{1}{\Delta} = \frac{1}{C}\left[ 1 + \left(\frac{r}{C}\right)\mu + \left(\frac{r}{C}\right)^2\left(-\frac{1}{2}+\frac{3}{2}\mu^2\right) \right.
$$

$$
\left. + \left(\frac{r}{C}\right)^3\left(-\frac{3}{2}\mu+\frac{5}{2}\mu^3\right) + \cdots \right],
\tag{4.47}
$$

and inspection of Eqs. (4.33)–(4.37) shows that Eq. (4.47) can be written as

$$
\frac{1}{\Delta} = \frac{1}{C}\sum_{n=0}^{\infty}\left(\frac{r}{C}\right)^n \mathcal{P}_n(\cos\psi) + \mathcal{O}(\epsilon_2).
\tag{4.48}
$$

Hence, the noncentral contribution to the total potential at $P$ is given by

$$
V_{\text{nc,int}} = -\gamma\mathcal{G}C^2\epsilon_2\sum_{n=0}^{\infty}\left(\frac{r}{C}\right)^2\int\int \mathcal{P}_2(\mu')\mathcal{P}_n(\cos\psi)\,\mathrm{d}\mu'\,\mathrm{d}\phi'.
\tag{4.49}
$$

From Eq. (4.43) we obtain

$$
\sum_{n=0}^{\infty}\left(\frac{r}{C}\right)^n\int\int \mathcal{P}_2(\mu')\mathcal{P}_n(\cos\psi)\,\mathrm{d}\mu'\,\mathrm{d}\phi' = \frac{4\pi}{5}\left(\frac{r}{C}\right)^2\mathcal{P}_2(\cos\theta).
\tag{4.50}
$$

Hence, the internal, noncentral contribution to the potential at $P$ is given by

$$
V_{\text{nc,int}} = -\frac{4\pi}{5}\gamma\mathcal{G}r^2\epsilon_2\mathcal{P}_2(\cos\theta),
\tag{4.51}
$$

and the total internal potential at $P$ is the sum of this term and that given by Eq. (4.22), that is,

$$
V_{\text{int}}(r,\theta) = -\frac{4}{3}\pi C^3\gamma\mathcal{G}\left[\frac{3C^2-r^2}{2C^3} + \frac{3}{5}\frac{r^2}{C^3}\epsilon_2\mathcal{P}_2(\cos\theta)\right].
\tag{4.52}
$$

Similarly, if $r > C$, then

$$
\frac{1}{\Delta} = \frac{1}{r}\sum_{n=0}^{\infty}\left(\frac{C}{r}\right)^n \mathcal{P}_n(\cos\psi) + \mathcal{O}(\epsilon_2)
\tag{4.53}
$$

and the total external potential is given by

$$
V_{\text{ext}}(r,\theta) = -\frac{4}{3}\pi C^3\gamma\mathcal{G}\left[\frac{1}{r} + \frac{3}{5}\frac{C^2}{r^3}\epsilon_2\mathcal{P}_2(\cos\theta)\right].
\tag{4.54}
$$

## 4.4 Tidal Deformation

If $h(\psi)$ is the local height of the equipotential surface, then given that the tidal potential at the surface of the planet is $-\zeta g\mathcal{P}_2(\cos\psi)$, the total potential is

$$
V_{\text{total}}(r,\psi) = -\frac{\mathcal{G}m_{\text{p}}}{B} + gh(\psi) - \zeta g\mathcal{P}_2(\cos\psi),
\tag{4.55}
$$

where $B$ is the mean radius of the planet. On an equipotential surface this must be independent of $\psi$; hence we must have $h(\psi) = \zeta \mathcal{P}_2(\cos \psi)$, where it is understood that $\psi$ is measured from the axis of symmetry of the tidal bulge. The equilibrium tide defines the shape of a shallow, zero-density ocean covering an inflexible, spherical planet. In practice, of course, no fluid has zero density and no solid is inflexible. We need to find the tidal deformation of a real solid or fluid body.

In order to show the factors involved, we consider the case of a simple, two-component model planet that has a mean radius $B$, a homogeneous, incompressible, fluid ocean of density $\sigma$, and a homogeneous, incompressible, solid core of mean radius $A$, density $\rho$, and rigidity $\mu$ (Street 1925, Dermott 1979a; see Fig. 4.6). The rigidity $\mu$ is a measure of the force needed to deform the elastic body.

The equilibrium tide is the second-order surface harmonic that describes the equipotential surface that would exist close to a completely rigid, ocean-free, spherical planet circled by a single, distant satellite. If the planet is not completely rigid and has an ocean, then the surfaces of the ocean and the equilibrium tide will *not* coincide. This is the case even if we neglect effects due to the kinetic energy of the ocean currents. To calculate the response of the ocean and core to the gravitational field of the satellite we must take into account the effect of the gravitational field of the tidal bulge itself (self-gravitation) and the effect of the elastic forces within the solid. We must also allow for the elastic yielding of the core under the action of all the forces that act on both the core and the ocean; that is, we must allow for the potentials of the deformed core and the ocean tide

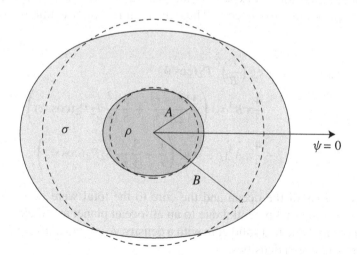

Fig. 4.6. Schematic diagram showing a model planet consisting of a deformed core of mean radius $A$ and density $\rho$, surrounded by a deformed ocean of mean radius $B$ and density $\sigma$. The circles of radii $A$ and $B$ are denoted by dashed lines.

itself, as well as the loading effects of the tides raised in the various parts of the planet (Street 1925).

Since the tide-raising potential is a solid spherical harmonic of the second degree, it follows that the deformation of the planet must be described in terms of the same harmonic function (Love 1911). If this were not the case, then it would not be possible for the surface of the ocean, for example, to be an equipotential surface. We have already seen in Sect. 4.2 that the tidal potential depends only on one angle, $\psi$, implying axial symmetry about the line joining the two centres. Thus, we can describe the deformed shapes of the core boundary and ocean surface by

$$R_{cb}(\psi) = A\left[1 + S_2 P_2(\cos\psi)\right] \tag{4.56}$$

and

$$R_{os}(\psi) = B\left[1 + T_2 P_2(\cos\psi)\right] \tag{4.57}$$

respectively, where $S_2$ and $T_2$ are constants. We will determine $S_2$ and $T_2$ from (i) the fact that the surface of the static ocean must be an equipotential and (ii) by considering the equilibrium of all the forces acting on the mean core boundary.

The potential within the ocean, $V_0(r, \psi)$, is the sum of three potentials: (i) the tidal potential due to the satellite,

$$V_3(r, \psi) = -\frac{\mathcal{G} m_s}{a^3} r^2 P_2(\cos\psi) = -\zeta g \left(\frac{r}{B}\right)^2 P_2(\cos\psi), \tag{4.58}$$

which is a generalisation of Eqs. (4.9) and (4.12), (ii) $V_{int}(r, \psi)$ due to the ocean, and (iii) $V_{ext}(r, \psi)$ due to the core. Thus, the total potential within the ocean is given by

$$V_0(r, \psi) = -\zeta g \left(\frac{r}{B}\right)^2 P_2(\cos\psi)$$
$$- \frac{4}{3}\pi B^3 \sigma \mathcal{G}\left(\frac{3B^2 - r^2}{2B^3} + \frac{3}{5}\frac{r^2}{B^3} T_2 P_2(\cos\psi)\right)$$
$$- \frac{4}{3}\pi A^3 (\rho - \sigma)\mathcal{G}\left(\frac{1}{r} + \frac{3}{5}\frac{A^2}{r^3} S_2 P_2(\cos\psi)\right). \tag{4.59}$$

The contributions of the ocean and the core to the total were derived by first calculating the internal potential due to an all-ocean planet and then adding the external potential due to a solid core with a density, $\rho - \sigma$, equal to the difference of the core and ocean densities.

We calculate the potential at the ocean surface, $V_{os}(r, \psi)$, by substituting the surface equation, $r = B\left[1 + T_2 P_2(\cos\psi)\right]$, into Eq. (4.59) and expanding the resulting expression, neglecting second and higher order terms in $\zeta/B$, $S_2$ and

$T_2$. This gives

$$V_{os}(r, \psi) = -\zeta g P_2(\cos \psi) - \frac{4}{3}\pi B^2 \sigma \mathcal{G} \left( \frac{1}{2} - \frac{2}{5}T_2 P_2(\cos \psi) \right)$$
$$- \frac{4}{3}\pi \frac{A^3}{B}(\rho - \sigma)\mathcal{G} \left( 1 - T_2 P_2(\cos \psi) + \frac{3}{5} \left( \frac{A}{B} \right)^2 S_2 P_2(\cos \psi) \right).$$

(4.60)

The terms in this equation that do not depend on $\psi$ give rise to compressive forces only; since we have assumed that the core and ocean are incompressible, these have no role in determining the shape of the planet and can be ignored. Given that the ocean surface is an equipotential, the sum of those terms that do depend on $\psi$ must be zero. Hence we must have

$$\frac{\zeta_c}{A} = \left[ \frac{2}{5}\frac{\sigma}{\rho} + \left( \frac{A}{B} \right)^3 \left( 1 - \frac{\sigma}{\rho} \right) \right] T_2 - \frac{3}{5} \left( \frac{A}{B} \right)^5 \left( 1 - \frac{\sigma}{\rho} \right) S_2.$$

(4.61)

Here we have introduced

$$\zeta_c = \frac{m_s}{m_c} \left( \frac{A}{a} \right)^3 A,$$

(4.62)

where $m_c$ is the mass of the core and

$$g_c = \frac{\mathcal{G}m_c}{A^2}$$

(4.63)

is the gravity at the core boundary. The quantity $\zeta_c$ is the amplitude of the "equilibrium tide" that would exist at the core boundary if the ocean were removed; it is related to $\zeta$ in Eq. (4.13) by

$$\zeta g A^2 = \zeta_c g_c B^2.$$

(4.64)

Equation (4.61) gives us our first relation between $S_2$ and $T_2$. A second relation is obtained by considering the equilibrium of all the forces acting on the mean core boundary.

Within the solid body of the core, the deforming potential, $V_c(r, \psi)$ is the sum of the tidal potential due to the satellite, $V_3(r, \psi)$, and the internal potentials due to the ocean and core. Thus,

$$V_c(r, \psi) = -\zeta_c g_c \left( \frac{r}{A} \right)^2 P_2(\cos \psi)$$
$$- \frac{4}{3}\pi B^3 \sigma \mathcal{G} \left( \frac{3B^2 - r^2}{2B^3} + \frac{3}{5}\frac{r^2}{B^3}T_2 P_2(\cos \psi) \right)$$
$$- \frac{4}{3}\pi A^3 (\rho - \sigma)\mathcal{G} \left( \frac{3A^2 - r^2}{2A^3} + \frac{3}{5}\frac{r^2}{A^3} S_2 P_2(\cos \psi) \right). \quad (4.65)$$

Consider a sphere within the core, concentric with the planet's centre. For a fixed value of $r$, the effective deforming potential is given by those terms in $V_c(r, \psi)$

that only depend on $P_2(\cos\psi)$. The other terms are compressive and can be ignored. Hence the effective deforming potential can be written as

$$V_c(r, \psi) = -Zr^2 P_2(\cos\psi), \qquad (4.66)$$

where

$$Z = \frac{g_c}{A}\left(\frac{\zeta_c}{A} + \frac{3}{5}\frac{\sigma}{\rho}(T_2 - S_2) + \frac{3}{5}S_2\right). \qquad (4.67)$$

Chree (1896a) showed that the yielding of the core under the force resulting from this deforming potential is the same as that which would be produced by an outward normal force per unit area of amount $\rho Z A^2 P_2(\cos\psi)$ acting at the mean core boundary, $r = A$.

Other pressures also act at this boundary. These arise from loading terms caused by the hydrostatic pressure in the ocean and the tide in the solid core. For example, in the case of a shallow ocean for which $(B - A) \ll B$ and there is no variation of $g$ within the ocean, these pressures arise from (i) the variation with $\psi$ of the ocean depth and (ii) the variation with $\psi$ of the distance of the core boundary from the centre of the planet. These pressures are given by the product of the local gravity, the density, and the height of the tide and can be written as

$$P_o(\psi) = g\sigma B(T_2 - S_2)P_2(\cos\psi) \qquad (4.68)$$

and

$$P_c(\psi) = g_c\rho A S_2 P_2(\cos\psi). \qquad (4.69)$$

For a deep ocean $g$ is not a constant and although the loading term due to the tide in the core remains the same, the angular variation of the hydrostatic pressure on the core is no longer determined by the height of the ocean tide alone. We must also take account of the angular variation of gravity within the ocean.

In the general case of a deep ocean, the pressure on the core boundary due to the ocean is given by

$$P_o(\psi) = \int_{R_{cb}}^{R_{os}} \sigma(r)\frac{\partial V_o(r, \psi)}{\partial r}dr, \qquad (4.70)$$

where the integral limits are from the core boundary to the ocean surface. As the ocean in our model is assumed to be incompressible and of uniform density, this reduces to

$$P_o(\psi) = \sigma\left[V_o(R_{os}, \psi) - V_o(R_{cb}, \psi)\right]. \qquad (4.71)$$

Furthermore, since only the variable part of the potential can contribute to the deforming forces (we are neglecting compression) and since the surface of the ocean is an equipotential surface, it follows that $V_o(R_{os}, \psi)$ is a constant and can be neglected.

We can obtain the potential at the core boundary, $V_{cb}(\psi)$, by substituting the equation for the core boundary, $R_{cb} = A\left[1 + S_2 P_2(\cos\psi)\right]$, into Eq. (4.60) or

Eq. (4.65) and expanding the resulting expression, neglecting second and higher order terms in $\zeta/B$, $S_2$, and $T_2$. This gives

$$V_{cb}(\psi) = \text{constant} - Ag_c \left( \frac{\zeta_c}{A} + \frac{3}{5}\frac{\sigma}{\rho}(T_2 - S_2) - \frac{2}{5}S_2 \right) P_2(\cos \psi), \qquad (4.72)$$

where the $\psi$ dependence is contained in the second term on the right-hand side of the equation; this is the part that will contribute to the variable part of the pressure at the core boundary, $P_o(\psi)_\psi$. Hence

$$P_o(\psi)_\psi = \sigma Ag_c \left( \frac{\zeta_c}{A} + \frac{3}{5}\frac{\sigma}{\rho}(T_2 - S_2) - \frac{2}{5}S_2 \right) P_2(\cos \psi). \qquad (4.73)$$

The total effective outward normal force per unit area at the mean core boundary is the sum of the elastic forces within the core and the loading terms due to the ocean and the core tides. We write this as $XP_2(\cos \psi)$, where

$$\begin{aligned} X &= \rho A^2 Z - P_o(\psi)_\psi - \rho g_c A S_2 \\ &= \frac{2}{5}\rho g_c A \left( 1 - \frac{\sigma}{\rho} \right) \left( \frac{5}{2}\frac{\zeta_c}{A} - S_2 + \frac{3}{2}\frac{\sigma}{\rho}(T_2 - S_2) \right). \end{aligned} \qquad (4.74)$$

Note that $X \to 0$ as $\sigma \to \rho$, which must be the case if the ocean is in hydrostatic equilibrium.

Love (1944) showed that the radial displacement of the solid core produced by this deforming pressure is

$$\Delta R(\psi) = \frac{5}{19}\frac{A}{\mu}XP_2(\cos \psi), \qquad (4.75)$$

which must be the same as $AS_2P_2(\cos \psi)$. This gives us our second relationship between $S_2$ and $T_2$:

$$S_2 = \frac{1}{\tilde{\mu}} \left( 1 - \frac{\sigma}{\rho} \right) \left( \frac{5}{2}\frac{\zeta_c}{A} - S_2 + \frac{3}{2}\frac{\sigma}{\rho}(T_2 - S_2) \right), \qquad (4.76)$$

where $\tilde{\mu}$, the *effective rigidity* of the solid core, is a dimensionless quantity defined by

$$\tilde{\mu} = \frac{19\mu}{2\rho g_c A}. \qquad (4.77)$$

It is a measure of the ratio of the elastic and gravitational forces acting at the core boundary.

If $\tilde{\mu} \ll 1$, then the core responds like a fluid, whereas for $\tilde{\mu} \gg 1$ elastic forces within the core dominate. Note that if $\sigma = \rho$, then $S_2 = 0$ and the elastic core is undeformed. If the planet is ocean free, then $\sigma = 0$ and, from Eq. (4.76), the amplitude of the tide on the isolated core is given by

$$AS_2 = \frac{(5/2)\zeta_c}{1 + \tilde{\mu}}. \qquad (4.78)$$

In the general case, we can write

$$AS_2 = F\frac{(5/2)\zeta_c}{1+\tilde{\mu}} \quad \text{and} \quad BT_2 = H\frac{5}{2}\zeta, \tag{4.79}$$

in which case $F$ is a dimensionless quantity that is a measure of the effect of the ocean on the amplitude of the core tide, and $H$ is a measure of the effect of the internal structure on the external shape of the planet. By eliminating $T_2$ from Eqs. (4.61) and (4.76), we obtain

$$F = \frac{(1+\tilde{\mu})(1-\sigma/\rho)(1+3/2\alpha)}{1+\tilde{\mu}-\sigma/\rho+(3\sigma/2\rho)(1-\sigma/\rho)-(9/4\alpha)(A/B)^5(1-\sigma/\rho)^2} \tag{4.80}$$

and

$$H = \frac{2\langle\rho\rangle}{5\rho}\left(\frac{1+\tilde{\mu}+(3/2)(A/B)^2 F\delta}{(1+\tilde{\mu})(\delta+2\sigma/5\rho)}\right), \tag{4.81}$$

where

$$\alpha = 1 + \frac{5}{2}\frac{\rho}{\sigma}\left(\frac{A}{B}\right)^3\left(1-\frac{\sigma}{\rho}\right) \quad \text{and} \quad \delta = \left(\frac{A}{B}\right)^3\left(1-\frac{\sigma}{\rho}\right) \tag{4.82}$$

and $\langle\rho\rangle$ is the mean density. If the planet is fluid throughout, or if because of thermal creep the solid core has relaxed to hydrostatic equilibrium (this has application to tidal bulges on satellites with spin rates equal to or *synchronous* with their orbital mean motions), then $\tilde{\mu} = 0$ and the hydrostatic value of $H$ is given by

$$H_h = \frac{2\langle\rho\rangle}{5\sigma}\left(\frac{1+(3\delta/5\gamma)(A/B)^2}{\delta+2\sigma/5\rho-(9\delta\sigma/25\gamma\rho)(A/B)^2}\right), \tag{4.83}$$

where

$$\gamma = \frac{2}{5} + \frac{3\sigma}{5\rho} \tag{4.84}$$

(Dermott 1979a).

We can now use these results to determine the tidal amplitudes in the limiting case of a planet with a shallow, uniform ocean. If $A = B$, $\zeta_c = \zeta$, and $\langle\rho\rangle = \rho$ the condition for the ocean surface to be an equipotential surface, Eq. (4.61), reduces to

$$\frac{\zeta}{A} = \frac{2}{5}S_2 + \left(1-\frac{3}{5}\frac{\sigma}{\rho}\right)(T_2-S_2). \tag{4.85}$$

Eliminating the terms containing $S_2$ alone from Eqs. (4.61) and (4.76) we obtain

$$A(T_2 - S_2) = \frac{\zeta\tilde{\mu}}{1-\sigma/\rho+\tilde{\mu}\left(1-3\sigma/5\rho\right)} \tag{4.86}$$

for the amplitude of the ocean tide. This agrees with that first given by Chree (1896b). We note that $A(T_2 - S_2) \to 0$ as $\tilde{\mu} \to 0$. Furthermore, if $\sigma = \rho$, then

$$A(T_2 - S_2) = \frac{5}{2}\zeta \tag{4.87}$$

and is independent of $\tilde{\mu}$. Thus, when the core and ocean have the same density, the core is undeformed and the amplitude of the ocean tide is a factor 5/2 larger than the "equilibrium" tide given by Eq. (4.13).

We can find the amplitude of the solid body tide in the case of a shallow ocean by substituting the expression for $T_2 - S_2$ from Eq. (4.86) into Eq. (4.61). This gives

$$AS_2 = \frac{5}{2}\zeta \left[ \frac{(1 - \sigma/\rho)}{1 - \sigma/\rho + \tilde{\mu}(1 - 3\sigma/5\rho)} \right]. \tag{4.88}$$

For an ocean-free planet, $\sigma = 0$ and

$$AS_2 = \frac{(5/2)\zeta}{1 + \tilde{\mu}}, \tag{4.89}$$

which agrees with the expression first given by Lord Kelvin (Thompson 1863). Kelvin applied this result to observations of the fortnightly tide for which $AS_2 \approx 0.6\zeta$ and hence deduced that the rigidity of the Earth is $\sim 1.2 \times 10^{11} \mathrm{Nm}^{-2}$ or $\sim 50\%$ greater than the rigidity of uncompressed steel. At the time, this was a surprising result because the interior of the Earth was then thought to be largely molten (Bullen 1975).

In the case of the Earth it is perhaps more realistic to consider the planet as being composed of a solid core of density $\rho$ and rigidity $\mu$ covered by a shallow ocean of density $\sigma$. In which case, the amplitudes of the tides in the ocean and the solid core are given by Eqs. (4.86) and (4.88). However, these amplitudes were calculated by assuming that the ocean is in hydrostatic equilibrium and thus that the tidal currents in the ocean have no role in determining the shape of the ocean surface. For the Earth, this is not a good assumption.

The effects of ocean currents depend on the natural oscillation frequency of the ocean basin and this is determined by the size, shape, and depth of the basin. For heuristic purposes we follow Proudman (1953) and consider a uniform, equatorial canal of depth $d$ bounded by two parallels of latitude. In a reference frame centred on the Earth and rotating with the mean motion of the Moon the tidal bulge is stationary. Hence the speed of the bulge with respect to the surface of the Earth is $U = 2\pi A / T_E \approx 500 \mathrm{\ ms}^{-1}$, where $T_E$ is the spin period of the Earth. However, $U$ is not the speed of the tidal current in the ocean. The fluid in the tidal bulge is supplied by a current of speed $u$ flowing uniformly throughout the depth of the ocean (see Fig. 4.7). If, at a given longitude, $\zeta$ denotes the height of the surface above the ocean floor, then from the equation of continuity we

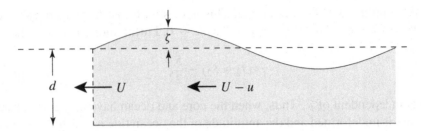

Fig. 4.7.  Schematic diagram of a tidal wave in a uniform equatorial canal of depth $d$ bounded by two parallels of latitude. The tidal bulge is stationary in a reference frame centred on the Earth and rotating with the Moon. The speed of the bulge with respect to the solid Earth is $U$. However, the fluid in the tidal bulge is supplied by a current of speed $u$, which, at a given longitude, flows uniformly throughout the depth of the ocean.

have

$$(U - u)(d + \zeta) = U d . \tag{4.90}$$

If $\zeta \ll d$, then

$$u = \frac{\zeta}{d} U . \tag{4.91}$$

The mean ocean depth on the Earth is $\sim 4$ km, and so $u \sim 0.1$ ms$^{-1}$. From Bernouilli's theorem, which relates work done by the hydrostatic pressure to the kinetic and potential energies of the fluid in streamline flow, we obtain

$$\frac{1}{2}(U - u)^2 + g\zeta + \Psi = \text{constant,} \tag{4.92}$$

where $\Psi$ is the tidal potential. If we assume that at the surface of the canal $\Psi = -g\bar{\zeta}$, then given that $U^2$ is a constant and $u^2 \ll uU$, we have

$$U u = g(\zeta - \bar{\zeta}) . \tag{4.93}$$

Substituting for $u$ from Eq. (4.91) we obtain

$$\zeta = \frac{\bar{\zeta}}{1 - U^2/gd} . \tag{4.94}$$

For the Earth resonance would occur in an equatorial canal of depth $d_{\text{res}} = U^2/g \approx 22$ km and given that the mean depth of the ocean is less than $d_{\text{res}}$ we would expect the tides to be inverted. This reasoning cannot be applied to the Earth as a whole because for near-global oceans we cannot assume that the total tidal potential is decoupled from the tidal currents. However, Eq. (4.94) provides a good indication that in calculating the shapes of the Earth's oceans we must include effects due to the ocean currents and consider the possibility of resonance in the various ocean basins.

## 4.5 Rotational Deformation

In Sect. 4.4 we showed how the tide raised by an orbiting satellite distorts the surface of a planet. By modelling the planet as consisting of a core and mantle we were able to derive expressions for the distortions of each component. The most important result is that the shape of the distorted planet (see Fig. 4.6) can be approximated by an *oblate spheroid* with long semi-axis $a$ lying along the planet–satellite line and with short semi-axes $b = c$ giving a circular cross section perpendicular to the axis of symmetry (the planet–satellite line). The spheroid is modelled using $P_2(\cos \psi)$, a Legendre polynomial of degree 2 where the angle $\psi$ is measured from the axis of symmetry. In this section we show that many of the analytical results derived for tidal distortion can be directly applied to rotational distortion.

Consider a spherical, rigid planet rotating at an angular rate $\Omega$ (see Fig. 4.8). A point $P$ on the surface experiences a centrifugal acceleration, $\mathbf{a}_{cf,x} = \Omega^2 r \sin\theta\, \hat{\mathbf{x}}$, or, since $x = r \sin\theta$, $\mathbf{a}_{cf,x} = \Omega^2 x\, \hat{\mathbf{x}}$. By symmetry a similar point located on the surface in the $y$–$z$ plane feels an acceleration $\mathbf{a}_{cf,y} = \Omega^2 y\, \hat{\mathbf{y}}$. The rotation of the planet produces no acceleration along the axis of rotation. Therefore, an arbitrary point on the surface at position $(x, y, z)$ experiences an acceleration

$$\mathbf{a}_{cf} = \Omega^2 (x\, \hat{\mathbf{x}}, y\, \hat{\mathbf{y}}) . \tag{4.95}$$

We can consider this centrifugal acceleration in terms of a centrifugal potential, $V_{cf}$, such that $\mathbf{a}_{cf} = -\nabla V_{cf}$, where, in polar coordinates,

$$V_{cf}(r, \theta) = -\frac{1}{2}\Omega^2 r^2 \sin^2\theta . \tag{4.96}$$

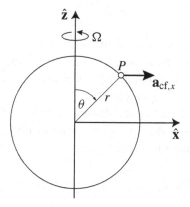

Fig. 4.8. The acceleration experienced at a point $P$ (in the $x$–$z$ plane) on the surface of a planet rotating at a rate $\Omega$. Here $\theta$ is the angle measure from the $z$ axis (the axis of rotation) and $r$ is the radial distance to the point.

Now consider a global ocean on the surface of the planet. The fluid experiences a total potential

$$V_{total}(r, \theta) = -\frac{\mathcal{G}m_p}{r} + V_{cf}(r, \theta). \tag{4.97}$$

We know that, in equilibrium, the surface of the fluid must lie on an equipotential surface where the surface is locally perpendicular to the net gravitational and centrifugal acceleration. Assuming that the distortion of the ocean surface from a sphere is small, we can write

$$r_{ocean} = a + \delta r(\theta), \tag{4.98}$$

where $a = r_{equatorial}$, the equatorial radius of the planet. Therefore the potential describing the surface is the constant given by

$$V_{total}(surface) \approx -\frac{\mathcal{G}m_p}{a} + \frac{\mathcal{G}m_p}{a^2}\delta r - \frac{1}{2}\Omega^2 a^2 \sin^2\theta - \Omega^2 a \sin^2\theta \delta r. \tag{4.99}$$

For most planets $\Omega^2 a \ll \mathcal{G}m_p/a^2$ (see below) and hence we can neglect the last term in Eq. (4.99). Hence

$$\delta r \approx constant + \frac{\Omega^2 a^4}{2\mathcal{G}m_p}\sin^2\theta. \tag{4.100}$$

Therefore the effect of rotation, as might be expected, is to cause flattening at the poles. We can quantify the extent of the deformation by defining the *oblateness* or *flattening* of the planet as

$$f = \frac{r_{equatorial} - r_{pole}}{r_{equatorial}}. \tag{4.101}$$

The analysis above suggests that for planets we should expect to find $f \approx q/2$, where

$$q = \frac{\Omega^2 a^3}{\mathcal{G}m_p} \tag{4.102}$$

is a dimensionless quantity equal to the ratio of the centrifugal acceleration at the equator to the gravitational acceleration. However, we have neglected the feedback effect that the rotational distortion has on the planet's gravity field given that the planet is no longer a sphere.

Before including the effect of the distortion, it is worthwhile to consider the extreme case, $q \to 1$, since this allows us to set an upper limit on the rotational velocity of a planet. Equation (4.102) gives

$$\Omega_{max} \approx \left(\frac{\mathcal{G}m_p}{a^3}\right)^{1/2} \approx a\,(\mathcal{G}\langle\rho\rangle)^{1/2}, \tag{4.103}$$

where $\langle\rho\rangle$ is the mean density of the planet. For the Earth, $\langle\rho\rangle = 5.52$ g cm$^{-3}$, and so $\Omega_{max} \approx 1.2 \times 10^{-3}$ rad s$^{-1}$ for a rotation period of $P_{min} = 1.4$ h. For Jupiter, $P_{min} \approx 2.9$ h compared with its current value of 9.9 h.

Because of rotational flattening, most planets (but *not* most satellites) may be treated to a good approximation as oblate spheroids (i.e., triaxial ellipsoids with two equal long axes ($a = b$) and one short axis ($c$)). A basic result of potential theory is that the *external* gravitational potential of any body with an axis of symmetry can be written in the form

$$V_{\text{gravity}}(r, \theta) = -\frac{\mathcal{G}m}{r} \left[ 1 - \sum_{n=2}^{\infty} J_n \left(\frac{R}{r}\right)^n \mathcal{P}_2(\cos\theta) \right], \qquad (4.104)$$

where $m$ is the total mass, $R$ ($= a$ for the case of rotational deformation) is the equatorial radius, $J_n$ are dimensionless constants, and, as before (see Sect. 4.3), $\mathcal{P}_n(\cos\theta)$ are Legendre polynomials of degree $n$. Note that there is no $n = 1$ term provided the origin of the coordinates is chosen as the body's centre of mass. The $J_n$s reflect the distribution of mass within the body and must be determined empirically for a planet. Of these quantities, by far the most important is $J_2$ and this has a simple physical interpretation in terms of the three moments of inertia, $\mathcal{A}$, $\mathcal{B}$, and $\mathcal{C}$, about the principal axes. MacCullagh's theorem (see Eq. (5.36) and the derivation by Cook 1973) allows us to write (Cook 1980)

$$J_2 = \frac{\mathcal{C} - \frac{1}{2}(\mathcal{A} + \mathcal{B})}{ma^2} \approx \frac{\mathcal{C} - \mathcal{A}}{ma^2}, \qquad (4.105)$$

where the approximation is valid when $\mathcal{A} \approx \mathcal{B}$, as is the case for rotational distortion. In general, $J_n$ is given by the integral

$$J_n = +\frac{1}{mR^n} \int_0^R \int_{-1}^{+1} r^n \mathcal{P}_n(\mu)\rho(r,\mu)\, 2\pi r^2 \, d\mu \, dr, \qquad (4.106)$$

where $\mu = \cos\theta$ and $\rho(r, \mu)$ is the internal density distribution. Since $\mathcal{P}_2(\mu)$ is an odd function for odd $n$, $J_3 = J_5 = J_7 = \cdots = 0$ for a planet whose northern and southern hemispheres are symmetric. In fact, only the Earth has a measured nonzero value of $J_3$. Provided that $q$ is small, it can be shown that $J_n \propto q^{n/2}$ where the constants of proportionality are usually of order unity. Therefore, since $q \ll 1$ generally, the higher-order $J_n$s rapidly become small. For a planet of uniform density it can be shown that $J_2 = q/2$. Values of $J_2$ and $J_4$ for the planets are listed in Table A.4.

We can now return to the problem of calculating the flattening of a rotating planet. We can write the centrifugal potential given in Eq. (4.96) as

$$V_{\text{cf}} = \frac{1}{3}\Omega^2 r^2 \left[ \mathcal{P}_2(\mu) - 1 \right]. \qquad (4.107)$$

Here we can ignore the term that is independent of $\mu$ so that $V_{\text{cf}}$ has the same angular dependence as the $J_2$ term in the planetary gravity field. The total potential experienced by an ocean on the surface of the planet is now

$$V_{\text{total}}(r, \theta) = -\frac{\mathcal{G}m_{\text{p}}}{r} + \left[ \frac{\mathcal{G}m_{\text{p}}a^2}{r^3} J_2 + \frac{1}{3}\Omega^2 r^2 \right] \mathcal{P}_2(\mu), \qquad (4.108)$$

where we have neglected $J_4$, $J_6$, etc. As before, we require the surface to be an equipotential and write $r = R + \delta r(\theta)$. On substituting into Eq. (4.108) and expanding this gives

$$\delta r = \text{constant} - \left[ J_2 + \frac{1}{3}q \right] R P_2(\mu). \qquad (4.109)$$

We can use the definition of $f$ in Eq. (4.101) with this new expression for $\delta r$ to obtain

$$f = \frac{3}{2}J_2 + \frac{1}{2}q, \qquad (4.110)$$

replacing our previous result, $f \approx q/2$. Using data from Yoder (1995) we can compare the calculated value of $f$ with its observed value. For the Earth $f_{\text{calc}} = 0.003349$ whereas $f_{\text{obs}} = 0.003353$. In the case of Jupiter $f_{\text{calc}} = 0.06670$ whereas $f_{\text{obs}} = 0.06487$. This value is sufficiently large that it is possible to see polar flattening in images of Jupiter's disc. From these comparisons it is clear that Eq. (4.110) provides a good estimate of the flattening.

The fact that both tidal deformation and rotational deformation give rise to a surface that can be modelled by means of a Legendre polynomial of degree two implies that the theory developed in Sect. 4.4 for the tidal deformation of a planet with a core and mantle can be directly applied to the case of rotational deformation. In both cases measurements of the extent of the deformation reveal information about the internal structure of the planet. Of course, the theory is equally applicable to the case of a satellite deformed by (i) the tides raised on it by the planet and (ii) the satellite's own rotation. In Sect. 4.7 we examine the particular case of the deformations on a satellite in synchronous rotation.

The $J_2$ of a planet modifies the gravitational field experienced by an orbiting object such as a satellite or ring particle. The main consequence is that the elliptical path of an orbiting object rotates or *precesses* in space. The dynamical consequences are discussed in more detail in Sects. 6.11, 7.7, 7.9, and 8.11. For our purposes it is sufficient to know that the precessional effect of $J_2$ can be observed directly by monitoring the orbits of satellites and narrow, eccentric rings. Therefore, $J_2$ is an observable quantity and Eq. (4.105) allows us to relate it to $C - A$, the difference in the two principal moments of inertia.

However, we still require an additional relation between $C$ and $A$ in order to calculate each separately and thereby constrain models of the interior. One such method for the case of the Earth is to observe the consequences of the torque exerted by the Sun and Moon on the rotationally flattened Earth. This causes the Earth's spin axis to rotate about the normal to Earth's orbit plane at a rate that is proportional to $(C - A)/C$ (Cook 1980), an effect called *luni-solar precession*. At the moment this technique for determining $C$ and $A$ can only be applied to the Earth–Moon system. For the other planets a different method must be employed.

## 4.6 The Darwin–Radau Relation

The Darwin–Radau relation (see, for example, Cook 1980) is an approximate equation relating the moment of inertia factor, $C/mR^2$ (where $m$ is the mass of the object and $R$ is its mean radius), to the values of $q$, $f$, and $J_2$ of the planet or satellite. The relation was first derived by Radau (1885) based on work by Clairaut (1743); Darwin (1899) also contributed to the problem. The underlying assumption is that the object concerned is in hydrostatic equilibrium. The relation can be expressed in a number of different forms but here we adopt the one used by Cook (1980). This gives

$$\frac{C}{mR^2} = \frac{2}{3}\left[1 - \frac{2}{5}\left(\frac{5}{2}\frac{q}{f} - 1\right)^{1/2}\right].$$

(4.112)

If we define the moment of inertia factor, $\bar{C}$, a dimensionless quantity, to be

$$\bar{C} = \frac{C}{mR^2}$$

(4.113)

and make use of the relationship between $J_2$, $q$, and $f$ given in Eq. (4.110) then we can rewrite the Darwin–Radau relation given in Eq. (4.112) as

$$\frac{J_2}{f} = -\frac{3}{10} + \frac{5}{2}\bar{C} - \frac{15}{8}\bar{C}^2.$$

(4.114)

However, the relationship between $\bar{C}$ and $J_2/f$ given by this form of the Darwin–Radau relation is just the limiting case of a more general result that can be derived using the distortion model for a more realistic planet derived in Sect. 4.4. For the core–mantle model derived there, Dermott (1979b) gives

$$\bar{C} = \frac{2}{5}\left[\frac{\sigma}{\langle\rho\rangle} + \left(1 - \frac{\sigma}{\langle\rho\rangle}\right)\left(\frac{A}{B}\right)^2\right].$$

(4.115)

This result can be derived from first principles using the definition of the moment of inertia and the known dimensions and distortions of the core and mantle. If the satellite surface and the core–mantle interface are in equilibrium then the factor $H$ has the hydrostatic value $H_h$ given by Eq. (4.83). Dermott (1979b) relates the value of $J_2/f$ to $H_h$ by means of the equation

$$\frac{J_2}{f} = \frac{2}{3}\left(1 - \frac{2}{5H_h}\right).$$

(4.116)

The formulae for $\delta$ and $\gamma$ allow us to relate $J_2/f$ to $\bar{C}$ for a given value of $A/B$. We obtain

$$\frac{J_2}{f} = \frac{2}{3} + \frac{\bar{C} - \frac{2}{5}(A/B)^2}{1 - (A/B)^2} + \frac{8 - 20(A/B)^5 + 10\bar{C}\left[5(A/B)^3 - 2\right]}{12\left[(A/B)^5 - 1\right] + 15\bar{C}\left[2 - 5(A/B)^3 + 3(A/B)^5\right]}.$$

(4.117)

There are two limiting cases of this formula to consider:

- In the case of a point core, $A/B \to 0$ and

$$J_2/f \to \bar{C}. \tag{4.118}$$

- In the case of the Darwin–Radau relation, $A/B \to 1$ and

$$J_2/f \to -\frac{3}{10} + \frac{5}{2}\bar{C} - \frac{15}{8}\bar{C}^2. \tag{4.119}$$

Figure 4.9 shows plots of $J_2/f$ as a function of $\bar{C}$ for the point core model, the Darwin–Radau model, and a general core–mantle model with $A/B = 0.5$. The known values of $J_2/f$ for Earth and the giant planets are indicated with horizontal lines denoting the limits on $\bar{C}$ for each planet. Note that the range of values of $\bar{C}$ for a given value of $J_2/f$ becomes smaller as $J_2/f$ increases. The key point is that measurements of $J_2/f$ allow limits to be placed on the moment of inertia factor. When this information is combined with measurements of $C - A$ from Eq. (4.105) we can obtain estimates for $A$ as well as $C$. Such estimates provide constraints on detailed models of planetary interiors.

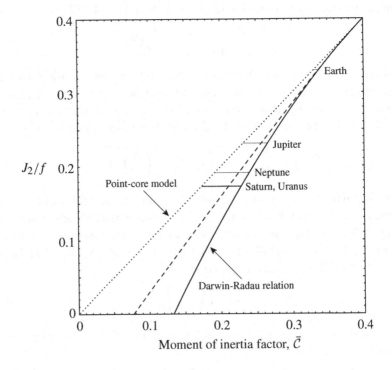

Fig. 4.9.  The value of $J_2/f$ as a function of the moment of inertia factor, $\bar{C}$, for the cases of (i) a point-core model (dotted line), (ii) a model with $A/B = 0.5$ (dashed line), and (iii) the Darwin–Radau relation (solid line). Values of $J_2/f$ for the Earth and the giant planets are indicated.

## 4.7 Shapes and Internal Structures of Satellites

Consider the case of a satellite in hydrostatic equilibrium. The satellite is assumed to exhibit synchronous rotation and to be moving in a near-equatorial, near-circular orbit about a planet. The satellite experiences tidal deformation due to the planet as well as rotational deformation due to its own spin. The theory developed above and the fact that the satellite's mean motion $n$ is equal to its spin frequency $\Omega$ allows us to show that the resulting shape of the satellite should be a triaxial ellipsoid. Indeed, because there are specific relationships between the semiaxes for a satellite in hydrostatic equilibrium, accurate measurements of a satellite's shape allow us to determine if it is in equilibrium and when combined with other data allow us to infer properties of the internal structure of the satellite.

The centrifugal potential at a point $(r, \theta, \psi)$ arising from the satellite's rotation is

$$V_{\text{rotational}} = \frac{1}{3}\Omega^2 r^2 \mathcal{P}_2(\cos\theta) \qquad (4.120)$$

(cf. Eq. (4.107)), where $\theta$ is the angle between the radius vector and the vertical axis, and $\mathcal{P}_2(\cos\theta) = (1/4)(3\cos 2\theta + 1)$ is the Legendre polynomial of degree two. Note that $V_{\text{rotational}}$ is independent of $\psi$, the angle between the projected radius vector and the $x$–$y$ plane; this accounts for the symmetry of the equipotential surface about the $z$ axis. The tidal potential due to the planet at the same point is

$$V_{\text{tidal}} = -\frac{\mathcal{G}m_{\text{p}}}{a^3}r^2 \mathcal{P}_2(\cos\psi) \qquad (4.121)$$

(cf. Eq. (4.9)), where $m_{\text{p}}$ is the mass of the planet. Note that $V_{\text{tidal}}$ is independent of $\theta$, the angle the radius vector makes with the $z$ axis; this results in the symmetry of the equipotential surface about the $x$ axis. However, by making use of Kepler's third law and noting that $n = \Omega$ we can write

$$V_{\text{tidal}} = -\Omega^2 r^2 \mathcal{P}_2(\cos\psi), \qquad (4.122)$$

and hence $V_{\text{rotational}}$ has exactly the same form as $V_{\text{tidal}}$, differing only by a factor 3 in magnitude and with different axes of symmetry. This means that we can apply the theory developed in Sect. 4.4 for tidal deformation directly to rotational deformation. Figure 4.10 shows a comparison of the resulting equipotential surfaces for each type of deformation. Note that in the case of rotational deformation (Fig. 4.10a) the shape is symmetric with respect to the $z$ axis whereas in the case of tidal deformation the $x$ axis (lying along the satellite–planet line) is the axis of symmetry.

Using the shape model for the surface of the mantle (i.e., the surface of the satellite) given by Eq. (4.57) we can now calculate the resulting shape for each form of deformation, recognising that an addition factor of $-1/3$ has to be

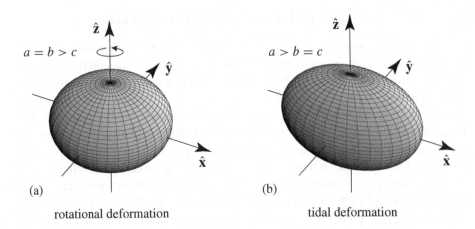

$a = b > c$ (a)
rotational deformation

$a > b = c$ (b)
tidal deformation

Fig. 4.10. Examples of the equipotential surfaces arising from (a) rotational deformation, where the $z$ axis is the axis of rotation, and (b) tidal deformation, where the tide-raising body lies along the direction of the $x$ axis.

introduced for the rotational case. It is easiest to define the shape in terms of the semiaxes $a$, $b$, and $c$ (along the $x$, $y$, and $z$ axes) of a general, triaxial ellipsoid. Each of these quantities can be calculated as a function of $B$ and $T_2$ alone by evaluating the Legendre polynomial for the appropriate values of $\theta$ and $\psi$.

For the rotational deformation we need only calculate $P_2(\cos\theta)$ at $\theta = \pi/2$ to give $a$ and $b$, and at $\theta = 0$ to give $c$. This gives

$$a_r = B(1 + T_2/6), \qquad b_r = B(1 + T_2/6), \qquad c_r = B(1 - T_2/3). \qquad (4.123)$$

For the tidal deformation we need only calculate $P_2(\cos\psi)$ at $\psi = 0$ to give $a$ and at $\psi = \pi/2$ to give $b$ and $c$. This gives

$$a_t = B(1 + T_2), \qquad b_t = B(1 - T_2/2), \qquad c_t = B(1 - T_2/2). \qquad (4.124)$$

If we assume that the spin axis is perpendicular to the orbit plane and that we can add the rotational and tidal contributions linearly, the resulting shape is a triaxial ellipsoid with semiaxes

$$a = B(1 + 7T_2/6), \qquad b = B(1 - T_2/3), \qquad c = B(1 - 5T_2/6). \qquad (4.125)$$

In particular, we have the result that the shape of a synchronously rotating satellite in hydrostatic equilibrium subjected to rotational and tidal deformations is such that

$$b - c = \frac{1}{4}(a - c) \qquad (4.126)$$

(Dermott 1979b). Furthermore, adapting Eq. (4.13) for the case of the tide raised by the planet on the satellite and combining it with Eq. (4.102) gives $\zeta/B = 3q/4$.

Hence, using the definition of $BT_2$ given in Eq. (4.79),

$$a - c = 2BT_2 = \frac{15}{4} H_h q B .$$ (4.127)

In this equation the quantities $a - c$, $B$, and $q$ can be measured from sufficiently high-resolution images of the satellite. Combining these with knowledge of the mass of the satellite gives its mean density, $\langle \rho \rangle$. From the definition of the factor $H_h$ in Eq. (4.83) we see that this now places constraints on the values of $A/B$ (where $A$ is the mean radius of the core) as well as $\sigma$ and $\rho$ (the densities of the core and mantle, respectively). This is the basis of a technique that can be used to determine the internal structure of satellites, particularly those close to a planet where the tidal and rotational distortions are large.

Dermott (1979b) used the theory presented here and in Sects. 4.5 and 4.6 to suggest that spacecraft determinations of (a) the gravitational moments of satellites such as Io, Ganymede, and Titan and (b) the shapes of satellites such as Mimas and Tethys could be used to provide evidence for internal differentiation. Dermott & Thomas (1988) applied a second-order version of the shape technique to Mimas using high-resolution images obtained by the *Voyager* spacecraft. They found that the shape of Mimas is a good approximation to a triaxial ellipsoid with measurements giving $(b-c)/(a-c) = 0.27 \pm 0.04$ compared with a predicted ratio of 0.25 from Eq. (4.126); this suggests that the satellite is close to hydrostatic equilibrium. Dermott & Thomas combined their estimated value of the mean radius, $B = 198.8$ km, with the mass of Mimas as determined by Kozai (1957) to obtain a mean density $\langle \rho \rangle = 1.137 \pm 0.018$ g cm$^{-3}$. They showed that $a - c = 16.9 \pm 0.7$ km compared with a predicted value of $20.3 \pm 0.3$ km for an undifferentiated satellite. The lower than expected bulge is strongly suggestive of a centrally condensed satellite. One model of its interior that is consistent with these observations predicts a rocky core of dimensions $A/B = 0.44 \pm 0.09$ with an ice mantle of density $\rho = 0.96 \pm 0.08$ g cm$^{-3}$. Another possibility is that Mimas has a deep regolith composed of porous (and hence significantly underdense) ice. It is intriguing to note that observations of the dynamical interaction of Janus and Epimetheus, the coorbital satellites of Saturn (see Sect. 3.12), suggest an icy composition of comparable porosity.

The orbital tour of the *Galileo* spacecraft around the jovian system has brought it to within 1,000 km or less of the surfaces of all four of the Galilean satellites. Using spacecraft tracking data Anderson et al. (1996a, 1996b, 1997a) have derived estimates of the moment of inertia factor for Io ($\bar{C} = 0.378 \pm 0.007$), Europa ($\bar{C} = 0.347 \pm 0.014$), Ganymede ($\bar{C} = 0.311 \pm 0.003$), and Callisto ($\bar{C} = 0.406 \pm 0.030$). Recall that if a satellite in hydrostatic equilibrium is homogeneous then we would expect to find $\bar{C} = 0.4$. Therefore, Io, Europa, and Ganymede are all centrally condensed. Indeed, Ganymede has the lowest measured value of $\bar{C}$ of any object in the solar system. However, the initial data

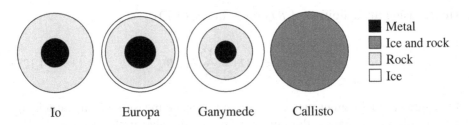

Fig. 4.11. Plausible models for the interiors of the Galilean satellites derived from spacecraft data. All the satellites are drawn to a common radial scale. The data are taken from Anderson et al. (1996a, 1996b, 1997a, 1997b).

for Callisto (Anderson et al. 1997b) suggested that it could be undifferentiated, with further encounters hinting at a partial separation of rock and ice (Anderson et al. 1998). Combined with data on the satellites' mean densities, it has been possible to derive models for the interiors. These are shown schematically in Fig. 4.11. The tidal heating of Io is discussed in Sect. 4.11, but here we note that one of the more intriguing possibilities arising from interpretation of the *Galileo* data is that the tidal heating of Europa has resulted in an ocean of liquid water below a crust of water ice.

Although we have assumed that the satellite is in a near-circular orbit, in reality the shape of a satellite changes with the varying tidal potential as the satellite moves around an orbit that is appreciably eccentric. In circumstances where a spacecraft has repeated, close approaches with a satellite, and measurements of $\bar{C}$ and $J_2$ can be derived at each encounter, it is possible to get extremely accurate measurements of the moments, as well as information about such physical properties as the satellite's rigidity. The *Cassini* spacecraft's use of repeated gravity assists from the saturnian satellite Titan should provide unprecedented information about its internal structure (Rappaport et al. 1997).

## 4.8 The Roche Zone

Consider a small, spherical satellite of mass $m_s$ and radius $R_s$ in synchronous rotation about a planet of mass $m_p$ and radius $R_p$. The semi-major axis of the satellite's circular orbit is taken to be $a$ and its mean motion is $n$. We have already seen in Sect. 3.6 that the unstable Lagrangian equilibrium points $L_1$ and $L_2$ lie on the line connecting the planet and satellite centres at a distance $d_L$ from the satellite, where

$$d_L = \left(\frac{m_s}{3m_p}\right)^{1/3} a. \tag{4.128}$$

The Jacobi constant associated with the critical zero-velocity curve that passes through these points can be found from Hill's equations (see Sect. 3.13) and is given by $C_H = 9(m_s/3m_p)^{2/3}$. The area enclosed by this curve is called the *Roche lobe*, and it has an important physical interpretation.

Consider the stability of a particle lying on both the equator of the satellite and the line connecting the planet and satellite centres. The problem is to calculate the semi-major axis $a_L$ at which the particle is no longer gravitationally bound to the satellite. For a particle at the centre of the satellite, the gravitational and centrifugal forces are in equilibrium and $\mathcal{G}m_p/a^2 = n^2 a$. However, the particle on the equator will experience (i) an excess gravitational or centrifugal force due to tidal shear, (ii) a centrifugal force due to the satellite's rotation, and (iii) a force due to the gravitational attraction of the satellite. If the particle is just in equilibrium, these forces will balance and

$$-\frac{\partial}{\partial a}\left(\frac{\mathcal{G}m_p}{a^2}\right)R_s + n^2 R_s = \frac{\mathcal{G}m_s}{R_s^2}, \tag{4.129}$$

which implies that

$$\frac{3m_p}{m_s} = \left(\frac{a}{R_s}\right)^3. \tag{4.130}$$

Thus, in the case of a small, spherical satellite, the *Roche limit*, $a_L$(spherical), is given by

$$a_L(\text{spherical}) = \left(\frac{3m_p}{m_s}\right)^{1/3} R_s = \left(\frac{3\rho_p}{\rho_s}\right)^{1/3} R_p, \tag{4.131}$$

where $\rho_p$ and $\rho_s$ are the mean densities of the planet and satellite, respectively. Comparison of Eqs. (4.128) and (4.131) shows that at the Roche limit, $L_1$ and $L_2$ just touch the surface of the satellite. If the two densities are approximately equal then $a_L$(spherical) $\approx 1.44 R_p$ and scales with the radius of the planet.

However, if the satellite is in hydrostatic equilibrium (or is a star), then at the Roche limit the satellite fills its Roche lobe. In this case Eq. (4.129) needs some modification. The tidal and centrifugal forces are greater, while the gravitational force due to the satellite is reduced and the term for this force in Eq. (4.129) becomes a poor approximation. A more complete analysis shows that $a_L$(hydrostatic) $= 2.46 R_p$ (Chandrasekhar 1987).

All planetary ring systems are found in the *Roche zone* between $a_L$(spherical) and $a_L$(hydrostatic) (see Fig. 10.1). Satellites in prograde orbits inside the synchronous orbit with masses greater than some lower limit will evolve towards the planet due to tidal friction on timescales less than the age of the solar system (see Sect. 4.9). If these satellites enter the Roche zone and are tidally disrupted, then this may help explain the origin and radial location of planetary rings. The only attractive forces in Eq. (4.129) are gravitational, but Aggarwal & Oberbeck (1974) calculate that a satellite with tensile strength

$$T \geq (8/57)\pi\mathcal{G}\rho_p\rho_s R_s^2 \tag{4.132}$$

could orbit at the surface of the planet. Thus, an icy satellite as large as $R_s \sim$ 200 km could orbit at the surface of Saturn, if the satellite had a tensile strength $T \geq 10^6$ Nm$^{-2}$. However, satellites close to the giant planets suffer intense cometary bombardment, leading, in the case of small satellites outside the Roche zone, to total disruption and then subsequent reaccretion, and in the case of small satellites inside the Roche zone, to total disruption, dispersal, and ring formation (Smith et al. 1981).

## 4.9 Tidal Torques

Consider the tide raised on a planet by a satellite moving in a circular, equatorial orbit with mean motion $n$ about a planet rotating with angular speed $\Omega$. If $\Omega \neq n$, then the planet experiences tidal oscillations. So far, we have assumed that the energy of the system is conserved. However, in practice, tidal oscillations always generate friction and this results in energy loss and a phase shift in the tidal response of the planet.

It is useful to compare the response of the planet to that of a forced harmonic oscillator. We write the equation of motion as

$$m\frac{d^2x}{dt^2} = -kx - \beta\frac{dx}{dt} + F_0 \cos \omega t, \qquad (4.133)$$

where $x$ is the displacement from the equilibrium configuration, $m$ is the inertia, and $kx$ is the restoring force, with $k(> 0)$ the *stiffness* parameter. The equation of motion can also be written as

$$\frac{d^2x}{dt^2} = -\omega_0^2 x - \frac{1}{\tau}\frac{dx}{dt} + \frac{F_0}{m} \cos \omega t, \qquad (4.134)$$

where $\omega_0$ is the natural frequency of the oscillator, $\tau(> 0)$ is the damping timescale, and $F_0/m$ and $\omega$ are, respectively, the amplitude and frequency of the external driving force. By substituting the trial solution

$$x = A \cos(\omega t + \delta) \qquad (4.135)$$

into Eq. (4.134), we can show that the steady-state response of the system is given by

$$A = (F_0/m)\left[\left(\omega_0^2 - \omega^2\right)^2 + (\omega/\tau)^2\right]^{-1/2}, \qquad (4.136)$$

where $A$ is a positive amplitude, and the phase shift $\delta$ is given by

$$\sin \delta = -(\omega/\tau)\left[\left(\omega_0^2 - \omega^2\right)^2 + (\omega/\tau)^2\right]^{-1/2}. \qquad (4.137)$$

Thus, $\delta$ depends on the frequency, $\omega$, of the driving force but is independent of the amplitude, $F_0/m$. Because the damping force always opposes the motion, $\delta$ is negative ($-\pi < \delta \leq 0$) for all values of $\omega$, and the response always lags the driving force (Baierlein 1983, Feynman et al. 1963).

We can relate the phase shift $\delta$ to the *specific dissipation function* $Q$ of the oscillator, which is defined by

$$Q = \frac{2\pi E_0}{\Delta E}, \tag{4.138}$$

where $\Delta E$ is the energy dissipated over one cycle and $E_0$ is the peak energy stored during the cycle. The work done by the restoring force over a displacement $\delta x$ is $kx\,\delta x$. Hence the peak energy stored in the oscillator is

$$E_0 = \int_0^A kx\,dx = \frac{1}{2}m\omega_0^2 A^2 = \frac{1}{2}\omega_0^2 \frac{F_0^2}{m}\left[\left(\omega_0^2 - \omega^2\right)^2 + (\omega/\tau)^2\right]^{-1}. \tag{4.139}$$

The work done by the drag force over a displacement $\delta x$ in a time $\delta t$ is $\beta\dot{x}\,\delta x$. Therefore, the rate of energy dissipation is $\dot{E} = -\beta\dot{x}^2$ and, using the time derivative of Eq. (4.135), the mean rate of energy dissipation is $\langle\dot{E}\rangle = (1/2)\beta(A\omega)^2$. The energy dissipated over one cycle is $\Delta E = \langle\dot{E}\rangle(2\pi/\omega)$. Hence,

$$\Delta E = \pi\left(F_0^2/m\right)(\omega/\tau)\left[\left(\omega_0^2 - \omega^2\right)^2 + (\omega/\tau)^2\right]^{-1}. \tag{4.140}$$

If $\omega_0^2 \gg \omega^2 \gg (\omega/\tau)^2$, that is, the system is not close to resonance and the damping is weak, then

$$\sin\delta = -1/Q. \tag{4.141}$$

Using the driven harmonic oscillator as a model, we argue that, in all cases, the effect of tidal friction is to cause a negative phase shift, $\delta$, in the tidal response of a planet. A satellite in a circular orbit raises a semidiurnal tide of frequency $2(\Omega-n)$ on a planet (see Sect. 4.2). If $\Omega > n$ (i.e., the satellite is above synchronous height), then the tidal bulge is carried ahead of the tide-raising satellite by an angle $\epsilon$ radians, where $2\epsilon = \delta = Q^{-1}$. Conversely, if $\Omega < n$ (i.e., the satellite is below synchronous height), and the tidal dissipation function $Q$ is independent of amplitude and frequency, then the tidal bulge lags behind the tide-raising body by the same angle $\epsilon$ (MacDonald 1964). Thus, because of tidal friction, the axis of the tidal bulge is not aligned with the line connecting the planet and satellite centres. This results in a *tidal torque* and a transfer of energy and angular momentum between the planet and the satellite.

The torque $\mathbf{\Gamma}$ on the satellite is determined by the gradient of the planet's external potential $V_{ext}$ at the position $\mathbf{r}$ of the satellite and is given by

$$\mathbf{\Gamma} = \mathbf{r} \times \mathbf{F}, \tag{4.142}$$

where

$$\mathbf{F} = -m_s\nabla V_{ext}. \tag{4.143}$$

Only the component of the force perpendicular to the line connecting the planet and the satellite centres, $F_\psi = -(m_s/r)\,(\partial V_{ext}/\partial\psi)$, contributes to the torque and

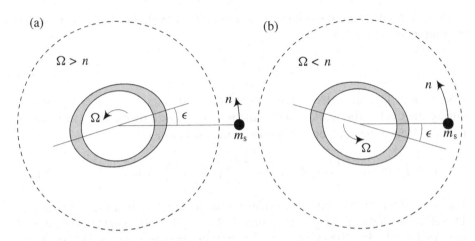

Fig. 4.12.   In all cases, the effect of tidal friction is to cause a negative phase shift, $\delta$, in the tidal response of a planet. Here we show a satellite in a circular orbit raising a semi-diurnal tide of frequency $2(\Omega - n)$ on a planet. (a) If $\Omega > n$ the satellite is above synchronous height (dashed circle) and the tidal bulge is carried ahead of the tide-raising satellite by an angle $\epsilon$ radians, where $2\epsilon = \delta = Q^{-1}$. (b) If $\Omega < n$ the satellite is below synchronous height and the tidal bulge lags behind the tide-raising body by the same angle $\epsilon$.

only the noncentral part of the planet's potential, $V_{nc,ext}$, contributes to that force component. Hence, the magnitude of the torque on the satellite is given by

$$\Gamma = -m_s \frac{\partial V_{nc,ext}}{\partial \psi} \tag{4.144}$$

and it follows from Newton's laws of motion that an equal and opposite torque acts on the planet.

If the satellite is above synchronous height ($\Omega > n$), the tidal bulge is carried ahead of the satellite (Fig. 4.12a) and the work done by this torque acts to increase the orbital energy of the system at a rate $\Gamma n$. At the same time, an equal and opposite torque works at a rate $\Gamma \Omega$ to decrease the rotational energy of the planet. Because $\Omega \neq n$, these rates of work are not equal and the total mechanical energy of the system, $E$, decreases at a rate

$$\dot{E} = -\Gamma(\Omega - n) < 0. \tag{4.145}$$

Conversely, if the satellite is below synchronous height ($\Omega < n$), the tidal bulge lags behind the satellite (Fig. 4.12b) and the signs of the torques are reversed. In this case, the orbital energy of the system decreases, while the rotational energy of the planet increases. However, the total mechanical energy of the system still decreases at a rate

$$\dot{E} = \Gamma(\Omega - n) < 0. \tag{4.146}$$

In both cases this energy is dissipated as heat in the planet and it is the rate of energy dissipation that determines the rate of orbital evolution.

The total energy of the system is the sum of the rotational energy of the planet, $\frac{1}{2}I\Omega^2$, where $I$ is the moment of inertia of the planet, and the orbital energy of the system, $-\mathcal{G}m_p m_s/2a$ (see Sect. 2.7). Hence, the rate of change of the total energy is given by

$$\dot{E} = \frac{d}{dt}\left(\frac{1}{2}I\Omega^2 - \mathcal{G}\frac{m_p m_s}{2a}\right) = I\Omega\dot{\Omega} + \mathcal{G}\frac{m_p m_s}{2a^2}\dot{a}. \tag{4.147}$$

Using Kepler's third law, $\mathcal{G}\left(m_p + m_s\right) = n^2 a^3$, we obtain

$$\dot{E} = I\Omega\dot{\Omega} + \frac{1}{2}\frac{m_p m_s}{\left(m_p + m_s\right)}n^2 a\dot{a}. \tag{4.148}$$

However, while some mechanical energy is dissipated as heat, the total angular momentum of the system,

$$L = I\Omega + \frac{m_p m_s}{\left(m_p + m_s\right)}a^2 n, \tag{4.149}$$

is conserved. Thus $\dot{L} = 0$ and

$$I\dot{\Omega} = -\frac{1}{2}\frac{m_p m_s}{\left(m_p + m_s\right)}na\dot{a}. \tag{4.150}$$

On substituting this expression for $I\dot{\Omega}$ into Eq. (4.148), we obtain

$$\dot{E} = -\frac{1}{2}\frac{m_p m_s}{\left(m_p + m_s\right)}na\dot{a}(\Omega - n), \tag{4.151}$$

and because $\dot{E} < 0$, we have

$$\text{sign}\,(\dot{a}) = -\text{sign}\,(\dot{\Omega}) = \text{sign}\,(\Omega - n). \tag{4.152}$$

Hence, if $\Omega > n$, then the semi-major axis of the satellite increases while the rate of rotation of the planet decreases. This is the case for the Earth–Moon system where the Moon is slowly receding from the Earth while the rotational period of the Earth is slowly increasing, thereby producing longer days. Conversely, if $\Omega < n$, then the semi-major axis of the satellite decreases while the rate of rotation of the planet increases. This is the case for the Martian satellite Phobos, which is now slowly spiraling in towards the planet. To calculate the timescale of orbital evolution, we need an expression for the magnitude of the tidal torque $\Gamma$ in terms of the tidal lag angle.

If a homogeneous planet is deformed by a tidal potential of magnitude $V_3(\psi) = -\zeta g P_2(\cos\psi)$, as in Eq. (4.12), then the noncentral part of the external potential of the deformed planet at some point $P(\psi)$ is given by Eq. (4.54) and can be written as

$$V_{\text{nc,ext}} = -\frac{3}{5}C\epsilon_2 g\left(\frac{C}{r}\right)^3 P_2(\cos\psi). \tag{4.153}$$

We have shown for the particular case of a homogeneous solid body that the elevation of the planetary surface is given by

$$C\epsilon_2 = \frac{(5/2)\zeta}{1 + \tilde{\mu}} \tag{4.154}$$

(see Eq. (4.89)). In the general case, we denote this elevation by $h_2\zeta$ and the noncentral part of the external potential is written as

$$V_{\text{nc,ext}} = -k_2\zeta g \left(\frac{C}{r}\right)^3 P_2(\cos\psi). \tag{4.155}$$

The coefficients $h_2$ and $k_2$ were introduced by A. E. H. Love and are known as *Love numbers*. They are mostly used as a convenient way of cloaking our ignorance of a body's internal structure.

However, if the internal structure of the planet is known, then the Love numbers can be calculated. For the simple case of a homogeneous solid body, we have

$$h_2 = \frac{5/2}{1 + \tilde{\mu}} \quad \text{and} \quad k_2 = \frac{3/2}{1 + \tilde{\mu}}. \tag{4.156}$$

For any other body whose outer surface is in hydrostatic equilibrium, the external potential is fully determined by the surface potential. For a point $P(r, \psi)$ on the surface, the total potential is the sum of the central potential, the noncentral potential given by Eq. (4.155) with $C = r$, and the tidal potential. Given that the surface is an equipotential, we have

$$-\frac{Gm_p}{r} - k_2\zeta g P_2(\cos\psi) - \zeta g P_2(\cos\psi) = \text{constant}. \tag{4.157}$$

Substituting the equation of the surface, $r = C\,[1 + \epsilon_2 P_2(\cos\psi)]$, into Eq. (4.157) and eliminating the dependence on $P_2(\cos\psi)$, we deduce that

$$k_2 = (C\epsilon_2/\zeta) - 1. \tag{4.158}$$

For the particular model planet described in Sect. 4.4, we have $C\epsilon_2/\zeta = \frac{5}{2}H$ (see Eq. (4.79)), where $2/5 \le H \le 1$, and hence $k_2 = (5/2)H - 1$.

From Eqs. (4.144), (4.155), and (4.13), given that $\partial P_2(\cos\psi)/\partial\psi = -\frac{3}{2}\sin 2\psi$, we obtain

$$\Gamma = \frac{3}{2}k_2\frac{Gm_s^2}{a^6}C^5\sin 2\epsilon, \tag{4.159}$$

where $\epsilon$ is the tidal lag angle. Hence, from Eqs. (4.145), (4.146), (4.151), and (4.159),

$$\dot{a} = \text{sign}\,(\Omega_p - n)\frac{3k_{2p}}{Q_p}\frac{m_s}{m_p}\left(\frac{C_p}{a}\right)^5 na, \tag{4.160}$$

where we have introduced the subscript p to emphasise that this is the case for a tide raised on a planet and the parameters $\Omega_p$, $k_{2p}$, $Q_p$, and $C_p$ refer to the planet.

If $I_p = \alpha_p m_p C_p^2$ (where $\alpha_p \le 2/5$), then

$$\dot{\Omega}_p = -\text{sign}\,(\Omega_p - n)\frac{3k_{2p}}{2\alpha_p Q_p}\frac{m_s^2}{m_p\,(m_p + m_s)}\left(\frac{C_p}{a}\right)^3 n^2 .$$ (4.161)

Just as the satellite raises a tide on the planet, the planet also raises a tide on the satellite. Using the same arguments we can show that, due to this tide,

$$\dot{a} = \text{sign}\,(\Omega_s - n)\frac{3k_{2s}}{Q_s}\frac{m_p}{m_s}\left(\frac{C_s}{a}\right)^5 na$$ (4.162)

and

$$\dot{\Omega}_s = -\text{sign}\,(\Omega_s - n)\frac{3k_{2s}}{2\alpha_s Q_s}\frac{m_p^2}{m_s\,(m_p + m_s)}\left(\frac{C_s}{a}\right)^3 n^2 .$$ (4.163)

The fact that Eqs. (4.162) and (4.163) can be obtained from Eqs. (4.160) and (4.161) by a simple change of subscript is a reflection of the basic symmetry of the motion of two bodies about their common centre of mass. However, we should note that if $m_s \ll m_p$, which would, for example, apply to the tide raised by the Moon on the Earth, and to the tide raised by the Earth on the Moon, then

$$\dot{\Omega}_p = -\text{sign}\,(\Omega_p - n)\frac{3k_{2p}}{2\alpha_p Q_p}\left(\frac{m_s}{m_p}\right)^2 \left(\frac{C_p}{a}\right)^3 n^2$$ (4.164)

and

$$\dot{\Omega}_s = -\text{sign}\,(\Omega_s - n)\frac{3k_{2s}}{2\alpha_s Q_s}\left(\frac{m_p}{m_s}\right)\left(\frac{C_s}{a}\right)^3 n^2 .$$ (4.165)

If $m_s \gg m_p$, which would apply to the tide raised by the Sun on the Earth, then

$$\dot{\Omega}_p = -\text{sign}\,(\Omega_p - n)\frac{3k_{2p}}{2\alpha_p Q_p}\left(\frac{m_s}{m_p}\right)\left(\frac{C_p}{a}\right)^3 n^2 .$$ (4.166)

In the above, we have assumed that the planet spins in the same sense as the orbital motion of the satellite. An important exception to this rule is the orbit of Triton, which is nearly circular but has an inclination with respect to Neptune's equator of 156.834°, that is, the orbit of the satellite is retrograde with respect to the spin of the planet. In this case we can deduce the sign of the orbital evolution by changing the direction of motion of the satellite in Fig. 4.12a (Goldreich & Soter 1966). We then see that the tide on the planet acts to retard both the spin of the planet and the orbit of the satellite with the result that, regardless of the position of the synchronous orbit, the satellite orbit decays at a rate given by

$$\dot{a} = -\frac{3k_{2p}}{Q_p}\frac{m_s}{m_p}\left(\frac{C_p}{a}\right)^5 an .$$ (4.167)

Table 4.1. Tidal despinning timescales, $\tau$, for a selection of planets and satellites. Values of $k_2$ and $Q$ in parentheses are estimates calculated using a mean rigidity $\langle \mu \rangle = 5 \times 10^{10}$ N m$^{-2}$ for a rocky body and $\langle \mu \rangle = 4 \times 10^9$ N m$^{-2}$ for an icy body. All other values are from Yoder (1995).

| Body | Type | Perturber | $k_2$ | $Q$ | $\tau$ (y) |
|------|------|-----------|-------|-----|------------|
| Mercury | rocky | Sun | (0.1) | (100) | $4 \times 10^9$ |
| Venus | rocky | Sun | 0.25 | (100) | $6 \times 10^{10}$ |
| Earth | rocky | Sun | 0.299 | 12 | $5 \times 10^{10}$ |
| Mars | rocky | Sun | 0.14 | 86 | $7 \times 10^{12}$ |
| Earth | rocky | Moon | 0.299 | 12 | $1 \times 10^{10}$ |
| Moon | rocky | Earth | 0.030 | 27 | $2 \times 10^7$ |
| Phobos | rocky | Mars | (0.0000004) | (100) | $3 \times 10^5$ |
| Io | rocky | Jupiter | (0.03) | (100) | $2 \times 10^3$ |
| Europa | rocky | Jupiter | (0.02) | (100) | $4 \times 10^4$ |
| Hyperion | icy | Saturn | (0.0003) | (100) | $1 \times 10^9$ |
| Miranda | icy | Uranus | (0.0009) | (100) | $8 \times 10^3$ |
| Ariel | icy | Uranus | (0.10) | (100) | $1 \times 10^4$ |
| Triton | icy | Neptune | (0.086) | (100) | $4 \times 10^4$ |
| Charon | icy | Pluto | (0.006) | (100) | $6 \times 10^5$ |
| Pluto | icy | Charon | (0.06) | (100) | $1 \times 10^7$ |

Thus, the rate of orbital evolution is a function of a number of constants. However, although $a$, $m_s$, $R_p$, $m_p$, and even $k_2$ are quite well determined, the planetary tidal dissipation functions, $Q$, are not well known (see Sect. 4.13). Some estimates of tidal despinning timescales for a number of important bodies in the solar system, assuming an initial period of 10 h, are shown in Table 4.1.

Furthermore, satellites in eccentric and inclined orbits raise tides with different frequencies and amplitudes on the planet and we need to consider the amplitude and frequency dependence of $Q$.

Lunar laser ranging experiments using the laser reflectors left by the *Apollo* and *Lunakhod* missions show that the Moon is receding from the Earth at a current rate of $\dot{a} \approx +10^{-9}$ ms$^{-1}$. This is consistent with independent measurements of the slowing down of the Earth's rotation.

## 4.10 Satellite Tides

So far we have assumed that the satellites move in circular, equatorial orbits. This approach is adequate for estimating tidal despinning timescales and the results shown in Table 4.1 indicate that, for satellites close to their primaries, these timescales are much less than the age of the solar system. However, if the orbit of a satellite is eccentric, then tidal evolution does not cease once the satellite has achieved synchronous rotation (i.e. $\Omega_s = n$). Tidal dissipation in the interior of the satellite, due to the tide raised on it by the planet, can act both to heat the satellite and circularise its orbit. In some cases this has resulted in widespread melting and the most spectacular volcanic activity in the solar system.

The eccentricity damping timescale can be estimated from the rate of dissipation of the total energy $E$, which is again given by $E = -\mathcal{G}m_p m_s/2a < 0$. In this case the planet can be treated as a point mass and the total angular momentum of the system, $L$, is the sum of the orbital angular momentum $L_{orbital}$ and the angular momentum associated with the spin of the synchronous satellite, that is,

$$L = \frac{m_p m_s}{(m_p + m_s)} a^2 n \left(1 - e^2\right)^{1/2} + \alpha_s m_s C_s^2 n.$$ (4.168)

However, if $a^2 \gg C_s^2$, then the angular momentum (and energy) associated with the satellite spin is negligible and we can write

$$e^2 = 1 + \frac{2E L_{orbital}^2}{\mathcal{G}^2} \frac{(m_p + m_s)}{(m_p m_s)^3}.$$ (4.169)

Because angular momentum is conserved, we have $\dot{L}_{orbital} = 0$ and hence

$$\dot{e} = -\frac{\dot{E}}{2eE} \left(1 - e^2\right) \approx -\frac{\dot{E}}{2eE}.$$ (4.170)

Because all tidal oscillations are dissipative, $\dot{E} < 0$ and thus the tides on the satellite always act to decrease the orbital eccentricity and the semi-major axis, while increasing the mean motion.

We now determine the nature of the tidal interaction between the planet and the satellite by considering the tidal potential experienced by a synchronous satellite moving in an eccentric orbit. In a frame centred on the planet, the satellite moves in an elliptical orbit with the planet at one focus (Fig. 4.13a) and we will assume that the spin axis of the satellite is normal to the orbital plane. Using the results from Sect. 2.5 we find that the orbital radius of the satellite at time $t$ after pericentre is given by

$$r = a(1 - e \cos E) \approx a(1 - e \cos nt).$$ (4.171)

If we let $\varphi$ denote the angle between the radius vector and the line joining the satellite to the empty focus (see Fig. 4.13a), then the sine rule gives

$$\frac{\sin \varphi}{2ae} = \frac{\sin f}{2a - r},$$ (4.172)

where $f$ is the true anomaly. Therefore,

$$\sin \varphi = \frac{2e}{1 + (1 - r/a)} \sin f \approx 2e \sin nt + \mathcal{O}(e^2).$$ (4.173)

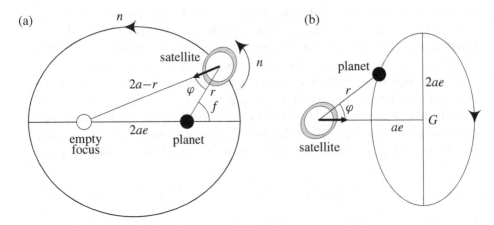

Fig. 4.13.    (a) The path of a satellite in an elliptical orbit in the frame centred on the planet. The satellite keeps one face (marked by an arrow) pointed toward the empty focus of its orbit. (b) The path of the planet in a frame centred on and rotating with the satellite. For small values of $e$ the planet moves about its guiding centre, $G$, on an ellipse with semi-major and semi-minor axes in the ratio 2:1.

We have shown in Sect. 2.6 that if terms of $\mathcal{O}(e^2)$ are neglected, then a satellite in synchronous rotation keeps one face pointed towards the empty focus of its orbit (see Fig. 4.13a). If we consider the same system, but now examine the path of the planet in a frame centred on and rotating with the solid body of the satellite, then we have the configuration shown in Fig. 4.13b. From this viewpoint the planet appears to execute an ellipse of semi-major axis $2ae$ and semi-minor axis $ae$ in the equatorial plane of the satellite. This is the guiding centre approximation (Sect. 2.6).

Using a spherical polar coordinate system centred on the satellite (Fig. 4.14), we now calculate the potential at some point $P$ on the surface of the satellite with spherical polar coordinates $(C_s, \theta, \phi)$, where $\theta$ is the colatitude measured from the spin axis of the satellite and $\phi$ is the longitude measured from the line $OX$ that connects the centre of the satellite to the guiding centre $G$ of the planet's orbit. Let $\alpha$ denote the angle between the position vector of $P$ and the satellite–planet line, and let $\Delta$ denote the distance between $P$ and the planet. The angle $\psi$ between the projection of the position vector of $P$ onto the $X$–$Y$ plane and the satellite–planet line is $\phi - \varphi = \phi - 2e \sin nt$, where $\phi$ is the longitude of $P$ (see Fig. 4.14). The angles $\alpha$, $\theta$, and $\psi$ are related by

$$\cos \alpha = \sin \theta \cos \psi \approx \sin \theta \cos(\phi - 2e \sin nt). \qquad (4.174)$$

Hence

$$\cos \alpha \approx \sin \theta (\cos \phi + 2e \sin \phi \sin nt). \qquad (4.175)$$

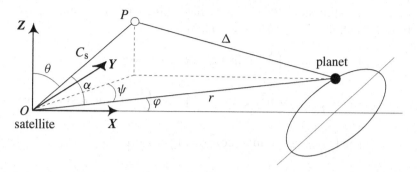

Fig. 4.14. The geometry used in the calculation of the tides raised on a synchronous satellite in an eccentric orbit. The coordinate frame is centred on and rotates with the satellite, while the path of the planet is an ellipse in the equatorial (and orbital) plane of the satellite.

We can now derive an expression for the potential at $P$ and extract from it those terms that generate the tidal forces on the satellite. The potential is given by

$$V = -\frac{\mathcal{G}m_{\mathrm{p}}}{\Delta}. \tag{4.176}$$

From Fig. 4.14, using the cosine rule and the approximation in Eq. (4.171), we have

$$\begin{aligned} \Delta^2 &= C_{\mathrm{s}}^2 + r^2 - 2rC_{\mathrm{s}}\cos\alpha \\ &= C_{\mathrm{s}}^2 + a^2(1 - 2e\cos nt) - 2aC_{\mathrm{s}}(1 - e\cos nt)\cos\alpha \end{aligned} \tag{4.177}$$

or

$$\Delta = a\left[1 - 2e\cos nt + \left(\frac{C_{\mathrm{s}}}{a}\right)^2 - 2\left(\frac{C_{\mathrm{s}}}{a}\right)(1 - e\cos nt)\cos\alpha\right]^{1/2}. \tag{4.178}$$

If we write $\Delta$ as

$$\Delta = a(1 - x)^{1/2}, \tag{4.179}$$

where

$$x = 2e\cos nt - \left(\frac{C_{\mathrm{s}}}{a}\right)^2 + 2\left(\frac{C_{\mathrm{s}}}{a}\right)(1 - e\cos nt)\cos\alpha \tag{4.180}$$

is a small quantity, then

$$\frac{1}{\Delta} = \frac{1}{a}\left(1 + \frac{1}{2}x + \frac{3}{8}x^2 + \frac{5}{16}x^3 + \cdots\right). \tag{4.181}$$

On calculating expressions for $x^2$ and $x^3$, neglecting terms in $e^2$ and $(C_s/a)^3$, and substituting back into Eq. (4.181) we obtain

$$
\frac{1}{\Delta} = \frac{1}{a} \left[ 1 + e \cos nt + \left( \frac{C_s}{a} \right) \sin \theta \left[ \cos \phi + 2e \cos(\phi - nt) \right] \right.
$$
$$
+ \left( \frac{C_s}{a} \right)^2 \frac{1}{2} \left( 3 \sin^2 \theta \cos^2 \phi - 1 \right) + 3e \left( \frac{C_s}{a} \right)^2 \sin^2 \theta \sin 2\phi \sin nt
$$
$$
\left. + 3e \left( \frac{C_s}{a} \right)^2 \frac{1}{2} \left( 3 \sin^2 \theta \cos^2 \phi - 1 \right) \cos nt \right]. \tag{4.182}
$$

The first two terms in Eq. (4.182) are independent of the position of $P$ and consequently make no contribution to the force. Using arguments similar to those in Sect. 4.2, we can show that the term in $C_s/a$ provides the mean force necessary for motion in an ellipse, while the terms in $(C_s/a)^2$ provide the tidal forces. If we write

$$
\cos \beta = \sin \theta \cos \phi, \tag{4.183}
$$

where $\beta$ is the angle between the position vector of $P$ and the $X$ axis or the line joining the centre of the satellite and the guiding centre, then the potential can be written as

$$
V_s = -\mathcal{G} \frac{m_p}{a} \left( \frac{C_s}{a} \right)^2 \left[ P_2(\cos \beta) + 3e P_2(\cos \beta) \cos nt + 3e \sin^2 \theta \sin 2\phi \sin nt \right]. \tag{4.184}
$$

The first term in Eq. (4.184) is independent of time and is equivalent to the tidal potential in the circular orbit case. This term gives rise to a fixed tidal bulge with axis of symmetry pointing towards the guiding centre of the planet's orbit. The additional terms in Eq. (4.184) are time dependent and a result of the orbital eccentricity. The second term gives rise to a time variation in the amplitude of the first term. This is the *radial tide*. The third term is the *librational tide*. The radial tide on the satellite has an obvious origin in the varying distance of the planet from the satellite. However, the equally important librational tide arises from the fact that while the satellite rotates to keep one face towards the empty focus of its orbit, the axis of the tidal bulge on the satellite always points towards the planet with the result that the tidal bulge oscillates or librates across the surface of the satellite. Both of these tides contribute to the dissipation of energy in the satellite at rates determined by their amplitudes.

If we introduce a new longitude, $\Phi = \phi - \pi/4$, by shifting the reference line by 45°, and a new angle $\beta'$, where $\beta'$ is the angle between the position vector of $P$ and the new reference line, then we can write the potential as

$$
V_s = -\mathcal{G} \frac{m_p}{a} \left( \frac{C_s}{a} \right)^2 \left[ P_2(\cos \beta) + 3e P_2(\cos \beta) \cos nt \right.
$$
$$
\left. + 4e P_2(\cos \beta') \sin nt - e \left( 3 \sin^2 \theta - 2 \right) \sin nt \right]. \tag{4.185}
$$

This shows that there are, in fact, three tides: a radial tide of amplitude $3e$, a librational tide of amplitude $4e$ with an axis of symmetry displaced by $45°$ from that of the radial tide, and a third tide that is symmetric about the spin axis (Longuet-Higgins 1950). The latter two tides act in phase and should be regarded as a single librational tide. However, we note that there is a phase shift of $\pi/2$ between the time variation of the librational tide and that of the radial tide. The time variations of these tidal distortions in the equatorial plane of the satellite are shown in Fig. 4.15.

Both the heating rate of a satellite and the timescale for the damping of its orbital eccentricity are completely determined by $\dot{E}$, the rate of loss of orbital energy due to tidal dissipation in the satellite. However, further progress with this problem can be made only by making some assumptions about the tidal dissipation mechanism. If we assume that the satellite's tidal dissipation function, $Q_s$, is independent of amplitude and frequency, which may be appropriate if, for example, the source of dissipation is weak friction in the solid body of the satellite, then the tidal oscillations of the satellite are linear and the effects of the radial and librational tides can be calculated separately and then simply added.

Thus, our first approach to estimating $\dot{E}$, which is valid only if the problem is linear, involves determining the energy stored in the radial and librational tides and then calculating $\dot{E}$ from the definition of $Q_s$. This is also the approach of choice if the nature of the dissipation mechanism is unknown, in which case all our uncertainties about the tidal dissipation mechanism are buried in the estimate of $Q_s$. From Eq. (4.138) we have

$$\dot{E} = \frac{n\,\Delta E}{2\pi} = \frac{nE_0}{Q_s}, \tag{4.186}$$

where, in this case, the tidal frequency is equal to the mean motion $n$. For solid satellites in the solar system that are either icy ($\mu = 4 \times 10^9$ Nm$^{-2}$ and density $\rho \approx 1$ g cm$^{-3}$) or rocky ($\mu = 5 \times 10^{10}$ Nm$^{-2}$ and density $\rho \approx 3$ g cm$^{-3}$) the ratio

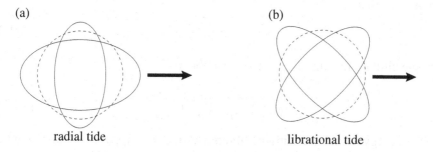

(a)

radial tide

(b)

librational tide

Fig. 4.15. The orientation of the equipotential curves in the equatorial plane ($\theta = \pi/2$) for the extremes of the (a) radial tide and (b) librational tide induced in a satellite due to its orbital eccentricity. In each case the arrows mark the direction of the planet.

of the elastic to gravitational forces $\tilde{\mu}_s \approx (10^4 \text{km}/C_s)^2 \gg 1$. Thus, the energy stored in the tidal deformation is mostly elastic and the gravitational component can be neglected. If we consider the deformation associated with the radial tide alone and write

$$R(\beta) = C_s \left[ 1 + \epsilon \mathcal{P}_2(\cos \beta) \right], \tag{4.187}$$

where $\epsilon \mathcal{P}_2(\cos \beta)$ is a measure of the strain, then the force per unit area at the surface of the satellite needed to maintain this strain is $(19/5)\mu \epsilon \mathcal{P}_2(\cos \beta)$ (see Eq. (4.75)). The work done per unit area against the elastic forces to increase the strain from zero to some maximum value $\epsilon_{\max}$ is given by

$$W = \int_0^{\epsilon_{\max}} \frac{19}{5} \mu \epsilon C_s \left[ \mathcal{P}_2(\cos \beta) \right]^2 d\epsilon = \frac{19}{10} \mu \epsilon_{\max}^2 C_s \left[ \mathcal{P}_2(\cos \beta) \right]^2 . \tag{4.188}$$

Hence, the peak stored energy associated with the radial tide is given by

$$E_{\text{radial}} = \frac{19}{10} \mu \epsilon_{\max}^2 C_s \int \left[ \mathcal{P}_2(\cos \beta) \right]^2 dA, \tag{4.189}$$

where the integral is over the sphere. Because the radial tide has an axis of symmetry, the element of area $dA$ can be written as

$$dA = 2\pi C_s^2 \sin \beta \, d\beta . \tag{4.190}$$

Hence,

$$E_{\text{radial}} = \frac{19}{5} \mu \epsilon_{\max}^2 \pi C_s^3 \int_0^\pi \left[ \mathcal{P}_2(\cos \beta) \right]^2 \sin \beta \, d\beta = \frac{57}{50} \mu \epsilon_{\max}^2 \mathcal{V}_s, \tag{4.191}$$

where $\mathcal{V}_s = (4/3)\pi C_s^3$ is the volume of the satellite. Even though the satellite is solid, we assume that either through the process of formation or relaxation due to solid-body creep there is no strain associated with the mean radial tide and that the strain is generated by the oscillations in the shape of the satellite about its mean configuration. In which case, from Eqs. (4.13), (4.77), (4.89), and (4.185) we deduce that

$$\epsilon_{\max} = 3e \frac{(5/2)\zeta}{\tilde{\mu}_s C_s} . \tag{4.192}$$

On substituting this into Eq. (4.191), we obtain

$$E_{\text{radial}} = \frac{27e^2}{4\tilde{\mu}_s} \left( \frac{C_s}{a} \right)^5 \frac{\mathcal{G}m_p^2}{a} . \tag{4.193}$$

The energy associated with the librational tide, $E_{\text{librational}}$, is calculated in the same way. From inspection of Eqs. (4.184) and (4.189), it follows that

$$E_{\text{librational}} = \frac{19}{10} \mu \epsilon_{\max}^2 C_s \int \left[ \sin^2 \theta \sin 2\phi \right]^2 dA, \tag{4.194}$$

where in this case the element of area $dA = C_s^2 \sin\theta \, d\theta \, d\phi$. Hence

$$
\begin{aligned}
E_{\text{librational}} &= \frac{19}{10} \mu \epsilon_{\text{max}}^2 C_s^3 \int \left[ \sin^2\theta \sin 2\phi \right]^2 \sin\theta \, d\theta \, d\phi \\
&= \frac{19}{10} \mu \epsilon_{\text{max}}^2 2\pi C_s^3 \int_0^{\pi/2} \sin^5\theta \, d\theta \\
&= \frac{4}{3} \frac{57}{50} \mu \epsilon_{\text{max}}^2 V_s
\end{aligned}
\tag{4.195}
$$

and it follows from Eq. (4.192) that

$$
E_{\text{librational}} = \frac{4}{3} \left[ \frac{27e^2}{4\tilde{\mu}_s} \left( \frac{C_s}{a} \right)^5 \frac{\mathcal{G}m_p^2}{a} \right].
\tag{4.196}
$$

If the tidal dissipation mechanism is a linear process, then the total peak energy stored in the tidal distortion, $E_0$, is simply the sum of the radial and librational components. Therefore

$$
\frac{dE}{dt} = -\frac{63e^2 n}{4\tilde{\mu}_s Q_s} \left( \frac{C_s}{a} \right)^5 \frac{\mathcal{G}m_p^2}{a}.
\tag{4.197}
$$

Substituting this into Eq. (4.170), we calculate that the eccentricity damping timescale is

$$
\tau_e = -\frac{e}{\dot{e}} = \frac{4}{63} \frac{m_s}{m_p} \left( \frac{a}{C_s} \right)^5 \frac{\tilde{\mu}_s Q_s}{n}
\tag{4.198}
$$

(Yoder & Peale 1981). Estimates of the energy dissipation rates and the eccentricity damping timescales for various satellites in the solar system are shown in Table 4.2.

Table 4.2. *Tidal dissipation rates and eccentricity damping timescales for a selection of satellites. Values of $\tilde{\mu}_s$ and $Q_s$ in parentheses are estimates calculated using the values given in Table 4.1. Values for the Moon are from Yoder (1995).*

| Satellite | Type | $\tilde{\mu}_s$ | $Q_s$ | $e$ | $\dot{E}$ (W) | $\tau_e$ (y) |
|-----------|------|------|------|------|------|------|
| Moon | rocky | 50 | 27 | 0.0549 | $3 \times 10^8$ | $2 \times 10^{10}$ |
| Io | rocky | (40) | (100) | 0.0043 | $3 \times 10^{12}$ | $6 \times 10^6$ |
| Europa | rocky | (80) | (100) | 0.0101 | $1 \times 10^{11}$ | $3 \times 10^8$ |
| Mimas | icy | (2700) | (100) | 0.0202 | $3 \times 10^8$ | $3 \times 10^8$ |
| Enceladus | icy | (2000) | (100) | 0.0045 | $1 \times 10^7$ | $7 \times 10^8$ |
| Titan | rocky/icy | (9) | (100) | 0.0289 | $4 \times 10^{10}$ | $2 \times 10^9$ |
| Miranda | icy | (1700) | (100) | 0.0027 | $3 \times 10^6$ | $3 \times 10^8$ |
| Ariel | icy | (1500) | (100) | 0.0034 | $2 \times 10^7$ | $6 \times 10^7$ |
| Triton | icy | (20) | (100) | 0.0000 | 0 | $9 \times 10^7$ |

In addition to changing the semi-major axis, the tide raised on a planet by a satellite also acts to increase the orbital eccentricity of the satellite. However, Jeffreys (1961) showed that the magnitude of the increase is negligible.

## 4.11 Tidal Heating of Io

Although all the tidal dissipation functions, $Q_s$, listed in Table 4.2, except that for the Moon, are mere estimates, it appears likely that some of the eccentricity damping timescales are considerably less than the age of the solar system and yet the observed orbital eccentricities are far from zero. This inconsistency has motivated much of the recent work on the dynamical evolution of the satellite systems of Jupiter, Saturn, Uranus, and Neptune. The Galilean satellite of Jupiter, Io, is of particular interest. Peale et al. (1979) were the first to realize the full implications of the above calculations. They pointed out that if $Q_s \approx 100$, then the tidal heating rate of Io is $\sim 3 \times 10^{12}$ W and probably three times greater than the radiogenic heating rate of the Moon. Moreover, the heating is not uniform throughout the body of the satellite; in the centre it is three times above the average. Given that seismic studies have shown the Moon to be near the melting point in its deep interior (Nakamura et al. 1976), Peale et al. (1979) argued that tidal heating may have melted Io's interior and that this would have led to a further increase in the heating rate and the *runaway melting* of most of the satellite's interior.

Some insight into this process can be obtained from Eq. (4.191). The stored elastic energy increases with rigidity, $\mu$, and the square of the strain. However, the strain increases with decreasing $\mu$. Hence, as long as $\tilde{\mu}_s \gg 1$ and the stored energy is mostly elastic, the stored energy must also increase with decreasing $\mu$. Thus the weaker the satellite, the greater the heating rate. Peale et al. (1979) showed that melting the interior effectively weakens the satellite. Even if $\mu$ and $Q_s$ are constant, melting the interior increases the strain in the remaining solid mantle with the result that the overall heating rate increases, thereby producing further melting. Equilibrium is reached when thermal conduction through the solid mantle is sufficient to remove the heat generated in and below the mantle. Peale et al. (1979) estimated that the solid mantle would have a mean thickness of $\sim 18$ km and on that basis they predicted that Io *"might currently be the most intensely heated terrestrial-type body in the solar system"* and that *"widespread and recurrent surface vulcanism would occur"*. Their highly prescient paper, which was published in *Science* on March 2, 1979, was spectacularly confirmed a few weeks later by Linda Morabito, a member of the *Voyager 1* optical navigation team, when she examined an image of Io and two background stars, taken by the *Voyager 1* camera for navigational purposes on March 8, 1979, and noticed two prominent volcanic plumes on the satellite surface (Morabito et al. 1979). During that first encounter, *Voyager 1* observed a total of nine volcanic eruptions,

Fig. 4.16. *Voyager* image of Io enhanced to show detail on the surface and in the plume of the erupting volcano, Prometheus, on the limb. The plume reaches a height of $\sim 50\,\mathrm{km}$ and is spread over a radius of $\sim 150\,\mathrm{km}$. *(Image courtesy of NASA/JPL.)*

some ejecting material with velocities of $1\,\mathrm{km\,s^{-1}}$ to heights of 250 km (Smith et al. 1979). Figure 4.16 shows a *Voyager* image of the eruption of the volcano Prometheus on Io's surface. Estimates of the resurfacing rate due to vulcanism range from 1 to $10\,\mathrm{cm\,y^{-1}}$ (Spohn 1997). To put this in perspective, consider that if this rate has been maintained over the age of the solar system, then a volume of material equal to $10^2$ to $10^3$ times the volume of the satellite has been recycled through the volcanic vents!

Ground-based observations of Io at infrared wavelengths show that the intrinsic luminosity of the satellite is $2.5\,\mathrm{Wm^{-2}}$, giving a total power of $10^{14}$ W (Veeder et al. 1994). Multiplying this by the age of the solar system (this, of course, needs some justification), we calculate a total energy loss of $1.4 \times 10^{31}$ J, which is comparable in magnitude to the present orbital energy, $\mathcal{G}m_p m_s/2a = 1.34 \times 10^{31}$ J.

## 4.12 Tides on Titan

In Sect. 4.11 we calculated $\dot{E}$ using an estimate of $Q_s$. However, if the tidal dissipation mechanism is known, then it may be possible to estimate $\dot{E}$ directly from first principles. This is the case, for example, for dissipation due to tidal currents in oceans and this could have application to Titan. Methane was detected spectroscopically in the atmosphere of this saturnian satellite by Kuiper (1944). *Voyager 1* observations have since revealed that the surface temperature is $\sim 95$ K, between the melting point (90.6 K) and the boiling point (118 K) of methane at the total surface pressure of 1.6 bar (Hanel et al. 1981). Thus, it is possible that methane and higher hydrocarbons exist in the liquid state on the surface of the satellite and this has raised some questions about the action of tides in the putative oceans (see Lunine 1993 for a review). The fact that Titan has a large orbital eccentricity is a problem because this requires that $Q_s > 200$ (see Table 4.2) and the only other object known to have a near-global ocean, the Earth, has $Q_p \approx 12$ (see Table 4.1). However, the dissipation rate depends on the depth of a global ocean (Sagan & Dermott 1982, Dermott & Sagan 1995).

The tidal dissipation rate can be determined by calculating the mean velocity of the tidal flow at all points on the ocean floor. Flow occurs because of the periodic changes in the height of the tide. If we consider the radial tide alone, then the time variation of the tide height at some point $P$ on the surface of the satellite (see Fig. 4.17) is given by

$$\zeta = 3eh\mathcal{P}_2(\cos \beta) \cos nt, \qquad (4.199)$$

where $h$, the mean amplitude of the ocean tide, is given by

$$h = \frac{(m_s/m_p)(C_s/a)^3 C_s}{[1 - (3\sigma/5\rho)] + [1 - \sigma/\rho]/\tilde{\mu}_s} \approx 120 \text{ m}, \qquad (4.200)$$

and where we have assumed that the ocean is predominantly ethane with a density $\sigma = 0.65$ g cm$^{-3}$ (see Eq. (4.86)). Because we are only interested in the eccentricity-driven oscillations of the shape of the satellite about its mean configuration, the above formula for $h$ applies even when the solid body of the satellite has relaxed to hydrostatic equilibrium.

Consider the change in the tide height as the satellite moves from apocentre to pericentre (Fig. 4.17). The increase in the tide height in the time interval d$t$ at

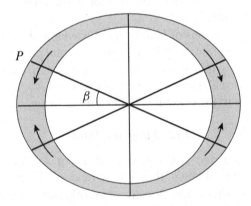

Fig. 4.17. Tidal flow due to a radial tide. The arrows mark the direction of the tidal flow due to the radial tide alone as the satellite moves from apocentre to pericentre and the amplitude of the tidal distortion increases. The tidal flow is symmetric about the line joining the centre of the satellite and the guiding centre of the planet's orbit. $P$ marks the surface of the cone of semiangle $\beta$. The flows across the plane through the poles perpendicular to the line of centres and across the equatorial plane of the satellite are always zero.

the point $P$ is

$$d\zeta = -3ehP_2(\cos\beta)n\sin nt\,dt\,. \tag{4.201}$$

Given that the increment in area is

$$dA = 2\pi C_s^2 \sin\beta\,d\beta \tag{4.202}$$

it follows that the increase in the volume of the ocean covering those parts of the surface inside the cone of semiangle $\beta$ is given by

$$\Delta V = -3ehn\,\sin nt\,dt\,2\pi C_s^2 \int_0^\beta P_2(\cos\beta)\sin\beta\,d\beta$$

$$= 3eh\pi C_s^2 n\,\sin nt\cos\beta\sin^2\beta\,dt\,. \tag{4.203}$$

If the tidal flow occurs in an ocean of depth $D$, then

$$\Delta V = 2\pi C_s \sin\beta Dv\,dt, \tag{4.204}$$

where $v$ is the velocity of the tidal current at $P$. Hence, from Eqs. (4.203) and (4.204), we have

$$v = \frac{1}{4}nC_s\left(\frac{3eh}{D}\right)\sin 2\beta \sin nt\,. \tag{4.205}$$

The energy dissipation rate per unit area due to boundary layer turbulence is

$$\dot{E} = f\sigma v^3, \tag{4.206}$$

where the dimensionless constant $f \approx 0.003$ (Sears 1995) is the coefficient of skin friction (Goldreich & Soter 1966). To find the average value of $\dot{E}$, the expression for $v$, Eq. (4.205), has to be averaged over time and space (weighted according to area). Given that

$$\langle\sin^3 nt\rangle = 4\pi/3 \tag{4.207}$$

and

$$\langle\sin^3 2\beta\rangle = \frac{2}{\pi}\int_0^{\pi/2}\sin^3 2\beta\sin\beta\,d\beta = 32\pi/35, \tag{4.208}$$

we calculate that

$$\langle v^3\rangle = \frac{2}{105\pi}\left[nC_s\left(\frac{3eh}{D}\right)\right]^3 \tag{4.209}$$

and that the rate of energy dissipation due to the radial tide alone is given by

$$\dot{E}_{\text{radial}} = 4\pi C_s^2 f\sigma\langle v^3\rangle\,. \tag{4.210}$$

We require the total rate of energy dissipation; this can be written as

$$\dot{E}_{\text{total}} = x\dot{E}_{\text{radial}}, \tag{4.211}$$

where $x$ is a constant factor to be determined. If the tidal dissipation mechanism were a linear process, then we could write

$$\dot{E}_{\text{total}} = \dot{E}_{\text{radial}} + \dot{E}_{\text{librational}}, \tag{4.212}$$

in which case, from Eqs. (4.191) and (4.195), we would have $x = 7/3$. However, if the tidal dissipation mechanism is due to boundary layer turbulence and is proportional to $v^3$, then the dissipation mechanism is certainly nonlinear and Eq. (4.212) is clearly inapplicable. Using a numerical hydrodynamical code, Sears (1995) has determined that $x \approx 6$ and Dermott & Sagan (1995) used this estimate to show that the present high orbital eccentricity of the satellite Titan ($e = 0.029$) can only be accounted for by a global hydrocarbon ocean deeper than 0.6 km or, if the ocean is not global, liquid hydrocarbons confined to a number of disconnected seas or crater lakes.

### 4.13 Tidal Evolution

Now consider the orbital evolution of a satellite due to the tide it raises on a planet. If $a_i$ and $a_0$ denote the initial and current values of the semi-major axis and $\Delta t$ denotes the corresponding time interval, then integration of Eq. (4.160) gives

$$\frac{2}{13}a_0^{13/2}\left[1 - (a_i/a_0)^{13/2}\right] = \frac{3k_{2p}}{Q_p}\left(\frac{\mathcal{G}}{m_p}\right)^{1/2} C_p^5 m_s \Delta t. \tag{4.213}$$

Here we have assumed that the tidal dissipation mechanism is linear and thus that the tidal dissipation function, $Q_p$, is independent of amplitude and frequency. It follows that if a system contains a number of satellites, then the orbital evolution of any one of these satellites can be treated without regard to the tides raised on the planet by the others. If the tidal dissipation mechanism is linear, then a satellite has a near-constant phase relationship with its own tide (small periodic variations in this phase may arise if the orbit is eccentric or inclined to the equator) and this results in a systematic exchange of energy and angular momentum between the spin of the planet and the orbit of the tide-raising satellite. However, the phases of the satellite with respect to the tides raised on the planet by the other satellites periodically change sign. Consequently, the effects of these indirect tidal interactions average to zero on comparatively short timescales. The latter statement needs some qualification if there is a resonant relationship between the mean motions of a pair of satellites, which, of course, is often the case, but the effects are small and we ignore them here.

   If the current semi-major axis is significantly different from the initial value and $(a_i/a_0)^{13/2} \ll 1$, then the second term on the left-hand side of Eq. (4.213) can be neglected and we can write

$$\frac{2}{13}a_0^{13/2} = \frac{3k_{2p}}{Q_p}\left(\frac{\mathcal{G}}{m_p}\right)^{1/2} C_p^5 m_s \Delta t, \tag{4.214}$$

implying that

$$\log a_0 = \frac{2}{13} \log m_\text{s} + \text{constant}. \tag{4.215}$$

Thus, if a satellite system is tidally evolved and $Q_\text{p}$ is independent of amplitude and frequency, we should expect to see a linear relationship between $\log a_0$ and $\log m_\text{s}$ of slope 2/13 (Allan 1969; Dermott 1972). The appropriate plots for the inner satellite systems of Saturn and Uranus are shown in Fig. 4.18. Taking $a_\text{i} = a_\text{sync}$, where $a_\text{sync}$ is the radius of the synchronous orbit, we have also plotted the curves defined by the full equation, Eq. (4.213). Since $Q_\text{p}$ is unknown, we have chosen the curves to pass through the points associated with Mimas and Ariel since these are the satellites for which $m_\text{s}/a^{13/2}$ is a maximum for each system. If the orbits of Mimas and Ariel are tidally evolved, then satellites above the synchronous orbit can exist above or close to the curve, but not below it. If a satellite is below synchronous orbit, as is the case for nine of the small satellites of Uranus discovered by *Voyager 2*, then the curve represents the value of $a_\text{i}$ of a satellite whose orbit would decay into the planet in a time interval $\Delta t$. If the tidal dissipation function varies with time, the above statements still hold, but $Q_\text{p}$ in Eq. (4.214) needs to be replaced by the average value, $\langle Q_\text{p} \rangle$, defined by

$$\Delta t = \langle Q_\text{p} \rangle \int_0^t \frac{\text{d}t}{Q_\text{p}(t)}. \tag{4.216}$$

The distributions of the satellites of Saturn and Uranus shown in Fig. 4.18 appear to favour the hypothesis that the orbits of a least the inner satellites of these

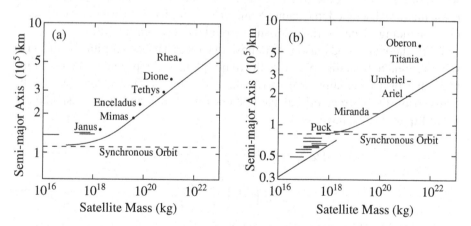

Fig. 4.18. Distribution of masses in the inner regions of the satellite systems of (a) Saturn and (b) Uranus. If tidal evolution has been appreciable, then the solid curve places bounds on the satellite distribution. The masses of the small satellites were estimated from their radii using a density of 1.1 g cm$^{-3}$ for the saturnian system and 1.6 g cm$^{-3}$ for the uranian system. The linear portions of the solid curves have a slope of 2/13.

planets are tidally evolved. The only anomalous satellite is the uranian satellite
Puck. However, this satellite is very close to the synchronous orbit and if $Q_p$ is a
function of the tidal frequency, $2(\Omega - n)$, we might expect these effects to be more
important for satellites with $\Omega \approx n$. Several other features of the satellite systems
offer further support for the tidal hypothesis. These are the existence of stable
orbit–orbit resonances, anomalously high orbital eccentricities and inclinations,
and satellites that have clearly experienced postformative vulcanism and surface
melting.

Tidal forces are clearly capable of changing the ratio of mean motions of a pair
of satellites. If we denote the ratio of a pair of the mean motions by $\mathcal{N} = n'/n$,
then

$$\frac{\dot{\mathcal{N}}}{\mathcal{N}} = \frac{\dot{n}'}{n'} - \frac{\dot{n}}{n}. \tag{4.217}$$

For tidally evolved satellites for which $(a_i/a_0)^{13/2} \ll 1$, we have

$$\frac{\dot{n}'}{n'} \approx \frac{\dot{n}}{n} \approx -\frac{3}{13\Delta t} \tag{4.218}$$

and hence $\dot{\mathcal{N}} \approx 0$. This indeed may now be the case for Miranda and Ariel
(Fig. 4.18b). However, during the early stages of orbital evolution there could
have been significant changes in $\mathcal{N}$ and pairs of satellites may have encountered
resonances for which the ratios of their mean motions were equal to ratios of
two small integers, $p$ and $q$, that is,

$$\frac{n'}{n} \approx \frac{p}{p+q}, \tag{4.219}$$

where the integer $q$ is the order of the resonance. The dynamics of orbit–orbit
resonance are discussed in Chapter 8 and in Sect. 8.15 we examine in more detail
the convincing evidence for resonant encounters between satellites.

While there are good reasons to believe that there has been orbital evolution of
satellites due to tidal dissipation in the major planets, the problem of calculating
the value of $Q_p$ remains. We can make estimates of $Q_{\text{Saturn}}$ and $Q_{\text{Uranus}}$ by
considering the current orbital parameters for Mimas and Ariel respectively.
From Eq. (4.24) we have

$$Q_p \geq \frac{39}{2} k_2 \left( \frac{\mathcal{G}}{m_p} \right)^{1/2} \frac{R_p^5}{a^{13/2}} m_s \, \Delta t$$

$$= \frac{39}{2} k_2 \left( \frac{R_p}{a} \right)^5 \frac{m_s}{m_p} \frac{2\pi \, \Delta t}{T}, \tag{4.220}$$

where $T$ is the orbital period of the satellite.

For Mimas, $a/R_p = 3.075$, $m_s/m_p = 6.6 \times 10^{-8}$, $T = 0.942$d, $k_2 = 0.35$, and
the time interval is taken to be the age of the solar system, that is, $\Delta t = 4.5 \times 10^9$ y.
This implies

$$Q_{\text{Saturn}} \geq 1.8 \times 10^4. \tag{4.221}$$

For Ariel, $a/R_p = 7.30$, $m_s/m_p = 1.7 \times 10^{-5}$, $T = 2.520d$, $k_2 = 0.32$, and again the time interval is taken to be $\Delta t = 4.5 \times 10^9 y$. This gives

$$Q_{Uranus} \geq 2.0 \times 10^4. \tag{4.222}$$

Work by Tittemore & Wisdom (1990) on the possible passage of uranian satellites through a variety of resonances suggests both a lower and an upper limit. They find

$$11,000 < Q_{Uranus} < 39,000. \tag{4.223}$$

The situation of Jupiter is complicated by the existence of the Laplace relation between three of the Galilean satellites, Io, Europa, and Ganymede (see Sect. 1.6.2 and Sect. 8.17). However, the dynamics of this system provide important information about tidal dissipation in both Jupiter and Io. The system is discussed in more detail in Sect. 8.17.

The whole problem of estimating values of $Q_p$ in the solar system was first discussed by Goldreich & Soter (1966). In the case of the Earth, tidal dissipation is due to friction between the tidally generated currents and the ocean floor and occurs mostly in the shallow seas. Lunar laser ranging (LLR) measurements of the position of the Moon show that its mean longitude $\lambda$ is decelerating at a rate

$$\frac{d^2\lambda}{dt^2} = \dot{n} = -25.3 \pm 1.2 \text{ arcsec century}^{-1}. \tag{4.224}$$

This is equivalent to $\dot{a} = -(2a/3n)\dot{n} = 3.74$ cm y$^{-1}$ and on substitution into Eq. (4.160) we calculate that

$$Q_{Earth} \approx 12. \tag{4.225}$$

This estimate is supported by analyses of the times of ancient eclipses and by calculations of the rate at which energy is dissipated in the oceans (Burns 1986). However, this low value of $Q_p$ does raise a problem in that if we use it to estimate the time of orbital evolution (Eq. (4.214)), we obtain the result that $\Delta t \approx 1.6 \times 10^9$ y, which is very much less than the age of the Earth–Moon system. This is similar to the problem we have with the orbital evolution of Io. Again, the point to emphasize is that if we use LLR measurements of $\dot{a}$ to estimate the age of the Moon we obtain a reasonably satisfying result, albeit one that is too low by a factor of 3. This obviously suggests that the rate of dissipation varies on geological timescales and given that the configuration of the oceans changes significantly on a timescale of $\sim 10^8$ y due to continental drift this is, perhaps, to be expected. Webb (1982) has calculated that the tidal oscillations of the Earth's oceans are, at present, near-resonant and that this accounts for the current high dissipation rate. In the past, when the Earth's spin period was shorter, the oceans would have been further from resonance and the rate of dissipation would have been correspondingly less.

In the case of the major planets, although we can make estimates of $Q_p$ by assuming that the orbits of some of the inner satellites are tidally evolved or, in the case of Jupiter, by measuring the heat output from Io, we do not know the sources of dissipation in these bodies. Turbulent viscosity in the fluid outer layers of the planets is inadequate by many orders of magnitude (Goldreich & Nicholson 1977a). Dermott (1979a) showed that solid friction in rocky cores could provide an adequate source of dissipation. The gravitational field of the tide in the underlying ocean enhances the core tide by a factor $\sim 2$, that is, $F$ in Eq. (4.80) is about 2 and because the elastic energy stored in the solid core increases as $F^2$, it follows that $\sim 4$ times more energy is stored in the core than would be the case without the overlying ocean. If $Q_{core} \sim 40$, which may be appropriate for rock close to its melting point, then in the cases of Jupiter and Uranus the cores would have to have volumes twice that of the Earth and in the case of Saturn the volume would have to be about eight times that of the Earth. The volumes of Jupiter, Saturn, and Uranus are 1,316, 763, and 63 Earth volumes, respectively. Whether small, solid, rocky cores exist at the centres of these giant planets is unknown. Stevenson (1983) has suggested that hysteresis associated with hydrogen–helium phase transitions may be an adequate source of the tidal dissipation in Jupiter.

Our closing remarks in this chapter concern the scales of satellite systems and the tidal evolution of small bodies in the solar system. The limiting distance from the centre of a body (or secondary) at which we might expect to find satellites in stable orbits is determined by the locations of the inner Lagrangian points, $L_1$ and $L_2$. We can write

$$a_L = \left( \frac{m_{sec}}{3 m_{prim}} \right)^{1/3} a_{sec} \qquad (4.226)$$

(see Eq. (3.75)). If the secondaries are planets or asteroids, then

$$a_L = \left( \frac{4\pi \rho_{sec}}{9 M_0} \right)^{1/3} a_{sec} C_{sec}, \qquad (4.227)$$

where $\rho_{sec}$ is the density of the secondary and $M_0$ is the mass of the Sun. Thus, $a_L$ scales with distance from the Sun and the radius of the secondary. For example, for asteroids of density $\rho_{sec} = 3 \mathrm{g\ cm^{-3}}$ we can calculate that

$$a_L \approx 200 a_{sec} C_{sec}, \qquad (4.228)$$

where the units of $a_{sec}$ are AU. Thus, $a_L / C_{sec} \gg 1$ and there is no lower limit to the size of an asteroid that could have a satellite in stable orbit. Indeed, we now know that asteroid (243) Ida, which has a mean radius of 15 km, has a satellite, Dactyl, of mean radius 0.7 km (see Fig. 2.6).

If we apply Eq. (4.226) to the satellites of planets, then we can write

$$a_L = \left( \frac{\rho_s}{3 \rho_p} \right)^{1/3} \frac{a_s}{C_p} C_s. \qquad (4.229)$$

Thus, $a_L$ scales with the radius of the satellite and the distance of the satellite from the planet. For satellites just outside the Roche limit (see Sect. 4.8), $a_L/C_s$ is close to unity. For more distant satellites such as, for example, Io or Titan, $a_L/C_s$ increases to 5.7 and 19.7 respectively. Thus, small, unseen satellites could exist close to these bodies. However, the action of tidal forces may have been sufficient to ensure the absence of all but the smallest satellites and the latter have probably been destroyed by cometary impacts (Smith et al. 1982).

Tidal forces are usually associated with large bodies, but it is possible that these forces have had a role in the spin–orbit evolution of some surprisingly small satellites and asteroids. The sign of orbital evolution is determined by the location of the satellite with respect to the synchronous orbit. For the synchronous orbit of a planet or asteroid, we can write

$$a_{sync} = \left(\frac{4\pi \mathcal{G} \rho_{sec}}{3\Omega_{sec}^2}\right)^{1/3} C_{sec},\qquad(4.230)$$

where $\Omega_{sec}$ is the spin frequency of the secondary, in this case the planet or asteroid. If we assume that, regardless of their size, all asteroids have spin periods of about 8 h (Alfvén 1964) and densities $\rho_{sec} = 3\text{g cm}^{-3}$, then $a_{sync}$ scales with the radius of the body and $a_{sync}/C_{sec} \approx 2.6$. For tidally despun planets with long spin periods, $a_{sync}/C_p$ can be much larger than 2.6. It is noticeable that Mercury and Venus, the only planets in the solar system without satellites, have been almost completely despun by solar tides. For Mercury, $a_{sync}/C_{sec} \approx 130$, while for Venus $a_{sync}/C_{sec} \approx 253$. Burns (1973) and Ward & Reid (1973) argued that the braking of spins of these planets by solar tides would have resulted in expansion of the synchronous orbits and that the subsequent tidal decay of the orbits of any subsynchronous satellites could have resulted in the loss of these satellites.

This mechanism may have also had a role in the loss of bodies orbiting satellites, that is, satellites of satellites (Reid 1973). For tidally despun satellites in synchronous rotation, the radius of the satellite's synchronous orbit is given by

$$a_{sync} = \left(\frac{4\pi \mathcal{G} \rho_s}{3n^2}\right)^{1/3} C_s .\qquad(4.231)$$

Hence, from Eq. (4.227) and Eq. (4.231), we deduce that

$$a_{sync}/a_L = 3^{1/3} .\qquad(4.232)$$

Thus, the possible stable orbits of satellites of tidally despun satellites are all subsynchronous.

## 4.14 The Double Synchronous State

It is highly likely that the orbit of the Moon and the spin of the Earth have changed considerably over the age of the solar system due, in particular, to the action of the semidiurnal tide raised by the Moon on the Earth (for a review see Burns 1986). Most of the angular momentum of the system now resides in the orbit of the Moon, but at the time of satellite formation the opposite was probably the case and most of the angular momentum probably resided in the spin of the Earth. We estimate that the maximum spin frequency that the Earth could have had, $\Omega_{max}$, is given by

$$\alpha m_p C_p^2 \Omega_{max} \approx m_s a^2 n + \alpha m_p C_p^2 \Omega, \qquad (4.233)$$

where $\Omega$ is the present spin frequency of the Earth and that the corresponding minimum spin period of the Earth could have been as small as $\sim 4$ h. However, the initial configuration of the system is not known and cannot be determined from contemporary measurements.

If, for heuristic purposes, we treat the orbit of the Moon as equatorial and neglect the influence of the solar tides, then the total angular momentum of the system, $L_{tot}$, is the sum of the orbital angular momentum $L_{orb}$ and the spin angular momentum of the Earth, $L_{spin}$. Given that $L_{tot}$ is conserved, we can divide the equation for the sum of $L_{orb}$ and $L_{spin}$ by $L_{tot}$ to obtain

$$1 = 0.832 \left(\frac{a}{60.4}\right)^{1/2} + 0.168 \left(\frac{24}{T}\right), \qquad (4.234)$$

where $a$ is the semi-major axis of the Moon in Earth radii and $T$ is the spin period of the Earth in hours. If we assume that, on average, the tidal dissipation function of the Earth is $Q_{Earth} \approx 34$, then we can use Eq. (4.234) to plot the semi-major axis of the Moon and the corresponding spin period of the Earth for various times in the evolution of the system (Fig. 4.19). Here, we are assuming that the Moon has evolved to its present orbit over $4.5 \times 10^9$ y. If the tidal dissipation function, $Q_{Earth}$, remains at $\sim 34$, then, after $\sim 5 \times 10^{10}$ y of evolution the transfer of angular momentum from the spin of the Earth to the orbit of the Moon will be nearly-complete and the system will be in a *double synchronous* state in which the spin periods of both the Earth and the Moon are the same and equal to the orbital period of the Moon. This will be achieved when the Moon has a semi-major axis of $\sim 87$ Earth radii and the spin period of the Earth is $\sim 47$ d. At this stage solar tides, which act to brake the spin of the Earth, cannot be neglected. The librations of any spin–orbit couplings (see Chap. 5) will grow in amplitude and any resonant locks will be released (Peale 1986). The synchronous orbit of the Earth will expand beyond the orbit of the Moon and then the semidiurnal tide raised by the Moon on the Earth will act to reduce the Moon's semi-major axis. After a further $\sim 5 \times 10^{10}$ y of evolution the Moon will approach the Earth, break up due to tidal stresses, and form a massive ring system (Jeffreys 1970).

The condition that a system evolves to the double synchronous state in a time less than the age of the solar system can be determined by assuming initially

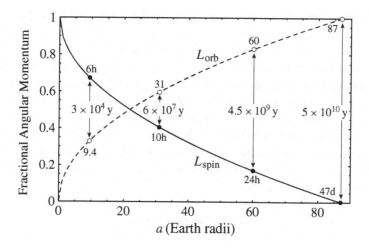

Fig. 4.19. Exchange of angular momentum between the orbit of the Moon, $L_{orb}$ (dashed curve), and the spin of the Earth, $L_{spin}$ (solid curve). The open circles and numbers on the curve for $L_{orb}$ are the distances of the Moon from the Earth expressed in Earth radii. The filled circles and numbers on the curve for $L_{spin}$ are the corresponding spin periods of the Earth. The times in years for each pair of numbers are the orbital evolution times calculated by assuming that the present configuration is the result of $4.5 \times 10^9$ y of evolution. Under various assumptions the double synchronous state of the Earth–Moon system will be reached after $5 \times 10^{10}$ y when the spin period of the Earth and the orbital period of the Moon will be 47 d and the semi-major axis of the Moon will be $\sim 87$ Earth radii.

$L_{tot} = L_{spin}$ and that finally $L_{tot} = L_{orb}$, in which case the final semi-major axis $a_f$ is related to the initial spin frequency $\Omega_i (= 2\pi/T_i)$ by

$$m_s \left(\mathcal{G}m_p\right)^{1/2} a_f^{1/2} = \alpha m_p C_p^2 \Omega_i . \tag{4.235}$$

Hence,

$$\left(\frac{a_f}{C_p}\right)^{1/2} = \alpha \frac{m_p}{m_s} \left(\frac{3\Omega_i^2}{4\pi \mathcal{G} \rho_p}\right)^{1/2} . \tag{4.236}$$

On substituting into Eq. (4.214) we deduce that, for the double synchronous state to be achieved, the ratio of the satellite and planet masses must exceed the critical value

$$\left(\frac{m_p}{m_s}\right)_{crit} = \left(\frac{2Q_p}{39\alpha k_{2p}\Omega_i \Delta t}\right)^{1/14} \left(\frac{3\alpha^2}{4\pi \mathcal{G} \rho_p}\right)^{1/2} \frac{2\pi}{T_i} . \tag{4.237}$$

This is determined largely by the initial spin period and is only very weakly dependent on any other properties of the system, including the size of the planet or primary. If we assume for the Earth–Moon system that $T_i \approx 4$ h, then $(m_s/m_p)_{crit} \approx 0.0147$ and is little greater than the observed ratio of 0.0123.

The system with the highest mass ratio in the solar system is that of Pluto and Charon. *Hubble Space Telescope* observations of the barycentric wobble of Pluto give a mass ratio of $0.0837 \pm 0.0147$ (Null et al. 1993), which is well above the critical value. At present Pluto and Charon are the only known bodies in a double synchronous state, but it is possible that small, unseen pairs of asteroids are also tidally despun. If we apply Eq. (4.237) to a small rocky body of density $\sim 3$ g cm$^{-3}$, rigidity $\mu \sim 5 \times 10^{10}$ N m$^{-2}$, and $Q_p \sim 100$ (none of these values is critical) and assume an initial spin period of $\sim 8$ h (Alfvén 1964), then the critical mass ratio for an asteroid of radius $\sim 100$ km is 0.034 and only increases to 0.065 for an asteroid as small as 10 m. Many very small bodies in the solar system could be tidally despun.

## Exercise Questions

**4.1**     In general, the $n$th harmonic coefficient of a planet's gravity field is given by the integral

$$J_n = -\frac{2\pi}{MR^n} \int_0^R \int_{-1}^1 P_n(\mu) r^{n+2} \rho(r, \mu) \, d\mu \, dr,$$

where $P_n$ is the Legendre polynomial of degree $n$, $\mu = \cos\theta$, $M$ and $R$ are the planet's mass and mean radius, and $\rho(r, \mu)$ is the interior density distribution. Consider a uniform density planet with oblateness $\epsilon = (a - b)/a$, where $a$ and $b$ are the equatorial and polar radii. By dividing the planet into a sphere of radius $b$, surrounded by a thin shell of variable thickness $h(\theta) = \epsilon R \sin^2\theta$, and evaluating the integral separately for the two pieces, show that $J_2 \approx 2\epsilon/5$, in the limit that $\epsilon \ll 1$. Why does this procedure fail for a planet with a radial density gradient? For a planet in hydrostatic equilibrium, $\epsilon = (3/2)J_2 + q/2$, where the quantity $q$ is related to the Love number $k_2$ by $k_2 = 3J_2/q$. Use the above expression for $J_2$ to show that $k_2 = 3/2$ for a planet of uniform density.

**4.2**     The moment of inertia of a planet, $C$, together with its mean density $\rho$, impose a strong constraint on its interior density distribution. Consider a spherical planet with a central core of radius $R_c$ and uniform density $\rho_c$, surrounded by a shell (the "mantle") of density $\rho_m$ and exterior radius $R$. The model thus has four free parameters, which may be adjusted (but not uniquely) to fit the observed values of $R$, $\rho$, and $C/MR^2$ (note that the total mass $M = (4/3)\pi\rho R^3$ is not independent of $R$ and $\rho$). (a) Derive analytic expressions for the fractional core radius, $x = r_c/R$, and core density $\rho_c$ for such a model in terms of $\rho$, $\rho_m$, and $\alpha = C/MR^2$. Sketch $x$ and $\rho_c$ as a function of $\rho_m$, and show graphically that there are two limiting cases corresponding to (i) $\rho_c \to \infty$ and $x \to 0$ and (ii) $\rho_m \to 0$. Use these limiting cases to determine the maximum mantle density, minimum core density, and maximum core radius compatible with specified values of $\rho$ and $\alpha$. (b) For the Earth, $\rho = 5.52$ g cm$^{-3}$, $\alpha = 0.332$, and $R = 6371$ km. What are the ranges of core and mantle densities, and core radii, permitted for the Earth by the two-layer model? Seismological data reveal that the core radius is $r_c = 3,480$ km. Use this information to deduce the average densities of the core and mantle. (c) For Mars, $\rho = 3.95$ g cm$^{-3}$ and $\alpha \simeq 0.375$. What are the ranges of core and mantle densities, and core radii, permitted for Mars by the two-layer model? A plausible mantle density is 3.5 g cm$^{-3}$; what are the corresponding fractional core radius and density?

**4.3**     Consider a damped simple harmonic oscillator as a one-dimensional analogue for tidal forcing in a planet:

$$\ddot{x} + \beta \dot{x} + \omega_0^2 x = F e^{i\omega t},$$

where $x$ is displacement, $\beta$ is a viscous damping factor, $\omega_0$ is the natural oscillation frequency, and the right-hand side is an external periodic forcing term that represents the tidal potential. (a) Show that the forced solution to the above differential equation, after any transients due to initial conditions have decayed away, is of the form $x(t) = A e^{i(\omega t - \epsilon)}$. Give expressions for the amplitude $A$ and phase delay $\epsilon$, in terms of $\omega_0$, $\omega$, $\beta$, and $F$. Sketch $A(\omega)$ and $\epsilon(\omega)$ for $\omega_0 = 1$ and $\beta = 0.1$, for $0.1 \leq \omega \leq 10$. (b) Show by direct integration that the work done by the damping force $-\beta \dot{x}$ over one complete cycle of the oscillation is $W = \pi \beta \omega A^2$. (c) Evaluate the maximum potential energy $E_{max}$ stored during a cycle due to work done against the restoring force $-\omega_0^2 x$, and combine these expressions to obtain an expression for $Q = 2\pi E_{max}/W$. Show that $\tan \epsilon \simeq Q^{-1}$ for $\omega \ll \omega_0$.

**4.4**     Write a computer program to track simultaneously the evolution of the Moon's semi-major axis $a$ and the Earth's spin rate $\omega$ under the *combined influence* of lunar and solar tides on the planet. (Neglect tides raised by the planet on the Moon and Sun, and assume zero-inclination, coplanar, circular orbits for the Earth and Moon.) (a) Apply your program to integrate the lunar orbit backwards in time, starting at $t = 0$ with the present conditions ($a = 3.84 \times 10^5$ km, $P = 2\pi/\omega = 24$ h), and using the constants $k_2 = 0.29$, $Q = 12$, and $C/MR^2 = 0.334$ for the Earth. Continue your integration until $\omega = n$ and thus $T_s = 0$. At what time in the past did this synchronous configuration occur, and what were the values of $a/R$ and the Earth's spin period $P$ at this time? (b) Now integrate the lunar orbit forwards in time, starting at $t = 0$ with the same initial conditions as in part (a), until a synchronous state is again reached. Answer the same three questions as in part (a). (c) Repeat the integration in part (b), but "turn off" the solar tides, and see how your answers change.

**4.5**     Consider the case of a satellite on a circular, inclined orbit. Use an argument based on conservation of energy and angular momentum to show that satellite tides will damp the inclination either to zero or to 180°, depending on the initial inclination. Derive an expression for the inclination damping time for a satellite in synchronous rotation. Do you expect the inclination time to be comparable to, or much longer than, the eccentricity damping time for typical satellites?

**4.6**      Because tides raised by satellites in their primaries generally result in the expansion of orbits initially outside the synchronous distance, the presence of a massive satellite in a close orbit sets an upper limit on either the age of the satellite or the planet's tidal dissipation function, $Q$. Use the orbital distances and masses of Deimos, Io, Mimas, Ariel, and Proteus to place lower limits on $Q$ for Mars, Jupiter, Saturn, Uranus, and Neptune. Why is the larger satellite Triton not suitable for this calculation for Neptune?

# 5

## Spin–Orbit Coupling

This common body,
Like to a vagabond flag upon the stream,
Goes to and back, lackeying the varying tide,
To rot itself with motion.

William Shakespeare, *Anthony and Cleopatra, I, iv*

### 5.1 Introduction

In the last chapter, we considered the effect of tides raised on a satellite by a planet where we assumed that the satellite was in a synchronous spin state (i.e., that the rotational period of the satellite was equal to its orbital period). As mentioned in Sect. 1.6, most of the major natural satellites in the solar system are observed to be rotating in the synchronous state. How did this situation arise and what determines the spin–orbit state of a given satellite or planet? In this chapter, we start by further examining the effects of a tidal torque on a satellite's rotation. This analysis reveals why, for example, in order to maintain its synchronous spin–orbit resonance, the Moon must have a permanent quadrupole moment. The consequences of this extra torque on the system are then examined and this leads to a general approach to the concept of spin–orbit resonance in the solar system. The origin and stability of these resonances are also discussed.

### 5.2 Tidal Despinning

Consider the case of a satellite orbiting a planet in an elliptical orbit. Those parts of the orbit in which the satellite's spin rate, which we denote by $\dot{\eta} + n$, is less (or greater) than its angular velocity or the rate of change of its true anomaly $\dot{f}$, are shown in Fig. 5.1a. If we transform to a reference frame that is centred on the satellite and rotates with the satellite's mean motion $n$, then in this rotating frame the planet moves about its guiding centre in a 2:1 ellipse (cf. Sect. 4.5) as

189

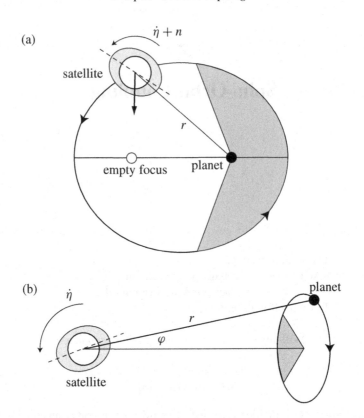

Fig. 5.1. (a) The path of a rotating satellite in an inertial reference frame centred on a planet. The nonshaded region shows that part of the orbit for which the spin rate of the satellite in inertial space, $\dot{\eta} + n > \dot{f}$. The dashed line denotes the axis of the tidal bulge. (b) The path of the planet in a reference frame centred on the satellite and rotating with its mean motion, $n$. The nonshaded region corresponds to the range of true anomaly for which $\dot{\eta} > \dot{\varphi}$.

shown in Fig. 5.1b. The rotation rate of the satellite in the rotating frame is $\dot{\eta}$ and the case $\langle \dot{\eta} \rangle = 0$ corresponds to the synchronous spin–orbit state.

For small values of the satellite's eccentricity $e$, the angle $\varphi$ shown in Fig. 5.1b is given by

$$\varphi \approx 2e \sin nt. \tag{5.1}$$

Thus $\dot{\varphi}$ is a function of time and changes sign as the planet moves around the 2:1 ellipse. If $\dot{\eta} < 2en$, then when the satellite is close to pericentre, it is possible that $\dot{\varphi} > \dot{\eta}$. In Fig. 5.2a, the angular range over which, for some value of $\dot{\eta}$, this applies is denoted by the shaded area. In this region, the tide raised on the satellite by the planet lags behind the satellite–planet line (cf. Fig. 4.6) and a couple acts on the satellite to increase $\dot{\eta}$ and spin up the satellite (see Fig. 5.2a). In the unshaded region where $\dot{\varphi} < \dot{\eta}$, the situation is reversed and the tide raised

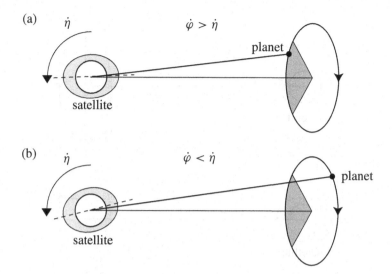

Fig. 5.2. (a) On the shaded part of orbit for which $\dot{\varphi} > \dot{\eta}$, the tide raised on the satellite by the planet lags behind the satellite–planet line (the dashed line denotes the axis of the tidal bulge) and a positive couple acts on the satellite to increase its spin rate. (b) On all other parts of the orbit, $\dot{\varphi} < \dot{\eta}$, the tidal bulge is carried ahead of the satellite–planet line and the spin of the satellite is braked.

on the satellite is carried ahead of the satellite–planet line (cf. Fig. 4.6). In this case, the resulting couple brakes the spin of the satellite and decreases $\dot{\eta}$ (see Fig. 5.2b):

By analogy with the situation examined in Sect. 4.3, the tidal torque acting to change the spin of the satellite is

$$N_s = -D \left(\frac{a}{r}\right)^6 \text{sign}\,(\dot{\eta} - \dot{\varphi}), \qquad (5.2)$$

where

$$D = \frac{3}{2} \frac{k_2}{Q_s} \frac{n^4}{G} R_s^5 \qquad (5.3)$$

and is a positive constant, and $Q_s$, $k_2$, and $R_s$ are the tidal dissipation function, Love number, and radius of the satellite, respectively. A positive torque will act to increase the spin of the satellite, $\dot{\eta}$. To find the mean torque, $\langle N_s \rangle$, we need to average $N_s$ over one orbital period of the satellite. If we consider the special case where the satellite is in synchronous rotation ($\dot{\eta} = 0$), then the torque is positive on the near side of the centred ellipse and negative on the far side (see Fig. 5.3a). The planet spends equal intervals of time on each half of a 2:1 ellipse. However, since the radial distance is smaller on the near side, the mean torque is positive and will act to spin up the satellite. For equilibrium and zero mean torque, we must have $\dot{\eta} > 0$. In this case, the sign of the torque does not reverse

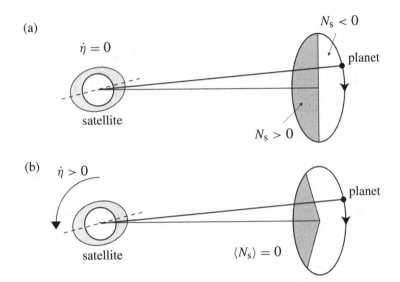

Fig. 5.3. (a) If $\dot{\eta} = 0$, then the tidal torque is positive and stronger on the near side of the planet's path and negative and weaker on the far side. Thus, the resultant net torque on the satellite is positive and the spin rate is increased. (b) In the equilibrium case, $\dot{\eta} > 0$ and the sign of the tidal torque on the satellite reverses closer to pericentre than apocentre. The stronger, positive torque (shaded area) now acts for less time than the weaker, negative torque.

at the midpoints of the 2:1 ellipse and equilibrium is achieved because the torque acting on the shaded part of the planet's path is stronger than that on the unshaded portion but acts for a shorter period of time (see Fig. 5.3b).

This argument suggests that the synchronous state is not stable and leads to a spinning up of the satellite. If this is the case, why are so many satellites observed to be in synchronous rotation? The answer is that other torques are acting, because, as in the case of the Moon, most satellites are at least partially solid and have permanent quadrupole moments, that is, permanent bulges or departures from sphericity. Before examining the effect of the quadrupole moment, we follow Goldreich (1966) and calculate the equilibrium spin rate in the absence of a permanent deformation.

The sign reversal in Eq. (5.2) occurs at those two points in orbit at which $\dot{\eta} = \dot{\varphi}$, or

$$\dot{f} = \dot{\eta} + n. \qquad (5.4)$$

If $t = 0$ is the time of pericentre passage and sign reversal occurs when $t = \pm T$, then, since

$$f \approx nt + 2e \sin nt + \frac{5}{4} e^2 \sin 2nt, \qquad (5.5)$$

sign reversal occurs when

$$\dot{\eta} = 2en \cos nT + \frac{5}{2}e^2n \cos 2nT. \tag{5.6}$$

At the time of sign reversal, let

$$f = \pm \left(\frac{\pi}{2} - \delta\right). \tag{5.7}$$

Then

$$\sin \delta = \cos f = \cos(nT + 2e \sin nT) = \cos nT - 2e \sin^2 nT. \tag{5.8}$$

From Eq. (5.6),

$$\cos nT = \frac{\dot{\eta}}{2en} - \frac{5}{4}e \cos 2nT. \tag{5.9}$$

Hence, to $\mathcal{O}(e)$,

$$\sin \delta = \frac{\dot{\eta}}{2en} - \frac{5}{4}e + \frac{1}{2}e \sin^2 nT. \tag{5.10}$$

The mean tidal torque acting on the satellite is given by

$$\langle N_s \rangle = -\frac{nD}{2\pi} \int_0^{2\pi/n} \left(\frac{a}{r}\right)^6 \mathrm{sign}(\dot{\eta} - \dot{\varphi}) \, dt$$

$$= -\frac{D}{2\pi} \int_0^{2\pi} \left(\frac{a}{r}\right)^4 \frac{\mathrm{sign}(\dot{\eta} - \dot{\varphi})}{(1 - e^2)^2} \, df. \tag{5.11}$$

Allowing for the change of sign, this reduces to

$$\langle N_s \rangle = +\frac{D}{\pi} \int_0^{(\pi/2)-\delta} (1 + 4e \cos f) \, df$$

$$- \frac{D}{\pi} \int_{(\pi/2)-\delta}^{\pi} (1 + 4e \cos f) \, df$$

$$= +\frac{2D}{\pi}(4e \cos \delta - \delta). \tag{5.12}$$

For equilibrium, we must have $\langle N_s \rangle = 0$ and this requires that

$$\delta = 4e \cos \delta \approx 4e \left(1 - \frac{1}{2}\delta^2\right) \approx 4e. \tag{5.13}$$

It follows from this and Eqs. (5.5) and (5.7) that we have the relationship

$$\sin nT \approx \sin f \approx \sin[\pm(\pi/2 - \delta)]$$

$$= \cos \delta \approx \cos 4e \approx 1 - 8e^2$$

$$\approx 1.$$

Substituting into Eq. (5.10), we obtain the result

$$\dot{\eta} = \frac{19}{2}e^2n. \tag{5.14}$$

Thus, in the absence of a permanent quadrupole moment, the Moon, for example, would rotate about 3% faster than the observed synchronous rate, and over a period of about 2.6 y, we would see both sides of the satellite.

Before closing this introductory section, we need to emphasize that in Eq. (5.2) we formulated the effects of tidal drag using the model due to MacDonald (1964) that assumes a constant lag angle for the tidal bulge. If we had chosen to use the alternative formulation of Darwin (1908) in which the tidal potential is expanded in a Fourier time series and each component of the tide is given a constant phase lag, then our conclusions would be substantially different. Goldreich & Peale (1966) discuss this in detail.

## 5.3 The Permanent Quadrupole Moment

To calculate the external gravitational field of a permanently deformed satellite at any distance from its centre of mass, we require a description of the distribution of mass within the satellite. At very large distances, the field is well represented by that of a point mass. At lesser, but still large, distances, the information on the mass distribution provided by the satellite's principal moments of inertia proves to be sufficient. The derivations that follow are based chiefly on those given by MacMillan (1936) and Ramsey (1937, 1940).

Consider an element of mass at a point $P$ within a body and let the position vector of this element with respect to an arbitrary origin at $O$ be $\mathbf{p} = (x, y, z)$ (see Fig. 5.4). We define the following moments of inertia with respect to the coordinate axes:

$$A = \sum \delta m (y^2 + z^2), \tag{5.15}$$

$$B = \sum \delta m (z^2 + x^2), \tag{5.16}$$

$$C = \sum \delta m (x^2 + y^2) \tag{5.17}$$

and the following products of inertia

$$D = \sum \delta m \, yz, \tag{5.18}$$

$$\mathcal{E} = \sum \delta m \, zx, \tag{5.19}$$

$$\mathcal{F} = \sum \delta m \, xy. \tag{5.20}$$

The moment of inertia $I_L$ about any line $OL$ can be expressed in terms of $A$, $B, C, D, \mathcal{E}$, and $\mathcal{F}$ and the direction cosines $l, m$, and $n$ of the line $OL$ with respect to the $x, y$, and $z$ axes. Let $PQ$ be the perpendicular from the point $P(x, y, z)$ to the line $OL$, where the position vector of $Q$ is given by $\mathbf{q} = (x', y', z')$. Hence $OP = p$ and $OQ = q$, the magnitudes of the vectors $\mathbf{p}$ and $\mathbf{q}$. Because $PQ$ is perpendicular to $OL$, we have

$$\mathbf{p} \cdot \mathbf{q} = q^2. \tag{5.21}$$

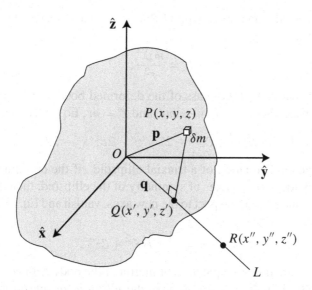

Fig. 5.4. The location of a mass element $\delta m$ at a point $P$ within the body with position vector $\mathbf{p} = (x, y, z)$. $OL$ is an arbitrary line from the origin of the coordinate system at $O$. $PQ$ is the perpendicular from $P$ to the point $Q$ with position vector $\mathbf{q} = (x', y', z')$ along the line $OL$. The arbitrary point $R$ with position vector $\mathbf{r} = (x'', y'', z'')$ also lies along the line $OL$.

However,

$$\mathbf{p} \cdot \mathbf{q} = xx' + yy' + zz' = x(lq) + y(mq) + z(nq) = q(lx + my + nz) \quad (5.22)$$

and hence

$$q = lx + my + nz. \quad (5.23)$$

The moment of inertia $I_L$ is given by

$$I_L = \sum \delta m (PQ)^2 = \sum \delta m \left[ x^2 + y^2 + z^2 - (lx + my + nz)^2 \right]. \quad (5.24)$$

Because $l^2 + m^2 + n^2 = 1$, we can write

$$I_L = \sum \delta m \left[ (x^2 + y^2 + z^2)(l^2 + m^2 + n^2) - (lx + my + nz)^2 \right], \quad (5.25)$$

which, on expanding and rearranging, gives

$$I_L = l^2 \sum \delta m (y^2 + z^2) + m^2 \sum \delta m (z^2 + x^2) + n^2 \sum \delta m (x^2 + y^2)$$
$$- 2mn \sum \delta m \, yz - 2nl \sum \delta m \, zx - 2lm \sum \delta m \, xy. \quad (5.26)$$

This can be written as

$$I_L = \mathcal{A}l^2 + \mathcal{B}m^2 + \mathcal{C}n^2 - 2\mathcal{D}mn - 2\mathcal{E}nl - 2\mathcal{F}lm. \quad (5.27)$$

If we now consider an arbitrary point $R(x'', y'', z'')$ a distance $r$ from $O$ on the line $OL$ and write

$$I_L = \frac{m_s \lambda^4}{r^2},\tag{5.28}$$

where $m_s = \sum \delta m$ is the total mass of the deformed body and $\lambda$ is an arbitrary length, then, given that $x'' = lr$, $y'' = mr$, and $z'' = nr$, Eq. (5.27) becomes

$$m_s \lambda^4 = \mathcal{A}x''^2 + \mathcal{B}y''^2 + \mathcal{C}z''^2 - 2\mathcal{D}y''z'' - 2\mathcal{E}z''x'' - 2\mathcal{F}x''y'',\tag{5.29}$$

which is the general equation of a triaxial ellipsoid. If the coordinate axes are chosen to coincide with the axes of symmetry of the ellipsoid, then the products of inertia $\mathcal{D}$, $\mathcal{E}$, and $\mathcal{F}$, with respect to the new axes, vanish and Eq. (5.29) reduces to

$$m_s \lambda^4 = \mathcal{A}x''^2 + \mathcal{B}y''^2 + \mathcal{C}z''^2.\tag{5.30}$$

These new axes are the principal axes of inertia of the body defined with respect to the point $O$. Equation (5.30) defines the *ellipsoid of inertia* (see Cauchy (1827) for details of the above calculation). This is an invariant of the body; it is independent of the orientation of the axes but varies with the position of the origin $O$. If $O$ is the centre of mass, then the ellipsoid is called the *central ellipsoid of inertia*. It follows from the properties of this ellipsoid that every body, regardless of its shape, possesses three mutually perpendicular axes, such that the moment of inertia about one of these axes is a maximum, while another is a minimum and the third is either intermediate or equal to one of the other two.

We now derive an expression for the external gravitational field of a permanently deformed satellite in terms of its principal moments of inertia, $A$, $B$, and $C$ defined with respect to the centre of mass. In our new system (see Fig. 5.5), we take the point $O$ to be at the centre of mass of the satellite and let $P$ be a point a distance $r$ from $O$. We will assume that $r$ is very much greater than the mean radius of the satellite. Let the coordinate system be defined such that the $x$, $y$, and $z$ directions lie along the principal axes of inertia of the satellite (see Fig. 5.5).

If $\delta m$ is the mass of a small mass element at the point $Q$, distance $R$ from $O$, then the potential of the satellite at $P$ is given by

$$V = -\sum \frac{\mathcal{G}\delta m}{\Delta} = -\sum \frac{\mathcal{G}\delta m}{(r^2 + R^2 - 2rR\cos\theta)^{1/2}},\tag{5.31}$$

where $\theta$ is the angle between $OP$ and $OQ$ and the summation is taken over all mass elements. Expanding Eq. (5.31) binomially and neglecting high-order terms ($r \gg R$ for all $Q$), we obtain

$$V \approx -\frac{\mathcal{G}m_s}{r} - \frac{\sum \mathcal{G}\delta m R\cos\theta}{r^2} - \frac{2\sum \mathcal{G}\delta m R^2 - 3\sum \mathcal{G}\delta m R^2 \sin^2\theta}{2r^3},\tag{5.32}$$

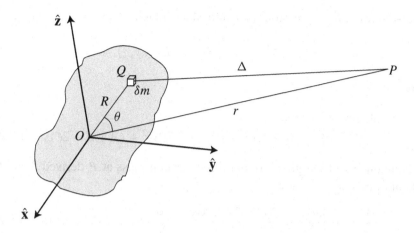

Fig. 5.5. A coordinate system with origin at the centre of mass $O$ of a satellite with axes aligned with its principal moments of inertia. The point $P$ is at a distance $r$ from $O$. The small mass element $\delta m$ at the point $Q$ is at a distance $R$ from $O$, and the line $OQ$ makes an angle $\theta$ with the line $OP$.

where, as before, $m_s = \sum \delta m$ is the mass of the satellite. Because $O$ is at the centre of mass of the satellite,

$$\sum \delta m \, R \cos \theta = 0. \tag{5.33}$$

We also have

$$
\begin{aligned}
2 \sum \delta m R^2 &= 2 \sum \delta m (x^2 + y^2 + z^2) \\
&= \sum \delta m (y^2 + z^2) + \sum \delta m (z^2 + x^2) + \sum \delta m (x^2 + y^2) \\
&= A + B + C. \tag{5.34}
\end{aligned}
$$

If we denote the moment of inertia of the body about the line $OP$ by $I$, then

$$I = \sum \delta m \, R^2 \sin^2 \theta \tag{5.35}$$

and, to the extent that (5.32) is a good approximation,

$$V = -\frac{\mathcal{G} m_s}{r} - \frac{\mathcal{G}(A + B + C - 3I)}{2r^3}. \tag{5.36}$$

This is MacCullagh's formula (MacCullagh 1844a,b; Haughton 1855).

If we now let $x$, $y$, and $z$ denote the coordinates of the point $P$, then $x/r$, $y/r$, and $z/r$ are the direction cosines of $P$ with respect to the principal axes of inertia and, from Eq. (5.27), we have

$$I = (A x^2 + B y^2 + C z^2)/r^2. \tag{5.37}$$

On substituting this expression for $I$ into MacCullagh's formula, we obtain

$$V = -\frac{Gm_s}{r} - \frac{G}{2r^5} f(A, B, C, x, y, z), \qquad (5.38)$$

where

$$f(A, B, C, x, y, z) = \\ (B + C - 2A)x^2 + (C + A - 2B)y^2 + (A + B - 2C)z^2. \quad (5.39)$$

The components of the gravitational force per unit mass at $P$ derived from the gradient of this potential are

$$F_x = -\frac{\partial V}{\partial x} = -\frac{Gm_s x}{r^3} + \frac{G(B + C - 2A)x}{r^5} - \frac{5Gx}{2r^7} f(A, B, C, x, y, z), \quad (5.40)$$

$$F_y = -\frac{\partial V}{\partial y} = -\frac{Gm_s y}{r^3} + \frac{G(C + A - 2B)y}{r^5} - \frac{5Gy}{2r^7} f(A, B, C, x, y, z), \quad (5.41)$$

$$F_z = -\frac{\partial V}{\partial z} = -\frac{Gm_s z}{r^3} + \frac{G(A + B - 2C)z}{r^5} - \frac{5Gz}{2r^7} f(A, B, C, x, y, z). \quad (5.42)$$

These forces exert a couple on a unit mass at $P$ and an equal and opposite couple acts on the deformed body about its centre of mass. The latter couple has components

$$N_x = z F_y - y F_z = +3G(C - B)yz/r^5, \qquad (5.43)$$

$$N_y = x F_z - z F_x = +3G(A - C)zx/r^5, \qquad (5.44)$$

$$N_z = y F_x - x F_y = +3G(B - A)xy/r^5. \qquad (5.45)$$

Euler's full equations of motion are

$$A\dot\omega_x - (B - C)\omega_y\omega_z = N_x, \qquad (5.46)$$

$$B\dot\omega_x - (C - A)\omega_z\omega_x = N_y, \qquad (5.47)$$

$$C\dot\omega_z - (A - B)\omega_x\omega_y = N_z, \qquad (5.48)$$

where $\omega_x$, $\omega_y$, and $\omega_z$ are the projections of the spin vector on the principal axes. In our problem, we wish to calculate the rotational motion of a satellite due to the torque exerted on its quadrupole moment by a distant planet. We will assume that the spin axis of the satellite is normal to its orbital plane and that $\omega_x$ and $\omega_y$ are zero. We now denote the direction cosines of the planet with respect to the $x$ axis and $y$ axis by $x/r = \cos\psi$ and $y/r = \sin\psi$, respectively (see Fig. 5.6). In this case, Euler's equations of motion reduce to Eq. (5.48), which can be written

$$C\ddot\theta - \frac{3}{2}(B - A)\frac{Gm_p}{r^3}\sin 2\psi = 0, \qquad (5.49)$$

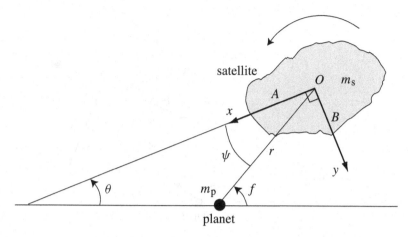

Fig. 5.6. Rotation of a satellite with its spin axis normal to the orbit plane. $\psi$ is the angle between the planet–satellite line and the principal axis $A$ associated with the minimum moment of inertia of the satellite. The angle $\theta$ is measured with respect to a direction fixed in inertial space.

where the angle $\theta$ is measured with respect to a direction fixed in inertial space. Note that, in some other formulations of the above equation of motion, for example, that given by Danby (1988), the sign of $\ddot{\theta}$ is negative rather than positive. This arises because Danby's choice of coordinate system implies that $\omega_z = -\dot{\theta}$.

We can obtain a simple, heuristic verification of Eq. (5.49) by considering the following. Represent a satellite with a permanent quadrupole moment by a spherical satellite with two equal, diametrically opposed point masses, $m$, embedded in its equatorial (and orbital) plane (see Fig. 5.7). Let the respective distances of these small masses from the planet be $r_1$ and $r_2$ and let $r$ denote the distance between the centre of the satellite and the centre of the planet. The line joining the planet and satellite centres makes an angle $\psi$ with the principal axis associated with $\mathcal{A}$, the minimum moment of inertia, that is, the line joining the two small masses, $m$.

If the satellite has a mean radius $R_s$, then the torque on the satellite due to the gravitational forces between the planet and the two small masses is $N_1 + N_2$, where

$$N_1 = +\mathcal{G}\frac{m_s m_p}{r_1^2} R_s \sin \alpha, \qquad N_2 = -\mathcal{G}\frac{m_s m_p}{r_2^2} R_s \sin \beta. \qquad (5.50)$$

The angles $\alpha$ and $\beta$ are defined in Fig. 5.7; the sign of $N_1$ is positive because it acts to increase $\theta$. Applying the cosine and sine rules, we have

$$\sin \alpha = \frac{r}{r_1} \sin \psi, \qquad \sin \beta = \frac{r}{r_2} \sin \psi, \qquad (5.51)$$

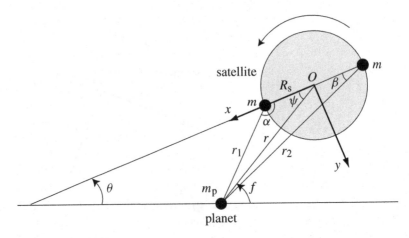

Fig. 5.7. Representation of a satellite with a quadrupole moment as a spherical object with two diametrically opposed small masses. The diameter joining the two masses defines the axis with the minimum moment of inertia and makes an angle $\psi$ with the planet–satellite line (cf. Fig. 5.6).

and

$$\frac{1}{r_1^3} \approx \frac{1}{r^3}\left[1 - \frac{3}{2}\left(\frac{R_s}{r}\right)^2 + 3\frac{R_s}{r}\cos\psi\right], \tag{5.52}$$

$$\frac{1}{r_2^3} \approx \frac{1}{r^3}\left[1 - \frac{3}{2}\left(\frac{R_s}{r}\right)^2 - 3\frac{R_s}{r}\cos\psi\right]. \tag{5.53}$$

Hence the equation of motion for $\theta$ reduces to

$$C\ddot{\theta} - \frac{3}{2}\left(2mR_s^2\right)\frac{\mathcal{G}m_p}{r^3}\sin 2\psi = 0 \tag{5.54}$$

and, given that $\mathcal{B} - \mathcal{A} = 2mR_s^2$, the parallel with Eq. (5.49) is complete.

## 5.4 Spin–Orbit Resonance

The gravitational interaction between the orbital motion of a planet and the quadrupole moment of an attendant satellite results in small, short-period oscillations in the rotation rate of the satellite that, usually, are of little consequence. However, there are circumstances in which this is not the case. These arise when there is a simple integer, or near-integer, relationship between the spin period of a satellite and its orbital period, in which case there may be significant spin–orbit coupling. The following discussion is based on the pioneering work of Goldreich & Peale (1966, 1968), Wisdom, Peale & Mignard (1984), and Wisdom (1987a,b).

Consider the motion of a small satellite whose spin axis is normal to the plane of its fixed elliptical orbit. Let the long axis of the satellite make an angle $\theta$ with

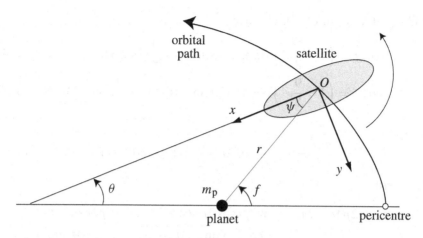

Fig. 5.8. The long axis of the satellite makes an angle $\theta$ with a reference axis that is fixed in inertial space, which we take to be the major axis of the satellite's fixed orbit.

a reference axis that is fixed in inertial space, which in this two-body, keplerian system we can take to be the major axis of the satellite's orbit. The long axis of the satellite makes an angle $\psi$ with the satellite–planet centre line. Hence

$$\psi = f - \theta, \tag{5.55}$$

where $f$ is the true anomaly (see Fig. 5.8). In the absence of tidal torques, the equation of motion for $\theta$ is (see Eq. (5.49))

$$\mathcal{C}\ddot{\theta} - \frac{3}{2}(\mathcal{B} - \mathcal{A})\frac{\mathcal{G}m_p}{r^3}\sin 2\psi = 0. \tag{5.56}$$

Because $r$ and $\psi$ vary with $f$, which is a nonlinear function of time, this equation is nonintegrable. However, in those cases of interest here, in which the spin rate $\dot{\theta}$ is commensurate with the mean motion $n$, we can derive an equation of motion that, although an approximation, is both useful and integrable.

Because we are interested in those cases for which $\dot{\theta}$ is a rational multiple of the mean motion, we introduce a new variable

$$\gamma = \theta - pM, \tag{5.57}$$

where $p$ is a rational and $M$ is the mean anomaly. Given that $n$ is a constant, $\ddot{\theta} = \ddot{\gamma}$ and the equation of motion for $\gamma$ is (cf. Eq. (5.56))

$$\ddot{\gamma} + \frac{3}{2}n^2\left(\frac{\mathcal{B} - \mathcal{A}}{\mathcal{C}}\right)\left(\frac{a}{r}\right)^3\sin(2\gamma + 2pM - 2f) = 0. \tag{5.58}$$

This equation can be expanded in a Fourier-like Poisson series in terms of $e$ and $M$ using standard expressions for $(a/r)^3$, $\sin f$, and $\cos f$ (see Sect. 2.5).

Including all terms of $\mathcal{O}(e^2)$, we have

$$\sin f = \left(1 - \frac{7}{8}e^2\right)\sin M + e\sin 2M + \frac{9}{8}e^2\sin 3M, \qquad (5.59)$$

$$\cos f = \left(1 - \frac{9}{8}e^2\right)\cos M + e(\cos 2M - 1) + \frac{9}{8}e^2\cos 3M, \qquad (5.60)$$

and

$$\left(\frac{a}{r}\right)^3 = 1 + 3e\cos M + \frac{3}{2}e^2(1 + 3\cos 2M). \qquad (5.61)$$

We can write

$$\sin(2\gamma + 2pM - 2f) = \sin 2\gamma(\cos 2pM\cos 2f + \sin 2pM\sin 2f)$$
$$+ \cos 2\gamma(\sin 2pM\cos 2f - \cos 2pM\sin 2f). \qquad (5.62)$$

Hence

$$\left(\frac{a}{r}\right)^3\sin(2\gamma + 2pM - 2f) = [S_1 + S_2]\sin 2\gamma + [S_3 - S_4]\cos 2\gamma, \qquad (5.63)$$

where

$$S_1 = \left(\frac{a}{r}\right)^3\cos 2pM\cos 2f, \qquad S_2 = \left(\frac{a}{r}\right)^3\sin 2pM\sin 2f,$$

$$S_3 = \left(\frac{a}{r}\right)^3\sin 2pM\cos 2f, \qquad S_4 = \left(\frac{a}{r}\right)^3\cos 2pM\sin 2f. \qquad (5.64)$$

To $\mathcal{O}(e^2)$, the $S_i$ are given by

$$S_1 = \frac{1}{2}[\cos 2(1-p)M + \cos 2(1+p)M]$$
$$+ \frac{1}{4}e[7\cos(3+2p)M + 7\cos(3-2p)M$$
$$- \cos(1+2p)M - \cos(1-2p)M]$$
$$+ \frac{1}{4}e^2[-5\cos 2(1+p)M - 5\cos 2(1-p)M$$
$$+ 17\cos 2(2+p)M + 17\cos 2(2-p)M], \qquad (5.65)$$

$$S_2 = \frac{1}{2}[\cos 2(1-p)M - \cos 2(1+p)M]$$
$$+ \frac{1}{4}e[-7\cos(3+2p)M + 7\cos(3-2p)M$$
$$- \cos(1-2p)M + \cos(1+2p)M]$$
$$+ \frac{1}{4}e^2[5\cos 2(1+p)M - 5\cos 2(1-p)M$$
$$- 17\cos 2(2+p)M + 17\cos 2(2-p)M], \qquad (5.66)$$

$$S_3 = \frac{1}{2}\left[\sin 2(1+p)M - \sin 2(1-p)M\right]$$

$$+ \frac{1}{4}e\left[7\sin(3+2p)M - 7\sin(3-2p)M\right.$$

$$\left. + \sin(1-2p)M - \sin(1+2p)M\right]$$

$$+ \frac{1}{4}e^2\left[-5\sin 2(1+p)M + 5\sin 2(1-p)M\right.$$

$$\left. +17\sin 2(2+p)M - 17\sin 2(2-p)M\right], \qquad (5.67)$$

$$S_4 = \frac{1}{2}\left[\sin 2(1+p)M + \sin 2(1-p)M\right]$$

$$+ \frac{1}{4}e\left[7\sin(3+2p)M + 7\sin(3-2p)M\right.$$

$$\left. - \sin(1-2p)M - \sin(1+2p)M\right]$$

$$+ \frac{1}{4}e^2\left[-5\sin 2(1+p)M - 5\sin 2(1-p)M\right.$$

$$\left. +17\sin 2(2+p)M + 17\sin 2(2-p)M\right]. \qquad (5.68)$$

Therefore, the equation of motion for $\gamma$, Eq. (5.58), can be written as

$$\ddot{\gamma} + \frac{3}{2}\frac{(\mathcal{B}-\mathcal{A})}{\mathcal{C}}n^2 \left([S_1 + S_2]\sin 2\gamma + [S_3 - S_4]\cos 2\gamma\right) = 0. \qquad (5.69)$$

Note that $S_1$ and $S_2$ only contain cosines, whereas $S_3$ and $S_4$ only contain sines. The equation is exact, but the $S_i$ are infinite series in $e$ and $M$ and thus the equation is still nonintegrable. To progress, we must resort to approximations.

If the spin rate of the satellite is close to a spin–orbit resonance, then $\dot{\theta} \approx pn$ and $\gamma$ is slowly varying, that is, $\dot{\gamma} \ll n$ and we can produce an approximate equation of motion by averaging all the terms in Eq. (5.69) over one orbital period while holding $\gamma$ fixed. We then obtain

$$\ddot{\gamma} + \frac{3}{2}\frac{(\mathcal{B}-\mathcal{A})}{\mathcal{C}}n^2 \left([\langle S_1\rangle + \langle S_2\rangle]\sin 2\gamma + [\langle S_3\rangle - \langle S_4\rangle]\cos 2\gamma\right) = 0, \qquad (5.70)$$

where

$$\langle S_i\rangle = \frac{1}{2\pi}\int_0^{2\pi} S_i \, dM, \qquad i = 1, 2, 3, 4 \qquad (5.71)$$

and it is now understood that $\gamma$ refers to the averaged value. The $S_i$ have to be evaluated for a particular value of the rational $p$ corresponding to the particular spin–orbit resonance under consideration. Because cosines and sines with arguments that are integer multiples of $M$ average out to zero over one orbital period, the only terms in $S_i$ that make a nonzero contribution to the equation of motion are those cosine terms with zero arguments. For example, in the synchronous case for which $p = 1$, only those cosine terms with arguments containing $p - 1$ as a factor contribute to the equation of motion. Inspection of Eqs. (5.81) to

(5.84) shows that in this case, to $\mathcal{O}(e^2)$, we have

$$\left([\langle S_1 \rangle + \langle S_2 \rangle]\sin 2\gamma + [\langle S_3 \rangle - \langle S_4 \rangle]\cos 2\gamma\right)_{p=1} = \left(1 - \frac{5}{2}e^2\right)\sin 2\gamma. \quad (5.72)$$

If we carry out the same procedure for other values of $p$, then inspection of the same equations (or the equivalent set that contains terms of higher order in $e$) shows that only values of $p$ that are an integer multiple of $1/2$ can contribute to the averaged equation of motion. In those cases, we can write

$$\ddot{\gamma} + \frac{3}{2}n^2\frac{(B-A)}{C}H(p,e)\sin 2\gamma = 0, \quad (5.73)$$

where, for example, to $\mathcal{O}(e^4)$,

$$H(-1, e) = +\frac{1}{24}e^4, \quad (5.74)$$

$$H(-1/2, e) = +\frac{1}{48}e^3, \quad (5.75)$$

$$H(0, e) = 0, \quad (5.76)$$

$$H(+1/2, e) = -\frac{1}{2}e + \frac{1}{16}e^3, \quad (5.77)$$

$$H(+1, e) = +1 - \frac{5}{2}e^2 + \frac{13}{16}e^4, \quad (5.78)$$

$$H(+3/2, e) = +\frac{7}{2}e - \frac{123}{16}e^3, \quad (5.79)$$

$$H(+2, e) = +\frac{17}{2}e^2 - \frac{115}{6}e^4, \quad (5.80)$$

$$H(+5/2, e) = +\frac{845}{48}e^3, \quad (5.81)$$

$$H(+3, e) = +\frac{533}{16}e^4. \quad (5.82)$$

Inspection of the eccentricity function, $G_{lpq}(e)$, defined by Kaula (1966) shows that

$$H(p, e) = G_{20(2p-2)}(e) \quad (5.83)$$

and that, except for the case $p = 0$, $H(p, e) = \mathcal{O}\left(e^{2|p-1|}\right)$.

Thus, by introducing an approximation, we have reduced the full equation of motion, Eq. (5.56), to the pendulum equation, which we can write as

$$\ddot{\gamma} = -[\text{sign } H(p, e)]\frac{1}{2}\omega_0^2\sin 2\gamma, \quad (5.84)$$

where

$$\omega_0 = n\left[3\left(\frac{B-A}{C}\right)|H(p, e)|\right]^{1/2} \quad (5.85)$$

is the libration frequency. In the presence of a tidal torque acting to brake the spin of the satellite, a term representing the mean tidal torque averaged over one orbital period, $\langle N_s \rangle$, can be added to the averaged equation of motion to give

$$\ddot{\gamma} = -[\text{sign } H(p, e)] \frac{1}{2} \omega_0^2 \sin 2\gamma + \langle N_s \rangle / C. \qquad (5.86)$$

If

$$|\langle N_s \rangle| / C < \frac{1}{2} \omega_0^2 \qquad (5.87)$$

then the sign of $\ddot{\gamma}$ must reverse periodically and thus it is possible for the satellite to be trapped in a spin–orbit resonance for which $\langle \dot{\theta} \rangle = pn$. If Eq. (5.87), the *strength criterion*, is satisfied, then the mean torque due to the resonant interaction between the planet and the quadrupole moment of the satellite compensates for the mean tidal torque acting to change the spin period of the satellite, $\langle \ddot{\gamma} \rangle = 0$, and $\gamma$ librates about an equilibrium value $\gamma_0$ given by

$$\gamma_0 = \frac{1}{2} \sin^{-1} \left[ \frac{2 \langle N_s \rangle}{-[\text{sign } H(p, e)] \omega_0^2 C} \right]. \qquad (5.88)$$

The equilibrium orientation of the satellite and the sign of $\gamma_0$ are determined by the sign of $H(p, e)$. For small displacements of $\gamma$ from $\gamma_0$, the sign of $\ddot{\gamma}$ must be such as to return $\gamma$ to the equilibrium displacement, $\gamma_0$. If the mean tidal torque is weak in comparison with the resonant torque, that is, if

$$|\langle N_s \rangle| / C \ll \frac{1}{2} \omega_0^2 \qquad (5.89)$$

then, if $H(p, e) > 0$, $\gamma_0 \approx 0$ or $\pi$ and the long axis of the satellite points towards the planet on passage of the satellite through pericentre. Conversely, if $H(p, e) < 0$, $\gamma_0 \approx \pi/2$ or $3\pi/2$ and the long axis of the satellite points in a direction perpendicular to the planet–satellite line on passage of the satellite through pericentre.

We now consider the rotation of Mercury and give a simple, physical interpretation of the averaged equation of motion. The case of Mercury is particularly interesting because it was only after radar observations revealed that the planet is trapped in a 3:2 spin–orbit resonance with the Sun, rather than the expected 1:1 synchronous state, that the dynamics of spin–orbit resonance was first investigated. A good history of these events has been given by Goldreich & Peale (1968). The rotational and orbital motions of Mercury in an inertial reference frame are shown in Fig. 5.9.

The spin period of the planet is 58.65 d, while its orbital period is 87.97 = $1.5 \times 58.65$ d. Thus, the planet rotates on its axis three times while it orbits the Sun twice and on successive passages of Mercury through perihelion, opposite faces of the planet are presented to the Sun. The physical meaning of the angle

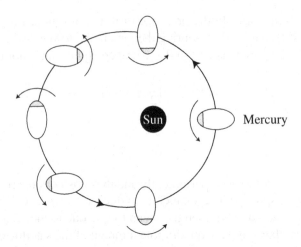

Fig. 5.9. In an inertial frame centred on the Sun, the planet Mercury completes 3/2 rotations each orbital period.

$\gamma$ is that it describes the orientation of the long axis of the satellite on passage of the satellite through pericentre, that is, it is a *stroboscopic angle* that is evaluated when $M = 0$. Given that $H(p, e) \approx +(7/2)e > 0$, we expect that $\gamma \approx 0$ and that at perihelion the long axis of the planet points towards the Sun (Fig. 5.10). However, it is possible for $\gamma$ to librate about the equilibrium value with an amplitude $\leq \pi/2$. If Mercury were trapped in the $p = +1/2$ resonance, for which $H(p, e) < 0$, then we would expect the orientation of the planet to be as shown in Fig. 5.11.

Figure 5.10a shows the motion of the Sun in a reference frame centred on Mercury and rotating with the planet's mean, resonant spin rate, $(3/2)n$, where $n$ is Mercury's mean motion. The points on the looped path indicate the position of the Sun at equal intervals of time. The path of the Sun in this rotating frame is closed only because the spin–orbit resonance exists and it is this crucial fact that validates our use of the averaging method. The average gravitational interaction between the quadrupole moment of the planet and the Sun can be modelled by spreading the mass of the Sun along this closed path in such a way that the local line density is proportional to the time spent by the Sun in that part of the path. This line density is inversely proportional to the spacing of the points shown in Fig. 5.10a. The angle $\gamma$ can now be interpreted as the deviation of the long axis of the planet from the planet–perihelion direction in the rotating frame (Fig. 5.10a). From the symmetry of this figure, we further deduce that the gravitational interaction could be modelled by replacing the mass distribution of the Sun with a circular distribution of uniform line density that does not contribute to the torque plus two point masses, $f m_{\text{sun}}$, positioned as shown in

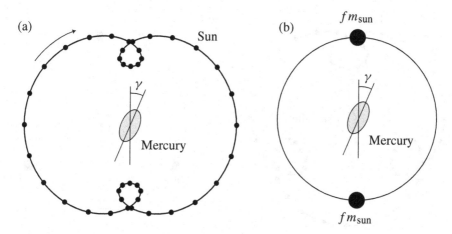

Fig. 5.10. (a) The motion of the Sun as seen in a reference frame centred on Mercury and rotating with Mercury's resonant spin rate, $(3/2)n$, where $n$ is Mercury's mean motion. The gravitational interaction between the quadrupole moment of Mercury and the Sun can be modelled by spreading the mass of the Sun around the closed path with a local line density proportional to the time spent in that part of the path (the points on the closed path show successive positions of the Sun at equal intervals of time) or by two point masses, $fm_{\text{sun}}$ where $f = (1/2)H(3/2, e)$, placed as shown in (b).

Fig. 5.10b, where

$$f = \frac{1}{2}H(p, e) \approx \frac{7}{4}e \tag{5.90}$$

and $m_{\text{sun}}$ is the mass of the Sun.

A closed path in a rotating reference frame, as shown in Fig. 5.11, is a necessary but not a sufficient condition for spin–orbit coupling. Equations (5.65) and

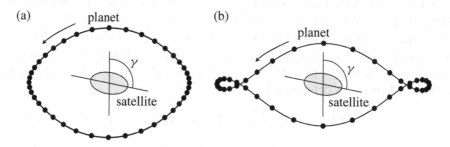

Fig. 5.11. The orientation of a satellite trapped in a $p = +1/2$ resonance is anomalous in that on passage of the satellite through pericentre its long axis points in a direction perpendicular to the planet–satellite line. The path of the planet in a rotating reference frame centred on the satellite is shown for (a) a small eccentricity and (b) a large eccentricity.

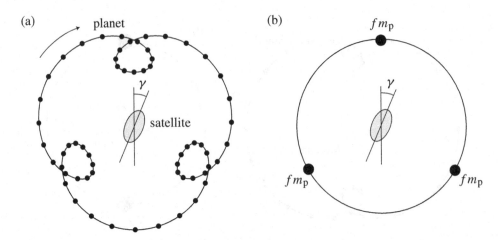

Fig. 5.12.   (a) The path of a planet in a reference frame centred on a satellite and rotating with the satellite's mean spin rate, in this case $(4/3)n$, where $n$ is the satellite's mean motion. The averaged gravitational torque on the planet due to the satellite's quadrupole moment can be modelled by replacing the looped path of the planet in the rotating reference frame by a circular distribution of mass plus three point masses placed as shown in (b).

(5.56) show that $p$ must also be an integer multiple of $1/2$ (this is determined by the twofold symmetry of the satellite's gravitational potential). It is worth considering why other values of $p$ do not contribute to any resonant interaction. In Fig. 5.12 we show the motion of a planet as seen in a reference frame centred on a satellite and rotating with the satellite's mean rotation rate, in this case $(4/3)n$, where $n$ is the satellite's mean motion. From the shape of the closed path in Fig. 5.12a, we can see that in this case the average gravitational force of the planet could be modelled by replacing the looped path by a circular distribution of mass and three equal point masses (see Fig. 5.12b). Now compare the configuration shown in Fig. 5.10b with that shown in Fig. 5.12b. In Fig. 5.10b, if $N \sin 2\gamma$ is the torque acting on the satellite due to one of the point masses, then the total torque on the satellite acting to restore the equilibrium configuration is $2N \sin 2\gamma$. However, in the case depicted in Fig. 5.12b for which $p = 4/3$, the total torque acting to change $\gamma$ is determined by

$$N \sin 2\gamma + N \sin 2\left(\frac{\pi}{3} + \gamma\right) - N \sin 2\left(\frac{\pi}{3} - \gamma\right) = 0 \qquad (5.91)$$

and the system is neutrally stable.

The special case of $p = 0$ is shown in Fig. 5.13. The torque $\delta N$ on the satellite due to the mass element $\delta m$ is given by

$$\delta N = \frac{3}{2}(\mathcal{B} - \mathcal{A})\frac{G\,\delta m}{r^3}\sin 2\psi. \qquad (5.92)$$

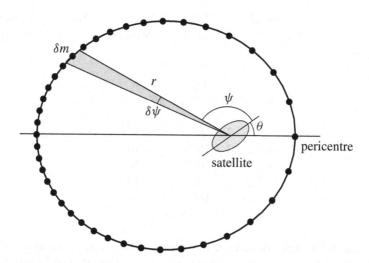

Fig. 5.13. The path of a planet in an inertial reference frame centred on the satellite. If $p = 0$, then the orientation of the satellite is fixed in inertial space and there is no resonant interaction with the planet.

From Kepler's law of areas, we obtain

$$\delta m = \frac{r^2 \delta \psi}{2\pi a^2 (1 - e^2)^{1/2}}.$$

(5.93)

Hence, the total torque on the satellite is given by

$$N = \frac{3(\mathcal{B} - \mathcal{A})\mathcal{G}}{4\pi a^3 (1 - e^2)^{3/2}} \int_0^{2\pi} [1 + e\cos(\psi + \theta)] \sin 2\psi \, d\psi,$$

(5.94)

and given that the satellite is not rotating in inertial space, $\theta$ is fixed and the integral is zero.

For a satellite to be trapped in a particular spin–orbit resonance, the torque on the satellite due to the resonance must exceed that due to tidal drag. From the strength criterion, Eq. (5.79) and Eq. (5.2), we calculate that $(\mathcal{B} - \mathcal{A})/\mathcal{C}$ must exceed a critical value given by

$$\left(\frac{\mathcal{B} - \mathcal{A}}{\mathcal{C}}\right)_{\text{critical}} = \frac{5}{2} \frac{k_2}{Q} \left(\frac{R_s}{a}\right)^3 \frac{m_p}{m_s} \frac{1}{|H(p, e)|},$$

(5.95)

where $m_p$ is the mass of the primary and we have assumed that $\mathcal{C} \approx (2/5)m_s R_s^2$. Critical values of $(\mathcal{B} - \mathcal{A})/\mathcal{C}$ for a series of spin–orbit resonances in the Sun–Mercury and the Earth–Moon systems are listed in Table 5.1. The orbital eccentricities of the Moon and Mercury are, respectively, 0.0549 and 0.206, while $(\mathcal{B} - \mathcal{A})/\mathcal{C} \approx 2.28 \times 10^{-4}$ for the Moon (Yoder 1995) and it is reasonable to assume that $(\mathcal{B} - \mathcal{A})/\mathcal{C}$ for Mercury is comparable. Inspection of Table 5.1 shows that there is certainly no problem in understanding the stability of Mercury's

Table 5.1. Critical values of $(B - A)/C$ for Mercury and the Moon. We assume for Mercury that $k_2 \approx 0.1$ and $Q = 100$ and for the Moon that $k_2 \approx 0.03$ and $Q = 27$ (Yoder 1995).

| | Mercury | Moon |
|---|---|---|
| $p$ | $(B - A)/C$ | $(B - A)/C$ |
| $+3$ | $2 \times 10^{-8}$ | $7 \times 10^{-5}$ |
| $+5/2$ | $7 \times 10^{-9}$ | $7 \times 10^{-6}$ |
| $+2$ | $3 \times 10^{-9}$ | $8 \times 10^{-7}$ |
| $+3/2$ | $2 \times 10^{-9}$ | $10^{-7}$ |
| $+1$ | $10^{-9}$ | $2 \times 10^{-8}$ |

present spin–orbit state, or that of the Moon. However, if we allow that the spins of both of these bodies have been tidally braked and that their initial orbital periods were short, then we do need to understand not only how these bodies came to be trapped in their present spin–orbit states but also how they were able to evolve through many other strong resonances without trapping.

## 5.5 Capture into Resonance

That there is a problem with understanding capture into resonance can be seen by following the evolution of a satellite's spin rate on encounter with a spin–orbit resonance. We assume that, initially, $\dot{\theta} > pn$ and that tides act to brake the spin of the satellite. Thus, initially, $\dot{\gamma} > 0$ and the resonance is approached from above (Fig. 5.14). The equation of motion of the resonant argument, $\gamma$, in the presence of drag, is

$$C\ddot{\gamma} + \frac{3}{2}(B - A)n^2 H(p, e) \sin 2\gamma = \langle N_s \rangle. \tag{5.96}$$

Integrating with respect to time, we obtain the *energy integral*

$$\frac{1}{2}C\dot{\gamma}^2 - \frac{3}{4}(B - A)n^2 H(p, e) \cos 2\gamma = E, \tag{5.97}$$

where $E$, the total energy, is given by

$$E = \langle N_s \rangle \gamma + E_0 \tag{5.98}$$

and $E_0$ is a constant determined by the initial conditions. For the energy equation, Eq. (5.97), to have physical solutions ($\dot{\gamma}^2 > 0$), we must have

$$E \geq -\frac{3}{4}(B - A)n^2 |H(p, e)|. \tag{5.99}$$

If $E > +(3/4)(B - A)n^2 |H(p, e)|$, then the sign of $\dot{\gamma}$ does not change and the motion of $\gamma$ is one of *circulation*. However, given that $\langle N_s \rangle < 0$, tidal forces act to reduce $E$ and resonance encounter occurs when $\dot{\gamma}$ is reduced to zero.

(a)

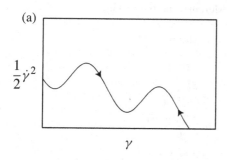

(b)

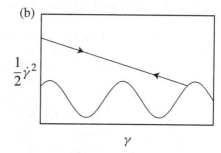

Fig. 5.14. (a) Variation of $(1/2)\dot{\gamma}^2$ with the resonant argument $\gamma$. (b) Separation of the variation of $(1/2)\dot{\gamma}^2$ with $\gamma$ into a potential term that varies sinusoidally with $\gamma$ and a term due to the tidal torque that decreases linearly with increasing $\gamma$. In the latter treatment, $(1/2)\dot{\gamma}^2$ is given by the *difference* of these two terms.

In Fig. 5.14a we show the total variation of $\dot{\gamma}^2$ with $\gamma$, while in Fig. 5.14b we separate the variation of $\dot{\gamma}^2$ into two components, one due to the potential term that varies sinusoidally with $\gamma$ and a drag term that decreases linearly with increasing $\gamma$ before resonance encounter and increases linearly with decreasing $\gamma$ after resonance encounter. In the latter representation, the variation of $\dot{\gamma}^2$ with $\gamma$ is given by the *difference* of the two terms. If $\langle N_s \rangle$ is constant, then the equation of motion is perfectly reversible: The sign of $\dot{\gamma}$ changes on resonance encounter, but the trajectory of the system in $(\gamma, \dot{\gamma})$ space after encounter duplicates that before encounter and capture into resonance cannot occur.

Goldreich & Peale (1968) explain this passage through resonance without capture using the following "pendulum" analogy. While the pendulum is circulating, a constant torque acts to brake its rotation. Thus, after a while, the pendulum will pass over its point of support for the last time (in the initial direction for which $\dot{\gamma} > 0$) and its rotation rate will be reduced to zero. The sense of rotation of the pendulum then reverses. However, both the magnitude and sign of the torque remain unchanged and thus the torque now acts to increase the rotation rate of the pendulum. Whatever energy was removed from the pendulum before it was braked is now resupplied and the pendulum swings back over its point of support and the magnitude of the rate of rotation (with $\dot{\gamma} < 0$) then continues to increase.

Given that the amplitude of the sinusoidal potential term in the energy equation is constant, it follows that for capture into resonance to occur, (a) $\langle N_s \rangle$ must somehow vary with $\dot{\gamma}$ and (b) during the last swing of the pendulum, during which the sign of $\dot{\gamma}$ reverses, the decrease in $E$ before resonance encounter ($\dot{\gamma} = 0$) must be greater than the increase in $E$ after resonance encounter, preventing the pendulum from swinging back over its point of support.

As we have already noted, the incorporation of energy dissipation into tidal theory is in many ways poorly developed. However, Goldreich & Peale (1966,

Table 5.2. *Physical and orbital quantities for Mercury and the Moon.*

| Quantity | Mercury | Moon |
|---|---|---|
| $k_2$ | 0.1 | 0.03 |
| $Q$ | 100 | 27 |
| $e$ | 0.206 | 0.0549 |
| $(\mathcal{B} - \mathcal{A})/\mathcal{C}$ | $10^{-4}$ | $2.28 \times 10^{-4}$ |
| $H(p, e)$ | 0.65 | 0.99 |
| $T_{\text{libration}}$ | 17 y | 2.88 y |
| $\gamma_0$ | 2 arcsec | 9.6 arcsec |
| $2\pi/\dot\theta_{\text{initial}}$ | 9 h | 9 h |
| $T_{\text{despin}}$ | $5 \times 10^9$ y | $3 \times 10^7$ y |
| $\langle N_s \rangle \pi / U$ | $10^{-4}$ | $6 \times 10^{-4}$ |

1968) give several plausible models of tidal dissipation that allow $\langle N_s \rangle$ to vary with $\dot\gamma$ and they have used these models to estimate the probability of capture into resonance. We will describe two of these models, both based on that due to Darwin (1908), but before doing so we need to estimate the magnitudes of the various terms in the energy equation. Tidal forces in the solar system are extremely weak and produce significant changes in some spin rates and orbital periods only because they act over billions of years. Using the values of $k_2$ and $Q$, etc., listed in Table 5.2, we calculate that if the initial spin periods ($2\pi/\dot\theta_{\text{initial}}$) of Mercury and the Moon were, say, 9 hours, then the times needed to brake these bodies were $5 \times 10^9$ and $3 \times 10^7$ years, respectively, implying that the Mercury–Sun spin–orbit resonance may be comparatively young. The lag angles, $\gamma_0$, given by Eq. (5.88) are only a few arcseconds and if we denote the amplitude of the potential term, $(3/4)(\mathcal{B} - \mathcal{A})n^2 H(p, e)\cos 2\gamma$, in the energy equation by $U$, then $\langle N_s \rangle \pi / U \ll 1$. We also note that the libration periods, $T_{\text{libration}} = 2\pi/\omega_0$ (using Eq. (5.85)), are greater than the orbital periods, but not by large factors.

Using Darwin's procedure for calculating the tidal torque on a satellite due to the tide raised on it by a planet, we expand the tidal potential in a Fourier time series and assume that each component raises an equilibrium tide on the satellite. The effects of tidal dissipation are then modelled by giving each component a phase shift such that the component either leads or lags the associated term in the potential. In our first model, we assume that the magnitudes, but not the signs, of the phase shifts are independent of the tidal frequencies. In this case, the mean tidal torque is given by

$$\langle N_s \rangle = -D \sum_{h=-\infty}^{\infty} [H(h, e)]^2 \, \text{sign}\,(\dot\theta - hn), \tag{5.100}$$

where $h$ is a half-integer, $D$ is a positive constant given by Eq. (5.3), and we have assumed that the spin axis of the satellite is normal to its orbital plane. For trapping in the $p$th resonance, $\dot{\theta} - pn = \dot{\gamma}$ and we can write

$$\langle N_s \rangle = -W - Z \, \mathrm{sign}\,(\dot{\gamma}), \tag{5.101}$$

where

$$W = +D \sum_{h \neq p} [H(h, e)]^2 \, \mathrm{sign}\,(p - h) \tag{5.102}$$

and

$$Z = +D \, [H(p, e)]^2 . \tag{5.103}$$

The rate of change of energy is given by

$$\frac{\mathrm{d}E}{\mathrm{d}t} = \langle N_s \rangle \dot{\gamma} . \tag{5.104}$$

Hence, the energy changes with $\gamma$ according to

$$\int \mathrm{d}E = \int_{\gamma_1}^{\gamma_2} \langle N_s \rangle \, \mathrm{d}\gamma . \tag{5.105}$$

Because $\pi \langle N_s \rangle \ll (3/4)(\mathcal{B} - \mathcal{A})n^2 H(p, e)$, the slopes of the straight lines in Fig. 5.15 are negligible and we can assume that in all integrations $\gamma_2 - \gamma_1 = \pi$.

Because the initial conditions are unspecified, the energy at the point $P$, where $\dot{\gamma}$ is first reduced to zero (Fig. 5.15), can lie anywhere in the range $E'_0$ to $E'_0 + \Delta E$, where

$$\Delta E = (W + Z)\pi . \tag{5.106}$$

On encounter with the $p$th resonance the sign of the $h = p$ term in Eq. (5.101) changes. Before resonance encounter, $\dot{\gamma} > 0$ and

$$\langle N_s \rangle = -W - Z, \tag{5.107}$$

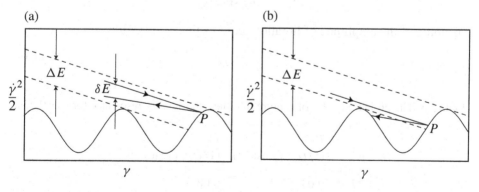

Fig. 5.15.  Capture into resonance depends on the magnitude and sign of $\delta E$ and the value of $\gamma$ at the location $P$ where $(1/2)\dot{\gamma}^2$ is first reduced to zero. The locations of $P$ show an encounter with a resonance (a) without capture and (b) with capture. The capture probability is given by $|\delta E / \Delta E|$ and capture is certain if $\delta E < 0$ and $|\delta E| > |\Delta E|$.

and after resonance encounter, $\dot{\gamma} < 0$ and

$$\langle N_s \rangle = -W + Z. \tag{5.108}$$

Thus, on resonance encounter, the magnitude of $\langle N_s \rangle$ decreases by $2Z$. It follows that the decrease in $E$ over one libration period is given by

$$\delta E = 2Z\pi \tag{5.109}$$

and that the probability $P_p$ of capture into the $p$th resonance is given by

$$P_p = \frac{2Z}{Z + W} = \frac{2\,[H(p, e)]^2}{[H(p, e)]^2 + \sum_{h \neq p} [H(p, e)]^2 \, \text{sign}\,(p - h)}. \tag{5.110}$$

In this case, the probability of capture does not depend on either $(\mathcal{B} - \mathcal{A})/\mathcal{C}$ or (assuming that the strength criterion, Eq. (5.87), is satisfied) on the magnitude of the tidal drag, that is, on $D$; It is determined by $p$ and $e$ alone.

Capture probabilities into the $p$th Mercury resonance calculated by Goldreich & Peale (1968) for $e = 0.2$ and $Q$ independent of frequency are shown in Table 5.3. The frequency-independent model gives a good explanation for both capture into the $p = +3/2$ resonance and avoidance of the other, higher-order resonances. However, it does not account for the damping of the amplitude of libration. For the case $H(p, e) > 0$, $\gamma$ librates about $\gamma_0 \approx 0$ with an amplitude $\gamma_{\max}$ obtained by solving Eq. (5.97) for $\dot{\gamma} = 0$. If we ignore the small displacement of $\gamma_0$ from zero, we get

$$\frac{1}{4} \omega_0^2 \cos 2\gamma_{\max} \approx -E/\mathcal{C} \tag{5.111}$$

and

$$\dot{\gamma}_{\max} \approx \frac{2\dot{E}}{\mathcal{C}\omega_0^2 \sin 2\gamma_{\max}}. \tag{5.112}$$

Damping of the amplitude of libration requires that $\dot{E} < 0$. However,

$$\dot{E} = \langle N_s \rangle \dot{\gamma} \tag{5.113}$$

Table 5.3. *The probability, $P_p$, of capture into the $p$th Mercury resonance for $e = 0.2$.*

| $p$ | $P_p$<br>$(1/Q \sim \text{Constant})$ | $P_p$<br>$(1/Q \sim \text{Frequency})$ |
|------|------|------|
| +5/2 | 0.03 | 0.007 |
| +2 | 0.15 | 0.016 |
| +3/2 | 0.73 | 0.067 |
| +1 | 1 | 0 |

and if $\langle N_s \rangle$ does not depend on $\dot{\gamma}$, then $\langle \dot{\gamma} \rangle = \langle \dot{E} \rangle = 0$ and the librations are undamped.

The second model considered by Goldreich & Peale (1968) allows that the tidal dissipation function is frequency dependent. In this case, they assumed that

$$\langle N_s \rangle = -K' \sum_{h=-\infty}^{\infty} [H(h, e)]^2 (\dot{\theta} - hn), \tag{5.114}$$

where $K'$ is a positive constant. Near a resonance $\dot{\theta} = pn + \dot{\gamma}$ and we can write

$$\langle N_s \rangle = -K \left( V + \frac{\dot{\gamma}}{n} \right), \tag{5.115}$$

where

$$K = K'n \sum_h [H(h, e)]^2 \tag{5.116}$$

and

$$V = \frac{\sum_h (p - h) [H(h, e)]^2}{\sum_h [H(h, e)]^2}. \tag{5.117}$$

The probability of capture into the $p$th resonance is given by

$$P_p = \frac{4(\omega_0/n)}{\pi V + 2(\omega_0/n)}. \tag{5.118}$$

For the frequency-dependent case, the probabilities depend on the low-amplitude libration frequency $(\omega_0/n)$ and the eccentricity $e$. The probabilities shown in Table 5.3 are low, but they are not negligible and, in this case, the dependence of $\langle N_s \rangle$ on $\dot{\gamma}$ does result in a term in the expression for $\dot{E}$ that is proportional to $\dot{\gamma}^2$. This term does not average to zero and thus the amplitude of libration is damped.

## 5.6 Forced Librations

For a satellite trapped in a spin–orbit resonance, for example, the synchronous, 1:1 resonance, analysis of the averaged equation of motion with the drag term included shows that any libration about the equilibrium configuration is damped to zero and the satellite then rotates uniformly with its long axis pointing exactly towards the planet on passage through pericentre. However, the full equation of motion contains short-period terms; consequently the rotational motion of the satellite has short-period librations about the equilibrium configuration.

Consider the rotation of a satellite in a reference frame centred on the satellite and rotating with the satellite's mean motion (see Fig. 5.16). The gravitational

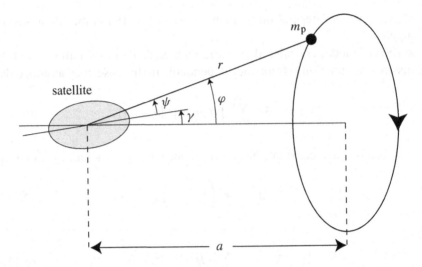

**Fig. 5.16.** Rotation of a satellite in a rotating reference frame centred on the satellite and rotating with the satellite's mean motion. The planet moves in a 2:1 ellipse about its guiding centre, which, in this frame, is stationary.

torque on the satellite due to the planet is determined by the angle $\psi$, which in this case is given by

$$\psi = \varphi - \gamma \approx 2e \sin nt - \gamma \tag{5.119}$$

to $\mathcal{O}(e)$, and the full equation of motion, Eq. (5.56), can be written

$$C\ddot{\gamma} - \frac{3}{2}(\mathcal{B} - \mathcal{A})n^2 \left(\frac{a}{r}\right)^3 \sin(4e \sin nt - 2\gamma) = 0. \tag{5.120}$$

For small deviations from the equilibrium configuration, given that the contribution due to the radial variation in distance $r$ is $(a/r)^3 \approx 1 + 3e \cos nt$, Eq. (5.120) reduces to

$$C\ddot{\gamma} = -\frac{3}{2}(\mathcal{B} - \mathcal{A})n^2(2\gamma - 4e \sin nt) \tag{5.121}$$

or

$$\ddot{\gamma} = -\omega_0^2 \gamma + 2\omega_0^2 e \sin nt, \tag{5.122}$$

where $\omega_0$ is the libration frequency (see Eq. (5.85)). Substituting $\gamma = \gamma_0 \sin nt$ into this equation and solving for $\gamma_0$, the amplitude of the *forced libration*, we obtain

$$\gamma = -\frac{2\omega_0^2 e}{n^2 - \omega_0^2} \sin nt. \tag{5.123}$$

Note that the variation in $r$ does not contribute to this equation. If the forcing frequency $n$ is less than the natural frequency $\omega_0$, then the librations are in phase with the force. However, if $n > \omega_0$, then the librations and the force are 180° out of phase.

For the Moon, the libration period is 2.86 y while the amplitude of forced libration is $\sim 15$ arcsec and is too small to detect. However, in the case of Phobos, the highly distorted, innermost satellite of Mars, Duxbury & Callahan (1982) using a control network of ninety-eight craters on forty-three *Viking Orbiter* Phobos pictures obtained a value of 0.8° ($\pm 0.2$°) for the amplitude of the forced libration. If we assume that the satellite is homogeneous, then the observed value of $\gamma_0$ requires that $(B - A)/C \approx 0.1$. However, if the satellite is homogeneous, then the observed shape of Phobos implies that $(B - A)/C \approx 0.2$. These two widely different values of $(B - A)/C$ could be reconciled if the satellite is not homogeneous but has a dense core surrounded by a deep, low-density regolith (Thomas et al. 1986).

## 5.7 Surface of Section

The method outlined in Sect. 5.4 suggests that, if $\dot{\theta} \approx pn$, we can analyse the motion in the vicinity of a resonance in terms of the slowly varying resonant argument $\gamma = \theta - pM$ using Eq. (5.73). In Fig. 5.17a, we show analytic solutions for the variation of $\dot{\theta}/n$ with $\theta$ using values of $\dot{\theta}$ obtained from the energy integral for the motion for $\gamma$, which is given by

$$\frac{1}{2}\dot{\gamma}^2 - \frac{1}{4}[\text{sign } H(p, e)]\omega_0^2 \cos 2\gamma = \frac{E_0}{C}, \tag{5.124}$$

where $E_0$ is a constant determined by the initial conditions. Analytic solutions are shown for the resonances corresponding to $p = +1/2, +1,$ and $+3/2$ and also for the nonresonant case associated with $p = 0$. For the cases corresponding to $p = +1, +3/2$, for which $H(p, e) > 0$, stable equilibrium points exist at $\theta = 0$ and $\pi$, whereas for the case corresponding to $p = +1/2$, for which $H(p, e) < 0$, a stable equilibrium point exists at $\theta = \pi/2$. In all cases, for $\dot{\gamma} = 0$ at the stable equilibrium points, we must have

$$E_0 = -\frac{1}{4}\omega_0^2 C \tag{5.125}$$

and $E_0$ is a local (remember that $\omega_0$ depends on $p$) minimum at these points. For values of $E_0$ in the range

$$-\frac{1}{4}\omega_0^2 C \le E_0 < +\frac{1}{4}\omega_0^2 C \tag{5.126}$$

$\gamma$ librates about the equilibrium position with an amplitude determined by $E_0$. The value of $E_0$ associated with the separatrix that separates regions of libration from those of circulation is given by

$$E_0 = +\frac{1}{4}\omega_0^2 C. \tag{5.127}$$

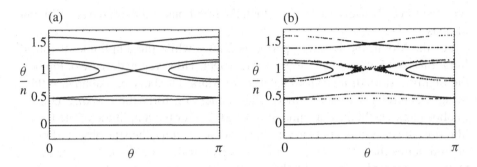

Fig. 5.17.   (a) Analytic variation of $\dot{\theta}/n$ with $\theta$ on the separatrices of the $p = +1/2$, $+1$, $+3/2$ resonances (and in the libration region of the $p = +1$ resonance) for $\alpha = \sqrt{3(\mathcal{B} - \mathcal{A})/\mathcal{C}} = 0.2$ and $e = 0.1$. (b) A surface of section plot for seven different trajectories in $(\theta, \dot{\theta}/n)$ phase space for the same values of $\alpha$ and $e$. Values of $\theta$ and $\dot{\theta}/n$ are plotted at every passage of the satellite through pericentre.

For motion on the separatrix, the maximum value of $|\dot{\gamma}|$ is given by

$$|\dot{\gamma}_{\mathrm{max}}| = \omega_0 \qquad\qquad (5.128)$$

and is a measure of the half-width of the resonance. In Fig. 5.17a we show the separatrices associated with various resonances for $e = 0.1$ and $\alpha = (3(\mathcal{B} - \mathcal{A})/\mathcal{C})^{1/2} = 0.2$. Note that, for small amplitudes of libration, the period of libration is $2\pi/\omega_0$, whereas the period of libration on the separatrix is infinite.

To investigate how good an approximation this is, we can integrate the full equation of motion, Eq. (5.56), numerically for fixed values of $e$ and the asphericity parameter $\alpha$. However, rather than trying to show the full variation of $\theta$ and $\dot{\theta}$ with time, we choose to produce a *surface of section* of the motion, which illustrates the fundamental properties of each solution. This involves solving the full equation of motion, Eq. (5.56), numerically and plotting the values of $\theta$ and $\dot{\theta}/n$ each time the satellite passes through pericentre. The choice of pericentre is arbitrary, but it is a natural choice in this case because at pericentre $M = 0$ and thus, in this case, the surface of section is equivalent to a plot of $\gamma$ against $\dot{\gamma}/n$.

In Fig. 5.17b we show surface of section plots for the range of initial conditions $(\theta, \dot{\theta}/n)$ corresponding to the values of $E_0/\mathcal{C}$ used to plot the analytic curves shown in Fig. 5.17a. In both cases, we use $e = 0.1$ and $\alpha = 0.2$. These large values of $e$ and $\alpha$ help to illustrate the sizes of the resonant regions. However, the resonant half-widths are large. For example, in the case of the $p = +1$ resonance, we have $\omega_0 = 0.2$ and it is questionable whether analytic solutions with large amplitudes of libration satisfy our requirement that they are in the "vicinity" of the resonance and that $\dot{\theta} \approx pn$. The surface-of-section method allows us to test the limitations of our analytical theory.

Numerical experiments show that in the heart of each resonant region, close to a stable equilibrium point where the amplitude of libration of $\theta$ is small, the surface of section trajectories in $(\theta, \dot{\theta}/n)$ space mimic the analytic curves with the exception that there is a displacement associated with the forced libration. For example, for the $p = +1$ resonance there is a contribution to $\dot{\gamma}/n$ from the forced libration of $-0.0083$ (see Eq. (5.115)). This small negative displacement is just about visible in Fig. 5.17b, as are the small positive displacements associated with the $p = +1/2$ and $p = +3/2$ resonances (the displacements are more obvious in Fig. 5.18b where we use even larger values of $e$ and $\alpha$). For small amplitudes of libration, successive points on the surface-of-section trajectory follow each other in a regular sequence, completing a closed path in a time given by the libration period. However, it is clear from Fig. 5.17b that motion at the separatrix has an additional property. This is particularly evident around the strong $p = +1$ resonance. Motion on the separatrix in our full integration is *chaotic* in the sense that although the evolution of $\theta$ follows deterministic laws, it is nevertheless unpredictable and wanders within some finite limits. This accounts for the "fuzzy" appearance of the separatrix trajectory at the $p = +1$ and the $p = +3/2$ resonances.

The trajectories shown in Fig. 5.17b are in very good agreement with the analytic curves shown in Fig. 5.17a, despite the fact that our analysis assumes that we can investigate each resonance in isolation and that the averaged effect of all the other resonances and forced librations is zero. However, it is obvious that as the half-widths of the resonances increase, we must reach a point where this assumption breaks down. In Fig. 5.18a, we show analytic solutions for the motion of $\theta$ for the case $e = 0.15$ and $\alpha = 0.3$. In this case, the curves associated with the $p = +1$ and the $p = +3/2$ separatrices overlap close to $\theta = 0$ and $\theta = \pi$, suggesting simultaneous libration in two spin–orbit states,

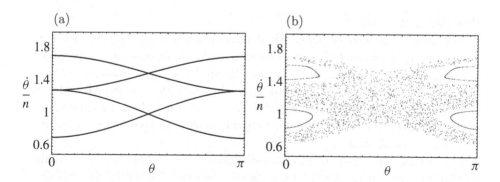

Fig. 5.18. (a) Analytic variation of $\dot{\theta}/n$ with $\theta$ on the separatrices and in the libration regions of the $p = +1, +3/2$ resonances for $\alpha = \sqrt{3(\mathcal{B} - \mathcal{A})/\mathcal{C}} = 0.3$ and $e = 0.15$. (b) A surface of section plot for three different trajectories in $(\theta, \dot{\theta}/n)$ phase space for the same values of $\alpha$ and $e$. Values of $\theta$ and $\dot{\theta}$ are plotted at every passage of the satellite through pericentre.

which is impossible. Chirikov's resonance overlap criterion states that when the sum of the two unperturbed half-widths equals the separation of the resonance centres, large-scale chaos ensues (Chirikov 1979). In the spin–orbit problem, the resonance overlap criterion for the two strongest resonances, the $p = +1$ and the $p = +3/2$, can be written

$$\omega_{0(p=+1)} + \omega_{0(p=+3/2)} \geq \frac{n}{2}, \tag{5.129}$$

which is satisfied if

$$\alpha \geq \frac{1}{2 + \sqrt{14e}} \tag{5.130}$$

(Wisdom et al. 1984).

Chirikov (1979) also estimated that the half-width of the chaotic separatrix, expressed in terms of the chaotic variations of the energy integral, is given by

$$\frac{\Delta E_0}{E_0} \approx 4\pi\varepsilon\lambda^3 \exp(-\pi\lambda/2), \tag{5.131}$$

where $\varepsilon$ is the ratio of the coefficient of the nearest perturbing high-frequency term to the coefficient of the perturbed term and $\lambda \, (= \Delta\Omega/\omega_0)$ is the ratio of the frequency difference, $\Delta\Omega$, between the resonant term and the nearest nonresonant term to the frequency of small-amplitude librations, $\omega_0$. For the synchronous spin–orbit state perturbed by the $p = +3/2$ term, $\varepsilon = H(+3/2, e)/H(+1, e) = 7e/2$ and $\lambda = \Delta\Omega/\omega_0 = n/n\alpha = 1/\alpha$. Hence

$$\frac{\Delta E_0}{E_0} \approx \frac{14\pi e}{\alpha^3} \exp(-\pi/2\alpha) \tag{5.132}$$

(Wisdom et al. 1984).

The width of the chaotic separatrix depends linearly on the eccentricity $e$ but exponentially on the asphericity parameter $\alpha$, and a small increase in $\alpha$ produces a dramatic increase in the width of the chaotic band. The quantity $\Delta E_0/E_0$ is extremely small for near-spherical bodies like Mercury and the Moon but is of the order of unity for irregular bodies like Hyperion and can be appreciable even for bodies like Mimas that have regular shapes but significant permanent distortions (see Table 5.4).

The chaotic zones associated with the separatrices may have had a role in the evolution of the spin rates and orientations of some satellites (Wisdom 1987a,b). The reduction in the energy integral on resonance encounter that must occur if permanent capture into resonance is to be achieved depends on the particular spin–orbit resonance encountered and on the nature of the tidal dissipation mechanism (see Sect. 5.5). However, capture into resonance will occur as described in Sect. 5.5 only if the energy reduction due to tidal forces that occurs in one

Table 5.4. Resonance half-widths.

| Body | $(\mathcal{B} - \mathcal{A})/\mathcal{C}$ | $e$ | $p$ | $H(p, e)$ | $\omega_0/n$ | $\Delta E_0/E_0$ | $\Delta E_{\text{tides}}/E_0$ |
|------|------|------|------|------|------|------|------|
| Mercury | 0.0001 | 0.2 | $+3/2$ | 0.64 | 0.014 | | |
| | | | $+1$ | 0.90 | 0.016 | $10^{-33}$ | $10^{-6}$ |
| | | | $+1/2$ | $-0.10$ | 0.006 | | |
| Moon | 0.000228 | 0.055 | $+3/2$ | 0.19 | 0.011 | | |
| | | | $+1$ | 0.99 | 0.026 | $10^{-21}$ | $10^{-5}$ |
| | | | $+1/2$ | $-0.03$ | 0.005 | | |
| Hyperion | 0.26 | 0.1 | $+3/2$ | 0.34 | 0.515 | | |
| | | | $+1$ | 0.97 | 0.870 | 1.1 | $10^{-9}$ |
| | | | $+1/2$ | $-0.05$ | 0.198 | | |
| Mimas | $0.06^{\dagger}$ | 0.02 | $+3/2$ | 0.07 | 0.112 | | |
| | | | $+1$ | 1.00 | 0.424 | 0.28 | $10^{-6}$ |
| | | | $+1/2$ | $-0.01$ | 0.042 | | |

$^{\dagger}$The value of $(\mathcal{B} - \mathcal{A})/\mathcal{C}$ for Mimas is taken from Dermott & Thomas (1988).

librational cycle is significantly greater than the width of the chaotic separatrix. In this regard, it is useful to compare the energy change

$$\delta E_{\text{tides}} \sim \pi \langle N_{\text{s}} \rangle \sim \pi \frac{3}{2} \frac{k_2}{Q_{\text{s}}} \frac{n^4}{\mathcal{G}} R_{\text{s}}^5 \qquad (5.133)$$

with the width of the chaotic separatrix, $2\Delta E_0$. Estimates of these quantities for the synchronous case are shown in Table 5.4. Note the marked contrast between the widths of the chaotic zones associated with Mercury and the Moon and those associated with Hyperion and Mimas.

Wisdom (1987a,b) has shown that the synchronous states of all small irregularly shaped satellites in the solar system, even some such as Deimos ($\alpha = 0.8$, $e = 0.0005$) with near-circular orbits, have significant chaotic zones. Furthermore, analysis by Wisdom et al. (1984) and Wisdom (1987a,b) of the stability of the spin axis orientation (we assume in this chapter that the spin axis is always perpendicular to the orbital plane) has shown that the chaotic zone of the synchronous state is attitude unstable. While the satellite is in the synchronous chaotic zone, the slightest deviation of the spin axis from the orbit normal grows exponentially on a very short timescale and the satellite tumbles chaotically in three-dimensional space. All synchronously rotating satellites in the solar system with irregular shapes may have spent a period of time comparable to the tidal despinning time tumbling chaotically before the spin evolved out of the chaotic zone. Wisdom (1987a,b) has speculated on some of the possible geophysical consequences of a prolonged period of tumbling. Of particular interest is the fact that the tumbling state would have involved a significant enhancement in the rate of tidal dissipation as compared to that in the regular synchronous state. This is because the amplitude of the variation of the tidal distortion in the tumbling state is a factor $1/e$ greater than that in the regular synchronous state (see Sect. 4.10) and the rate of tidal heating of the satellite is increased by a factor $1/e^2$ while the eccentricity damping timescale is decreased by the same factor. Wisdom (1987a,b) has given an interesting discussion of the possible role of chaotic tumbling in the heating of Miranda and the damping of the orbital eccentricity of Deimos.

## Exercise Questions

**5.1**     Where necessary use the data in Appendix A to answer the following questions. (a) Estimate $a - c$ for Jupiter's satellite Io, due to the combined effects of rotation and tidal distortion. (Give your answer in kilometres, and in fractions of the mean radius; assume uniform density.) (b) Repeat part (a) for the Earth's moon, in its current orbit. (c) At an early stage in the history of the Earth–Moon system, the Moon was probably at a distance of only 10 Earth radii, while the Earth's rotation period was about 10 h. Estimate the oblateness, $\epsilon$, and $J_2$ of the Earth at this time, as well as $a - c$ for the Moon. (d) Mercury rotates with a period of 56 days, or 2/3 of its orbital period of 88 days. Calculate its hydrostatic oblateness, assuming a Love number $k_2 = 3/2$. Do you think $\epsilon$ is detectable by current techniques of measurement? Is there any other practicable way to obtain an estimate of the planet's moment of inertia? (e) Both Pluto and its moon Charon are believed to be in states of synchronous rotation. Estimate $a - c$ for each object, assuming that they have equal densities. (You will have to estimate the density first, from Kepler's third law.)

**5.2**     The current semi-major axis of Charon's orbit is 19,636 km, while the orbital period (and Pluto's spin period) is 6.3872 days. The radius of Pluto is $\sim 1{,}137$ km and that of Charon approximately 600 km. Charon is assumed to be synchronously rotating also, so that the system has reached a stable end point of tidal evolution. (a) Calculate the total mass of the system and the average density of the two bodies. What measurements would be necessary to ascertain the individual masses and densities of Pluto and Charon? (b) Show that the time taken by tides in Pluto to synchronize its spin is

$$\tau_P = \frac{2\gamma_P Q_P}{3\,(k_2)_P} \frac{m_P(m_P + m_C)}{m_C^2} \left(\frac{a}{R_P}\right)^3 \frac{\omega_P}{n^2},$$

where $\gamma = C/mR^2$, the initial spin period of Pluto is $2\pi/\omega_P$, and the subscripts P and C refer to Pluto and Charon respectively. Estimate $\tau_P$, assuming reasonable values for the unknown physical parameters $Q$, $\gamma$, and $k_2$, and $2\pi/\omega_P = 10$ h. (c) Write down (by inspection) the analogous expression for Charon's despinning timescale, $\tau_C$, and show that $\tau_C/\tau_P \approx (R_C/R_P)^6 = 1/50$. (d) What was the initial semi-major axis of the system and the corresponding orbital period if both Pluto and Charon started with spin periods of 10 h?

**5.3** Show from first principles that the approximate potential, $V$, experienced at a point $P$ at a distance $r$ from the centre of mass, $O$, of a body of mass $m$ is given by MacCullagh's formula,

$$V = -\frac{\mathcal{G}m}{r} - \frac{\mathcal{G}(\mathcal{A} + \mathcal{B} + \mathcal{C} - 3I)}{2r^3},$$

where $\mathcal{A}$, $\mathcal{B}$, and $\mathcal{C}$ are the principal moments of inertia of the body defined with respect to $O$, and $I$ is the moment of inertia of the body along the line from $O$ to $P$.

**5.4** Consider the rotation of a satellite as it moves in an equatorial orbit around a planet. Let the satellite's orbital radius, semi-major axis, eccentricity, mean anomaly, and true anomaly be $r$, $a$, $e$, $M$, and $f$, respectively. An analytical approach to the study of spin–orbit resonance, where the spin rate is approximately $p$ times the mean motion, requires averaging the expression $(a/r)^3 \sin(2\gamma + 2pM - 2f)$ over one orbital period (see Sect. 5.4); here $\gamma$ is an orientation angle. Use the fourth-order expansions of $(a/r)^3$, $\cos f$, and $\sin f$ given in Sect. 2.5 to derive expressions (to fourth-order in the eccentricity $e$) for the quantities $S_1$, $S_2$, $S_3$, and $S_4$ in the equation

$$\left(\frac{a}{r}\right)^3 \sin(2\gamma + 2pM - 2f) = (S_1 + S_2) \sin 2\gamma + (S_3 - S_4) \cos 2\gamma.$$

By finding the time-averaged values, $\langle S_i \rangle = (1/2\pi) \int_0^{2\pi} S_i \, dM$, show that

$$\left\langle \left(\frac{a}{r}\right)^3 \sin(2\gamma + 2pM - 2f) \right\rangle = H(p, e) \sin 2\gamma$$

and verify the expressions for $H(p, e)$ given in Eqs. (5.74)–(5.82). Given that $H(p, e)$ is related to the strength of a spin–orbit resonance, find the smallest value of $e$ for which the $p = 1$ resonance is weaker than the $p = 3/2$ resonance.

**5.5** The long axis of a small, ellipsoidal satellite makes an angle $\theta$ with the pericentre direction of its orbit. The differential equation describing the evolution of $\theta$ is

$$\mathcal{C}\ddot{\theta} - \frac{3}{2}(\mathcal{B} - \mathcal{A})\frac{\mathcal{G}m_p}{r^3} \sin 2\psi = 0,$$

where $\mathcal{A}$, $\mathcal{B}$, and $\mathcal{C}$ are the (constant) moments of inertia about the different axes, $m_p$ is the mass of the planet, $r$ is the radial distance of the satellite from the planet, and $\psi = f - \theta$ where $f$ is the true anomaly of the satellite in its orbit. Theoretically, the maximum variation in $\theta$ for a satellite trapped in the 1:1 spin–orbit resonance is $\pm 90°$ (see Fig. 5.17a). However, the effect of other resonances reduces this amplitude. Assuming that the satellite is on a fixed orbit such that $r$

can be found at any time by solving Kepler's equation, write a computer program to solve this differential equation. Taking $(3(\mathcal{B} - \mathcal{A})/\mathcal{C})^{1/2} = 0.25$ and $e = 0.2$, start the satellite at its pericentre with $\theta = 0$. Use different initial values of $\dot{\theta}/n$ (where $n$ is the satellite's mean motion) to determine the maximum variation of $\theta$ in the 1:1 spin–orbit resonance.

**5.6**     Use the analytical theory given in Sect. 5.7 to derive an expression for the minimum and maximum values of $\dot{\theta}/n$ on a surface of section for the retrograde rotation of a satellite trapped in the $p = -1/2$ and the $p = -1$ spin–orbit resonances. If the eccentricity $e$ of the satellite's orbit is fixed, derive an expression for the value of $\alpha = (3(\mathcal{B} - \mathcal{A})/\mathcal{C})^{1/2}$ at which you expect the libration "islands" from these two resonances to intersect. Similarly, if $\alpha$ is fixed, derive an expression for the value of $e$ at which intersection occurs. Use the program written for Question 5.5 to produce a surface of section for $\alpha = 0.25$ and $e = 0.15$ showing clear examples of libration, circulation, and motion near the separatrix for the $p = -1/2$ and $p = -1$ spin–orbit resonances.

# 6

# The Disturbing Function

O polished perturbation!

William Shakespeare, *Henry IV (2), IV, v*

## 6.1 Introduction

In Chapter 3 we approached the three-body problem from the point of view of the location and stability of equilibrium points in the restricted problem. However, we made no attempt to tackle the more general problem of the motion of a third body under the gravitational effects of the two other bodies for arbitrary initial conditions. This problem is nonintegrable, but we can make some progress by analysing the accelerations experienced by the three bodies. If their motions are dominated by a central or primary body, then the orbits of the secondary bodies are conic sections with small deviations due to their mutual gravitational perturbations. In this chapter, we show how these deviations can be calculated by defining and analysing the *disturbing function*.

Consider a mass $m_i$ orbiting a primary of mass $m_c$ in an elliptical path. As we have seen in Chapter 2, this problem is integrable and the orbital elements $a_i$, $e_i$, $I_i$, $\varpi_i$, and $\Omega_i$ of the mass $m_i$ are constant, provided the gravitational effect of the central body can be treated as arising from a point mass. If we now introduce a third mass, $m_j$, then the mutual gravitational force between the masses $m_i$ and $m_j$ results in accelerations in addition to the standard two-body accelerations due to $m_c$ (see Fig. 6.1). These additional accelerations of the secondary masses *relative* to the primary can be obtained from the gradient of the perturbing potential, also called the *disturbing function*.

This chapter is concerned with a mathematical analysis of the properties of a Fourier series expansion of the disturbing function. We show how particular problems in solar system dynamics can be tackled by isolating the appropriate

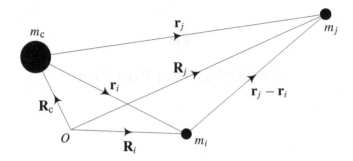

Fig. 6.1. The position vectors $\mathbf{r}_i$ and $\mathbf{r}_j$, of two masses $m_i$ and $m_j$, with respect to the central mass $m_c$. The three masses have position vectors $\mathbf{R}$, $\mathbf{R}'$, and $\mathbf{R}_c$ with respect to an arbitrary, fixed origin $O$.

terms in the expansion of the disturbing function and by assuming that the time-averaged contributions to the equations of motion of all the other terms are negligible. An understanding of the properties of the disturbing function is the key to understanding the dynamics of resonance and other long-period motions in the solar system.

## 6.2 The Disturbing Function

Let the position vectors with respect to a fixed origin $O$ of the three bodies of masses $m_c$, $m_i$, and $m_j$ be $\mathbf{R}_c$, $\mathbf{R}_i$, and $\mathbf{R}_j$ respectively. Let $\mathbf{r}_i$ and $\mathbf{r}_j$ denote the position vectors of the secondary masses $m_i$ and $m_j$ relative to the primary, where

$$|\mathbf{r}_i| = r_i = \left(x_i^2 + y_i^2 + z_i^2\right)^{1/2}, \qquad |\mathbf{r}_j| = r_j = \left(x_j^2 + y_j^2 + z_j^2\right)^{1/2}, \qquad (6.1)$$

and

$$|\mathbf{r}_j - \mathbf{r}_i| = \left[(x_j - x_i)^2 + (y_j - y_i)^2 + (z_j - z_i)^2\right]^{1/2} \qquad (6.2)$$

and the primary is the origin of the coordinate system (see Fig. 6.1).

From Newton's laws of motion and the law of gravitation we obtain the equations of motion of the three masses in the inertial reference frame:

$$m_c \ddot{\mathbf{R}}_c = \mathcal{G} m_c m_i \frac{\mathbf{r}_i}{r_i^3} + \mathcal{G} m_c m_j \frac{\mathbf{r}_j}{r_j^3}, \qquad (6.3)$$

$$m_i \ddot{\mathbf{R}}_i = \mathcal{G} m_i m_j \frac{(\mathbf{r}_j - \mathbf{r}_i)}{|\mathbf{r}_j - \mathbf{r}_i|^3} - \mathcal{G} m_i m_c \frac{\mathbf{r}_i}{r_i^3}, \qquad (6.4)$$

$$m_j \ddot{\mathbf{R}}_j = \mathcal{G} m_j m_i \frac{(\mathbf{r}_i - \mathbf{r}_j)}{|\mathbf{r}_i - \mathbf{r}_j|^3} - \mathcal{G} m_j m_c \frac{\mathbf{r}_j}{r_j^3}. \qquad (6.5)$$

The accelerations of the secondaries relative to the primary are given by

$$\ddot{\mathbf{r}}_i = \ddot{\mathbf{R}}_i - \ddot{\mathbf{R}}_c, \tag{6.6}$$

$$\ddot{\mathbf{r}}_j = \ddot{\mathbf{R}}_j - \ddot{\mathbf{R}}_c. \tag{6.7}$$

Substituting the expressions for $\ddot{\mathbf{R}}_c$, $\ddot{\mathbf{R}}_i$, and $\ddot{\mathbf{R}}_j$ from Eqs. (6.3)–(6.5) we get

$$\ddot{\mathbf{r}}_i + \mathcal{G}\left(m_c + m_i\right)\frac{\mathbf{r}_i}{r_i^3} = \mathcal{G}m_j\left(\frac{\mathbf{r}_j - \mathbf{r}_i}{\left|\mathbf{r}_j - \mathbf{r}_i\right|^3} - \frac{\mathbf{r}_j}{r_j^3}\right), \tag{6.8}$$

$$\ddot{\mathbf{r}}_j + \mathcal{G}\left(m_c + m_j\right)\frac{\mathbf{r}_j}{r_j^3} = \mathcal{G}m_i\left(\frac{\mathbf{r}_i - \mathbf{r}_j}{\left|\mathbf{r}_i - \mathbf{r}_j\right|^3} - \frac{\mathbf{r}_i}{r_i^3}\right). \tag{6.9}$$

These relative accelerations can be written as gradients of scalar functions, that is, we can write

$$\ddot{\mathbf{r}}_i = \nabla_i\left(U_i + \mathcal{R}_i\right) = \left(\hat{\mathbf{i}}\frac{\partial}{\partial x_i} + \hat{\mathbf{j}}\frac{\partial}{\partial y_i} + \hat{\mathbf{k}}\frac{\partial}{\partial z_i}\right)\left(U_i + \mathcal{R}_i\right) \tag{6.10}$$

and

$$\ddot{\mathbf{r}}_j = \nabla_j\left(U_j + \mathcal{R}_j\right) = \left(\hat{\mathbf{i}}\frac{\partial}{\partial x_j} + \hat{\mathbf{j}}\frac{\partial}{\partial y_j} + \hat{\mathbf{k}}\frac{\partial}{\partial z_j}\right)\left(U_j + \mathcal{R}_j\right), \tag{6.11}$$

where

$$U_i = \mathcal{G}\frac{\left(m_c + m_i\right)}{r_i} \quad \text{and} \quad U_j = \mathcal{G}\frac{\left(m_c + m_j\right)}{r_j} \tag{6.12}$$

are the central, or two-body, parts of the total potential. The subscript $i$ or $j$ is included in the $\nabla$ operator to emphasise that the gradient is with respect to the coordinates of the mass $m_i$ or $m_j$. The $\mathcal{R}$ term in the potential is the *disturbing function*, which represents the potential that arises from the other secondary mass. Since $\mathbf{r}_i$ is not a function of $x_j$, $y_j$, and $z_j$, and $\mathbf{r}_j$ is not a function of $x_i$, $y_i$, and $z_i$, we can write

$$\mathcal{R}_i = \frac{\mathcal{G}m_j}{\left|\mathbf{r}_j - \mathbf{r}_i\right|} - \mathcal{G}m_j\frac{\mathbf{r}_i \cdot \mathbf{r}_j}{r_j^3}, \tag{6.13}$$

$$\mathcal{R}_j = \frac{\mathcal{G}m_i}{\left|\mathbf{r}_i - \mathbf{r}_j\right|} - \mathcal{G}m_i\frac{\mathbf{r}_i \cdot \mathbf{r}_j}{r_i^3}. \tag{6.14}$$

The leading terms in these expressions are called the *direct terms* while the other terms that arise from the choice of the origin of the coordinate system are called the *indirect terms*. If the origin of the coordinate system was at the centre of mass, then these indirect terms would not appear.

The above analysis can be extended to any number of bodies. In addition, the accelerations associated with the disturbing function can arise from any source and not just from point-mass gravitational forces. They could, for example, arise from a potential associated with the oblateness of the central mass (see Sect. 6.11). However, in what follows in this chapter we are mostly concerned with the particular case of two point-mass secondaries of masses $m$ and $m'$ and position vectors $\mathbf{r}$ and $\mathbf{r}'$ relative to the central mass, where $r < r'$ always. With this notation, the equation of motion of the inner secondary is

$$\ddot{\mathbf{r}} + \mathcal{G}\left(m_c + m\right)\frac{\mathbf{r}}{r^3} = \mathcal{G}m'\left(\frac{\mathbf{r}' - \mathbf{r}}{|\mathbf{r}' - \mathbf{r}|^3} - \frac{\mathbf{r}'}{r'^3}\right) \tag{6.15}$$

and its disturbing function can be written

$$\mathcal{R} = \frac{\mu'}{|\mathbf{r}' - \mathbf{r}|} - \mu'\frac{\mathbf{r} \cdot \mathbf{r}'}{r'^3}, \tag{6.16}$$

where $\mu' = \mathcal{G}m'$ and the associated reference orbit has osculating elements $n^2 a^3 = \mathcal{G}\left(m_c + m\right)$. Similar equations can be written for the outer secondary giving

$$\ddot{\mathbf{r}}' + \mathcal{G}\left(m_c + m'\right)\frac{\mathbf{r}'}{r'^3} = \mathcal{G}m\left(\frac{\mathbf{r} - \mathbf{r}'}{|\mathbf{r} - \mathbf{r}'|^3} - \frac{\mathbf{r}}{r^3}\right). \tag{6.17}$$

The corresponding disturbing function for the outer secondary is then

$$\mathcal{R}' = \frac{\mu}{|\mathbf{r} - \mathbf{r}'|} - \mu\frac{\mathbf{r} \cdot \mathbf{r}'}{r^3}, \tag{6.18}$$

where $\mu = \mathcal{G}m$ and the associated reference orbit has osculating elements $n'^2 a'^3 = \mathcal{G}\left(m_c + m'\right)$.

Although this is the most straightforward way to derive expressions for $\mathcal{R}$ and $\mathcal{R}'$, it is worth pointing out that this procedure and the resulting expressions are not unique. For example, it is possible to add an additional term, $\mathcal{G}m\mathbf{r}'/r'^3$, to each side of the equation of motion for the mass $m'$, Eq. (6.17), resulting in an additional term $-\mu/r'$ in the expression for $\mathcal{R}'$; however, this requires that the associated reference orbit for $m'$ has osculating elements $n'^2 a'^3 = \mathcal{G}\left(m_c + m + m'\right)$.

## 6.3 Expansion Using Legendre Polynomials

Consider the configuration shown in Fig. 6.2 where $\mathbf{r}$ and $\mathbf{r}'$ denote the position vectors of the masses $m$ and $m'$ respectively. Let $\psi$ denote the angle between the two position vectors. From the cosine rule we have

$$\left|\mathbf{r}' - \mathbf{r}\right|^2 = r^2 + r'^2 - 2rr'\cos\psi, \tag{6.19}$$

or, alternatively,

$$\frac{1}{|\mathbf{r}' - \mathbf{r}|} = \frac{1}{r'}\left[1 - 2\frac{r}{r'}\cos\psi + \left(\frac{r}{r'}\right)^2\right]^{-\frac{1}{2}}. \tag{6.20}$$

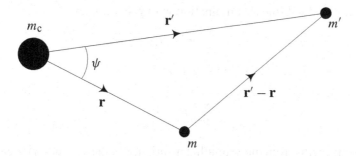

Fig. 6.2.  The position vectors $\mathbf{r}$ and $\mathbf{r}'$ of two masses $m$ and $m'$, with respect to a central mass $m_c$. The angle between the position vectors is $\psi$.

This can be expanded in Legendre polynomials to give

$$\frac{1}{|\mathbf{r}' - \mathbf{r}|} = \frac{1}{r'} \sum_{l=0}^{\infty} \left(\frac{r}{r'}\right)^l P_l(\cos\psi), \qquad (6.21)$$

where $P_0(\cos\psi) = 1$, $P_1(\cos\psi) = \cos\psi$, $P_2(\cos\psi) = \frac{1}{2}(3\cos^2\psi - 1)$, etc. (see Sect. 4.2).

Since $\mathbf{r} \cdot \mathbf{r}' = rr'\cos\psi = rr'P_1(\cos\psi)$, the disturbing function for the inner secondary can be written

$$\mathcal{R} = \frac{\mu'}{r'} \sum_{l=2}^{\infty} \left(\frac{r}{r'}\right)^l P_l(\cos\psi), \qquad (6.22)$$

where the $P_0(\cos\psi)$ term has been omitted because it does not depend on $r$ and, ultimately, we are only interested in the gradient of $\mathcal{R}$ with respect to the coordinates of the inner secondary.  Similarly, the disturbing function for the outer secondary can be written

$$\mathcal{R}' = \frac{\mu}{r'} \sum_{l=2}^{\infty} \left(\frac{r}{r'}\right)^l P_l(\cos\psi) + \mu \frac{r}{r'^2} \cos\psi - \mu \frac{r'}{r^2} \cos\psi. \qquad (6.23)$$

Thus, apart from two extra terms (that are actually unimportant for the applications discussed in the book), the expressions for $\mathcal{R}$ and $\mathcal{R}'$ are very similar.

This chapter is concerned with the series expansion of the disturbing functions $\mathcal{R}$ and $\mathcal{R}'$ in terms of the orbital elements (as opposed to the Cartesian coordinates) of $m$ and $m'$. We use the standard orbital elements $a$, $e$, $I$, $\varpi$, $\Omega$, and $\lambda$ to denote the semi-major axis, eccentricity, inclination, longitude of pericentre, longitude of ascending node, and mean longitude, respectively, of the mass $m$, with similar primed quantities for the mass $m'$. We show that the expansion of $\mathcal{R}$ has the form

$$\mathcal{R} = \mu' \sum S(a, a', e, e', I, I') \cos\varphi. \qquad (6.24)$$

Here $\varphi$ is a permitted linear combination with general form

$$\varphi = j_1\lambda' + j_2\lambda + j_3\varpi' + j_4\varpi + j_5\Omega' + j_6\Omega, \tag{6.25}$$

where the $j_i$ $(i = 1, 2, \ldots, 6)$ are integers and

$$\sum_{i=1}^{6} j_i = 0. \tag{6.26}$$

This property stems from the azimuthal invariance of the primary's potential. By knowing the explicit form of the function $S$ and the permissible combinations of the angles in $\varphi$, we can identify those terms that make the dominant contributions to the equations of motion and, conversely, those that can be neglected.

To illustrate the nature of this expansion let us consider the special case where the orbits of the two masses $m$ and $m'$ lie in the same plane and we can ignore any terms arising from the inclination. In this case we can write the angle $\psi$ as the difference of the true longitudes,

$$\psi = (f' + \varpi') - (f + \varpi), \tag{6.27}$$

where $f$ and $f'$ denote the true anomalies of $m$ and $m'$. Hence,

$$\cos\psi = (\cos f' \cos\varpi' - \sin f' \sin\varpi')(\cos f \cos\varpi - \sin f \sin\varpi)$$
$$+ (\sin f' \cos\varpi' + \cos f' \sin\varpi')(\sin f \cos\varpi + \cos f \sin\varpi). \tag{6.28}$$

We have already given series expansions for $\cos f$ and $\sin f$ in Sect. 2.5 and we can find similar series for $\cos f'$ and $\sin f'$ by substituting $M'$ for $M$ and $e'$ for $e$. Taking these expansions to second degree in $e$ and $e'$ we find

$$\cos\psi = \left(1 - e^2 - e'^2\right)\cos[M - M' + \varpi - \varpi']$$
$$- e\cos[M' - \varpi + \varpi'] - e'\cos[M + \varpi - \varpi']$$
$$+ e\cos[2M - M' + \varpi - \varpi'] + e'\cos[M - 2M' + \varpi - \varpi']$$
$$- \frac{1}{8}e^2\cos[M + M' - \varpi + \varpi'] - \frac{1}{8}e'^2\cos[M + M' + \varpi - \varpi']$$
$$+ \frac{9}{8}e^2\cos[3M - M' + \varpi - \varpi'] + \frac{9}{8}e'^2\cos[M - 3M' + \varpi - \varpi']$$
$$+ ee'\cos[\varpi - \varpi'] + ee'\cos[2M - 2M' + \varpi - \varpi']$$
$$- ee'\cos[2M + \varpi - \varpi'] - ee'\cos[2M' - \varpi + \varpi']. \tag{6.29}$$

Even at this stage some properties of the expression for $\cos\psi$ are evident. It is clear that the degree of the eccentricity term associated with each cosine argument is at least the modulus of the sum of the coefficients of the mean anomalies in the argument. Another property shows up if we express the angles in terms of the mean longitudes rather than the mean anomalies using the substitutions

$M = \lambda - \varpi$ and $M' = \lambda' - \varpi'$. This gives

$$
\begin{aligned}
\cos \psi = {} & \left(1 - e^2 - e'^2\right) \cos[\lambda - \lambda'] - e \cos[\lambda' - \varpi] - e' \cos[\lambda - \varpi'] \\
& + e \cos[2\lambda - \lambda' - \varpi] + e' \cos[\lambda - 2\lambda' + \varpi'] \\
& - \frac{1}{8} e^2 \cos[\lambda + \lambda' - 2\varpi] - \frac{1}{8} e'^2 \cos[\lambda + \lambda' - 2\varpi'] \\
& + \frac{9}{8} e^2 \cos[3\lambda - \lambda' - 2\varpi] + \frac{9}{8} e'^2 \cos[\lambda - 3\lambda' + 2\varpi'] \\
& + e e' \cos[\varpi - \varpi'] + e e' \cos[2\lambda - 2\lambda' - \varpi + \varpi'] \\
& - e e' \cos[2\lambda - \varpi - \varpi'] - e e' \cos[2\lambda' - \varpi - \varpi'].
\end{aligned}
\tag{6.30}
$$

With this choice of angles it is clear that the sum of the integer coefficients of the longitudes in each argument is zero. This particular property is also true of the final expansion when the angles are expressed in terms of longitudes and it allows us to determine the permissible arguments.

If we now turn our attention to the radially dependent parts of the disturbing function Eq. (6.22), we can write

$$
\mathcal{R} = \frac{\mu'}{a'} \sum_{l=2}^{\infty} \alpha^l \left(\frac{a'}{r'}\right)^{l+1} \left(\frac{r}{a}\right)^l P_l(\cos \psi),
\tag{6.31}
$$

where

$$
\alpha = \frac{a}{a'} < 1
\tag{6.32}
$$

is the ratio of the semi-major axes of the masses $m$ and $m'$.

If we consider the terms with $l = 2$ then the series expansion for $r/a$ given in Sect. 2.5 gives

$$
\left(\frac{r}{a}\right)^2 \approx 1 - 2e \cos M + \left(\frac{1}{2}\right) e^2 (3 - \cos 2M),
\tag{6.33}
$$

$$
\left(\frac{a'}{r'}\right)^3 \approx 1 + 3e' \cos M' + \left(\frac{3}{2}\right) e'^2 (1 + 3 \cos 2M'),
\tag{6.34}
$$

with

$$
\begin{aligned}
\left(\frac{r}{a}\right)^2 \left(\frac{a'}{r'}\right)^3 \approx {} & 1 + \frac{3}{2} e^2 + \frac{3}{2} e'^2 - 2e \cos M + 3e' \cos M' \\
& - \frac{1}{2} e^2 \cos 2M + \frac{9}{2} e'^2 \cos 2M' \\
& - 3e e' \cos[M - M'] - 3e e' \cos[M + M'].
\end{aligned}
\tag{6.35}
$$

Since $P_2(x) = (1/2)(3x^2 - 1)$, $P_3(x) = (1/2)(5x^3 - 3x)$, etc., considerable effort is required to calculate the $P_l(\cos \psi)$ given the complexity of our expression for $\cos \psi$. In fact, for $l = 2$ there are fourteen separate arguments, while for $l = 3$ there are thirty-six arguments. However, since the series for $(a'/r')^{l+1}(r/a)^l$ only involves sums and differences of the mean anomalies, this means that their product with the terms in the $P_l(\cos \psi)$ series will always preserve the property that the sum of the coefficients of the longitudes in any cosine argument is zero.

It is clear, even from this simple analysis, that the expansion of the disturbing function is a nontrivial task best undertaken with the assistance of computer algebra systems. The end result is a series in $\alpha$ involving a large number of different arguments. Before considering how best to deal with this series, it is essential to generalise the expansion to three dimensions and introduce the inclinations and nodes of the two orbits.

The disturbing function $\mathcal{R}$ can be expanded in terms of standard orbital elements using the method developed by Kaula (1961, 1962), in which the disturbing function for an inner secondary is expanded in an infinite series in the osculating (i.e., instantaneous) elliptic elements referred to the equator of the primary. The expression for $\mathcal{R}$ in Eq. (6.16) can be written

$$
\mathcal{R} = \frac{\mu'}{a'} \sum_{l=2}^{\infty} \alpha^l \sum_{m=0}^{l} (-1)^{l-m} \kappa_m \frac{(l-m)!}{(l+m)!}
$$

$$
\times \sum_{p,p'=0}^{l} F_{lmp}(I) F_{lmp'}(I') \sum_{q,q'=-\infty}^{\infty} X_{l-2p+q}^{l,l-2p}(e) X_{l-2p'+q'}^{-l-1,l-2p'}(e')
$$

$$
\times \cos[(l-2p'+q')\lambda' - (l-2p+q)\lambda - q'\varpi' + q\varpi
$$

$$
+ (m-l+2p')\Omega' - (m-l+2p)\Omega], \tag{6.36}
$$

where $\alpha = a/a'$, $\lambda$ and $\lambda'$ are mean longitudes, $\varpi$ and $\varpi'$ are the longitudes of pericentre, and $\kappa_0 = 1$ and $\kappa_m = 2$ for $m \neq 0$.

The $F_{lmp}(I)$ are the inclination functions defined as

$$
F_{lmp}(I) = \frac{i^{l-m}(l+m)!}{2^l p!(l-p)!}
$$

$$
\times \sum_k (-1)^k \binom{2l-2p}{k} \binom{2p}{l-m-k} c^{3l-m-2p-2k} s^{m-l+2p+2k}, \tag{6.37}
$$

where $i = \sqrt{-1}$, $k$ is summed from $k = \max(0, l-m-2p)$ to $k = \min(l-m, 2l-2p)$, $s = \sin \frac{1}{2}I$, and $c = \cos \frac{1}{2}I$.

The quantities $X_c^{a,b}(e)$ are *Hansen coefficients*, which can be defined by

$$
X_c^{a,b}(e) = e^{|c-b|} \sum_{\sigma=0}^{\infty} X_{\sigma+\alpha,\sigma+\beta}^{a,b} e^{2\sigma}. \tag{6.38}
$$

In this context $\alpha = \max(0, c-b)$, $\beta = \max(0, b-c)$, and the $X_{c,d}^{a,b}$ are *Newcomb operators*, which can be defined recursively by

$$
X_{0,0}^{a,b} = 1, \tag{6.39}
$$

$$
X_{1,0}^{a,b} = b - a/2, \tag{6.40}
$$

and, for $d = 0$,

$$4cX_{c,0}^{a,b} = 2(2b-a)X_{c-1,0}^{a,b+1} + (b-a)X_{c-2,0}^{a,b+2}, \tag{6.41}$$

or, for $d \neq 0$,

$$4dX_{c,d}^{a,b} = -2(2b+a)X_{c,d-1}^{a,b-1} - (b+a)X_{c,d-2}^{a,b-2}$$

$$- (c-5d+4+4b+a)X_{c-1,d-1}^{a,b}$$

$$+ 2(c-d+b)\sum_{j\geq2}(-1)^j\binom{3/2}{j}X_{c-j,d-j}^{a,b}. \tag{6.42}$$

Also, $X_{c,d}^{a,b} = 0$ if $c < 0$ or $d < 0$. If $d > c$ then $X_{c,d}^{a,b} = X_{d,c}^{a,-b}$.

Additional information concerning Hansen coefficients and Newcomb operators can be found in Plummer (1918) and Hughes (1981). In particular, Hughes (1981) describes the properties of Hansen coefficients and their various recursive relations.

We must also consider the expansion of $\mathcal{R}'$. It is interesting to note that this expansion is curiously absent from the literature. It can only be assumed that since this form of expansion was developed for handling the perturbations on artificial satellites due to the exterior orbits of the Moon and Sun, the need for a similar expansion for $\mathcal{R}'$ never arose.

The expression for $\mathcal{R}'$ is

$$\mathcal{R}' = \frac{\mu}{a'}\sum_{l=1}^{\infty}\alpha^l\sum_{m=0}^{l}\kappa_m\frac{(l-m)!}{(l+m)!}$$

$$\times \sum_{p,p'=0}^{l}F_{lmp}(I)F_{lmp'}(I')\sum_{q,q'=-\infty}^{\infty}X_{l-2p+q}^{l,l-2p}(e)X_{l-2p'+q'}^{-(l+1),l-2p'}(e')$$

$$\times \cos\left[(l-2p'+q')\lambda' - (l-2p+q)\lambda - q'\varpi' + q\varpi\right.$$

$$\left. + (m-l+2p')\Omega' - (m-l+2p)\Omega\right]$$

$$- \frac{\mu a'}{a^2}\sum_{m=0}^{1}\kappa_m\frac{(1-m)!}{(1+m)!}$$

$$\times \sum_{p,p'=0}^{1}F_{1mp}(I)F_{1mp'}(I')\sum_{q,q'=-\infty}^{\infty}X_{1-2p+q}^{-2,1-2p}(e)X_{1-2p'+q'}^{1,1-2p'}(e')$$

$$\times \cos\left[(1-2p'+q')\lambda' - (1-2p+q)\lambda - q'\varpi' + q\varpi\right.$$

$$\left. + (m-1+2p')\Omega' - (m-1+2p)\Omega\right]. \tag{6.43}$$

## 6.4 Literal Expansion in Orbital Elements

Given the importance of the disturbing function in solar system dynamics, a number of authors have derived high-order expansions. Peirce (1849) derived an expansion to sixth order in the eccentricities and mutual inclination. One of the major expansions of the disturbing function, and one of the most commonly used, is due to Le Verrier (1855), who published a seventh-order expansion; Boquet (1889) extended Le Verrier's expansion to eighth order. Although Le Verrier's expansion contains a number of trivial errors, most of which were corrected in subsequent volumes of the *Annals of the Paris Observatory*, a single nontrivial mistake was found by Murray (1985). Other expansions include the symbolic development to sixth order by Newcomb (1895) and the low-order expansions by Brown & Shook (1933) and Brouwer & Clemence (1961). Although all these expansions were carried out in terms of the individual eccentricities and longitudes of pericentre of the two orbiting bodies, they all made use of a mutual inclination and a mutual ascending node. The reason for this was probably to reduce the amount of calculation required, but in the era of computer algebra such restrictions no longer apply. An expansion complete to second order in the individual eccentricities and inclinations is derived in Sect. 6.5, while Appendix B contains a literal expansion, which is complete to fourth order. Both expansions were derived using the method outlined below.

Given the complexity of the expansion, it is customary to distinguish between the direct and indirect parts of the disturbing function. Using the definitions in Eqs. (6.16) and (6.18), we can write

$$R = \frac{\mu'}{a'}\mathcal{R}_D + \frac{\mu'}{a'}\alpha\mathcal{R}_E \tag{6.44}$$

and

$$\mathcal{R}' = \frac{\mu}{a'}\mathcal{R}_D + \frac{\mu}{a'}\frac{1}{\alpha^2}\mathcal{R}_I, \tag{6.45}$$

where

$$\mathcal{R}_D = \frac{a'}{|\mathbf{r}' - \mathbf{r}|} \tag{6.46}$$

and

$$\mathcal{R}_E = -\left(\frac{r}{a}\right)\left(\frac{a'}{r'}\right)^2 \cos\psi, \tag{6.47}$$

$$\mathcal{R}_I = -\left(\frac{r'}{a'}\right)\left(\frac{a}{r}\right)^2 \cos\psi. \tag{6.48}$$

In these expressions $\mathcal{R}_D$ is derived from the direct part of the disturbing function, $\mathcal{R}_E$ comes from the indirect part due to an external perturber, and $\mathcal{R}_I$ comes from the indirect part for an internal perturber. It is clear from Eqs. (6.44)–(6.46) that we can use an expansion of $\mathcal{R}_D$ to obtain the direct part of either $\mathcal{R}$ or $\mathcal{R}'$.

To isolate the appropriate terms in the disturbing function for any particular problem in solar system dynamics, we need to obtain a series expansion of $\mathcal{R}$ or $\mathcal{R}'$ in terms of the individual orbital elements of the two orbiting bodies. This requires separate expansions of the direct part $\mathcal{R}_D$ defined in Eq. (6.46) and the indirect parts $\mathcal{R}_E$ and $\mathcal{R}_I$ defined in Eqs. (6.47) and (6.48) respectively. The different cosine arguments in the expansion given in Appendix B are labelled D, E, or I according to which part of the disturbing function they are derived from.

Using Eq. (6.19) we can write

$$\frac{1}{\Delta} = \frac{\mathcal{R}_D}{a'} = \left[r^2 + r'^2 - 2rr' \cos \psi\right]^{-1/2}, \tag{6.49}$$

where $\Delta = |\mathbf{r}' - \mathbf{r}|$ is the separation of the two masses and $\psi$ is the angle between the two radius vectors (see Fig. 6.2). Since $\mathbf{r} \cdot \mathbf{r}' = rr' \cos \psi$ we can write

$$\cos \psi = \frac{xx' + yy' + zz'}{rr'}. \tag{6.50}$$

From Eq. (2.122) we have

$$\frac{x}{r} = \cos \Omega \cos(\omega + f) - \sin \Omega \sin(\omega + f) \cos I, \tag{6.51}$$

$$\frac{y}{r} = \sin \Omega \cos(\omega + f) + \cos \Omega \sin(\omega + f) \cos I, \tag{6.52}$$

$$\frac{z}{r} = \sin(\omega + f) \sin I, \tag{6.53}$$

with similar expressions for $x'/r'$, $y'/r'$, and $z'/r'$.

Each of the above equations can be expanded as a series in $M$ and $M'$ using the series expansions for $\cos f$ and $\sin f$ given in Sect. 2.5 and hence we can derive a series expansion for $\cos \psi$. If we define

$$\Psi = \cos \psi - \cos(\theta - \theta'), \tag{6.54}$$

where $\theta = \varpi + f$ and $\theta' = \varpi' + f'$ are the true longitudes of the inner and outer bodies respectively, then, as we shall see later, the resulting series for $\Psi$ is of second order in $\sin I$ and $\sin I'$ and the expression for $\Delta^{-1}$ can be expanded as a Taylor series in $\Psi$. We have

$$\begin{aligned}
\frac{1}{\Delta} &= \left[r^2 + r'^2 - 2rr' \left(\cos(\theta - \theta') + \Psi\right)\right]^{-1/2} \\
&= \frac{1}{\Delta_0} + rr'\Psi \cdot \frac{1}{\Delta_0^3} + \frac{3}{2}(rr'\Psi)^2 \cdot \frac{1}{\Delta_0^5} + \cdots \\
&= \sum_{i=0}^{\infty} \frac{(2i)!}{(i!)^2} \cdot \left(\frac{1}{2}rr'\Psi\right)^i \cdot \frac{1}{\Delta_0^{2i+1}},
\end{aligned} \tag{6.55}$$

where

$$\frac{1}{\Delta_0} = \left[r^2 + r'^2 - 2rr' \cos(\theta - \theta')\right]^{-1/2}. \tag{6.56}$$

Let

$$\rho_0 = \left[a^2 + a'^2 - 2aa' \cos(\theta - \theta')\right]^{1/2}. \tag{6.57}$$

Using a Taylor series expansion in $\rho_0$, we can write

$$\frac{1}{\Delta_0^{2i+1}} = \frac{1}{\rho_0^{2i+1}} + (r-a)\frac{\partial}{\partial a}\left(\frac{1}{\rho_0^{2i+1}}\right) + (r'-a')\frac{\partial}{\partial a'}\left(\frac{1}{\rho_0^{2i+1}}\right) + \cdots. \quad (6.58)$$

Let $D_{m,n}$ denote the differential operator

$$D_{m,n} = a^m a'^n \frac{\partial^{m+n}}{\partial a^m \partial a'^n}, \quad (6.59)$$

and let

$$\varepsilon = \frac{r}{a} - 1, \qquad \varepsilon' = \frac{r'}{a'} - 1. \quad (6.60)$$

From the expansion of $r/a$ given in Eq. (2.81), it is clear that $\varepsilon$ is of $\mathcal{O}(e)$ and $\varepsilon'$ is of $\mathcal{O}(e')$. Hence we have

$$\frac{1}{\Delta_0^{2i+1}} = \left[1 + \varepsilon D_{1,0} + \varepsilon' D_{0,1} + \right.$$
$$\left. \frac{1}{2!}\left(\varepsilon^2 D_{2,0} + 2\varepsilon\varepsilon' D_{1,1} + \varepsilon'^2 D_{0,2}\right) + \cdots \right]\frac{1}{\rho_0^{2i+1}}. \quad (6.61)$$

However, from Eq. (6.57),

$$\frac{1}{\rho_0^{2i+1}} = \left[a^2 + a'^2 - 2aa'\cos(\theta - \theta')\right]^{-\left(i+\frac{1}{2}\right)}$$

$$= a'^{-(2i+1)}\left[1 + \alpha^2 - 2\alpha\cos(\theta - \theta')\right]^{-\left(i+\frac{1}{2}\right)}$$

$$= a'^{-(2i+1)}\frac{1}{2}\sum_{j=-\infty}^{\infty} b_{i+\frac{1}{2}}^{(j)}(\alpha)\cos j(\theta - \theta'), \quad (6.62)$$

where the $b_s^{(j)}(\alpha)$ are *Laplace coefficients*, each of which can be expressed as a uniformly convergent series in $\alpha$ for all $\alpha < 1$. Since the $D_{m,n}$ operators act only on the Laplace coefficients, we can define functions $A_{i,j,m,n}$ by

$$A_{i,j,m,n} = D_{m,n}\left(a'^{-(2i+1)}b_{i+\frac{1}{2}}^{(j)}(\alpha)\right) = a^m a'^n \frac{\partial^{m+n}}{\partial a^m \partial a'^n}\left(a'^{-(2i+1)}b_{i+\frac{1}{2}}^{(j)}(\alpha)\right), \quad (6.63)$$

and we can now write

$$\frac{1}{\Delta_0^{2i+1}} = \frac{1}{2}\sum_{j=-\infty}^{\infty}\left[A_{i,j,0,0} + \varepsilon A_{i,j,1,0} + \varepsilon' A_{i,j,0,1} + \cdots\right]\cos j(\theta - \theta'). \quad (6.64)$$

If we generalise this expression we obtain

$$\frac{1}{\Delta_0^{2i+1}} = \frac{1}{2}\sum_{j=-\infty}^{\infty}\left[\sum_{l=0}^{\infty}\frac{1}{l!}\sum_{k=0}^{l}\binom{l}{k}\varepsilon^k\varepsilon'^{l-k}A_{i,j,k,l-k}\right]\cos j(\theta - \theta'). \quad (6.65)$$

Care must be taken in the calculation of the partial derivatives with respect to $a$ and $a'$ in the $A_{i,j,k,l-k}$ since $a$ and $a'$ are still contained implicitly within the Laplace coefficients $b^{(j)}_{i+\frac{1}{2}}(a/a')$.

Substituting Eq. (6.65) in Eq. (6.55) we obtain

$$\mathcal{R}_D = \sum_{i=0}^{\infty} \frac{(2i)!}{(i!)^2} \left(\frac{1}{2} \frac{r}{a} \frac{r'}{a'} \Psi\right)^i \frac{a^i a'^{i+1}}{2}$$

$$\times \sum_{j=-\infty}^{\infty} \left[ \sum_{l=0}^{\infty} \frac{1}{l!} \sum_{k=0}^{l} \binom{l}{k} \varepsilon^k \varepsilon'^{l-k} A_{i,j,k,l-k} \right] \cos j(\theta - \theta'). \qquad (6.66)$$

It is worthwhile noting that the inclinations, $I$ and $I'$, are only contained in $\Psi$ and the eccentricities are only contained in the $\varepsilon$ and $\varepsilon'$ terms in Eq. (6.66).

The expansion of the indirect parts, $\mathcal{R}_E$ and $\mathcal{R}_I$, is more straightforward using the series obtained from expanding $\cos \psi$ in Eq. (6.50) and the series given in Sect. 2.5. Note that the expansion of these terms does not involve Laplace coefficients.

The literal expansion makes use of Laplace coefficients, which are explicit functions of $\alpha$ rather than the individual coefficients of powers of $\alpha$ that we encountered in Kaula's expansion. The Laplace coefficient $b^{(j)}_{i+\frac{1}{2}}(\alpha)$ in Eq. (6.62) is defined by

$$\frac{1}{2} b_s^{(j)}(\alpha) = \frac{1}{2\pi} \int_0^{2\pi} \frac{\cos j\psi \, d\psi}{(1 - 2\alpha \cos \psi + \alpha^2)^s}, \qquad (6.67)$$

where $s = i + 1/2$ is a half-integer (i.e., $s = 1/2, 3/2, 5/2, \ldots$) and $\alpha = a/a'$. Alternatively we can write this in series form as

$$\frac{1}{2} b_s^{(j)}(\alpha) = \frac{s(s+1)\ldots(s+j-1)}{1 \cdot 2 \cdot 3 \ldots j} \alpha^j$$

$$\times \left[ 1 + \frac{s(s+j)}{1(j+1)} \alpha^2 + \frac{s(s+1)(s+j)(s+j+1)}{1 \cdot 2(j+1)(j+2)} \alpha^4 + \ldots \right]. \qquad (6.68)$$

In the case where $j = 0$ the factor outside the brackets is equal to unity. It can be shown that the series definition of the Laplace coefficient is always convergent for $\alpha < 1$.

Useful relations between Laplace coefficients and their derivatives are given in Brouwer & Clemence (1961). These include

$$b_s^{(-j)} = b_s^{(j)}, \qquad (6.69)$$

$$Db_s^{(j)} = s \left( b_{s+1}^{(j-1)} - 2\alpha b_{s+1}^{(j)} + b_{s+1}^{(j+1)} \right), \qquad (6.70)$$

$$D^n b_s^{(j)} = s \left( D^{n-1} b_{s+1}^{(j-1)} - 2\alpha D^{n-1} b_{s+1}^{(j)} \right.$$

$$\left. + D^{n-1} b_{s+1}^{(j+1)} - 2(n-1) D^{n-2} b_{s+1}^{(j)} \right), \qquad (6.71)$$

and

$$\alpha^n \left( D^n b_s^{(j)} - D^n b_s^{(j-2)} \right) =$$
$$- (j + n - 1)\alpha^{n-1} D^{n-1} b_s^{(j)} - (j - n - 1)\alpha^{n-1} D^{n-1} b_s^{(j-2)}$$
$$+ 2(j-1) \left[ \alpha^n D^{n-1} b_s^{(j-1)} + (n-1)\alpha^{n-1} D^{n-2} b_s^{(j-1)} \right], \quad (6.72)$$

where $n \geq 2$ in the last two relations and $D \equiv d/d\alpha$ is a differential operator.

## 6.5 Literal Expansion to Second Order

As an illustration of the techniques outlined in Sect. 6.4, we will now derive an expansion of the disturbing function complete to second order in the eccentricities and inclinations.

To derive a series expansion for $\cos \psi$ we first need to make use of the expansions for $\sin f$ and $\cos f$ in terms of the mean anomaly $M$, given in Eqs. (2.84) and (2.85) respectively. To second order we have

$$\sin f = \sin M + e \sin 2M + e^2 \left( \frac{9}{8} \sin 3M - \frac{7}{8} \sin M \right), \quad (6.73)$$

$$\cos f = \cos M + e (\cos 2M - 1) + e^2 \left( \frac{9}{8} \cos 3M - \frac{9}{8} \cos M \right). \quad (6.74)$$

Hence

$$\cos[\omega + f] = \cos \omega \cos f - \sin \omega \sin f$$
$$\approx \cos[\omega + M] + e \left( \cos[\omega + 2M] - \cos \omega \right)$$
$$+ e^2 \left( -\cos[\omega + M] - \frac{1}{8} \cos[\omega - M] + \frac{9}{8} \cos[\omega + 3M] \right) \quad (6.75)$$

and

$$\sin[\omega + f] = \sin \omega \cos f + \cos \omega \sin f$$
$$\approx \sin[\omega + M] + e \left( \sin[\omega + 2M] - \sin \omega \right)$$
$$+ e^2 \left( -\sin[\omega + M] + \frac{1}{8} \sin[\omega - M] + \frac{9}{8} \sin[\omega + 3M] \right). \quad (6.76)$$

In keeping with a number of previous expansions (including that by Kaula discussed in Sect. 6.3) we wish to express the disturbing function in terms of powers of $\sin \frac{1}{2}I$ and $\sin \frac{1}{2}I'$ rather than sines and cosines of the inclinations. Therefore we make use of the relations

$$\cos I = 1 - 2 \sin^2 \frac{1}{2}I = 1 - 2s^2 \quad (6.77)$$

and

$$\sin I = 2 \sin \frac{1}{2}I \left( 1 - \sin^2 \frac{1}{2}I \right)^{\frac{1}{2}} = 2s + \mathcal{O}(s^3), \quad (6.78)$$

where $s = \sin \frac{1}{2} I$. Substitution of these expressions and our expansions of $\cos[\omega + f]$ and $\sin[\omega + f]$ in Eqs. (6.51)–(6.53) gives

$$\frac{x}{r} \approx \cos[\omega + \Omega + M] + e \left(\cos[\omega + \Omega + 2M] - \cos[\omega + \Omega]\right)$$
$$+ e^2 \left(\frac{9}{8} \cos[\omega + \Omega + 3M] - \frac{1}{8} \cos[\omega + \Omega - M] - \cos[\omega + \Omega + M]\right)$$
$$+ s^2 \left(\cos[\omega - \Omega + M] - \cos[\omega + \Omega + M]\right), \qquad (6.79)$$

$$\frac{y}{r} \approx \sin[\omega + \Omega + M] + e \left(\sin[\omega + \Omega + 2M] - \sin[\omega + \Omega]\right)$$
$$+ e^2 \left(\frac{9}{8} \sin[\omega + \Omega + 3M] - \frac{1}{8} \sin[\omega + \Omega - M] - \sin[\omega + \Omega + M]\right)$$
$$- s^2 \left(\sin[\omega - \Omega + M] + \sin[\omega + \Omega + M]\right), \qquad (6.80)$$

and

$$\frac{z}{r} \approx 2s \sin[\omega + M] + 2es \left(\sin[\omega + 2M] - \sin \omega\right). \qquad (6.81)$$

Similar expressions can be obtained for $x'/r'$, $y'/r'$, and $z'/r'$ by replacing unprimed quantities by primed quantities in the above equations. Hence we can derive an expression for $\cos \psi$ using Eq. (6.50). At the same time we can use the relations $M = \lambda - \varpi$ and $\omega = \varpi - \Omega$ to express the expansion in terms of longitudes. We get

$$\cos \psi \approx$$
$$(1 - e^2 - e'^2 - s^2 - s'^2) \cos[\lambda - \lambda'] + ee' \cos[2\lambda - 2\lambda' - \varpi + \varpi']$$
$$+ ee' \cos[\varpi - \varpi'] + 2ss' \cos[\lambda - \lambda' - \Omega + \Omega']$$
$$+ e \cos[2\lambda - \lambda' - \varpi] - e \cos[\lambda' - \varpi]$$
$$+ e' \cos[\lambda - 2\lambda' + \varpi'] - e' \cos[\lambda - \varpi']$$
$$+ \frac{9}{8} e^2 \cos[3\lambda - \lambda' - 2\varpi] - \frac{1}{8} e^2 \cos[\lambda + \lambda' - 2\varpi]$$
$$+ \frac{9}{8} e'^2 \cos[\lambda - 3\lambda' + 2\varpi'] - \frac{1}{8} e'^2 \cos[\lambda + \lambda' - 2\varpi']$$
$$- ee' \cos[2\lambda - \varpi - \varpi'] - ee' \cos[2\lambda' - \varpi - \varpi']$$
$$+ s^2 \cos[\lambda + \lambda' - 2\Omega] + s'^2 \cos[\lambda + \lambda' - 2\Omega']$$
$$- 2ss' \cos[\lambda + \lambda' - \Omega - \Omega']. \qquad (6.82)$$

Since $\theta = \omega + \Omega + f$ we have

$$\cos[\theta - \theta'] = (\cos \Omega \cos[\omega + f] - \sin \Omega \sin[\omega + f])$$
$$\times (\cos \Omega' \cos[\omega' + f'] - \sin \Omega' \sin[\omega' + f'])$$
$$+ (\sin \Omega \cos[\omega + f] + \cos \Omega \sin[\omega + f])$$
$$\times (\sin \Omega' \cos[\omega' + f'] + \cos \Omega' \sin[\omega' + f']). \qquad (6.83)$$

By comparing this with Eqs. (6.51)–(6.53) we see that the expansion for $\cos[\theta - \theta']$ can be obtained from the expansion for $\cos \psi$ by setting $I = I' = 0$. Since $\Psi = \cos \psi - \cos[\theta - \theta']$, the expansion of $\cos \psi$ shows that $\Psi$ is the inclination-dependent part of $\cos \psi$ and

$$
\begin{aligned}
\Psi = {} & s^2 \left(\cos[\lambda + \lambda' - 2\Omega] - \cos[\lambda - \lambda']\right) \\
& + 2ss' \left(\cos[\lambda - \lambda' - \Omega + \Omega'] - \cos[\lambda + \lambda' - \Omega - \Omega']\right) \\
& + s'^2 \left(\cos[\lambda + \lambda' - 2\Omega'] - \cos[\lambda - \lambda']\right).
\end{aligned}
\tag{6.84}
$$

Note that $\Psi$ is of second order in the inclinations.

Since $r/a = 1 + \mathcal{O}(e)$ and $r'/a' = 1 + \mathcal{O}(e')$, it is clear that, to second order in the eccentricities and inclinations, we can write

$$
\begin{aligned}
\left(\frac{1}{2} \frac{r}{a} \frac{r'}{a'} \Psi\right) = {} & \frac{1}{2} s^2 \left(\cos[\lambda + \lambda' - 2\Omega] - \cos[\lambda - \lambda']\right) \\
& + ss' \left(\cos[\lambda - \lambda' - \Omega + \Omega'] - \cos[\lambda + \lambda' - \Omega - \Omega']\right) \\
& + \frac{1}{2} s'^2 \left(\cos[\lambda + \lambda' - 2\Omega'] - \cos[\lambda - \lambda']\right),
\end{aligned}
\tag{6.85}
$$

which is independent of $e$ to this order. Since we are only interested in a second-order expansion, and since $\Psi$ is already of second order, we can ignore second and higher powers of $\Psi$.

We have now obtained the first of the two major terms required for the series for $\mathcal{R}_D$ (see Eq. (6.66)). We need to derive an expression for $\cos j[\theta - \theta']$, where $j$ is an arbitrary integer. We start by noting that

$$
\begin{aligned}
\cos j[\theta - \theta'] = {} & \cos j[\omega + \Omega + f] \cos j[\omega' + \Omega' + f'] \\
& + \sin j[\omega + \Omega + f] \sin j[\omega' + \Omega' + f'].
\end{aligned}
\tag{6.86}
$$

From Eq. (2.88) we have

$$
f = M + 2e \sin M + \frac{5}{4} e^2 \sin 2M + \mathcal{O}(e^3).
\tag{6.87}
$$

If we substitute this expression in $\cos j[\omega + \Omega + f]$ and $\sin j[\omega + \Omega + f]$, transform to longitudes as before, and carry out a Taylor series expansion we obtain

$$
\begin{aligned}
\cos j\theta \approx {} & (1 - j^2 e^2) \cos[j\lambda] \\
& + \left(\frac{1}{2} j^2 e^2 - \frac{5}{8} je^2\right) \cos[(2 - j)\lambda - 2\varpi] \\
& + \left(\frac{1}{2} j^2 e^2 + \frac{5}{8} je^2\right) \cos[(2 + j)\lambda - 2\varpi] \\
& - je \cos[(1 - j)\lambda - \varpi] + je \cos[(1 + j)\lambda - \varpi]
\end{aligned}
\tag{6.88}
$$

and

$$\sin j\theta \approx (1 - j^2 e^2) \sin[j\lambda]$$
$$+ \left(\frac{5}{8} j e^2 - \frac{1}{2} j^2 e^2\right) \sin[(2 - j)\lambda - 2\varpi]$$
$$+ \left(\frac{5}{8} j e^2 + \frac{1}{2} j^2 e^2\right) \sin[(2 + j)\lambda - 2\varpi]$$
$$+ j e \sin[(1 - j)\lambda - \varpi] + j e \sin[(1 + j)\lambda - \varpi]. \qquad (6.89)$$

By substituting unprimed quantities for primed ones we can easily obtain similar expressions for $\cos j\theta'$ and $\sin j\theta'$. The resulting expression for $\cos j[\theta - \theta']$ is

$$\cos j[\theta - \theta'] \approx$$
$$(1 - j^2 e^2 - j^2 e'^2) \cos[j(\lambda - \lambda')]$$
$$+ \left(\frac{5}{8} j e^2 + \frac{1}{2} j^2 e^2\right) \cos[(2 + j)\lambda - j\lambda' - 2\varpi]$$
$$+ \left(\frac{1}{2} j^2 e^2 - \frac{5}{8} j e^2\right) \cos[(2 - j)\lambda + j\lambda' - 2\varpi]$$
$$+ j e \cos[(1 + j)\lambda - j\lambda' - \varpi] - j e \cos[(1 - j)\lambda + j\lambda' - \varpi]$$
$$+ \left(\frac{1}{2} j^2 e'^2 - \frac{5}{8} j e'^2\right) \cos[j\lambda + (2 - j)\lambda' - 2\varpi']$$
$$+ \left(\frac{5}{8} j e'^2 + \frac{1}{2} j^2 e'^2\right) \cos[j\lambda - (2 + j)\lambda' + 2\varpi']$$
$$- j e' \cos[j\lambda + (1 - j)\lambda' - \varpi'] + j e' \cos[j\lambda - (1 + j)\lambda' + \varpi']$$
$$- j^2 e e' \cos[(1 + j)\lambda + (1 - j)\lambda' - \varpi - \varpi']$$
$$- j^2 e e' \cos[(1 - j)\lambda + (1 + j)\lambda' - \varpi - \varpi']$$
$$+ j^2 e e' \cos[(1 + j)\lambda - (1 + j)\lambda' - \varpi + \varpi']$$
$$+ j^2 e e' \cos[(1 - j)\lambda - (1 - j)\lambda' - \varpi + \varpi']. \qquad (6.90)$$

Although the summation over $j$ in Eq. (6.66) is over all values, in practice we do not need to carry out this summation (see Sect. 6.9 for an example).

From Eqs. (6.60) and (2.81) we have

$$\varepsilon = \frac{r}{a} - 1 \approx -e \cos M + \frac{1}{2} e^2 (1 - \cos 2M)$$
$$= -e \cos[\lambda - \varpi] + \frac{1}{2} e^2 \left(1 - \cos[2\lambda - 2\varpi]\right) \qquad (6.91)$$

and hence

$$\varepsilon^2 \approx \frac{1}{2} e^2 + \frac{1}{2} e^2 \cos 2M = \frac{1}{2} e^2 + \frac{1}{2} e^2 \cos[2\lambda - 2\varpi], \qquad (6.92)$$

with similar expressions for $\varepsilon'$ and $\varepsilon'^2$. No powers beyond the second are necessary for this expansion since $\varepsilon$ is of $\mathcal{O}(e)$.

Finally, before carrying out the summation in Eq. (6.66), we need to calculate the derivatives of the Laplace coefficients given by the $A_{i,j,m,n}$ function. This task can be simplified by noting that, for a given value of $i$ in the summation, we need to calculate

$$a^i a'^{i+1} A_{i,j,m,n} = a^{i+m} a'^{i+n+1} \frac{\partial^{m+n}}{\partial a^m \partial a'^m} \left( a'^{-(2i+1)} b^{(j)}_{i+\frac{1}{2}} (a/a') \right). \tag{6.93}$$

The result of the differentiation is to leave a function of $a/a'$ alone. In our case the required values of $A_{i,j,m,n}$ are:

$$a^i a'^{i+1} A_{i,j,0,0} = \alpha^i b^{(j)}_{i+\frac{1}{2}} (\alpha), \tag{6.94}$$

$$a^i a'^{i+1} A_{i,j,1,0} = \alpha^{i+1} D b^{(j)}_{i+\frac{1}{2}} (\alpha), \tag{6.95}$$

$$a^i a'^{i+1} A_{i,j,0,1} = -\alpha^{i+1} D b^{(j)}_{i+\frac{1}{2}} (\alpha) - (2i+1)\alpha^i b^{(j)}_{i+\frac{1}{2}} (\alpha), \tag{6.96}$$

$$a^i a'^{i+1} A_{i,j,2,0} = \alpha^{i+2} D^2 b^{(j)}_{i+\frac{1}{2}} (\alpha), \tag{6.97}$$

$$a^i a'^{i+1} A_{i,j,1,1} = -\alpha^{i+2} D^2 b^{(j)}_{i+\frac{1}{2}} (\alpha) - 2\alpha^{i+1}(i+1) D b^{(j)}_{i+\frac{1}{2}} (\alpha), \tag{6.98}$$

and

$$a^i a'^{i+1} A_{i,j,2,2} = \alpha^{i+2} D^2 b^{(j)}_{i+\frac{1}{2}} (\alpha) + 4\alpha^{i+1}(i+1) D b^{(j)}_{i+\frac{1}{2}} (\alpha)$$
$$+ 2\alpha^i (2i^2 + 3i + 2) b^{(j)}_{i+\frac{1}{2}} (\alpha), \tag{6.99}$$

where $i$ will take the values 0 and 1; higher values can be ignored because of the presence of the $\Psi^i$ term in Eq. (6.66).

We are now in a position to carry out the summation over $i$. To second order in the eccentricity and inclination we have

$$\mathcal{R}_D = \left( \frac{1}{2} [a' A_{0,j,0,0} + \varepsilon a' A_{0,j,1,0} + \varepsilon' a' A_{0,j,0,1} \right.$$
$$+ \varepsilon^2 a' A_{0,j,2,0} + \varepsilon\varepsilon' a' A_{0,j,1,1} + \varepsilon'^2 a' A_{0,j,0,2}]$$
$$\left. + \left( \frac{1}{2} \frac{r}{a} \frac{r'}{a'} \Psi \right) aa'^2 A_{1,j,0,0} \right) \cos j[\theta - \theta']. \tag{6.100}$$

Using the series that we have already derived for the quantities in this equation, we get an expansion with twenty-three cosine arguments. These can be categorised by the *order* of the argument, which is simply the sum of the coefficients of $\lambda$ and $\lambda'$. If we write the second-order expansion as

$$\mathcal{R}_D = \mathcal{R}_D^{(0)} + \mathcal{R}_D^{(1)} + \mathcal{R}_D^{(2)}, \tag{6.101}$$

where $\mathcal{R}^{(i)}$ denotes the part of the expansion containing the arguments of order

$i$, then

$$
\begin{aligned}
\mathcal{R}_D^{(0)} &= \left( \frac{1}{2} b_{\frac{1}{2}}^{(j)} + \frac{1}{8} (e^2 + e'^2) \left[ -4j^2 + 2\alpha D + \alpha^2 D^2 \right] b_{\frac{1}{2}}^{(j)} \right) \cos[j\lambda - j\lambda'] \\
&\quad + \left( \frac{1}{8} ee' \left[ 2j + 4j^2 - 2\alpha D - \alpha^2 D^2 \right] b_{\frac{1}{2}}^{(j)} \right) \\
&\qquad \times \cos[(1+j)\lambda - (1+j)\lambda' - \varpi + \varpi'] \\
&\quad + \left( \frac{1}{8} ee' \left[ -2j + 4j^2 - 2\alpha D - \alpha^2 D^2 \right] b_{\frac{1}{2}}^{(j)} \right) \\
&\qquad \times \cos[(1-j)\lambda - (1-j)\lambda' - \varpi + \varpi'] \\
&\quad + \left( \frac{1}{4} (s^2 + s'^2)[-\alpha] b_{\frac{3}{2}}^{(j)} \right) \cos[(1+j)\lambda - (1+j)\lambda'] \\
&\quad + \left( \frac{1}{4} (s^2 + s'^2)[-\alpha] b_{\frac{3}{2}}^{(j)} \right) \cos[(1-j)\lambda - (1-j)\lambda'] \\
&\quad + \left( \frac{1}{2} ss'[\alpha] b_{\frac{3}{2}}^{(j)} \right) \cos[(1+j)\lambda - (1+j)\lambda' - \Omega + \Omega'] \\
&\quad + \left( \frac{1}{2} ss'[\alpha] b_{\frac{3}{2}}^{(j)} \right) \cos[(1-j)\lambda - (1-j)\lambda' - \Omega + \Omega'], \qquad (6.102)
\end{aligned}
$$

$$
\begin{aligned}
\mathcal{R}_D^{(1)} &= \left( \frac{1}{4} e[2j - \alpha D] b_{\frac{1}{2}}^{(j)} \right) \cos[(1+j)\lambda - j\lambda' - \varpi] \\
&\quad + \left( \frac{1}{4} e[-2j - \alpha D] b_{\frac{1}{2}}^{(j)} \right) \cos[(1-j)\lambda + j\lambda' - \varpi] \\
&\quad + \left( \frac{1}{4} e'[1 + 2j + \alpha D] b_{\frac{1}{2}}^{(j)} \right) \cos[j\lambda - (1+j)\lambda' + \varpi'] \\
&\quad + \left( \frac{1}{4} e'[1 - 2j + \alpha D] b_{\frac{1}{2}}^{(j)} \right) \cos[j\lambda + (1-j)\lambda' - \varpi'], \qquad (6.103)
\end{aligned}
$$

$$
\begin{aligned}
\mathcal{R}_D^{(2)} &= \left( \frac{1}{16} e^2 \left[ 5j + 4j^2 - 2\alpha D - 4j\alpha D + \alpha^2 D^2 \right] b_{\frac{1}{2}}^{(j)} \right) \\
&\qquad \times \cos[(2+j)\lambda - j\lambda - 2\varpi] \\
&\quad + \left( \frac{1}{16} e^2 \left[ -5j + 4j^2 - 2\alpha D + 4j\alpha D + \alpha^2 D^2 \right] b_{\frac{1}{2}}^{(j)} \right) \\
&\qquad \times \cos[(2-j)\lambda + j\lambda - 2\varpi] \\
&\quad + \left( \frac{1}{8} ee' \left[ 2j - 4j^2 - 2\alpha D + 2j\alpha D - \alpha^2 D^2 \right] b_{\frac{1}{2}}^{(j)} \right) \\
&\qquad \times \cos[(1+j)\lambda + (1-j)\lambda' - \varpi - \varpi'] \\
&\quad + \left( \frac{1}{8} ee' \left[ -2j - 4j^2 - 2\alpha D - 2j\alpha D - \alpha^2 D^2 \right] b_{\frac{1}{2}}^{(j)} \right) \\
&\qquad \times \cos[(1-j)\lambda + (1+j)\lambda' - \varpi - \varpi'] \\
&\quad + \left( \frac{1}{16} e'^2 \left[ 4 + 9j + 4j^2 + 6\alpha D + 4j\alpha D + \alpha^2 D^2 \right] b_{\frac{1}{2}}^{(j)} \right) \\
&\qquad \times \cos[j\lambda - (2+j)\lambda' + 2\varpi']
\end{aligned}
$$

$$+ \left( \frac{1}{16} e'^2 \left[ 4 - 9j + 4j^2 + 6\alpha D - 4j\alpha D + \alpha^2 D^2 \right] b_{\frac{1}{2}}^{(j)} \right)$$
$$\times \cos[j\lambda + (2 - j)\lambda' - 2\varpi']$$
$$+ \left( \frac{1}{4} s^2 [\alpha] b_{\frac{3}{2}}^{(j)} \right) \cos[(1 - j)\lambda + (1 + j)\lambda' - 2\Omega]$$
$$+ \left( \frac{1}{4} s^2 [\alpha] b_{\frac{3}{2}}^{(j)} \right) \cos[(1 + j)\lambda + (1 - j)\lambda' - 2\Omega]$$
$$+ \left( \frac{1}{2} ss' [-\alpha] b_{\frac{3}{2}}^{(j)} \right) \cos[(1 - j)\lambda + (1 + j)\lambda' - \Omega - \Omega']$$
$$+ \left( \frac{1}{2} ss' [-\alpha] b_{\frac{3}{2}}^{(j)} \right) \cos[(1 + j)\lambda + (1 - j)\lambda' - \Omega - \Omega']$$
$$+ \left( \frac{1}{4} s'^2 [\alpha] b_{\frac{3}{2}}^{(j)} \right) \cos[(1 - j)\lambda + (1 + j)\lambda' - 2\Omega']$$
$$+ \left( \frac{1}{4} s'^2 [\alpha] b_{\frac{3}{2}}^{(j)} \right) \cos[(1 + j)\lambda + (1 - j)\lambda' - 2\Omega']. \tag{6.104}$$

The arguments in this expansion are not all unique and further simplification is possible. This is clear from an inspection of the different terms since, apart from the first term in $\mathcal{R}_D^{(0)}$, they occur in pairs with similar form. Because the summation in $j$ in Eq. (6.66) is over all values, we can always carry out a transformation of $j$ of the form $j \to \pm j + k$ where $k$ is an integer, provided that we apply it to the argument and its associated term. Also, since only cosines appear in the expansion we can always change the sign of the argument. We can then use these procedures to reduce the arguments in the expansion to some arbitrary standard form. In our case we have decided to make $j$ the coefficient of $\lambda'$ in each argument.

As an example, consider the two terms in $ee'$ in $\mathcal{R}_D^{(0)}$. These can be transformed to the same cosine argument,

$$j\lambda' - j\lambda + \varpi' - \varpi, \tag{6.105}$$

by changing $j$ to $-j$ in the first, and then applying the transformation $j \to j + 1$ in each. The resulting term associated with this argument is then

$$\frac{1}{4} ee' \left[ 2 + 6j + 4j^2 - 2\alpha D - \alpha^2 D^2 \right] b_{\frac{1}{2}}^{(j+1)}. \tag{6.106}$$

Similar procedures can be carried out for the other arguments and the total number of arguments can be reduced from twenty-three to eleven. In such transformations we have made use of the fact that $b_s^{(-j)} = b_s^{(j)}$, as given in Eq. (6.69). We also point out that, even with our decision to express all the arguments in a form where the coefficient of $\lambda'$ is $j$, the final form of our expansion is not unique and transformations of the form $j \to -j$ followed by reversal of the argument produce arguments of a different form.

The final form of our second-order expansion of the direct part is

$$
\begin{aligned}
\mathcal{R}_\mathrm{D} = {} & \left( \frac{1}{2} b_{\frac{1}{2}}^{(j)} + \frac{1}{8} \left( e^2 + e'^2 \right) \left[ -4j^2 + 2\alpha D + \alpha^2 D^2 \right] b_{\frac{1}{2}}^{(j)} \right. \\
& \left. + \frac{1}{4} \left( s^2 + s'^2 \right) \left( [-\alpha] b_{\frac{3}{2}}^{(j-1)} + [-\alpha] b_{\frac{3}{2}}^{(j+1)} \right) \right) \\
& \times \cos[j\lambda' - j\lambda] \\
& + \left( \frac{1}{4} ee' \left[ 2 + 6j + 4j^2 - 2\alpha D - \alpha^2 D^2 \right] b_{\frac{1}{2}}^{(j+1)} \right) \\
& \times \cos[j\lambda' - j\lambda + \varpi' - \varpi] \\
& + \left( ss' [\alpha] b_{\frac{3}{2}}^{(j+1)} \right) \cos[j\lambda' - j\lambda + \Omega' - \Omega] \\
& + \left( \frac{1}{2} e [-2j - \alpha D] b_{\frac{1}{2}}^{(j)} \right) \cos[j\lambda' + (1-j)\lambda - \varpi] \\
& + \left( \frac{1}{2} e' [-1 + 2j + \alpha D] b_{\frac{1}{2}}^{(j-1)} \right) \cos[j\lambda' + (1-j)\lambda - \varpi'] \\
& + \left( \frac{1}{8} e^2 \left[ -5j + 4j^2 - 2\alpha D + 4j\alpha D + \alpha^2 D^2 \right] b_{\frac{1}{2}}^{(j)} \right) \\
& \times \cos[j\lambda' + (2-j)\lambda - 2\varpi] \\
& + \left( \frac{1}{4} ee' \left[ -2 + 6j - 4j^2 + 2\alpha D - 4j\alpha D - \alpha^2 D^2 \right] b_{\frac{1}{2}}^{(j-1)} \right) \\
& \times \cos[j\lambda' + (2-j)\lambda - \varpi' - \varpi] \\
& + \left( \frac{1}{8} e'^2 \left[ 2 - 7j + 4j^2 - 2\alpha D + 4j\alpha D + \alpha^2 D^2 \right] b_{\frac{1}{2}}^{(j-2)} \right) \\
& \times \cos[j\lambda' + (2-j)\lambda - 2\varpi'] \\
& + \left( \frac{1}{2} s^2 [\alpha] b_{\frac{3}{2}}^{(j-1)} \right) \\
& \times \cos[j\lambda' + (2-j)\lambda - 2\Omega] \\
& + \left( ss' [-\alpha] b_{\frac{3}{2}}^{(j-1)} \right) \cos[j\lambda' + (2-j)\lambda - \Omega' - \Omega] \\
& + \left( \frac{1}{2} s'^2 [\alpha] b_{\frac{3}{2}}^{(j-1)} \right) \cos[j\lambda' + (2-j)\lambda - 2\Omega'].
\end{aligned}
\tag{6.107}
$$

Generating the indirect parts of the disturbing function defined in Eqs. (6.47) and (6.48) is relatively simple since we have already derived an expression for $\cos \psi$. Expressions for $r/a$ and $(a'/r')^2$ can be obtained from the elliptical expansions given in Sect. 2.5. To second order,

$$
\frac{r}{a} = 1 - e \cos[\lambda - \varpi] + \frac{1}{2} e^2 (1 - \cos[2\lambda - 2\varpi])
\tag{6.108}
$$

and

$$
\left( \frac{a'}{r'} \right) = 1 + 2e' \cos[\lambda' - \varpi'] + \frac{1}{2} e'^2 \left( 1 + 5 \cos[2\lambda' - 2\varpi'] \right)
\tag{6.109}
$$

and hence

$$\mathcal{R}_E = -\frac{r}{a}\left(\frac{a'}{r'}\right)^2 \cos\psi$$

$$\approx \left(-1 + \frac{1}{2}e^2 + \frac{1}{2}e'^2 + s^2 + s'^2\right)\cos[\lambda' - \lambda]$$

$$- ee'\cos[2\lambda' - 2\lambda - \varpi' + \varpi] - 2ss'\cos[\lambda' - \lambda - \Omega' + \Omega]$$

$$- \frac{1}{2}e\cos[\lambda' - 2\lambda + \varpi] + \frac{3}{2}e\cos[\lambda' - \varpi] - 2e'\cos[2\lambda' - \lambda - \varpi']$$

$$- \frac{3}{8}e^2\cos[\lambda' - 3\lambda + 2\varpi] - \frac{1}{8}e^2\cos[\lambda' + \lambda - 2\varpi]$$

$$+ 3ee'\cos[2\lambda - \varpi' - \varpi] - \frac{1}{8}e'^2\cos[\lambda' + \lambda - 2\varpi']$$

$$- \frac{27}{8}e'^2\cos[3\lambda' - \lambda - 2\varpi'] - s^2\cos[\lambda' + \lambda - 2\Omega]$$

$$+ 2ss'\cos[\lambda' + \lambda - \Omega' - \Omega] - s'^2\cos[\lambda' + \lambda - 2\Omega'], \tag{6.110}$$

where we have changed the sign of the argument in some cases in order to adopt the same convention used to derive $\mathcal{R}_D$.

We can use similar methods to derive an expression for $\mathcal{R}_I$. Alternatively we can reverse the primed and unprimed quantities in our expression for $\mathcal{R}_E$. We obtain

$$\mathcal{R}_I = -\frac{r'}{a'}\left(\frac{a}{r}\right)^2 \cos\psi$$

$$\approx \left(-1 + \frac{1}{2}e^2 + \frac{1}{2}e'^2 + s^2 + s'^2\right)\cos[\lambda' - \lambda]$$

$$- ee'\cos[2\lambda' - 2\lambda - \varpi' + \varpi] - 2ss'\cos[\lambda' - \lambda - \Omega' + \Omega]$$

$$- 2e\cos[\lambda' - 2\lambda + \varpi] + \frac{3}{2}e'\cos[\lambda - \varpi'] - \frac{1}{2}e'\cos[2\lambda' - \lambda - \varpi']$$

$$- \frac{27}{8}e^2\cos[\lambda' - 3\lambda + 2\varpi] - \frac{1}{8}e^2\cos[\lambda' + \lambda - 2\varpi]$$

$$+ 3ee'\cos[2\lambda - \varpi' - \varpi] - \frac{1}{8}e'^2\cos[\lambda' + \lambda - 2\varpi']$$

$$- \frac{3}{8}e'^2\cos[3\lambda' - \lambda - 2\varpi'] - s^2\cos[\lambda' + \lambda - 2\Omega]$$

$$+ 2ss'\cos[\lambda' + \lambda - \Omega' - \Omega] - s'^2\cos[\lambda' + \lambda - 2\Omega']. \tag{6.111}$$

## 6.6 Terms Associated with a Specific Argument

The method outlined in Sect. 6.5 permits the calculation of a full expansion of the planetary disturbing function to any specified order in the eccentricities and inclinations. Its major disadvantage is that to find the terms associated with a specific argument one has to carry out a complete expansion to the order of that argument. However, in Sect. 6.3 we showed that in this respect there are distinct advantages to using Kaula's form of the expansion. The major disadvantage of Kaula's formulae is the lack of Laplace coefficients.

Ellis & Murray (1999) derived a variation on Kaula's expansion that incorporates the best features of both approaches. Furthermore, they give explicit formulae for the *finite* series associated with a specific argument expanded to a specific order. Let the argument have the form

$$\varphi = j_1\lambda' + j_2\lambda + j_3\varpi' + j_4\varpi + j_5\Omega' + j_6\Omega \qquad (6.112)$$

and let $N_{max}$ be the maximum order of the expansion. Ellis & Murray showed that the expression for $\mathcal{R}_D$ associated with $\varphi$ is

$$\mathcal{R}_D = \sum_{i=0}^{i_{max}} \frac{(2i)!}{i!} \frac{(-1)^i}{2^{2i+1}} \alpha^i$$

$$\times \sum_{s=s_{min}}^{i} \sum_{n=0}^{n_{max}} \frac{(2s-4n+1)(s-n)!}{2^{2n}n!(2s-2n+1)!} \sum_{m=0}^{s-2n} \kappa_m \frac{(s-2n-m)!}{(s-2n+m)!}$$

$$\times (-1)^{s-2n-m} F_{s-2n,m,p}(I) F_{s-2n,m,p'}(I') \sum_{l=0}^{i-s} \frac{(-1)^s 2^{2s}}{(i-s-l)!l!}$$

$$\times \sum_{\ell=0}^{\ell_{max}} \frac{(-1)^\ell}{\ell!} \sum_{k=0}^{\ell} \binom{\ell}{k} (-1)^k \alpha^\ell \frac{d^\ell}{d\alpha^\ell} b_{i+\frac{1}{2}}^{(j)}(\alpha)$$

$$\times X_{-j_2}^{i+k,\,-j_2-j_4}(e) X_{j_1}^{-(i+k+1),\,j_1+j_3}(e')$$

$$\times \cos\left[j_1\lambda' + j_2\lambda + j_3\varpi' + j_4\varpi + j_5\Omega' + j_6\Omega\right] \qquad (6.113)$$

where, as before, $\kappa_0 = 1$ and $\kappa_m = 2$ for $m \neq 0$.

The following relationships hold throughout the calculation:

$$q = j_4, \qquad (6.114)$$

$$q' = -j_3, \qquad (6.115)$$

$$\ell_{max} = N_{max} - |j_5| - |j_6|, \qquad (6.116)$$

$$p_{min} = -(j_5 + j_6)/2, \quad p'_{min} = 0 \quad \text{if} \quad j_5 + j_6 < 0, \qquad (6.117)$$

$$p_{min} = 0, \quad p'_{min} = (j_5 + j_6)/2 \quad \text{if} \quad j_5 + j_6 \geq 0, \qquad (6.118)$$

$$s_{min} = \max\left(p_{min}, p'_{min}, j_6 + 2p_{min}, -j_5 + 2p'_{min}\right), \qquad (6.119)$$

$$i_{max} = \left[(N_{max} - |j_3| - |j_4|)/2\right], \qquad (6.120)$$

where the square brackets in Eq. (6.120) denote the integer part of the expression. A number of intermediate definitions are required for the summation. These are:

$$n_{max} = \left[(s - s_{min})/2\right], \qquad (6.121)$$

$$m_{min} = 0 \quad \text{if} \quad s, j_5 \quad \text{are both even or both odd,} \qquad (6.122)$$

$$m_{min} = 1 \quad \text{if} \quad s, j_5 \quad \text{are neither both even nor both odd,} \qquad (6.123)$$

$$p = (-j_6 - m + s - 2n)/2 \quad \text{with } p \le s - 2n \text{ and } p \ge p_{\min}, \quad (6.124)$$

$$p' = (j_5 - m + s - 2n)/2 \quad \text{with } p' \le s - 2n \text{ and } p' \ge p'_{\min}, \quad (6.125)$$

$$j = |j_2 + i - 2\bar{a} - 2n - 2p + q|. \quad (6.126)$$

Note that $q$ and $q'$ are determined directly from $\varphi$ and remain fixed over all the summations. However, $p$ and $p'$ change with $s$, $n$, and $m$ but the relationships given in Eqs. (6.124) and (6.125) always hold.

Ellis & Murray (1999) showed that the summations involved in the definitions of the functions of eccentricity and inclination in Eq. (6.113) need only be evaluated to a finite order that is at most equal to $N_{\max}$. The Hansen coefficient in $e$ need only include terms up to order $N_{\max} - |j_3| - |j_5| - |j_6|$ in $e$; similarly the Hansen coefficient in $e'$ need only include terms up to order $N_{\max} - |j_4| - |j_5| - |j_6|$ in $e'$. The $F$ inclination function in $I$ need only include terms up to order $N_{\max} - |j_3| - |j_4| - |j_5|$ in $I$; similarly the $F$ function in $I'$ need only include terms up to order $N_{\max} - |j_3| - |j_4| - |j_6|$ in $I'$.

For the indirect parts we have

$$\mathcal{R}_E = -\kappa_m \frac{(1-m)!}{(1+m)!} F_{1,m,p}(I) F_{1,m,p'}(I') X_{-j_2}^{1,-j_2-j_4}(e) X_{j_1}^{-2,j_1+j_3}(e')$$
$$\times \cos\left[j_1\lambda' + j_2\lambda + j_3\varpi' + j_4\varpi + j_5\Omega' + j_6\Omega\right] \quad (6.127)$$

and

$$\mathcal{R}_I = -\kappa_m \frac{(1-m)!}{(1+m)!} F_{1,m,p}(I) F_{1,m,p'}(I') X_{-j_2}^{-2,-j_2-j_4}(e) X_{j_1}^{1,j_1+j_3}(e')$$
$$\times \cos\left[j_1\lambda' + j_2\lambda + j_3\varpi' + j_4\varpi + j_5\Omega' + j_6\Omega\right], \quad (6.128)$$

where each of the quantities $p$, $p'$, and $m$ must be integers and equal to 0 or 1. If these conditions are not satisfied then the given argument does not appear in the expansion of the indirect part. As with $\mathcal{R}_D$ we can reduce the extent of the series expansions in powers of the eccentricity and inclination, and the same modifications apply. An analysis of the integers involved in the expansion of this indirect part gives the following relationships:

$$q = j_4, \quad (6.129)$$

$$q' = -j_3, \quad (6.130)$$

$$p = (j_2 + j_4 + 1)/2, \quad (6.131)$$

$$p' = -(j_1 + j_3 - 1)/2, \quad (6.132)$$

$$m = j_5 - 2p' + 1. \quad (6.133)$$

## 6.7 Use of the Disturbing Function

The complete expansions for $\mathcal{R}_D$, $\mathcal{R}_E$, and $\mathcal{R}_I$ contain an infinite number of cosine arguments. However, in practice we are only interested in certain cosine arguments and we can neglect all others. Our basis for doing this is the *averaging principle* whereby we assume (with some justification) that all the unimportant terms will be of short period and therefore their effects will average out to zero over the longer-period motion. This concept is illustrated in Sect. 6.9. All that concerns us here is that the averaging principle allows us to isolate those terms in the disturbing function that are appropriate for a particular problem and to ignore the infinite number of remaining terms. Effectively we move from a consideration of the infinite series of the full disturbing function, $\mathcal{R}$, to a finite series of the averaged disturbing function, $\langle\mathcal{R}\rangle$. This concept is the basis of our analysis of secular perturbations in Chapter 7, resonant perturbations in Chapter 8, and their applications to chaotic motion in Chapter 9 and planetary rings in Chapter 10. This approach to the use of the planetary disturbing function permits us to carry out analytical studies when we move beyond the simplicity of the two-body problem.

The procedure for determining the appropriate term, $\langle\mathcal{R}\rangle$ or $\langle\mathcal{R}'\rangle$, in the disturbing function is as follows:

1) Decide which combination of angles, $\varphi$, is applicable to the problem at hand. This requires knowledge of the physical problem and will be discussed in Sect. 6.9.
2) Determine the "order", $N$, of the argument. This is equal to the absolute value of the sum of the coefficients of $\lambda$ and $\lambda'$ in $\varphi$.
3) By looking at the appropriate order terms in the expansion of $\mathcal{R}_D$, determine the value of the integer $j$ that gives agreement with the desired argument, $\varphi$.
4) Calculate the combination of Laplace coefficients for that value of $j$ to give the explicit form of the term of interest, $\langle\mathcal{R}_D\rangle$ say.
5) Decide whether an external or an internal perturbation is being considered. This is determined by the nature of the problem.
6) If the perturbation is external, then look at the appropriate order terms in the expansion of the indirect part, $\mathcal{R}_E$, and isolate a matching argument, if it exists, and read off the corresponding indirect term $\langle\mathcal{R}_E\rangle$.
7) If the perturbation is internal, then look at the appropriate order terms in the expansion of the indirect part, $\mathcal{R}_I$, and isolate a matching argument, if it exists, and read off the corresponding indirect term $\langle\mathcal{R}_I\rangle$.
8) If the perturbation is external then

$$\langle\mathcal{R}\rangle = \frac{\mu'}{a'}\left(\langle\mathcal{R}_D\rangle + \alpha\langle\mathcal{R}_E\rangle\right). \tag{6.134}$$

9) If the perturbation is internal then

$$\langle\mathcal{R}'\rangle = \frac{\mu}{a}\left(\alpha\langle\mathcal{R}_D\rangle + \frac{1}{\alpha}\langle\mathcal{R}_I\rangle\right). \tag{6.135}$$

Use of the explicit expansion of the indirect part in steps 6 and 7 can be avoided altogether since the averaged indirect part of the disturbing function, $\langle\mathcal{R}_E\rangle$ or

$\langle \mathcal{R}_I \rangle$, can be obtained from the averaged direct part, $\langle \mathcal{R}_D \rangle$. The procedure is as follows: To obtain $\langle \mathcal{R}_E \rangle$ replace every occurrence of $\alpha^n D^n A_1$ ($n = 0, 1$) in $\langle \mathcal{R}_D \rangle$ by $-1$ and replace every occurrence of $\alpha^n D^n B_0$ ($n = 0, 1$) in $\langle \mathcal{R}_D \rangle$ by $-2$; all other terms are ignored. To obtain $\langle \mathcal{R}_I \rangle$ replace every occurrence of $\alpha^n D^n A_1$ ($n = 0, 1, 2, \ldots$) in $\langle \mathcal{R}_D \rangle$ by $(-1)^{n+1}(n + 1)!$ and replace every occurrence of $\alpha^n D^n B_0$ ($n = 0, 1, 2, \ldots$) in $\langle \mathcal{R}_D \rangle$ by $(-1)^{n+1}(2n + 2)n!$, ignoring all other terms.

Throughout this analysis we have assumed that $r' > r$ (i.e., that the orbits do not intersect). The convergence of the resulting series will therefore depend on how close the orbits are to intersection. Obviously if the orbits intersect there will be a singularity since $\mathbf{r} = \mathbf{r}'$ at some longitude and the first term in Eq. (6.16) or (6.18) becomes undefined. Hence an approximate condition for convergence is

$$a(1 + e) < a'(1 - e'), \qquad (6.136)$$

or that the apocentric distance of the inner orbit has to be less than the pericentric distance of the outer orbit.

Another advantage of the Legendre-type expansion given in Eq. (6.113) is that it is easy to see the form of the lowest order terms in the expansion. We have already stated in Sect. 6.3 (and shown in Sect. 6.5) that the potential associated with the perturbations of the orbit of the mass $m$ by the mass $m'$ can be written as

$$\mathcal{R} = \mu' \sum S \cos \varphi, \qquad (6.137)$$

where $S$ is a function of the semi-major axes, eccentricities, and inclinations of $m$ and $m'$. From the definitions of mean longitude and longitude of pericentre, the general form of the argument $\varphi$ can be written as

$$\varphi = (l - 2p' + q')\lambda' - (l - 2p + q)\lambda - q'\varpi' + q\varpi$$
$$+ (m - l + 2p')\Omega' - (m - l + 2p)\Omega, \qquad (6.138)$$

where, in this case, $l$, $m$, $p$, $p'$, $q$, and $q'$ are all integers. We can calculate the valid arguments by using the property that the sum of the integer coefficients of the angle variables in each argument is zero. If we write the general form of an argument as

$$\varphi = j_1\lambda' + j_2\lambda + j_3\varpi' + j_4\varpi + j_5\Omega' + j_6\Omega \qquad (6.139)$$

then our condition on the coefficients implies that

$$\sum_{i=1}^{6} j_i = 0. \qquad (6.140)$$

This is the *d'Alembert relation* and it does *not* apply to any choice of angles – we must use angles that are referred to a fixed direction (i.e., longitudes rather than anomalies). The longitudes $\lambda$, $\lambda'$, $\varpi$, $\varpi'$, $\Omega$, and $\Omega'$ form an appropriate set of angles. Hamilton (1994) provides an overview of the d'Alembert rules that determine such relationships.

Now consider the form of $S$, the "strength" of an individual term. From the properties of $X_{l-2p+q}^{l,l-2p}(e)$ and $F_{lmp}(I)$ we can calculate the lowest order terms in the eccentricities and inclinations. Use of Eqs. (6.37)–(6.42) gives

$$X_{l-2p+q}^{l,l-2p}(e) = \mathcal{O}(e^{|q|}), \qquad X_{l-2p'+q'}^{-l-1,l-2p'}(e') = \mathcal{O}(e'^{|q'|}) \qquad (6.141)$$

and

$$F_{lmp}(I) = \mathcal{O}(s^{|m-l+2p|}), \qquad F_{lmp'}(I') = \mathcal{O}(s'^{|m-l+2p'|}), \qquad (6.142)$$

where $s = \sin \frac{1}{2}I$ and $s' = \sin \frac{1}{2}I'$. Therefore we can write

$$S \approx \frac{f(\alpha)}{a'} e^{|q|} e'^{|q'|} s^{|m-l+2p|} s'^{|m-l+2p'|} = \frac{f(\alpha)}{a'} e^{|j_4|} e'^{|j_3'|} s^{|j_6|} s'^{|j_5|}, \qquad (6.143)$$

where $f(\alpha)$ can be expressed as a function of Laplace coefficients. Hence the lowest power of $e$, for example, in a given term is at least equal to the absolute value of the coefficient of $\varpi$. Similarly the lowest powers of $e'$, $\sin \frac{1}{2}I$, and $\sin \frac{1}{2}I'$ are greater than or equal to the absolute value of the coefficients of $\varpi'$, $\Omega$, and $\Omega'$ respectively in $\varphi$. This property is clear from the second-order expansion given in Sect. 6.5 and the fourth-order expansion in Appendix B.

## 6.8 Lagrange's Planetary Equations

The expansion of the disturbing function gives us the dependence of the perturbing potential on the orbital elements. Now we need to quantify the resulting orbital variations of the perturbed body. To do this we make use of *Lagrange's planetary equations*. These are best derived using a Hamiltonian formulation (see Sect. 2.10). Here we confine ourselves to a statement of the equations. Full derivations can be found in Brouwer & Clemence (1961) and Roy (1988).

The use of Lagrange's equations require the introduction of an additional angle. If we write

$$\lambda = M + \varpi = n(t - \tau) + \varpi = nt + \epsilon, \qquad (6.144)$$

where $\lambda$ is the mean longitude, $M$ is the mean anomaly, $\varpi$ is the longitude of pericentre, $t$ is time, and $\tau$ is the time of pericentre passage, then the new angle $\epsilon$ denotes the *mean longitude at epoch* (i.e., the mean longitude of the mass $m$ at the moment from which time is measured). Lagrange's equations for the variations of the orbital elements are

$$\frac{da}{dt} = \frac{2}{na} \frac{\partial \mathcal{R}}{\partial \epsilon}, \qquad (6.145)$$

$$\frac{de}{dt} = -\frac{\sqrt{1-e^2}}{na^2 e}(1 - \sqrt{1-e^2})\frac{\partial \mathcal{R}}{\partial \epsilon} - \frac{\sqrt{1-e^2}}{na^2 e}\frac{\partial \mathcal{R}}{\partial \varpi}, \qquad (6.146)$$

$$\frac{d\epsilon}{dt} = -\frac{2}{na} \frac{\partial \mathcal{R}}{\partial a} + \frac{\sqrt{1-e^2}(1 - \sqrt{1-e^2})}{na^2 e}\frac{\partial \mathcal{R}}{\partial e} + \frac{\tan \frac{1}{2}I}{na^2 \sqrt{1-e^2}}\frac{\partial \mathcal{R}}{\partial I}, \qquad (6.147)$$

$$\frac{d\Omega}{dt} = \frac{1}{na^2\sqrt{1-e^2}\sin I}\frac{\partial \mathcal{R}}{\partial I}, \tag{6.148}$$

$$\frac{d\varpi}{dt} = \frac{\sqrt{1-e^2}}{na^2 e}\frac{\partial \mathcal{R}}{\partial e} + \frac{\tan\frac{1}{2}I}{na^2\sqrt{1-e^2}}\frac{\partial \mathcal{R}}{\partial I}, \tag{6.149}$$

$$\frac{dI}{dt} = \frac{-\tan\frac{1}{2}I}{na^2\sqrt{1-e^2}}\left(\frac{\partial \mathcal{R}}{\partial \epsilon} + \frac{\partial \mathcal{R}}{\partial \varpi}\right) - \frac{1}{na^2\sqrt{1-e^2}\sin I}\frac{\partial \mathcal{R}}{\partial \Omega}. \tag{6.150}$$

A problem arises if we consider the expression for $\dot{\epsilon}$ given in Eq. (6.147) (see, e.g., Brouwer & Clemence 1961). Since the right-hand side of the equation contains a factor $\partial \mathcal{R}/\partial a$ we have to be aware that the semi-major axis occurs explicitly in the Laplace coefficients of the disturbing function and implicitly in the arguments of the cosine terms as the mean motion since $\lambda = nt + \epsilon$. This gives rise to the time occurring as a factor when the partial derivative is taken. The problem can be overcome if we define a new mean longitude at epoch, $\epsilon^*$, by

$$\frac{d\epsilon^*}{dt} = \frac{d\epsilon}{dt} + t\frac{dn}{dt}. \tag{6.151}$$

Hence

$$\frac{d\lambda}{dt} = n + \frac{d\epsilon^*}{dt} \tag{6.152}$$

and

$$\lambda = \int n\,dt + \epsilon^*. \tag{6.153}$$

This can also be written as

$$\lambda = \rho + \epsilon^*, \tag{6.154}$$

where

$$\frac{d\rho}{dt} = n, \qquad \frac{d^2\rho}{dt^2} = \frac{dn}{dt} = -\frac{3}{2}\frac{n}{a}\frac{da}{dt} \tag{6.155}$$

or

$$\frac{d^2\rho}{dt^2} = -\frac{3}{a^2}\frac{\partial \mathcal{R}}{\partial \epsilon}. \tag{6.156}$$

In this case we should consider any derivatives of $\partial/\partial\epsilon$, such as those that occur in the expressions for $\dot{a}$, $\dot{e}$, and $\dot{I}$, to mean $\partial/\partial\lambda$. In practice the variation of $\epsilon$ can usually be neglected since it is a small effect.

The variation of the orbital elements of the mass $m'$ can be expressed by equations similar to Eqs. (6.145)–(6.150), with $\mathcal{R}$ replaced by $\mathcal{R}'$ and all unprimed variables exchanged for primed ones. The derivation of Lagrange's planetary equations does not assume that $\mathcal{R}$ arises from perturbations by an external mass. Therefore we can equally well use these equations to study the perturbations on the mass $m$ due to, for example, a nonspherical central mass. This will be considered in Sect. 6.11. Similarly the equations are equally applicable if we use the averaged disturbing functions $\langle \mathcal{R} \rangle$ and $\langle \mathcal{R}' \rangle$.

We have already seen in Sect. 2.9 that the variations in the orbital elements can be expressed in terms of the radial, tangential, and orthogonal forces acting on an orbiting object. However, Lagrange's equations allow us to derive similar variations but based on the Fourier series expansion of the disturbing function discussed in this chapter. As such they provide the basis for most of the perturbation calculations that follow.

## 6.9 Classification of Arguments in the Disturbing Function

We can now approach the subject of the physical significance of the expansion of the disturbing function. So far we have expressed the perturbing potential as a series involving an infinite number of permissible combinations of angles. But which angles are important in any given problem? In other words, which of the infinite terms in the expansion are important and which can be ignored? To a large extent the answers to these questions depend on the semi-major axis of the perturbed orbit. We can classify all arguments by considering the frequencies or periods associated with the cosine arguments in the expansion.

Each cosine argument contains a linear combination of the angles $\lambda'$, $\lambda$, $\varpi'$, $\varpi$, $\Omega'$, and $\Omega$. We know that in the unperturbed problem the mean longitudes, $\lambda'$ and $\lambda$, increase linearly at rates $n'$ and $n$ respectively. In contrast, all the other angles are constant in the unperturbed problem. Therefore, when we consider the perturbed system $\lambda'$ and $\lambda$ are rapidly varying quantities, whereas all the other angles undergo slow variations. Therefore, any valid arguments that do not involve mean longitudes are slowly varying. These give rise to *secular* terms, from the Latin verb *saeculum* meaning century, or long period. This does not imply that all other arguments are of short period. Consider a general argument of the form $\varphi = j_1 \lambda' + j_2 \lambda + j_3 \varpi' + j_4 \varpi + j_5 \Omega' + j_6 \Omega$ with

$$\lambda' \approx n't + \epsilon' \quad \text{and} \quad \lambda \approx nt + \epsilon \qquad (6.157)$$

(see Eq. (6.144)). Therefore $j_1 \lambda' + j_2 \lambda \approx (j_1 n' + j_2 n)t + \text{constant}$ and so, if the semi-major axes are such that

$$j_1 n' + j_2 n \approx 0, \qquad (6.158)$$

then this argument also has a period longer than either orbital period. Equation (6.158) is satisfied when there is a commensurability between the two mean motions or orbital periods (see Sect. 1.7). We classify such arguments as giving rise to *resonant* terms in the expansion. If we consider the semi-major axes, the equivalent condition is

$$a \approx (|j_1|/|j_2|)^{\frac{2}{3}} a'. \qquad (6.159)$$

Because of the dependence on semi-major axis, resonant terms are localised. Whereas a particular combination of angles may be slowly varying at one semi-major axis of the perturbed body, the same combination would be varying rapidly at another. In contrast the secular terms can be considered as global.

Any argument that is neither secular nor resonant is considered to give rise to a *short-period* term. In practise the application of the averaging principle mentioned in Sect. 6.7 allows us to ignore the infinite number of short-period terms in the expansion and accept that the dynamics is dominated by the appropriate secular and resonant terms.

Below we provide predictions of motion under secular and resonant terms in the context of the elliptical restricted three-body problem with small inclination, and we compare the answers with the results of numerical integrations. Here we assume that the mass $m$ is negligible and that the orbit of $m'$ is a fixed ellipse in the reference plane. Our starting point is a set of the lowest order form of Lagrange's equations for $\dot{a}$, $\dot{e}$, $\dot{\varpi}$, and $\dot{\Omega}$ derived from inspection of Eqs. (6.145), (6.146), (6.149), and (6.148). The equations of motion are

$$\frac{da}{dt} = \frac{2}{na}\frac{\partial\langle\mathcal{R}\rangle}{\partial\lambda}, \tag{6.160}$$

$$\frac{de}{dt} = -\frac{1}{na^2e}\frac{\partial\langle\mathcal{R}\rangle}{\partial\varpi}, \tag{6.161}$$

$$\frac{d\varpi}{dt} = +\frac{1}{na^2e}\frac{\partial\langle\mathcal{R}\rangle}{\partial e}, \tag{6.162}$$

$$\frac{d\Omega}{dt} = +\frac{1}{na^2\sin I}\frac{\partial\langle\mathcal{R}\rangle}{\partial I}, \tag{6.163}$$

where $\langle\mathcal{R}\rangle$ is the averaged part of the disturbing function for an external perturber.

### 6.9.1 Secular Terms

Secular terms arise from those arguments that do not contain the mean longitudes. Inspection of the direct part of the second-order expansion in Eq. (6.107) shows that secular terms are obtained by setting $j = 0$ in those cosine arguments containing $j\lambda' - j\lambda$. This gives

$$\langle\mathcal{R}_{\mathrm{D}}\rangle = C_0 + C_1(e^2 + e'^2) + C_2 s^2 + C_3 ee'\cos(\varpi' - \varpi), \tag{6.164}$$

where

$$C_0 = \frac{1}{2}b_{\frac{1}{2}}^{(0)}(\alpha), \tag{6.165}$$

$$C_1 = \frac{1}{8}\left[2\alpha D + \alpha^2 D^2\right]b_{\frac{1}{2}}^{(0)}(\alpha), \tag{6.166}$$

$$C_2 = -\frac{1}{2}\alpha b_{\frac{3}{2}}^{(1)}(\alpha), \tag{6.167}$$

$$C_3 = \frac{1}{4}\left[2 - 2\alpha D - \alpha^2 D^2\right]b_{\frac{1}{2}}^{(1)}(\alpha). \tag{6.168}$$

Note that there are no $ss'$ or $s'^2$ terms in $\langle \mathcal{R}_{\mathrm{D}} \rangle$ because we are taking $s' = 0$ and that $C_0$ is a function of $\alpha$ only. Furthermore, inspection of the terms in $\mathcal{R}_{\mathrm{E}}$ (Eq. (6.110)) shows that all the arguments contain at least one mean longitude and so there are no secular contributions from the indirect part of the disturbing function. Hence the low-order version of Lagrange's equations becomes

$$\left( \frac{da}{dt} \right)_{\mathrm{sec}} = 0, \tag{6.169}$$

$$\left( \frac{de}{dt} \right)_{\mathrm{sec}} = n\alpha(m'/m_{\mathrm{c}})C_3 e' \sin(\varpi - \varpi'), \tag{6.170}$$

$$\left( \frac{d\varpi}{dt} \right)_{\mathrm{sec}} = n\alpha(m'/m_{\mathrm{c}}) \left[ 2C_1 + C_3(e'/e) \cos(\varpi - \varpi') \right], \tag{6.171}$$

$$\left( \frac{d\Omega}{dt} \right)_{\mathrm{sec}} = n\alpha(m'/m_{\mathrm{c}})(C_2/2), \tag{6.172}$$

where we have used the fact that $\mu' = \mathcal{G}m' \approx n^2 a^3(m'/m_{\mathrm{c}})$, where $m_{\mathrm{c}}$ is the mass of the central object. If we assume that $e \gg e'$ then the approximate solutions to these equations are

$$a = a_0, \tag{6.173}$$

$$e = e_0 - \frac{n\alpha}{\varpi}(m'/m_{\mathrm{c}})C_3 e' \left[ \cos \varpi_0 - \cos \varpi \right], \tag{6.174}$$

$$\varpi = \varpi_0 + n\alpha(m'/m_{\mathrm{c}})2C_1 t, \tag{6.175}$$

$$\Omega = \Omega_0 + n\alpha(m'/m_{\mathrm{c}})(C_2/2)t, \tag{6.176}$$

where the subscript 0 denotes the initial ($t = 0$) value of a quantity, and we have taken $\varpi' = 0$. These solutions predict that there is no secular change in $a$, that $e$ varies sinusoidally with an amplitude of

$$(\Delta e)_{\mathrm{sec}} = \left| (n\alpha/\varpi)(m'/m_{\mathrm{c}})C_3 e' \right|, \tag{6.177}$$

and that $\varpi$ and $\Omega$ will either increase or decrease linearly with time depending on the signs of $C_1$ and $C_2$.

Figures 6.3a–d show the results of a numerical integration of the full equations of motion of the elliptical restricted three-body problem with $a' = 1$, $e' = 0.048$, $\varpi' = 0$, $I' = 0$, and $m'/m_{\mathrm{c}} = 1/1047.355$ with starting conditions $a_0 = 0.192$, $e_0 = 0.1$, $\varpi_0 = 130°$, $\Omega_0 = 200°$, $\lambda_0 = 300°$, and $\lambda' = 0°$. Substitution of $\alpha = a/a' = 0.192$ in Eqs. (6.166)–(6.168) gives $C_1 = 0.0148335$, $C_2 = -0.0593339$, and $C_3 = -0.00708688$; note that $2C_1 = -C_2/2$. Since the mass ratio is that of the Jupiter–Sun ratio, the integration was designed to mimic the motion of an asteroid perturbed by Jupiter, and so the time units in the plots are given as Jupiter periods. However, the semi-major axis was deliberately chosen to be far away from Jupiter in order to avoid proximity to strong resonances. In these circumstances the secular perturbations alone should provide a good approximation to the motion.

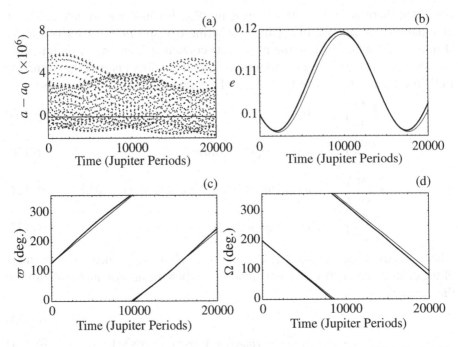

Fig. 6.3.   A comparison of the results of a full numerical integration (thick line) with predictions from analytical theory (thin line) for the variation of (a) semi-major axis, (b) eccentricity, (c) longitude of perihelion, and (d) longitude of ascending node for a test particle undergoing predominantly secular perturbations from Jupiter.

The results show that the agreement is excellent over the 20,000 Jupiter periods of the integration. There are variations in $a$ but these are extremely small; note that the scale in Fig. 6.3a is enlarged. The fact that the semi-major axis is almost constant justifies the evaluation of the Laplace coefficients for a fixed value of $\alpha$. The eccentricity does vary as predicted, and while $\varpi$ increases linearly with time (since $C_1 > 0$), $\Omega$ is decreasing linearly at the same rate (since $2C_1 = -C_2/2$; cf. Eqs. (6.175) and (6.176)). Prograde motion of the pericentre (or node) is called *precession* and retrograde motion is called *regression*. The behaviour of $\varpi$ and $\Omega$ is a natural consequence of the secular terms in the disturbing function.

Because of the infinite number of short-period terms in the disturbing function, which we have neglected, there should be differences between the results of a full integration and the predictions of our analytical theory. We can see this already in the Fig. 6.3a, where there are small, but detectable short-period changes in the semi-major axis from the constant value predicted by theory. Figure 6.4 shows the difference between the "observed" eccentricity (i.e., the one determined by the numerical integration) and the calculated value from theory, as a function of time for the first 1,000 Jupiter periods of the integration. Here again we can see the effect of the short-period terms inherently included in any full integration.

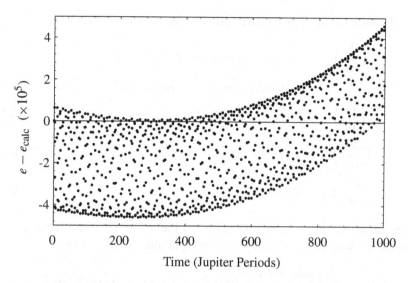

Fig. 6.4.  Differences between the observed and calculated values of the test particle's eccentricity as a function of time. The data are sampled every Jupiter period and show the short-period variations in $e$.

### 6.9.2 Resonant Terms

Now suppose, for example, that we want to study an asteroid's motion at 3.27 AU, under the perturbing effect of Jupiter. Since Jupiter's semi-major axis is 5.20 AU we have, using Kepler's third law, that the ratio of their periods is $(3.27/5.20)^{3/2} \approx 0.499$. Hence, we have the relation $2n' \approx n$ and we would expect resonant terms to be important. Therefore, in the vicinity of the 2:1 resonance, as well as the secular terms discussed above, we also need to consider those terms in the expansion of the disturbing function that contain $2\lambda' - \lambda$ (i.e., the resonant terms for this location).

Inspection of Eq. (6.107) shows that in a second-order expansion there are two terms in $\langle \mathcal{R}_\mathrm{D} \rangle / a'$ that have a cosine argument containing $2\lambda' - \lambda$ for specific values of $j$. The relevant direct part of the averaged disturbing function is

$$\langle \mathcal{R}_\mathrm{D} \rangle = C_0 + C_1(e^2 + e'^2) + C_2(s^2 + s'^2) + C_3 ee' \cos(\varpi - \varpi')$$
$$+ C_4 e \cos(2\lambda' - \lambda - \varpi) + C_5 e' \cos(2\lambda' - \lambda - \varpi'), \quad (6.178)$$

where the additional constants $C_4$ and $C_5$ are given by

$$C_4 = \frac{1}{2}[-4 - \alpha D] b_{\frac{1}{2}}^{(2)}(\alpha), \quad (6.179)$$

$$C_5 = \frac{1}{2}[3 + \alpha D] b_{\frac{1}{2}}^{(1)}(\alpha). \quad (6.180)$$

The second of these two resonant arguments makes no contribution to $\dot{e}$, $\dot{\varpi}$, and $\dot{\Omega}$ but does contribute a term to $\dot{a}$. Inspection of Eq. (6.110) shows that there is also a $-2\alpha e'$ contribution to the same argument from the indirect part.

Application of the approximate form of Lagrange's equations gives

$$\left(\frac{da}{dt}\right)_{\text{res}} = 2n\alpha a(m'/m_c)C_4 e \sin(2\lambda' - \lambda - \varpi)$$
$$+ 2n\alpha a(m'/m_c)(C_5 - 2\alpha)e' \sin(2\lambda' - \lambda - \varpi'), \quad (6.181)$$

$$\left(\frac{de}{dt}\right)_{\text{res}} = n\alpha(m'/m_c)C_4 \sin(2\lambda' - \lambda - \varpi), \quad (6.182)$$

$$\left(\frac{d\varpi}{dt}\right)_{\text{res}} = n\alpha(m'/m_c)(C_4/e) \cos(2\lambda' - \lambda - \varpi), \quad (6.183)$$

$$\left(\frac{d\Omega}{dt}\right)_{\text{res}} = 0 \quad (6.184)$$

for the variations in $a$, $e$, $\varpi$, and $\Omega$ due to the 2:1 resonance. If we consider approximate solutions for these resonant equations alone we obtain

$$a = a_0 - \frac{2n\alpha a(m'/m_c)C_4 e}{2n' - n - \dot{\varpi}} \left[\cos(2\lambda' - \lambda - \varpi) - \cos(\lambda_0 + \omega_0)\right]$$
$$- \frac{2n\alpha a(m'/m_c)(C_5 - 2\alpha)e'}{2n' - n} \left[\cos(2\lambda' - \lambda - \varpi') - \cos\lambda_0\right], \quad (6.185)$$

$$e = e_0 + \frac{n\alpha(m'/m_c)C_4}{2n' - n - \dot{\varpi}} \left[\cos(2\lambda' - \lambda - \varpi) - \cos(\lambda_0 + \omega_0)\right], \quad (6.186)$$

$$\varpi = \varpi_0 + \frac{n\alpha(m'/m_c)(C_4/e)}{2n' - n - \dot{\varpi}} \left[\sin(2\lambda' - \lambda - \varpi) + \sin(\lambda_0 + \omega_0)\right], \quad (6.187)$$

$$\Omega = \Omega_0. \quad (6.188)$$

To derive these solutions we have assumed that the only time-varying quantities on the right-hand side of the equations for $\dot{a}$, $\dot{e}$, and $\dot{\varpi}$ are in the cosine arguments and that $\varpi$ increases linearly with time at a constant rate $\dot{\varpi}$ determined by secular theory. These equations suggest that $a$, $e$, and $\varpi$ will experience sinusoidal variations with maximum amplitudes

$$(\Delta a)_{\text{res}} = 2n\alpha a(m'/m_c)\left(\left|\frac{C_4 e}{2n' - n - \dot{\varpi}}\right| + \left|\frac{(C_5 - 2\alpha)e'}{2n' - n}\right|\right), \quad (6.189)$$

$$(\Delta e)_{\text{res}} = \left|\frac{n\alpha(m'/m_c)C_4}{2n' - n - \dot{\varpi}}\right|, \quad (6.190)$$

$$(\Delta\varpi)_{\text{res}} = \left|\frac{n\alpha(m'/m_c)(C_4/e)}{2n' - n - \dot{\varpi}}\right|. \quad (6.191)$$

These are only approximate solutions, particularly in the case of the semi-major axis where we have just combined the amplitudes of the terms associated with each of the two resonant arguments.

Figures 6.5a–d show the result of a full integration of the equations of motion and a comparison with the predicted variations from the combined secular and resonant theory outlined above. The calculations were done with the same

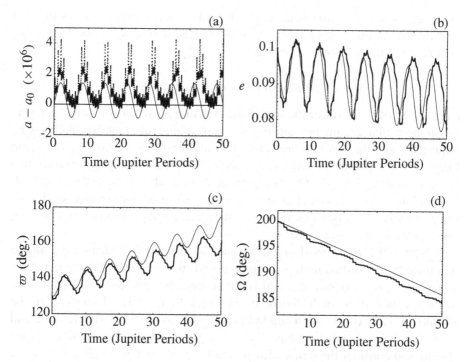

**Fig. 6.5.** A comparison of the results of a full numerical integration (points or thick line) with predictions from analytical theory (thin line) for the variation of (a) semi-major axis, (b) eccentricity, (c) longitude of perihelion, and (d) longitude of ascending node for a test particle near the 2:1 resonance undergoing resonant and secular perturbations from Jupiter.

starting values as in Sect. 6.9.1, but with $a = 0.6$ in order to place the test particle close to (but not in) the 2:1 jovian resonance. The relevant constants are now $C_1 = 0.314001$, $C_2 = -1.25600$, $C_3 = -0.447005$, $C_4 = -1.04332$, and $C_5 = 1.55230$. Note that the magnitudes of $C_1$ and $C_2$ have increased by a factor of $\sim 20$ over those in our secular example with $a = 0.192$. This is because the separation from Jupiter has decreased, thereby increasing the size of the secular effects. Examination of Fig. 6.5 shows that there is good agreement between the predictions and the numerical results, with the amplitudes and frequencies of the variations in $a$, $e$, and $\varpi$ being close to their predicted values. We would expect there to be some differences, partly due to our approximate form of Lagrange's equations and partly due to the fact that in order to integrate the differential equations we took the quantities $a$ and $e$ on the right-hand side of Eqs. (6.181)–(6.183) to be constant, whereas clearly they are varying due to the resonance.

Note from Eqs. (6.189)–(6.191) that all the amplitudes contain a divisor of the form $2n' - n - \dot{\varpi}$ (i.e., the time derivative of the resonant argument $2\lambda' - \lambda - \varpi$). This implies that the changes in the elements will become even larger as the exact resonance is approached. However, it is in such circumstances that the assumptions in our simple analytical model break down. We consider a more complete model of resonance in Sect. 8.

### 6.9.3 Short-Period and Small-Amplitude Terms

Knowledge of the form of the disturbing function allows us to isolate the permissible secular arguments and the resonant arguments that are likely to be important. In effect, we are assuming that all other terms involving the mean longitudes are of short period and that their effect will average out to zero; this is the averaging principle. Our comparison of the analytical theory with the full integrations in Figs. 6.4 and 6.5 shows that this is a good approximation. Therefore, although short-period terms exist, their effects appear to be negligible, at least for the examples we chose.

In Sect. 6.9.2 we showed that if we want to know which terms are going to dominate the perturbed motion of the asteroid, we should find those terms for which $j_1 n' + j_2 n \approx 0$, where $j_1$ and $j_2$ are integers, because these will contribute to the creation of a small divisor in Eqs. (6.189)–(6.191). Therefore, in the vicinity of 3.27 AU the dominant terms are likely to be those with $j_1 = \pm 2$ and $j_2 = \mp 1$ since we will then have $2n' - n \approx 0$. However, this implies that we should also consider the terms with $j_1 = \pm 4, \pm 6, \ldots$ and $j_2 = \mp 2, \mp 3, \ldots$ since these will also give rise to small divisors. Can there be an infinite number of such terms, all contributing to motion at this resonance? Simple number theory tells us that we can always approximate the ratio of two real numbers (in our case the two mean motions) by a rational number, to arbitrary precision. Ought there be an infinite number of resonances that could contribute large-amplitude terms to the disturbing function at any semi-major axis?

We can resolve these paradoxes by considering our expression for $S$, the "strength" of the disturbing function (see Eq. (6.143)). For simplicity consider the case of a near commensurability of mean motions in the planar, circular, restricted, three-body problem. Let the resonant argument be

$$\varphi = (j + k)\lambda' - j\lambda - k\varpi \qquad (6.192)$$

and let us assume that there is a near commensurability such that

$$(j + k)n' - jn \approx 0, \qquad (6.193)$$

where $j$ and $k$ are integers. This means that arguments which contain expressions of the form $(j + k)\lambda' - j\lambda - k\varpi$ can vary slowly and produce long-period, large-amplitude perturbations. For example, in the case of the 2:1 resonance we have $j = \pm 1, \pm 2, \pm 3, \ldots$ and $k = \pm 1, \pm 2, \pm 3, \ldots$. However, although there is an infinite number of possible resonances for each pair of $j$ and $k$, most of them are weak. This is because $S \propto e^{|k|}$ (see Eq. (6.143)) and $e < 1$. Therefore, as $k$ increases the strength decreases. In effect these other terms exist but they are of small amplitude. By a similar argument we can overcome the difficulty of always being arbitrarily close to a resonance. For example, the 21:10 resonance is close to the 2:1 resonance, yet in this case $S \propto e^{11}$ and so the resonance is weak. Therefore the "nearly resonant" terms corresponding to higher orders in the eccentricities and inclinations can be discarded in the same way as all the other short-period terms.

The *order k* of a resonant term is identical to the order $N = |j_1 + j_2|$ of an argument in the disturbing function. Thus, if we require all the arguments that could contribute to a given second-order resonance, we should look at the arguments labelled D2 and E2 (or I2) in the expansion of the disturbing function given in Appendix B. We may need to consider other arguments as well, because if we require a fourth-order expansion in the orbital elements we should also look at the D4 and E4 (or I4) arguments. Similarly, because the secular terms do not contain mean longitudes, we only need to consider arguments labelled D0 in Appendix B.

## 6.10 Sample Calculations of the Averaged Disturbing Function

Here we consider the calculation of the appropriate terms in the disturbing function for two commensurabilities. In the first case, that of a second-order commensurability, we make use of the literal expansion given in Appendix B. In the second case the commensurability is of eleventh order and we resort to the form of the expansion for explicit arguments derived by Ellis & Murray (1999) and given in Sect. 6.6.

### 6.10.1 Terms Associated with the 3:1 Commensurability

Here we derive the terms required for a study of asteroid motion at 2.50 AU, close to the 3:1 commensurability with Jupiter. This will be used in our study of this resonance in Sect. 9.5.2. If we assume that the asteroid mass in negligible ($m \ll m'$) and that its eccentricity and inclination are small enough to allow us to use a second-order expansion of the disturbing function (i.e., we can ignore higher order terms in the fourth-order expansion given in Appendix B), then the necessary secular terms in the expansion are 4D0.1, 4D0.2, and 4D0.3 with $j = 0$, while the resonant terms are 4D2.1, 4D2.2, 4D2.3, 4D2.4, 4D2.5, 4D2.6 with $j = 3$, and 4E2.5. This gives an expression for the averaged disturbing function of the form

$$\frac{a'}{\mu'}\langle\mathcal{R}\rangle = A_0 + A_1 e^2 + A_2 s^2 + A_3 ee' \cos(\varpi' - \varpi) + A_4 ss' \cos(\Omega' - \Omega)$$

$$+ A_5 e^2 \cos(3\lambda' - \lambda - 2\varpi) + A_6 ee' \cos(3\lambda' - \lambda - \varpi' - \varpi)$$

$$+ A_7 e'^2 \cos(3\lambda' - \lambda - 2\varpi') + A_8 s^2 \cos(3\lambda' - \lambda - 2\Omega)$$

$$+ A_9 ss' \cos(3\lambda' - \lambda - \Omega' - \Omega) + A_{10} s'^2 \cos(3\lambda' - \lambda - 2\Omega'), \quad (6.194)$$

where the $A_i$ $(i = 0, 1, \ldots, 10)$ now denote combinations of Laplace coefficients and their derivatives. Note that there are other terms in the secular part of the expansion that contain expressions of second order in $e'$ and $s'$. However, since we are interested in studying the motion of the asteroid (not Jupiter) we can take the orbital elements of Jupiter to be fixed and hence these expressions are effectively constants. The explicit forms of the constants $A_i$ are

$$A_0 = \frac{1}{2} b_{\frac{1}{2}}^{(0)}(\alpha), \tag{6.195}$$

$$A_1 = \frac{1}{8} \left[ 2\alpha D + \alpha^2 D^2 \right] b_{\frac{1}{2}}^{(0)}(\alpha), \tag{6.196}$$

$$A_2 = \frac{1}{2} \left[ -\alpha \right] b_{\frac{3}{2}}^{(1)}(\alpha), \tag{6.197}$$

$$A_3 = \frac{1}{4} \left[ 2 - 2\alpha D - \alpha^2 D^2 \right] b_{\frac{1}{2}}^{(1)}(\alpha), \tag{6.198}$$

$$A_4 = [\alpha] b_{\frac{3}{2}}^{(1)}(\alpha), \tag{6.199}$$

$$A_5 = \frac{1}{8} \left[ 21 + 10\alpha D + \alpha^2 D^2 \right] b_{\frac{1}{2}}^{(3)}(\alpha), \tag{6.200}$$

$$A_6 = \frac{1}{4} \left[ -20 - 10\alpha D - \alpha^2 D^2 \right] b_{\frac{1}{2}}^{(2)}(\alpha), \tag{6.201}$$

$$A_7 = \frac{1}{8} \left[ 17 + 10\alpha D + \alpha^2 D^2 \right] b_{\frac{1}{2}}^{(1)}(\alpha) - \frac{27}{8}\alpha, \tag{6.202}$$

$$A_8 = \frac{1}{2} [\alpha] b_{\frac{3}{2}}^{(2)}(\alpha), \tag{6.203}$$

$$A_9 = [-\alpha] b_{\frac{3}{2}}^{(2)}(\alpha), \tag{6.204}$$

$$A_{10} = \frac{1}{2} [\alpha] b_{\frac{3}{2}}^{(2)}(\alpha). \tag{6.205}$$

Note that, for the reasons given above, we have excluded the terms in 4D0.1 that only contained primed quantities. The $-(27/8)\alpha$ term in $A_7$ comes from the indirect term 4E2.5.

A numerical value for each of the $A_i$ shown above can be calculated at a given value of $\alpha$, the ratio of the semi-major axes. It is customary to fix this value of $\alpha$ when the asteroid is known to be in close proximity to the resonance such that the resonant terms in the expansion dominate. This is a good approximation, especially when the asteroid is actually inside the resonance. We can find the value of $\alpha$ for the nominal location of the resonance from the formula

$$\alpha_{3:1} = \frac{a_{3:1}}{a'} = \left( \frac{1}{3} \right)^{2/3} \left( \frac{m_c}{m_c + m'} \right)^{1/3}, \tag{6.206}$$

where $m_c$ is the mass of the Sun and $m'$ is the mass of Jupiter. This gives $\alpha_{3:1} \approx 0.480597$. The values of the constants $A_i$ in this case are given in Table 6.1.

Table 6.1. The values of the constants $A_i$ for the 3:1 jovian commensurability.

| $i$ | $A_i$ |
|---|---|
| 0 | 1.06671 |
| 1 | 0.142097 |
| 2 | −0.568387 |
| 3 | −0.165406 |
| 4 | 1.13677 |
| 5 | 0.598100 |
| 6 | −2.21124 |
| 7 | 0.362954 |
| 8 | 0.330812 |
| 9 | −0.661625 |
| 10 | 0.330812 |

By fixing $a$ and $a'$, the term in our expression for $\langle \mathcal{R} \rangle$ associated with $A_0$ becomes effectively a constant and can be neglected since ultimately we will be taking partial derivatives of $\langle \mathcal{R} \rangle$.

### 6.10.2 Terms Associated with the 18:7 Commensurability

Consider one of the terms relevant to a study of the motion of minor planet (2) Pallas. If $n'$ and $n$ denote the mean motions of Jupiter and Pallas respectively, then observations show that

$$18n' - 7n = -0.45°\text{y}^{-1}. \tag{6.207}$$

Therefore Jupiter and Pallas are close to a 18:7 resonance. In an eleventh-order expansion of the disturbing function there are 182 arguments associated with this resonance. Here we follow the example of Ellis & Murray (1999) and derive the terms associated with one of these arguments, namely

$$\varphi = 18\lambda' - 7\lambda - 5\varpi - 6\Omega. \tag{6.208}$$

Applying the definitions given in Eq. (6.114)–(6.120) gives $q = -5$, $q' = 0$, $\ell_{max} = 5$, $p_{min} = 3$, $p'_{min} = 0$, $s_{min} = 3$, and $i_{max} = 3$. Since $s_{min} = i_{max}$ the only contribution will come from $i = s = 3$ and hence $l = 0$. Similarly, since $n_{max} = [(s - 3)/2] = 0$ we must have $n = 0$. Hence, from Eq. (6.124) the only valid value of $p$ is $p = 3$; hence $m = 3$ and so from Eq. (6.125) $p' = 0$; we also have $j = 15$. We can now write the simplified form of Eq. (6.113) as

$$\langle \mathcal{R}_D \rangle_+ = \frac{\alpha^3}{720} \sum_{\ell=0}^{5} \frac{(-1)^\ell}{\ell!} \sum_{k=0}^{\ell} \binom{\ell}{k} (-1)^k \alpha^\ell D^\ell b_{\frac{7}{2}}^{(15)}(\alpha) F_{3,3,3}(I) F_{3,3,0}(I')$$

$$\times X_7^{3+k,\,12}(e) X_{18}^{-(4+k),\,18}(e') \cos\left[18\lambda' - 7\lambda - 5\varpi - 6\Omega\right]. \tag{6.209}$$

To complete the calculation we need to investigate the possibility that there are terms associated with the negative of our original argument, namely $\varphi = -(18\lambda' - 7\lambda - 5\varpi - 6\Omega)$. In this case inspection of Eq. (6.114)–(6.126) shows that there are no contributions and $\langle \mathcal{R}_D \rangle_- = 0$.

We only require two evaluations of the inclination function and twelve evaluations of Hansen coefficients. Although our expansion is to eleventh order, according to the approximations given in Eq. (6.141)–(6.142) the function $F_{3,3,3}(I)$ will produce terms of $\mathcal{O}(I^6)$ and $X_7^{3+k,\,12}(e)$ will produce terms of $\mathcal{O}(e^5)$. Thus we are only concerned with the lowest order terms in all function evaluations. This means we can ignore the higher order terms in $F_{3,3,0}(I') = 15 + \mathcal{O}(I'^2)$ and $X_{18}^{-(4+k),\,18}(e') = 1 + \mathcal{O}(e'^2)$ for $k = 0, 1, \ldots 5$. We have

$$F_{3,3,3}(I) = 15s^6, \tag{6.210}$$

$$X_7^{3,12}(e) = -\frac{1577149}{1280}e^5, \tag{6.211}$$

$$X_7^{4,12}(e) = -\frac{1473703}{960}e^5, \tag{6.212}$$

$$X_7^{5,12}(e) = -\frac{7280077}{3840}e^5, \tag{6.213}$$

$$X_7^{6,12}(e) = -\frac{1486337}{640}e^5, \tag{6.214}$$

$$X_7^{7,12}(e) = -\frac{10842187}{3840}e^5, \tag{6.215}$$

$$X_7^{8,12}(e) = -\frac{409031}{120}e^5, \tag{6.216}$$

and the resulting expression for $\langle \mathcal{R}_D \rangle$ is

$$\langle \mathcal{R}_D \rangle = -\frac{e^5 s^6}{12288} \Big[ 4731447\alpha^3 + 1163365\alpha^4 D + 110950\alpha^5 D^2$$
$$+ 5130\alpha^6 D^3 + 115\alpha^7 D^4 + \alpha^8 D^5 \Big] b_{\frac{15}{2}}^{(15)}(\alpha)$$
$$\times \cos\left[18\lambda' - 7\lambda - 5\varpi - 6\Omega\right]. \tag{6.217}$$

Application of the algorithm given in Sect. 6.6 for the indirect parts shows that none exist in this case; furthermore, there are no indirect terms associated with any of the 182 possible arguments to eleventh order at this resonance. Therefore the averaged part of the disturbing function associated with this argument is given by $\langle \mathcal{R} \rangle = (\mathcal{G}m'/a') \langle \mathcal{R}_D \rangle$.

## 6.11 The Effect of Planetary Oblateness

When the disturbing function was introduced in Sect. 6.2 it was in the context of the perturbing potential experienced by an orbiting mass due to the gravitational effect of another body. More generally it can be thought of as the terms in the potential over and above those associated with the simple $1/r$ term arising from treating the central object as a point mass. Therefore Lagrange's equations are equally valid in cases where the disturbing function arises from a nonspherical or *oblate* central object. In this section we demonstrate how such additional terms lead to changes in the keplerian ellipse of the two-body orbit. This has particular importance for the dynamical study of planetary satellites (see Sects. 7.7, 7.9, and 8.11) and their smaller counterparts, planetary ring particles (see Chapter 10).

Only a perfectly rigid planet could be spherical, and so any consideration of the motion of planetary satellites has to take account of the nonpoint mass terms in the planet's potential. Consider a satellite orbiting a planet of mass $m_p$ and mean radius $R_p$. Let the coordinates of the satellite be $(r, \phi, \alpha)$ in a coordinate system centred on the planet, where $r$ is the radial distance, $\phi$ is the longitude, and $\alpha$ now denotes the latitude of the satellite (see Fig. 6.6). Here $\alpha = 0$ corresponds to a position in the planet's equatorial plane, and $\phi = 0$ denotes a position on the zero longitude (or prime meridian) of the planet. It can be shown from potential theory that the gravitational potential experienced by the satellite can be written as

$$V = -\frac{\mathcal{G}m_c}{r}\left[1 - \sum_{i=2}^{\infty} J_i (R_p/r)^i P_i(\sin\alpha)\right], \qquad (6.218)$$

where $P_i(\sin\alpha)$ is the Legendre polynomial of degree $i$ in $\sin\alpha$ and the $J_i$ are dimensionless coefficients that characterise the size of the nonspherical com-

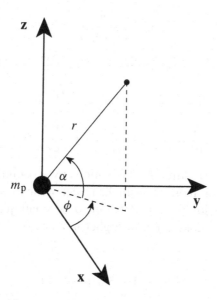

Fig. 6.6. The relationships among the radius $r$, longitude $\phi$, and latitude $\alpha$, and the standard Cartesian coordinates for an object in orbit around a mass $m_p$.

ponents of the potential. If $i$ is even then the $J_i$ are called the zonal harmonic coefficients. Known values of $J_2$ and $J_4$ for the planets are listed in Table A1.4. The letter $J$ is used in honour of Sir Harold Jeffreys (1891–1989), the English geophysicist. Note that we are assuming that $V$ is a function of $r$ and $\alpha$ only and that we have reverted to the standard practice of having a negative potential. The absence of $\phi$ in $V$ means that the planet's potential is taken to be axisymmetric. This is a reasonable approximation because the nonsphericity of the planet is predominantly due to its rotational flattening along the spin axis.

In this coordinate system the velocity of the satellite is given by

$$\mathbf{v} = \dot{r}\hat{\mathbf{r}} + r\dot{\alpha}\hat{\boldsymbol{\alpha}} + r\cos\alpha\dot{\phi}\hat{\boldsymbol{\phi}}, \tag{6.219}$$

where $\hat{\mathbf{r}}, \hat{\boldsymbol{\alpha}}$, an $\hat{\boldsymbol{\phi}}$ are the standard mutually orthogonal unit vectors. The equations of motion can be derived by considering a generalisation of the formulae used in Eq. (2.7). By comparing the individual components we obtain

$$\ddot{r} - r\dot{\alpha}^2 - r\cos^2\alpha\dot{\phi}^2 + \frac{\partial V}{\partial r} = 0, \tag{6.220}$$

$$\frac{\mathrm{d}}{\mathrm{d}t}\left(r^2\dot{\alpha}\right) + r^2\sin\alpha\cos\alpha\dot{\phi}^2 + \frac{\partial V}{\partial\alpha} = 0, \tag{6.221}$$

$$\frac{\mathrm{d}}{\mathrm{d}t}\left(r^2\cos^2\alpha\dot{\phi}\right) = 0. \tag{6.222}$$

Note that the last equation implies that the vertical component of the angular momentum is conserved and we can write

$$\dot{\phi} = \frac{h}{r^2\cos^2\alpha}, \tag{6.223}$$

where $h$ is a constant. Hence, Eqs. (6.220) and (6.221) can be written as

$$\ddot{r} - r\dot{\alpha}^2 - \frac{h^2}{r^3\cos^2\alpha} + \frac{\partial V}{\partial r} = 0, \tag{6.224}$$

$$\frac{\mathrm{d}}{\mathrm{d}t}\left(r^2\dot{\alpha}\right) + \frac{h^2\sin\alpha}{r^2\cos^3\alpha} + \frac{\partial V}{\partial\alpha} = 0. \tag{6.225}$$

We can linearise these differential equations by assuming that the deviations from a circular, equatorial orbit are small. Hence we let $r = a(1 + \varepsilon)$, where $a$ is the satellite's semi-major axis, and let $\varepsilon$ and $\alpha$ be small quantities such that we can neglect terms of second order or higher in $\varepsilon$ and $\alpha$. Then the equations of motion become

$$a\ddot{\varepsilon} - \frac{h^2}{a^3}(1 - 3\varepsilon) + \frac{\partial V}{\partial r} = 0, \tag{6.226}$$

$$a\ddot{\alpha} + \frac{h^2}{a^3}\alpha + \frac{1}{a}\frac{\partial V}{\partial\alpha} = 0. \tag{6.227}$$

Now we perform Taylor series expansions of $\partial V/\partial r$ and $\partial V/\partial \alpha$ in the vicinity of $r = a$ and $\alpha = 0$. This gives

$$\frac{\partial V}{\partial r}(r, \alpha) \approx a(A + B\varepsilon + C\alpha), \tag{6.228}$$

$$\frac{\partial V}{\partial \alpha}(r, \alpha) \approx a^2(D + E\varepsilon + F\alpha), \tag{6.229}$$

where

$$A = \frac{1}{a}\left(\frac{\partial V}{\partial r}\right)_0, \qquad B = \left(\frac{\partial^2 V}{\partial r^2}\right)_0, \qquad C = \frac{1}{a}\left(\frac{\partial^2 V}{\partial r \partial \alpha}\right)_0 \tag{6.230}$$

and

$$D = \frac{1}{a^2}\left(\frac{\partial V}{\partial \alpha}\right)_0, \qquad E = \frac{1}{a}\left(\frac{\partial^2 V}{\partial r \partial \alpha}\right)_0, \qquad F = \frac{1}{a^2}\left(\frac{\partial^2 V}{\partial \alpha^2}\right)_0 \tag{6.231}$$

are constants and the subscript 0 denotes evaluation of the partial derivative at $r = a$ and $\alpha = 0$. Therefore the linearised equations of motion can be written as

$$\ddot{\varepsilon} + 3(h^2/a^4)\varepsilon + B\varepsilon + C\alpha = (h^2/a^4) - A, \tag{6.232}$$

$$\ddot{\alpha} + (h^2/a^4)\alpha + E\varepsilon + F\alpha = -D. \tag{6.233}$$

In order to simplify the problem we assume that all the $J_3$, $J_5$, etc. coefficients are zero. These are small quantities and, in the case of all but a few planets, are difficult to measure. If we consider only those terms with even values of $i$ in Eq. (6.218), then it can easily be shown that $(\partial^2 V/\partial r \partial \alpha)_0 = 0$. Hence $C = E = 0$ and the equations of motion for $\varepsilon$ and $\alpha$ become uncoupled. If we assume that the time-averaged radial excursion is zero (i.e., that $\langle \varepsilon \rangle = 0$) then Eq. (6.232) gives $A = h^2/a^4$ and hence the variation in the radial direction is given by

$$\ddot{\varepsilon} + (3A + B)\varepsilon = 0 \tag{6.234}$$

with solution

$$\varepsilon = e \cos \kappa t, \tag{6.235}$$

where $\kappa^2 = 3A + B$ and $e$, from our knowledge of the two-body problem, is the eccentricity of the orbit. Similarly the vertical variation is given by

$$\ddot{\alpha} + (A + F)\alpha = -D \tag{6.236}$$

with solution

$$\alpha = -\frac{D}{A + F} + I \cos(vt + \delta), \tag{6.237}$$

where $v^2 = A + F$ and $I$, again from our knowledge of the two-body problem, is the inclination of the orbit. Finally, we can return to Eq. (6.223) to consider the variation in $\phi$. Using the same approximations as before, we can write

$$\dot{\phi} \approx (h/a^2)(1 - 2\varepsilon) = (h/a^2)(1 - 2e \cos \kappa t) \tag{6.238}$$

with solution

$$\phi = A^{\frac{1}{2}}t - (2A^{\frac{1}{2}}/\kappa)e\sin\kappa t. \tag{6.239}$$

Note that the satellite's mean motion is just the average value of $\dot{\phi}$ and hence

$$n^2 = \langle\dot{\phi}\rangle^2 = A. \tag{6.240}$$

The solutions we have obtained show that the motion under the point mass and $J_{2i}$ terms in the potential gives rise to three frequencies: the mean motion, $n$, the radial frequency, $\kappa$, and the vertical frequency, $\nu$. According to our first-order theory in $e$ and $I$, these are given by

$$n^2 = \frac{1}{a}\left(\frac{\partial V}{\partial r}\right)_0, \tag{6.241}$$

$$\kappa^2 = \frac{3}{a}\left(\frac{\partial V}{\partial r}\right)_0 + \left(\frac{\partial^2 V}{\partial r^2}\right)_0, \tag{6.242}$$

$$\nu^2 = \frac{1}{a}\left(\frac{\partial V}{\partial r}\right)_0 + \frac{1}{a^2}\left(\frac{\partial^2 V}{\partial \alpha^2}\right)_0. \tag{6.243}$$

As stated above, the radial and vertical motions are now uncoupled. It should also be noted that the sinsusoidal component of the $\phi$ motion is $\pi/2$ out of phase with the $\varepsilon$ motion and has approximately twice the amplitude if $n \approx \kappa$. Expressions for these frequencies as series in $J_{2i}$ and $(R_p/a)^{2i}$ can now be calculated by evaluating the various partial derivatives. Considering terms up to and including $J_4$ we have

$$n^2 = \frac{\mathcal{G}m_p}{a^3}\left[1 + \frac{3}{2}J_2\left(\frac{R_p}{a}\right)^2 - \frac{15}{8}J_4\left(\frac{R_p}{a}\right)^4\right], \tag{6.244}$$

$$\kappa^2 = \frac{\mathcal{G}m_p}{a^3}\left[1 - \frac{3}{2}J_2\left(\frac{R_p}{a}\right)^2 + \frac{45}{8}J_4\left(\frac{R_p}{a}\right)^4\right], \tag{6.245}$$

$$\nu^2 = \frac{\mathcal{G}m_p}{a^3}\left[1 + \frac{9}{2}J_2\left(\frac{R_p}{a}\right)^2 - \frac{75}{8}J_4\left(\frac{R_p}{a}\right)^4\right]. \tag{6.246}$$

Note that if $J_2 = J_4 = 0$ then $n^2 = \kappa^2 = \nu^2 = n_0^2$, where $n_0 = (\mathcal{G}m_p/a^3)^{1/2}$ is the keplerian mean motion of the satellite around a point-mass planet. In the case of the mean motion, $n$, the inclusion of the additional terms means that for a given semi-major axis the satellite moves faster than the rate expected at that location if the motion was purely keplerian. Therefore, since the observable quantity for a satellite is usually $n$, the semi-major axis is not that determined from Kepler's third law. Instead it is necessary to solve Eq. (6.244), a nonlinear equation in $a$.

The major geometrical consequence of the inclusion of the $J_{2i}$ terms and the resulting small differences in the three frequencies is that the orbit is no longer closed or, in other words, that there is motion of the pericentre and the node. We can see this by noting that the extent to which the rates of the radial and vertical excursions (i.e., those due to the eccentricity and the inclination) differ from $n$ is a measure of the rate of change of the pericentre and node respectively. Hence

$$\dot\varpi = n - \kappa, \tag{6.247}$$

$$\dot\Omega = n - \nu \tag{6.248}$$

and, to $\mathcal{O}(R_p/a)^4$, we have

$$\dot\varpi = +n_0 \left[ \frac{3}{2} J_2 \left( \frac{R_p}{a} \right)^2 - \frac{15}{4} J_4 \left( \frac{R_p}{a} \right)^4 \right], \tag{6.249}$$

$$\dot\Omega = -n_0 \left[ \frac{3}{2} J_2 \left( \frac{R_p}{a} \right)^2 - \frac{9}{4} J_2^2 \left( \frac{R_p}{a} \right)^4 - \frac{15}{4} J_4 \left( \frac{R_p}{a} \right)^4 \right]. \tag{6.250}$$

The approach given above is that used in the discussion of planetary rings in Sect. 10.3. However, similar results can be obtained by making use of Lagrange's equations. Roy (1988) uses this technique to study the oblateness problem, including the effects of the $J_{2i+1}$ coefficients.

Here we follow the approach of Roy (1988) but restrict ourselves to an analysis of the long-term effects of $J_2$ and $J_4$ and include terms up to second order in $e$ and $I$. The disturbing function can be written as (cf. Eq. (6.218))

$$\mathcal{R} = -\frac{\mathcal{G}m_p}{r} \left[ J_2 \left( \frac{R_p}{r} \right)^2 P_2(\sin \alpha) + J_4 \left( \frac{R_p}{r} \right)^4 P_4(\sin \alpha) \right], \tag{6.251}$$

where

$$P_2(\sin \alpha) = (1/2)(3 \sin^2 \alpha - 1) \tag{6.252}$$

and

$$P_4(\sin \alpha) = (1/8)(35 \sin^4 \alpha - 30 \sin^2 \alpha + 3). \tag{6.253}$$

From Fig. 6.6 and Eq. (2.122) we can relate the latitude to the inclination $I$, the true anomaly $f$, and the argument of pericentre $\omega$ using the equation

$$\sin \alpha = \sin I \sin(f + \omega). \tag{6.254}$$

Hence, using $r = a(1 - e^2)/(1 + e \cos f)$, the relevant averaged terms in $\mathcal{R}$ to second order are given by

$$\langle \mathcal{R} \rangle = \frac{1}{2} n^2 a^2 \left[ \frac{3}{2} J_2 \left( \frac{R_p}{a} \right)^2 - \frac{9}{8} J_2^2 \left( \frac{R_p}{a} \right)^4 - \frac{15}{4} J_4 \left( \frac{R_p}{a} \right)^4 \right] e^2$$
$$- \frac{1}{2} n^2 a^2 \left[ \frac{3}{2} J_2 \left( \frac{R_p}{a} \right)^2 - \frac{27}{8} J_2^2 \left( \frac{R_p}{a} \right)^4 - \frac{15}{4} J_4 \left( \frac{R_p}{a} \right)^4 \right] \sin^2 I, \tag{6.255}$$

where we have made use of the expressions in Sect. 2.5 relating $\sin f$ and $\cos f$ to power series in the mean anomaly, $M$, and then averaged the resulting series from $M = 0$ to $M = 2\pi$, ignoring constant terms. We will use this form of the oblateness contribution when we consider its incorporation in a secular perturbation theory in Sect. 7.7.

Now we are in a position to use Lagrange's equations to derive expressions for $\dot{\varpi}$ and $\dot{\Omega}$. To second order in $e$ and $I$ this gives

$$\dot{\varpi} = +n \left[ \frac{3}{2} J_2 \left( \frac{R_p}{a} \right)^2 - \frac{9}{8} J_2^2 \left( \frac{R_p}{a} \right)^4 - \frac{15}{4} J_4 \left( \frac{R_p}{a} \right)^4 \right], \qquad (6.256)$$

$$\dot{\Omega} = -n \left[ \frac{3}{2} J_2 \left( \frac{R_p}{a} \right)^2 - \frac{27}{8} J_2^2 \left( \frac{R_p}{a} \right)^4 - \frac{15}{4} J_4 \left( \frac{R_p}{a} \right)^4 \right]. \qquad (6.257)$$

These differ from the expressions in Eqs. (6.249) and (6.250) because we have chosen to express the rates in terms of $n$ rather than $n_0$. To convert between the expressions we make use of Eq. (6.244). As noted by Greenberg (1981), care must be taken to distinguish between osculating and mean elements in the use of expressions for $\dot{\varpi}$ and $\dot{\Omega}$ (see Elliot & Nicholson (1984) for a discussion on this subject).

## Exercise Questions

**6.1**    Asteroid (3805) Goldreich is close to the 8:3 resonance with Jupiter. In December 1997 Goldreich had a semi-major axis $a = 2.68463$ AU while Jupiter's semi-major axis was $a' = 5.20335$ AU. Calculate the mean motion (in $°\mathrm{d}^{-1}$) of the asteroid ($n$) and Jupiter ($n'$) and hence the value of $8n' - 3n$ (i.e., the separation in $°\mathrm{d}^{-1}$ of the asteroid from the nominal location of the 8:3 resonance). Use the d'Alembert rules to write down the twenty-eight possible arguments for this resonance in an expansion of the disturbing function complete to fifth order in the eccentricities and inclinations of both bodies. For each argument give the relevant powers of the eccentricities and inclinations in the associated term. Use the method given in Sect. 6.6 to find an explicit expression for the term associated with the single resonant argument $8\lambda' - 3\lambda - \varpi' - 2\varpi - \Omega' - \Omega$; your answer should include terms up to fifth order in the eccentricities and inclinations. Evaluate the resulting Laplace coefficients using $\alpha = a/a'$. Find the smallest values of the positive integers $p$ and $q > 5$ such that $|(p + q)n' - pn| < |8n' - 3n|$.

**6.2**    (*Due to Marcus Ansorg*) The Laplace coefficient, $b_s^{(j)}$, can be written in terms of a hypergeometric series as

$$\frac{1}{2} b_s^{(j)} = \frac{s(s + 1) \dots (s + j - 1)}{j!} \alpha^j F(\alpha^2),$$

where $\alpha = a/a' < 1$ is the ratio of the semi-major axes and

$$F(\alpha^2) = 1 + \frac{s(s+j)}{1!(j+1)}\alpha^2 + \frac{s(s+1)(s+j)(s+j+1)}{2!(j+1)(j+2)}\alpha^4 + \dots.$$

As $\alpha \to 1$ the series is slow to converge making numerical evaluation difficult. However, by a judicious change of variable the convergence properties can be improved. It can be shown that $F$ is a solution of the ordinary differential equation

$$x(x-1)\frac{d^2 F}{dx^2} + \frac{dF}{dx}[(2s+j+1)x - (j+1)] + s(s+j)F = 0.$$

With the substitutions $y = 1 - x$, $G(y) = F(1-y) = F(x)$, derive a differential equation for $G$ and use direct substitution to show that this equation has a solution

$$G(y) = \sum_{l=0}^{\infty} A_l y^{l-2s+1} + \ln y \sum_{l=0}^{\infty} B_l y^l,$$

where $A_l$ and $B_l$ are constants. Derive expressions for $A_l$ and $B_l$. Write a program to calculate Laplace coefficients using (i) the series for $F(\alpha^2)$ given above (suitable for small $\alpha$) and (ii) the series

$$F(\alpha^2) = \sum_{l=0}^{\infty} a_l(1-\alpha^2)^{l-2s+1} + \ln(1-\alpha^2) \sum_{l=0}^{\infty} b_l(1-\alpha^2)^l.$$

Use your two programs to compare the time taken to calculate $b_{3/2}^{(2)}(0.999)$ correct to seven decimal places using each method.

**6.3**     The precession of Saturn's orbit due to Jupiter can be roughly estimated by approximating Jupiter's time-averaged influence by that of a ring (or torus) of mass $m_J$ and radius $a_J$, and then treating this ring as producing an effective "$J_2$" for the Sun. It can be shown that $J_2^{\text{eff}} = (C - A)/(MR^2)$, where $C$ and $A$ are the polar and equatorial moments of inertia of the ring and $M$ and $R$ are the mass and radius of the Sun. Show that $J_2^{\text{eff}} = m_J a_J^2/(2MR^2)$ and then derive expressions for the pericentre and node precession rates of Saturn's orbit using the formulae derived in Sect. 6.11. (You may assume that the eccentricity and inclination of Saturn are small quantities.) Evaluate the period of Saturn's pericentre precession in years.

**6.4**     The gravity field of the Earth (mass $m_E$) has a small third-harmonic component, due to an asymmetry between the northern and southern hemispheres, which may be written:

$$R_3 = -\frac{\mu J_3 R^3}{2r^4} \cos\theta \left(5\cos^2\theta - 3\right),$$

where $\mu = Gm_{\mathrm{E}}$ and $\theta$ is colatitude. Evaluate the average value of $R_3$ with respect to the mean anomaly $M$ and show that

$$\langle R_3 \rangle = -\frac{3\mu J_3 R^3}{2a^4} \frac{e}{(1-e^2)^{5/2}} \sin I \left( \frac{5}{4} \sin^2 I - 1 \right) \sin \omega,$$

where $\omega$ is the argument of periapse. Note that the precession of $\omega$ due to $J_2$ implies that the *long-term* average is zero for any $e$ and $I$. Moreover, the dependence of $\langle R_3 \rangle$ on $\omega$ means that, unlike the case for $J_2$, there are now perturbations of $e$ and $I$. Use Lagrange's perturbation equations to derive an expression for $\mathrm{d}I/\mathrm{d}t$. Combine this expression with the expression for $\mathrm{d}\varpi/\mathrm{d}t$ due to $J_2$ derived in the notes to show that the inclination varies with $\omega$ according to the equation

$$\frac{\mathrm{d}I}{\mathrm{d}\omega} \simeq \frac{J_3 R}{2 J_2 a} \frac{e}{(1-e^2)} \cos I \cos \omega,$$

where we assume $J_3 \ll J_2$. Use this expression to find the approximate variation in $I$ over the course of one precessional cycle of $\omega$, assuming that the variations in $I$ are small.

**6.5**    The lowest order effect of General Relativity (GR) on planetary orbits is to induce a precession of the pericentres beyond that due to inter planet perturbations. The effect is largest for Mercury, and its confirmation was one of the earliest successful tests of GR. The dominant GR effects can be modelled by adding an additional potential term to the Sun's gravitational potential, of the form

$$V_{\mathrm{GR}} = -\frac{\mathcal{G} M h^2}{c^2 r^3},$$

where $M$ is the mass of the Sun, $c$ is the speed of light, $r$ is the radial distance of the planet (with semi-major axis $a$ and eccentricity $e$), and $h = r^2 \dot{\theta} = [\mathcal{G} M a (1 - e^2)]^{1/2}$ is the orbital angular momentum of the planet per unit mass. (a) Use the epicyclic theory developed in Sect. 6.11 to evaluate the perturbed mean motion $n$ and the epicyclic frequency $\kappa$ for a near-circular orbit with mean radius $a$. Show that

$$\dot{\varpi}_{\mathrm{GR}} \simeq \frac{6\pi \mathcal{G} M}{a c^2 T} \simeq 3 (v_{\mathrm{orb}}/c)^2 n,$$

where $T = 2\pi/n$ is the orbital period and $v_{\mathrm{orb}}$ is the average orbital velocity. (b) Evaluate $\dot{\varpi}$ for Mercury and for the Earth, and compare the GR-induced precession rates with their precession rates due to planetary secular perturbations, which are of order 10 arcsec y$^{-1}$, or $\sim 1.5 \times 10^{-12}$ rad s$^{-1}$.

**6.6**    A simple description of the interaction of satellite precessional motions due to planetary and solar perturbations can often be obtained by comparing the precession *rates*. (a) The nodal precession rate of a satellite's orbit due to solar perturbations may be written

$$\dot{\Omega}_{\text{solar}} = -(3/4)(n_{\text{p}}^2/n)\cos\beta,$$

where $\beta$ is the inclination of the satellite orbit relative to the planet's orbital plane and $n_{\text{p}}$ is the mean motion of the planet around the Sun. By comparing this rate to the nodal precessional rate, $\dot{\Omega}$, of the satellite orbit due to the planet's $J_2$ (see Sect. 6.11), show that there exists a critical semi-major axis, $a_c$, for which these two rates are equal, and derive an expression for $a_c$ in terms of $M_{\text{p}}/M_{\text{Sun}}$, $R$, $a_{\text{p}}$, $J_2$, and $\beta$. (b) Evaluate $a_c$ for the Earth, Saturn, and Uranus, in units of planetary radii and for near-equatorial orbits ($I \ll \pi/2$), and determine in each case which satellites (if any) lie exterior to $a_c$. (Note that for Uranus $\beta = 98°$.) (c) Describe, in qualitative terms, the nodal precession of satellites with $a \ll a_c$ and $a \gg a_c$. Can you explain the *absence* of any equatorial ($I \simeq 0$) satellites with $a > a_c$? (d) Calculate $\dot{\Omega}$ and the corresponding precessional periods (in years) for the Moon, Mimas, Titan, Miranda, and Oberon.

# 7

# Secular Perturbations

Past and to come seem best, things present worst.

William Shakespeare, *Henry IV, (2), I, iii*

## 7.1 Introduction

In the last chapter we saw how the disturbing function can be expanded in an infinite series where the individual terms can be classified as secular, resonant, or short period, according to the given physical problem. We have already stated in Sect. 3 that the $N$-body problem (for $N \geq 3$) is nonintegrable. However, in this chapter we will show how, with suitable approximations, it is possible to find an analytical solution to a particular form of the $N$-body problem that can be applied to the motion of solar system bodies. We can do this by considering the effects of the purely secular terms in the disturbing function for a system of $N$ masses orbiting a central body. The resulting theory can be applied to satellites orbiting a planet, or planets orbiting the Sun, and then used to study the motion of small objects orbiting in either of these systems. This is the subject of *secular perturbation theory*.

## 7.2 Secular Perturbations for Two Planets

Consider the motion of two planets of mass $m_1$ and $m_2$ moving under their mutual gravitational effects and the attraction of a point-mass central body of mass $m_c$ where $m_1 \ll m_c$ and $m_2 \ll m_c$. Let $\mathcal{R}_1$ and $\mathcal{R}_2$ be the disturbing functions describing the perturbations on the orbit of the masses $m_1$ and $m_2$ respectively, where $\mathcal{R}_1$ and $\mathcal{R}_2$ are functions of the standard *osculating* orbital elements of both bodies. Osculating elements, from the Latin verb *osculare* meaning "to kiss", are instantaneous elements derived from the values of the position and velocity of an object assuming an unperturbed keplerian orbit. The perturbations on the orbital elements are given by Lagrange's equations, Eqs. (6.145)–(6.150).

In the absence of any mean motion commensurabilities between two masses, the secular perturbations arising from the gravitational perturbations between $m_1$, $m_2$, and $m_c$ are obtained by isolating the terms in the disturbing function that are independent of the mean longitudes. We can also exclude any terms that depend only on the semi-major axis since, from Eq. (6.145), these will not make any contribution to secular evolution. To second order in the eccentricities and inclinations (and first order in the masses), the only terms in the expansion of the disturbing function that do not contain the mean longitudes are, from Appendix B, the terms 4D0.1, 4D0.2, and 4D0.3 with $j = 0$. Hence the general, averaged, secular, direct part of the disturbing function is

$$\mathcal{R}_D^{(\text{sec})} = \frac{1}{8}\left[2\alpha_{12}D + \alpha_{12}^2 D^2\right]b_{\frac{1}{2}}^{(0)}(e_1^2 + e_2^2) - \frac{1}{2}\alpha_{12}b_{3/2}^{(1)}(s_1^2 + s_2^2)$$

$$+ \frac{1}{4}\left[2 - 2\alpha_{12}D - \alpha_{12}^2 D^2\right]b_{\frac{1}{2}}^{(1)}e_1 e_2 \cos(\varpi_1 - \varpi_2)$$

$$+ \alpha_{12}b_{3/2}^{(1)}s_1 s_2 \cos(\Omega_1 - \Omega_2), \tag{7.1}$$

where the subscripts 1 and 2 refer to the inner and outer body respectively and $\alpha_{12} = a_1/a_2$ where $a_1 < a_2$. There is no indirect part. In fact, as can be seen from Appendix B, all the indirect terms involve at least one mean longitude and hence will never contribute purely secular terms (see Brouwer & Clemence 1961).

When calculating $\mathcal{R}_1$ and $\mathcal{R}_2$ from $\mathcal{R}_D^{(\text{sec})}$ we have to take account of the fact that $\mathcal{R}_1$ arises from an external perturbation by $m_2$ whereas $\mathcal{R}_2$ comes from an internal perturbation by $m_1$. Hence, from Eqs. (6.134) and (6.135), $\mathcal{R}_1$ and $\mathcal{R}_2$ can be written as

$$\mathcal{R}_1 = \frac{\mathcal{G}m_2}{a_2}\mathcal{R}_D^{(\text{sec})} = \frac{\mathcal{G}m_2}{a_1}\alpha_{12}\mathcal{R}_D^{(\text{sec})} \tag{7.2}$$

and

$$\mathcal{R}_2 = \frac{\mathcal{G}m_1}{a_1}\alpha_{12}\mathcal{R}_D^{(\text{sec})} = \frac{\mathcal{G}m_1}{a_2}\mathcal{R}_D^{(\text{sec})}. \tag{7.3}$$

Using the following relationships between the Laplace coefficients and their derivatives,

$$2\alpha\frac{db_{1/2}^{(0)}}{d\alpha} + \alpha^2\frac{d^2 b_{1/2}^{(0)}}{d\alpha^2} = \alpha b_{3/2}^{(1)}, \tag{7.4}$$

$$2b_{1/2}^{(1)} - 2\alpha\frac{db_{1/2}^{(1)}}{d\alpha} - \alpha^2\frac{d^2 b_{1/2}^{(1)}}{d\alpha^2} = -\alpha b_{3/2}^{(2)}, \tag{7.5}$$

and the approximations $\mathcal{G}m_c \approx n_1^2 a_1^3 \approx n_2^2 a_2^3$, we can write

$$\mathcal{R}_1 = n_1^2 a_1^2 \frac{m_2}{m_c + m_1} \left[ \frac{1}{8} \alpha_{12}^2 b_{3/2}^{(1)} e_1^2 - \frac{1}{8} \alpha_{12}^2 b_{3/2}^{(1)} I_1^2 \right.$$

$$- \frac{1}{4} \alpha_{12}^2 b_{3/2}^{(2)} e_1 e_2 \cos(\varpi_1 - \varpi_2)$$

$$\left. + \frac{1}{4} \alpha_{12}^2 b_{3/2}^{(1)} I_1 I_2 \cos(\Omega_1 - \Omega_2) \right] \tag{7.6}$$

and

$$\mathcal{R}_2 = n_2^2 a_2^2 \frac{m_1}{m_c + m_2} \left[ \frac{1}{8} \alpha_{12} b_{3/2}^{(1)} e_2^2 - \frac{1}{8} \alpha_{12} b_{3/2}^{(1)} I_2^2 \right.$$

$$- \frac{1}{4} \alpha_{12} b_{3/2}^{(2)} e_1 e_2 \cos(\varpi_1 - \varpi_2)$$

$$\left. + \frac{1}{4} \alpha_{12} b_{3/2}^{(1)} I_1 I_2 \cos(\Omega_1 - \Omega_2) \right], \tag{7.7}$$

where we have assumed that $I_1$ and $I_2$ are small enough so that the approximations $s_1 = \sin \frac{1}{2} I_1 \approx \frac{1}{2} I_1$ and $s_2 = \sin \frac{1}{2} I_2 \approx \frac{1}{2} I_2$ are valid.

The equations for $\mathcal{R}_1$ and $\mathcal{R}_2$ given in Eqs. (7.6) and (7.7) can be combined to give

$$\mathcal{R}_j = n_j a_j^2 \left[ \frac{1}{2} A_{jj} e_j^2 + A_{jk} e_1 e_2 \cos(\varpi_1 - \varpi_2) \right.$$

$$\left. \frac{1}{2} B_{jj} I_j^2 + B_{jk} I_1 I_2 \cos(\Omega_1 - \Omega_2) \right], \tag{7.8}$$

where $j = 1, 2$; $k = 2, 1$ ($j \neq k$); and

$$A_{jj} = +n_j \frac{1}{4} \frac{m_k}{m_c + m_j} \alpha_{12} \bar{\alpha}_{12} b_{3/2}^{(1)}(\alpha_{12}), \tag{7.9}$$

$$A_{jk} = -n_j \frac{1}{4} \frac{m_k}{m_c + m_j} \alpha_{12} \bar{\alpha}_{12} b_{3/2}^{(2)}(\alpha_{12}), \tag{7.10}$$

$$B_{jj} = -n_j \frac{1}{4} \frac{m_k}{m_c + m_j} \alpha_{12} \bar{\alpha}_{12} b_{3/2}^{(1)}(\alpha_{12}), \tag{7.11}$$

$$B_{jk} = +n_j \frac{1}{4} \frac{m_k}{m_c + m_j} \alpha_{12} \bar{\alpha}_{12} b_{3/2}^{(1)}(\alpha_{12}), \tag{7.12}$$

where $\bar{\alpha}_{12} = \alpha_{12}$ if $j = 1$ (an external perturber) and $\bar{\alpha}_{12} = 1$ if $j = 2$ (an internal perturber). From the definition of the Laplace coefficients given in Sect. 6.4 we have

$$b_{3/2}^{(1)}(\alpha) = \frac{1}{\pi} \int_0^{2\pi} \frac{\cos \psi \, d\psi}{(1 - 2\alpha \cos \psi + \alpha^2)^{\frac{3}{2}}}, \tag{7.13}$$

$$b_{3/2}^{(2)}(\alpha) = \frac{1}{\pi} \int_0^{2\pi} \frac{\cos 2\psi \, d\psi}{(1 - 2\alpha \cos \psi + \alpha^2)^{\frac{3}{2}}}. \tag{7.14}$$

Note that in this case $B_{11} = -B_{12}$ and $B_{21} = -B_{22}$. However, the situation is different when we have to take account of terms due to the oblateness of the central body (see Sect. 7.7). All these quantities are frequencies that can be thought of as the constant elements of two matrices $\mathbf{A}$ and $\mathbf{B}$ given by

$$\mathbf{A} = \begin{pmatrix} A_{11} & A_{12} \\ A_{21} & A_{22} \end{pmatrix} \quad \text{and} \quad \mathbf{B} = \begin{pmatrix} B_{11} & B_{12} \\ B_{21} & B_{22} \end{pmatrix}. \quad (7.15)$$

Note that the elements of these matrices are only functions of the masses and the (fixed) semi-major axes of the two bodies and that the rows (or columns) of the matrix $\mathbf{B}$ are not linearly independent.

Taking the lowest order terms in $e$ and $I$ in Eqs. (6.146), (6.148), (6.149), and (6.150) we can easily derive an approximate form of Lagrange's equations for the time variation of the original orbital elements:

$$\dot{e}_j = -\frac{1}{n_j a_j^2 e_j} \frac{\partial \mathcal{R}_j}{\partial \varpi_j}, \qquad \dot{\varpi}_j = +\frac{1}{n_j a_j^2 e_j} \frac{\partial \mathcal{R}_j}{\partial e_j}, \qquad (7.16)$$

$$\dot{I}_j = -\frac{1}{n_j a_j^2 I_j} \frac{\partial \mathcal{R}_j}{\partial \Omega_j}, \qquad \dot{\Omega}_j = +\frac{1}{n_j a_j^2 I_j} \frac{\partial \mathcal{R}_j}{\partial I_j}. \qquad (7.17)$$

Given the form of the equations above, it is convenient to define the vertical and horizontal components of eccentricity and inclination "vectors" by:

$$h_j = e_j \sin \varpi_j, \qquad k_j = e_j \cos \varpi_j \qquad (7.18)$$

and

$$p_j = I_j \sin \Omega_j, \qquad q_j = I_j \cos \Omega_j. \qquad (7.19)$$

These variables have the advantage that they avoid the singularities inherent in Eqs. (7.16) and (7.17) for low $e$ and $I$. The general secular part of the disturbing function can now be written

$$\mathcal{R}_j = n_j a_j^2 \left[ \frac{1}{2} A_{jj} (h_j^2 + k_j^2) + A_{jk} (h_j h_k + k_j k_k) \right.$$

$$\left. + \frac{1}{2} B_{jj} (p_j^2 + q_j^2) + B_{jk} (p_j p_k + q_j q_k) \right]. \qquad (7.20)$$

Note that when $k$ is used as a subscript it is always equal to either 1 or 2, denoting the interior or exterior body; this should not be confused with the use of $k$ as the horizontal component of the eccentricity vector.

Since each of the $h_j$, $k_j$, $p_j$, and $q_j$ is a function of two variables we can write

$$\frac{dh_j}{dt} = \frac{\partial h_j}{\partial e_j} \frac{de_j}{dt} + \frac{\partial h_j}{\partial \varpi_j} \frac{d\varpi_j}{dt}, \qquad \frac{dk_j}{dt} = \frac{\partial k_j}{\partial e_j} \frac{de_j}{dt} + \frac{\partial k_j}{\partial \varpi_j} \frac{d\varpi_j}{dt}, \qquad (7.21)$$

$$\frac{dp_j}{dt} = \frac{\partial p_j}{\partial I_j} \frac{dI_j}{dt} + \frac{\partial p_j}{\partial \Omega_j} \frac{d\Omega_j}{dt}, \qquad \frac{dq_j}{dt} = \frac{\partial q_j}{\partial I_j} \frac{dI_j}{dt} + \frac{\partial q_j}{\partial \Omega_j} \frac{d\Omega_j}{dt}, \qquad (7.22)$$

where, from the definitions given above, the partial derivatives are given by

$$\frac{\partial h_j}{\partial e_j} = \frac{h_j}{e_j}, \qquad \frac{\partial k_j}{\partial e_j} = \frac{k_j}{e_j}, \qquad \frac{\partial h_j}{\partial \varpi_j} = +k_j, \qquad \frac{\partial k_j}{\partial \varpi_j} = -h_j \qquad (7.23)$$

and

$$\frac{\partial p_j}{\partial I_j} = \frac{p_j}{I_j}, \qquad \frac{\partial q_j}{\partial I_j} = \frac{q_j}{I_j}, \qquad \frac{\partial p_j}{\partial \Omega_j} = +q_j, \qquad \frac{\partial q_j}{\partial \Omega_j} = -p_j. \qquad (7.24)$$

After some calculation it can be shown that the perturbation equations can be written as

$$\dot{h}_j = +\frac{1}{n_j a_j^2} \frac{\partial \mathcal{R}_j}{\partial k_j}, \qquad \dot{k}_j = -\frac{1}{n_j a_j^2} \frac{\partial \mathcal{R}_j}{\partial h_j}, \qquad (7.25)$$

$$\dot{p}_j = +\frac{1}{n_j a_j^2} \frac{\partial \mathcal{R}_j}{\partial q_j}, \qquad \dot{q}_j = -\frac{1}{n_j a_j^2} \frac{\partial \mathcal{R}_j}{\partial p_j}, \qquad (7.26)$$

where $\mathcal{R}_j$ is as given in Eq. (7.20).

The full equations for the variation of $h_j, k_j, p_j$, and $q_j$ ($j = 1, 2$) then become

$$\begin{aligned}
\dot{h}_1 &= +A_{11}k_1 + A_{12}k_2, & \dot{k}_1 &= -A_{11}h_1 - A_{12}h_2, \\
\dot{h}_2 &= +A_{21}k_1 + A_{22}k_2, & \dot{k}_2 &= -A_{21}h_1 - A_{22}h_2, \\
\dot{p}_1 &= +B_{11}q_1 + B_{12}q_2, & \dot{q}_1 &= -B_{11}p_1 - B_{12}p_2, \\
\dot{p}_2 &= +B_{21}q_1 + B_{22}q_2, & \dot{q}_2 &= -B_{21}p_1 - B_{22}p_2.
\end{aligned} \qquad (7.27)$$

Thus, to lowest order, the equations for the time variation of $\{h_j, k_j\}$ are decoupled from those of $\{p_j, q_j\}$. Furthermore, these are linear differential equations with constant coefficients, and hence the problem of secular perturbations reduces to two sets of eigenvalue problems. The solutions are given by

$$h_j = \sum_{i=1}^{2} e_{ji} \sin(g_i t + \beta_i), \qquad k_j = \sum_{i=1}^{2} e_{ji} \cos(g_i t + \beta_i), \qquad (7.28)$$

$$p_j = \sum_{i=1}^{2} I_{ji} \sin(f_i t + \gamma_i), \qquad q_j = \sum_{i=1}^{2} I_{ji} \cos(f_i t + \gamma_i), \qquad (7.29)$$

where the frequencies $g_i$ ($i = 1, 2$) are the eigenvalues of the matrix $\mathbf{A}$, with $e_{ji}$ the components of the two corresponding eigenvectors, and $f_i$ ($i = 1, 2$) are the eigenvalues of the matrix $\mathbf{B}$, with $I_{ji}$ the components of the corresponding eigenvectors. The phases $\beta_i$ and $\gamma_i$, as well as the amplitudes of the eigenvectors, are determined by the initial conditions. This would correspond to making observations of the osculating eccentricities and inclinations at some time. The solution described by Eqs. (7.28) and (7.29) is the classical *Laplace–Lagrange secular solution* of the secular problem.

With the introduction of the solution to the eigenvalue problem it is easy to confuse the quantities associated with the two bodies and those associated with the two eigenmodes of the system. In our notation the subscript $j$ always denotes the planet number while the subscript $i$ always denotes the mode number.

It is interesting to note that in our case the characteristic equation for **B** is

$$\begin{vmatrix} B_{11} - f & B_{12} \\ B_{21} & B_{22} - f \end{vmatrix} = 0, \tag{7.30}$$

which reduces to

$$f\,[f - (B_{11} + B_{22})] = 0 \tag{7.31}$$

since $B_{11}B_{22} - B_{12}B_{21} = 0$ from the definitions given in Eqs. (7.11) and (7.12). Thus one of the roots of the characteristic equation is $f_1 = 0$ and there is a degeneracy in the problem. This highlights a subtle difference between the $\{h, k\}$ and the $\{p, q\}$ solutions. Whereas an eccentric orbit introduces an asymmetry and a reference line into the problem, a spherical or point-mass central body has no natural reference plane. Physically it is only meaningful to talk about a mutual inclination and hence the choice of a reference plane is arbitrary. For example, it is customary to refer satellite orbits to the equatorial plane of the planet (i.e., the plane perpendicular to its spin vector). However, as we shall see, the introduction of a nonspherical planet adds terms to the diagonal elements of **B** and removes the degeneracy problem.

Another point concerning our solution is that it is independent of the mean longitudes because these have been deliberately excluded from the averaged part of the disturbing function. Therefore, although we are able to predict the variations in the eccentricities, inclinations, pericentres, and nodes of the two bodies, we have no information about their positions in space.

The solution given in Eqs. (7.28) and (7.29) implies that the resulting motion of all the masses is stable *for all time*. However, it is important to remember the assumptions under which this result was derived: (i) no mean motion commensurabilities, (ii) $\mathbf{r}_1 < \mathbf{r}_2$, and (iii) the $e$s and $I$s are small enough that a second-order expansion of the disturbing function is sufficient to describe the motion. But the amplitudes of the eccentricity eigenvectors, for example, could be large enough for the orbits to intersect, violating conditions (ii) and (iii). As we shall see there may be situations where no mean motion commensurabilities exist, but where "small divisor" terms are still important. We have derived a theory that is correct only to the first order in the masses and so it is important to realise that there could be significant contributions from a second-order theory.

## 7.3 Jupiter and Saturn

We will now apply the theory given above to the case of Jupiter (mass $m_1$) and Saturn (mass $m_2$) orbiting the Sun (mass $m_c$). In 1983 the system had the following parameters:

$$m_1/m_c = 9.54786 \times 10^{-4}, \qquad m_2/m_c = 2.85837 \times 10^{-4},$$
$$a_1 = 5.202545 \text{ AU}, \qquad a_2 = 9.554841 \text{ AU},$$
$$n_1 = 30.3374°\text{y}^{-1}, \qquad n_2 = 12.1890°\text{y}^{-1},$$
$$e_1 = 0.0474622, \qquad e_2 = 0.0575481,$$
$$\varpi_1 = 13.983865°, \qquad \varpi_2 = 88.719425°,$$
$$I_1 = 1.30667°, \qquad I_2 = 2.48795°,$$
$$\Omega_1 = 100.0381°, \qquad \Omega_2 = 113.1334°. \qquad (7.32)$$

Since $\alpha = a_1/a_2 = 0.544493$, we can use the definition of Laplace coefficients given in Eqs. (7.13) and (7.14) to get

$$b_{3/2}^{(1)} = 3.17296, \qquad b_{3/2}^{(2)} = 2.07110. \qquad (7.33)$$

Using the definitions of the matrix elements given in Eqs. (7.9)–(7.12) we have

$$\mathbf{A} = \begin{pmatrix} +0.00203738 & -0.00132987 \\ -0.00328007 & +0.00502513 \end{pmatrix} °\text{y}^{-1} \qquad (7.34)$$

and

$$\mathbf{B} = \begin{pmatrix} -0.00203738 & +0.00203738 \\ +0.00502513 & -0.00502513 \end{pmatrix} °\text{y}^{-1}. \qquad (7.35)$$

We can now find the eigenvalues of $\mathbf{A}$ and $\mathbf{B}$ by solving the respective characteristic equations:

$$\begin{vmatrix} A_{11} - g & A_{12} \\ A_{21} & A_{22} - g \end{vmatrix} = g^2 - (A_{11} + A_{22})g + (A_{11}A_{22} - A_{21}A_{12}) = 0 \quad (7.36)$$

and

$$\begin{vmatrix} B_{11} - f & B_{12} \\ B_{21} & B_{22} - f \end{vmatrix} = f^2 - (B_{11} + B_{22})f + (B_{11}B_{22} - B_{21}B_{12}) = 0. \quad (7.37)$$

The solutions of the resulting quadratic equations are

$$g_1 = 9.63435 \times 10^{-4} °\text{y}^{-1}, \qquad g_2 = 6.09908 \times 10^{-3} °\text{y}^{-1} \qquad (7.38)$$

and

$$f_1 = 0, \qquad f_2 = -7.06251 \times 10^{-3} °\text{y}^{-1}. \qquad (7.39)$$

The eigenvectors of $\mathbf{A}$ and $\mathbf{B}$ are the four vectors $\mathbf{x}_1, \mathbf{x}_2, \mathbf{y}_1,$ and $\mathbf{y}_2$ that satisfy the equations

$$\mathbf{A}\mathbf{x}_i = g_i\mathbf{x}_i \qquad \text{and} \qquad \mathbf{B}\mathbf{y}_i = f_i\mathbf{y}_i \qquad (i = 1, 2). \qquad (7.40)$$

However, it is clear from these definitions that if $\mathbf{x}_i$ is an eigenvector of the matrix $\mathbf{A}$ then so is $c\mathbf{x}_i$, where $c$ is a constant. Therefore each eigenvector is only determined up to some arbitrary scaling constant. If we let $\bar{e}_{ji}$ and $\bar{I}_{ji}$ denote the components of these unscaled eigenvectors and let $S_i$ and $T_i$ denote the scaling constant (or magnitude) of each eigenvector, then

$$S_i \bar{e}_{ji} = e_{ji} \quad \text{and} \quad T_i \bar{I}_{ji} = I_{ji} \quad (i = 1, 2). \tag{7.41}$$

The values of $\bar{e}_{ji}$ and $\bar{I}_{ji}$ are obtained by solving four sets of two simultaneous linear equations in two unknowns. The four resulting (unscaled) eigenvectors are

$$\begin{pmatrix} \bar{e}_{11} \\ \bar{e}_{21} \end{pmatrix} = \begin{pmatrix} -0.777991 \\ -0.628275 \end{pmatrix}, \quad \begin{pmatrix} \bar{e}_{12} \\ \bar{e}_{22} \end{pmatrix} = \begin{pmatrix} 0.332842 \\ -1.01657 \end{pmatrix},$$

$$\begin{pmatrix} \bar{I}_{11} \\ \bar{I}_{21} \end{pmatrix} = \begin{pmatrix} 0.707107 \\ 0.707107 \end{pmatrix}, \quad \begin{pmatrix} \bar{I}_{12} \\ \bar{I}_{22} \end{pmatrix} = \begin{pmatrix} -0.40797 \\ 1.00624 \end{pmatrix}. \tag{7.42}$$

The scaling factors $S_i$ and $T_i$ are determined from the boundary conditions. At time $t = 0$ we have

$$h_1 = 0.0114692, \quad h_2 = 0.0575337, \quad k_1 = 0.0460556, \quad k_2 = 0.00128611 \tag{7.43}$$

and

$$p_1 = 0.0224566, \quad p_2 = 0.0399314, \quad q_1 = -0.00397510, \quad q_2 = -0.0170597, \tag{7.44}$$

where we have converted the inclinations from degrees to radians. Substituting $t = 0$ in our general solution given in Eqs. (7.28) and (7.29) we get

$$h_j = S_1 \bar{e}_{j1} \sin \beta_1 + S_2 \bar{e}_{j2} \sin \beta_2, \quad k_j = S_1 \bar{e}_{j1} \cos \beta_1 + S_2 \bar{e}_{j2} \cos \beta_2 \tag{7.45}$$

and

$$p_j = T_1 \bar{I}_{j1} \sin \gamma_1 + T_2 \bar{I}_{j2} \sin \gamma_2, \quad q_j = T_1 \bar{I}_{j1} \cos \gamma_1 + T_2 \bar{I}_{j2} \cos \gamma_2, \tag{7.46}$$

where the subscript $j$ ($= 1, 2$) denotes the planet (Jupiter or Saturn). These can be considered as four sets of two simultaneous linear equations in the eight unknowns $S_i \sin \beta_i$, $S_i \cos \beta_i$, $T_i \sin \gamma_i$, and $T_i \cos \gamma_i$ with ($i = 1, 2$). In our case the solutions are

$$\begin{pmatrix} S_1 \sin \beta_1 \\ S_2 \sin \beta_2 \end{pmatrix} = \begin{pmatrix} -0.0308089 \\ -0.375549 \end{pmatrix}, \quad \begin{pmatrix} S_1 \cos \beta_1 \\ S_2 \cos \beta_2 \end{pmatrix} = \begin{pmatrix} -0.0472469 \\ 0.027935 \end{pmatrix},$$

$$\begin{pmatrix} T_1 \sin \gamma_1 \\ T_2 \sin \gamma_2 \end{pmatrix} = \begin{pmatrix} 0.0388876 \\ 0.0123566 \end{pmatrix}, \quad \begin{pmatrix} T_1 \cos \gamma_1 \\ T_2 \cos \gamma_2 \end{pmatrix} = \begin{pmatrix} -0.0109598 \\ -0.00925221 \end{pmatrix}. \tag{7.47}$$

These give

$$\beta_1 = -146.892°, \quad \beta_2 = -53.3565,° \quad \gamma_1 = 105.74° \quad \gamma_2 = 126.825° \tag{7.48}$$

and

$$S_1 = 0.0564044, \quad S_2 = 0.0468053, \quad T_1 = 0.0404025, \quad T_2 = 0.0154366. \tag{7.49}$$

The resulting, scaled eigenvectors are

$$\begin{pmatrix} e_{11} \\ e_{21} \end{pmatrix} = \begin{pmatrix} -0.0438821 \\ -0.0354375 \end{pmatrix}, \quad \begin{pmatrix} e_{12} \\ e_{22} \end{pmatrix} = \begin{pmatrix} 0.0155788 \\ -0.047581 \end{pmatrix},$$

$$\begin{pmatrix} I_{11} \\ I_{21} \end{pmatrix} = \begin{pmatrix} 0.0285689 \\ 0.0285689 \end{pmatrix}, \quad \begin{pmatrix} I_{12} \\ I_{22} \end{pmatrix} = \begin{pmatrix} -0.00629766 \\ 0.015533 \end{pmatrix}, \quad (7.50)$$

where the $I_{ji}$ are expressed in radians.

We have now determined all the constants in Eqs. (7.28) and (7.29). Therefore we can obtain $h$, $k$, $p$, and $q$ for Jupiter and Saturn at any time $t$. The solution is of the form

$$h_j = e_{j1} \sin(g_1 t + \beta_1) + e_{j2} \sin(g_2 t + \beta_2),$$

$$k_j = e_{j1} \cos(g_1 t + \beta_1) + e_{j2} \cos(g_2 t + \beta_2),$$

$$p_j = I_{j1} \sin(f_1 t + \gamma_1) + I_{j2} \sin(f_2 t + \gamma_2),$$

$$q_j = I_{j1} \cos(f_1 t + \gamma_1) + I_{j2} \cos(f_2 t + \gamma_2), \quad (7.51)$$

where $j = 1$ for Jupiter and $j = 2$ for Saturn. From these solutions we can derive the orbital elements of the two planets at any time $t$. For example, the relation $e_j(t) = (h_j^2 + k_j^2)^{1/2}$ is used to calculate the eccentricity of planet $j$. Using our results we obtain

$$e_1(t) = \sqrt{0.00217 - 0.00137 \cos(93.5° + 0.005514\,t)},$$

$$e_2(t) = \sqrt{0.00352 + 0.00337 \cos(93.5° + 0.005514\,t)}, \quad (7.52)$$

where the phases are in degrees and the frequencies in degrees per year. This implies a fixed periodicity of $\sim 70,100$ y in the variation of the eccentricity of each planet. Figure 7.1a shows the evolution of the eccentricities of the two planets over a time span of 200,000 y derived from our secular solution; the periodicity in the variation is clear. The different signs in the magnitude of the cosine imply that a maximum in Jupiter's eccentricity coincides with a minimum in Saturn's eccentricity and vice versa.

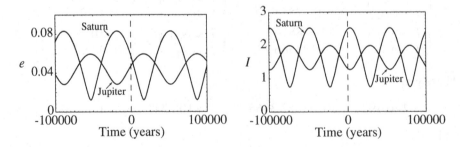

Fig. 7.1. The (a) eccentricities and (b) inclinations of Jupiter and Saturn derived from a secular perturbation theory calculated over a time span of 200,000 y centred on 1983.

Similarly the relation $I_j = (p_j^2 + q_j^2)^{1/2}$ is used to calculate the inclination of planet $j$ at any time $t$. Our results (in radians) give

$$I_1(t) = \sqrt{0.000856 - 0.00360\cos(21.1° - 0.00706\,t)},$$
$$I_2(t) = \sqrt{0.00106 + 0.000888\cos(21.1° - 0.00706\,t)}. \tag{7.53}$$

In this case the associated period of the secular variation in each planet is $\sim 51{,}000$ y; since $f_1 = 0$, this period is just $360°/f_2$. The variation for each planet is shown in Fig. 7.1b.

The secular solution that we have derived for Jupiter and Saturn is only an approximation to the actual variations in their orbital elements. In reality the perturbations from the planets Uranus and Neptune exert considerable influence on their orbits. A further complication is that the orbits of Jupiter and Saturn are close to a 5:2 commensurability. This introduces additional perturbations on timescales that are shorter than those associated with the secular variation.

## 7.4 Free and Forced Elements

We have shown that under certain conditions we can construct a secular solution to the motion of two orbiting bodies moving under their mutual gravitational effects; at any time we can obtain the eccentricities, longitudes of pericentre, inclinations, and longitudes of ascending node of both bodies. We can make use of this solution to study the motion of an additional body, of negligible mass, moving under the influence of the central body and perturbed by the other two bodies.

Following the example given in Sect. 7.2 for the secular theory for two bodies, the disturbing function $\mathcal{R}$ for a test particle with orbital elements $a$, $n$, $e$, $I$, $\varpi$, and $\Omega$ is given by

$$\mathcal{R} = na^2 \left[ \frac{1}{2} A e^2 + \frac{1}{2} B I^2 \right.$$

$$\left. + \sum_{j=1}^{2} A_j e e_j \cos(\varpi - \varpi_j) + \sum_{j=1}^{2} B_j I I_j \cos(\Omega - \Omega_j) \right], \tag{7.54}$$

where

$$A = +n\frac{1}{4}\sum_{j=1}^{2} \frac{m_j}{m_c} \alpha_j \bar{\alpha}_j b_{3/2}^{(1)}(\alpha_j), \tag{7.55}$$

$$A_j = -n\frac{1}{4}\frac{m_j}{m_c} \alpha_j \bar{\alpha}_j b_{3/2}^{(2)}(\alpha_j), \tag{7.56}$$

$$B = -n\frac{1}{4}\sum_{j=1}^{2} \frac{m_j}{m_c} \alpha_j \bar{\alpha}_j b_{3/2}^{(1)}(\alpha_j), \tag{7.57}$$

$$B_j = +n\frac{1}{4}\frac{m_j}{m_c} \alpha_j \bar{\alpha}_j b_{3/2}^{(1)}(\alpha_j) \tag{7.58}$$

and

$$\alpha_j = \begin{cases} a_j/a & \text{if} & a_j < a, \\ a/a_j & \text{if} & a_j > a, \end{cases} \tag{7.59}$$

$$\bar{\alpha}_j = \begin{cases} 1 & \text{if} & a_j < a, \\ a/a_j & \text{if} & a_j > a. \end{cases} \tag{7.60}$$

If we now transform to a new set of variables $h$, $k$, $p$, and $q$ for the test particle and $h_j$, $k_j$, $p_j$, and $q_j$ $(j = 1, 2)$ for each perturbing body, where

$$h = e \sin \varpi, \qquad k = e \cos \varpi \tag{7.61}$$

and

$$p = I \sin \Omega, \qquad q = I \cos \Omega \tag{7.62}$$

and the other elements are already defined in Eqs. (7.18) and (7.19), we have

$$\mathcal{R} = na^2 \left[ \frac{1}{2} A(h^2 + k^2) + \frac{1}{2} B(p^2 + q^2) \right.$$
$$\left. + \sum_{j=1}^{2} A_j(hh_j + kk_j) + \sum_{j=1}^{2} B_j(pp_j + qq_j) \right]. \tag{7.63}$$

The equations of motion are

$$\dot{h} = +\frac{1}{na^2} \frac{\partial \mathcal{R}}{\partial k}, \qquad \dot{k} = -\frac{1}{na^2} \frac{\partial \mathcal{R}}{\partial h}, \tag{7.64}$$

$$\dot{p} = +\frac{1}{na^2} \frac{\partial \mathcal{R}}{\partial q}, \qquad \dot{q}_j = -\frac{1}{na^2} \frac{\partial \mathcal{R}}{\partial p}. \tag{7.65}$$

Substituting for $\mathcal{R}$ from Eq. (7.63) we can write the equations of motion as

$$\dot{h} = +Ak + \sum_{j=1}^{2} A_j k_j, \qquad \dot{k} = -Ah - \sum_{j=1}^{2} A_j h_j, \tag{7.66}$$

$$\dot{p} = +Bq + \sum_{j=1}^{2} B_j q_j, \qquad \dot{q} = -Bp - \sum_{j=1}^{2} B_j p_j, \tag{7.67}$$

where the values of $h_j$, $k_j$, $p_j$, and $q_j$ are derived from the secular solution given in Eqs. (7.28) and (7.29). Substituting from these equations we get

$$\dot{h} = +Ak + \sum_{j=1}^{2} A_j \sum_{i=1}^{2} e_{ji} \cos(g_i t + \beta_i), \tag{7.68}$$

$$\dot{k} = -Ah - \sum_{j=1}^{2} A_j \sum_{i=1}^{2} e_{ji} \sin(g_i t + \beta_i), \tag{7.69}$$

$$\dot{p} = +Bq + \sum_{j=1}^{2} B_j \sum_{i=1}^{2} I_{ji} \cos(f_i t + \gamma_i), \tag{7.70}$$

$$\dot{q} = -Bp - \sum_{j=1}^{2} B_j \sum_{i=1}^{2} I_{ji} \sin(f_i t + \gamma_i). \tag{7.71}$$

By taking another time derivative of each equation and using Eq. (7.26) again we have

$$\ddot{h} = -A^2 h - \sum_{i=1}^{2} v_i (A + g_i) \sin(g_i t + \beta_i), \tag{7.72}$$

$$\ddot{k} = -A^2 k - \sum_{i=1}^{2} v_i (A + g_i) \cos(g_i t + \beta_i), \tag{7.73}$$

$$\ddot{p} = -B^2 p - \sum_{i=1}^{2} \mu_i (B + f_i) \sin(f_i t + \gamma_i), \tag{7.74}$$

$$\ddot{q} = -B^2 q - \sum_{i=1}^{2} \mu_i (B + f_i) \cos(f_i t + \gamma_i), \tag{7.75}$$

where

$$v_i = \sum_{j=1}^{2} A_j e_{ji} \quad \text{and} \quad \mu_i = \sum_{j=1}^{2} B_j I_{ji}. \tag{7.76}$$

The solutions to the uncoupled differential equations in Eqs. (7.72)–(7.75) are

$$\begin{aligned} h &= e_{\text{free}} \sin(At + \beta) + h_0(t), \qquad k = e_{\text{free}} \cos(At + \beta) + k_0(t), \\ p &= I_{\text{free}} \sin(Bt + \gamma) + p_0(t), \qquad q = I_{\text{free}} \cos(Bt + \gamma) + q_0(t), \end{aligned} \tag{7.77}$$

where $e_{\text{free}}$, $I_{\text{free}}$, $\beta$, and $\gamma$ are constants determined from the boundary conditions and

$$h_0(t) = -\sum_{i=1}^{2} \frac{v_i}{A - g_i} \sin(g_i t + \beta_i), \tag{7.78}$$

$$k_0(t) = -\sum_{i=1}^{2} \frac{v_i}{A - g_i} \cos(g_i t + \beta_i), \tag{7.79}$$

$$p_0(t) = -\sum_{i=1}^{2} \frac{\mu_i}{B - f_i} \sin(f_i t + \gamma_i), \tag{7.80}$$

$$q_0(t) = -\sum_{i=1}^{2} \frac{\mu_i}{B - f_i} \cos(f_i t + \gamma_i). \tag{7.81}$$

Note that $h_0$, $k_0$, $p_0$, and $q_0$ are only functions of the (constant) semi-major axis of the particle and do not involve any of its other orbital elements. However, since the values of $h_0$, $k_0$, $p_0$, and $q_0$ also depend on the secular solution for the two perturbing bodies, they will vary with time.

If we define the quantities

$$e_{\text{forced}} = \sqrt{h_0^2 + k_0^2}, \qquad I_{\text{forced}} = \sqrt{p_0^2 + q_0^2} \qquad (7.82)$$

then the solutions given in Eq. (7.77) have a simple geometrical interpretation (Figs. 7.2 and 7.3). In the case of the $h$–$k$ solution, the values of $k$ and $h$ for the particle define a point in the $k$–$h$ plane. The vector from the origin to this point has a length $e$ and it makes an angle $\varpi$ with the $k$ axis. In the light of our solution given above, this vector can also be thought of as the vector sum of two other vectors: The first goes from the origin to the point $(k_0, h_0)$; it has a length $e_{\text{forced}}$ and makes an angle $\varpi_{\text{forced}}$ with the $k$ axis. The second goes from $(k_0, h_0)$ to the point $(k, h)$; it has a length $e_{\text{free}}$ and makes an angle $\varpi_{\text{free}} = At + \beta$ with the $k$ axis. This implies that the particle's motion can be thought of as motion around a circle with centre $(k_0, h_0)$ at a constant rate $A$ while this point itself moves in some complicated path determined by the secular solution for the two perturbing bodies. This is illustrated in Fig. 7.2. The quantities $e_{\text{forced}}$ and $\varpi_{\text{forced}}$ are derived from $h_0$ and $k_0$ and they are called the *forced eccentricity* and *forced longitude of pericentre* of the particle. Their values are determined solely by the semi-major axis of the particle and the secular solution for the two perturbing bodies. In contrast, $e_{\text{free}}$ and $\varpi_{\text{free}}$, the *free eccentricity* and *free longitude of pericentre* of the particle, are derived from the boundary conditions and denote fundamental orbital parameters of the particle. These quantities are also referred to as the *proper eccentricity* and *proper longitude of pericentre* of the particle's orbit.

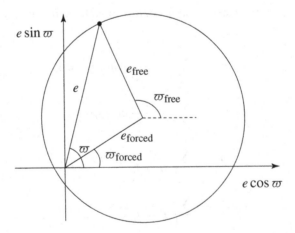

Fig. 7.2.   The geometrical relationship among the osculating, free, and forced eccentricities and longitudes of pericentre for the case $e_{\text{free}} > e_{\text{forced}}$.

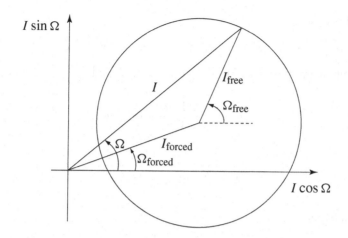

Fig. 7.3. The geometrical relationship among the osculating, free, and forced inclinations and longitudes of ascending node for the case $I_{\text{free}} < I_{\text{forced}}$.

It is important to note that the circle shown in Fig. 7.2 need not encompass the origin. If $e_{\text{free}}$ is small enough or the value of $e_{\text{forced}}$ for the particle's semi-major axis is large enough, the motion of the particle around the circle might be such that $\varpi$ or $\Omega$ (the osculating longitudes of pericentre or ascending node) vary over some fixed range of angles.

The $p$–$q$ solution is shown in Fig. 7.3, where $I_{\text{forced}}$, $\Omega_{\text{forced}}$, $I_{\text{free}}$, and $\Omega_{\text{free}}$ denote the forced and free inclinations and nodes of the particle. Here we illustrate a situation where the forced inclination is larger than the free one, such that the circle does not enclose the origin. As stated above, this is an outcome of the boundary conditions.

Because the expressions for $e_{\text{forced}}$ and $I_{\text{forced}}$ given in Eq. (7.82) depend on the definitions of $h_0$, $k_0$, $p_0$, and $q_0$ given in Eqs. (7.78)–(7.81), it is clear that potentially large values of $e_{\text{forced}}$ or $I_{\text{forced}}$ can arise if either of the conditions $A - g_i \approx 0$ or $B - f_i \approx 0$ is satisfied. The $g_i$ and $f_i$ are the eigenfrequencies of the system of two interacting bodies whereas, from Eqs. (7.55) and (7.57), the quantities $A$ and $B$ are functions of the semi-major axis of the test particle. This implies that at certain locations in semi-major axis there will be singularities in the forced eccentricities or inclinations. We will consider a specific example of this in Sect. 7.5.

Another important point concerns the limiting values of $e_{\text{forced}}$ and $I_{\text{forced}}$ as the orbit of either of the two perturbers is approached. Since the $A$, $A_j$, $B$, and $B_j$ in Eqs. (7.55)–(7.58) as well as the definitions of the $\nu_i$ and $\mu_i$ in Eq. (7.76) involve the Laplace coefficients $b_{3/2}^{(1)}(\alpha_j)$ or $b_{3/2}^{(2)}(\alpha_j)$, which all approach infinity as $\alpha_j \to 1$, it is not obvious that there are finite limiting values for $e_{\text{forced}}$ and $I_{\text{forced}}$ at the orbits of the perturbers. Let us assume that we are considering the behaviour of $e_{\text{forced}}$ in the vicinity of a perturbing body denoted by subscript

$j = l$. In this case

$$b_{3/2}^{(1)}(\alpha_l) \to \infty \quad \text{and} \quad b_{3/2}^{(2)}(\alpha_l) \to \infty \quad \text{as} \quad \alpha_l \to 1. \qquad (7.83)$$

At the orbit of the perturber $A \gg g_i$ for $i = 1, 2$ and $A_l \gg A_i$, where $i \neq l$. Therefore

$$A - g_i \approx A \approx \frac{1}{4} n \frac{m_l}{m_c} \alpha_l \bar{\alpha}_l b_{3/2}^{(1)}(\alpha_l) \qquad (7.84)$$

and

$$v_i = \sum_{j=1}^{2} A_j e_{ji} \approx A_l e_{li}. \qquad (7.85)$$

This implies that

$$h_0(t) \approx -\sum_{i=1}^{2} \frac{A_l e_{li}}{A} \sin(g_i t + \beta_i) = +\frac{b_{3/2}^{(2)}(\alpha_l)}{b_{3/2}^{(1)}(\alpha_l)} \sum_{i=1}^{2} e_{li} \sin(g_i t + \beta_i). \qquad (7.86)$$

Since the definition of the Laplace coefficient given in Eqs. (6.67) and (6.68) can also be written as

$$\frac{1}{2} b_s^{(j)}(\alpha) = \frac{s(s+1)\dots(s+j-1)}{j!} \alpha^j F(s, s+j, j+1; \alpha^2), \qquad (7.87)$$

where $F(a, b, c; d)$ is the standard hypergeometric function, we have

$$\lim_{\alpha_l \to 1} \frac{b_{3/2}^{(2)}(\alpha_l)}{b_{3/2}^{(1)}(\alpha_l)} = \lim_{\alpha_l \to 1} \left[ \frac{5}{4} \alpha_l \frac{F(\frac{3}{2}, \frac{7}{2}, 3, \alpha_l^2)}{F(\frac{3}{2}, \frac{5}{2}, 2, \alpha_l^2)} \right] = 1 \qquad (7.88)$$

from the properties of hypergeometric series and their relationship with elliptical integrals. Therefore

$$\lim_{\alpha_l \to 1} h_0(t) = \sum_{i=1}^{2} e_{li} \sin(g_i t + \beta_i) = h_l. \qquad (7.89)$$

Similarly

$$\lim_{\alpha_l \to 1} k_0(t) = \sum_{i=1}^{2} e_{li} \cos(g_i t + \beta_i) = k_l. \qquad (7.90)$$

Therefore, as the orbit of perturbing body $l$ is approached,

$$e_{\text{forced}} = \sqrt{h_0^2 + k_0^2} \to \sqrt{h_l^2 + k_l^2} = e_l. \qquad (7.91)$$

This implies that the forced values of the eccentricity and longitude of pericentre at the orbit of the perturber are equal to the equivalent osculating values of these elements for the perturber. A similar result for the forced values of the inclination and longitude of ascending node can be shown using the same method as given above. However, we have to be careful not to make too many generalisations about the nature of the particle's orbit in the vicinity of the orbit of a perturber. We have already seen in Fig. 3.30 that particles near the $L_1$ and $L_2$ points acquire an eccentricity from their encounter with the satellite, despite the fact that the satellite is moving in a circular orbit. Therefore our result concerning the forced eccentricity and inclination is really only valid for particle orbits far from $L_1$ and $L_2$ with low values of $e$ and $I$.

## 7.5 Jupiter, Saturn, and a Test Particle

We can illustrate the results derived in Sect. 7.4 by calculating the forced orbital elements on a test particle moving under the effects of secular perturbations from Jupiter and Saturn. In Sect. 7.3 we obtained the secular solution for the Jupiter–Saturn system and so we have values of $h_j$, $k_j$, $p_j$, and $q_j$ ($j = 1, 2$) at any time $t$. This allows us to calculate the forced elements of a test particle at any location in the solar system.

Before proceeding it is important to point out that this entire analysis ignores the effect of mean motion resonances. We have already seen in Sect. 6.9.2 that at certain locations the particle is subjected to additional perturbations over an above those due to the secular terms in the disturbing function. A more complete analysis of the resonant terms is given in Sect. 8. Here we content ourselves with the warning that our results on forced elements do not take account of the effects of mean motion resonances.

The first step in the secular theory for the particle's motion is to calculate the value of the frequency, $A$, given in Eq. (7.55). Note that in the case of two bodies orbiting a spherical or point-mass central object, we have $B = -A$. The value of $A$ depends on (i) the masses of the perturbers and (ii) the semi-major axis of the perturbed particle.

Figure 7.4 shows the variation in $A$ from 0 to 30 AU. The singularities close to 5 and 10 AU arise from the fact that the Laplace coefficients tend to infinity as $\alpha_j \rightarrow 1$. On the same diagram the three, nonzero eigenfrequencies of the system are denoted by the solid and dashed horizontal lines. Two of them, namely $g_1 = 0.00096°\mathrm{y}^{-1}$ and $g_2 = 0.0061°\mathrm{y}^{-1}$, are the eccentricity–pericentre eigenfrequencies while the third one, $f_2 = -0.0071°\mathrm{y}^{-1}$, is the single nonzero inclination–node eigenfrequency. The intersection of these lines with the curve showing the variation of $A$ identify the semi-major axes where large forced eccentricities or inclinations can be expected.

Since the value of $A$ only depends on the semi-major axes and masses of the planets and the particle, and since all these quantities are constant (recall that there is no secular change in the semi-major axes), the value of $A$ is constant at any given semi-major axis. It is also independent of time.

The values of the forced eccentricity and longitude of pericentre as a function of semi-major axis for a given time can be calculated using Eqs. (7.76) and

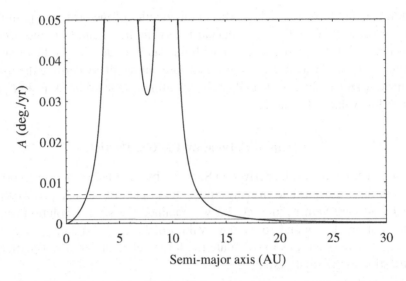

Fig. 7.4. The variation of the frequency $A$, defined in Eq. (7.55), as a function of semi-major axis, derived from perturbations by Jupiter and Saturn. The horizontal solid lines denote the values of the two eccentricity–pericentre eigenfrequencies, $A = g_1 = 0.00096°\mathrm{y}^{-1}$ and $A = g_2 = 0.0061°\mathrm{y}^{-1}$; the dashed line denotes the value of the nonzero inclination–node eigenfrequency, $A = -B = -f_2 = 0.0071°\mathrm{y}^{-1}$. Singularities in the plot correspond to the orbital semi-major axes of Jupiter and Saturn.

(7.78)–(7.82) where the values of $e_{ji}$ and $I_{ji}$ are given in Sect. 7.3. Figure 7.5 shows the variation of $e_{\mathrm{forced}}$ and $\varpi_{\mathrm{forced}}$ in the range 0 to 30 AU. Note that the values of $e_{\mathrm{forced}}$ and $\varpi_{\mathrm{forced}}$ at the orbits of Jupiter and Saturn are equal to the osculating values of the two planets at this time. This phenomenon was discussed in Sect. 7.4. It is important to point out that the curves shown in Fig. 7.5 are for one specific time ($t = 0$). Since the values of the osculating elements for the planets vary according to the secular solution, so too will the shape of the curves.

The singularities in the $e_{\mathrm{forced}}$ plot (see Fig. 7.5) close to 0.5, 2, 12.5, and 17.5 AU are a consequence of the small divisors inherent in Eqs. (7.78) and (7.79). At each of these locations the value of $A$ is equal to one of the $g_i$ eigenfrequencies of the perturbing system. The equivalent effect in the forced longitudes of perihelion, $\varpi_{\mathrm{forced}}$, is to cause a sudden shift in the longitudes of 180° (see Fig. 7.5b). Although the shape of both curves will alter with time as the osculating elements of Jupiter and Saturn vary, the singularities will remain at the same locations.

The equivalent set of plots for the forced inclination and forced longitude of ascending node are shown in Fig. 7.6. Since there is only one nonzero eigenfrequency in the system, there will only be two singularities (in this case near 2

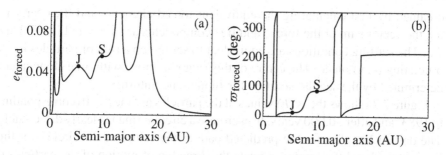

Fig. 7.5. The variation in (a) the forced eccentricity and (b) the forced longitude of perihelion as a function of semi-major axis at time $t = 0$. The letters J and S denote the osculating values for Jupiter and Saturn respectively at their semi-major axes. The singularities near 0.5, 2, 12.5, and 17.5 AU arise from the small divisors $A - g_i$ in Eqs. (7.78)–(7.79).

and 12 AU) in the plots for $I_{forced}$ and $\Omega_{forced}$ corresponding to the two points of intersection of $A$ with the dashed line shown in Fig. 7.4.

We can provide an additional illustration of the results derived in Sect. 7.4 by considering the orbital evolution of a number of test particles moving under the effects of perturbations from Jupiter and Saturn. A numerical integration of the full equations of motion was carried out of a system consisting of 250 test particles, Jupiter, and Saturn. The planets perturbed each other and the particles but the test particles were assumed to have negligible mass. The test particles were started with the same semi-major axis (1.8 AU), free eccentricity (0.049), and free inclination (2.12°), but with random initial free longitudes of pericentre and ascending nodes. These values were chosen to minimise any resonant effects (see Sect. 6.9). The initial osculating values of $e$, $I$, $\varpi$, and $\Omega$ for each particle were

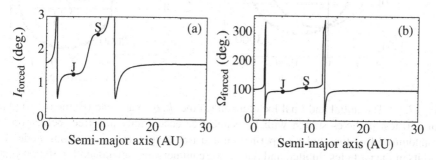

Fig. 7.6. The variation in (a) the forced inclination and (b) the forced longitude of ascending node as a function of semi-major axis at time $t = 0$. The letters J and S denote the osculating values Jupiter and Saturn respectively at their semi-major axes. The singularities near 2 and 12 AU arise from the small divisor $B - f_2$ in Eqs. (7.80) and (7.81).

obtained by first calculating the equivalent forced elements and then carrying out the vector sum of the forced and free components as shown in Figs. 7.2 and 7.3. The starting conditions are equivalent to setting up a ring of particles in the osculating $(k, h)$ plane. The centre of this ring will move with time along a path determined by the secular solution for Jupiter and Saturn.

Figure 7.7a shows the $(k, h)$ values of the particles at time $t = 0$ (corresponding to 1983) and after 30,000 years. In each case a circular ring is clearly discernable. The dashed line denotes the predicted path of the centre of the circle over this time interval and the arrows indicate the direction of motion of the particles in the circle and of the centre of the circle itself. Note that the circle of particles in the $(k, h)$ plane is maintained over the timescale of the integration and that the predicted centre of the circle is in excellent agreement with its observed location. This implies that the results of the numerical integration agree with the predictions of secular perturbation theory and that the motion of a test particle can be described as a combination of uniform rotation in a circle where the centre of the circle is executing some well-defined path in the $(k, h)$ plane. It is clear from Fig. 7.7a that there has been some movement of the particles along the circle and in the radial direction. This can probably be accounted for by the effects of the short-period terms neglected in the secular calculation.

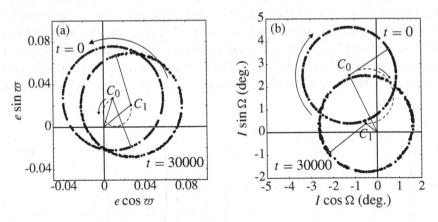

Fig. 7.7.  The initial and final locations in (a) the $(k, h)$ plane and (b) the $(q, p)$ plane of 250 test particles started with the same free eccentricity and free inclination but randomised free longitudes of perihelion and free longitudes of ascending node. The orbits of the particles, Jupiter, and Saturn were numerically integrated for 30,000 years. The points $C_0$ and $C_1$ denote the coordinates of $(k_0, h_0)$ (and $(q_0, p_0)$) at times $t = 0$ and $t = 30,000$ y respectively. For each circle there is a line (of length equal to the forced eccentricity or inclination) joining the origin to the centre and another line (of length equal to the free eccentricity or inclination) from there to the first test particle, denoted by a small white circle.

The equivalent results for the $(q, p)$ values of the same particles are shown in Fig. 7.7b. Once again the preservation of the circle with time is evident, as well as the fact that the centre of the circle lies close to the calculated location, $(q_0, p_0)$, determined from secular perturbation theory. As with Fig. 7.7a there is also some movement along the circle and in the radial direction caused by the effects of short-period terms.

A further subtlety arises when we consider direct comparisons between secular perturbation theory and full numerical integrations. Any integration of the full equations of motion must, by necessity, include all the infinite number of short-period terms in the expansion of the disturbing function. However, the secular perturbation theory we have derived and the resulting theory of free and forced elements are based on the truncation of the disturbing function and the averaging principle. This difference causes a problem when we have to consider starting values for the orbital elements of Jupiter and Saturn. In order to make the comparison as fair as possible we should numerically integrate the orbits of Jupiter and Saturn using the full equations of motion. The results should then be Fourier analysed to determine an approximation to the secular solution at time $t = 0$. These orbital elements should be used in the secular perturbation theory, which leads to a determination of the forced elements. In fact this procedure was adopted for the comparison shown in Fig. 7.7. This method is also used in the example given in Sect. 8.19 where we have to take account of resonant and secular terms.

## 7.6 Gauss's Averaging Method

We introduced the concept of secular perturbations by considering a mathematical approach using the disturbing function to isolate those terms that were independent of the mean longitudes. However, there is also a more physical approach that gives the same results and provides an insight into the disturbing function method and the averaging principle.

Consider the *averaged* precessional effect of an external perturber of mass $m'$ on the longitude of pericentre of the orbit of an internal body of mass $m$. Both objects are assumed to orbit in the same plane. Let $r$, $f$, $a$, $e$, and $\varpi$ denote the orbital radius, true anomaly, semi-major axis, eccentricity (assumed small), and longitude of pericentre of the internal mass with similar primed quantities for the external mass.

There are two approaches to the problem. The first approach is to make use of Lagrange's equations and an expansion of the disturbing function to isolate those terms (the secular terms) that will be important; this is the method we have already adopted. The alternative approach is to assume that the perturbational effect of the external body is equivalent to spreading its mass around its orbit so that the internal body moves under the force exerted by a "ring" of material. The density of this ring is nonuniform when $e' \neq 0$. The radial and tangential components of the perturbing force are then calculated and used with Gauss's form of the perturbational equations to derive the precession rate. This is *Gauss's method*. As we shall see, both approaches produce identical results.

To lowest order Lagrange's equations give

$$\langle \dot{\varpi} \rangle = \frac{1}{na^2 e} \frac{\partial \langle \mathcal{R} \rangle}{\partial e} \tag{7.92}$$

(cf. Eq. (7.16)), where the relevant parts of $\langle \mathcal{R} \rangle$ are given by the terms 4D0.1 and 4D0.2 with $j = 0$. There are no indirect terms. Neglecting the term in 4D0.1 that is independent of the eccentricities, we have

$$\langle \mathcal{R} \rangle = \frac{\mathcal{G}m'}{a'} \left[ \frac{1}{8}(e^2 + e'^2) \left( 2\alpha D + \alpha^2 D^2 \right) b_{\frac{1}{2}}^{(0)} \right.$$
$$\left. + \frac{1}{4} ee' \left( 2 - 2\alpha D - \alpha^2 D^2 \right) b_{\frac{1}{2}}^{(1)} \cos[\varpi' - \varpi] \right]. \tag{7.93}$$

Therefore, assuming $e \neq 0$, substitution of Eq. (7.93) in Eq. (7.92) and the use of Kepler's third law gives

$$\langle \dot{\varpi} \rangle = +\frac{1}{4} \frac{m'}{m_c} n\alpha^2 \left[ \left( 2D + \alpha D^2 \right) b_{\frac{1}{2}}^{(0)} \right.$$
$$\left. + \frac{e'}{e} \frac{1}{\alpha} \left( 2 - 2\alpha D - \alpha^2 D^2 \right) b_{\frac{1}{2}}^{(1)} \cos(\varpi' - \varpi) \right] \tag{7.94}$$

for the general case of the perturber moving in an elliptical orbit, where $m_c$ is the mass of the central body.

According to Gauss's method the averaged effect of the perturbations from the external mass can be calculated by smearing out the mass of the perturber around the orbit and then determining the potential or force that this exerts on the orbiting object. For a perturber moving in an eccentric orbit the line density is derived by smearing out the mass in such a way that the ring mass in a fixed time interval is the same all along the orbit. Therefore the resulting ring is densest at apocentre and least dense at pericentre. By Kepler's second law this is equivalent to a line density

$$\rho = \frac{m'r'}{2\pi a'b'}, \tag{7.95}$$

where $b' = a'\sqrt{1 - e'^2}$ is the semi-minor axis of the orbit.

Consider the mass $m$ at a particular point in its orbit given by the polar coordinates $(r, f)$ referred to the pericentre of its orbit. Consider a line element at position $(r', f')$ on the external ring of material. The mass of this element is $\rho r' d\phi$, where

$$\phi = f' + \varpi' - f - \varpi \tag{7.96}$$

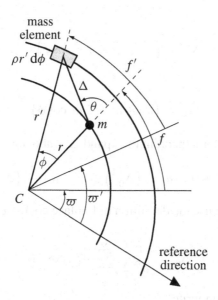

Fig. 7.8. The relationship between the orbital elements of the mass $m$ and mass element of the smeared ring of material, separated by a distance $\Delta$. The central mass is located at the point $C$.

is the angle between the two position vectors (the difference between the two true longitudes). If $\Delta$ is the distance between the mass $m$ and the ring element, then simple trigonometry gives

$$\Delta \sin \theta = r' \sin \phi \quad \text{and} \quad \Delta \cos \theta = r' \cos \phi - r. \quad (7.97)$$

The relationships between the various angles are shown in Fig. 7.8.

The mass element exerts a force of attraction on $m$ and this can be divided into components directed along the radius vector from the focus to the mass $m$ and at right angles to it in the positive sense. The radial and tangential components of the force are given by

$$d\bar{R} = \frac{\mathcal{G}\rho r'}{\Delta^2} \cos \theta \, d\phi \quad \text{and} \quad d\bar{T} = \frac{\mathcal{G}\rho r'}{\Delta^2} \sin \theta \, d\phi, \quad (7.98)$$

where $\theta$ is the angle that the line joining $m$ to the mass element makes with the extended radius vector of the mass $m$. Note that although we can also express the mass element as $(\rho \Delta / \cos(\theta - \phi)) d\theta$, and integrate over $\theta$, the integration over $\phi$ is easier to accomplish. Substituting for $\cos \theta$ and $\sin \theta$ from Eq. (7.97) we have

$$d\bar{R} = \mathcal{G}\rho r' \frac{(r' \cos \phi - r)}{\Delta^3} \, d\phi \quad \text{and} \quad d\bar{T} = \frac{\mathcal{G}\rho r'^2 \sin \phi}{\Delta^3} \, d\phi. \quad (7.99)$$

We can also express $\Delta$ in terms of $r$, $r'$, and $\phi$. Using the cosine rule we have

$$\Delta^2 = r'^2 + r^2 - 2rr' \cos \phi. \quad (7.100)$$

Therefore,

$$\Delta^{-3} = \left( r'^2 + r^2 - 2rr' \cos\phi \right)^{-\frac{3}{2}}. \tag{7.101}$$

If we write

$$\Delta_0^{-3} = \left( a'^2 + a^2 - 2aa' \cos\phi \right)^{-\frac{3}{2}} \tag{7.102}$$

then we can use Taylor's theorem to expand $\Delta^{-3}$ about $r = a$. This gives

$$\Delta^{-3} \approx \Delta_0^{-3} + (r - a)\frac{\partial}{\partial a}\left(\Delta_0^{-3}\right) + (r' - a')\frac{\partial}{\partial a'}\left(\Delta_0^{-3}\right). \tag{7.103}$$

Furthermore, we can use the definition of Laplace coefficients given in Sect. 6.4 to write

$$\Delta_0^{-3} = \frac{1}{a'^3}\frac{1}{2}\sum_{j=-\infty}^{\infty} b_{3/2}^{(j)} \cos j\phi. \tag{7.104}$$

Since $\alpha = a/a'$ we can write

$$\frac{\partial}{\partial a} \equiv \frac{1}{a'}\frac{\partial}{\partial \alpha} \qquad \text{and} \qquad \frac{\partial}{\partial a'} \equiv -\frac{\alpha}{a'}\frac{\partial}{\partial \alpha}. \tag{7.105}$$

Therefore

$$\frac{\partial}{\partial a}\left(\Delta_0^{-3}\right) = \frac{1}{a'^4}\frac{1}{2}\sum_{j=-\infty}^{\infty} \frac{db_{3/2}^{(j)}}{d\alpha} \cos j\phi \tag{7.106}$$

and

$$\frac{\partial}{\partial a'}\left(\Delta_0^{-3}\right) = -\frac{1}{a'^4}\frac{1}{2}\sum_{j=-\infty}^{\infty} \left( 3b_{3/2}^{(j)} + \alpha\frac{db_{3/2}^{(j)}}{d\alpha} \right) \cos j\phi. \tag{7.107}$$

For motion in an ellipse we have

$$r = \frac{a(1 - e^2)}{1 + e\cos f} \tag{7.108}$$

and hence

$$r - a \approx -ae\cos f + \mathcal{O}(e^2). \tag{7.109}$$

Therefore, to first order in $e$ we have

$$\Delta^{-3} = \frac{1}{2}\frac{1}{a'^3}\sum_{j=-\infty}^{\infty} \left[ b_{3/2}^{(j)} - \alpha e\frac{db_{3/2}^{(j)}}{d\alpha} \cos f \right.$$
$$\left. - e'\left( 3b_{3/2}^{(j)} + \alpha\frac{db_{3/2}^{(j)}}{d\alpha} \right) \cos f' \right] \cos j\phi. \tag{7.110}$$

In order to calculate the magnitude of the total radial force $\bar{R}$ acting on the fixed mass $m$, we need to find

$$\bar{R} = \oint d\bar{R} = \int_{\phi=0}^{2\pi} \mathcal{G}\rho r'(r'\cos\phi - r)\Delta^{-3}\,d\phi$$

$$= \int_{f'=0}^{2\pi} \mathcal{G}\rho r'(r'\cos[f' + \varpi' - f - \varpi] - r)\Delta^{-3}\,df', \qquad (7.11)$$

where the substitution $\phi = f' + \varpi' - f - \varpi$ has been made in the expression for $\Delta^{-3}$ given in Eq. (7.110), and it is understood that

$$r' \approx a'(1 - e'\cos f'). \qquad (7.112)$$

The magnitude of the total tangential force is given by

$$\bar{T} = \oint d\bar{T} = \int_{\phi=0}^{2\pi} \mathcal{G}\rho r'^2 \sin\phi\,\Delta^{-3}\,d\phi$$

$$= \int_{f'=0}^{2\pi} \mathcal{G}\rho r'^2 \sin[f' + \varpi' - f - \varpi]\Delta^{-3}\,df'. \qquad (7.113)$$

At this stage it is important to realise that the infinite summation in the expansion of $\Delta^{-3}$ reduces to a finite summation after the integration over $f'$ in Eqs. (7.111) and (7.113). For example,

$$\cos f' \cos[j(f' + \beta)] = \frac{1}{2}(\cos[f' - j(\beta + f')] + \cos[f' + j(\beta + f')]), \quad (7.114)$$

where $\beta$ is an angle independent of $f'$, and hence

$$\int_0^{2\pi} \cos f' \cos[j(f' + \beta)]\,df' = \begin{cases} 0 & \text{if } j \neq \pm 1, \\ \pi\cos\beta & \text{if } j = \pm 1. \end{cases} \qquad (7.115)$$

Therefore only a finite number of terms in the summation over $j$ are required. Since we only require terms to the first order in the eccentricity we only need to take, at most, the summation from $j = -2$ to $j = +2$. We can also make use of the fact that $b_s^{(-j)} = b_s^{(j)}$ to simplify the resulting expressions.

To find the associated precession rate due to $\bar{R}$ and $\bar{T}$ we use Gauss's form of the perturbation equations. We have, from Eq. (2.165),

$$\dot{\varpi} = \frac{1}{nae}\left[-\bar{R}\cos f + \bar{T}\left(\frac{2 + e\cos f}{1 + e\cos f}\right)\sin f\right], \qquad (7.116)$$

where terms of $\mathcal{O}(e^2)$ have been neglected.

Using the definition of $\rho$ from Eq. (7.95) and the approximations

$$r'^2 \approx a'^2(1 - 2e'\cos f') \quad \text{and} \quad r'^3 \approx a'^3(1 - 3e'\cos f') \qquad (7.117)$$

we get

$$\bar{R} = \frac{\mathcal{G}m'}{2\pi a'b'} \int_0^{2\pi} [a'^3(1 - 3e'\cos f')\cos[f' + \varpi' - f - \varpi]$$
$$- ra'^2(1 - 2e'\cos f')]\Delta^{-3}\,df' \qquad (7.118)$$

for the radial component of the force and

$$\bar{T} = \frac{\mathcal{G}m'}{2\pi a'b'} \int_0^{2\pi} a'^3(1 - 3e'\cos f')\sin[f' + \varpi' - f - \varpi]\Delta^{-3}\,df' \qquad (7.119)$$

for the tangential component, where $\Delta^{-3}$ is given by Eq. (7.110) with $\phi = f' + \varpi' - f - \varpi$.

In the calculation of $\bar{R}$ and $\bar{T}$ the only contributing values of $j$ in the summation that defines $\Delta^{-3}$ are $j = 0, \pm 1, \pm 2$. The resulting expressions for $\bar{R}$ and $\bar{T}$ are

$$\bar{R} = \frac{\mathcal{G}m'}{2a'^2}\left[ b_{3/2}^{(1)} - \alpha b_{3/2}^{(0)} + \alpha e\left( b_{3/2}^{(0)} - \frac{db_{3/2}^{(1)}}{d\alpha} + \alpha\frac{db_{3/2}^{(0)}}{d\alpha} \right)\cos f \right.$$
$$\left. - \alpha e'\left( b_{3/2}^{(1)} - \frac{1}{2}\frac{db_{3/2}^{(0)}}{d\alpha} - \frac{1}{2}\frac{db_{3/2}^{(2)}}{d\alpha} + \alpha\frac{db_{3/2}^{(1)}}{d\alpha} \right)\cos(\varpi' - \varpi - f) \right] \qquad (7.120)$$

and

$$\bar{T} = \frac{\mathcal{G}m'}{4a'^2}\alpha e'\left( \frac{db_{3/2}^{(2)}}{d\alpha} - \frac{db_{3/2}^{(0)}}{d\alpha} \right)\sin(\varpi' - \varpi - f), \qquad (7.121)$$

where we have used the fact that $b' = a' + \mathcal{O}(e^2)$. Note that when $e' = 0$ there is no tangential component to the force.

Using the substitutions from Eqs. (7.120) and (7.121) we can now obtain time-averaged values of the two terms in Eq. (7.116) that are functions of $f$. This is done by writing

$$\langle F(f) \rangle = \frac{1}{2\pi} \int_0^{2\pi} F(f)\,dM, \qquad (7.122)$$

where $F(f)$ is a general function of $f$ and $M$ is the true anomaly. Here we can make use of the expressions for $\cos f$ and $\sin f$ in terms of $M$ given in Sect. 2.5. This gives the averaged precession rate in the eccentric case as

$$\langle \dot{\varpi} \rangle =$$
$$+ \frac{1}{4}\frac{m'}{m_c}n\alpha^2\left[ -3\alpha b_{3/2}^{(0)} + 2b_{3/2}^{(1)} + \alpha\frac{db_{3/2}^{(1)}}{d\alpha} - \alpha^2\frac{db_{3/2}^{(0)}}{d\alpha} \right.$$
$$\left. + \frac{e'}{e}\frac{\alpha}{2}\left( -3\frac{db_{3/2}^{(0)}}{d\alpha} + 2b_{3/2}^{(1)} + 2\alpha\frac{db_{3/2}^{(1)}}{d\alpha} + \frac{db_{3/2}^{(2)}}{d\alpha} \right)\cos(\varpi' - \varpi) \right]. \qquad (7.123)$$

To compare this result with that derived from using the disturbing function method (see Eq. (7.94)), we need to rewrite the various Laplace coefficients in Eq. (7.123) using the relationships given in Eqs. (6.70)–(6.72). Alternatively each combination of Laplace coefficients can be expressed as a series in $\alpha$. Either method shows that the expression in Eq. (7.123) is identical to that given in Eq. (7.94). Therefore, using Gauss's method, we have shown that the secular change in the pericentre due to an external perturber with its mass distributed around its orbit is identical to that obtained by ignoring all but the most important secular terms in the planetary disturbing function. This provides a more rigorous justification of our use of the averaging principle.

## 7.7 Generalised Secular Perturbations

Now let us consider a more generalised form of the secular theory applied to the orbits of $N$ bodies moving around a nonspherical central mass. We can take account of the effects of an oblate central body when considering the perturbing potential on object $j$ by adding a term of the form

$$\langle \mathcal{R}_j^{(\text{obl})} \rangle =$$

$$\frac{1}{2} n_j a_j^2 n_j \left[ \frac{3}{2} J_2 \left( \frac{R_c}{a_j} \right)^2 - \frac{9}{8} J_2^2 \left( \frac{R_c}{a_j} \right)^4 - \frac{15}{4} J_4 \left( \frac{R_c}{a_j} \right)^4 \right] e_j^2$$

$$- \frac{1}{2} n_j a_j^2 n_j \left[ \frac{3}{2} J_2 \left( \frac{R_c}{a_j} \right)^2 - \frac{27}{8} J_2^2 \left( \frac{R_c}{a_j} \right)^4 - \frac{15}{4} J_4 \left( \frac{R_c}{a_j} \right)^4 \right] I_j^2, \quad (7.124)$$

where $R_c$ is the radius of the central body and $J_2$ and $J_4$ are its first two zonal gravity coefficients (cf. Eq. (6.255)).

Note that this expression is only applicable to situations where the perturbing and perturbed bodies are isolated, single bodies that have well-defined osculating elements. In the case of planetary rings the observed quantity is normally the geometrical semi-major axis of the ring (i.e., the best-fit ellipse traced by the ring particles). This requires changes in the definition of the mean motion of the ring because in this case the semi-major axis is not the same as the quantity $a$ used in the definition of the mean motion in Eq. (6.244).

The orbiting mass will experience the usual secular perturbations from other orbiting bodies. Consider the secular effect of a mass $m_k$ on the mass $m_j$. If we write

$$\langle \mathcal{R}_D(\alpha_{jk}) \rangle = \frac{1}{8} \alpha_{jk} b_{3/2}^{(1)} e_j^2 - \frac{1}{4} \alpha_{jk} b_{3/2}^{(2)} e_j e_k \cos(\varpi_j - \varpi_k)$$

$$- \frac{1}{8} \alpha_{jk} b_{3/2}^{(1)} I_j^2 + \frac{1}{4} \alpha_{jk} b_{3/2}^{(1)} I_j I_k \cos(\Omega_j - \Omega_k), \quad (7.125)$$

where $\alpha_{jk}$ is the ratio of the semi-major axes of the two objects then for an external perturbation we have

$$\langle \mathcal{R}_j^{(\text{sec})} \rangle = \frac{\mathcal{G} m_k}{a_k} \langle \mathcal{R}_D(a_j/a_k) \rangle \qquad (a_j < a_k), \tag{7.126}$$

whereas for an internal perturbation we get

$$\langle \mathcal{R}_j^{(\text{sec})} \rangle = \frac{\mathcal{G} m_k}{a_k} \frac{a_k}{a_j} \langle \mathcal{R}_D(a_k/a_j) \rangle \qquad (a_j > a_k). \tag{7.127}$$

The inconvenience of having to specify whether the perturbation is internal or external can be overcome if we define the quantities $\alpha_{jk}$ and $\bar{\alpha}_{jk}$ by

$$\alpha_{jk} = \begin{cases} a_k/a_j & \text{if} \quad a_j > a_k \quad \text{(internal perturber)}, \\ a_j/a_k & \text{if} \quad a_j < a_k \quad \text{(external perturber)} \end{cases} \tag{7.128}$$

and

$$\bar{\alpha}_{jk} = \begin{cases} 1 & \text{if} \quad a_j > a_k \quad \text{(internal perturber)}, \\ a_j/a_k & \text{if} \quad a_j < a_k \quad \text{(external perturber)}. \end{cases} \tag{7.129}$$

This notation was first used by Dermott & Nicholson (1986) in their secular perturbation theory for the uranian satellites. Using our usual approximation, $\mathcal{G} \approx n_j^2 a_j^3/m_c$, we can now write the secular part of the disturbing function experienced by the mass $m_j$ due to all the other $N-1$ masses. We have

$$\langle \mathcal{R}_j^{(\text{sec})} \rangle = n_j a_j^2 n_j \sum_{k=1, k \neq j}^{N} \frac{m_k}{m_c + m_j} \left[ \frac{1}{8} \alpha_{jk} \bar{\alpha}_{jk} b_{3/2}^{(1)} e_j^2 - \frac{1}{8} \alpha_{jk} \bar{\alpha}_{jk} b_{3/2}^{(1)} I_j^2 \right.$$
$$\left. - \frac{1}{4} \alpha_{jk} \bar{\alpha}_{jk} b_{3/2}^{(2)} e_j e_k \cos(\varpi_j - \varpi_k) + \frac{1}{4} \alpha_{jk} \bar{\alpha}_{jk} b_{3/2}^{(1)} I_j I_k \cos(\Omega_j - \Omega_k) \right]. \tag{7.130}$$

We can combine the $\mathcal{R}^{(\text{obl})}$ and $\mathcal{R}^{(\text{sec})}$ terms and write down an expression for the perturbing function for the mass $m_j$:

$$\mathcal{R}_j = n_j a_j^2 \left[ \frac{1}{2} A_{jj} e_j^2 + \frac{1}{2} B_{jj} I_j^2 \right.$$
$$\left. + \sum_{k=1, k \neq j}^{N} A_{jk} e_j e_k \cos(\varpi_j - \varpi_k) + \sum_{k=1, k \neq j}^{N} B_{jk} I_j I_k \cos(\Omega_j - \Omega_k) \right], \tag{7.131}$$

where

$$A_{jj} = +n_j \left[ \frac{3}{2} J_2 \left( \frac{R_c}{a_j} \right)^2 - \frac{9}{8} J_2^2 \left( \frac{R_c}{a_j} \right)^4 - \frac{15}{4} J_4 \left( \frac{R_c}{a_j} \right)^4 \right.$$
$$\left. + \frac{1}{4} \sum_{k=1, k \neq j}^{N} \frac{m_k}{m_c + m_j} \alpha_{jk} \bar{\alpha}_{jk} b_{3/2}^{(1)}(\alpha_{jk}) \right], \tag{7.132}$$

$$A_{jk} = -\frac{1}{4}\frac{m_k}{m_c + m_j}n_j\alpha_{jk}\bar{\alpha}_{jk}b_{3/2}^{(2)}(\alpha_{jk}) \qquad (j \neq k), \qquad (7.133)$$

$$B_{jj} = -n_j\left[\frac{3}{2}J_2\left(\frac{R_c}{a_j}\right)^2 - \frac{27}{8}J_2^2\left(\frac{R_c}{a_j}\right)^4 - \frac{15}{4}J_4\left(\frac{R_c}{a_j}\right)^4\right.$$

$$\left. + \frac{1}{4}\sum_{k=1,k\neq j}^{N}\frac{m_k}{m_c + m_j}\alpha_{jk}\bar{\alpha}_{jk}b_{3/2}^{(1)}(\alpha_{jk})\right], \qquad (7.134)$$

$$B_{jk} = +\frac{1}{4}\frac{m_k}{m_c + m_j}n_j\alpha_{jk}\bar{\alpha}_{jk}b_{3/2}^{(1)}(\alpha_{jk}) \qquad (j \neq k). \qquad (7.135)$$

The quantities $A_{jj}$, $A_{jk}$, $B_{jj}$, and $B_{jk}$ can be thought of as the constant elements of two $N \times N$ matrices, $\mathbf{A}$ and $\mathbf{B}$.

If we now transform to the new variables $h_j$, $k_j$, $p_j$, and $q_j$ as given in Eqs. (7.18) and (7.19), we obtain the generalised form of Eq. (7.20):

$$\langle\mathcal{R}_j\rangle = n_j a_j^2\left[\frac{1}{2}A_{jj}(h_j^2 + k_j^2) + \frac{1}{2}B_{jj}(p_j^2 + q_j^2)\right.$$

$$\left. + \sum_{k=1,k\neq j}^{N}A_{jk}(h_j h_k + k_j k_k) + \sum_{k=1,k\neq j}^{N}B_{jk}(p_j p_k + q_j q_k)\right]. \qquad (7.136)$$

The resulting equations of motion are as given in Eqs. (7.25) and (7.26) and their solution is

$$h_j = \sum_{i=1}^{N}e_{ji}\sin(g_i t + \beta_i), \qquad k_j = \sum_{i=1}^{N}e_{ji}\cos(g_i t + \beta_i), \qquad (7.137)$$

$$p_j = \sum_{i=1}^{N}I_{ji}\sin(f_i t + \gamma_i), \qquad q_j = \sum_{i=1}^{N}I_{ji}\cos(f_i t + \gamma_i), \qquad (7.138)$$

where the $g_i$ and the $f_i$ are the two sets of $N$ eigenvalues of the matrices $\mathbf{A}$ and $\mathbf{B}$. As before, the phases $\beta_i$ and $\gamma_i$, as well as the magnitudes of the eigenvectors $e_{ji}$ and $I_{ji}$, are determined from the boundary values. At any time $t$, the squares of the eccentricity and inclination of the mass $m_j$ are given by

$$e_j^2 = \left[\sum_{i=1}^{N}e_{ji}\sin(g_i t + \beta_i)\right]^2 + \left[\sum_{i=1}^{N}e_{ji}\cos(g_i t + \beta_i)\right]^2, \qquad (7.139)$$

$$I_j^2 = \left[\sum_{i=1}^{N}I_{ji}\sin(f_i t + \gamma_i)\right]^2 + \left[\sum_{i=1}^{N}I_{ji}\cos(f_i t + \gamma_i)\right]^2. \qquad (7.140)$$

Similarly the longitude of pericentre and the longitude of ascending node of the mass $m_j$ at any time can be found from the values of $h_j$, $k_j$, $p_j$, and $q_j$.

One important difference that arises from the introduction of the oblateness terms is that there is no longer a degeneracy in the inclination–node eigenvalues.

Mathematically this is because the $J_2$ and $J_4$ terms imply that the rows or columns of the matrix **B** are no longer linearly dependent; therefore **B** has a rank equal to $N$, the number of rows or columns. Physically the degeneracy is removed because there is now a well-defined reference plane introduced by the nonspherical shape of the planet.

## 7.8 Secular Theory for the Solar System

In the case of the motion of the planets the oblateness of the Sun has negligible effect, although it is interesting to note that the small discrepancies in the precession of the orbit of Mercury, which provided confirmation of the theory of general relativity, were once thought to be due to solar oblateness. The first application of the Laplace–Lagrange secular theory resulted in Laplace's claim to have proved the stability of the solar system, although, as we have seen, no such claim can be made without a more detailed study. We shall return to the whole question of the stability of the solar system in Chapter 9.

A number of secular theories have been derived to describe the long-term behaviour of the planets. In this century the most widely used is due to Brouwer & van Woerkom (1950). They analysed the problem of the secular motion of all the planets, except Pluto, using a modified classical theory to take account of a near-resonance between Jupiter and Saturn. The resulting theory has ten rather than eight eigenfrequencies and phases for the $e$–$\varpi$ solution. Brouwer & van Woerkom's derived values of the eigenfrequencies and phases are given in Table 7.1. The extra two frequencies arise from dominant terms in a higher order (in the masses) secular theory (see Sect. 7.11) and are given by $g_9 = 2g_5 - g_6$ and $g_{10} = 2g_6 - g_5$. The elements of all the corresponding eigenvectors are given in Tables 7.2 and 7.3. In all these tables $g_i$ and $\beta_i$ denote the eigenfrequencies and phases associated with the $\{h, k\}$ solution and $f_i$ and $\gamma_i$ denote those associated with the $\{p, q\}$ solution. Time $t = 0$ corresponds to the year A.D. 1900. As before the subscripts $i$ and $j$ denote the eigenmode and the planet number respectively.

A study of Table 7.1 shows that $f_5 \approx 0$. This results from the degeneracy in the $I$–$\Omega$ solution in the absence of oblateness terms (cf. Sect. 7.3). It should be noted that to obtain the variation of the orbital elements of planet $j$ we need to use formulae such as those given in Eqs. (7.139) and (7.140). Although it is tempting to associate each $i$ subscript with a planet, this would be incorrect. The best that can be said is that subscript $i$ denotes the mode that is likely to be dominated by planet $j = i$. The existence of analytical solutions such as that derived by Brouwer & van Woerkom (1950) allows us to investigate the long-term variations in the orbital elements of the planets due to their mutual gravitational

Table 7.1. Brouwer & van Woerkom's (1950) secular eigenfrequencies $g_i$ and $f_i$ (in arcseconds per year) and the associated phases $\beta_i$ and $\gamma_i$ (in degrees) for the planetary system.

| $i$ | $g_i$ ($''\,\mathrm{y}^{-1}$) | $f_i$ ($''\,\mathrm{y}^{-1}$) | $\beta_i$ (deg.) | $\gamma_i$ (deg.) |
|---|---|---|---|---|
| 1 | 5.46326 | −5.20154 | 92.182 | 19.433 |
| 2 | 7.34474 | −6.57080 | 196.881 | 318.057 |
| 3 | 17.32832 | −18.74359 | 335.224 | 255.031 |
| 4 | 18.00233 | −17.63331 | 317.948 | 296.541 |
| 5 | 4.29591 | 0.00000 | 29.550 | 107.102 |
| 6 | 27.77406 | −25.73355 | 125.120 | 127.367 |
| 7 | 2.71931 | −2.90266 | 131.944 | 315.063 |
| 8 | 0.63332 | −0.67752 | 69.021 | 202.293 |
| 9 | −19.18225 | — | 293.979 | — |
| 10 | 51.25222 | — | 220.691 | — |

attractions, although we have to be aware that these are only approximations to the true behaviour.

Figure 7.9 shows plots of the variation of the orbital eccentricity and inclination of Mercury, Venus, Earth, and Mars as a function of time. These show that there is significant variation in the orbital elements of the inner planets as a result of secular perturbations. In the case of Mercury there is a large-amplitude, long-period variation in its eccentricity and inclination. Note that Mercury's

Table 7.2. The components $e_{ij}$ of the eigenvectors for the $e$–$\varpi$ solution taken from Brouwer & van Woerkom's (1950) secular theory for the planetary system. All quantities have been multiplied by a factor of $10^5$.

| | $j = 1$ | 2 | 3 | 4 | 5 | 6 | 7 | 8 |
|---|---|---|---|---|---|---|---|---|
| $i = 1$ | 17454 | 610 | 392 | 64 | −1 | −1 | 0 | 0 |
| 2 | −2550 | 2094 | 1634 | 290 | −1 | −1 | 0 | 0 |
| 3 | 154 | −1255 | 1043 | 2972 | 0 | −1 | 0 | 0 |
| 4 | −169 | 1483 | −1483 | 7237 | 0 | −1 | 0 | 0 |
| 5 | 3571 | 1909 | 1834 | 1861 | 4482 | 3291 | −3188 | 63 |
| 6 | 96 | 6 | 283 | 1502 | −1535 | 4863 | −248 | −11 |
| 7 | 43 | 42 | 44 | 59 | 147 | 148 | 3043 | −323 |
| 8 | 1 | 1 | 1 | 2 | 5 | 6 | 143 | 938 |
| 9 | −21 | −11 | −12 | −20 | −16 | −49 | 20 | 0 |
| 10 | 10 | 4 | 15 | 64 | −54 | 202 | −16 | 0 |

Table 7.3. The components $I_{ij}$ of the eigenvectors for the $I$–$\Omega$ solution taken from Brouwer & van Woerkom's (1950) secular theory for the planetary system. All quantities have been multiplied by a factor of $10^5$. Note that the resulting planetary inclinations are in radians.

| | $j = 1$ | 2 | 3 | 4 | 5 | 6 | 7 | 8 |
|---|---|---|---|---|---|---|---|---|
| $i = 1$ | 12449 | 1178 | 849 | 180 | −2 | −3 | 2 | 0 |
| 2 | −3545 | 1004 | 810 | 180 | −1 | −2 | 1 | 0 |
| 3 | 409 | −2680 | 2448 | −3589 | 0 | 0 | 0 | 0 |
| 4 | 116 | −685 | 453 | 5025 | 0 | −2 | 0 | 0 |
| 5 | 2757 | 2757 | 2757 | 2757 | 2757 | 2757 | 2757 | 2757 |
| 6 | 28 | 12 | 281 | 965 | −631 | 1572 | −69 | −8 |
| 7 | −332 | −192 | −173 | −126 | −96 | −78 | 1760 | −207 |
| 8 | −145 | −132 | −130 | −123 | −117 | −113 | 110 | 1175 |

eccentricity does not appear to be periodic and yet it must be because it is derived from a periodic solution. Note too that there is good evidence of coupling between Earth and Venus, with each showing similar variations in their eccentricities and inclinations. Mars's orbit changes from near circular to an eccentricity of 0.14 on a timescale of approximately 1 My.

Figure 7.10 shows similar plots for Jupiter, Saturn, Uranus, and Neptune. Here one of the most striking features is the high frequency of the variations in $e$ and $I$ of Jupiter and Saturn. We have already seen in Sect. 7.3 that a simplified secular theory based on just these two planets leads to a coupling with periodic variations on a timescale of $\sim 51{,}000$ y. This variation persists in the complete secular theory, although the effect of the 5:2 near-resonance between the two planets leads to the introduction of additional eigenfrequencies in the $e$–$\varpi$ solution. This can be seen by plotting the variation in $e$ and $I$ over the same interval as shown in Fig. 7.1 (see Fig. 7.11). While there is good agreement between the variation in inclination, the period of the eccentricity variation has changed.

The quantities given in Tables 7.1–7.3 can be checked against the values obtained by integrating the full equations of motion and identifying the frequencies, phases, and amplitudes of the secular solution. Such a procedure is called *synthetic secular perturbation theory* and it is frequently used in studies of long-term stability. Again, we stress that the existence of the types of analytical solutions we have discussed does not imply that the dynamical system is stable over infinite timescales – this can only be resolved by other means, usually involving numerical integration of the equations of motion.

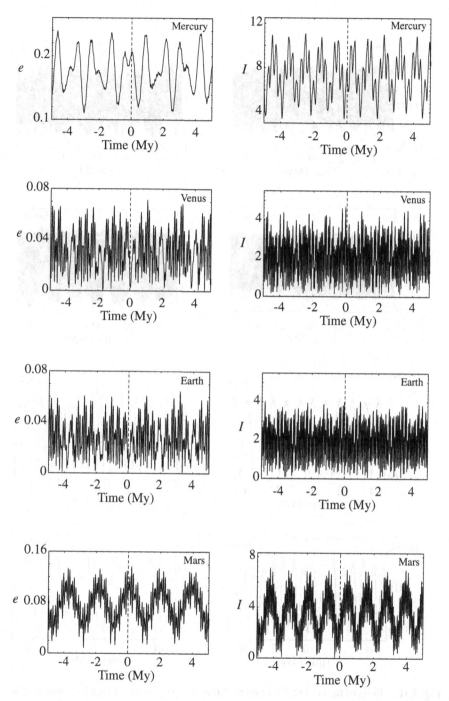

Fig. 7.9. Plots of the eccentricity and inclination (in degrees) of Mercury, Venus, Earth, and Mars over a period of 10 million years centred on AD 1900, according to the secular theory of Brouwer & van Woerkom (1950).

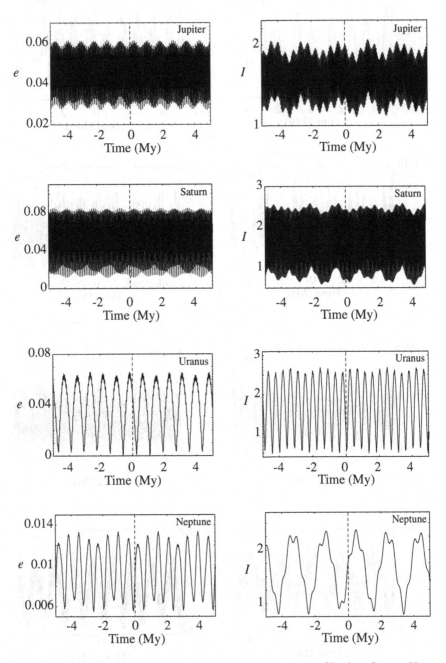

Fig. 7.10.   Plots of the eccentricity and inclination (in degrees) of Jupiter, Saturn, Uranus, and Neptune over a period of 10 million years centred on AD 1900, according to the secular theory of Brouwer & van Woerkom (1950).

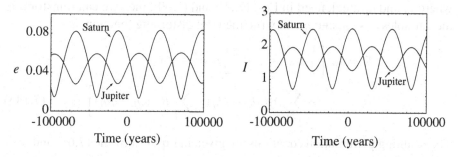

Fig. 7.11. Plots of the eccentricity and inclination (in degrees) of Jupiter and Saturn over a period of 200,000 years centred on AD 1983, according to the secular theory of Brouwer & van Woerkom (1950).

## 7.9 Generalised Free and Forced Elements

For completeness we now proceed to derive a theory for the free and forced elements of a test particle moving in a system of $N$ bodies orbiting an oblate central mass. In this case we assume that we already have a secular theory for the motion of the $N$ bodies along the lines of that derived in Sect. 7.7 above. In the following analysis we will assume that the particle has orbital elements $a$, $e$, $I$, $\varpi$, and $\Omega$.

The relevant oblateness and secular terms in the disturbing function are given by Eqs. (7.124) and (7.130) with (i) the subscript $j$ removed and (ii) the subscript $k$ replaced by the subscript $j$. The condition $j \neq k$ in the summation no longer applies.

If we define

$$A = +n \left[ \frac{3}{2} J_2 \left( \frac{R_c}{a} \right)^2 - \frac{9}{8} J_2^2 \left( \frac{R_c}{a} \right)^4 - \frac{15}{4} J_4 \left( \frac{R_c}{a} \right)^4 \right. $$
$$\left. + \frac{1}{4} \sum_{j=1}^{N} \frac{m_j}{m_c} \alpha_j \bar{\alpha}_j b_{3/2}^{(1)}(\alpha_j) \right], \tag{7.141}$$

$$A_j = -\frac{1}{4} \frac{m_j}{m_c} n \alpha_j \bar{\alpha}_j b_{3/2}^{(2)}(\alpha_j), \tag{7.142}$$

$$B = -n \left[ \frac{3}{2} J_2 \left( \frac{R_c}{a} \right)^2 - \frac{27}{8} J_2^2 \left( \frac{R_c}{a} \right)^4 - \frac{15}{4} J_4 \left( \frac{R_c}{a} \right)^4 \right. $$
$$\left. + \frac{1}{4} \sum_{j=1}^{N} \frac{m_j}{m_c} \alpha_j \bar{\alpha}_j b_{3/2}^{(1)}(\alpha_j) \right], \tag{7.143}$$

$$B_j = +\frac{1}{4} \frac{m_j}{m_c} n \alpha_j \bar{\alpha}_j b_{3/2}^{(1)}(\alpha_j), \tag{7.144}$$

where $\alpha_j$ and $\bar{\alpha}_j$ are defined in Eqs. (7.59) and (7.60), then we can transform to the variables $h$, $k$, $p$, and $q$ and write the total disturbing function as

$$\mathcal{R} = na^2 \left[ \frac{1}{2} A(h^2 + k^2) + \frac{1}{2} B(p^2 + q^2) \right.$$

$$\left. + \sum_{j=1}^{N} A_j (hh_j + kk_j) + \sum_{j=1}^{N} B_j (pp_j + qq_j) \right]. \tag{7.145}$$

The resulting perturbation equations are given in Eqs. (7.66) and (7.67) and their solutions are

$$h = e_{\text{free}} \sin(At + \beta) + h_0(t), \qquad k = e_{\text{free}} \cos(At + \beta) + k_0(t),$$
$$p = I_{\text{free}} \sin(Bt + \gamma) + p_0(t), \qquad q = I_{\text{free}} \cos(Bt + \gamma) + q_0(t), \tag{7.146}$$

which is identical in form to Eq. (7.77), where $e_{\text{free}}$, $I_{\text{free}}$, $\beta$, and $\gamma$ are constants determined from the boundary conditions and

$$h_0(t) = -\sum_{i=1}^{N} \frac{\nu_i}{A - g_i} \sin(g_i t + \beta_i), \tag{7.147}$$

$$k_0(t) = -\sum_{i=1}^{N} \frac{\nu_i}{A - g_i} \cos(g_i t + \beta_i), \tag{7.148}$$

$$p_0(t) = -\sum_{i=1}^{N} \frac{\mu_i}{B - f_i} \sin(f_i t + \gamma_i), \tag{7.149}$$

$$q_0(t) = -\sum_{i=1}^{N} \frac{\mu_i}{B - f_i} \cos(f_i t + \gamma_i). \tag{7.150}$$

The equations for the resulting forced eccentricity, $e_{\text{forced}}$, and forced inclination, $I_{\text{forced}}$, are given in Eq. (7.82).

We have already noted in Sect. 7.5 that singularities will arise at semi-major axes where the value of $A$ or $B$ match one of the $g_i$ or $f_i$ eigenfrequencies of the system. Naturally, the effect of small divisors also applies when we extend the analysis to include more bodies and the effects of oblateness. Figure 7.12 shows a plot of the proper precession and node rates as a function of semi-major axis in the inner (Fig. 7.12a) and outer (Fig. 7.12b) parts of the solar system, ignoring any oblateness; in this case $A = -B$. Superimposed on each plot are horizontal lines denoting the values of the $g_i$ and $-f_i$ eigenfrequencies based on the theory of Brouwer & van Woerkom (1950) detailed in Sect. 7.8. At the points of intersection of the lines and the curves we would expect there to be large values of the forced elements.

The existence of these small divisors is connected with the phenomenon of secular resonance, which we discuss in more detail in Sect. 7.11. At this stage we content ourselves with noting that in the region between Mars and Jupiter

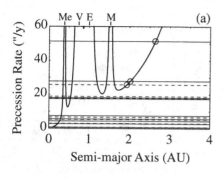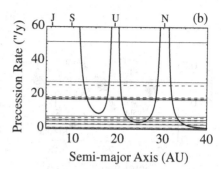

Fig. 7.12. The precession rate of a test particle as a function of semi-major axis in the (a) inner and (b) outer parts of the solar system according to the theory of Brouwer & van Woerkom (1950). The solid and dashed horizontal lines denote the $g_i$ and $-f_i$ eigenfrequencies of the system taken from Table 7.1. The intersections of three of these lines with the precession rate in the region of the asteroid belt are indicated by circles. The letters above each plot denote the locations of the eight planets and their associated singularities.

(the location of the asteroid belt) there are three values of the semi-major axis where such intersections occur: Two of them are close to 2 AU where the $g_6$ and $-f_6$ frequencies match the rate, and another is near 2.6 AU where the $g_{10}$ frequency matches. These are indicated in Fig. 7.12.

## 7.10 Hirayama Families and the IRAS Dust Bands

The analysis given in Sect. 7.4 and Sect. 7.9 implies that the osculating eccentricity of a small mass object, such as an asteroid or a dust particle, moving under the gravitational effects of the planets can be thought of as having two components: (i) a free or proper eccentricity, which reflects the "inherent" eccentricity of the object, and (ii) a forced eccentricity, which is a result of its current semi-major axis and the relative positions of its perturbers. The same arguments apply to the inclination of the object. Therefore the free or proper elements provide information on the body's natural elements rather than those that are matters of its circumstances.

Hirayama (1918) derived proper elements for the limited number of asteroids known at the time. He showed that some asteroids tended to cluster in groups in $(a, e)$ and $(a, I)$ space and that the clustering was more pronounced when plotted using free rather than osculating elements. He proposed that each cluster, or *family*, represented objects that had a common dynamical origin and were the remnants of the breakup of a parent body. It is customary to name an identified family after its largest member. Today, soon after a valid orbit has been determined, an asteroid's proper elements are calculated and family associations are examined. On this basis almost half of all asteroids are thought to be family members.

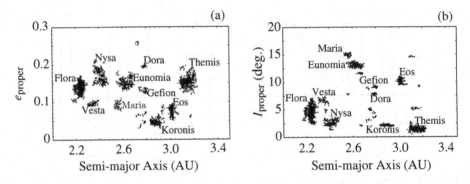

Fig. 7.13. A plot of (a) the proper (or free) eccentricity and (b) the proper inclination of asteroids in the range $2.0 \leq a \leq 3.5$ AU that have been identified as family members.

Figure 7.13a shows the proper eccentricity as a function of semi-major axis for those asteroids that have been identified as belonging to families. The equivalent diagram for proper inclination is shown in Fig. 7.13b. There are a number of obvious clusterings of asteroids and it is important to realise that these are groupings in three-dimensional $(a, e, I)$ space.

We have already shown that the variation of the orbital elements can be thought of as motion around a circle with a centre determined by the forced component (see Figs. 7.2 and 7.3). In Fig. 7.14 we illustrate this by plotting osculating values of $k = e \cos \varpi$ versus $h = e \sin \varpi$ and $q = I \cos \Omega$ versus $p = I \sin \Omega$ for the

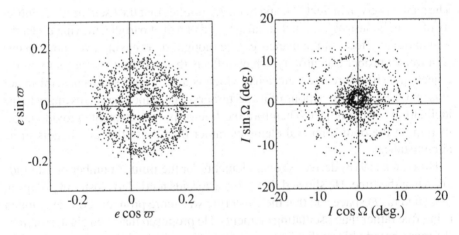

Fig. 7.14. (a) A plot of the osculating values of $k = e \cos \varpi$ and $h = e \sin \varpi$ for all the asteroids shown in Fig. 7.13a. A number of "rings" are clearly visible. (b) A plot of the osculating values of $q = I \cos \Omega$ and $p = I \sin \Omega$ for all the asteroids shown in Fig. 7.13b. A number of "rings" are clearly visible, although most have a small radius (i.e., a low inclination).

Table 7.4. The proper and forced orbital elements of the Koronis, Eos, and Themis asteroid families.

| Family | $a$ (AU) | $e_{proper}(°)$ | $I_{proper}(°)$ | $e_{forced}$ | $\varpi_{forced}(°)$ | $I_{forced}(°)$ | $\Omega_{forced}(°)$ |
|---|---|---|---|---|---|---|---|
| Koronis | 2.875 | 0.049 | 2.12 | 0.037 | 6.2 | 1.16 | 96.1 |
| Eos | 3.015 | 0.071 | 10.20 | 0.037 | 7.6 | 1.19 | 97.1 |
| Themis | 3.136 | 0.152 | 1.42 | 0.038 | 8.7 | 1.22 | 97.8 |

family members in our sample. In each case the presence of "circles" of asteroids is clearly visible. This implies that members of a given asteroid family do indeed have a common forced eccentricity and inclination (determined by their semi-major axis) and a common free eccentricity and inclination with randomised free pericentres and nodes. Note that, as predicted from secular theory, the centres of these circles are not identical.

Three of the largest clusterings are associated with the Koronis, Eos, and Themis families, all of which are thought to be the fragments of larger bodies that suffered catastrophic collisions at least $10^7$ years ago. The proper and forced orbital elements of these families are listed in Table 7.4. The data used in this table are taken from Dermott *et al.* (1985). If these families were formed by the collisional breakup of a larger body then such an event would have produced a large amount of asteroidal dust. Dramatic evidence for the collisional theory for the origin of the Hirayama families came with the discovery by the Infrared Astronomical Satellite (IRAS) of dust bands in the solar system (Low et al. 1984, Neugebauer et al. 1984).

IRAS carried out an all-sky survey at wavelengths of 12, 25, 60, and 100 $\mu$m. Surveys at these wavelengths are particularly adept at detecting infrared radiation from dust in the solar system. The background flux detected by IRAS at $25\mu$m is shown in Fig. 7.15a. In this plot the IRAS dust bands are barely detectable as small "bumps" in the distribution close to ecliptic latitudes of $0°$ and $\pm 10°$. However, if the background component of the curve is removed and the smoothed residuals are plotted (Fig. 7.15b), then the bands are clearly visible – there is a central band that appears to be split and two side bands at $\pm 10°$. But how do we know that these bands originate from the collisions that formed the main families? The answer lies in an understanding of secular perturbations and orbital geometry.

We have already noted that a test particle's forced elements are a function of its semi-major axis alone. Figure 7.16 shows the values of the forced elements in the region of the asteroid belt according to the theory of Brouwer & van Woerkom (1950). Fragments from the breakup of an asteroid would have approximately the same free eccentricity and free inclination but would quickly have acquired

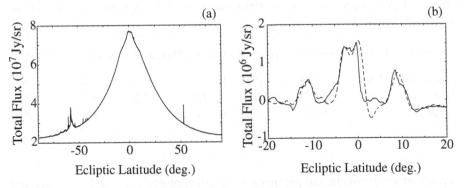

Fig. 7.15. (a) The IRAS background infrared flux at a wavelength of 25 μm. The spikes close to ecliptic latitudes −60° and +50° are caused by residual contributions from the galactic plane. (b) The smoothed, residual data showing the flux associated with the dust bands (solid line) together with the model data (dashed line) obtained by assuming that the signal comes from dust in the same distribution of orbits as the Themis, Eos, and Koronis families of asteroids.

randomised free pericentres and free nodes. Having the same forced inclination means that all the fragments precess about a common mean plane determined by the forced inclination and forced node. Because the vertical component of the their motion with respect to this plane is simple harmonic in form, the asteroids will spend most of their time at the extremes of their motion, giving rise to a bunching at these locations. The result is that, viewed from the Sun, the asteroids will give the appearance of lying in two bands separated by $2I_{\text{forced}}$ in latitude (see Fig. 7.17).

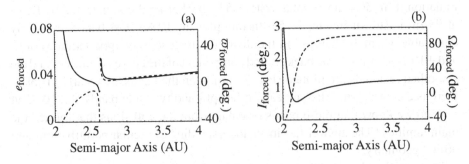

Fig. 7.16. Forced orbital elements as a function of semi-major axis using the secular theory of Brouwer & van Woerkom (1950). (a) The forced eccentricity (solid curve, left-hand axis) and forced longitude of perihelion (dashed curve, right-hand axis). (b) The forced inclination (solid curve, left-hand axis) and forced longitude of ascending node (dashed curve, right-hand axis).

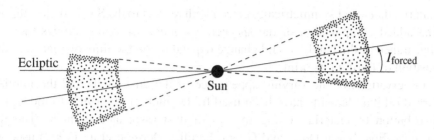

Fig. 7.17. A vertical cross-sectional view of a distribution of asteroids with the same free and forced inclination but randomised free nodes. The radial extent of the cross section is due to the eccentricity of the asteroids.

However, there is also an effect due to the eccentricity. This is shown in Fig. 7.18. Orbits with the same forced and free eccentricities and the same forced pericentre but randomised free pericentres generate an orbital distribution that is symmetrical with respect to a point C that is not the Sun (denoted by the point S in Fig. 7.18). If we envisage asteroidal dust moving in these orbits then the combined effect of the inclination and eccentricity is to produce a cloud of

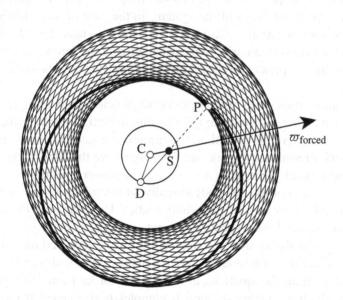

Fig. 7.18. The distribution of eccentric orbits having the same semi-major axis $a$, the same forced eccentricity $e_{forced}$ and proper eccentricity $e_{proper}$, the same forced pericentre $\varpi_{forced}$, but randomised proper longitudes of pericentre. S denotes the position of the Sun and C is the centre of symmetry. An individual orbit with pericentre at P and centre at D is highlighted. In the diagram the lines CS, DS, and DC have lengths $ae_{forced}$, $ae$, and $ae_{proper}$ respectively, where $e$ is the eccentricity of the orbit. Note that the centre of symmetry is not the Sun.

material that is not symmetrically placed with respect to the Sun and the ecliptic. The added complication of the spacecraft's geocentric view implies that the appearance of the bands should change depending on the time of year (i.e., the position of Earth in its orbit).

Observations of the varying appearance of the dust bands and the implied forced orbital elements have been used by Dermott et al. (1992) to model the distribution of material. Using sources of dust associated with the Themis, Koronis, Eos, Nysa, Dora, and Gefion families Dermott et al. (1992) used an iterative procedure to produce a model profile (dashed line in Fig. 7.15b) that gives excellent agreement with the observations. The implied association of the dust with the collisions that formed the major Hirayama families appears to be confirmed.

## 7.11 Secular Resonance

In our study of Brouwer & van Woerkom's (1950) secular perturbation theory for the solar system (Sect. 7.9) we noted that there are problems in calculating the forced elements of test particles at those locations in semi-major axis where either of the proper precession rates (denoted by $A$ or $B$) of the particle equals one of the eigenfrequencies of the system. In the case of the asteroid belt we noted three such locations: two near 2 AU and one near 2.6 AU. The latter explains the obvious singularity in the calculation of the forced eccentricity and forced longitude of pericentre shown in Fig. 7.16a. This is an example of *secular resonance*.

A resonance arises when two periods or frequencies are in a simple numerical ratio. We have already seen examples of this in Sect. 5.4 where the frequencies in question were the orbital and spin rates of a satellite (or planet). In the case of secular resonance the relevant frequencies are the rates of change of the proper longitude of pericentre ($A = \dot{\varpi}_{\text{proper}}$) or proper longitude of ascending node ($B = \dot{\Omega}_{\text{proper}}$) of the test body (usually an asteroid) and one of the eigenfrequencies of the system of perturbing bodies. Unfortunately the techniques required to analyse these resonances are not as simple as in the spin–orbit case. The basic secular theory used throughout this chapter is based on an expansion of the disturbing function to second degree in the eccentricities and inclinations and use of Lagrange's equations in their lowest order form. This produces a system in which the $(e, \varpi)$ solution is completely decoupled from the $(I, \Omega)$ solution. Although this is sufficient to allow us to suggest where secular resonances might occur, a more complete theory requires that higher order terms be taken into account. Furthermore, it is also necessary to take into account terms of the second order in the masses, whereas we have only considered a first-order theory. This makes the mathematics more difficult and introduces a coupling between the eccentricity and inclination terms. For further details see the review articles by Knežević & Milani (1994) and Froeschlé & Morbidelli (1994).

Williams (1969) derived a semianalytical secular theory without using an expansion of the disturbing function. His calculation of proper elements and subsequent identification of asteroid families (Williams 1979) were fundamental advances in modern studies of asteroid dynamics. Because of coupling the location of secular resonances actually correspond to surfaces in $(a, e, I)$ space rather than being centred on a single semi-major axis. The location of these surfaces for the asteroid belt has been calculated by Williams (1969) and Williams & Faulkner (1981). They studied the secular resonances where the frequencies $A - g_5$, $A - g_6$, and $B - f_6$ are all approximately zero. These are referred to as the *linear secular resonances* and, as we have noted, their existence is suggested by the secular perturbation theory developed in this chapter. These are also known as the $\nu_5$, $\nu_6$, and $\nu_{16}$ secular resonances, where the suffix denotes the $i$th value of the eigenfrequency involved ($\nu_1 = g_1, \ldots, \nu_{10} = g_{10}, \nu_{11} = f_1, \ldots, \nu_{18} = f_8$).

Figure 7.19 shows the location in proper semi-major axis–proper inclination space of the linear secular resonances in the asteroid belt, calculated for $e_{\text{proper}} = 0.1$ (after Milani & Knežević 1990). These are superimposed on the

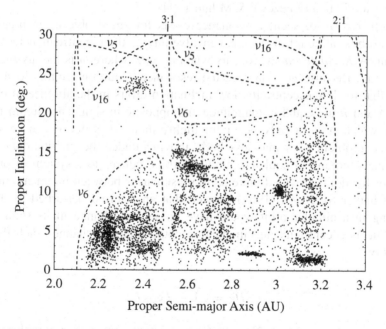

Fig. 7.19. The location of the important $\nu_5$, $\nu_6$, and $\nu_{16}$ linear secular resonances (calculated using $e_{\text{proper}} = 0.1$) and the numbered asteroids' $I_{\text{proper}}$ as a function of proper semi-major axis.

actual distribution of the proper elements for asteroids in the main belt. For $\sin I_{proper} < 0.3$ the elements were calculated by Milani & Knežević (1990), using a theory developed by Yuasa (1973), while the elements of Lemaître & Morbidelli (1994) were used for larger inclinations. Note the singularities in Fig. 7.19 close to 2.5 and 3.3 AU. These are the locations of the 3:1 and 2:1 mean motion resonances (see Chapter 8) where the orbital period of the asteroid is a simple fraction of Jupiter's period. At these locations it is necessary to use a secular theory that incorporates the effect of mean motion resonance. The additional singularity close to 2 AU has already been mentioned (see above and Fig. 7.12a). The distribution of asteroids is clearly nonrandom, partly due to the existence of the Kirkwood gaps at a number of resonant locations (see Fig. 1.7 and Sect. 9.8). However, it is also clear that the inner edge of the main belt in $a$–$I$ space is correlated with the location of the $\nu_6$ secular resonance for $I < 15°$. At higher inclinations there appears to be a group isolated by the three secular resonances and the 3:1 Kirkwood gap.

Other secular resonances are possible, subject to a d'Alembert-like relationship on the permitted combinations of frequencies. These are the *nonlinear secular resonances* and all involve higher powers of eccentricity and/or inclination in the equations of motion. Nine of them ($A + B - g_5 - f_6$, $A + B - g_6 - f_6$, $A + B - g_5 - f_7$, $A - 2g_6 + g_5$, $A - 2g_6 + g_7$, $A - 3g_6 + 2g_5$, $B - f_6 - g_5 + g_6$, $2A + B - 2g_6 - f_6$, and $3A + B - 3g_6 - f_6$) give rise to important secular resonances in the asteroid belt (Knežević & Milani 1994).

Another form of secular resonance exists for small objects on highly inclined orbits, although it does not involve any of the eigenfrequencies of the system. A *Kozai resonance* occurs when $\dot\omega = 0$, where $\omega$ is the argument of pericentre. Because $\varpi = \omega + \Omega$, the resonance condition reduces to $A = B$. Note that for low-eccentricity, low-inclination orbits in the absence of oblateness $A$ and $B$ are equal in magnitude and opposite in sign. The circumstances under which $A = B$ only occur for highly inclined orbits. It can be shown that the problem of a massless body moving under the gravitational effect of planets moving in coplanar, circular orbits reduces to a system of one degree of freedom, provided there are no resonances between the mean motions (see Chapter 8). Kozai (1962) showed that an asteroid perturbed by Jupiter moving in a circular orbit would have no secular change in its semi-major axis but its eccentricity and inclination could undergo changes such that the quantity

$$H_K = \sqrt{1 - e^2}\,\cos I \qquad\qquad (7.151)$$

always remains constant. Note that for constant semi-major axis this is just another way of stating that the third Delaunay momentum, $H$, is constant (see Eq. (2.176)). This is also related to the Tisserand relation discussed in Sect. 3.4.

A consequence of this constant is that the eccentricity and inclination of a small object's orbit are coupled such that $e$ is a maximum when $I$ is a minimum and *vice versa*.

Kozai's theory was extended by Michel & Thomas (1995) to include the four giant planets. The theory shows that for low inclinations it is possible for $\omega$ to librate about stable points at $\omega = 0°$ and $\omega = 180°$. For inclinations greater that $\sim 30°$ these points become unstable and new stable equilibrium points appear at $\omega = 90°$ and $\omega = 270°$. Thomas & Morbidelli (1996) have shown that the Kozai resonance can only affect orbits with large $e$ and $I$. They also confirmed the conclusion of Bailey et al. (1992) that the Kozai resonance is the mechanism by which some long-period comets can become sungrazing.

## 7.12 Higher Order Secular Theory

In this chapter we have used the secular theory of Brouwer & van Woerkom (1950) as our adopted theory for the long-term variation of the planetary orbits. This is referred to a linear theory and for the most part it follows the methods of Sect. 7.7 in that it is only to first order in the masses and is based on an expansion of the disturbing function to second degree in the eccentricities and inclinations. Using a treatment due to Hill (1897), Brouwer & van Woerkom made some attempt to incorporate a higher order theory to account for the Jupiter–Saturn interactions. These interactions give rise to the $g_9$ and $g_{10}$ eigenfrequencies and their associated eigenvectors listed in Tables 7.1–7.3.

Incorporating higher degree terms in the orbital elements, Bretagnon (1974) developed a secular theory for all the planets except planets to second order in the masses. A later version incorporated the effects of relativity and lunar perturbations (Bretagnon 1982). Following the methods of Duriez (1979), a new secular theory was devised by Laskar (1985, 1986a). This included the same perturbations as Bretagnon's later theory, but with the addition of even higher order terms in the eccentricities and inclinations. The resulting theory was numerically integrated over a time span of 30 My and Fourier analysed (Laskar 1988). Around the same time a number of researchers undertook extensive numerical integrations of the outer solar system in studies of long-term stability of planetary orbits (Kinoshita & Nakai 1984, Applegate et al. 1986, Carpino et al. 1987). There have also been integrations of the orbits of the inner planets. For example, Quinn et al. (1991) investigated the Earth's orbit over a timescale of 3 My. Their results, subsequently extended to 6 My, were compared with the semianalytical secular theory of Laskar (1989) and were found to be in good agreement (Laskar et al. 1992). All of these studies provided deep insights into the secular interactions of the planets. These are examined in more detail in Sect. 9.10.

### Exercise Questions

**7.1**    Consider the secular interactions of two planets orbiting a star. Let $\alpha = a_1/a_2$, $q = m_2/m_1$, $v = n_2/n_1 = \alpha^{3/2}$, and $\beta = b_{3/2}^{(2)}(\alpha)/b_{3/2}^{(1)}(\alpha)$, where subscripts 1 and 2 refer to the inner and outer planet, respectively. By writing out explicit expressions for the four matrix coefficients $A_{ij}$ in the secular theory, show that the two eccentricity eigenfrequencies are given by

$$g_{\pm} = (\sigma/2)\left\{q\alpha + v \pm \sqrt{(q\alpha - v)^2 + 4qv\alpha\beta^2}\right\},$$

where $\sigma = (1/4)n_1\mu_1\alpha b_{3/2}^{(1)}(\alpha)$. Show further that the eigenvector amplitude ratios are given by

$$\left(\frac{e_2}{e_1}\right)_{\pm} = \frac{q\alpha - g_{\pm}/\sigma}{q\alpha\beta} = \frac{v\beta}{v - g_{\pm}/\sigma}.$$

These results take on a particularly simple form when $\alpha \ll 1$. Using the approximations $b_{3/2}^{(1)}(\alpha) \simeq 3\alpha$, $b_{3/2}^{(2)}(\alpha) \simeq \frac{15}{4}\alpha^2$ valid for $\alpha \ll 1$, show that the eccentricity eigenfrequencies are given approximately by

$$g_{+} \simeq \frac{3}{4}\mu_2 n_1 \alpha^3, \qquad g_{-} \simeq \frac{3}{4}\mu_1 n_2 \alpha^2.$$

Show further that $|(e_1/e_2)_{+}| \gg 1$ and $(e_1/e_2)_{-} \ll 1$. Sketch the motion in the $(h, k)$ plane for each body, and use this to argue that the $g_{+}$ mode is dominated by simple pericentre precession of the inner planet while the $g_{-}$ mode corresponds to pericentre precession of the outer planet.

**7.2**    The formulae derived in Question 7.1 are quite useful in many different contexts. Use the simplified expression for $g_{-}$ to estimate the pericentre precession period of Saturn's orbit due to perturbations by Jupiter. Compare your answer with that obtained in Question 6.3. Use the simplified expression for $g_{+}$ to estimate the precession rate of Venus's orbit due to the Earth. Your results will be only roughly correct; can you suggest two reasons why? The simplified expression for $g_{+}$ can also be used to calculate the precession rate of a satellite's orbit due to solar perturbations by treating the Sun as a very distant outer satellite of the planet, with mass ratio $\mu_2 = M_{sun}/M_p$, where $M_p$ is the mass of the planet. Show that, in this case, the satellite's precession rate is $\dot{\varpi} \simeq (3/4)n_p^2/n$, where $n_p$ is the mean motion of the planet's orbit about the Sun. Use this result to calculate the precession period of the lunar orbit due to the Sun, in years. Compare the solar-induced precession rate with that due to the Earth's oblateness, taking $J_2 = 1.08 \times 10^{-3}$. Which perturbation is dominant for the lunar orbit?

**7.3**    A plot of the distribution of the longitudes of perihelia of the asteroids shows a clustering around the longitude of Jupiter's perihelion. By isolating the appropriate *lowest order* secular terms in the expansion of the disturbing function and assuming that all motion is in the plane of Jupiter's orbit (semi-major axis 5.20 AU), derive an expression for $\dot{\varpi}$, the rate of change of the longitude of perihelion of an asteroid orbiting interior to Jupiter. Taking the Jupiter/Sun mass ratio to be $10^{-3}$ and the eccentricity of Jupiter to be 0.048, and using $a = 2.86$ AU and $e = 0.15$ as typical values for an asteroid, calculate the numerical value of $\dot{\varpi}$ in units of degrees/century in the cases (i) where the perihelia of the asteroid and Jupiter are aligned and (ii) where the perihelia differ by $180°$. Use your results to explain the observed distribution of asteroid perihelia.

**7.4** Use the data in Tables A.2 and A.3 to determine the semi-major axis, eccentricity, inclination, longitude of perihelion, and longitude of ascending node of Jupiter and Saturn at JD 2450800.5. Hence, using the theory for two-planet secular perturbations given in Sect. 7.2, derive a full secular solution for the variation of the orbits of Jupiter and Saturn. Find the free and forced orbital elements of asteroid (243) Ida under the secular effects of Jupiter and Saturn using the data given in Table A.16 and the secular theory already derived. Compare your answers for Ida with those obtained using Brouwer & van Woerkom's secular theory for the solar system (see Sect. 7.8).

**7.5** *Voyager* observations in 1980 and 1981 led to the determination of orbital elements for the Saturnian satellites Prometheus ($a_1 = 139{,}377$ km, $n_1 = 587.2890°\mathrm{d}^{-1}, e_1 = 0.0024, \varpi_1 = 173°\mathrm{d}^{-1}$) and Pandora ($a_2 = 141{,}712$ km, $n_2 = 572.7891°\mathrm{d}^{-1}$, $e_2 = 0.0042$, $\varpi_2 = 22°$) and the F ring ($a_3 = 140{,}175$ km, $n_3 = 582.27°\mathrm{d}^{-1}$, $e_3 = 0.0026$, $\varpi_3 = 230°$), where $a$ is the semi-major axis, $n$ the mean motion, $e$ the eccentricity, and $\varpi$ the longitude of pericentre; there were no detectable inclinations. Taking the $J_2$ of Saturn to be 0.0163, construct a secular theory for the interaction of all three objects taking the Prometheus/Saturn and Pandora/Saturn mass ratios to be $1.15 \times 10^{-9}$ and $7.66 \times 10^{-10}$ respectively and allowing the mass of the F ring to take values in steps of 0.1 between zero and 3 times the mass of Prometheus. For each of the assumed masses of the F ring use your secular solution to determine the minimum separation of Prometheus and the F ring over a 20,000 day period starting from 23 August 1981 (the epoch of the quoted elements). For what values of the mass of the F ring is the minimum separation less than 70 km, the longest radial dimension of Prometheus? The quoted masses of the two satellites are based on assumed densities of $1.2$ g cm$^{-3}$. There is some evidence that the densities could be as low as $0.6$ g cm$^{-3}$. Repeat your calculations using the lower estimate for the density. Comment on the ability of the secular perturbation theory to model the interactions between the F ring and the two satellites.

**7.6** Use the theory given in Sect. 7.9 and the data given in Table A.9 to calculate the $g_i$ (eccentricity/pericentre) and $f_i$ (inclination/node) eigenfrequencies in units of $°d^{-1}$ for the Saturn system. Your model should include all the eight major ($\langle R \rangle > 100$ km) prograde satellites as well as the effect of Saturn's $J_2$ and $J_4$; the effects of resonance should be ignored. Calculate all values of the semi-major axis where the value of the natural pericentre precession rate ($A$) or nodal regression rate ($B$) equals one of the $g_i$ or $f_i$, respectively.

# 8

# Resonant Perturbations

The heavens themselves, the planets, and this centre
Observe degree, priority, and place,
Insisture, course, proportion, season, form,
Office, and custom, in all line of order.

William Shakespeare, *Troilus and Cressida, I, iii*

## 8.1 Introduction

We saw in Chapter 6 how resonant effects arise in the small divisor problem
when we considered the motion of an asteroid whose orbital period was a simple
fraction of Jupiter's period. The naïve theory predicted that, as the ratio of mean
motions approached the exact resonant value, the small divisor approached zero
and large-amplitude variations in the elements would result. In this chapter we
examine the theory of resonance in more detail. Starting from simple geometrical
and physical approaches, we go on to show how the simple model breaks down.
In order to understand the basic dynamics of resonance we start by using the
pendulum approach valid for resonant phenomena in the asteroid belt. We then
give a complete and detailed model of resonance using a Hamiltonian approach.
Throughout this chapter we develop a variety of approaches to handle the problem
of orbit–orbit resonance in the solar system and beyond.

Although there is an extensive range of literature on celestial mechanics, there
is little devoted specifically to the theory of resonance. Useful reviews of the
subject, particularly in the context of orbital evolution through resonance, have
been given by Greenberg (1977), Peale (1986), and Malhotra (1988).

## 8.2 The Geometry of Resonance

We consider the geometrical importance of resonant mechanisms. Consider an
asteroid in a 2:1 resonance with Jupiter. For simplicity we assume that Jupiter is

321

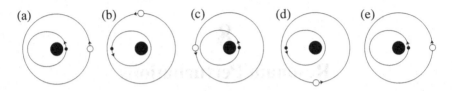

Fig. 8.1.   The relative positions of Jupiter (white circle) and an asteroid (small filled circle) for the stable configuration when their orbital periods are in a ratio of 2:1. If $T_J$ is the period of Jupiter's orbit then the diagrams illustrate the configurations at times (a) $t = 0$, (b) $t = \frac{1}{4}T_J$, (c) $t = \frac{1}{2}T_J$, (d) $t = \frac{3}{4}T_J$, and (e) $t = T_J$.

in a circular orbit and that all motion takes place in the plane of Jupiter's orbit. In this situation we are ignoring any perturbations between the two objects as we are only interested in how resonant relationships lead to repeated encounters of "good" and "bad" kinds.

Figure 8.1 illustrates the relative configurations of the asteroid and Jupiter such that, at some time $t = 0$, Jupiter and the asteroid are at conjunction and the asteroid is at the perihelion of its orbit. This is illustrated in Fig. 8.1a. Since the objects are in a 2:1 resonance, the asteroid will complete two periods for every one period of Jupiter. We consider time steps of one quarter Jupiter periods. Neglecting any actual perturbations between the objects, the asteroid and Jupiter will have moved to relative positions shown in Fig. 8.1b at time $t = \frac{1}{4}T_J$. The asteroid is now at the aphelion of its orbit, and Jupiter has completed 1/4 of an orbit. Although the closest approach between the two orbits is at the asteroid's aphelion, Jupiter is not nearby when the asteroid is at this position. Similarly, when Jupiter reaches this position at time $t = \frac{1}{2}T_J$, the asteroid is back at perihelion (Fig. 8.1c). At time $t = \frac{3}{4}T_J$ the asteroid returns to the danger point but Jupiter is not nearby (Fig. 8.1d), while at time $t = T_J$ (Fig. 8.1e) the original configuration shown in Fig. 8.1a is repeated. Thus, although there appears to be the potential for close approaches and large perturbations from Jupiter at the asteroid's aphelion, such approaches are avoided by the resonant mechanism. This would be an example of a stable equilibrium configuration between Jupiter and the asteroid.

Conversely, if we start Jupiter and the asteroid at conjunction at the asteroid's aphelion (Fig. 8.2), we would have an unstable equilibrium configuration, where the damaging close approaches would be repeated every Jupiter period.

We can examine the geometry of resonance for a more general case by first considering two satellites moving around a central planet in circular, coplanar orbits. Let us assume that

$$\frac{n'}{n} = \frac{p}{p+q}, \tag{8.1}$$

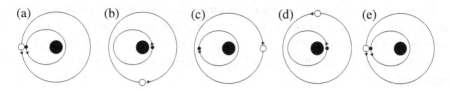

Fig. 8.2. The relative positions of Jupiter (white circle) and an asteroid (small filled circle) for the unstable configuration when their orbital periods are in a ratio of 2:1. If $T_J$ is the period of Jupiter's orbit then the diagrams illustrate the configurations at times (a) $t = 0$, (b) $t = \frac{1}{4}T_J$, (c) $t = \frac{1}{2}T_J$, (d) $t = \frac{3}{4}T_J$, and (e) $t = T_J$.

where $n$ and $n'$ are the mean motions of the inner and outer satellites respectively and $p$ and $q$ are integers. If the two satellites are in conjunction at time $t = 0$ then the next conjunction will occur when

$$nt - n't = 2\pi \tag{8.2}$$

and the period, $T_{con}$, between successive conjunctions is given by

$$T_{con} = \frac{2\pi}{n - n'}. \tag{8.3}$$

But $p(n - n') = qn'$ and therefore

$$T_{con} = \frac{p}{q}\frac{2\pi}{n'} = \frac{p}{q}T' = \frac{p+q}{q}T, \tag{8.4}$$

where $T$ and $T'$ are the orbital periods of the two satellites. Hence

$$qT_{con} = pT' = (p+q)T. \tag{8.5}$$

If $q = 1$, then each satellite completes a whole number of orbits between successive conjunctions and every conjunction occurs at the same longitude in inertial space. If $q = 2$, then only every second conjunction occurs at the same longitude, etc.

Now consider the case when $e = 0$, $e' \neq 0$, and $\varpi' \neq 0$, where $e$ denotes the eccentricity and $\varpi$ the longitude of pericentre. If the resonant relation

$$(p+q)n' - pn - q\varpi' = 0 \tag{8.6}$$

is satisfied, then we can rewrite this as

$$\frac{n' - \varpi'}{n - \varpi'} = \frac{p}{p+q}, \tag{8.7}$$

where $n' - \varpi'$ and $n - \varpi'$ are relative motions; these can be considered as the mean motions in a reference frame corotating with the pericentre of the outer satellite. From the viewpoint of this reference frame the *orbit* of the outer satellite is fixed or stationary.

Now if $q = 1$, then every conjunction takes place at the same point in the orbit of the outer satellite, but not at the same longitude in inertial space. If $q = 2$, then every second conjunction takes place at the same point in the orbit, etc.

If the resonant relation given in Eq. (8.6) holds, the corresponding resonant argument is

$$\varphi = (p + q)\lambda' - p\lambda - q\varpi'. \tag{8.8}$$

At a conjunction of the two satellites, $\lambda = \lambda'$ and we have

$$\varphi = q(\varpi' - \lambda') = q(\varpi' - \lambda). \tag{8.9}$$

Thus $\varphi$ is a measure of the displacement of the longitude of conjunction from pericentre of the outer satellite. For example, observations of the saturnian satellites Titan and Hyperion show that the resonant angle $\varphi = 4\lambda' - 3\lambda - \varpi'$ oscillates or librates about the apocentre of Hyperion's orbit with an amplitude of $14.0°$ and a period of $18.75$ y.

At this stage we reintroduce the concept of the rotating reference frame used throughout Chapter 3 and mentioned in the context of spin–orbit coupling in Chapter 5. Consider the orbits of an asteroid and Jupiter where there exists a 2:1 commensurate relationship between the orbital periods and Jupiter is assumed to move in a circular orbit (Fig. 8.3). Again we are ignoring the actual perturbations between these objects since we are only interested in the relative geometry of their orbits.

The path in the rotating frame (Fig. 8.3b) approximates that of a centred ellipse for low eccentricities of the keplerian orbit. As the orbital periods are related by

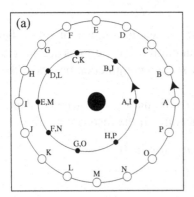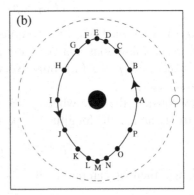

Fig. 8.3. (a) Points along the orbit of an asteroid (small black circles) and Jupiter (white circles) at fixed time intervals of $1/16$ Jupiter period in a nonrotating reference frame for the 2:1 resonance. The letters at each point on one orbit match up with an equivalent point on the other orbit. (b) The path of the asteroid in a rotating frame moving with the mean motion of Jupiter. The letters denote the same points as shown in Fig. 8.3a. The eccentricity of the asteroid's orbit is 0.2.

the 2:1 commensurability, the asteroid will orbit twice for every single orbit of Jupiter. At each of its two aphelion passages (points E and M) Jupiter is ±90° away in longitude. At aphelion the angular velocity of the asteroid is a minimum and close to that of Jupiter. Therefore the points become closer together as seen in the rotating frame. For sufficiently large values of the eccentricity the angular velocity of the asteroid at aphelion is smaller than the fixed angular velocity of Jupiter and so in the rotating frame the asteroid will appear to be moving backwards.

We can illustrate the behaviour shown in Fig. 8.3 for a variety of eccentricities and resonances. Figure 8.4 shows the paths in the rotating frame for particles at a variety of first-order interior, second-order interior, first-order exterior, and second-order exterior resonances respectively. In each case there are no mutual perturbations.

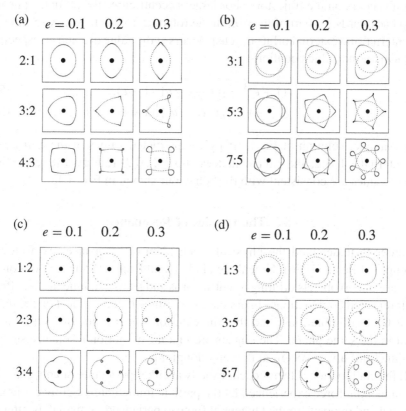

Fig. 8.4. Paths in the rotating frame for a test particle at (a) the 2:1, 3:2, and 4:3 first-order, interior resonances, (b) the 3:1, 5:3, and 7:5 second-order, interior resonances, (c) the 1:2, 2:3, and 3:4 first-order, exterior resonances, and (d) the 1:3, 3:5, and 5:7 second-order, exterior resonances for values of the eccentricity $e = 0.1$, 0.2, and 0.3. The positions of the particle along each path are drawn at equal time intervals.

The particle paths in the rotating frame illustrate the relationship between the resonance and the frequency of conjunctions with the internal or external object. In the case where the particle is in a $p+q : p$ interior resonance with the external object the configuration repeats itself every $p+q$ orbits of the particle. However, when the particle is in a $p : p+q$ exterior resonance with the internal object there is a repeated configuration every $p$ orbits of the particle.

One obvious feature of the plots shown in Fig. 8.4 is the formation of "loops" in the rotating frame trajectories. When they occur it is always at the apocentre of internal particle orbits and the pericentre of external particle orbits. Because it takes $p + q$ and $p$ orbits for particle's configuration to repeat itself in the $p+q : p$ interior and $p : p+q$ exterior resonances respectively, we would expect the respective number of loops to be $p + q$ and $p$ in each case. At a unique, critical value of the particle's eccentricity the angular velocity of the particle on its orbit matches the (constant) angular velocity of the perturber; in this case a "cusp" occurs on the trajectory. For larger eccentricities the particle appears to loop backwards for part of its orbit in the rotating frame. It is easy to show that for internal and external orbits the cusp occurs at the value of $e$ and $e'$ respectively that satisfy the cubic equations

$$(1 + e)^3 = [(p + q)/p]^2 (1 - e), \tag{8.10}$$
$$(1 - e')^3 = [p/(p + q)]^2 (1 + e'). \tag{8.11}$$

For example, solution of Eq. (8.10) gives predicted values of the critical $e$ for the internal 2:1, 3:2, and 4:3 resonances of 0.365, 0.211, and 0.148 respectively. These values are consistent with the behaviour shown in Fig. 8.4a.

## 8.3 The Physics of Resonance

In order to understand the physical mechanism of resonance, we follow the example of Peale (1976), Greenberg (1977), and Peale (1986). Let us consider the case of two objects orbiting a central mass and let us examine the net effect of their repeated conjunctions. In our case we assume that the object on the interior orbit has negligible mass and that the exterior object is moving on a circular orbit in the same plane. The objects are in a mean motion resonance such that conjunctions always occur at the same longitude.

If the conjunctions always occur exactly at either pericentre or apocentre, then the tangential force experienced by the particle immediately before conjunction is equal and opposite to the tangential force experienced immediately after conjunction. Thus, there is no net tangential force. From the formulae derived in Sect. 2.9, we know that the angular momentum can change only under the effects of a tangential force (see Eq. (2.149)). Similarly, if conjunctions occur at pericentre or apocentre (i.e., $f = 0$ or $f = 180°$ respectively in Eq. (2.145)) then

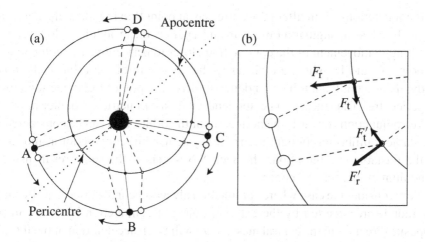

Fig. 8.5. Schematic diagrams illustrating resonant encounters between an object of negligible mass on an internal, elliptical orbit and a massive object on an external, circular orbit. (a) The geometry of encounters at four different points of conjunction, labelled A, B, C, and D. The dotted line denotes the major axis of the internal object. The arrows denote the directions towards which each conjunction should be moving; the stable conjunction is at the pericentre of the orbit. (b) A close-up of conjunction A showing the radial (subscript r) and tangential (subscript t) forces acting on the internal object immediately before (unprimed quantities) and after (primed quantities) the conjunction.

we also require a tangential force to change the semi-major axis. This symmetry is destroyed if conjunctions occur at any other point on the orbit.

We can examine the effects of other types of encounters by considering conjunctions at the points labelled A, B, C, and D in Fig. 8.5a. Here we are assuming that the longitude of pericentre of the interior object is fixed and that the resonance relationship implies that conjunctions will occur at the same point if there is no gravitational interaction between the objects.

If conjunction occurs at point A, close to the pericentre of the interior object, then the objects on their orbits are in the process of diverging from one another (see Fig. 8.5b). Therefore the tangential force, $F_t$, experienced by the interior object immediately prior to conjunction is larger than the tangential force, $F_t'$, it experiences immediately after conjunction. Furthermore, the angular velocity of the inner object is closer to the (constant) angular velocity of the outer object just before conjunction than just after. Therefore the larger tangential force in the direction of motion acts for a longer time than the smaller tangential force, which acts in the opposite after conjunction has occurred. Hence the net result of the encounter is an increase in angular momentum of the inner object and a decrease in its mean angular velocity; this means that subsequent conjunctions will occur closer to pericentre.

If conjunctions occur after pericentre, say at point B in Fig. 8.5a, the net result would be a loss of angular momentum and a gain in mean angular velocity, again resulting in movement of the conjunction closer to pericentre. For conjunctions at points C and D, which are closer to the apocentre of the orbit of the inner body, there would again be a tendency to drive the point of conjunction closer to pericentre. Thus there is a net tendency at all asymmetric encounter points to drive conjunctions towards the pericentre of the orbit. This can be considered as the stable equilibrium configuration of the system, while conjunction at apocentre is the unstable configuration. In Sect. 8.6 we show the same result using a pendulum model for resonance.

Now consider the case where conjunction always occur at pericentre. Because the radial force exerted by the external object at conjunction is always in the opposite direction to the central mass, there will be an acceleration of the interior mass away from the central mass and therefore it will move on an orbit that is slightly larger than it would have done if it had not received the perturbation (cf. Eq. (2.145)). This means that it will have reached its pericentre slightly later than it would have done and hence its longitude of pericentre has moved in the prograde sense (cf. Eq. (2.165)). Given that natural satellite orbits usually undergo rapid precession due to the oblateness of the planet, this implies that the effect would be enhanced for those satellites trapped in a resonance with an exterior satellite.

In the case where the object of negligible mass is on an external, eccentric orbit (this is the case examined by Peale (1976, 1986) and Greenberg (1977)), it can easily be shown that conjunctions at apocentre give the stable configuration. Furthermore, the resonance mechanism in these cases results in a regression of the pericentre of the small mass on the exterior orbit; in the case of satellite orbits, especially those that are far from the central planet, the pericentre regression effect can dominate over the pericentre precession introduced by the $J_2$ of the planet. For example, the saturnian satellite Hyperion is in a 3:4 external resonance with Titan; this is sufficient to introduce a net regression of $19°$ $y^{-1}$ in Hyperion's orbit. The dynamics of this resonance will be discussed later in this chapter.

## 8.4 Variation of Orbital Elements

In Chapter 6 we derived Lagrange's equations, which determine the variation of the orbital elements for a given perturbing potential. If we consider only the lowest order terms in $e$ and $I$ then the approximate form of Eqs. (6.145)–(6.150) is given by

$$\dot{n} = -\frac{3}{a^2}\frac{\partial \mathcal{R}}{\partial \lambda}, \tag{8.12}$$

$$\dot{e} = -\frac{1}{na^2 e} \frac{\partial \mathcal{R}}{\partial \varpi}, \tag{8.13}$$

$$\dot{i} = -\frac{1}{na^2 \sin I} \frac{\partial \mathcal{R}}{\partial \Omega}, \tag{8.14}$$

$$\dot{\varpi} = \frac{1}{na^2 e} \frac{\partial \mathcal{R}}{\partial e} + \frac{\sin \frac{1}{2} I}{na^2} \frac{\partial \mathcal{R}}{\partial I}, \tag{8.15}$$

$$\dot{\Omega} = \frac{1}{na^2 \sin I} \frac{\partial \mathcal{R}}{\partial I}, \tag{8.16}$$

$$\dot{\epsilon} = \frac{e}{2na^2} \frac{\partial \mathcal{R}}{\partial e}, \tag{8.17}$$

where we have chosen to write the equation for $\dot{n}$ rather than $\dot{a}$ (using $\dot{n} = -\frac{3}{2} n\dot{a}/a$), and we have used the usual procedure of replacing partial derivatives involving the mean longitude of epoch by those involving the mean longitude $\lambda$ (see Sect. 6.7). Note that in the case of $\dot{\epsilon}$ we have ignored terms involving partial derivatives with respect to $a$; this is equivalent to assuming that the terms involving Laplace coefficients are constant.

From Eq. (6.139) the general form of an appropriate argument in the expansion of the disturbing function is

$$\varphi = j_1 \lambda' + j_2 \lambda + j_3 \varpi' + j_4 \varpi + j_5 \Omega' + j_6 \Omega \tag{8.18}$$

and to lowest order the general term in the averaged expansion is

$$\langle \mathcal{R} \rangle = \frac{\mathcal{G}m'}{a'} \left[ \mathcal{R}_{\mathrm{D}}^{(\mathrm{sec})} + e^{|j_4|} e'^{|j_3|} s^{|j_6|} s'^{|j_5|} \left[ f_{\mathrm{d}}(\alpha) + f_{\mathrm{e}}(\alpha) \right] \cos \varphi \right], \tag{8.19}$$

$$\langle \mathcal{R}' \rangle = \frac{\mathcal{G}m}{a} \left[ \alpha \mathcal{R}_{\mathrm{D}}^{(\mathrm{sec})} + e^{|j_4|} e'^{|j_3|} s^{|j_6|} s'^{|j_5|} \left[ \alpha f_{\mathrm{d}}(\alpha) + f_{\mathrm{i}}(\alpha) \right] \cos \varphi \right] \tag{8.20}$$

for internal and external resonances respectively where $s = \sin \frac{1}{2} I$, $s' = \sin \frac{1}{2} I'$, and we have included the secular contribution given by

$$\mathcal{R}_{\mathrm{D}}^{(\mathrm{sec})} = (e^2 + e'^2) f_{\mathrm{s},1}(\alpha) + ee' f_{\mathrm{s},2}(\alpha) \cos(\varpi' - \varpi)$$
$$+ (s^2 + s'^2) f_{\mathrm{s},3}(\alpha) + ss' f_{\mathrm{s},4}(\alpha) \cos(\Omega' - \Omega) \tag{8.21}$$

(cf. Eq. (7.1)). We can use Appendix B to obtain expressions for $f_{\mathrm{d}}(\alpha)$, $f_{\mathrm{e}}(\alpha)$, $f_{\mathrm{i}}(\alpha)$, and $f_{\mathrm{s},i}(\alpha)$ ($i = 1, 2, 3, 4$). For convenience these are given explicitly in Tables 8.1–8.4.

Note that the secular terms given in Eq. (8.21) will make no contribution if we restrict our expansions to terms of the first order in the eccentricities and inclinations. Furthermore, in these cases there will be no contribution from the inclination terms, because $s$ and $s'$ only occur when terms of order two or higher are considered.

Table 8.1. Direct terms in the expansion of the disturbing function.

| $j_1$ | $j_2$ | $j_3$ | $j_4$ | $j_5$ | $j_6$ | $f_d(\alpha)$ |
|---|---|---|---|---|---|---|
| $j$ | $1-j$ | $0$ | $-1$ | $0$ | $0$ | $\frac{1}{2}[-2j-\alpha D]b_{1/2}^{(j)}$ |
| $j$ | $1-j$ | $-1$ | $0$ | $0$ | $0$ | $\frac{1}{2}[-1+2j+\alpha D]b_{1/2}^{(j-1)}$ |
| $j$ | $2-j$ | $0$ | $-2$ | $0$ | $0$ | $\frac{1}{8}[-5j+4j^2-2\alpha D$ $+4j\alpha D+\alpha^2 D^2]b_{1/2}^{(j)}$ |
| $j$ | $2-j$ | $-1$ | $-1$ | $0$ | $0$ | $\frac{1}{4}[-2+6j-4j^2+2\alpha D$ $-4j\alpha D-\alpha^2 D^2]b_{1/2}^{(j-1)}$ |
| $j$ | $2-j$ | $-2$ | $0$ | $0$ | $0$ | $\frac{1}{8}[2-7j+4j^2-2\alpha D$ $+4j\alpha D+\alpha^2 D^2]b_{1/2}^{(j-2)}$ |
| $j$ | $2-j$ | $0$ | $0$ | $0$ | $-2$ | $\frac{1}{2}\alpha b_{3/2}^{(j-1)}$ |
| $j$ | $2-j$ | $0$ | $0$ | $-1$ | $-1$ | $-\alpha b_{3/2}^{(j-1)}$ |
| $j$ | $2-j$ | $0$ | $0$ | $-2$ | $0$ | $\frac{1}{2}\alpha b_{3/2}^{(j-1)}$ |

Table 8.2. Indirect terms in the expansion of the disturbing function for an external perturber.

| $j_1$ | $j_2$ | $j_3$ | $j_4$ | $j_5$ | $j_6$ | $f_e(\alpha)$ |
|---|---|---|---|---|---|---|
| $2$ | $-1$ | $-1$ | $0$ | $0$ | $0$ | $-2\alpha$ |
| $3$ | $-1$ | $-2$ | $0$ | $0$ | $0$ | $-\frac{27}{8}\alpha$ |

Table 8.3. Indirect terms in the expansion of the disturbing function for an internal perturber.

| $j_1$ | $j_2$ | $j_3$ | $j_4$ | $j_5$ | $j_6$ | $f_i(\alpha)$ |
|---|---|---|---|---|---|---|
| $2$ | $-1$ | $-1$ | $0$ | $0$ | $0$ | $-\frac{1}{2}/\alpha$ |
| $3$ | $-1$ | $-2$ | $0$ | $0$ | $0$ | $-\frac{3}{8}/\alpha$ |

Table 8.4. Secular terms in the expansion of the disturbing function.

| $i$ | $f_{s,i}(\alpha)$ |
|---|---|
| 1 | $\frac{1}{8}\left[2\alpha D + \alpha^2 D^2\right]b_{1/2}^{(0)}$ |
| 2 | $\frac{1}{4}\left[2 - 2\alpha D - \alpha^2 D^2\right]b_{1/2}^{(1)}$ |
| 3 | $-\frac{1}{2}\alpha b_{3/2}^{(1)}$ |
| 4 | $\alpha b_{3/2}^{(1)}$ |

The equations for the time variation of the orbital elements are required when we have to calculate the time derivative of the general angle $\varphi$ given in Eq. (8.18). This is given by

$$\dot{\varphi} = j_1(n' + \dot{\epsilon}') + j_2(n + \dot{\epsilon}) + j_3\dot{\varpi}' + j_4\dot{\varpi} + j_5\dot{\Omega}' + j_6\dot{\Omega} = 0. \qquad (8.22)$$

We say that the perturbed body is in *exact resonance* when the time variation of a particular resonant argument, $\dot{\varphi}$, is exactly zero. This implies that in this case there exists a particular linear combination of mean motions and precession rates (i.e., a particular set of values of the $j_1, j_2, \ldots, j_6$) such that $\dot{\varphi} = 0$.

It is clear from Eq. (8.23) that if we neglect the contributions from all the variations of the longitudes then "exact" resonance occurs when

$$j_1 n' + j_2 n \approx 0. \qquad (8.23)$$

Let $j_1 = p + q$ and $j_2 = -p$, where $p$ and $q$ are positive integers and $q$ is the *order of the resonance*. We define the *nominal resonance location*, $a_n$, of the $p + q : p$ resonance to be the semi-major axis of the inner body that satisfies the relation

$$a_n = \left(\frac{p}{p+q}\right)^{2/3} a', \qquad (8.24)$$

with a similar definition for external resonances. However, this relationship provides only an approximate location for the resonance. In cases where there are significant contributions to the precession rates the locations in semi-major axis of the different exact resonances associated with a $p + q : p$ commensurability are separate and distinct. For example, in Sect. 6.10.1 we noted that to second order there are six possible resonant arguments associated with the basic 3:1 commensurability. If we ignore the variations in mean longitude of epoch the six time derivatives of the resonant angles are given by

$$\dot{\varphi}_1 = 3n' - n - 2\dot{\varpi}, \quad \dot{\varphi}_2 = 3n' - n - \dot{\varpi} - \dot{\varpi}', \quad \dot{\varphi}_3 = 3n' - n - 2\dot{\varpi}',$$
$$\dot{\varphi}_4 = 3n' - n - 2\dot{\Omega}, \quad \dot{\varphi}_5 = 3n' - n - \dot{\Omega} - \dot{\Omega}', \quad \dot{\varphi}_6 = 3n' - n - 2\dot{\Omega}'. \qquad (8.25)$$

If the bodies involved in these resonances are orbiting an oblate planet then the values of $\varpi$, $\varpi'$, $\Omega$, and $\Omega'$ would be principally determined by the value of $J_2$ and the distance of the relevant body from the planet. Therefore, it is likely that $|\dot{\varpi}| > |\dot{\varpi}'|$ and $|\dot{\Omega}| > |\dot{\Omega}'|$. Furthermore, the nodal rates are usually opposite in sign to the pericentre precession rates. This combination leads to a "splitting" of the basic 3:1 resonance over a range of semi-major axes. The relative extent of this splitting at each planet is thought to provide an explanation for the abundance of mean motion commensurabilities in the jovian and saturnian systems and a lack of such phenomena in the uranian system. This is discussed further in the Sect. 8.15.

## 8.5  Resonance in the Circular Restricted Three-Body Problem

Although we can proceed to investigate the properties of resonant behaviour in a variety of circumstances it proves instructive to concentrate on perhaps the simplest case of interest. This is the planar, circular, restricted problem where a body on an external orbit perturbs an inner body of negligible mass, with both objects moving in the reference plane. The general term in the averaged expansion reduces to

$$\langle \mathcal{R} \rangle = \frac{\mathcal{G}m'}{a'} \left[ f_{s,1}(\alpha)e^2 + f_d(\alpha)e^{|j_4|} \cos\varphi \right], \tag{8.26}$$

where

$$\varphi = j_1\lambda' + j_2\lambda + j_4\varpi \tag{8.27}$$

and the corresponding equations of motion become

$$\dot{n} = 3j_2 C_r n e^{|j_4|} \sin\varphi, \tag{8.28}$$

$$\dot{e} = j_4 C_r e^{|j_4|-1} \sin\varphi, \tag{8.29}$$

$$\dot{\varpi} = 2C_s + |j_4|C_r e^{|j_4|-2} \cos\varphi, \tag{8.30}$$

$$\dot{\epsilon} = C_s e^2 + \frac{1}{2}|j_4|C_r e^{|j_4|} \cos\varphi. \tag{8.31}$$

Here the constants arising from the resonant and secular parts of the disturbing function are given by

$$C_r = \frac{\mathcal{G}m'}{na^2a'} f_d(\alpha) = \left(\frac{m'}{m_c}\right) n\alpha f_d(\alpha) \tag{8.32}$$

and

$$C_s = \frac{\mathcal{G}m'}{na^2a'} f_{s,1}(\alpha) = \left(\frac{m'}{m_c}\right) n\alpha f_{s,1}(\alpha) \tag{8.33}$$

respectively, where $m_c$ is the mass of the central body and we have used Kepler's third law to write $\mathcal{G} = n^2a^3/m_c$. Note that $f_d$ comes from either the first or third entry in Table 8.1 and $f_{s,1}$ is given in Table 8.4

Examining Eqs. (8.28)–(8.31) allows us to deduce some basic properties of the variation of the orbital elements. Given that $|j_4| \geq 1$, the mean motion and semi-major axis are almost constant, because of the $e^{|j_4|}$ term on the right-hand side of Eq. (8.28). The variation of $e$ is proportional to one smaller power of $e$ (see Eq. (8.29)) and so the eccentricity will undergo larger variations. The equation for $\varpi$ shows that a near-circular orbit will undergo rapid changes in the longitude of pericentre as the second (resonant) term on the right-hand side of Eq. (8.30) dominates over the first (secular) term. Similarly the variation of $\epsilon$ will be dominated by the resonant contribution for first-order resonances, but the secular term will be comparable to the resonant term for second-order resonances. Since $\dot{n} = -(3/2)(n/a)\dot{a}$ we have that

$$\dot{a} = -2j_2 C_r a e^{|j_4|} \sin \varphi \tag{8.34}$$

and hence

$$\frac{da}{de} = -2(j_2/j_4)ae . \tag{8.35}$$

This implies that the variations of $a$ and $e$ are always correlated. If $j_2$ and $j_4$ have the same sign (the case for an interior resonance), $da/de < 0$ and a maximum of $a$ corresponds to a minimum of $e$ and vice versa. For opposite signs of $j_2$ and $j_4$ (the case for an exterior resonance) $da/de > 0$ and $a$ and $e$ are at maximum or minimum together.

It is interesting to compare Eq. (8.35) with a similar expression derived from the Tisserand relation. In Sect. 3.4 we showed that

$$\frac{1}{2a} + \sqrt{a(1 - e^2)} \cos I = \text{constant} \tag{8.36}$$

for the restricted three-body problem, provided $a$, $e$, and $I$ are calculated far from the perturber. Recall that in this expression $a$ is the semi-major axis of the perturber. Setting $I = 0$ in Eq. (8.36) and differentiating with respect to $e$ gives

$$\frac{da}{de} = \frac{2a^{\frac{5}{2}}e}{a^{\frac{3}{2}} - 1} \tag{8.37}$$

to lowest order in $e$. Therefore, for orbits interior to that of the perturber (i.e., $a < 1$ in these units), $da/de < 0$ whereas for exterior orbits $da/de > 0$. This gives the same result regarding maxima and minima as stated above for the resonance case. In Sect. 8.8 we use a Hamiltonian approach to show that the Tisserand relation is a constant of the motion.

To complete the resonant equations of motions we have to derive an equation for the variation of the resonant angle $\varphi$. Recalling that the orbit of the external body is fixed we have

$$\dot{\varphi} = j_1 n' + j_2(n + \dot{\epsilon}) + j_4 \dot{\varpi}. \tag{8.38}$$

Table 8.5. *The numerical values of $\alpha$ and the coefficients $\alpha f_{s,1}(\alpha)$ and $\alpha f_d(\alpha)$ for a selection of first- and second-order internal resonances of the form $p+q : p$. The value of $\alpha$ is the nominal value for each resonance.*

| $p+q:p$ | $\alpha$ | $\alpha f_{s,1}(\alpha)$ | $\alpha f_d(\alpha)$ |
|---|---|---|---|
| 2 : 1 | 0.629961 | 0.244190 | −0.749964 |
| 3 : 2 | 0.763143 | 0.879751 | −1.54553 |
| 4 : 3 | 0.825482 | 1.88147 | −2.34472 |
| 5 : 4 | 0.861774 | 3.24494 | −3.14515 |
| 6 : 5 | 0.885549 | 4.96857 | −3.94613 |
| 3 : 1 | 0.480750 | 0.0683812 | 0.287852 |
| 5 : 3 | 0.711379 | 0.515657 | 2.32892 |
| 7 : 5 | 0.799064 | 1.33523 | 6.28903 |
| 9 : 7 | 0.845740 | 2.51812 | 12.1673 |
| 11 : 9 | 0.874782 | 4.06179 | 19.9639 |

Table 8.5 gives the values of $\alpha = a/a'$, $\alpha f_{s,1}(\alpha)$, and $\alpha f_d(\alpha)$ for a number of first- and second-order internal resonances of the form $p + q : p$ using the nominal value of $\alpha$ in each case. The quantities have been derived from the definitions given in Tables 8.1 and 8.4. Note that the magnitude of the $\alpha f_{s,1}(\alpha)$ and $\alpha f_d(\alpha)$ terms increases as $\alpha$ increases.

The set of formulae given in Eqs. (8.28)–(8.31) and (8.38) describe the variation of the orbital elements in the vicinity of a resonance in the planar, circular restricted problem. These can easily be extended to cover the variation of orbital elements for other cases. For example, if we consider motion in circular inclined orbits there is a parallel between the behaviour of the inclination and node and that of the eccentricity and pericentre seen above.

## 8.6 The Pendulum Model

We now examine the parallels between the simple resonance model outlined above and the motion of a pendulum. Consider the second time derivative of the argument $\varphi$ given in Eq. (8.27). Given that we are assuming $\dot{n}' = 0$, the time derivative of Eq. (8.38) gives

$$\ddot{\varphi} = j_2 \dot{n} + j_2 \dot{\epsilon} + j_4 \dot{\varpi} .  \tag{8.39}$$

If we write

$$\dot{\epsilon} = C_s e^2 + |j_4| C_r F(e) \cos \varphi,  \tag{8.40}$$

$$\dot{\varpi} = 2 C_s + |j_4| C_r G(e) \cos \varphi,  \tag{8.41}$$

where

$$F(e) = \frac{1}{2} e^{|j_4|} \quad \text{and} \quad G(e) = e^{|j_4|-2},  \tag{8.42}$$

then the second time derivatives of $\epsilon$ and $\varpi$ are given by

$$\ddot{\epsilon} = 2C_s e \dot{e} + |j_4| C_r \left( \frac{dF(e)}{de} \dot{e} \cos\varphi - F(e)\dot{\varphi} \sin\varphi \right), \tag{8.43}$$

$$\ddot{\varpi} = |j_4| C_r \left( \frac{dG(e)}{de} \dot{e} \cos\varphi - G(e)\dot{\varphi} \sin\varphi \right). \tag{8.44}$$

Recalling that the constants $C_r$ and $C_s$ already contain a factor of $m'/m_c$, which is usually a small quantity, the presence of $\dot{e}$ and $\dot{\varphi}$ in Eqs. (8.43) and (8.44) introduces another factor of $m'/m_c$ and so the contribution of $\ddot{\epsilon}$ and $\ddot{\varpi}$ to $\ddot{\varphi}$ can be neglected in most circumstances. However, an inspection of Eq. (8.44) shows that in the case of first-order resonances (in our model those for which $|j_4| = 1$) there can be a significant contribution to the precession of $\varpi$ because $G(e) = 1/e$ and $e$ is a small quantity. In these circumstances the $\ddot{\varpi}$ term can contribute to $\ddot{\varphi}$. Note that in either case there is no longer a contribution from the secular part of the disturbing function.

If we neglect the contributions from $\ddot{\epsilon}$ and $\ddot{\varpi}$, and make use of the expression for $\dot{n}$ in Eq. (8.28), then the equation for $\ddot{\varphi}$ becomes

$$\ddot{\varphi} = 3 j_2^2 C_r n e^{|j_4|} \sin\varphi. \tag{8.45}$$

An inspection of the last column of Table 8.5 suggests that the constant $C_r$ is negative when $q$, the order of the resonance, is odd and positive when $q$ is even. If we restrict ourselves to odd-order resonances then the equation for $\ddot{\varphi}$ has the form

$$\ddot{\varphi} = -\omega_0^2 \sin\varphi, \tag{8.46}$$

where we take

$$\omega_0^2 = -3 j_2^2 C_r n e^{|j_4|} \tag{8.47}$$

and we assume that $n$, $e$, and hence $\omega_0$ are approximately constant. Note that $\omega_0^2$ is always a positive quantity. Therefore in the case of odd-order resonances the equation for $\ddot{\varphi}$ is identical to that of a simple pendulum with stable motion about $\varphi = 0$.

In the case of even-order resonances the motion is still similar to that described by a simple pendulum equation but in this case the stable point is about $\varphi = \pi$ rather than $\varphi = 0$. This can easily be seen by making the substitution $\varphi' = \varphi + \pi$ in Eq. (8.46), giving $\ddot{\varphi}' = -\omega_0^2 \sin\varphi'$. In further discussions of librational motion for all resonances we will refer to Eq. (8.46) as the equation of a simple pendulum, even though technically there are minor differences in the case of even-order resonances. Note that when $\varphi$ is small the differential equation can be written as $\ddot{\varphi} = -\omega_0^2 \varphi$, the solution of which describes simple harmonic motion with a period independent of amplitude.

The differential equation given in Eq. (8.46) has some well-known properties. The solution can be described either as a circulation or a libration of the argument

$\varphi$, with the type of motion depending on the energy of the system, and hence on the starting conditions. The total energy $E$ of the system is the sum of the kinetic energy per unit mass ($T = \frac{1}{2}\dot{\varphi}^2$) and potential energy per unit mass ($U = 2\omega_0^2 \sin^2 \frac{1}{2}\varphi$). Hence

$$E = \frac{1}{2}\dot{\varphi}^2 + 2\omega_0^2 \sin^2 \frac{1}{2}\varphi. \qquad (8.48)$$

We can now classify the various types of motion by considering different values of the energy $E$. This is illustrated in Fig. 8.6a where we have plotted the potential energy and three possible values of the total energy, each corresponding to a different behaviour of the angle $\varphi$.

1) If $E > E_3$ (as in the case $E = E_1$) then the motion of $\varphi$ is *unbounded* and this corresponds to *circulation* of the angle $\varphi$. In the case of a pendulum this implies a 360° motion of the bob about the point of suspension.
2) If $E < E_3$ (as in the case $E = E_2$) then the motion of $\varphi$ is *bounded* and this corresponds to *oscillation* or *libration* of the angle $\varphi$. For a pendulum this implies a backwards and forwards motion about the point of suspension.
3) In the special case when $E = E_3$ motion occurs on the *separatrix*, which divides the circulation regime from the libration regime. This corresponds to the suspension of the pendulum bob in the vertical upward position.

The different possible trajectories are shown in Fig. 8.6b. For $E = E_1$ there is variation of $\dot{\varphi}$ with $\varphi$ but $\dot{\varphi}$ is never zero. However, for $E = E_2$, $\dot{\varphi}$ does take zero values, each of which corresponds to an extreme of the motion of the angle $\varphi$. In the special case where $E = E_3$, the energy associated with the separatrix, we have a peculiar type of motion where the period is infinite.

Note from Fig. 8.6b that there are two singular points at $\dot{\varphi} = 0$. The first is at the origin, which corresponds to an *elliptic* fixed point, which is *stable* to small displacements. The second is at $\varphi = \pm\pi$, which corresponds to a *hyperbolic*

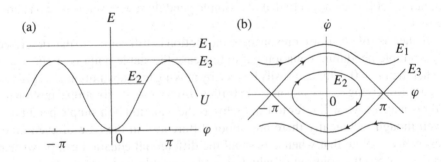

Fig. 8.6. (a) The potential energy $U$ as a function of $\varphi$ compared with three possible values $E_1$, $E_2$ and $E_3$ of the total energy $E$. (b) Three trajectories of the pendulum in $(\varphi, \dot{\varphi})$ space corresponding to each of the total energies shown in (a).

fixed point, which is *unstable* to small displacements. Thus we have provided the mathematical reasoning behind the physical arguments given in Sect. 8.3 for stable and unstable resonant configurations.

The time taken for $\varphi$ to move from an angle $\varphi = 0$ to an angle $\varphi = \varphi_0$ is given by

$$t = \int_0^{\varphi_0} \frac{d\varphi}{\dot{\varphi}}. \tag{8.49}$$

If $\dot{\varphi} = 0$ when $\varphi = \varphi_0$ the expression for the librational period of the motion is

$$T_{\text{lib}} = 4 \int_0^{\varphi_0} \frac{d\varphi}{\dot{\varphi}}. \tag{8.50}$$

Hence, because $E = 2\omega_0^2 \sin^2 \varphi_0/2$,

$$T_{\text{lib}} = \frac{1}{\omega_0} \int_0^{2\pi} \frac{d\theta}{\left(1 - (E/2\omega_0^2) \sin^2 \theta\right)^{1/2}} = \frac{4}{\omega_0} K\left(\frac{E}{2\omega_0^2}\right), \tag{8.51}$$

where we have set $\sin(\varphi/2) = \sin(\varphi_0/2) \sin \theta$. Here $K(x)$ is the complete elliptical integral of the first kind. This has the property that $K(0) = \pi/2$. Therefore, when $E$ is small corresponding to small amplitude librations about the equilibrium point, we have

$$\lim_{\varphi_0 \to 0} T_{\text{lib}} = \frac{2\pi}{\omega_0}. \tag{8.52}$$

This approximation is equivalent to assuming that $\sin \varphi \approx \varphi$. In this case the pendulum equation reduces to that for simple harmonic motion, $\ddot{\varphi} = -\omega_0^2 \varphi$ where the period of the libration is independent of the amplitude. In the more general case the period *is* a function of the amplitude and, in fact, the nature of the function $K(x)$ is such that $T_{\text{lib}} \to \infty$ as the separatrix is approached.

## 8.7 Libration Width

The principal advantage of deriving an analytical model of resonance is that we can provide estimates of the variation in orbital parameters caused by individual resonances. In the context of solar system dynamics the most important such estimate is the extent of the libration in semi-major axis (or mean motion) for an object in resonance. This is because we can then try to relate the known libration width with an observed phenomenon such as a ring feature or a gap in the asteroid distribution. We can accomplish this by using our simple pendulum model and making modifications incorporating additional terms that arise in the case of low eccentricity orbits near first-order resonances.

Consider the relationship between the kinetic and potential energies given by Eq. (8.48) where we assume that we are dealing with the equation of a simple pendulum. It is clear from Fig. 8.6 that the energy associated with maximum libration occurs when $\dot\varphi = 0$ at $\varphi = \pm\pi$. This implies

$$E_{\max} = -6j_2^2 C_r n e^{|j_4|}. \tag{8.53}$$

We now set the value of $E$ equal to $E_{\max}$ and consider the variation of $\varphi$ given by

$$\dot\varphi = \pm j_2 \left(12|C_r|ne^{|j_4|}\right)^{1/2} \cos\frac{1}{2}\varphi. \tag{8.54}$$

We can relate the variation of $\varphi$ to the variation in $n$ by means of Eq. (8.28). This gives

$$\begin{aligned} dn &= 3j_2 C_r n e^{|j_4|} \frac{\sin\varphi}{\dot\varphi}\, d\varphi \\ &= \pm\left(3|C_r|ne^{|j_4|}\right)^{1/2} \sin\frac{1}{2}\varphi\, d\varphi \end{aligned} \tag{8.55}$$

making use of Eq. (8.54). Integration gives

$$n = n_0 \pm \left(12|C_r|ne^{|j_4|}\right)^{1/2} \cos\frac{1}{2}\varphi, \tag{8.56}$$

and so the maximum change in the mean motion is

$$\delta n_{\max} = \pm\left(12|C_r|ne^{|j_4|}\right)^{1/2}, \tag{8.57}$$

which occurs when $\varphi = 0$. The equivalent maximum change in semi-major axis can be easily derived from Kepler's third law. It is

$$\delta a_{\max} = \pm\left(\frac{16}{3}\frac{|C_r|}{n}e^{|j_4|}\right)^{1/2} a. \tag{8.58}$$

In the derivation of the pendulum-type equation in Sect. 8.6 we noted that in the case of first-order resonances ($|j_4| = 1$) we cannot neglect the contribution to $\ddot\varphi$ from significant terms in $\varpi$. Now we consider this in more detail and determine the modifications to the maximum libration width. Using $j_4 = -1$ we take the general argument of a first-order resonance to be

$$\varphi = j_1\lambda' + j_2\lambda - \varpi \tag{8.59}$$

with

$$\dot\varphi = j_1 n' + j_2 n - \dot\varpi, \tag{8.60}$$

$$\ddot\varphi = j_2\dot n - \ddot\varpi, \tag{8.61}$$

where, from Eqs. (8.29) and (8.44),

$$\ddot\varpi = \frac{C_r^2}{e^2}\sin 2\varphi - \frac{C_r}{e}\left(j_1 n' + j_2 n\right)\sin\varphi. \tag{8.62}$$

This gives

$$\ddot{\varphi} = \left[3j_2^2 C_r n e + \frac{C_r}{e}(j_1 n' + j_2 n)\right] \sin\varphi - \frac{C_r^2}{e^2}\sin 2\varphi. \tag{8.63}$$

We can construct a solution to this modified form of the pendulum equation by assuming a form similar to the previous solution given in Eq. (8.56). We write

$$n = n_0 + k\cos\frac{1}{2}\varphi, \tag{8.64}$$

where $n_0$ is the mean motion associated with the nominal value of the resonance and $k$ is a constant that has yet to be determined. Note that $j_1 n' + j_2 n_0$ is also a constant. Ignoring the smaller, secular term in Eq. (8.31), we can now rewrite Eq. (8.60) as

$$j_1 n' + j_2 n_0 + j_2 k\cos\frac{1}{2}\varphi = \dot{\varphi} + \frac{C_r}{e}\cos\varphi. \tag{8.65}$$

Taking $\dot{\varphi} = 0$ at $\varphi = \pi$ gives

$$j_1 n' + j_2 n_0 = -\frac{C_r}{e}. \tag{8.66}$$

Hence

$$j_1 n' + j_2 n = -\frac{C_r}{e} + j_2 k\cos\frac{1}{2}\varphi, \tag{8.67}$$

and the equation for $\ddot{\varphi}$ becomes

$$\ddot{\varphi} = \left[3j_2^2 C_r n e - \frac{C_r^2}{e^2}\right]\sin\varphi + j_2 k\frac{C_r}{e}\cos\frac{1}{2}\varphi\sin\varphi - \frac{C_r^2}{e^2}\sin 2\varphi, \tag{8.68}$$

with the energy integral

$$\frac{1}{2}\dot{\varphi}^2 = 2\left[3j_2^2 C_r n e - \frac{C_r^2}{e^2}\right]\sin^2\frac{1}{2}\varphi - \frac{4}{3}j_2 k\frac{C_r}{e}\cos^3\frac{1}{2}\varphi - \frac{C_r^2}{e^2}\sin^2\varphi + E. \tag{8.69}$$

As before we find $E_{\max}$ by setting $\dot{\varphi} = 0$ and $\varphi = \pi$. With $E = E_{\max}$ we can find an equation for $\dot{\phi}$ when $\varphi = 0$. This procedure yields

$$\dot{\varphi}^2\Big|_{\varphi=0} = -12j_2^2 C_r n e + 4\frac{C_r}{e^2} - \frac{8}{3}j_2 k\frac{C_r}{e}. \tag{8.70}$$

Taking $\varphi = 0$ in Eqs. (8.65) and (8.67) and squaring the resulting expression for $\dot{\varphi}$ gives

$$\dot{\varphi}^2\Big|_{\varphi=0} = 4\frac{C_r}{e^2} + j_2^2 k^2 - 4j_2 k\frac{C_r}{e}. \tag{8.71}$$

Equating this expression with Eq. (8.70) gives the following quadratic in the unknown constant $k$:

$$k^2 - \frac{4}{3}\frac{1}{j_2}\frac{C_r}{e}k + 12C_r n e = 0, \tag{8.72}$$

with the solution

$$k = \frac{2}{3}\frac{1}{j_2}\frac{C_r}{e} \pm \left(12|C_r|ne\right)^{1/2}\left(1 + \frac{1}{27 j_2^2 e^3}\frac{|C_r|}{n}\right)^{1/2}, \tag{8.73}$$

where we have made use of the fact that $C_r < 0$ for first-order resonances. Substituting $\varphi = 0$ in Eq. (8.64) and making use of Eq. (8.66) gives

$$\delta n_{\max} = k + \frac{|C_r|}{j_2 e} \tag{8.74}$$

and hence, using our solution for $k$, we have

$$\delta n_{\max} = \pm \left(12|C_r|ne\right)^{1/2}\left(1 + \frac{1}{27 j_2^2 e^3}\frac{|C_r|}{n}\right)^{1/2} + \frac{|C_r|}{3 j_2 e}. \tag{8.75}$$

Note that for moderate values of $e$ this is consistent with the approximate expression given in Eq. (8.57) with $j_4 = -1$. These expressions are equivalent to those given by Dermott & Murray (1983). In terms of semi-major axis we can write

$$\frac{\delta a_{\max}}{a} = \pm \left(\frac{16}{3}\frac{|C_r|}{n}e\right)^{1/2}\left(1 + \frac{1}{27 j_2^2 e^3}\frac{|C_r|}{n}\right)^{1/2} - \frac{2}{9 j_2 e}\frac{|C_r|}{n}. \tag{8.76}$$

In Sect. 8.8.1 we use a Hamiltonian approach to derive a similar expression for the case of a first-order resonance.

These results imply that in the case of first-order resonances the maximum libration width in terms of mean motion or semi-major axis decreases as the eccentricity decreases until, for a low enough values of $e$, the width starts to increase again. We illustrate this behaviour in Fig. 8.7 where we have plotted the maximum libration width for the 3:1, 2:1, 5:3, 3:2, and 4:3 jovian internal resonances. The plots were produced by taking $m'/m_c = 9.54 \times 10^{-4}$ and $a' = 5.2033$ with values of $|C_r|/n = (m'/m_c)\alpha f_d(\alpha)$ taken from Table 8.5.

The plots shown in Fig. 8.7 have to be treated with some caution. It has to be remembered that the expansion of the disturbing function on which they are based has been truncated to lowest order in the eccentricity. Therefore the libration widths will not be accurate at large eccentricity. Furthermore, an object with an eccentricity of 0.25 at the 4:3 resonance would have an orbit that intersected the orbit of Jupiter and so some of the convergence conditions of the series expansion of the disturbing function would not hold. However, the plots illustrate the difference between first- and second-order resonances at low eccentricity. As $e \to 0$ there is an enhanced precessional effect for first-order

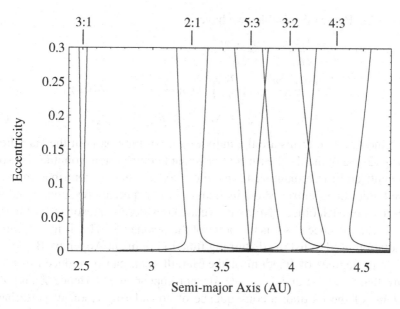

Fig. 8.7. Maximum libration zones as a function of semi-major axis and eccentricity for a selection of internal jovian resonances using an analytical model based on the circular restricted three-body problem. The nominal resonance locations are indicated above the plot.

resonances that increases the libration width, although this has to be treated with some caution (see Sect. 8.9 for a discussion of this phenomenon).

## 8.8 The Hamiltonian Approach

Given the similarity between the equations of motion of an object in resonance and those of a simple pendulum, it is not surprising that many celestial mechanicians have used a generalisation of the Hamiltonian for a pendulum as their fundamental model of resonance. This formulation was introduced by Poincaré (1902, 1905) and used by Message (1966), Yoder (1973), Peale (1976), Wisdom (1980), and Henrard & Lemaître (1983) and resulted in a powerful new Hamiltonian method for studying resonance. Their approach proves particularly useful when considering dissipative systems (such as tidally evolving satellites) and allows a new understanding of resonance capture and the effects of resonance passage.

The approach given below follows that of Peale (1986). From the definitions of the Poincaré variables (see Sect. 2.10) we note that $e^2 \approx 2\Gamma/\Lambda$ for moderate eccentricities. Hence we can derive the Hamiltonian for our system by making use of our expression for the disturbing function (the perturbing potential per unit mass) given in Eq. (8.26). To begin with we consider the effect of the $e^k$ term (where $k = 1, 2, 3, \ldots$) associated with a single, $k$th order resonant argument. In terms of the Poincaré variables we have

$$
\begin{aligned}
\mathcal{H} = &-\frac{\mathcal{G}^2 m_c^2 m^3}{2\Lambda^2} - \frac{\mathcal{G}^2 m_c^2 m'^3}{2\Lambda'^2} \\
&- \frac{\mathcal{G}^2 m_c m m'^3}{\Lambda'^2} f_d \left(\frac{2\Gamma}{\Lambda}\right)^{\frac{k}{2}} \cos[j\lambda' + (k-j)\lambda + k\gamma] \\
&- \Gamma \dot{\varpi}_{\text{sec}} - Z \dot{\Omega}_{\text{sec}} + \Lambda \dot{\lambda}_{\text{sec}} - \Gamma' \dot{\varpi}'_{\text{sec}} - Z' \dot{\Omega}'_{\text{sec}} + \Lambda' \dot{\lambda}'_{\text{sec}},
\end{aligned} \qquad (8.77)
$$

where the first two terms are the individual two-body parts of the Hamiltonian (cf. Eq. (2.182)); the third term is the resonant contribution from the $e^k$ resonant argument; and the remaining terms allow for all secular contributions (to the second order in eccentricity and inclination) to the pericentres, nodes, and mean longitudes of each object. Note that we are considering a resonance of the form $j : (j-k)$ and hence $k > 0$ is the order of the resonance. This is in keeping with the notation used to list the disturbing function terms in Appendix B.

The six degrees of freedom of the Hamiltonian can be reduced to four by noting that $z = -\Omega$ and $z' = -\Omega'$ do not appear in $\mathcal{H}$. Hence $Z$ and $Z'$ are constants of the motion, a consequence of considering resonant perturbations dominated by a planar term. We can now introduce a new set of four variables, $\theta_i$, defined by

$$
\begin{aligned}
\theta_1 &= j\lambda' + (k-j)\lambda + k\gamma, & (8.78) \\
\theta_2 &= j\lambda' + (k-j)\lambda + k\gamma', & (8.79) \\
\theta_3 &= \lambda, & (8.80) \\
\theta_4 &= \lambda'. & (8.81)
\end{aligned}
$$

The relationships between the corresponding momenta, $\Theta_i$, and the original momenta can be obtained by solving the simple system of equations

$$
\sum_{i=1}^{4} \Theta_i \, d\theta_i = \Lambda \, d\lambda + \Lambda' \, d\lambda' + \Gamma \, d\gamma + \Gamma' \, d\gamma' \qquad (8.82)
$$

to give

$$
\begin{aligned}
(k-j)(\Theta_1 + \Theta_2) + \Theta_3 &= \Lambda, & (8.83) \\
j(\Theta_1 + \Theta_2) + \Theta_4 &= \Lambda', & (8.84) \\
k\Theta_1 &= \Gamma, & (8.85) \\
k\Theta_2 &= \Gamma', & (8.86)
\end{aligned}
$$

where $\Theta_2$, $\Theta_3$, and $\Theta_4$ are all constants since $\theta_2$, $\theta_3$, and $\theta_4$ no longer appear in $\mathcal{H}$. At this stage the problem has been reduced to a one degree of freedom system in the variables $\theta_1$ and $\Theta_1$. Before proceeding it is worthwhile examining these constants in more detail.

From Eqs. (8.83)–(8.86) we have

$$
\Theta_4 = \Lambda' - j(\Theta_1 + \Theta_2) \approx m' \sqrt{\mathcal{G}(m_c + m')a'} \qquad (8.87)
$$

to lowest order. Therefore $\Theta_4$ is related to $a'$, a constant in the absence of a perturbation from the mass $m$. From the definitions we have

$$\Theta_2 = \frac{\Gamma'}{k} = \frac{m'}{k}\sqrt{\mathcal{G}(m_c + m)a'}\left(1 - \sqrt{1 - e'^2}\right) \approx \frac{m'}{2k}\sqrt{\mathcal{G}(m_c + m)a'}\,e'^2. \quad (8.88)$$

Therefore $\Theta_2$ is related to the eccentricity of the external body. Because there is no perturbation from the inner body included in $\mathcal{H}$, the quantities $a'$ and $e'$ are fixed and hence $\Theta_2$ must be a constant.

The third constant, $\Theta_3$, is another form of the Tisserand relation, itself derived from the Jacobi constant (see Sect. 3.4). We can demonstrate this by using Eq. (8.83) to write

$$\Theta_3 = \Gamma - (k - j)(\Theta_1 + \Theta_2) = \text{constant}. \quad (8.89)$$

Because $\Theta_2$ is also a constant we have

$$\Lambda - (k - j)\frac{\Gamma}{k} = \text{constant} \quad (8.90)$$

and hence

$$\dot{\Lambda} = \frac{k - j}{k}\dot{\Gamma}. \quad (8.91)$$

Expressed in terms of $a$ and $e$ the previous relation can be rewritten as

$$\frac{\dot{a}}{\dot{e}} = 2\frac{k - j}{k}ae, \quad (8.92)$$

which is equivalent to the previous result given in Eq. (8.35).

Returning to our expression for $\dot{\Lambda}$ in Eq. (8.91) we can find another expression for $(k - j)/k$ by making use of the fact that at the $j : (j - k)$ resonance location we have the approximate relation $jn' - (j - k)n \approx 0$ and hence

$$j + (k - j)m^3\Lambda^{-3} \approx 0, \quad (8.93)$$

with a suitable choice of units such that $n' = 1$. Therefore

$$\frac{k}{k - j} = 1 - m^3\Lambda^{-3} \quad (8.94)$$

and we can write

$$\dot{\Lambda}\left(1 - \frac{m^3}{\Lambda^3}\right) = \dot{\Gamma}, \quad (8.95)$$

which integrates to give

$$\Lambda + \frac{m^3}{2\Lambda^2} = \Gamma + \text{constant}. \quad (8.96)$$

In our choice of units this can be rewritten as

$$\frac{1}{2a} + \sqrt{a(1 - e^2)} = \text{constant} \quad (8.97)$$

(cf. Eq. (3.46) with $I = 0$).

In terms of the new coordinates and momenta the Hamiltonian given in Eq. (8.77) can be written as

$$
\mathcal{H} = -\frac{\mathcal{G}^2 m_c^2 m^3}{2[\Theta_3 + (k-j)\Theta_1]^2} - \frac{\mathcal{G}^2 m_c^2 m'^3}{2(\Theta_4 + j\Theta_1)^2}
$$
$$
- \frac{\mathcal{G}^2 m_c m m'^3}{(\Theta_4 + j\Theta_1)^2} \frac{(2k\Theta_1)^{k/2}}{[\Theta_3 + (k-j)\Theta_1]^{k/2}} f_{\mathrm{d}} \cos\theta_1
$$
$$
- k\Theta_1 \varpi_{\mathrm{sec}} + [\Theta_3 + (k-j)\Theta_1]\dot\lambda_{\mathrm{sec}} - k\Theta_2 \varpi'_{\mathrm{sec}} + (j\Theta_1 + \Theta_4)\dot\lambda'_{\mathrm{sec}}, \quad (8.98)
$$

where $\Theta_2$ has been set to zero or incorporated with $\Theta_3$ and $\Theta_4$ where it occurs in $\Lambda$ and $\Lambda'$; this has little effect on the dynamics of the resonance. The Hamiltonian in Eq. (8.98) can be used to study the dynamics for arbitrarily large eccentricities (see Wisdom 1980). Except for these cases, we can legitimately carry out the expansions given below. From the definitions we see that $|\Theta_1| \ll \Theta_3$ and $|\Theta_1| \ll \Theta_4$, so that we can make use of the approximations

$$
[\Theta_3 + (k-j)\Theta_1]^{-2} \approx \frac{1}{\Theta_3^2} - 2(k-j)\frac{\Theta_1}{\Theta_3^3} + 3(k-j)^2 \frac{\Theta_1^2}{\Theta_3^4}, \quad (8.99)
$$

$$
[\Theta_4 + j\Theta_1]^{-2} \approx \frac{1}{\Theta_4^2} - 2j\frac{\Theta_1}{\Theta_4^3} + 3j^2 \frac{\Theta_1^2}{\Theta_4^4}, \quad (8.100)
$$

where the first term on the right-hand side of each equation is a constant and will therefore be ignored when we substitute back in $\mathcal{H}$. Similarly we can ignore the $\Theta_1$ and $\Theta_2$ contributions to the divisors in the third term in $\mathcal{H}$ given in Eq. (8.98) since there is already a small factor in the numerator. Taking $\Theta_3 \approx \Lambda$ and $\Theta_4 \approx \Lambda'$, and substituting our definitions of $\Theta_1, \Theta_2, \Lambda$, and $\Lambda'$, we can now write the approximate Hamiltonian as

$$
\mathcal{H} = \left[ j(n' + \dot\lambda'_{\mathrm{sec}}) + (k-j)(n + \dot\lambda_{\mathrm{sec}}) - k\varpi_{\mathrm{sec}} \right]\left(\frac{\Gamma}{k}\right)
$$
$$
- \frac{3}{2}\left[ \frac{j^2}{m'a'^2} + \frac{(k-j)^2}{ma^2} \right]\left(\frac{\Gamma}{k}\right)^2
$$
$$
- (n^2)^{1-k/4} f_{\mathrm{d}} \frac{a^{3-k}}{a'} \frac{m'}{m_c} m^{1-k/2} (2\Gamma)^{k/2} \cos\theta_1, \quad (8.101)
$$

where again we have ignored constant terms in the Hamiltonian and made use of the relationships $n^2 \approx \mathcal{G}m_c/a^3$ and $n'^2 \approx \mathcal{G}m_c/a'^3$ (i.e., we have assumed that $m$ and $m'$ are much smaller than $m_c$).

Now we make a small change to the resonant variable, $\theta_1$, by defining $\theta'_1 = \theta_1/k$ and dropping the prime. The reasons for this step are outlined by Peale (1986) and involve the need to maintain a proper choice of canonical variables when $k > 1$. At the same time we change the sign of the Hamiltonian by setting $\mathcal{H}^\dagger = -\mathcal{H}$. This gives rise to a Hamiltonian of the form

$$
\mathcal{H}^\dagger = \bar\alpha\Gamma + \bar\beta\Gamma^2 + \bar\epsilon(2\Gamma)^{k/2}\cos k\theta_1, \quad (8.102)
$$

where

$$\bar{\alpha} = [(j - k)n^* - jn'^* + k\dot{\varpi}_{\text{sec}}]/k, \tag{8.103}$$

$$\bar{\beta} = \frac{3}{2k^2} \left[ \frac{(j - k)^2}{ma^2} + \frac{j^2}{m'a'^2} \right], \tag{8.104}$$

$$\bar{\epsilon} = (n^2)^{1-\frac{k}{4}} f_{\text{d}} \frac{a^{3-k}}{a'} \frac{m'}{m_{\text{c}}} m^{1-k/2}, \tag{8.105}$$

with $n^* = n + \dot{\lambda}_{\text{sec}}$ and $n'^* = n' + \dot{\lambda}'_{\text{sec}}$. Note that the quantity $\bar{\alpha}$ is a measure of the proximity to the resonance, with $\bar{\alpha} = 0$ denoting the location of exact resonance (i.e., the place where the time derivative of the resonant argument is zero).

So far in the analysis we have only considered a single dominant resonant term, which corresponds to that of an internal, $k$th-order resonance in the restricted problem. However, it is relatively easy to extend this to an external resonance. In this case the Hamiltonian becomes

$$\mathcal{H}' = -\frac{\mathcal{G}^2 m_{\text{c}}^2 m^3}{2\Lambda^2} - \frac{\mathcal{G}^2 m_{\text{c}}^2 m'^3}{2\Lambda'^2}$$
$$- \frac{\mathcal{G}^2 m_{\text{c}} m m'^3}{\Lambda'^2} f_{\text{d}} \left( \frac{2\Gamma'}{\Lambda'} \right)^{k/2} \cos[j\lambda' + (k - j)\lambda + k\gamma']$$
$$- \Gamma \dot{\varpi}_{\text{sec}} - Z\dot{\Omega}_{\text{sec}} + \Lambda\dot{\lambda}_{\text{sec}} - \Gamma'\dot{\varpi}'_{\text{sec}} - Z'\dot{\Omega}'_{\text{sec}} + \Lambda'\dot{\lambda}'_{\text{sec}}, \tag{8.106}$$

which differs from the original $\mathcal{H}$ in only one term, and $f_{\text{d}}$ now incorporates any relevant indirect terms. The transformed Hamiltonian is

$$\mathcal{H}'^{\dagger} = \bar{\alpha}'\Gamma' + \bar{\beta}'\Gamma'^2 + \bar{\epsilon}'(2\Gamma')^{k/2} \cos k\theta_2, \tag{8.107}$$

where

$$\bar{\alpha}' = [(j - k)n^* - jn'^* + \dot{\varpi}'_{\text{sec}}]/k, \tag{8.108}$$

$$\bar{\beta}' = \frac{3}{2k^2} \left[ \frac{(j - k)^2}{ma^2} + \frac{j^2}{m'a'^2} \right] = \bar{\beta}, \tag{8.109}$$

$$\bar{\epsilon}' = (n'^2)^{1-\frac{k}{4}} f_{\text{d}} a'^{2-k} \frac{m}{m_{\text{c}}} m'^{1-k/2}. \tag{8.110}$$

Note that $\mathcal{H}'^{\dagger}$ has a form that is identical to that of $\mathcal{H}^{\dagger}$ and so we can proceed with an analysis of the internal resonance case knowing that it also applies to the external case with slight modifications to the constants.

We have made the implicit assumption that $f_{\text{d}}$ is treated as a constant throughout, even though in reality it is a function of $\alpha = a/a'$. However, it can easily be shown that the size of the cosine term in Eq. (8.107) is dominated by the variation in eccentricity rather than semi-major axis. Similarly the effects of the variation of $a$ on the parameter $\bar{\beta}$ are negligible in most circumstances (Peale 1976). Under this assumption it is possible further to simplify the Hamiltonian

in Eq. (8.107) by a scale transformation designed to introduce a single free parameter related to $\bar{\alpha}$, the measure of the proximity to resonance. If we scale the
momentum such that

$$\Phi = \frac{\Gamma}{\eta}, \tag{8.111}$$

where $\eta > 0$ is a constant, the Hamiltonian becomes

$$\mathcal{H} = \bar{\alpha}\eta\Phi + \bar{\beta}\eta^2\Phi^2 + \bar{\epsilon}\eta^{k/2}(2\Phi)^{k/2}\cos k\phi, \tag{8.112}$$

where the coordinate conjugate to $\Phi$ is given by

$$\phi = \begin{cases} \theta_1 + \pi, & \text{if } \bar{\epsilon} > 0, \\ \theta_1, & \text{if } \bar{\epsilon} < 0. \end{cases} \tag{8.113}$$

Note that the definition of $\phi$ depends on the sign of $\bar{\epsilon}$, which in turn depends on
the sign of $f_d$, the constant involving Laplace coefficients. It can be shown that
this is positive or negative depending on whether the resonance has an even or
odd order, respectively (see, for example, Table 8.5).

The requirement that the coefficients of the last two terms in $\mathcal{H}$ are equal
implies that

$$\bar{\epsilon}\eta^{k/2} = (-1)^k 2\bar{\beta}\eta^2, \tag{8.114}$$

and hence the scaling parameter is given by

$$\eta = \left[ \frac{(-1)^k 2\bar{\beta}}{\bar{\epsilon}} \right]^{\frac{2}{k-4}}, \tag{8.115}$$

where the factor $(-1)^k$ is needed to take account of both possible signs of $\bar{\epsilon}$.
Scaling $\mathcal{H}$ by the factor $\bar{\beta}\eta^2$ gives

$$\mathcal{H} = \bar{\delta}\Phi + \Phi^2 + 2(-1)^k(2\Phi)^{k/2}\cos k\phi, \tag{8.116}$$

where the Hamiltonian is now parameterised by

$$\bar{\delta} = \frac{\bar{\alpha}}{\bar{\beta}\eta} = \bar{\alpha} \left[ \frac{4}{\bar{\epsilon}^2 \bar{\beta}^{2-k}} \right]^{\frac{1}{4-k}}. \tag{8.117}$$

Note that if $\bar{\alpha} = 0$ in Eq. (8.117) then $\bar{\delta} = 0$ and the perturbed object is at exact
resonance.

Now we can investigate some properties of the equilibrium points of the new
Hamiltonian. Because the time derivatives of $\phi$ and $\Phi$ are derived from the partial
derivatives of $\mathcal{H}$ with respect to $\Phi$ and $\phi$, the equilibrium points are simply the
solutions of $\partial\mathcal{H}/\partial\phi = \partial\mathcal{H}/\partial\Phi = 0$. The resulting simultaneous equations are

$$\bar{\delta} + 2\Phi + 2(-1)^k k(2\Phi)^{\frac{k-2}{2}}\cos k\phi = 0, \tag{8.118}$$

$$2(-1)^{k+1}k(2\Phi)^{k/2}\sin k\phi = 0. \tag{8.119}$$

For $k \geq 2$ there is always an equilibrium point at $\Phi = 0$. For nonzero $\Phi$ the second equation can only be satisfied when $k\phi = i\pi$, where $i$ is an integer. The first equation then gives the value of $\Phi$ at the equilibrium points as the solution of

$$\bar{\delta} + 2\Phi + 2(-1)^{k+i} k (2\Phi)^{\frac{k-2}{2}} = 0. \tag{8.120}$$

We can rewrite this in terms of a new variable, $R = \sqrt{2\Phi}$, where $R$ is the radial location of the equilibrium point and is always positive. We there by obtain

$$\bar{\delta} + R^2 = (-1)^{k+i+1} 2k R^{k-2}. \tag{8.121}$$

We can illustrate the solutions by defining two functions,

$$f(\bar{\delta}, R) = \bar{\delta} + R^2 \quad \text{and} \quad g(k, R) = (-1)^{k+i+1} 2k R^{k-2}. \tag{8.122}$$

The first function always describes a parabola with a minimum at $(0, \bar{\delta})$. The nature of $g(k, R)$ depends on the value of $k$. For example, for $k = 1$ the resulting curves are two hyperbolas, for $k = 2$ they are two straight lines parallel to the $R$ axis and for $k = 3$ they are two straight lines through the origin with equal and opposite gradients. The values of $R$ at the points of intersection of $f(\bar{\delta}, R)$ and $g(k, R)$ give the location of the equilibrium points for a particular value of $\bar{\delta}$. Families of such curves and their intersection points for the cases $k = 1, 2$, and 3 are shown in Fig. 8.8.

It is clear from Fig. 8.8 that a double root of Eq. (8.121) (and hence a bifurcation point) will occur where the two functions meet tangentially. This occurs where $f'(\bar{\delta}, R) = g'(k, R)$, where the prime denotes partial differentiation with respect

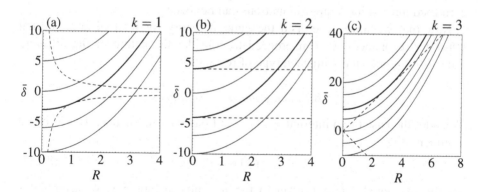

Fig. 8.8. Plots of the parabolas $f(\bar{\delta}, R)$ for various values of $\bar{\delta}$ (solid curves) and $g(k, R)$ (dashed curves and lines) for $k = 1, 2$, and 3. In each case the parabola defined by $f(\bar{\delta}_b, R)$ is shown as a thicker curve. (a) The curves for $k = 1$ and $\bar{\delta} = -10, -7, 0,$ 5 and the bifurcation value of $\bar{\delta}_b = -3$. (b) The curves for $k = 2$ and $\bar{\delta} = -10, -7, 0,$ 7 and the two bifurcation values of $\bar{\delta}_b = \pm 4$. (c) The curves for $k = 3$ and $\bar{\delta} = -10,$ $-5, 0, 5, 15, 20$ and the bifurcation value of $\bar{\delta}_b = 9$.

to $R$. The resulting bifurcation value, $R_b$, is given by

$$R_b = \left[(-1)^{k+i+1}k(k-2)\right]^{\frac{1}{4-k}}. \tag{8.123}$$

Substituting $R = R_b$ in Eq. (8.121) and solving for $\bar{\delta}$ gives the corresponding value of $\bar{\delta}$ at the bifurcation point as

$$\bar{\delta}_b = (4-k)\left[(-1)^{k+i+1}k\right]^{\frac{2}{4-k}}(k-2)^{\frac{k-2}{4-k}}. \tag{8.124}$$

With the appropriate signs this equation gives $\bar{\delta}_b = -3, \pm 4$, and 9 for $k = 1, 2$, and 3 respectively.

Before proceeding let us examine the scaling that we have introduced. After so many transformations it is easy to lose track of the relationship between the current set of variables and the original orbital elements; this is one of the disadvantages of working with canonical elements. However, from the definitions of $\bar{\beta}$ and $\bar{\epsilon}$, and making use of the approximation $\alpha \approx [(j-1)/j]^{2/3}$ derived from Kepler's third law, the relationship between $R$ and $e$ for the case of an external perturber can be written

$$R = \left\{\frac{3(-1)^k}{k^2 f_d}\left[(j-k)^{\frac{4}{3}}j^{\frac{2}{3}}\frac{m_c}{m'} + j^{\frac{4}{3}}(j-k)^{\frac{2}{3}}\frac{m}{m'}\frac{m_c}{m'}\right]\right\}^{\frac{1}{4-k}}e. \tag{8.125}$$

The equivalent expression for the case of an internal perturber is

$$R = \left\{\frac{3(-1)^k}{k^2 f_d}\left[(j-k)^{\frac{4}{3}}j^{\frac{2}{3}}\frac{m'}{m}\frac{m_c}{m} + j^2\frac{m_c}{m}\right]\right\}^{\frac{1}{4-k}}e'. \tag{8.126}$$

Note that in the case where the perturbing mass is significantly larger than the perturbed mass suitable approximations can be made.

Now that we have developed the general theory for internal and external $k$th-order resonances we can proceed to consider the nature of the different trajectories associated with each type of resonance.

### 8.8.1 The e and e' Resonances

Consider the case of an internal, $j : (j-1)$ first-order resonance. The resonant argument is

$$\theta = j\lambda' + (1-j)\lambda - \varpi \tag{8.127}$$

and the associated $f_d$ term is given by the first entry in Table 8.1. We refer to this as the *e resonance* because the associated term in the disturbing function is of $\mathcal{O}(e)$. The relevant values of the single-parameter Hamiltonian and the constants $\bar{\delta}, \bar{\alpha}, \bar{\beta}$, and $\bar{\epsilon}$ can easily be obtained by substituting $k = 1$ in Eq. (8.116) and Eqs. (8.103)–(8.105). The final Hamiltonian is

$$\mathcal{H}_1 = \bar{\delta}\Phi + \Phi^2 - 2\sqrt{2\Phi}\cos\phi, \tag{8.128}$$

where $\phi$ is given by Eq. (8.113) with $\theta_1 = \theta$ and

$$\bar{\delta} = \bar{\alpha} \left( \frac{4}{\bar{\epsilon}^2 \bar{\beta}} \right)^{\frac{1}{3}}.$$
(8.129)

To make it easier to study the types of motion we introduce a mixed canonical transformation (see Sect. 2.10) given by

$$x = \sqrt{2\Phi} \cos \phi \qquad \text{and} \qquad y = \sqrt{2\Phi} \sin \phi,$$
(8.130)

where, as above, $\sqrt{2\Phi}$ can be thought of as the radial distance, $R$, from the origin. The Hamiltonian becomes

$$\mathcal{H}_1(x, y) = \frac{1}{2} \bar{\delta}(x^2 + y^2) + \frac{1}{4}(x^2 + y^2)^2 - 2x.$$
(8.131)

From Eq. (8.125) the approximate relationships between $R$ and $e$ is

$$R \approx \left[ \frac{-3}{f_{\mathrm{d}}}(j - 1)^{\frac{4}{3}} j^{\frac{2}{3}} \frac{m_{\mathrm{c}}}{m'} \right]^{\frac{1}{3}} e,$$
(8.132)

and the radial distance from the origin in $(x, y)$ space is simply a scaled value of the eccentricity of the perturbed object.

Now consider the case of an external, $j : (j - 1)$ first-order resonance. The resonant argument is

$$\theta = j\lambda' + (1 - j)\lambda - \varpi'$$
(8.133)

and the associated $f_{\mathrm{d}}$ term is given by the second entry in Table 8.1. We refer to this as the $e'$ *resonance* because the associated term in the disturbing function is of $\mathcal{O}(e')$. In the special case of $j = 2$ we need to include the indirect term $f_{\mathrm{e}}$ (the first entry Table 8.3); this is best done by absorbing it in $f_{\mathrm{d}}$. We have already noted that the relevant Hamiltonian, $\mathcal{H}'$, has a form identical to $\mathcal{H}$, with suitable, minor modifications to the constants. The resulting approximate relationship between $R$ and $e'$ for the case where $m' \ll m$ is

$$R \approx \left[ \frac{-3}{f_{\mathrm{d}}} j^2 \frac{m_{\mathrm{c}}}{m} \right]^{\frac{1}{3}} e'.$$
(8.134)

Therefore the theory developed for the $e$ resonance can also be applied to the $e'$ resonance with a suitable change of resonant angle and scaling.

In the previous section we showed how the radial locations of the equilibrium points of the system can be found for a $k$th-order resonance. By solving $\partial \mathcal{H}_1 / \partial x = 0$ and $\partial \mathcal{H}_1 / \partial y = 0$ it is easy to show that the equilibrium points lie along the $x$ axis and are given by the solution of the cubic equation

$$x^3 + \bar{\delta}x - 2 = 0.$$
(8.135)

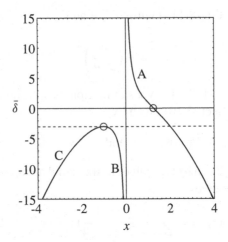

Fig. 8.9. A plot of $\bar{\delta} = (2-x^3)/x$ illustrating the relationship between the value of $\bar{\delta}$ and the position on the $x$ axis of the equilibrium points for $e$ and $e'$ resonances. Branches A and B give rise to stable equilibrium points; branch C gives rise to an unstable equilibrium point. The intersection of the left-hand branch of the curves and the dashed line given by $\bar{\delta} = -3$ denotes the location of the bifurcation point (circled) where the stable and unstable branches meet. The intersection point associated with exact resonance at $\bar{\delta} = 0$ is also circled.

Here, of course, $x$ can be positive or negative whereas before we were interested in the radial distance, $R$, of the equilibrium points. The roots of this cubic give the $x$ values of equilibrium points of $\mathcal{H}_1$ for a given value of $\bar{\delta}$. Figure 8.9 shows $\bar{\delta}$ as a function of $x$. Note that for small $|x|$ this equation becomes $\bar{\delta}x - 2 \approx 0$ whereas for large $|x|$ we have $x^2 + \bar{\delta} \approx 0$. The solutions of Eq. (8.135) are

$$x_1 = \frac{-3^{\frac{1}{3}}\bar{\delta} + \Delta^{\frac{2}{3}}}{3^{\frac{2}{3}}\Delta^{\frac{1}{3}}}, \tag{8.136}$$

$$x_{2,3} = \frac{(3^{\frac{1}{3}} \pm 3^{\frac{5}{6}}\,\mathrm{i})\bar{\delta} + (-1 \pm \sqrt{3}\,\mathrm{i})\Delta^{\frac{2}{3}}}{3^{\frac{2}{3}}2\Delta}, \tag{8.137}$$

where $\mathrm{i} = \sqrt{-1}$ and

$$\Delta = 9 + \sqrt{3}\sqrt{27 + \bar{\delta}^3}. \tag{8.138}$$

Substituting $k = 1$ in Eq. (8.124) gives $\bar{\delta}_b = -3$ for the value of $\bar{\delta}$ at which the bifurcation occurs. This point and the equilibrium point associated with exact resonance are circled in Fig. 8.9. It is clear from its definition that $\Delta$ is always real for $\bar{\delta} \geq -3$ and this always gives one real root, $x_1$. For $\bar{\delta} < -3$ there are always three real roots, even though the quantity $\Delta$ becomes complex.

We can illustrate the dependence on $\bar{\delta}$ by plotting the curves of constant Hamiltonian for various values of $\bar{\delta}$. The initial values of the semi-major axis, eccentricity, and $\phi$ determine the Hamiltonian (a constant of the system) and

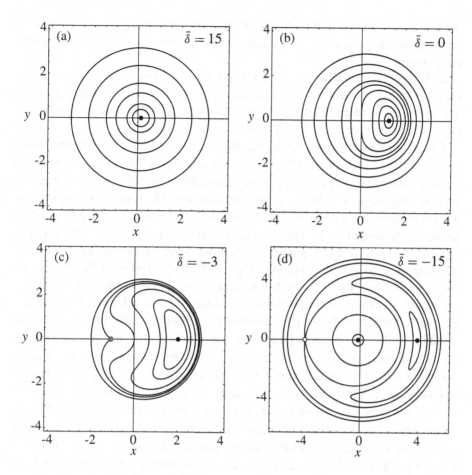

Fig. 8.10. Curves of constant Hamiltonian for the $e$ and $e'$ resonances. (a) $\bar{\delta} = 15$ with $\mathcal{H}_1 = 1, 3, 10, 20, 50, 100$. (b) $\bar{\delta} = 0$ (exact resonance) with $\mathcal{H}_1 = -9/5, -3/2, -1, 0, 1, 2, 5, 10, 20$. (c) $\bar{\delta} = -3$ (the bifurcation value) with $\mathcal{H}_1 = -5, -4, -2, 0, 3/4, 2$. (d) $\bar{\delta} = -15$ with $\mathcal{H}_1 = -63, -55, -48.57189, -20, -1$.

the particular value of $\bar{\delta}$ for an individual orbit. The orbit will then remain on that curve for all time. In Fig. 8.10 the locations of the stable and unstable equilibrium points are marked by small, filled and unfilled circles, respectively. For $\bar{\delta} = 15$ (i.e., large and positive), there is only one equilibrium point situated close to, but just to the right of, the origin in the positive $x$ direction (Fig. 8.10a). In this case there is only one type of motion for most eccentricities (recalling that the radial coordinate in Fig. 8.10 is a scaled eccentricity), although for very small eccentricities it is possible that the angle $\phi$, directly related to the resonant angle, is librating about the stable equilibrium point at $\phi = 0$.

The situation at exact resonance ($\bar{\delta} = 0$) is shown in Fig. 8.10b. Here the equilibrium point is well to the right of the origin. Note that although some of the trajectories in this plot represent libration in resonance (i.e., the curve does not enclose the origin) the larger curves of constant $\mathcal{H}_1$ are circulating rather than librating.

In the special case where $\bar{\delta} = -3$ a cusp forms (see Fig. 8.10c), indicating the presence of a double root at $x = -1$ in addition to the single root with positive $x$. The small, crossed circle in Fig. 8.10c indicates the location of the bifurcation point where the stable and unstable branching occurs.

As $\bar{\delta}$ is decreased further we see that the existence of three equilibrium points is now obvious (see Fig. 8.10d). The first, corresponding to the C branch of the left-hand curve in Fig. 8.9, is always unstable while the second (which approaches the origin as $\bar{\delta}$ decreases) is always stable. The third point, the one that always exists for any value of $\bar{\delta}$, now moves in the positive $x$ direction and the stable orbits around it describe a characteristic "banana" shape, all with large eccentricity and, for large-amplitude librations, large excursions in eccentricity. It is important to note that although Figs. 8.10a and 8.10d show the trajectories for large $|\bar{\delta}|$ values that are equidistant from exact resonance, they illustrate very different types of resonant phenomena. At this stage it is helpful to recall that the radial distance in Fig. 8.10 is a scaled eccentricity and the radial location of a stable equilibrium point in these diagrams allows us to calculate the value of the forced eccentricity at an equilibrium point. An interesting fact is that the curves of constant Hamiltonian derived directly from the unexpanded Hamiltonian given in Eq. (8.98) are identical in form to those shown in Fig. 8.10. However, a new bifurcation occurs for large negative values of $\bar{\delta}$ resulting in (i) the appearance of an unstable equilibrium point at $\theta = 0$ close to the origin and (ii) the disappearance of the stable equilibrium point at $\theta = \pi$. This was found by Wisdom (1980) and is relevant to the motion of objects with large eccentricity.

This work has direct applications to the study of planetary rings, where eccentricities are typically very small ($e \sim 10^{-6}$). Figure 8.10 shows that the shape of the curves depends on the value of $\bar{\delta}$. Although there are fundamental differences between Figs. 8.10a and 8.10d there are similarities in the vicinity of the origin. For small $x$ and large $|\bar{\delta}|$ in Eq. (8.135) the $x^3$ term can be neglected. Therefore, for a given $\bar{\delta}$ the $x$ value of the equilibrium point in the vicinity of the origin is given by $x \approx 2/\bar{\delta}$. Figure 8.11 shows additional curves of constant Hamiltonian for $\bar{\delta} = -15$ (Fig. 8.11a) and $\bar{\delta} = +15$ (Fig. 8.11b). Note that both show concentric circles displaced to either side of the origin. The centre of each circle is at $x \approx \mp 2/15$ in these cases. Therefore the traversal of the resonance introduces a shift of 180° in $\phi$. Note that the forced eccentricity is a simple function of the distance from resonance. The mechanism illustrated in Fig. 8.11, and the phase shift on either side of exact resonance, are the basis of our understanding of the first-order Lindblad resonances that occur in planetary ring systems (see Sect. 10.3.2).

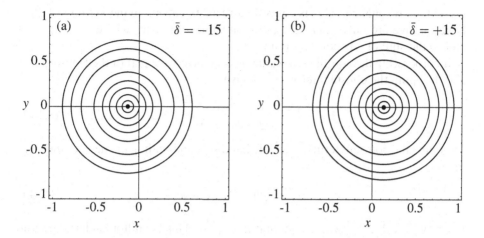

Fig. 8.11. The curves of constant Hamiltonian for the $e$ and $e'$ resonances in the case where the eccentricity is small and the system is not close to exact resonance (i.e., large values of $|\bar{\delta}|$). (a) The curves for $\bar{\delta} = -15$ with $\mathcal{H}_1 = -4, -3, -2, -1, -0.5, -0.2, 0,$ $0.1$. (b) The curves for $\bar{\delta} = 15$ with $\mathcal{H}_1 = -0.1, 0, 0.2, 0.5, 1, 2, 3, 4, 5$. Note the difference in scale from Figs. 8.10a and 8.10d.

The same resonant equations, but in a different context, are also applicable to the study of the resonant dynamics of asteroids, where $e \approx 0.1$. In fact, as we show below, the pendulum model provides an adequate representation of the dynamics. We can consider the Hamiltonian for the first-order resonance (Eq. (8.128)) as being composed of two parts. We write

$$\mathcal{H}_1 = \mathcal{H}_0 + \mathcal{H}_{\text{res}}, \tag{8.139}$$

where

$$\mathcal{H}_0 = \bar{\delta}\Phi + \Phi^2 \quad \text{and} \quad \mathcal{H}_{\text{res}} = -2\sqrt{2\Phi}\cos\phi. \tag{8.140}$$

Here we are treating the resonant term, $\mathcal{H}_{\text{res}}$, as a perturbation to $\mathcal{H}_0$. The resonant value of the eccentricity given by the $\mathcal{H}_0$ system alone has only small variations. We can find the value of $\Phi$ (and hence the eccentricity) at exact resonance by solving for $\Phi$ when $\dot{\phi} = 0$. Setting $\partial\mathcal{H}_1/\partial\Phi = 0$ gives

$$\bar{\delta} + 2\Phi_{\text{res}} - \frac{2}{\sqrt{2\Phi_{\text{res}}}}\cos\phi = 0. \tag{8.141}$$

However, we are assuming that $e$ is large and hence we can ignore the $\cos\phi$ term in this equation, obtaining

$$\Phi_{\text{res}} = \frac{1}{2}|\bar{\delta}| \tag{8.142}$$

for the resonant value. Because the resonant term does not vary significantly with respect to $\Phi$ in the vicinity of the resonance, we replace $\Phi$ in $\mathcal{H}_{\text{res}}$ by its value,

$\Phi_{\text{res}}$, at the resonance. This is equivalent to an expansion in the vicinity of $\Phi_{\text{res}}$. Note that we are only interested in the $\Phi$ values for cases where $\bar{\delta} < \bar{\delta}_b = -3$ and so the sign of $\bar{\delta}$ is not important.

Incorporating the equilibrium value of $\Phi$ in the cosine part of the Hamiltonian and dropping the constant term $-\bar{\delta}^2/4$ gives

$$\mathcal{H}_1 = \left(\Phi + \frac{1}{2}|\bar{\delta}|\right)^2 - 2\sqrt{|\bar{\delta}|}\cos\phi. \tag{8.143}$$

Dropping the constant term on the left-hand side and scaling the new momentum gives

$$\tilde{\mathcal{H}}_1 = \frac{1}{2}\Psi^2 - 2K\cos\psi, \tag{8.144}$$

where $\Psi = \sqrt{2}(\Phi + \frac{1}{2}|\bar{\delta}|)$, $\psi = \phi$, and $K = \sqrt{|\bar{\delta}|}$. This Hamiltonian has the same form as that of a simple pendulum. Therefore we can use the method shown in Sect. 8.6 to derive an expression for the change in $\Psi$ for the Hamiltonian associated with maximum libration. This gives $\Delta\Psi_{\text{max}} = 2\sqrt{2K}$ or, equivalently, $\Delta\Phi_{\text{max}} = 2\sqrt{K} = 2\left(|\bar{\delta}|\right)^{\frac{1}{4}}$. But $e$ is related to $\Phi$ by

$$S^2 e^2 = 2\Phi, \tag{8.145}$$

where

$$S = \left[\frac{-3}{f_d}(j-1)^{\frac{4}{3}}j^{\frac{2}{3}}\frac{m_c}{m'}\right]^{\frac{1}{3}}. \tag{8.146}$$

Differentiating and setting $\Delta\Phi$ equal to its maximum value gives

$$2e\,\Delta e = (4/S^2)\left(|\bar{\delta}|\right)^{\frac{1}{4}}. \tag{8.147}$$

We can use the same scaling relationship between $e$ and $\Phi$ to relate $|\bar{\delta}|$ to $e_{\text{res}}$. We have

$$\sqrt{2\Phi_{\text{res}}} = S\,e_{\text{res}} = \sqrt{|\bar{\delta}|}. \tag{8.148}$$

Substituting for $|\bar{\delta}|$ in Eq. (8.147) yields

$$2e\,\Delta e = 4S^{-\frac{3}{2}}\sqrt{e_{\text{res}}}. \tag{8.149}$$

From our discussion of the relationship between $\Theta_3$ and the Jacobi constant we showed that $\dot{a}$ and $\dot{e}$ are related by Eq. (8.92). Taking $k = 1$ in this equation (we are only considering first-order resonances) gives

$$\frac{\delta a}{a} = 2(1-j)e\,\Delta e \tag{8.150}$$

and hence

$$\Delta a_{\text{max}} = \pm\left(\frac{16}{3}\frac{|\mathcal{C}_r|}{n}e_{\text{res}}\right)^{\frac{1}{2}}a, \tag{8.151}$$

where we have used the relations $C_r = (m'/m_c)n\alpha f_d$ and $\alpha = [(j-1)/j]^{\frac{2}{3}}$. The resulting expression for $\Delta a_{max}$ is identical to that given in Eq. (8.58) where we first considered the pendulum model.

The pendulum approach can be used to understand the dynamics of asteroids at resonance. In the context of planetary ring dynamics the $e'$ resonance is called a *first-order corotation resonance* (see Sect. 10.3.1), while the $e$ resonance is called a *first-order Lindblad resonance* (see Sect. 10.3.2). Although we have restricted our analysis to first-order resonances, with minor modifications the pendulum model is equally applicable to all subsequent resonances discussed in this chapter.

### 8.8.2 The $e^2$, $e'^2$, $I^2$, and $I'^2$ Resonances

Consider the case of an internal, $j : (j-2)$ second-order resonance where the single resonant argument is now

$$\theta = j\lambda' + (2-j)\lambda - 2\varpi \tag{8.152}$$

and the associated $f_d$ term is given by the third term in Table 8.1. We refer to this as the $e^2$ *resonance* because the associated term in the disturbing function is of $\mathcal{O}(e^2)$. From Eq. (8.116) with $k = 2$ the Hamiltonian of the system is given by

$$\mathcal{H}_2 = \delta\Phi + \Phi^2 + 4\Phi \cos 2\phi, \tag{8.153}$$

where $\phi$ is given by Eq. (8.113) with $\theta_1 = \theta$ and

$$\bar\delta = 2\bar\alpha/\bar\epsilon \tag{8.154}$$

is a measure of the proximity to exact resonance. The definitions of the constants $\bar\alpha$ and $\bar\epsilon$ are given in Eqs. (8.103) and (8.105). Note that in this case the expression for $\bar\delta$ does not involve $\bar\beta$. In terms of $x$ and $y$ defined in Eq. (8.130) we have

$$\mathcal{H}_2 = \frac{1}{2}\bar\delta\left(x^2 + y^2\right) + \frac{1}{4}\left(x^2 + y^2\right)^2 + 2\left(x^2 - y^2\right). \tag{8.155}$$

Substitution of $k = 2$ in Eq. (8.125) gives the approximate scaling relationship between $R = \sqrt{2\Phi}$ and $e$ as

$$R \approx \left[\frac{3}{4f_d}(j-2)^{\frac{4}{3}}j^{\frac{2}{3}}\frac{m_c}{m'}\right]^{\frac{1}{2}} e. \tag{8.156}$$

As with first-order resonances we can also consider the case of an external, $j : (j-2)$ second-order resonance. Here the single resonant argument is

$$\theta = j\lambda' + (2-j)\lambda - 2\varpi' \tag{8.157}$$

and the corresponding expression for $f_d$ is the fifth entry in Table 8.1. We refer to this as the $e'^2$ *resonance* because the associated term in the disturbing function is of $\mathcal{O}(e'^2)$. In the case when $j = 2$ an indirect term given by the second entry in Table 8.3 has to be incorporated in $f_d$. The scaling relationship is

$$R \approx \left[ \frac{3}{4 f_d} j^2 \frac{m_c}{m} \right]^{\frac{1}{2}} e'. \tag{8.158}$$

Therefore the theory developed for the $e^2$ resonance can also be applied to the $e'^2$ resonance with a suitable change of resonant angle and scaling.

Although so far we have only considered resonances involving the eccentricity of either body, we can now extend the work to include inclination resonances. Because inclinations always occur in even powers in the expansion of the disturbing function there is no such thing as a first-order inclination resonance. However, when we consider an expansion to second order we have already noted (see, for example, Sect. 6.10.1) that there are three possible resonant arguments involving ascending nodes. Two of these can be investigated using the Hamiltonian theory for the $e^2$ and $e'^2$ resonances.

Consider the case of an internal, $j : (j - 2)$ second-order resonance where the single resonant argument is now

$$\theta = j\lambda' + (2 - j)\lambda - 2\Omega \tag{8.159}$$

and the expression for $f_d$ is given by the sixth term in Table 8.1. We refer to this as the $I^2$ *resonance*. From the definitions of the Poincaré variables in Eq. (2.176) we have

$$Z = m\sqrt{\mathcal{G}m_c a(1 - e^2)}(1 - \cos I) \approx m\sqrt{\mathcal{G}m_c a}\, 2s^2, \tag{8.160}$$

where $s = \sin \frac{1}{2} I$. Hence

$$I^2 \approx \frac{2Z}{\Lambda}, \tag{8.161}$$

and so the Hamiltonian has the same form as the general one given in Eq. (8.153), which was derived for the purely eccentric resonances. Consequently, the scaling developed for second-order eccentricity resonances based on that Hamiltonian is equally applicable to the $I^2$ resonance provided we replace $e$ by $I$ in Eq. (8.156) and it is understood that the constant $f_d$ now refers to the sixth term in Table 8.1.

The other inclination resonance that can be handled by the same method has the argument

$$\theta = j\lambda' + (2 - j)\lambda - 2\Omega', \tag{8.162}$$

and the expression for $f_d$ is given by the eighth term in Table 8.1. We refer to this as the $I'^2$ *resonance*. The scaling law is that given in Eq. (8.158) with $e'$ replaced by $I'$ and $f_d$ now referring to its $I'^2$ value in Table 8.1.

We are now in a position to examine the dynamics of the $e^2$, $e'^2$, $I^2$, and $I'^2$ resonances. From Eq. (8.121) with $k = 2$ we know that the radial distances of the equilibrium points (apart from the origin itself) are given by the real, positive values of

$$R = \sqrt{\pm 4 - \bar{\delta}} \qquad (8.163)$$

and that the bifurcation points occur when $\bar{\delta} = \pm 4$. A selection of curves of constant Hamiltonian for six different values of $\bar{\delta}$ are shown in Fig. 8.12. As with Fig. 8.10 the locations of the stable and unstable equilibrium points are marked by small, filled and unfilled circles, respectively. Bifurcation points are marked with a crossed circle.

For $\bar{\delta} > 4$ (see Fig. 8.12a) there is a single stable equilibrium point at the origin. As $\bar{\delta}$ decreases the curves narrow until at $\bar{\delta} = 4$ the first bifurcation point is reached (see Fig. 8.12b). The bifurcation leads to the formation of an unstable point at the origin and two stable equilibrium points equidistant from the origin along the $y$ axis. This is illustrated in Fig. 8.12c for the case of exact resonance. A further bifurcation occurs when $\bar{\delta} = -4$ (see Fig. 8.12d). At this stage the origin becomes a stable point and two unstable points are created along the $x$ axis. This is illustrated in Fig. 8.12e for $\bar{\delta} = -7$. There are no more bifurcations

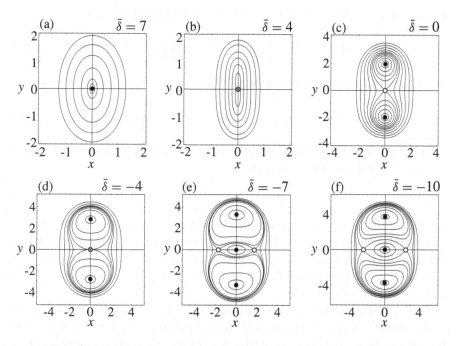

Fig. 8.12. The curves of constant Hamiltonian for the $e^2$, $e'^2$, $I^2$, and $I'^2$, resonances. (a) $\bar{\delta} = 7$ with $\mathcal{H}_2 = 0.05, 0.2, 1, 3, 6, 10$. (b) $\bar{\delta} = 4$ (a bifurcation value) with $\mathcal{H}_2 = 0.04, 0.2, 0.5, 1, 2, 3$. (c) $\bar{\delta} = 0$ (exact resonance) with $\mathcal{H}_2 = -3.7, -3, -2, -1, 0, 1, 3, 6, 10, 15$. (d) $\bar{\delta} = -4$ (a bifurcation value) with $\mathcal{H}_2 = -15, -8, -3, 0, 2, 4, 10, 20$. (e) $\bar{\delta} = -7$ with $\mathcal{H}_2 = -28, -20, -8, -3, -9/4, -1, 4, 10$. (f) $\bar{\delta} = -10$ with $\mathcal{H}_2 = -48, -40, -30, -20, -15, -9, -4, -1, 10$.

and so further reductions in $\bar{\delta}$ do not change the fundamental character of the system (see Fig. 8.12f).

### 8.8.3 The $e^3$ and $e'^3$ Resonances

It is relatively easy to extend our analysis to selected third-order resonant terms. Here we outline the method involved for two third-order arguments. For the case of the $e^3$ *and* $e'^3$ *resonances* the resonant arguments are

$$\theta = j\lambda' + (3 - j)\lambda - 3\varpi \quad \text{and} \quad \theta = j\lambda' + (3 - j)\lambda - 3\varpi' \tag{8.164}$$

respectively. These can be identified as the arguments 4D3.1 and 4D3.4 in Appendix B, and the relevant $f_d$ terms can be calculated using the definitions of $f_{82}$ and $f_{85}$ in Table B.12. The Hamiltonian of the system is given by

$$\mathcal{H}_3 = \bar{\delta}\Phi + \Phi^2 - 2(2\Phi)^{\frac{3}{2}} \cos 3\phi, \tag{8.165}$$

where

$$\bar{\delta} = 4\bar{\alpha}\bar{\beta}/\bar{\epsilon}^2. \tag{8.166}$$

In terms of $x = \sqrt{2\Phi}\cos\phi$ and $y = \sqrt{2\Phi}\sin\phi$ the Hamiltonian is

$$\mathcal{H}_3 = \frac{1}{2}\bar{\delta}\left(x^2 + y^2\right) + \frac{1}{4}\left(x^2 + y^2\right)^2 - 2x\left(x^2 - 3y^2\right). \tag{8.167}$$

The approximate scaling relationships between $R = \sqrt{2\Phi}$ and $e$ and $e'$ are

$$R \approx \left[\frac{-1}{3f_d}(j - 3)^{\frac{4}{3}}j^{\frac{2}{3}}\frac{m_c}{m'}\right]e \quad \text{and} \quad R \approx \left[\frac{-1}{3f_d}j^2\frac{m_c}{m}\right]e'. \tag{8.168}$$

The radial distances of the equilibrium points (apart from the origin itself) are given by the real, positive values of

$$R = \pm 3 \pm \sqrt{9 - \bar{\delta}}, \tag{8.169}$$

and the bifurcation points occur when $\bar{\delta} = 0$ and $9$. A selection of curves of constant Hamiltonian for six different values of $\bar{\delta}$ are shown in Fig. 8.13. As with Fig. 8.10 the locations of the stable and unstable equilibrium points are marked by small, filled and unfilled circles, respectively. Bifurcation points are marked with a crossed circle.

For $\bar{\delta} > 9$ (see Fig. 8.13a) there is a single stable equilibrium point at the origin with the curves following a triangular pattern around it. As $\bar{\delta}$ decreases the sides of the triangle narrow until at $\bar{\delta} = 9$ three bifurcation points occur (see Fig. 8.13b). The bifurcation leads to the formation of three pairs of stable and unstable equilibrium points at two distances from the origin, each pair 120° from the next. As $\bar{\delta}$ decreases the unstable points move towards the origin, reaching it at $\bar{\delta} = 0$. For this value of $\bar{\delta}$ (exact resonance; see Fig. 8.13c) a further bifurcation occurs. The origin remains a stable point and three unstable points are created

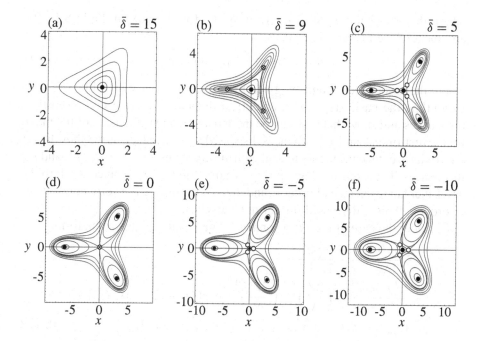

Fig. 8.13. The curves of constant Hamiltonian for the $e^3$ and $e'^3$ resonances. (a) $\bar{\delta} = 15$ with $\mathcal{H}_3 = 1, 5, 10, 20, 40$. (b) $\bar{\delta} = 9$ (a bifurcation value) with $\mathcal{H}_3 = 1, 4, 27/4, 15/2, 10, 15, 25, 40$. (c) $\bar{\delta} = 5$ with $\mathcal{H}_3 = -25, -10, 3/4, 10, 25, 50$. (d) $\bar{\delta} = 0$ (exact resonance and a bifurcation value) with $\mathcal{H}_3 = -100, -50, -10, 0, 20, 50, 100$. (e) $\bar{\delta} = -5$ with $\mathcal{H}_3 = -200, -100, -30, -10, \sqrt{14} - 3, 50, 100$. (f) $\bar{\delta} = -10$ with $\mathcal{H}_3 = -300, -200, -50, -10, \sqrt{19} - 3, 100, 300, 500$.

at intervals of $120°$. This is illustrated in Fig. 8.13e for $\bar{\delta} = -7$. Note that the small "triangles" near the origin are convex for $\bar{\delta} > 0$ and concave for $\bar{\delta} < 0$, with the unstable point moving from negative to positive values of $x$. There are no more bifurcations and so further reductions in $\bar{\delta}$ do not change the fundamental character of the system (see Fig. 8.13f).

### 8.8.4 *The ee' and II' Resonances*

So far we have avoided any resonances that have an argument involving the pericentres or nodes of both masses. We refer to these as *mixed resonances*. If we restrict ourselves to second-order resonances then there are two mixed resonant arguments to consider. The first of these is the *ee' resonance* with argument

$$\theta = j\lambda' + (2 - j)\lambda - \varpi' - \varpi, \tag{8.170}$$

where the associated $f_d$ term is given by the fourth entry in Table 8.1. The argument of the $II'$ *resonance* is

$$\theta = j\lambda' + (2 - j)\lambda - \Omega' - \Omega, \tag{8.171}$$

where the associated $f_d$ term is given by the seventh entry in Table 8.1. Here we outline the basic steps in an analytical theory to describe motion in these resonances and show that they can be treated as a simple extension of the theory for the first-order $e$ and $e'$ resonances discussed above. In fact, as we show below, it is trivial to extend the theory to deal with the higher-order $ee'^k$, $e^k e'$, $II'^k$, and $I^k I'$ resonances. In the terminology of ring dynamics the $ee'^k$ resonance is referred to as a Lindblad resonance of order $k$, and the $II'^k$ resonance is a vertical resonance of order $k$. This is discussed further in Chapter 10.

Consider the case of the $II'$ resonance. The Hamiltonian has the form

$$
\begin{aligned}
\mathcal{H} = &-\frac{\mathcal{G}^2 m_c^2 m^3}{2\Lambda^2} - \frac{\mathcal{G}^2 m_c^2 m'^3}{2\Lambda'^2} \\
&- \frac{\mathcal{G}^2 m_c m m'^3}{\Lambda'^2} f_d \left(\frac{2Z}{\Lambda}\right)^{1/2} \left(\frac{2Z'}{\Lambda'}\right)^{1/2} \cos[j\lambda' + (2-j)\lambda + z + z'] \\
&- \Gamma \dot{\varpi}_{\text{sec}} - Z \dot{\Omega}_{\text{sec}} + \Lambda \dot{\lambda}_{\text{sec}} - \Gamma' \dot{\varpi}'_{\text{sec}} - Z' \dot{\Omega}'_{\text{sec}} + \Lambda' \dot{\lambda}'_{\text{sec}}
\end{aligned} \tag{8.172}
$$

(cf. Eq. (8.77)), where $z = -\Omega$ and $z' = -\Omega'$. In this case $\Gamma$ and $\Gamma'$ do not appear in $\mathcal{H}$ and so they (and hence $e$ and $e'$) are constants of the motion. The new set of four variables is given by

$$\theta_1 = j\lambda' + (2 - j)\lambda + z' + z, \tag{8.173}$$
$$\theta_2 = j\lambda' + (2 - j)\lambda + 2z', \tag{8.174}$$
$$\theta_3 = \lambda, \tag{8.175}$$
$$\theta_4 = \lambda'. \tag{8.176}$$

The conjugate momenta, $\Theta_i$, can be found by equating coefficients in

$$\sum_{i=1}^{4} \Theta_i \, d\theta_i = \Lambda \, d\lambda + \Lambda' \, d\lambda' + Z \, dz + Z' \, dz'. \tag{8.177}$$

This gives

$$(2 - j)(\Theta_1 + \Theta_2) + \Theta_3 = \Lambda, \tag{8.178}$$
$$j(\Theta_1 + \Theta_2) + \Theta_4 = \Lambda', \tag{8.179}$$
$$\Theta_1 = Z, \tag{8.180}$$
$$\Theta_1 + 2\Theta_2 = Z', \tag{8.181}$$

where $\Theta_2$, $\Theta_3$, and $\Theta_4$ are all constants since $\theta_2$, $\theta_3$, and $\theta_4$ no longer appear in $\mathcal{H}$. The constant of interest is

$$\Theta_2 = \frac{1}{2}(Z' - Z). \tag{8.182}$$

Because $I^2 \approx 2Z/\Lambda$, the existence of $\Theta_2$ implies that the scaled *difference* in the inclinations is a constant of the motion. It is easy to show that the approximate Hamiltonian can now be written as

$$\mathcal{H} = \left[ j(n' + \dot{\lambda}'_{\text{sec}}) + (2-j)(n + \dot{\lambda}_{\text{sec}}) - 2\dot{\Omega}_{\text{sec}} \right] Z$$
$$- \frac{3}{2} \left[ \frac{j^2}{m'a'^2} + \frac{(2-j)^2}{ma^2} \right] Z^2$$
$$- n \, f_{\text{d}} \alpha \frac{m'}{m_{\text{c}}} (2Z)^{1/2} (2Z+C)^{1/2} \cos \theta_1, \tag{8.183}$$

where $C = 4\Theta_2$ is a constant. By carrying out a simple scale transformation with $\bar{\alpha} \to 2\bar{\alpha}$ and $\bar{\beta} \to 4\bar{\beta}$ we obtain

$$\mathcal{H} = \bar{\delta}\Phi + \Phi^2 + 2\sqrt{2\Phi}\sqrt{2\Phi + c} \cos \phi, \tag{8.184}$$

where $c = C/\eta$ and $\phi = \theta_1$.

Comparing this Hamiltonian with that given in Eq. (8.116) we see that it has characteristics of second-order resonances (the powers of $\Phi$ in the resonant term) as well as first-order resonances (the resonant angle occurs in the form $\phi$ and not $2\phi$). Furthermore, our new Hamiltonian is a function of two parameters: $\bar{\delta}$ is a measure of the distance to exact resonance and $c$ is related to the constant difference in inclination.

The preceding analysis can equally be applied to the $ee'$ resonance if we replace $Z$, $Z'$, $z$, and $z'$ by $\Gamma$, $\Gamma'$, $\gamma$, and $\gamma'$, respectively. In this case it is the difference in eccentricities that is a constant. Note too that we can easily modify the analysis to deal with the case where $I$ or $e$ rather than $I'$ or $e'$ is fixed.

The remaining analysis of the Hamiltonian (locating the equilibrium and bifurcation points) is similar to that carried out for the first-order resonances. Taking $R = \sqrt{2\Phi}$ the equilibrium points are located at the intersection of the two functions

$$f(\bar{\delta}, R) = \bar{\delta} + R^2 \quad \text{and} \quad g(R, c) = 2(-1)^{i+1} \left[ \frac{\sqrt{R^2+c}}{R} + \frac{R}{\sqrt{R^2+c}} \right], \tag{8.185}$$

where $i = 0, 1, 2, \dots$. Similarly, the bifurcation points occur where $f'(\bar{\delta}, R) = g'(R, c)$. If we make the substitutions $x = \sqrt{2\Phi} \cos \phi$ and $y = \sqrt{2\Phi} \sin \phi$ the Hamiltonian can be written as

$$\mathcal{H} = \frac{1}{2} \left( x^2 + y^2 \right) \bar{\delta} + \frac{1}{4} \left( x^2 + y^2 \right)^2 + 2x \left( x^2 + y^2 + c \right)^{1/2}, \tag{8.186}$$

with equilibrium points along the $x$ axis. Because the location of the equilibrium points depends on $c$ as well as $\bar{\delta}$, we have to consider plots of

$$\bar{\delta} = \frac{1}{x} \left( -x^3 - 2(x^2+c)^{\frac{1}{2}} - 2x^2(x^2+c)^{-1/2} \right) \tag{8.187}$$

and the curves of constant Hamiltonian for each individual value of $c$. In Fig. 8.14 we show plots of (i) $f(\bar{\delta}, R)$ and $g(R, c)$, (ii) $\bar{\delta}$ as a function of the $x$ value of the

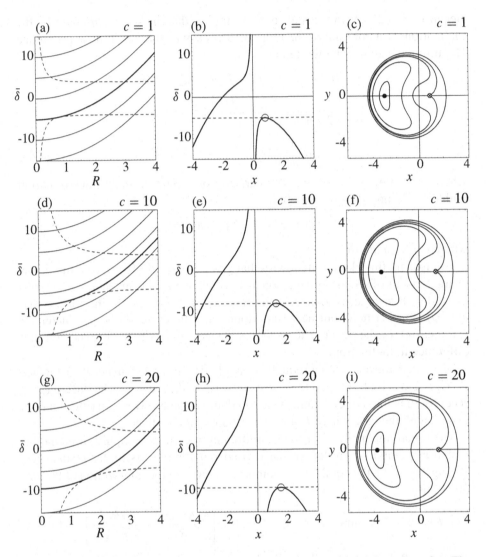

Fig. 8.14. The dynamics of the mixed resonance as a function of the value of $c$. The conventions used in Figs. 8.8, 8.9, and 8.10 apply. (a) Plots of $f(\bar{\delta}, R)$ (parabolas) and $g(R, c)$ (dashed curves) for $c = 1$. The thicker parabola denotes $f$ for the bifurcation value $\bar{\delta} = -5.09017$. (b) The relationship between $\bar{\delta}$ and the value of $x$ associated with the equilibrium points for $c = 1$. (c) Curves of constant Hamiltonian in the $x$–$y$ plane for $\bar{\delta} = -5.09017$ and $c = 1$. (d) Plots of $f(\bar{\delta}, R)$ and $g(R, c)$ for $c = 10$. The thicker parabola denotes $f$ for the bifurcation value $\bar{\delta} = -7.70166$. (e) The relationship between $\bar{\delta}$ and the value of $x$ associated with the equilibrium points for $c = 10$. (f) Curves of constant Hamiltonian for $\bar{\delta} = -7.70166$ and $c = 10$. (g) Plots of $f(\bar{\delta}, R)$ and $g(R, c)$ for $c = 20$. The thicker parabola denotes $f$ for the bifurcation value $\bar{\delta} = -9.1646$. (h) The relationship between $\bar{\delta}$ and the value of $x$ associated with the equilibrium points for $c = 20$. (i) Curves of constant Hamiltonian for $\bar{\delta} = -9.1646$ and $c = 20$.

equilibrium point (from Eq. (8.187)), and (iii) a selection of curves of constant Hamiltonian for the bifurcation value of $\bar{\delta}$, all for three different values of $c$. Note the general similarity to the plots shown in Figs. 8.8, 8.9, and 8.10. As the value of $c$ increases the bifurcation value of $\bar{\delta}$ becomes more negative (Figs. 8.14a,d,g), the equilibrium points move further from the origin (Figs. 8.14b,e,h), and the area enclosed increases (Figs. 8.14c,f,i).

So far we have only considered positive values of $c$. For negative values of $c$ an interesting phenomenon occurs. The presence of the $\sqrt{x^2 + y^2 + c}$ term in the Hamiltonian in Eq. (8.186) means that for negative values of $c$ the condition $R \geq \sqrt{-c}$ must hold. This places restrictions on the location of the equilibrium points and trajectories. This is illustrated in Fig. 8.15 for the case when $c = -5$. The general form of the various curves are the same as those shown in Fig. 8.14 but equilibrium points and trajectories cannot occur within a distance $R \leq \sqrt{5}$ of the origin.

The foregoing analysis can be extended to deal with the $ee'^k$, $e^k e'$, $II'^k$, and $I^k I'$ resonances. Consider the case of the $II'^k$ resonance. The angular variables are

$$\theta_1 = j\lambda' + (k + 1 - j)\lambda + kz' + z, \tag{8.188}$$

$$\theta_2 = j\lambda' + ((k + 1 - j)\lambda + (k + 1)z,' \tag{8.189}$$

$$\theta_3 = \lambda, \tag{8.190}$$

$$\theta_4 = \lambda'. \tag{8.191}$$

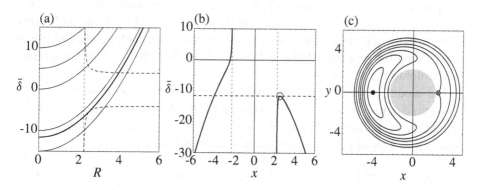

Fig. 8.15. The dynamics of the mixed resonance for $c = -5$. The conventions used in Fig. 8.14 apply. (a) Plots of $f(\bar{\delta}, R)$ (parabolas) and $g(R, c)$ (dashed curves). The thicker parabola denotes $f$ for the bifurcation value $\bar{\delta} = -11.6098$. The vertical dashed line denotes the asymptote at $R = \sqrt{5}$. (b) The relationship between $\bar{\delta}$ and the value of $x$ associated with the equilibrium points. The vertical dashed lines denote the asymptotes at $x = \pm\sqrt{5}$. (c) Curves of constant Hamiltonian in the $x$–$y$ plane for $\bar{\delta} = -11.6098$. The shaded area denotes the excluded region defined by the circle $x^2 + y^2 = 5$.

The conjugate momenta, $\Theta_i$, are

$$(k + 1 - j)(\Theta_1 + \Theta_2) + \Theta_3 = \Lambda, \tag{8.192}$$

$$j(\Theta_1 + \Theta_2) + \Theta_4 = \Lambda', \tag{8.193}$$

$$\Theta_1 = Z, \tag{8.194}$$

$$\Theta_1 + (k + 1)\Theta_2 = Z', \tag{8.195}$$

where the new constant of interest,

$$\Theta_2 = \frac{1}{k + 1}\left(Z' - Z\right), \tag{8.196}$$

again depends on the difference in the inclinations. The final Hamiltonian is

$$\mathcal{H} = \bar{\delta}\Phi + \Phi^2 + 2(-1)^k (2\Phi)^{1/2} (2\Phi + c)^{k/2} \cos\phi, \tag{8.197}$$

where $\phi = \theta_1$. It is clear that the analysis we have developed for the $II'$ resonance can be easily extended to the case of the $II'^k$ and other mixed resonances. However, in cases where $k$ is odd and $c$ is negative, there is always the possibility of excluded regions.

## 8.9 The 2:1 Resonance

We now consider some numerical examples that illustrate some of the concepts discussed in this chapter. We have carried out full numerical integrations of the equations of motion of the planar, circular, restricted, three-body problem for test particle orbits in the vicinity of the internal 2:1 resonance using a mass ratio of $m'/(m_c + m') = 0.001$. The results are based on the work of Winter & Murray (1997). In the regime we have selected the effects of the additional equilibrium point found by Wisdom (1980) and discussed in Sect. 8.8.1 are not important (see Colombo et al. 1968).

The model can be considered analogous to an investigation of the motion of asteroids close to 3.29 AU, the location of the strong, first-order, 2:1 jovian resonance in the asteroid belt. We have chosen a system of units where the semi-major axis of Jupiter is unity. From Table 8.5 we have

$$\alpha = 0.629961, \qquad \alpha f_{s,1} = 0.244190, \qquad \alpha f_d = -0.749964, \tag{8.198}$$

where $\alpha$ is the nominal value of $a/a'$ for the 2:1 resonance. We are dealing with a first-order, internal resonance and so Eq. (8.132) with $j = 2$ gives the appropriate scaling of the eccentricity. We have

$$\sqrt{2\Phi} = R = 15.874\,e, \tag{8.199}$$

where $e$ is the eccentricity of the test particle (or asteroid), and the only resonant

Table 8.6. *The initial values of semi-major axis* ($a_0$), *eccentricity* ($e_0$), *and longitude of perihelion* ($\varpi_0$) *for the numerical integrations shown in Fig. 8.16 and Fig. 8.17. The value of* $\lambda_0$ *was zero in each case. The initial orbital elements were calculated using the position and velocity with respect to an origin at the centre of mass of the system.*

| Plot | $a_0$ | $e_0$ | $\varpi_0$ | Description |
|------|-------|-------|------------|-------------|
| a | 0.625277 | 0.128386 | 0° | exact resonance |
| b | 0.633424 | 0.0725011 | 0° | medium-amplitude libration |
| c | 0.637837 | 0.0122862 | 0° | large-amplitude libration |
| d | 0.636705 | 0.060146 | 180° | apocentric libration |
| e | 0.638222 | 0.0184545 | 180° | inner circulation |
| f | 0.610592 | 0.1975 | 0° | outer circulation |

angle is

$$\phi = 2\lambda' - \lambda - \varpi . \tag{8.200}$$

For each of the six starting conditions given in Table 8.6 we have integrated the full equations of motion for 100 Jupiter periods and calculated the semi-major axis, eccentricity, longitude of pericentre, and the resonant argument as a function of time. The results are displayed in Fig. 8.16 where the data are plotted six times per Jupiter period. Note that in Fig. 8.16 the dashed line in each of the semi-major axis plots denotes the location of the nominal resonance. The equivalent plots of $x = \sqrt{2\Phi}\cos\phi$ and $y = \sqrt{2\Phi}\sin\phi$ are shown in Fig. 8.17. Again a point is plotted six times per Jupiter period. Note that we have chosen the starting conditions such that all orbits have the same value of the Jacobi constant, $C_J = 3.163$.

We now examine each of the plots in detail and comment on the various resonant and near-resonant phenomena that are illustrated.

### 8.9.1 Exact Resonance

Figures 8.16a and 8.17a show the trajectory of a test particle started close to the exact resonance. In the plot of $a$ as a function of time it is clear that the particle is staying close to the resonant value, while its eccentricity also maintains a constant value. Although the pericentre is regressing at a near-uniform rate (due to the resonant effects), the resonant angle is fixed close to $\phi = 0$. This is more clearly seen in Fig. 8.17a where the amplitude of libration is small. Note that although the test particle is close to exact resonance, the value of the semi-major axis given in Table 8.6, 0.625277, differs from the resonant value of 0.629961 because the former is calculated in the centre of mass frame whereas the latter is in the heliocentric frame.

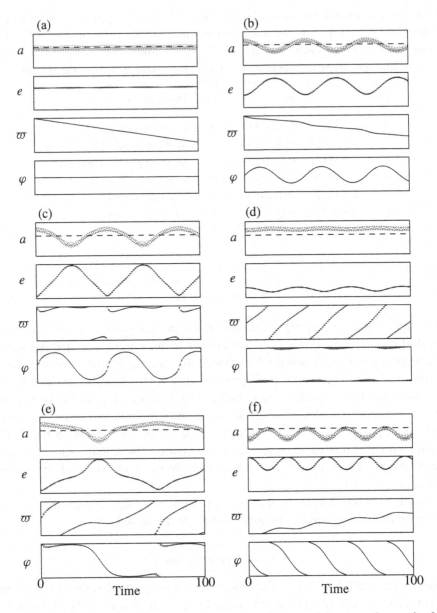

Fig. 8.16.   The variation of the orbital elements of test particles started near the 2:1 jovian resonance and integrated for 100 Jupiter periods. In each plot $a$ is the semi-major axis (on a scale from 0.6 to 0.65), $e$ is the eccentricity (on a scale from 0 to 0.2), $\varpi$ is the longitude of perihelion (on a scale from $0°$ to $360°$), and $\varphi = 2\lambda' - \lambda - \varpi$ is the resonant argument (on a scale from $-180°$ to $180°$). The starting values are given in Table 8.6. The plots illustrate (a) exact resonance, (b) medium-amplitude libration, (c) large-amplitude libration, (d) apocentric libration, (e) inner circulation, and (f) outer circulation.

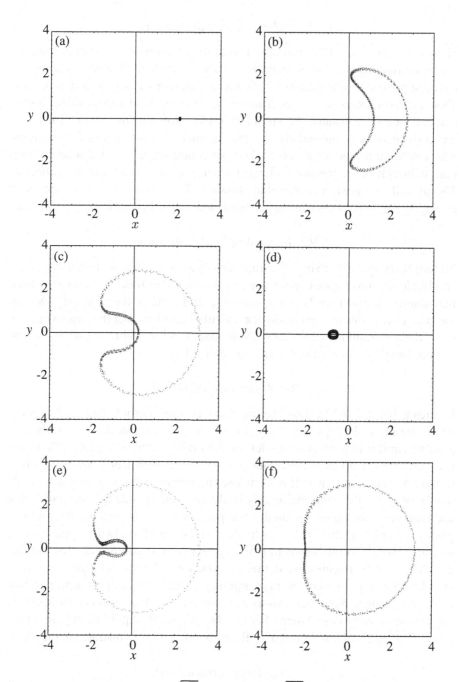

Fig. 8.17. The variation of $x = \sqrt{2\Phi}\cos\phi$ and $y = \sqrt{2\Phi}\sin\phi$ for the same trajectories that are shown in Fig. 8.16. The quantity $\sqrt{2\Phi}$ is defined in Eq. (8.199) and $\phi = 2\lambda' - \lambda - \varpi$ is the resonant angle. The plots illustrate (a) exact resonance, (b) medium-amplitude libration, (c) large-amplitude libration, (d) apocentric libration, (e) inner circulation, and (f) outer circulation.

### 8.9.2 Medium-Amplitude Libration

Figures 8.16b and 8.17b show the trajectory of a test particle undergoing a medium-amplitude libration about the exact resonance. The semi-major axis has an approximate amplitude of 0.0068 as it undergoes a long-period oscillation about the resonant value. The eccentricity also varies about a mean value close to that at the exact resonance shown in Fig. 8.16a. Note that the $a$ and $e$ variations are correlated as suggested by Eq. (8.92), such that when $a$ is a maximum, $e$ is a minimum, and vice versa. The pericentre regresses at a near-uniform rate, although the regression is largest when $e$ is at a minimum, as expected. The overall rate is slower than that shown in Fig. 8.16a. The resonant angle undergoes sinusoidal oscillations, as predicted by the pendulum approximation.

### 8.9.3 Large-Amplitude Libration

Figures 8.16c and 8.17c show the trajectory of a test particle undergoing a large-amplitude libration about the exact resonance. The libration is close to the maximum value and so the excursions in $a$ and $e$ are at their largest. Despite the large excursions in $e$ our numerical calculations show that the approximation $e^2 \approx 2\Gamma/\Lambda$ is valid to within 20%. Note that $\varpi$ is itself librating about 0. The resonant angle $\phi$ takes values in the range of $-146°$ to $+146°$.

### 8.9.4 Apocentric Libration

Figures 8.16d and 8.17d show the trajectory of a test particle undergoing apocentric libration. In Figs. 8.9 and 8.10 we see that, provided $\bar{\delta} < -3$, there is a stable equilibrium point to the left of the origin in the $x$–$y$ plots. Therefore it is possible for orbits started with $\phi = 180°$ and sufficiently low values of $e$ to have a trajectory that will not enclose the origin in the $x$–$y$ diagram. Note that in our case the semi-major axis is always well outside the resonant value and that the variation of $e$ is small. The pericentre was started at $180°$ and it is precessing quickly; this explains why the resonant angle $\phi$ always remains close to $180°$ – the $\varpi$ contribution to $\phi$ is large enough to compensate for the fact that $2n' - n \not\approx 0$. Because we define a resonance to be the libration rather than circulation of a particular resonant argument, a particle in apocentric libration is in a resonance. Note that a similar situation arises for small eccentricities at large positive values of $\bar{\delta}$ where there is always a stable equilibrium point to the right of the origin at low values of $e$. Both phenomena are illustrated in Fig. 8.11.

### 8.9.5 Inner Circulation

Figures 8.16e and 8.17e show the trajectory of a test particle undergoing inner circulation. It is clear from Fig. 8.17e that the trajectory encloses the origin and hence there is circulation of the resonant argument. However, the trajectory is still inside the critical curve that passes through the unstable equilibrium point close to $(-1.5, 0)$. Note that although the motion is circulatory, there are still large variations in $a$ and $e$, with magnitudes comparable to those at maximum libration of the resonant argument. The minimum $a$ and maximum $e$ occur when $\phi = 0$. When $\phi \approx 180°$ the trajectory is close to apocentric libration and for a while aspects of the plots resemble those seen in Fig. 8.16d and Fig. 8.17d.

### 8.9.6 Outer Circulation

Figures 8.16f and 8.17f show the trajectory of a test particle undergoing outer circulation. From the semi-major axis plot we see that $a$ is almost always less than the resonant value. Note that although the semi-major axis can take values on either side of the resonant value, it does not follow that the particle is in resonance; that must be determined from other considerations, the most important of which is the behaviour of the resonant angle. The variations in $a$ and $e$ are much smaller than those shown for inner circulation, primarily because the particle does not undergo a "detour" around the apocentric libration point. The pericentre is precessing and the resonant angle is clearly circulating.

### 8.9.7 Other Types of Motion

The orbits discussed above illustrate the basic types of motion to be found at the 2:1 and other first-order resonances. Other forms of motion include circulation about the single equilibrium point for large, negative values of $\bar{\delta}$ and circulation about the stable equilibrium point to the left of the origin for large, positive values of $\bar{\delta}$ (see Fig. 8.11) These and other types of orbits are not discussed here although their properties can easily be deduced from the information above.

### 8.9.8 Comparison with Analytical Theory

The validity of the Hamiltonian approach is illustrated in Fig. 8.18 where we compare the trajectories shown in Fig. 8.17 with those calculated analytically. One of the drawbacks of the Hamiltonian approach is that quantities such as $\bar{\delta}$, a measure of the proximity to exact resonance, have to be determined from observation. It is important to note that the analytical theory is an approximation to the motion, albeit a good one in certain circumstances. Numerical integration of the full equations of motion incorporates all the short-period effects and their influence is clearly seen in the numerical integrations; in contrast, the averaged Hamiltonian theory, by its very nature, ignores all short-period effects. This explains the basic differences between Fig. 8.18a and Fig. 8.18b. Each of the trajectories shown in Fig. 8.17 crosses the $y = 0$ line at two points, $x_1$ and $x_2$, which must have the same value of both $\bar{\delta}$ and $\mathcal{H}_1$. Setting $\mathcal{H}_1(x_1, 0) = \mathcal{H}_1(x_2, 0)$ in Eq. (8.131) gives

$$\bar{\delta} = \frac{1}{2}(x_2^2 + x_1^2) + 4(x_1 + x_2)^{-1} \tag{8.201}$$

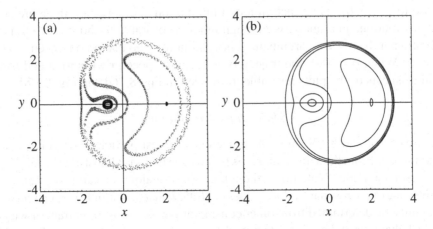

Fig. 8.18.   A comparison of the results of the numerical integrations with analytical theory. (a) The accumulated trajectories in $x$–$y$ space taken from Fig. 8.17. (b) The equivalent curves of constant Hamiltonian based on the analytical theory. The values of $\bar{\delta}$ and $\mathcal{H}_1$ for each curve are given in Table 8.7. In the case of the apocentric libration, the outer branch of the $\mathcal{H}_1$ curve has not been plotted. Note from Table 8.7 that all the trajectories have different values of $\bar{\delta}$ and $\mathcal{H}_1$.

for the approximate value of $\bar{\delta}$ for the trajectory. We can then find $\mathcal{H}_1$ by calculating $\mathcal{H}_1(x_1, 0)$ or $\mathcal{H}_1(x_2, 0)$. The resulting values of $\bar{\delta}$ and $\mathcal{H}_1$ are given in Table 8.7.

Note that the match between Figs. 8.18a and 8.18b is by no means perfect. There are two notable differences. The extent of the medium-amplitude libration has been underestimated by the analytical theory. The numerical results for the inner circulation show a single branch for the trajectory, whereas the analytical theory predicts motion on one of two branches, either a variation on apocentric libration or an outer circulation. Exact correspondence between the results from

Table 8.7. *The values of $\bar{\delta}$ and $\mathcal{H}_1$ that were used to produce the theoretical trajectories shown in Fig. 8.18b. The plot identification refers to Fig. 8.17.*

| Plot | $\bar{\delta}$ | $\mathcal{H}_1$ | Description |
|------|------|------|-------------|
| a | −3.62568 | −7.31775 | exact resonance |
| b | −3.57917 | −4.45861 | medium-amplitude libration |
| c | −3.69155 | −0.685164 | large-amplitude libration |
| d | −3.44688 | 0.562329 | apocentric libration |
| e | −2.38857 | 0.271609 | inner circulation |
| f | −3.58826 | 0.692563 | outer circulation |

different methods should not be expected, especially when the motion is close to a critical curve.

## 8.10 The 3:1 and 7:4 Resonances

To illustrate motion at other resonances we have carried out numerical integrations of the planar, circular, restricted three-body problem close to two other resonances.

At the 3:1 resonance the expected resonant argument is $\theta = 3\lambda' - \lambda - 2\varpi$. In units in which the perturber has unit semi-major axis the nominal resonance location is at $a_{\text{res}} = (1/3)^{2/3} = 0.48075$. Figure 8.19 shows the results of a single numerical integration of the equations of motion for starting conditions giving rise to librational motion in the resonance. The heliocentric starting values are $a_0 = 0.4809$, $e_0 = 0.13$, $\varpi_0 = 0$, and $\lambda_0 = 180°$ with $\lambda'_0 = 0$ and $\mu_2 = 0.001$. The value of $a_0$ is just greater than the nominal resonant value. Note from Fig. 8.19a that $\varpi$ is always increasing. The positive pericentre rate compensates for the fact that $3n' - n$ is also positive, such that the combination $\dot\theta = 3n' - n - 2\dot\varpi$ is close to zero. As expected from theory, stable libration is seen to occur around $\theta = 180°$. For these starting conditions the amplitude of libration is large ($\sim 130°$).

According to the Hamiltonian approach, the scaling factor $R$ for the $e^2$ resonance is given by Eq. (8.156) with $j = 3$. Using the value of $f_{\text{d}} = 0.598756$ for the 3:1 resonance derived from Table 8.5 gives $R = 51.044e$. A plot of the values

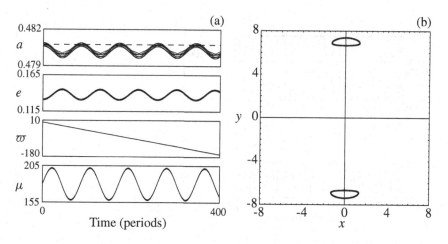

Fig. 8.19. Libration of a test particle in the 3:1 $e^2$ resonance in the planar circular restricted three-body problem. (a) Plots of the semi-major axis $a$, the eccentricity $e$, the longitude of pericentre $\varpi$, and the resonant angle $\theta = 3\lambda' - \lambda - 2\varpi$ as a function of time. (b) A plot of $x = \sqrt{2R}\cos\phi$ and $y = \sqrt{2R}\sin\phi$ (where $\phi = \theta/2$) for the same integration showing the two branches of the stable libration.

of $x = \sqrt{2R}\cos\phi$ and $y = \sqrt{2R}\sin\phi$ (where $\phi = \theta/2$) is shown in Fig. 8.19b. This shows the type of librational behaviour already seen in Fig. 8.12 for values of $\bar\delta < -4$

At the 7:4 resonance in the planar circular problem we expect the argument $\theta = 7\lambda' - 4\lambda - 3\varpi$ to be librating. The scaling relationship is given by the first part of Eq. (8.168) with $j = 5$ and, from Tables B.12 and B.14,

$$f_d = \frac{1}{48}\left[-1456 - 408\alpha D - 36\alpha^2 D^2 - \alpha^3 D^3\right]b_{1/2}^{(7)} = -3.86673. \qquad (8.202)$$

Using $\alpha = (4/7)^{\frac{2}{3}} = 0.688612$ gives $R = 2003.0e$. Figure 8.20 shows the evolution of the test particle's orbital elements and $(x, y)$ values for librational motion at the 7:4 resonance for a mass value of $\mu_2 = 0.001$. The heliocentric starting values are $a_0 = 0.691249$, $e_0 = 0.204339$, $\varpi_0 = 0.0$, $\lambda_0 = 0$, with $\lambda'_0 = 0$. It is clear from Fig. 8.20 that there are a number of additional short-period terms present in the orbital elements. Furthermore, although there are three clear centres of libration in Fig. 8.20b, the presence of the short-period terms causes significant departures from the expected trajectories shown in Fig. 8.13. This is due to the fact that the resonance lies between the 2:1 and 3:2 resonances where perturbations from other resonances cannot be avoided.

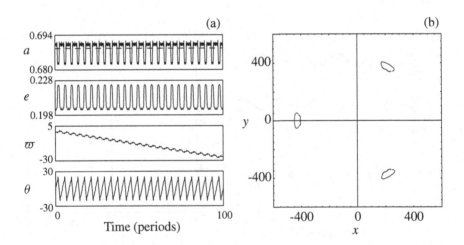

Fig. 8.20.   Libration of a test particle in the 7:4 $e^3$ resonance in the planar circular restricted three-body problem. (a) Plots of the semi-major axis $a$, the eccentricity $e$, the longitude of pericentre $\varpi$, and the resonant angle $\theta = 7\lambda' - 4\lambda - 3\varpi$ as a function of time. (b) Plot of $x = \sqrt{2R}\cos\phi$ and $y = \sqrt{2R}\sin\phi$ (where $\phi = \theta/3 + \pi$) for the same integration showing the three branches of the stable libration.

## 8.11 Additional Resonances and Resonance Splitting

In all of the numerical integrations carried out in Sects. 8.9 and 8.10 there was good agreement with the theoretical models derived from a Hamiltonian approach based on a truncated form of the disturbing function. This is exemplified by Fig. 8.18 where a direct comparison was made. The unavoidable presence of short-period terms in the full integrations (Fig. 8.18a) gives rise to the "fuzziness" of the $(x, y)$ curves when compared to those of constant Hamiltonian (Fig. 8.18b), but otherwise the comparison is good. However, there are circumstances where use of just a single resonant term from the disturbing function may not be appropriate.

So far we have considered resonances in isolation and yet at an arbitrary separation in semi-major axis from the perturber, basic number theory tells us that the mean motion can always be approximated by a rational number. Hence the perturbed object is always close to *some* resonance although, as we have seen, the strength depends on the order of the resonance. For example, if we restrict ourselves to first-order resonances then the separation in semi-major axis between the nominal locations of the consecutive $p+1:p$ and $p+2:p+1$ resonances is

$$\Delta a = \left[\left(\frac{p+1}{p+2}\right)^{2/3} - \left(\frac{p}{p+1}\right)^{2/3}\right] a' \approx \frac{2}{3}\frac{1}{p^2}a', \qquad (8.203)$$

where $a'$ is the semi-major axis of the perturber. Therefore, $\Delta a$ decreases as $p$ increases and the orbit of the external perturber is approached. Therefore, where the perturbed mass lies close to the perturbing mass, it may no longer be valid to consider just a single resonance. Given that each resonance has a finite libration width there will come a point where adjacent first-order resonances overlap. The consequences of this phenomenon are discussed in the next chapter. Note that if there are additional perturbing masses then we may also have to consider additional resonances.

Now consider the case where the particle's eccentricity is moderate to large. In such cases we would have to use expansions to higher order in $e$. Because of the relationship between the order of an expansion and the order of the resonant argument, the inclusion of higher-order terms has implications for the number of arguments to be considered. For example, at the 2:1 resonance we have considered the single resonant argument $\theta_1 = 2\lambda' - \lambda - \varpi$ and its associated term of first order in $e$. Although there *are* higher order terms in $e$ associated with this argument (see the entry for 4D1.1 in Table B.4) if we include them then for consistency we would need to include second- and higher-order arguments such as $\theta_2 = 4\lambda' - 2\lambda - 2\varpi$, $\theta_3 = 6\lambda' - 3\lambda - 3\varpi$, and $\theta_4 = 8\lambda' - 4\lambda - 4\varpi$ and their associated terms of order 2, 3, and 4 respectively. The exact locations of these resonances are determined by values of the semi-major axes for which the time derivative of the resonant argument is exactly zero. Because $\theta_2$, $\theta_3$, and $\theta_4$ are all simple multiples of $\theta_1$, exact resonance occurs at the same semi-major axis. The effect on the curves of constant Hamiltonian is to introduce additional equilibrium points as we change from first- to second- and higher-order resonances.

Further complications arise if we allow the perturber to move in an elliptical orbit in a different plane to the perturbed object. Many additional resonant arguments are now possible, with the location of each exact resonance depending on the particular combination of angles. For example, if our analysis is to second order in the orbital elements then at the 2:1 resonance we must consider the additional first-order argument $2\lambda' - \lambda - \varpi'$ as well as the additional second-order arguments $4\lambda' - 2\lambda - \varpi' - \varpi$, $4\lambda' - 2\lambda - 2\varpi'$, $4\lambda' - 2\lambda - 2\Omega$, $4\lambda' - 2\lambda - \Omega' - \Omega$, and $4\lambda' - 2\lambda - 2\Omega'$. Given that the exact locations of these resonances depend on the values of the quantities $\dot{\varpi}$, $\dot{\varpi}'$, $\dot{\Omega}$, and $\dot{\Omega}'$ it is clear that the resonances can be widely separated where these values are large. This is the phenomenon of *resonance splitting* and it is particularly important in satellite systems. Although it increases the number of resonances to be considered, if the resonances are sufficiently well separated then each can be treated individually, irrespective of the perturbing effects of the others.

Saturn's oblateness guarantees that the rates of pericentre precession and nodal regression are large for objects orbiting close to the planet. Figure 8.21 shows the location of the Mimas 6:4 and Tethys 3:1 resonances in the saturnian system, with $e$ and $I$ denoting the eccentricity and inclination of the object in resonance, with single and double primes denoting the equivalent values for Mimas and Tethys respectively. The resonant locations were calculated using the formulae

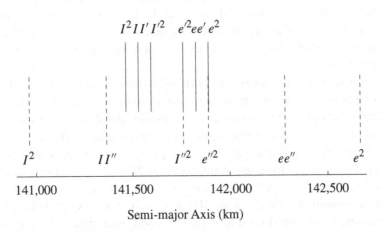

Fig. 8.21. Resonance splitting in the saturnian system illustrated by the locations in semi-major axis of the Mimas 6:4 (solid lines) and Tethys 3:1 (dashed lines) exact resonances. The symbols above and below each set of lines denote the appropriate terms in eccentricity and inclination associated with each argument.

in Sect. 6.11 and the values of $\varpi$, $\Omega$, $\varpi'$, and $\Omega'$ given by Harper & Taylor (1993). The Mimas 3:2 $e$ and $e'$ resonances are coincident with the $e^2$ and $e'^2$ resonances for the reasons explained above. Note that the resonances associated with each satellite are arranged in groups of three because of the resonance splitting. The rates of change of the longitudes of pericentre and node are always of opposite sign under the effect of the planet's $J_2$, resulting in the inclination resonances always lying interior to the eccentricity ones. The fact that Tethys is more distant than Mimas results in rates that are a factor five smaller ($\sim \pm 0.19°\text{d}^{-1}$ versus $\sim \pm 1°\text{d}^{-1}$). Note the near coincidence of the Tethys $I''^2$ and the Mimas $e'^2$ resonances, as well as the Tethys $e''^2$ and Mimas $e^2$ resonances. The reason for the general proximity of the resonances of these satellites is the fact that Mimas and Tethys are involved in a 4:2 resonance with each other (see Sect. 8.14.2 below). Some of the resonances shown in Fig. 8.21 may also have played a role in the orbital evolution of the satellite Pandora (see Sect. 10.11).

## 8.12 Resonant Encounters

In Chapter 4 we saw how the effect of tides raised on a planet by a satellite leads to an increase in the satellite's semi-major axis, provided that it is in a prograde orbit outside the synchronous location. We also showed that there is some evidence that the satellite systems of Saturn and Uranus have undergone significant tidal evolution over the age of the solar system. Indeed, such evolution is thought to be the reason why there are so many mean motion resonances between the satellites of Saturn (Goldreich 1965), although a primordial origin is more likely in some cases. The fact that there are no known resonances between the satellites of Uranus does not imply that there was no tidal evolution, since this absence can be explained by fundamental differences between the nature of resonances at Saturn and Uranus (see Chapter 9). Other uncertainties exist. For example, we do not know the rate of tidal evolution, because it depends on the tidal dissipation function of the planet, a poorly determined quantity for most solar system objects.

If significant tidal evolution has occurred in satellite systems, then we expect that there would have been occasions in the past when a resonant encounter occurred between a pair of satellites. Indeed, such encounters may have led to the observed preference for commensurabilities in some systems. Not all encounters would have resulted in resonant capture, although, as we shall see, the outcome would always have led to some change in the characteristics of the orbit. In Chapter 4 we saw how the major satellites of Saturn and Uranus are thought to have encountered a number of first- and second-order resonances during their dynamical evolution. It is important to note that although all these satellites are evolving outwards (i.e., their semi-major axes are increasing), relatively speaking some pairs are approaching one another and some are retreating. As we shall

see it is the relative direction of motion that determines whether or not capture
is possible in the classical problem.

One of the major advantages of the Hamiltonian approach to resonance introduced in Sect. 8.8 is that it improves our knowledge of the dynamics of resonance
encounter and provides a unified approach to the problem. It also demonstrates
that the *action* of an orbit can be used to predict the outcome of a resonant
encounter. For a given value of $\bar{\delta}$ the action of the orbit is defined as

$$J = \oint \Phi \, d\phi = \oint x \, dy \tag{8.204}$$

and this is just the area enclosed by the trajectory in the $(\Phi, \phi)$ or $(x, y)$ plots.
Provided the orbit is expanding or contracting sufficiently slowly (i.e., the change
in $\bar{\delta}$ over one cycle is negligible) the action is an *adiabatic invariant* of the
system and its value is a conserved quantity. Note that the curves of constant
Hamiltonian (i.e., the trajectories) far from the librational region for the first-
and second-order resonances can be closely approximated to circles centred on
the origin (see Figs. 8.10 and 8.11). Therefore the initial action, $J_0$, is simply
the area of that circle and hence

$$J_{\text{init}} = \pi (x_{\text{init}}^2 + y_{\text{init}}^2) = 2\pi \, \Phi_{\text{init}}, \tag{8.205}$$

where $x_{\text{init}}$, $y_{\text{init}}$, and $\Phi_{\text{init}}$ are the initial values of $x$, $y$, and $\Phi$, which are related
by Eqs. (8.130).

However, the adiabatic condition is not satisfied when the orbit evolves to
a position close to the resonant separatrix (see, for example, Lichtenberg &
Lieberman 1983). This is because the period of the librational or circulatory
motion goes to infinity as the separatrix is approached. After the separatrix has
been crossed the final action can once again become an approximate adiabatic
invariant of the motion. For example, if the object was captured into resonance
then the final action will be the area enclosed by the separatrix when it was
encountered. We can make use of these properties to predict the outcome of
resonant encounters and hence understand some of the orbital histories of natural
satellites and other solar system objects that have undergone significant orbital
evolution over the past $4.6 \times 10^9$ y.

Using this approach it is convenient to view the phenomenon as that of the
resonance encountering the orbit rather than the other way around. The orbit
and the area enclosed by it remain fixed as the curves of constant Hamiltonian
change around it following the increase or decrease in $\bar{\delta}$. Eventually a separatrix
encounters the orbit with the outcome depending on a number of factors; the end
result is a change in the nature of the orbit, in terms of its eccentricity and/or its
librational or circulatory character.

### 8.12.1 Encounters with First-Order Resonances

In order to understand the possible outcomes of resonant encounters we consider the evolution of several orbits as they evolve towards resonance. Following Peale (1986) we concentrate on encounters with first-order resonances. Here we change from the use of $\bar{\delta}$ (where $\bar{\delta} = 0$ denotes exact resonance) to $\delta$, where, for first-order resonances, the two quantities are related by

$$\delta = -\frac{1}{3}\bar{\delta} - 1 \tag{8.206}$$

and $\delta = 0$ now denotes the bifurcation value. This is in keeping with the notation of Peale (1986). If we consider the orbit of the external body to be fixed then the relationship between $\bar{\delta}$ and $\bar{\alpha}$ given in Eq. (8.117) implies that for $\delta < 0$ the inner body has a semi-major axis less than the resonant value. Similarly $\delta > 0$ implies a semi-major axis greater than the resonant value. Whether $\delta$ increases or decreases with time depends on how fast each mass is evolving in semi-major axis. As we shall see below, the key quantities are the initial value of $\delta$ and the sign of $\dot{\delta}$, that is, whether the two masses are approaching exact resonance at $\delta = 0$ from above ($\delta > 0$ and $\dot{\delta} < 0$) or below ($\delta < 0$ and $\dot{\delta} > 0$).

Figure 8.22 shows the evolution of an orbit that starts with a large negative value of $\delta$ (a large positive value of $\bar{\delta}$) and a relatively small value of the eccentricity, which, we recall, is proportional to the radial distance from the origin. The area enclosed by the trajectory is constant and equal to the initial action of the orbit (Figs. 8.22a,b). We see that when $\delta = 0$ (Fig. 8.22c, when the critical curve first exists) this area is less than the area of the critical curve (drawn for comparison); this is the condition for certain, but smooth, capture into resonant libration and it obviously depends on the initial area and hence on the initial eccentricity of the orbit. As $\delta$ increases beyond 0 (Figs. 8.22d,e) the orbit continues to evolve but now it is captured within the resonance region. In this case the action is still approximately equal to the initial value. However, as the stable equilibrium point moves further to the right with increasing $\delta$ (cf. Fig. 8.9), so the eccentricity of the captured orbit increases. From the properties of the curves

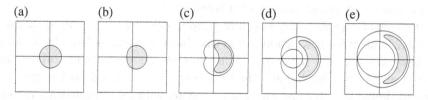

Fig. 8.22. Resonance encounter for increasing $\delta$ for a trajectory with an initial eccentricity smaller than the critical value. (a) $\delta = -6$, (b) $\delta = -1$, (c) $\delta = 0$, (d) $\delta = 1$, and (e) $\delta = 3$. The separatrix is shown for all values of $\delta \geq 0$.

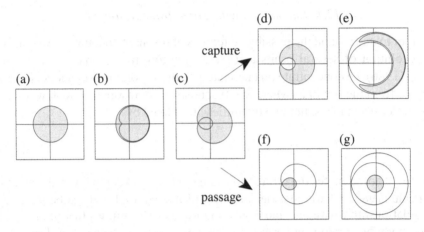

Fig. 8.23. Resonance encounter for increasing $\delta$ for a trajectory with an initial eccentricity larger than the critical value. In (a) $\delta = -6$ and in (b) $\delta = 0$. In (c), (d), and (f) $\delta = 0.485584$, the critical value, associated with the initial area enclosed by the trajectory. In (e) and (g) $\delta = 4$. The separatrix is shown for all values of $\delta \geq 0$.

shown in Fig. 8.10d we see that this evolution coincides with a decrease in the libration amplitude.

In Fig. 8.23 we show the evolution of an orbit with increasing $\delta$ for an initial enclosed area (see Fig. 8.23a) that is larger than that associated with the critical curve at $\delta = 0$. This orbit has an initial eccentricity that is larger than that shown in Fig. 8.22. Again the action is conserved but in this case when the resonant separatrix is finally encountered beyond the $\delta = 0$ value at $\delta = 0.485584$, there is a discontinuous change in the action. There are two possible outcomes, each depending critically on the exact conditions at separatrix crossing, and there is only a finite probability of capture. In the first case (Figs. 8.23d,e) capture into resonance occurs and the action changes to the area enclosed between the two branches of the encountered separatrix (Fig. 8.23d); there is no change in the eccentricity associated with the capture process. As $\delta$ increases further the resonance continues and a process similar to that shown in Fig. 8.22d,e occurs; conservation of the new action implies a gradual increase in eccentricity and a corresponding decrease in libration amplitude.

If resonant capture had not occurred during the stage of evolution shown in Fig. 8.23c, then the new action would have been the area enclosed by the inner defining curve of the separatrix. Since this always encloses the origin there would always be circulation. However, since the enclosed area is now less than the initial area, the instantaneous encounter with the separatrix has reduced the eccentricity of the orbit. Therefore, failure to be captured into resonance with increasing $\delta$ always implies an outcome that leads to a reduction in the eccentricity of the orbit as the separatrix is encountered (see Fig. 8.23f). As

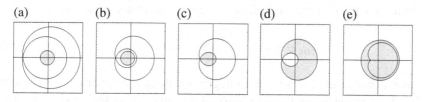

Fig. 8.24. Resonance encounter for decreasing $\delta$. In (a) $\delta = 4$ and in (b) $\delta = 1$. In (c) and (d) $\delta = 0.581777$, the critical value, associated with the initial area enclosed by the trajectory. In (e) $\delta = 0$. The separatrix is shown for all values of $\delta \geq 0$.

$\delta$ increases further the trajectory continues to circulate in a near-circular path about the origin (see Fig. 8.23g).

Figure 8.24 shows the outcome of a resonant encounter in the case of decreasing $\delta$. The initial orbit has a relatively small eccentricity enclosing the origin (see Fig. 8.24a). As $\delta$ decreases, circulation can continue (Fig. 8.24b) until the resonant separatrix is encountered (Fig. 8.24c) and the area enclosed by the orbit must expand to fill the area enclosed by the outer branch of the separatrix (Fig. 8.24d). In these circumstances capture into resonance cannot occur. However, since the resulting action is larger than the initial value, the encounter will always lead to an increase in the eccentricity. Subsequent evolution still conserves the action and the new eccentricity maintains its value. Note that in the case of decreasing $\delta$ it is also possible for the orbit to exhibit apocentric libration (cf. Fig. 8.17d).

To provide a quantitative understanding of these evolutionary histories we have to study the properties of the various curves. In particular we must examine the separatrices as functions of $\delta$ and the critical curve at $\delta = 0$.

In our discussion of the Hamiltonian approach above we showed that the locations of the equilibrium points (in terms of $\Phi$) for first-order resonances are given by Eqs. (8.136)–(8.137). Because the unstable equilibrium point at $\Phi_3$ (or $x_3$) lies on the critical curve that encloses the resonant region around the stable equilibrium point at $\Phi_1$, we can use information about the location of $\Phi_3$ to obtain the area of the critical curve associated with the formation of the resonant separatrix at $\delta = 0$ (see Fig. 8.10c). From Eq. (8.128) and transforming to $\delta$, the equation describing the critical curve at $\delta = 0$ is given by

$$-3\Phi + \Phi^2 - 2\sqrt{2\Phi} \cos \phi = \frac{3}{4}. \qquad (8.207)$$

Following Peale (1986) and Malhotra (1988), the area enclosed by this curve can be calculated by deriving an expression for $d\phi$ in terms of $\Phi$ and $d\Phi$, thereby changing the integration variable from $\phi$ to $\Phi$. By deriving expressions for $d\phi$ from $d\phi/d\tau = \partial\mathcal{H}_1/\partial\Phi$ and $d\Phi/d\tau = -\partial\mathcal{H}_1/\partial\phi$, where $\tau$ is the scaled time, the action associated with the critical curve can be calculated. This gives

$$J_{\text{crit}} = 2 \int_{1/2}^{9/2} \frac{\Phi \dot\phi}{\dot\Phi} \, d\Phi = 6\pi = 2\pi \Phi_{\text{crit}}, \tag{8.208}$$

where it is understood that we are integrating along the upper part of the curve and we have made use of the symmetry of the curve about the $x$ axis. The lower and upper limits on the integral are derived from the values of $\Phi$ on the critical curve at $\phi = 180°$ and $\phi = 0°$ respectively.

Therefore, provided $J_{\text{init}} \leq J_{\text{crit}}$ (i.e., $\Phi_{\text{init}} \leq \Phi_{\text{crit}} = 3$), the orbit will be captured into resonance as $\delta$ increases. Using Eqs. (8.132) this reduces to the condition $e_{\text{init}} \leq e_{\text{crit}}$, where $e_{\text{init}}$ is the initial eccentricity of the interior object (assumed to have negligible mass) and

$$e_{\text{crit}} = \sqrt{6} \left[ \frac{3}{f_{\text{d}}} (1-j)^{\frac{4}{3}} j^{\frac{2}{3}} \frac{m_{\text{c}}}{m'} \right]^{-1/3}. \tag{8.209}$$

In the case of an exterior object of negligible mass we use Eq. (8.134) and the condition is $e'_{\text{init}} \leq e'_{\text{crit}}$, where

$$e'_{\text{crit}} = \sqrt{6} \left[ \frac{3}{f_{\text{d}}} j^2 \frac{m_{\text{c}}}{m} \right]^{-1/3}. \tag{8.210}$$

Note that we can also modify the values of $e_{\text{crit}}$ and $e'_{\text{crit}}$ by using Eqs. (8.125) and (8.126) with $k = 1$ if we wish to consider situations where neither mass is negligible.

To calculate the changes in the eccentricity when a resonance is encountered we need to calculate the corresponding change in the action, which is related to the area enclosed by the outer and inner branches of the separatrix that starts at the unstable equilibrium point at $\Phi_3$. Above we outlined the procedure to be adopted for the critical curve at $\delta = 0$; now we need to generalise this for arbitrary, positive values of $\delta$.

To do the integral we need to make use of the equation for the value of $\Phi_3$ as a function of $\delta$ derived from Eq. (8.137) and then calculate the other two values of $\Phi$ where the separatrix crosses at $\phi = 0$. Malhotra (1988) gives the following exact formulae for these two crossing points, $\Phi_{\text{min}}$ and $\Phi_{\text{max}}$, as well as expressions for the areas $|A_{\text{inner}}|$ and $|A_{\text{outer}}|$ enclosed by the inner and outer branches of the separatrix respectively. Using our notation these expressions are

$$\Phi_{\text{min}} = 3(\delta + 1) - \Phi_3 - 2 \left( 2\Phi_3 \right)^{1/4}, \tag{8.211}$$

$$\Phi_{\text{max}} = 3(\delta + 1) - \Phi_3 + 2 \left( 2\Phi_3 \right)^{1/4}, \tag{8.212}$$

$$A_{\text{inner}} = 3 \left[ \frac{1}{2} \left[ \Phi_{\text{max}} + \Phi_{\text{min}} + 2\Phi_3 - 2(\delta + 1) \right] \left( \frac{\pi}{2} - \gamma \right) \right.$$
$$\left. - \sqrt{(\Phi_{\text{max}} - \Phi_3)(\Phi_3 - \Phi_{\text{min}})} \right], \tag{8.213}$$

$$A_{\text{outer}} = -3 \left[ \frac{1}{2} [\Phi_{\max} + \Phi_{\min} + 2\Phi_3 - 2(\delta + 1)] \left( \frac{\pi}{2} + \gamma \right) \right.$$
$$\left. + \sqrt{(\Phi_{\max} - \Phi_3)(\Phi_3 - \Phi_{\min})} \right], \tag{8.214}$$

where

$$\gamma = \sin^{-1} \left( \frac{\Phi_{\max} + \Phi_{\min} - 2\Phi_3}{\Phi_{\max} - \Phi_{\min}} \right). \tag{8.215}$$

Although the expressions for $A_{\text{inner}}$ and $A_{\text{outer}}$ are complicated, it can be shown (see, for example, Peale 1986) that their sum is given by the simpler expression

$$A_{\text{inner}} + A_{\text{outer}} = 6\pi(\delta_t + 1), \tag{8.216}$$

where $\delta_t$ is the value of $\delta \geq 0$ when transition occurs.

Provided that (i) we know $\delta_t$ and (ii) we can approximate the initial and final trajectories as circles centred on the origin in the $x$–$y$ plane, we have the following relationship between the initial and final values of $\Phi$:

$$\Phi_{\text{init}} + \Phi_{\text{final}} = 3(\delta_t + 1). \tag{8.217}$$

Hence the initial and final values of the eccentricity are related by

$$e_{\text{init}}^2 + e_{\text{final}}^2 = 6(\delta_t + 1) \left[ \frac{3}{f_d}(1 - j)^{\frac{4}{3}} j^{\frac{2}{3}} \frac{m_c}{m'} \right]^{-\frac{2}{3}} \tag{8.218}$$

for internal resonances and by

$$e_{\text{init}}'^2 + e_{\text{final}}'^2 = 6(\delta_t + 1) \left[ \frac{3}{f_d} j^2 \frac{m_c}{m} \right]^{-\frac{2}{3}} \tag{8.219}$$

for external resonances.

These formulae cover the case when capture does not occur either because $\delta$ is always decreasing (and so capture is impossible) or $\delta$ is always increasing but the initial eccentricity is greater than the critical value and resonant capture (a probabilistic event in such circumstances) does not occur.

We can quantify the expected changes in $\Phi$ when capture does not occur by plotting values of $|A_{\text{inner}}|/2\pi$ and $|A_{\text{outer}}|/2\pi$ as a function of $\delta_t$. This is shown in Fig. 8.25. If we know $\Phi_{\text{init}}$, we can use the plot to calculate $\Phi_{\text{final}}$ in the case when capture into resonance has not occurred. This can happen in two ways. The first case occurs when $\delta$ is increasing and $\Phi_{\text{init}} > \Phi_{\text{crit}}$ (see Fig. 8.23). We find the value of $|A_{\text{outer}}|/2\pi = \Phi_{\text{init}}$ and read off the corresponding value of $|A_{\text{inner}}|/2\pi = \Phi_{\text{final}}$. In the second case $\delta$ is decreasing (see Fig. 8.24). We find the value of $|A_{\text{inner}}|/2\pi = \Phi_{\text{init}}$ and read off the corresponding value of $|A_{\text{outer}}|/2\pi = \Phi_{\text{final}}$. In each case we can calculate the changes in the eccentricity from the scaling relations given in Eqs. (8.132) and (8.134).

We have already noted that in the case of decreasing $\delta$ (i.e., orbits that are receding from one another) the probability of capture into resonance is zero. In

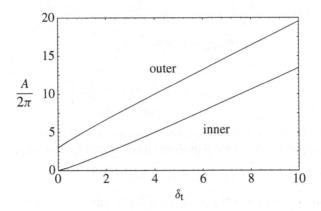

Fig. 8.25.   Plots of the scaled action, $A/2\pi$, for the inner and outer branches of the separatrix as a function of $\delta_t$, the value of $\delta$ at transition, for the case of first-order resonances.

the case of increasing $\delta$ (i.e., orbits that are approaching one another) capture into resonance is certain if $\Phi_{init} \leq \Phi_{crit} = 3$. In situations where $\Phi_{init} > \Phi_{crit}$, the probability, $P_{cap}$, of capture into resonance can be quantified by means of formulae devised by Henrard (1982) and then greatly simplified by Borderies & Goldreich (1984). By introducing the concept of "balance of energy integrals" these authors have shown that

$$P_{cap} = \frac{4\gamma}{\pi + 2\gamma},$$   (8.220)

where $\gamma$, as defined in Eq. (8.215), is evaluated using the value of $\delta = \delta_t$. Figure 8.26 shows the value of $P_{cap}$ as a function of $\delta_t$.

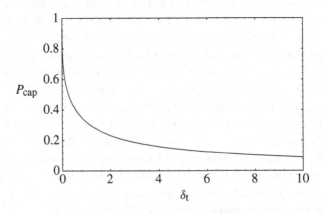

Fig. 8.26.   The probability of resonance capture as a function of $\delta_t$, the value of $\delta$ at transition, for the case of first-order resonances.

### 8.12.2  *Encounters with Second-Order Resonances*

In the case of second-order resonances the approach is similar to that used above for first-order resonances. Again there is a critical separatrix at $\delta = 0$ and it is easy to show that for this value of $\delta$ there are five equilibrium points at $\Phi_1 = 1/2$ (with $x = \pm 1$, $y = 0$) and $\Phi_2 = \Phi_3 = 0$ (three points coincident at the origin). Therefore the critical curve passes through the origin; in common with all curves in the $x$–$y$ plane for second-order resonances, the critical curve will be symmetric about the $x$ axis and $y$ axis. It is described by

$$2\Phi^2 - \Phi(1 + \cos 2\phi) = 0. \tag{8.221}$$

The area enclosed by this curve, the critical action, is given by

$$J_{\text{crit}} = 4 \int_{1/2}^{0} \frac{\Phi\dot{\phi}}{\dot{\Phi}} \, d\Phi = \pi = 2\pi \Phi_{\text{crit}}, \tag{8.222}$$

where the lower and upper limits on the integral are derived from the values of $\Phi$ on the critical curve when $\phi = 0$ and $\phi = \pi/2$.

As before, we are primarily interested in the case when $\delta$ is always increasing, since this can lead to capture if $\Phi_{\text{init}}$ is less than a critical value. Therefore, in the case of second-order resonance, provided $J_{\text{init}} < J_{\text{crit}}$ (i.e., $\Phi_{\text{init}} < \Phi_{\text{crit}} = 1/2$), the orbit will be captured into resonance as $\delta$ increases. Using Eqs. (8.156) and (8.158) the condition for certain capture reduces to $e_{\text{init}} < e_{\text{crit}}$, where

$$e_{\text{crit}} = \left[ \frac{3}{16 f_{\text{d}}} (2 - j)^{\frac{4}{3}} j^{\frac{2}{3}} \frac{m_{\text{c}}}{m'} \right]^{-1/2} \tag{8.223}$$

in the case of an internal resonance and

$$e'_{\text{crit}} = \left[ \frac{3 j^2}{16 f_{\text{d}}} \frac{m_{\text{c}}}{m} \right]^{-1/2} \tag{8.224}$$

in the case of external resonances. Using Eqs. (8.125) and (8.126) we can also make use of more accurate scaling relationships in cases where the two masses are comparable.

In order to calculate the outcome of the resonant encounters we need to calculate the areas enclosed by the inner and outer branches of the separatrix that has an unstable equilibrium point at $\Phi_3$ corresponding to the $x = 0$, $y = \pm\sqrt{\delta_{\text{t}}}$, where $\delta_{\text{t}} \geq 0$ is the value of $\delta$ when the transition occurs. Malhotra (1988) gives the formulae for the minimum and maximum values of $\Phi$ along the separatrix and the area enclosed by their inner and outer boundaries of the separatrix. These are

$$\Phi_{\text{min}} = \frac{1}{2}(1 + \delta) - \frac{1}{2}\sqrt{1 + 2\delta}, \tag{8.225}$$

$$\Phi_{\text{max}} = \frac{1}{2}(1 + \delta) + \frac{1}{2}\sqrt{1 + 2\delta}, \tag{8.226}$$

$$A_{\text{inner}} = 2\left[\frac{1}{2}\left[\Phi_{\max} + \Phi_{\min} + 2\Phi_3\right]\left(\frac{\pi}{2} - \gamma\right)\right.$$
$$\left. - \sqrt{(\Phi_{\max} - \Phi_3)(\Phi_3 - \Phi_{\min})}\right], \tag{8.227}$$

$$A_{\text{outer}} = -2\left[\frac{1}{2}\left[\Phi_{\max} + \Phi_{\min} + 2\Phi_3\right]\left(\frac{\pi}{2} + \gamma\right)\right.$$
$$\left. + \sqrt{(\Phi_{\max} - \Phi_3)(\Phi_3 - \Phi_{\min})}\right], \tag{8.228}$$

where, as before,

$$\gamma = \sin^{-1}\left(\frac{\Phi_{\max} + \Phi_{\min} - 2\Phi_3}{\Phi_{\max} - \Phi_{\min}}\right). \tag{8.229}$$

In this case,

$$A_{\text{inner}} + A_{\text{outer}} = \pi(1 + 2\delta_t). \tag{8.230}$$

Provided we can approximate the initial and final trajectories far from resonance as circles in the $x$–$y$ plane, we can write

$$\Phi_{\text{init}} + \Phi_{\text{final}} = \frac{\pi}{2}, \tag{8.231}$$

and hence the initial and final values of the eccentricity are related by

$$e_{\text{init}}^2 + e_{\text{final}}^2 = (1 + 2\delta_t)\left[\frac{3}{16f_d}(2 - j)^{\frac{4}{3}}j^{\frac{2}{3}}\frac{m_c}{m'}\right]^{-1} \tag{8.232}$$

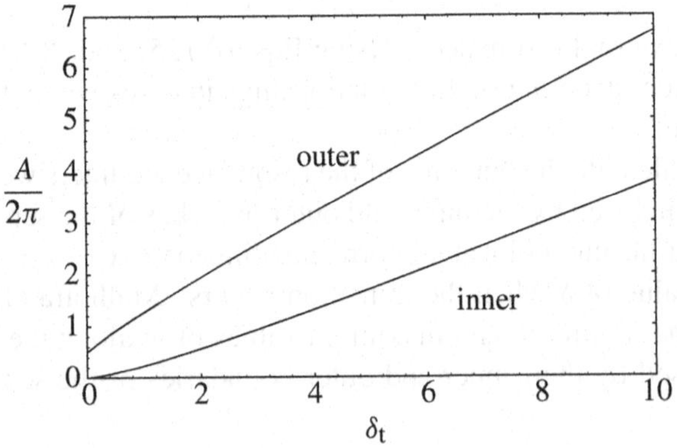

Fig. 8.27. Plots of the scaled action, $A/2\pi$, for the inner and outer branches of the separatrix as a function of $\delta_t$, the value of $\delta$ at transition, for the case of second-order resonances.

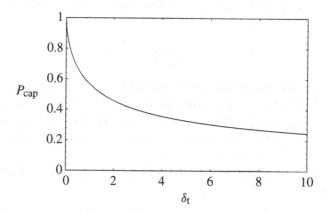

Fig. 8.28. The probability of resonance capture as a function of $\delta_t$, the value of $\delta$ at transition, for the case of second-order resonances.

for internal resonances and by

$$e_{init}'^2 + e_{final}'^2 = (1 + 2\delta_t)\left[\frac{3j^2}{16f_d}\frac{m_c}{m}\right]^{-1} \tag{8.233}$$

for external resonances.

Following the example shown above for the case of first-order resonances, we can plot $A/2\pi$ for the inner and outer branches of the separatrix as a function of $\delta_t$, the value of $\delta$ at the transition. These are shown in Fig. 8.27. The formula for the probability, $P_{cap}$, of capture into a second-order resonance in the case of increasing $\delta$ is identical to that given in Eq. (8.220), where the values of $\Phi_3 = \delta/2$, $\Phi_{min}$, and $\Phi_{max}$ are given by Eqs. (8.225) and (8.226). Figure 8.28 shows the resulting probability as a function of $\delta_t$.

## 8.13 The Dynamics of Capture and Evolution in Resonance

We have already discussed the dynamics of capture in the context of spin–orbit resonance (see Sect. 5.5) and there are a number of similarities when we consider the orbit–orbit case. In order for resonance capture to occur there must be dissipation in the system. In Sect. 8.12 above we used the conservation of the action to analyse the effect of resonant encounters. We noted that capture is impossible in the classical system if the resonance is approached from above with decreasing $\delta$. When approached from below, capture is certain for an initial eccentricity below a critical value, while it is a probabilistic event for larger initial values.

To illustrate the dynamics of capture and the subsequent evolution, consider the case of two satellites of masses $m$ and $m'$ evolving due to tides. The satellites are assumed to be in prograde orbits outside the synchronous orbit. Using Kepler's third law and Eq. (4.160) we can write the rate of change of the mean motion due to tides as

$$\dot{n}_t = -\frac{9}{2}\frac{n^2}{Q_p}k_2\frac{m}{m_p}\left(\frac{R_p}{a}\right)^5, \qquad (8.234)$$

where $a$ is the semi-major axis of the satellite; $n$ is its mean motion, and $k_2$, $m_p$, and $Q_p$ denote the Love number, mass, and the tidal dissipation function, respectively, of the planet. There is a similar equation for $\dot{n}'_t$ with $m$, $n$, and $a$ replaced by $m'$, $n'$, and $a'$. Here unprimed and primed quantities refer to the inner and outer satellites respectively.

Now consider the effect of a resonance. Let the resonant variable have the general form

$$\varphi = j_1\lambda' + j_2\lambda + j_3\varpi' + j_4\varpi + j_5\Omega' + j_6\Omega, \qquad (8.235)$$

and let the averaged disturbing functions experienced by the inner and outer satellites be

$$\mathcal{R} = \mu'S\cos\varphi, \qquad \mathcal{R}' = \mu S\cos\varphi, \qquad (8.236)$$

where $\mu = \mathcal{G}m$, $\mu' = \mathcal{G}m'$ and, from the d'Alembert properties,

$$S = e'^{|j_3|}e^{|j_4|}s'^{|j_5|}s^{|j_6|}\frac{f_d(\alpha)}{a'}. \qquad (8.237)$$

Here $s = \sin\frac{1}{2}I$ and $s' = \sin\frac{1}{2}I'$, where $I$ denotes inclination, and $f_d(\alpha)$ denotes the contribution of the direct part of the disturbing function as a combination of Laplace coefficients in $\alpha = a/a'$ (see Sect. 6.7).

From the lowest-order form of Lagrange's equation for $\dot{n}$ (see Eq. (8.12)) we can write the second time derivative of $\varphi$ as

$$\ddot{\varphi} = 3gS\sin\varphi + F, \qquad (8.238)$$

where

$$g = \frac{j_1^2\mu}{a'^2} + \frac{j_2^2\mu'}{a^2} \qquad \text{and} \qquad F = j_1\dot{n}'_t + j_2\dot{n}_t. \qquad (8.239)$$

Here we have ignored any contributions to $\ddot{\varphi}$ from the second time derivatives of the pericentres and longitudes at epoch. Note that $j_1$ and $j_2$ have opposite signs for a standard resonant argument. Treating $a$, $a'$, $e$, $e'$, $s$, and $s'$ in $S$ as constants, we can integrate Eq. (8.238) to obtain the energy equation

$$\frac{1}{2}\dot{\varphi}^2 + 3gS\cos\varphi = E + F\varphi. \qquad (8.240)$$

As pointed out by Allan (1969), the dynamics of this system correspond to those of a particle moving with velocity $\dot{\varphi}$ in a one-dimensional potential that is a function only of $\varphi$. As with the case of spin–orbit resonance, (i) the variation of $\frac{1}{2}\dot{\varphi}^2$ with $\varphi$ can be thought of as comprising a sinusoidal part and a linear part (cf. Fig. 5.14) and (ii) capture into the resonance depends on the magnitude and sign of the change in energy over one cycle (cf. Fig. 5.15).

Once the two satellites are trapped in resonance the mean value of the second derivative of $\varphi$, $\langle \ddot{\varphi} \rangle$, is zero and hence

$$\langle S \sin \varphi \rangle = -\frac{F}{3g}. \tag{8.241}$$

Allan (1969) shows that once capture occurs the two satellites will remain in resonance even though each continues to evolve away from the planet. We can make use of Eq. (8.241) and our lowest-order form of Lagrange's equations given in Eqs. (8.12)–(8.17) to derive expressions for the time-averaged variation in the orbital elements of the satellites in resonance. For the eccentricities these are

$$\frac{\langle \dot{e} \rangle}{e} = -\frac{j_4}{e^2} \frac{m'}{m_{\mathrm{p}}} na \frac{F}{3g}, \qquad \frac{\langle \dot{e}' \rangle}{e'} = -\frac{j_3}{e'^2} \frac{m}{m_{\mathrm{p}}} n'a' \frac{F}{3g}, \tag{8.242}$$

with similar equations for the inclinations.

Dermott et al. (1988) point out that the tidal torque on the inner satellite is usually larger than that on the outer satellite because of the extreme dependence on semi-major axis. However, as it is the perturbation of the outer satellite on the inner that dominates their gravitational interaction, we can often neglect the first terms in $g$ and $F$. Using Kepler's third law the variation in $e$ can be written as

$$\frac{\langle \dot{e} \rangle}{e} = +\frac{1}{e^2} \frac{j_4}{2j_2} \left( \frac{\dot{a}}{a} \right)_{\mathrm{t}}. \tag{8.243}$$

Because $j_2$ and $j_4$ have the same sign this implies that the eccentricity will always grow as the satellite evolves outwards. There is a similar result for the inclination, although, provided there is significant resonance splitting, the satellite should enter the inclination resonance at a different time. These phenomena were described analytically and investigated numerically by Dermott et al. (1988). They point out that in cases where the individual resonances at a given commensurability are not well separated, the evolution within a given resonance can cause the effects of a neighbouring resonance to become important (see Sect. 9.6).

## 8.14 Two-Body Resonance in the Solar System

The solar system has many examples of two objects orbiting a central mass where both objects are involved in a mean motion resonance. We refer to these as *two-body resonances* and in each case one (or more) arguments in the expansion of the disturbing function is librating. In Chapter 3 examples of the special case of objects librating in a 1:1 resonances (e.g., the Trojan asteroids and the coorbital satellites of Saturn) were discussed. Table 8.8 gives examples of the known first- and second-order two-body resonances involving planets and satellites in the solar system. In addition to these resonances listed in Table 8.8 several asteroids

Table 8.8. Known first- and second-order mean motion resonances involving planets or satellites in the solar system. In each case the unprimed and primed quantities refer to the inner and outer bodies respectively. All known planetary and satellite resonances are included.

| System | Resonant Argument | Amplitude | Period (y) |
|--------|-------------------|-----------|------------|
| *Planets* | | | |
| Neptune–Pluto | $3\lambda' - 2\lambda - \varpi'$ | 76° | 19,670 |
| *Jupiter* | | | |
| Io–Europa | $2\lambda' - \lambda - \varpi$ | 1° | — |
| Io–Europa | $2\lambda' - \lambda - \varpi'$ | 3° | — |
| Europa–Ganymede | $2\lambda' - \lambda - \varpi$ | 3° | — |
| *Saturn* | | | |
| Mimas–Tethys | $4\lambda' - 2\lambda - \Omega' - \Omega$ | 43.6° | 71.8 |
| Enceladus–Dione | $2\lambda' - \lambda - \varpi$ | 0.297° | 11.1 |
| Titan–Hyperion | $4\lambda' - 3\lambda - \varpi'$ | 36.0° | 1.75 |

are known to be involved in mean motion resonances with Jupiter. These include (279) Thule, (153) Hilda, (1362) Griqua, and (887) Alinda, which are in 4:3, 3:2, 2:1, and 3:1 resonances respectively (see Yoshikawa (1989) for a study of the dynamics of a number of these resonances). The ring systems of Saturn, Uranus, and Neptune contain many resonant features due to perturbing satellites; these will be discussed in Chapter 10. Resonances involving three gravitationally interacting objects orbiting a central mass will be discussed in Sect. 8.15.

To gain some insight into resonance mechanisms, we provide a study of two of the satellite resonances listed in Table 8.8. The Neptune–Pluto resonance is discussed in Sect. 9.9.

### 8.14.1 The Titan–Hyperion Resonance

Titan ($e = 0.0292$, $I = 0.33°$) and Hyperion ($e' = 0.1042$, $I' = 0.43°$) are trapped in a 4:3 resonance. The mass of Hyperion is negligible compared to that of Titan and the resonance can be considered in the context of the planar, circular, restricted problem of three bodies where Hyperion is in an external 3:4 resonance with Titan. From observations the argument

$$\varphi = 4\lambda' - 3\lambda - \varpi' \qquad (8.244)$$

librates about 180° so that conjunction of the two satellites librates about the apocentre of Hyperion. From the orbital periods given in Table A1.8 in Appendix A, we have

$$4n' - 3n = -18.679°\text{y}^{-1}. \qquad (8.245)$$

The observed motion of Hyperion's pericentre is

$$\dot{\varpi}' = -18.663\,°\mathrm{y}^{-1}. \tag{8.246}$$

Recall that the effect of Saturn's $J_2$ should cause a prograde motion of Hyperion's pericentre (see Sect. 6.11), and therefore the observed regression must be due to Titan. With these values

$$4n' - 3n - \dot{\varpi}' \approx 0 \tag{8.247}$$

to within observational error, where we have used averaged rates throughout.

The dynamics of the resonance are difficult to study because Hyperion's large eccentricity causes convergence problems if standard expansions of the disturbing function are used. An alternative method developed by Woltjer (1928) and used by Message (1989, 1993) is to determine the location of the periodic orbit associated with exact resonance and investigate the perturbed motion in its vicinity.

It is unlikely that the Titan–Hyperion resonance resulted from tidal evolution, even though both satellites would have been approaching each other as they receded from Saturn. Sinclair (1972) showed that the critical eccentricity below which Hyperion's capture was guaranteed is $e' = 0.068$, implying significant evolution of $e'$ within the resonance to reach its present value. However, to be captured into the 4:3 resonance Hyperion likely would have come close to Titan's orbit without the resonance mechanism allowing it to avoid dangerous conjunctions (Sinclair 1972). Hyperion may well be the last example of a number of objects that once orbited in the region close to Titan. Without the protection of a resonance the other moons would have been scattered by Titan. If Hyperion is the sole survivor of such a group then this would imply that the resonance is primordial.

### 8.14.2 The Mimas–Tethys Resonance

Mimas ($e = 0.02$, $I = 1.53°$) and Tethys ($e' = 0.0$, $I' = 1.09°$) are trapped in 4:2 resonance, which, since it is of second order, can involve the inclinations of the satellites as well as their eccentricities. Of the six known resonances between pairs of planets or satellites this is the only example of a second-order resonance. From observations the particular librating argument is

$$\varphi = 4\lambda' - 2\lambda - \Omega' - \Omega. \tag{8.248}$$

This implies that conjunction of the two satellites librates about the midpoint of the nodes. The observed amplitude of libration is $43.6°$. Our knowledge of the disturbing function tells us that the term associated with $\varphi$ will involve a factor $II'$ as well as the usual combination of Laplace coefficients. Yet this is one of three possible arguments at the 4:2 commensurability that involve inclinations, the others having factors $I^2$ and $I'^2$. There will also be eccentricity terms associated with the same commensurability. Hence, at the 4:2 Mimas–Tethys commensurability there are six possible resonant terms:

$$\langle \mathcal{R} \rangle_1 = f_1 e^2 \cos(4\lambda' - 2\lambda - 2\varpi), \qquad \langle \mathcal{R} \rangle_2 = f_2 e e' \cos(4\lambda' - 2\lambda - \varpi' - \varpi),$$

$$\langle \mathcal{R} \rangle_3 = f_3 e'^2 \cos(4\lambda' - 2\lambda - 2\varpi'), \qquad \langle \mathcal{R} \rangle_4 = f_4 I^2 \cos(4\lambda' - 2\lambda - 2\Omega),$$

$$\langle \mathcal{R} \rangle_5 = f_5 I I' \cos(4\lambda' - 2\lambda - \Omega' - \Omega), \quad \langle \mathcal{R} \rangle_6 = f_6 I'^2 \cos(4\lambda' - 2\lambda - 2\Omega'),$$
$$\tag{8.249}$$

where the $f_i = f_i(\alpha)$ are different combinations of Laplace coefficients that can be derived from our expansion of the disturbing function. Each resonance has a different location in semi-major axis determined by knowledge of the mean motions, pericentre, and node rates (cf. Fig. 8.21).

The question of why the Mimas–Tethys system is actually librating in the $II'$ resonance rather than any other can only be answered by considering the origin of such resonances and their link with the tidal evolution of satellite systems. Allan (1969), Sinclair (1972), and Greenberg (1973) have investigated this question in detail. Our analysis of the mechanism of resonance capture shows that the satellites have to be approaching each other as both recede from the planet under the effects of tides. Therefore it is likely that the $I^2$ resonance was encountered first and then the $II'$ resonance (see Fig. 8.21). However, Sinclair (1972) showed that the probability of capture into each of these resonances was as low as 0.07 and 0.04 respectively.

At this stage it is worthwhile making a comparison with resonance in the asteroid belt. In the case of the asteroid belt resonances, where tidal effects are unimportant, if an asteroid is trapped in a resonance it is usually the argument with the largest associated coefficient that is seen to be librating. In a sense the asteroid will be found to be librating in the strongest resonance. For example, if we consider a simple 3:2 first-order resonance (such as found in the Hilda group of asteroids), then a resonant asteroid could be in either the $e$ resonance with argument $3\lambda' - 2\lambda - \varpi$ or the $e'$ resonance with argument $3\lambda' - 2\lambda - \varpi'$. However, Jupiter's eccentricity of 0.048 is smaller than the average asteroid eccentricity of 0.15. Therefore, because usually $e' \ll e$ and the Laplace coefficient terms are comparable, the $e$ resonance will be stronger, although the relative masses of the objects are also important.

## 8.15 Resonant Encounters in Satellite Systems

Tidal evolution provides a mechanism for pairs of satellites to encounter a resonance. Consider a satellite of mass $m$ and semi-major axis $a$ moving in a prograde orbit about a planet. In Sect. 4.9 we showed that the tides raised on the planet by the satellite cause $a$ to change at a rate given by

$$\dot{a}_t = \frac{3k_2}{Q_p} \left( \frac{\mathcal{G}}{m_p} \right)^{1/2} R_p^5 \frac{m}{a^{11/2}} \tag{8.250}$$

(cf. Eq. (4.160)), where $m_p$, $R_p$, $Q_p$, and $k_2$ denote the mass, radius, tidal dissipation function, and Love number of the planet, respectively. If the physical properties of the planet remain constant over time, then $\dot{a}_t$ is purely a function of $a$ and $m$. Therefore, if we consider another satellite with semi-major axis $a'$ and mass $m'$ on an exterior orbit, the two satellites will be approaching one another if at a given time the ratio

$$(\dot{a}/\dot{a}') = (m/m')(a'/a)^{11/2} \tag{8.251}$$

is greater than one. In terms of the mean motions, using the ratio $\mathcal{N} = n'/n$, the condition for capture can be written as $\dot{\mathcal{N}} = d(n'/n)/dt > 1$.

The equation for $\dot{a}_t$ can be solved to give

$$a(t) = \left[ a_0^{13/2} - \frac{13}{2} Km(t - t_0) \right]^{2/13}, \tag{8.252}$$

where $a_0 = a(t_0)$ (the current value of $a$) and $K$ (assumed to be constant) is a function of $m_p$, $R_p$, $Q_p$, and $k_2$. Although $m_p$ and $R_p$ are well known (see Table A.4), it is more difficult to estimate $Q_p$ and $k_2$, let alone their primordial values (see Sect. 4.13). Nevertheless, Kepler's third law and the fact that the semi-major axes of satellites vary with time means that resonances can be encountered.

For capture to take place between satellites above the synchronous orbit, it is required that the orbit of the inner satellite must be expanding faster than the orbit of the outer satellite and thus that the satellites are on converging orbits (Dermott et al. 1988). On capture into a stable orbit–orbit resonance, $\dot{\mathcal{N}} \approx 0$ and the orbits of the satellites expand at the same rate. Angular momentum is lost from the spin of the planet at the same rate as before the resonance encounter, but now gravitational forces between the satellites due to the resonance act to transfer angular momentum from the inner to the outer satellite allowing their orbits to expand together, thus maintaining the resonance and keeping $\dot{\mathcal{N}} \approx 0$. But these same resonant forces also act to increase the eccentricities or inclinations of the satellites involved in the resonance at rates determined by the resonant argument (see Sect. 8.13), and it is these increases that provide the evidence for orbital evolution.

For example, as we have seen, the satellites Mimas and Tethys are trapped in an inclination-type resonance for which

$$4n' - 2n - \dot{\Omega}' - \dot{\Omega} = 0, \tag{8.253}$$

where the primed and unprimed quantities refer, respectively, to Tethys and Mimas. To a good approximation, we have

$$\frac{\langle \dot{I} \rangle}{I} \approx \frac{1}{4I^2} \left( \frac{\dot{a}}{a} \right)_{\text{tides}} \tag{8.254}$$

and

$$\frac{\langle \dot{I}' \rangle}{I'} \approx \frac{1}{4I'^2} \left( \frac{\dot{a}'}{a'} \right)_{\text{tides}}. \qquad (8.255)$$

Thus, the timescales for the increase of the orbital inclinations are considerably less than those for the increase of the semi-major axes. The present orbital inclinations of Mimas and Tethys are $1.53°$ and $1.09°$, respectively, and are significantly greater than their near neighbours. Allan (1969) argued that these anomalously high values are evidence for significant orbital evolution. Furthermore, if $Q_p$ is independent of amplitude and frequency, then, for this resonance, the condition for stability, $\mathcal{N} > 1$, is only marginally satisfied. For Mimas $m/a^{13/2} = 2.19 \times 10^{-38}$ while for Tethys $m'/a'^{13/2} = 1.75 \times 10^{-38}$ in cgs units. It follows that we might expect the resonance to be young and the amplitude of libration to be still high. This is indeed the case. The observed amplitude of libration of the resonant argument is $97.040°$ and Allan (1969) calculated that the resonance was formed $2.4 \times 10^8$ years ago, an age that is only 5% of the age of the solar system. An interesting extension of the work of Allan (1969) has recently been given by Champenois & Vienne (1999a,b)

The orbital inclinations of the major uranian satellites (see Table A.11) are particularly interesting. The anomalously high inclination of Miranda ($4.22°$) clearly suggests the action of resonance. However, unlike the satellite systems of Jupiter and Saturn, which are nearly saturated with orbit–orbit resonances, the uranian system does not contain a single resonance involving any of the five major satellites, Miranda, Ariel, Umbriel, Titania, and Oberon. Dermott (1984) suggested that this difference arises from the fact that the dynamical oblateness, $J_2$, of Uranus is small and consequently that any orbit–orbit resonances in the uranian system would be chaotic and unstable. Therefore, although the uranian satellite system is devoid of resonances now, the satellites may have been trapped in resonances at some time in the past and these resonances may have been responsible for significant changes in both the orbital inclinations and eccentricities. This problem has been considered by Dermott et al. (1988) for the uranian and saturnian satellites, by Peale (1988) for the uranian satellites, and in more detail by Tittemore & Wisdom (1988, 1989, 1990) also for the uranian satellites.

We can provide examples of possible evolutionary paths by using Eq. (8.252) (with suitable values of $K$) to describe the semi-major axis of each satellite as a function of time over the same fixed interval. This follows the example of Dermott et al. (1988) who carried out a study of tidal evolution in the saturnian and uranian satellite systems, both of which are thought to be tidally evolved. Figure 8.29 shows possible relative changes in semi-major axis over time for each system. In these plots the time is to be regarded as an integral involving

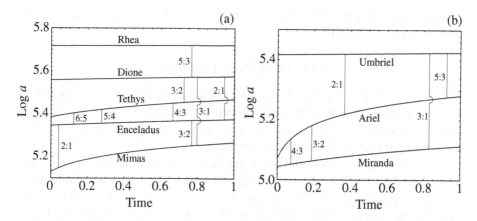

Fig. 8.29. Sample changes in the semi-major axes (measured in km) of satellites in (a) the saturnian and (b) the uranian system. A selection of first- and second-order resonances between pairs of satellites is indicated on each plot.

averaging of $Q_p$ (see Dermott et al. 1988). Consequently, the zero point on the time axis should not be considered to be fixed.

Recall that in the saturnian system (Fig. 8.29a) we know that there are currently two resonances: the Mimas–Tethys 4:2 and the Enceladus–Dione 2:1. The plots show that currently each pair of satellites is evolving outwards at almost the same rate although they approached one another at a higher rate in the past. Consequently, the necessary condition for capture occurs in each case. In contrast Enceladus and Tethys have always evolved on diverging paths and could have passed through a number of first-order resonances without capture. Curiously Mimas and Enceladus are on clearly converging paths and yet have managed to escape capture into a resonance.

There are no known resonances between pairs of satellites in the uranian system and yet the paths shown in Fig. 8.29b suggest that the pairs Miranda–Ariel and Ariel–Umbriel are on converging paths. In particular, for reasonable values of $Q_p$, it is likely that Miranda and Umbriel once encountered the 3:1 resonance in a direction that could permit capture. In light of this it is reasonable to ask the question: Why did resonances arise in the saturnian system and not in the uranian one? One possible explanation may be related to the fact that because the $J_2$ of Uranus is a factor ten smaller than that of Saturn, the resonances in the uranian system are less well separated than those in the saturnian system. Therefore, for resonances of similar strengths (i.e., similar widths in semi-major axis) overlap of adjacent resonances will occur for lower values of eccentricity and inclination. Furthermore, the individual resonances cannot be treated in isolation. Because resonance overlap leads to chaotic motion (see Sect. 9.6) we would expect satellites at Uranus entering resonances to be more likely to exhibit chaotic motion and less likely to conform to the classical model of capture than those at Jupiter or Saturn.

## 8.16 Three-Body Resonance

We have already investigated the dynamics of a two-body resonance where there is a simple numerical relationship between the mean motions of two orbiting bodies (see Eq. (8.1)). A *three-body resonance* occurs when there is a simple relationship involving three orbiting bodies.

Let $n_1$, $n_2$, and $n_3$ denote the mean motions of three objects of mass $m_1$, $m_2$, and $m_3$ that lie on near-circular, near-coplanar orbits at increasing distances from a central object of mass $m_c$. We assume that $n_1 > n_2 > n_3$ and hence that the first mass has the smallest semi-major axis of the three, and the third mass the largest semi-major axis. A *Laplace relation* is said to exist between the three orbiting masses if there is a numerical relationship among the masses of the form

$$qn_1 - (p+q)n_2 + pn_3 \approx 0, \tag{8.256}$$

where $p$ and $q$ are positive integers. The existence of a Laplace relation is a special example of a three-body resonance. If $T_{rep}$ is the time between successive repetitions of the initial configuration, we can write

$$\frac{p}{n_1 - n_2} \approx \frac{q}{n_2 - n_3} \approx \frac{p+q}{n_1 - n_3} \approx \frac{T_{rep}}{2\pi}, \tag{8.257}$$

where the mean motions are measured in radians per time unit. It is important to note that the existence of a Laplace relation does not necessarily imply that there is a resonant relationship between any pair of satellites, such as that given in Eq. (8.1).

To study the dynamics of three-body resonances we have to isolate from the disturbing function all those arguments that give rise to the resonant combination of mean longitudes given by

$$\varphi = q\lambda_1 - (p+q)\lambda_2 + p\lambda_3. \tag{8.258}$$

Note that this angle is independent of the longitudes of pericentre of any of the bodies. Having isolated the relevant terms we must then find expressions for the three components on the right-hand side of

$$\ddot{\varphi} = q\frac{d^2\lambda_1}{dt^2} - (p+q)\frac{d^2\lambda_2}{dt^2} + p\frac{d^2\lambda_3}{dt^2} \tag{8.259}$$

and show that $\ddot{\varphi}$ has the form of a pendulum equation,

$$\ddot{\varphi} = c\sin\varphi, \tag{8.260}$$

where $c$ is a function of $p$, $q$, and the masses, semi-major axes, and mean motions of the three bodies involved in the resonance.

The most comprehensive work on this subject in recent years is that by Aksnes (1988). He derived detailed expressions for the value of $c$ in Eq. (8.260) and applied the theory of three-body resonances to the Galilean satellites of Jupiter and the uranian satellites. Here we outline briefly the method of Aksnes.

The starting point of the theory of three-body resonances is the realisation that resonant arguments of the form given in Eq. (8.258) above arise from specific combinations of arguments of the form $\beta_i - \alpha_i$ ($i = 1, 2, 3$), where the $\alpha_i$ and $\beta_i$ are taken from one of the following combinations:

$$\alpha_1^{(k)} = (p+q)\lambda_2 - (p+q-k)\lambda_1 - k\varpi_1, \tag{8.261}$$

$$\beta_1^{(k)} = p\lambda_3 - (p-k)\lambda_1 - k\varpi_1, \tag{8.262}$$

$$\alpha_2^{(k)} = (q+k)\lambda_2 - q\lambda_1 - k\varpi_2, \tag{8.264}$$

$$\beta_2^{(k)} = p\lambda_3 - (p-k)\lambda_2 - k\varpi_2, \tag{8.264}$$

$$\alpha_3^{(k)} = (q+k)\lambda_3 - q\lambda_1 - k\varpi_3, \tag{8.265}$$

$$\beta_3^{(k)} = (p+q+k)\lambda_3 - (p+q)\lambda_2 - k\varpi_3, \tag{8.266}$$

where $\varpi_1$, $\varpi_2$, and $\varpi_3$ are the longitudes of pericentre of the three bodies and

$$k = 0, \pm1, \pm2 \pm 3, \text{ etc.} \tag{8.267}$$

Using the properties of the disturbing function (see Chapter 6 and Appendix B) we note that $|k|$ is simply the order of the argument. Aksnes (1988) showed the derivation of the terms arising from taking $k = 0$ (the zeroth-order terms) and $k = \pm1$ (the first-order terms). The total contribution of the zeroth- and first-order terms is given by

$$c = c^{(0)} + c^{(1)}, \tag{8.268}$$

where each term on the right-hand side provides three contributions to $\ddot{\varphi}$ written as

$$c^{(0)} = c_1^{(0)} + c_2^{(0)} + c_3^{(0)}, \qquad c^{(1)} = c_1^{(1)} + c_2^{(1)} + c_3^{(1)}, \tag{8.269}$$

where in each case the first term is from $q\ddot{\lambda}_1$, the second is from $-(p+q)\ddot{\lambda}_2$, and the third from $p\ddot{\lambda}_3$. Explicit expressions for each of these terms are given by Aksnes (1988).

Treating Eq. (8.260) as a pendulum equation we can calculate the libration period of the resonant argument and the maximum libration width of the three-body resonance using the techniques developed in Sects. 8.6 and 8.7. This gives

$$T_{\text{lib}} = \frac{2\pi}{\sqrt{c}}, \tag{8.270}$$

$$\delta a_{\text{max}} = \frac{4a}{3n}\sqrt{c}. \tag{8.271}$$

The only known example of a three-body resonance in the solar system is the Laplace resonance between three of the Galilean moons of Jupiter. It is important to note that a three-body resonance can exist even if there are no individual two-body resonant pairs involved in the dynamics. For example, it had been thought that Miranda, Ariel, and Umbriel, three of the moons of Uranus, were involved in a Laplace relation (see Greenberg 1975, 1976) although no resonant pairs exist. However, improved determinations of the orbital elements of the uranian satellites have shown that the associated argument is circulating and not librating.

Three-body resonances were also invoked by Dermott & Gold (1977) to account for early observations of the narrow rings of Uranus. They proposed that it was possible for arcs of material to be maintained at the locations of three-body resonances with pairs of uranian satellites. Goldreich & Nicholson (1977b) showed that even if this mechanism worked, the three-body resonances involving Miranda, the smallest of the classical uranian satellites, were stronger than those considered by Dermott & Gold. The paper by Goldreich & Nicholson included expressions for strengths of the three-body resonances.

### 8.17 The Laplace Resonance

The most well known Laplace relation involves three of the Galilean satellites of Jupiter. Io, Europa, and Ganymede are involved in the most wonderfully intricate set of resonances in the solar system. If we use subscripts 1, 2, and 3 to refer, respectively, to Io, Europa, and Ganymede, then we have

$$\lambda_1 - 2\lambda_2 + \varpi_1 = 0°, \quad \lambda_1 - 2\lambda_2 + \varpi_2 = 180°, \quad \lambda_2 - 2\lambda_3 + \varpi_2 = 0°,$$

$$n_1 - 2n_2 + \dot\varpi_1 = 0, \quad n_1 - 2n_2 + \dot\varpi_2 = 0, \quad n_2 - 2n_3 + \dot\varpi_2 = 0 \quad (8.272)$$

and, finally, the Laplace relation

$$\phi_L = \lambda_1 - 3\lambda_2 + 2\lambda_3 = 180°,$$

$$n_1 - 3n_2 + 2n_3 = 0. \quad (8.273)$$

Although the above relation between the mean motions is exact, the resonant argument, $\phi_L$, librates about 180° with an amplitude of 0.064° and a period of 2,071 d (Lieske 1998). The fact that the condition $\phi_L = 180°$ always holds implies that there can never be a triple conjunction of all three satellites involved in the resonance.

We illustrate the sequence of possible conjunctions in Fig. 8.30, where we follow the positions of the three satellites over the interval $T_{rep}$ (see Eq. (8.257)), the time taken for the cycle of conjunctions to repeat itself. The initial configuration shows a conjunction between Europa and Ganymede; at this time Io is 180° away (Fig. 8.30a). When Io and Ganymede have their first conjunction Europa is 60° away (Fig. 8.30b). Ganymede is 90° behind Io when it has its first conjunction

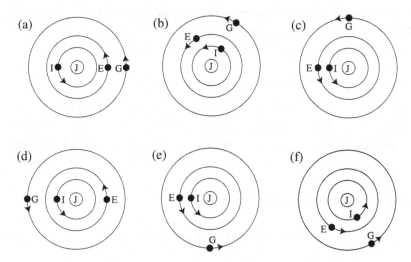

Fig. 8.30. The sequence of conjunctions for the Galilean satellites. The configurations at times (a) $t = 0$, (b) $t = T_{rep}/6$, (c) $t = T_{rep}/4$, (d) $t = T_{rep}/2$, (e) $t = 3T_{rep}/4$, and (f) $t = 5T_{rep}/6$. The letters J, I, E, and G denote Jupiter, Io, Europa, and Ganymede respectively.

with Europa (Fig. 8.30c), and then 180° and 90° at the next two conjunctions (Figs. 8.30d,e). The second conjunction between Io and Ganymede occurs with Europa trailing by 60°.

The full dynamics of this resonance are too complex to describe here and the reader is referred to the paper by Yoder & Peale (1981) for a more complete analysis. However, because a study of the dynamics provides vital information on tidal dissipation in both Jupiter and Io, we need to state some of the more important results.

Yoder & Peale (1981) argue convincingly that the observed near-zero amplitude of libration of the resonant argument, $\phi_L$, can be accounted for if the system of three satellites is now in a near-equilibrium state in which the effects of tidal dissipation in Jupiter are balanced chiefly by those in Io. The rate of tidal dissipation in Io is the key controlling factor in this process. Substituting the values in Table 8.9 into Eq. (8.272), we calculate that the rate of change of Io's longitude of pericentre is given by

$$\dot{\varpi}_1 = 2n_2 - n_1 = -0.7396° \, \text{d}^{-1}. \tag{8.274}$$

This precession rate has a secular contribution due to the oblateness of Jupiter and a resonant contribution due to the gravitational interaction between Io and Europa. Thus,

$$\dot{\varpi}_1 = \dot{\varpi}_{\text{sec}} + \dot{\varpi}_{\text{res}}, \tag{8.275}$$

Table 8.9. Physical and orbital parameters of the satellites of Jupiter involved in the Laplace resonance. The mass, semi-major axis, mean motion, orbital period, and eccentricity of the $i$th satellite are denoted by $m_i$, $a_i$, $n_i$, $T_i$, and $e_i$ respectively.

| Satellite | $i$ | $m_i$ $(10^{23} \mathrm{g})$ | $a_i$ $(10^3 \mathrm{km})$ | $n_i$ $(^\circ \mathrm{d}^{-1})$ | $T_i$ (d) | $e_i$ |
|---|---|---|---|---|---|---|
| Io | 1 | 893.3 | 421.67 | 203.4890 | 1.769137 | 0.0041 |
| Europa | 2 | 479.7 | 670.9 | 101.3747 | 3.551182 | 0.0101 |
| Ganymede | 3 | 1482 | 1070 | 50.3176 | 7.154554 | 0.0006 |

where the secular contribution is given by

$$\dot{\varpi}_{\mathrm{sec}} = \frac{3}{2} n_1 J_2 \left( \frac{C_\mathrm{p}}{a_1} \right)^2 = 0.1290^\circ \, \mathrm{d}^{-1}. \tag{8.276}$$

Io's eccentricity is purely a forced eccentricity due to the resonant interaction between Io and Europa, and its magnitude determines the pericentre rate, $\dot{\varpi}_{\mathrm{res}}$. In Chap. 6 we show that

$$\dot{\varpi}_{\mathrm{res}} = \frac{n_1}{e_1} \frac{m_2}{m_\mathrm{p}} \alpha F(\alpha) = -0.8686^\circ \, \mathrm{d}^{-1}, \tag{8.277}$$

where $\alpha = a_1/a_2 = 0.630$ and $F(\alpha)$ is given by

$$F(\alpha) = \frac{1}{2} \left[ -4 - \alpha \frac{\mathrm{d}}{\mathrm{d}\alpha} \right] b_{1/2}^{(2)}(\alpha) = -1.19049, \tag{8.278}$$

where $b_{1/2}^{(2)}(\alpha)$ is a Laplace coefficient (see Eq. (6.179)). It follows from this analysis that we should expect $e_1 = 0.0044$ and this is in fact close to the observed value of 0.0041 (see Table 8.9).

While $e_1$ is maintained at this value by the action of the Io–Europa resonance, the satellite is heated at a rate given by

$$\frac{\mathrm{d}E}{\mathrm{d}t} = -f \left\{ \frac{21}{2} \frac{k_{21}}{Q_1} e_1^2 n_1 \left( \frac{C_1}{a_1} \right)^5 \frac{\mathcal{G} m_\mathrm{p}^2}{a_1} \right\}, \tag{8.279}$$

where the expression in braces applies to a completely solid Io (Eq. (4.197)) for which Io's Love number is given by $k_{21} = 3/2\bar{\mu}_1$ and the factor $f$ allows for the increase in heating due to Io's molten interior (Peale et al. 1979). The energy dissipated as heat in Io is derived from the satellite's orbit and results in the mean motion increasing at a rate

$$\frac{\dot{n}_1}{n_1} = \frac{3}{2} \frac{\dot{E}}{E}, \tag{8.280}$$

where $E = -\mathcal{G} m_\mathrm{p} m_1/2a_1$ (see Sect. 4.10). The equilibrium between the effects of tidal dissipation in Jupiter that acts to decrease $n_1$ and tidal dissipation in Io that acts to increase $n_1$ allowed Yoder & Peale (1981) to deduce that

$$f \frac{k_{21}}{k_{2p}} \left( \frac{m_p}{m_1} \right)^2 \left( \frac{C_1}{C_p} \right)^5 \frac{Q_p}{Q_1} \approx \frac{1}{13e_1^2} \approx 4 \times 10^4, \tag{8.281}$$

where $k_{2p}$ is the Love number of the planet Jupiter. There are a number of unknowns in this equation. However, the energy dissipated in Io is radiated into space at a rate, $\dot{E}$, that can be measured from Earth. By eliminating the factors $e_1^2 f k_{21}/Q_1$ associated with Io from Eqs. (8.279) and (8.281), we obtain

$$\dot{E} \approx -\frac{21}{26} \frac{k_{2p}}{Q_p} \left( \frac{C_1}{a_1} \right)^2 \frac{\mathcal{G}m_1^2}{a_1} n_1, \tag{8.282}$$

showing that the rate of dissipation of energy in Io is determined by the rate of dissipation of tidal energy in Jupiter and thus that the tidal dissipation function, $Q_p$, of Jupiter can be determined from measurements of $\dot{E}$. From the dynamical oblateness, $J_2$, of Jupiter, we estimate that $k_{2p} \approx 0.53$. Hence we calculate from Eq. (8.282) that

$$\dot{E} \approx \frac{3 \times 10^{18}}{Q_p} \text{W}. \tag{8.283}$$

From ground-based infrared observations of Io, Veeder et al. (1994) estimate that $\dot{E} = 10^{14}$ W from which we deduce that $Q_p \approx 3 \times 10^4$.

The first point to make about this result is that it is remarkably close to the independent estimate $Q_p \geq 8 \times 10^4$ obtained by Goldreich & Soter (1966) from Eq. (8.252) using the assumption that the Laplace resonance is as old as the solar system. (With this assumption the mass of Io in Eq. (8.252) is effectively reduced by a factor 4.2 that allows for the transfer of angular momentum from the orbit of Io to the orbits of Europa and Ganymede.) The fact that the first estimate of $Q_p$ is actually less than Goldreich and Soter's lower bound is apparently discordant, but that may only mean either that the Veeder et al. estimate of $\dot{E}$ is a factor of $\sim 3$ too large or that the rates of tidal dissipation in either Io or Jupiter or both have, over the age of the solar system, varied with time.

From Eq. (8.279) we also calculate that if $\dot{E} \approx 10^{14}$W and $k_{21} \approx 0.03$, then $Q_1 \approx 2f$ and given that $f$ is unlikely to exceed $\sim 10$ (Peale et al. 1979), we must have $Q_1 \leq 20$. This upper bound on $Q_1$ is very low and suggests that the Veeder et al. estimate of $\dot{E}$ may in fact be too high.

## 8.18 Secular and Resonant Motion

In Chapter 7 we showed how the dynamical evolution of the orbits of $N$ gravitationally interacting masses could be modelled by means of secular perturbation theory. One of the early assumptions in the theory was that there were no mean motion resonances between pairs of masses. However, it is possible to adapt the standard secular perturbation theory to include the averaged effect of resonances or near-resonances. This method relies on the fact that the three timescales involved in the motion of an object (the orbital, resonant or circulation, and precessional periods) tend to be very different, thereby allowing certain approximations to be made in the theory. This *separation of timescale* technique was first used by Wisdom (1985a) in his study of chaotic motion at the 3:1 jovian resonance in the asteroid belt (see Chapter 9). It was subsequently used by Malhotra et al. (1989) in their study of the uranian satellite system. In this section we develop the theory of secular and resonant motion in the context of two masses orbiting an oblate central mass. The resulting theory is then incorporated into the full secular theory developed in Chapter 7.

Consider two masses $m$ and $m'$ orbiting an oblate central mass $m_c$. The two masses experience a gravitational potential due to the nonspherical potential of the central body, as well as their mutual perturbations. It is assumed that the orbiting masses are close to a first-order, $j : j - 1$ mean motion resonance. To lowest order the appropriate secular and resonant terms in the disturbing function for inner and outer masses respectively are:

$$\mathcal{R} = \frac{\mathcal{G}m_c}{2a} J_2 \left( \frac{R}{a} \right)^2$$
$$+ \frac{\mathcal{G}m'}{a'} \left( \frac{1}{2} b_{1/2}^{(0)} + Ce \cos[\theta - \varpi] + C'e' \cos[\theta - \varpi'] \right), \quad (8.284)$$

$$\mathcal{R}' = \frac{\mathcal{G}m_c}{2a'} J_2 \left( \frac{R}{a'} \right)^2$$
$$+ \frac{\mathcal{G}m}{a'} \left( \frac{1}{2} b_{1/2}^{(0)} + Ce \cos[\theta - \varpi] + C'e' \cos[\theta - \varpi'] \right), \quad (8.285)$$

where $R$ is the radius of the central mass and

$$\theta = j\lambda' + (1 - j)\lambda, \quad (8.286)$$

$$C = \frac{1}{2} [-2j - \alpha D] b_{1/2}^{(j)}, \quad (8.287)$$

$$C' = \frac{1}{2} [-1 + 2j + \alpha D] b_{1/2}^{(j-1)} - \frac{1}{2\alpha^2} \delta_{j,2}, \quad (8.288)$$

where $\alpha = a/a'$ and $\delta$ is the Krönecker delta function. The first terms in $\mathcal{R}$ and $\mathcal{R}'$ are due to the oblateness of the central body. The remaining terms are derived from the disturbing function expansion given in Appendix B. The $\frac{1}{2} b_{1/2}^{(0)}$ term is the lowest-order part of the secular term from 4D0.1 ($j = 0$). The terms in $C$ and $C'$ are derived from the lowest-order parts of 4D1.1, 4D1.2, and 4I1.3. Note that because we have restricted our attention to lowest-order terms the inclinations of both orbits do not occur.

Malhotra et al. (1989) showed that it is possible to derive a pendulum equation that describes the behaviour of the near-resonance. This involves the derivation of expressions for $j\dot{\lambda}'$ and $(1 - j)\dot{\lambda}$. Neglecting the resonant contribution of the variation of the mean longitude at epoch we have

$$\dot{\lambda} = n \left[ 1 + 3 J_2 \left( \frac{R}{a} \right)^2 - \frac{m'}{m_c} \alpha^2 \frac{\mathrm{d}}{\mathrm{d}\alpha} b_{1/2}^{(0)} \right], \tag{8.289}$$

$$\dot{\lambda}' = n' \left[ 1 + 3 J_2 \left( \frac{R}{a'} \right)^2 + \frac{m}{m_c} \left( 1 + \alpha \frac{\mathrm{d}}{\mathrm{d}\alpha} \right) b_{1/2}^{(0)} \right]. \tag{8.290}$$

A further differentiation with respect to time gives

$$\ddot{\lambda} = F \dot{n} + F' \dot{n}', \tag{8.291}$$
$$\ddot{\lambda}' = G \dot{n} + G' \dot{n}', \tag{8.292}$$

where

$$F = 1 + 7 J_2 \left( \frac{R}{a} \right)^2 + \frac{m'}{m_c} \alpha \left( \frac{1}{3} \alpha \frac{\mathrm{d}}{\mathrm{d}\alpha} + \frac{2}{3} \alpha^2 \frac{\mathrm{d}^2}{\mathrm{d}\alpha^2} \right) b_{1/2}^{(0)}, \tag{8.293}$$

$$F' = -\frac{2}{3} \frac{m'}{m_c} \alpha^{-\frac{1}{2}} \left( 2\alpha \frac{\mathrm{d}}{\mathrm{d}\alpha} + \alpha^2 \frac{\mathrm{d}^2}{\mathrm{d}\alpha^2} \right) b_{1/2}^{(0)}, \tag{8.294}$$

$$G = -\frac{2}{3} \frac{m}{m_c} \alpha^{\frac{3}{2}} \left( 2\alpha \frac{\mathrm{d}}{\mathrm{d}\alpha} + \alpha^2 \frac{\mathrm{d}^2}{\mathrm{d}\alpha^2} \right) b_{1/2}^{(0)}, \tag{8.295}$$

$$G' = 1 + 7 J_2 \left( \frac{R}{a'} \right)^2 + \frac{m}{m_c} \alpha \left( 1 + \frac{7}{3} \alpha \frac{\mathrm{d}}{\mathrm{d}\alpha} + \frac{2}{3} \alpha^2 \frac{\mathrm{d}^2}{\mathrm{d}\alpha^2} \right) b_{1/2}^{(0)}. \tag{8.296}$$

Malhotra et al. (1989) used the separation of timescale technique to show that the time-averaged values of $\cos\theta$ and $\sin\theta$ over the period of circulation are given by

$$\langle \cos\theta \rangle_T = \varepsilon (Ck + C'k') + \mathcal{O}(\varepsilon^2), \tag{8.297}$$
$$\langle \sin\theta \rangle_T = \varepsilon (Ch + C'h') + \mathcal{O}(\varepsilon^2), \tag{8.298}$$

where

$$\varepsilon = \frac{3}{2} \frac{n'^2}{\omega^2} \left[ (j-1) \frac{m'}{m_c} [(j-1)F - jG] \alpha^{-2} + j \frac{m}{m_c} [jG' - (j-1)F'] \right] \tag{8.299}$$

and we have used the standard notation $h = e \sin\varpi$, $k = e \cos\varpi$, $h' = e' \sin\varpi'$, and $k' = e' \cos\varpi'$ (see Chapter 7).

It can be shown that the averaged variation in $\dot{n}$ and $\dot{n}'$ is zero and therefore there is no secular change in $a$ or $a'$ due to the resonant perturbations. This means that we can use $a = \langle a \rangle$ and $a' = \langle a' \rangle$ in our perturbation equations.

The resonant contributions to the perturbation equations given in Eqs. (7.25) and (7.26) are given by

$$\dot{h} = +\frac{m'}{m_c} n \frac{a}{a'} C \cos\theta, \qquad \dot{k} = -\frac{m'}{m_c} n \frac{a}{a'} C \sin\theta,$$
$$\dot{h}' = +\frac{m}{m_c} n' C' \cos\theta, \qquad \dot{k}' = -\frac{m}{m_c} n' C' \sin\theta. \tag{8.300}$$

Taking the averaged value of each side we get

$$\langle \dot{h} \rangle = +\frac{m'}{m_c}n\frac{a}{a'}C\varepsilon(Ck + C'k'), \qquad \langle \dot{k} \rangle = -\frac{m'}{m_c}n\frac{a}{a'}C\varepsilon(Ch + C'h'),$$

$$\langle \dot{h}' \rangle = +\frac{m}{m_c}n'C'\varepsilon(Ck + C'k'), \qquad \langle \dot{k}' \rangle = -\frac{m}{m_c}n'C'\varepsilon(Ch + C'h'). \tag{8.301}$$

Using a simple change of notation and dropping the averaging symbols used above, we can adapt the standard equations of secular perturbation theory to include the effects of near-resonance between any pair of $N$ gravitationally interacting masses. This gives

$$\dot{h}_j = +\sum_{k=1}^{N}\left(A_{jk} + \varepsilon\bar{A}_{jk}\right)k_k, \qquad \dot{k}_j = -\sum_{k=1}^{N}\left(A_{jk} + \varepsilon\bar{A}_{jk}\right)h_k, \tag{8.302}$$

where the $A_{jk}$ are as defined in Eqs. (7.132) and (7.133), and where

$$\bar{A}_{jl} = n_j\frac{m_l}{m_c}\bar{\alpha}_{jl}C_jC_l \tag{8.303}$$

in the case of a near-resonance between the mass $m_j$ and $m_l$. In all other cases $\bar{A}_{jk} = 0$. We also have

$$\bar{\alpha}_{jk} = \begin{cases} a_j/a_k & \text{if } j < k, \\ 1 & \text{if } k < j, \end{cases} \tag{8.304}$$

where we are assuming that $a_1 < a_2 < a_3 \cdots$. In the new notation $C_j = C$, $C_l = C'$ if $j < l$ and $C_j = C'$, $C_l = C$ if $j > l$, with similar definitions for the other quantities associated with the bodies involved in the near-resonance.

Inspection of Eqs. (8.302) shows that the effect of each near-resonance is to modify one or more of the off-diagonal terms in the matrix composed of the $A_{jk}$. This implies that the qualitative nature of the secular solution given in Sect. 7.7 is unaffected – the near-resonance will only affect the quantitative aspects of the solution.

## 8.19 LONGSTOP Uranus

Immediately prior to the *Voyager 2* encounter with Uranus in January 1986, Dermott & Nicholson (1986) pointed out that previous determinations of the masses of the five uranian satellites known at that time were likely to be incorrect as they had been derived without making proper allowance for secular perturbations. In order to provide a numerical test of the validity of the secular theory as applied to the uranian satellites, a long-term numerical integration of the full equations of motion was carried out on the Cray 1S computer at the University of London Computer Centre. This integration was an extension of Project LONGSTOP (LONG-term Gravitational Stability Test of the Outer Planets), which involved a numerical integration of the five outer planets for $10^8$

Table 8.10. Initial conditions for the LONGSTOP Uranus numerical integration.

| j | Satellite | $m_j/m_c$ $\times 10^5$ | $a_j$ (km) | $e_j$ | $\varpi_j$ (deg.) | $I_j$ (deg.) | $\Omega_j$ (deg.) |
|---|-----------|------------------------|-----------|-------|-------------------|--------------|-------------------|
| 1 | Miranda | 0.1 | 129,775 | 0.0027 | 111 | 4.22 | 21 |
| 2 | Ariel | 1.8 | 190,822 | 0.0034 | 120 | 0.31 | 263 |
| 3 | Umbriel | 1.1 | 265,832 | 0.0050 | 193 | 0.36 | 279 |
| 4 | Titania | 3.2 | 436,035 | 0.0022 | 147 | 0.142 | 311 |
| 5 | Oberon | 3.4 | 583,117 | 0.0008 | 212 | 0.101 | 234 |

years (see Roy et al. 1988 and Chapter 9). The details of the application to the uranian satellites (LONGSTOP Uranus) are given in Malhotra et al. (1989).

The initial conditions of the LONGSTOP Uranus integration are given in Table 8.10. The values $\mathcal{G}m_c = 5.784184 \times 10^6 \text{ km}^3 \text{ s}^{-2}$, $J_2 = 3.3450 \times 10^{-3}$, $J_4 = -3.21 \times 10^{-5}$, and $R_c = 26,200$ km were also used. The full equations of motion were integrated for 3,790 y, or approximately 980,000 orbital periods of Miranda.

The $h$ and $p$ time series were Fourier analysed to provide approximate frequencies, amplitudes, and phases for the dominant spectral lines for each satellite. The tentative identifications of the fundamental modes were then used as input for a program that performed a least squares fit on either the $\{h, k\}$ or the $\{p, q\}$ data for all the satellites simultaneously. This provided best fits to the frequencies $\{f_i, g_i\}$ and phases $\{\beta_i, \gamma_i\}$ of the fundamental modes. Tables 8.11 and 8.12 give a comparison between the frequencies calculated from the numerical integration and those derived from the classical Laplace–Lagrange theory. Using the fitted frequencies and phases the program also determined the amplitudes $E_j^{(i)}$ and $I_j^{(i)}$ of the components of each mode identified in the data for each satellite. This technique is known as *synthetic secular perturbation theory*.

Before the derived values could be compared with those from the classical theory it was necessary to derive a set of "averaged" starting conditions, since the

Table 8.11. Comparison of secular eccentricity/pericentre frequencies, $g_i$ (° y$^{-1}$).

| Mode, $i$ | Numerical Integration | Classical Theory | Error | Modified Theory | Error |
|-----------|----------------------|------------------|-------|-----------------|-------|
| 1 | 20.299 | 20.589 | +1.4% | 20.289 | −0.05% |
| 2 | 6.000 | 5.965 | −0.6% | 5.965 | −0.6% |
| 3 | 2.909 | 2.856 | −1.8% | 2.874 | −1.2% |
| 4 | 1.924 | 1.608 | −16.4% | 1.874 | −2.6% |
| 5 | 0.367 | 0.352 | −4.1% | 0.367 | −0.0% |

Table 8.12. Comparison of secular inclination/node frequencies, $f_i$ ($^\circ$ y$^{-1}$).

| Mode, $i$ | Numerical Integration | Classical Theory | Error | Modified Theory | Error |
|---|---|---|---|---|---|
| 1 | −20.495 | −20.587 | −0.4% | −20.524 | −0.1% |
| 2 | −6.013 | −6.018 | −0.1% | | |
| 3 | −2.815 | −2.819 | −0.1% | | |
| 4 | −1.676 | −1.693 | −1.0% | | |
| 5 | −0.248 | −0.249 | −0.2% | | |

numerical integration had used osculating orbital elements while the classical secular theory requires the corresponding averaged elements. Consequently, using the fitted values of $f_i$, $g_i$, $\beta_i$, $\gamma_i$, $E_j^{(i)}$, and $I_j^{(i)}$ ($i, j = 1, 2, 3, 4, 5$), the synthetic solution was evaluated at $t = 0$ to provide an initial set of $\{h_j, k_j, p_j, q_j\}$ that would be independent of short-period effects.

The most striking agreement between the results of the numerical integration and the classical theory is found in the $I$–$\Omega$ eigenfrequencies (Table 8.12). If the numerical integration results are taken to be the "true" values then the largest error in the theoretical frequencies is 1%, and this occurs for $f_4$. This spectacular agreement implies that the standard secular theory based on a second-order (in $e$ and $I$) expansion of the disturbing function is capable of providing a perfectly adequate theory for the long-term dynamics of the uranian satellite system.

When the $e$–$\varpi$ eigenfrequencies are compared in Table 8.11, it is obvious that the results are quite different. Indeed, the smallest error for the $e$–$\varpi$ eigenfrequencies is comparable to the greatest error in the $I$–$\Omega$ eigenfrequencies, while the largest error exceeds 15%. Although the largest discrepancy is in the $g_4$ eigenfrequency, the errors in $g_5$ (4%) and $g_3$ (2%) are still bigger than the "acceptable" values exemplified by the $f_i$ agreements. It is clear that the full integration includes effects that have not been modelled by the classical theory. A prime source for such discrepancies is the effect of near-resonant interactions between pairs of satellites.

An examination of the dominant periodic terms in the $\{h, k\}$ theory for Umbriel, Titania, and Oberon in Laskar's theory of the Uranian satellites (Laskar 1986b) reveals large contributions from the 3:2 Titania–Oberon and 2:1 Umbriel–Titania near-resonances with periods of 144.9 d and 86.4 d respectively. It was clear that the near-resonances have indirect effects on the secular solution itself. Since the two near-resonances are both of the form $j : j - 1$ (i.e., both are first-order, near-resonances) it is to be expected from the form of the disturbing function that the eccentricities, but *not* the inclinations, would be affected. This is consistent with the numerical results. In fact, as we have seen in the previous section, the secular effects of these near-resonant terms can be computed analytically, and we can derive expressions for the modifications they introduce in the secular solution for the eccentricities and pericentres. Malhotra et al. (1989) also included a small correction to the inclination/node solution by including higher-order terms in Miranda's inclination.

A comparison of the error columns in Table 8.11 shows that the modified secular theory gives values for the eigenfrequencies that greatly improve upon those derived from the classical Laplace–Lagrange theory. The residuals in the eccentricity eigenfrequencies are now at most 2%, as compared with the 16% and 4% residuals for $g_4$ and $g_5$, respectively, obtained with the classical theory. Therefore, the modified secular theory gives excellent agreement with the results of a numerical integration of the full equations of motion.

Christou & Murray (1997) used a higher-order (in the masses) secular perturbation theory to provide further improvements on the method presented by Malhotra et al. (1989) for LONGSTOP Uranus. They incorporated the effects of more near-resonances and used a comparison of their higher-order theory with the results of a numerical integration. The improved theory reduced the largest error to less than 0.5%.

## 8.20 Pulsar Planets

A *pulsar* is a rapidly rotating neutron star that emits regular "pulses" of radiation at radio wavelengths. One class of pulsars, the *millisecond pulsars,* rotate more rapidly than the others and are believed to be composed of older pulsars that have been spun up by the accretion of matter from a companion star.

The pulsar PSR1257+12 emits pulses of radio waves every 6.2 ms and was discovered in February 1990. Monitoring of the time of arrival of these pulses took place over more than a year using the Arecibo radio telescope and the Very Large Array. Wolszczan & Frail (1992) reported that delays in the times of arrival of the pulses had distinct periodicities, which they attributed to the effects of two planet-sized objects orbiting the pulsar. The effect they claimed to have detected would be a consequence of the neutron star's movement about the common centre of mass of the system.

In Sect. 2.7 we noted how in the closed two-body system the masses orbit in ellipses about the centre of mass. The addition of a third mass complicates the motion of all three objects, but it is easy to show that the centre of mass of the system will still move with uniform velocity (or will be stationary) and the large central mass will orbit around it. In this case there should be two obvious periods in the system – one associated with each of the two orbiting masses.

By analyzing the data from the pulsar Wolszczan & Frail (1992) deduced the approximate orbital elements of two such planets and provided estimates of their masses. There was also preliminary evidence for the existence of a third planet. The original data taken from their paper are given in Table 8.13. Because

Table 8.13. Parameters of the two pulsar planets proposed by Wolszczan & Frail (1992). $m$ is the mass in units of Earth masses, $a \sin I$ is the projected semi-major axis, $I$ is the (unknown) inclination, $e$ is the eccentricity, $T$ is the time of pericentre passage, $P$ is the orbital period, and $\omega$ is the argument of pericentre.

| Parameter | Planet 1 | Planet 2 |
|---|---|---|
| $m$ ($M_E$) | $3.4/\sin I$ | $2.8/\sin I$ |
| $a \sin I$ (light ms) | $1.31 \pm 0.01$ | $1.41 \pm 0.01$ |
| $e$ | $0.022 \pm 0.007$ | $0.020 \pm 0.006$ |
| $T$ (JD) | $2448105.3 \pm 1.0$ | $2447998.6 \pm 1.0$ |
| $P$ (s) | $5751011.0 \pm 800.0$ | $8487388.0 \pm 1800.0$ |
| $\omega$ (deg.) | $252 \pm 20$ | $107 \pm 20$ |

the elements of the planets are deduced from timing measurements there is no information about the inclinations and nodes of their orbital planes. This is allowed for by including a common $\sin I$ projection factor in some of the orbital parameters shown in Table 8.13.

Wolszczan & Frail (1992) pointed out that the two planets were close to a 3:2 orbital resonance and the $\sim 180°$ difference in their pericentres was to be expected from secular perturbations. Rasio et al. (1992) realised that the secular and resonant effects of perturbations by one planet on the other would lead to long-period changes in their orbital elements, along the lines of those investigated in Sect. 8.17; this would provide a means of confirming the original interpretation of Wolszczan & Frail (1992).

The dynamics of the putative planetary system were subsequently investigated by several other authors including Malhotra et al. (1992) and Peale (1993a). By analysing three years of timing observations obtained using the Arecibo radio telescope, Wolszczan (1994) finally announced the confirmation of the existence of three planets around the millisecond pulsar, now renamed PSR B1257+12.

Questions still remain about the origin of such a resonant configuration. Did the planets evolve into the resonance before the supernova explosion that produced the pulsar, or is the resonance a consequence of some form of drag resulting from material produced after the explosion? An improved understanding of the first planetary resonance discovered outside our solar system may require advances in the theory of stellar evolution as well as dissipative dynamics.

## Exercise Questions

**8.1**    Prior to the advent of interplanetary spacecraft, the only method of determining satellite masses was to use observations of their gravitational interactions, particularly if resonances existed. Enceladus and Dione, two satellites of Saturn, are involved in a 2:1 mean motion resonance. The observed orbital mean motions of Enceladus (the inner and less massive satellite) and Dione are respectively $n_E = 262.732° \, d^{-1}$ and $n_D = 131.535° \, d^{-1}$; Enceladus's eccentricity is $e = 0.0048$. The resonance variable $\phi = 2\lambda_D - \lambda_E - \varpi_E$ is observed to librate about an equilibrium value of zero with an amplitude of $0.297°$. Taking the secular pericentre precession rate of Enceladus to be $\dot\varpi_E = 0.415° \, d^{-1}$ and dominated by Saturn's $J_2$, use the theory developed in Sect. 8.8 to derive expressions for $\bar\delta$ and $R$ as functions of $m_D/m_{Saturn}$ for this resonance. If $e$ is forced by the resonance, use the relationship between $\bar\delta$ and $R$ at an equilibrium point to calculate the value of $m_D/m_{Saturn}$ and hence find the numerical value of $\bar\delta$ and $R$ for the resonance. If Dione's radius is 560 km and the mass of Saturn is $5.685 \times 10^{29}$ g, use your mass determination to estimate Dione's mean density. Is your result physically plausible?

**8.2**    The results of a long-term numerical integration of the motion of three small satellites (with equal masses, low eccentricities ($\sim 0.001$), and low inclinations) around a spherical planet are compared with the analytical solution of the classical Laplace–Lagrange theory. The satellites have semi-major axes of 7.31, 9.31, and 11.60 in units of the radius of the planet. The comparison reveals a large discrepancy between the eccentricity–pericentre secular solution derived from the integration and that derived from theory for the first and third satellites. However, comparison of the inclination–node solution using both methods shows excellent agreement for all satellites. Calculate expressions for the *approximate* amplitudes of the variations in eccentricity due to possible 2:1 near-resonances between every pair of satellites (i.e., the 2:1 inner and the 1:2 outer resonance between the pairs (1,2), (1,3), (2,3)) and use your answers to suggest an explanation for the discrepancies between the secular theory and the results of the numerical integration.

**8.3**    Use Kepler's third law to calculate the semi-major axis (in AU) of the nominal location of all first- and second-order interior mean motion resonances with Jupiter that lie between 2.0 and 4.5 AU. Regions of resonance overlap are associated with chaotic motion (see Sect. 9.6). Using the pendulum model for resonant motion in the planar, circular, restricted three-body problem and neglecting contributions from $\varpi$, calculate the approximate width in semi-major axis for each of these resonances, taking the typical eccentricity of an asteroid to be 0.15 and the Jupiter/Sun mass ratio to be $9.54 \times 10^{-4}$. Identify those ranges of semi-major axis where any of the resonances overlap.

**8.4**    Consider the region of the solar system between 5.9 and 8.8 AU. Following the example of the previous question, calculate the nominal location of all first- and second-order exterior mean motion resonances with Jupiter that lie in this region and calculate their widths using the same parameters as in the previous question. Now find the location and width of all first- and second-order interior mean motion resonances with Saturn in the same region, taking the Saturn/Sun mass ratio to be $2.86 \times 10^{-4}$. Identify those ranges of semi-major axis where any of the resonances overlap and comment on the implications of your results for the long-term stability of objects between the two planets.

**8.5**    Consider the case of a tidally evolving satellite on a circular, prograde orbit around a planet. As the semi-major axis of the satellite increases, a test particle on a coplanar, interior orbit passes without capture through a succession of first- and second-order resonances with the satellite, leading to discrete increases in its eccentricity. The satellite/planet mass ratio is $10^{-6}$ and the initial values of the satellite's semi-major axis and orbital period are $a' = 1.5 \times 10^5$ km and 17 h respectively, and $a'$ increases at a constant rate $\dot{a}' = 3 \times 10^{-4}$ km y$^{-1}$. Given that the test particle's semi-major axis is constant and that it has an initial orbital period of 14.4 h, use the theory developed in Sect. 8.12 to predict the approximate times at which the particle encounters all $e$ and $e^2$ mean motion resonances; you may ignore the effect of the planet's oblateness. Calculate the predicted change in the particle's eccentricity at each resonant encounter.

**8.6**    In 1991 timing data obtained from radio observations of the pulsar PSR1257+12 suggested that at least two planets were in orbit around it. It was also observed that these planets appeared to be close to a 3:2 mean motion resonance. The orbital elements were deduced to be $a_1 = 0.36$ AU, $a_2 = 0.47$ AU, $e_1 = 0.022$, $e_2 = 0.020$, $\varpi_1 = 252°$, $\varpi_2 = 107°$, $P_1 = 66.563$ d, and $P_2 = 98.234$ d, where $a$, $e$, $\varpi$, and $P$ denote semi-major axis, eccentricity, longitude of pericentre, and orbital period, respectively. The corresponding mass ratios were $m_1/M = 7.1 \times 10^{-6}$ and $m_2/M = 5.9 \times 10^{-6}$, where $m$ is the mass of the planet and $M$ is the mass of the pulsar. The existence of the planets was confirmed by observations of variations in the orbital elements. Use the data given above to calculate the amplitude and period of the expected variation in $a$, $e$, and $\varpi$ due to the effects of the near 3:2 resonance between the planets. What other effects could be important in determining the short- and long-term variation in the orbital elements?

# 9

# Chaos and Long-Term Evolution

Take but degree away, untune that string,
And hark what discord follows. Each thing meets
In mere oppugnancy.

William Shakespeare, *Troilus and Cressida, I, iii*

## 9.1 Introduction

In this book we have derived a number of equations of motion to study the
rotational and orbital motion of solar system objects. These equations have
described either conservative systems, such as the two- and three-body problems,
or dissipative systems, such as the equations governing tidal evolution or the
dynamical effects of drag forces. However, all have a common characteristic:
They describe systems that are *deterministic*. This means that the current state
of the system allows us to calculate its past and future state providing we know
all the forces that are acting on it. In the case of the two-body problem we were
able to solve the equations of motion and calculate the behaviour of the system
at all past and future times. A complete analytical solution was not possible in
the case of the three-body problem and we had to rely on numerical solutions
if we wanted to follow the orbital evolution of a test particle. However, there
was an implicit assumption that, given the initial state of the system, we should
be able to calculate its future state by obtaining solutions of the equations of
motion. Unfortunately this assumption is not valid for some of the systems we
have investigated and this is because of the phenomenon called *chaos*.

In Chapter 7 we saw how the simple assumptions of secular perturbation theory allowed us to derive an analytical solution to the problem of the mutual
interaction of $N$ masses. Although we gained no information about the positions
of bodies on their orbits we could at least calculate the other orbital elements at
any time, past or present, from knowledge of the current values. The solution
of the secular problem by Laplace allowed him to make assertions about the

409

long-term stability of the solar system. Laplace himself believed in a deterministic universe where once the laws of nature had been discovered, it would only be a matter of knowing all the initial conditions and solving the appropriate equations; then everything about the system would be known. We now know that this view is false.

In the late nineteenth century Henri Poincaré, now acknowledged as the founder of the science of nonlinear dynamics, began a study of the mathematics of the three-body problem. This seminal work, eventually published in his *Les Méthodes Nouvelles de la Mécanique Céleste,* (Poincaré 1892, 1893, 1899), hinted at the complicated nature of the motion that can arise in solutions to the problem. Poincaré had realized that some starting conditions can give rise to unusual trajectories and laid a solid mathematical foundation for the study of chaos.

The major drawback to numerical studies in Poincaré's time was the lack of an efficient means of obtaining solutions to the equations of motion. The advent of digital computers with ever-increasing performances has added an experimental approach to the whole study of nonlinear dynamics in general, and solar system dynamics in particular. Combined with new observations and advances in theory, this means that we can now recognize the important role that chaos has played in determining the dynamical structure and evolution of the solar system.

There is still no universally accepted definition of chaos, even though it can be detected in a variety of dynamical systems. However, for our purposes we can make use of the following definition: An object in the solar system can be said to exhibit chaotic motion if its final dynamical state is sensitively dependent on its initial state. Given that the measure of any physical quantity has a built-in error, the lack of precision in starting conditions is transformed into an uncertainty in final conditions. In this chapter we examine the nature and consequences of chaotic phenomena in the context of long-term dynamical evolution. The article by Murray (1998) provides a summary of the major results in this field.

## 9.2 Sensitive Dependence on Initial Conditions

Consider the motion of a test particle in the vicinity of a planet orbiting the Sun. If the planet is in a circular orbit then this is an example of the circular restricted three-body problem. In the planar case the components of the particle's osculating position and velocity vectors at any time define four quantities, $x$, $y$, $\dot{x}$, and $\dot{y}$, which give the particle a unique "position" in a four-dimensional space referred to as *phase space*. As the particle evolves it follows a trajectory in phase space as well as the usual *configuration space* consisting of its motion in the $x$–$y$ plane. If the particle experiences only the gravitational field of the Sun then its motion would be entirely predictable. However, the perturbations from the planet cause certain regions of the phase space to become chaotic; the orbital evolution of test particles placed in such regions will take place in an unpredictable fashion.

Figure 9.1 illustrates this phenomenon for two nearby starting conditions for the orbit of a test particle perturbed by Jupiter in the planar, circular restricted problem. Both orbits take the test particle close to Jupiter and yet a difference of only 0.3° in initial longitude is sufficient to produce a dramatically different outcome. This is an example of what is called the *butterfly effect*. This was first mentioned in the context of chaotic weather systems, whereby it was suggested that under suitable conditions the flapping of a butterfly's wings in one part of the world could ultimately lead to a hurricane elsewhere in the world. In the context of solar system dynamics the test particle can be thought of as a spacecraft receiving a gravity assist from a Jupiter flyby. In such examples of chaos a small change in starting values changes the geometry of the encounter and hence the size of the direct perturbation received from the planet.

Comets and asteroids with sufficiently large eccentricities can intersect planetary orbits. With a semi-major axis $a = 13.75$ AU and an eccentricity $e = 0.385$, asteroid (2060) Chiron (see entry in Table A.17) has a perihelion distance inside the orbit of Saturn and an aphelion distance close to the orbit of Uranus. Using the best available orbital elements Oikawa & Everhart (1979) carried out several numerical integrations of Chiron's orbit, all based on starting conditions close to the accepted values. These showed that Chiron will undergo several close approaches to planets. Based on the fact that different starting conditions showed different outcomes, a characteristic of chaotic motion, Oikawa & Everhart were only able to take a probabilistic approach to determining the ultimate fate of

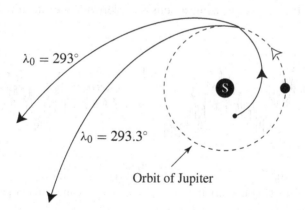

Fig. 9.1. The trajectories of two test particles started with the same semi-major axis ($a_0 = 0.8$), eccentricity ($e_0 = 0.4$), and longitude of pericentre ($\varpi = 295°$) but with slightly different initial mean longitudes ($\lambda_0 = 293°$ and $\lambda_0 = 293.3°$). Jupiter's orbit is denoted by a dashed circle and it has zero initial longitude. The numerical integration lasted one Jupiter period.

Chiron. They estimated a 1 in 8 chance that Saturn will place Chiron on to a hyperbolic orbit that ejects it from the solar system. The more likely outcome (a 7 in 8 chance) is that the close approaches to Saturn will cause Chiron's orbit to evolve towards the inner solar system where it would come under gravitational perturbations from Jupiter. The classification of Chiron as an asteroid is now debatable with the discovery of large, nonperiodic changes in its brightness (Tholen et al. 1988) and a coma (Meech & Belton 1989). Chiron is now classified as a *Centaur*, one of a number of giant, planet-crossing, cometlike objects that may represent Edgeworth–Kuiper belt objects on their way to becoming short-period comets (see, for example, Luu 1994).

Another example of the butterfly effect in the context of solar system dynamics is the orbital history of Comet Shoemaker-Levy 9. Numerical integrations of the comet's orbit show that prior to its impact with Jupiter in July 1994, it was disrupted by a close approach to Jupiter in 1992. Work by Chodas & Yeomans (1996) suggests that the comet was originally captured into orbit around Jupiter in $1929 \pm 9$ y. Prior to capture its orbit was probably similar to that of many of the Jupiter family of comets, with a low-eccentricity orbit that was likely to have been interior to Jupiter's orbit. Further information about the history of the comet is impossible, because (i) the orbit is chaotic and (ii) there will be no new astrometric data to enable an improved initial orbit to be determined.

However, chaos can be more subtle. Figure 9.2 shows the results of two numerical integrations of the equations of motion in the circular restricted three-body problem. In this case the orbital evolution is almost exactly the same for approximately 150 Jupiter periods, after which they gradually differ. There is no obvious reason for the drift (for example, there are no close approaches to Jupiter during the integration period) but the behaviour is again characteristic of chaotic motion with the orbits showing a sensitive dependence on initial conditions.

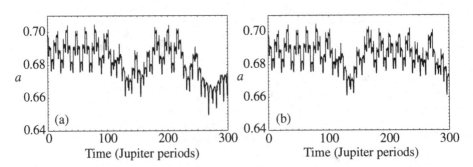

Fig. 9.2. The variation in semi-major axis of two test particles perturbed by Jupiter ($\mu_2 = 0.001$) in the planar, circular restricted problem. The particles were started with the same semi-major axis ($a_0 = 0.6984$), eccentricity ($e_0 = 0.1967$), and longitude of pericentre ($\varpi = 0°$) but with slightly different initial mean longitudes ((a) $\lambda_0 = 0°$ and (b) $\lambda_0 = 10^{-6°}$). Jupiter was started with zero initial longitude.

The type of chaotic behaviour contrasts with the more dramatic example shown above (see Fig. 9.1) and yet both are valid illustrations of chaos in the circular restricted problem.

## 9.3 Regular and Chaotic Orbits

We have already seen that the two-body problem is integrable (see Chapter 2) whereas the three-body problem is not (see Chapter 3). In fact there are chaotic solutions to the restricted three-body problem for certain starting conditions but we have deliberately avoided discussing them up to this point. We shall carry out a more thorough examination of chaos in the context of the circular, restricted problem in Sect. 9.4. Before proceeding with these and other examples of chaotic motion in the solar system, we demonstrate the differences between regular (i.e., nonchaotic) motion and chaotic motion, showing the characteristics of each and how some of the properties of chaotic orbits can be quantified.

### 9.3.1 The Poincaré Surface of Section

We met the concept of a surface of section in Sect. 5.4 when we discussed spin–orbit resonance. In that case we were representing the solution of a second-order, nonlinear differential equation for the time variation of $\theta$, the orientation angle of a nonspherical satellite. In Fig. 5.17b and 5.18b we plotted points denoting the values of $\dot{\theta}/n$ and $\theta$ at every pericentre passage, where $n$ was the mean motion of the satellite.

In the case of the planar, circular, restricted three-body problem the situation is more complicated. The equations of motion given in Eqs. (3.16) and (3.17) consist of two, simultaneous, nonlinear, second-order differential equations. The equations are repeated here for convenience:

$$\ddot{x} - 2n\dot{y} - n^2 x = -\mu_1 \frac{x + \mu_2}{r_1^3} + \mu_2 \frac{x - \mu_1}{r_2^3}, \qquad (9.1)$$

$$\ddot{y} + 2n\dot{x} - n^2 y = -\left(\frac{\mu_1}{r_1^3} + \frac{\mu_2}{r_2^3}\right) y, \qquad (9.2)$$

where $\mu_1 = m_1/(m_1 + m_2)$, $\mu_2 = m_2/(m_1 + m_2)$, and

$$r_1^2 = (x + \mu_2)^2 + y^2, \qquad (9.3)$$
$$r_2^2 = (x - \mu_1)^2 + y^2, \qquad (9.4)$$

where $r_1$ and $r_2$ are the distances of the test particle from the masses $m_1$ and $m_2$ respectively.

The solution consists of sets of values of $x$, $y$, $\dot{x}$, and $\dot{y}$ at a sequence of times, where these quantities denote the components of the position and velocity vectors in the rotating reference frame. Recall that apart from special starting conditions the restricted problem is nonintegrable and we have to resort to numerical solutions of the equations of motion.

We showed in Chapter 3 that there exists a constant of the motion, the Jacobi constant, defined by

$$C_J = n^2 \left( x^2 + y^2 \right) + 2 \left( \frac{\mu_1}{r_1} + \frac{\mu_2}{r_2} \right) - \dot{x}^2 - \dot{y}^2. \tag{9.5}$$

We have already discussed how the values of $x$, $y$, $\dot{x}$, and $\dot{y}$ at any given time correspond to a single point in a four-dimensional phase space. Because of the existence of the Jacobi constant in the circular restricted problem the path of the particle in phase space is confined to a surface (see Fig. 9.3a). Therefore, for a given, fixed value of the Jacobi constant we only require three out of the four quantities to define the osculating orbit at that time uniquely. This is because the other quantity can be determined, at least up to a sign change, by the equation defining the Jacobi constant. For example, if we take $x$, $y$, and $\dot{x}$ as our three quantities, the other one, $\dot{y}$, is determined by Eq. (9.5) (see Fig. 9.3b), provided we have noted the value of the Jacobi constant at time $t = 0$. If we now define a plane, say $y = 0$, in the resulting three-dimensional space, the values of $x$ and $\dot{x}$ can be plotted every time the particle has $y = 0$ (see Fig. 9.3c). The ambiguity in the sign of $\dot{y}$ is removed by considering only those crossings with a fixed sign of $\dot{y}$. This is the method of the *Poincaré surface of section* or the *Poincaré map* and this is the technique we have used to illustrate the regular and chaotic regions in the circular restricted problem. The section is obtained by fixing a plane in the phase space and plotting the points when the trajectory intersects this plane in a particular direction. Note that as a result we do not plot the points at equal time intervals; we only plot a point when an intersection takes place.

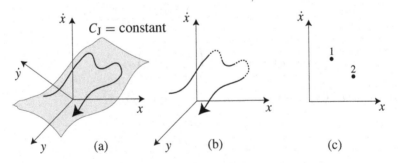

Fig. 9.3. The production of a Poincaré surface of section. (a) The position and velocity of the particle define a point in a four-dimensional phase space at any particular time. The existence of the Jacobi constant means that the test particle's path is moving on a particular surface in this space. (b) We can consider just the $x$, $y$, and $\dot{x}$ values, since the value of $\dot{y}$ is always determined up to an arbitrary sign change by Eq. (9.5). The dashed line denotes the part of the trajectory where $y < 0$. (c) A point is plotted every time $y = 0$ with $\dot{y} > 0$, resulting in a sequence of points in a two-dimensional space.

At first glance we seem to have removed, or at least hidden, a lot of information about the orbit. However, the main advantage of the surface of section method is this very sparseness. Instead of plotting the trajectory in $x-y$ space and following the test particle around numerous orbits, which may only change slightly, with the surface of section method we get an overview of changes in the orbit. All the information is still contained within those points, even though some effort may be required to extract it. There are other advantages to the method.

We are able to use this technique because we are considering the circular, planar, restricted problem. If we now assume that the eccentricity of the secondary mass is nonzero then we can no longer make use of a Poincaré surface of section for the problem. However, we can make use of the averaging principle and isolate only those terms in an expansion of the disturbing function that are likely to dominate. In some circumstances it is possible to obtain analytical solutions of the averaged system and this was one of the techniques employed in our study of resonant perturbations in the previous chapter.

Now we consider numerical solutions to our system of equations using $\mu_2 = 10^{-3}$, a value comparable to the Jupiter/Sun mass ratio.

### 9.3.2 Regular Orbits

Consider the trajectory of a test particle with starting values $x_0 = 0.55$, $y_0 = 0$, and $\dot{x}_0 = 0$, with $\dot{y} > 0$ determined from the solution of Eq. (9.5) with $C_J = 3.07$. We can convert these values to their equivalent semi-major axis $a$ and eccentricity $e$ using the formulae given in Chapter 2. This gives $a_0 = 0.6944$ and $e_0 = 0.2065$ for the initial values. In the Sun–Jupiter system this value of $a_0$ would correspond to a test particle at 3.612 AU, well beyond the main belt of asteroids.

In Fig. 9.4a we show the evolution of $e$ as a function of time. Although there are obvious variations, the plot shows a regular variation in the eccentricity in the

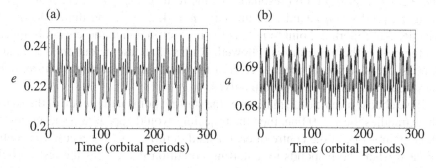

Fig. 9.4. The time variability of (a) eccentricity $e$ and (b) semi-major axis $a$ for initial values $a_0 = 0.6944$ and $e_0 = 0.2065$. The plots show a behaviour characteristic of regular orbits. (Adapted from Murray 1998.)

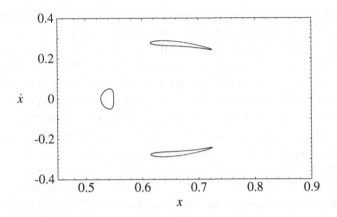

Fig. 9.5.   A surface of section for the regular trajectory shown in Fig. 9.4. The points are plotted whenever $y = 0$ with $\dot{y} > 0$. (Adapted from Murray 1998.)

range 0.206–0.248 over the course of the integration. At this particular location we would expect to see resonant perturbations since $(a/a_J)^{3/2} = 0.564 \approx 4/7$ and the orbit of the test particle is close to a 7:4 resonance with Jupiter. The equivalent plot for the variation of the semi-major axis is shown in Fig. 9.4b. Note that the changes in $a$ are correlated with those in $e$, as we would expect from some of the simple relations discussed in Sect. 8.5. The value of $a$ appears to be librating close to the location of the nominal resonance at $a = (4/7)^{2/3} \approx 0.689$. Figure 9.4 only shows the variations in $e$ and $a$. A complete description of the trajectory should also involve plots of $\varpi$ and another angle, such as the mean longitude $\lambda$.

Figure 9.5 shows the same trajectory displayed as a surface of section, obtained by plotting the values of $x$ and $\dot{x}$ whenever $y = 0$ with $\dot{y} > 0$. The pattern that emerges shows three, distinct "islands". The appearance of such islands is one characteristic of resonant motion when displayed as a surface of section. In these cases a mean motion resonance of the form $p + q : p$, where $p$ and $q$ are integers, produces $q$ islands. In this case $p = 4$, $q = 3$, and three islands are visible. An important point to note is the manner in which these islands appear on the plot as the trajectory is followed. Rather than tracing out one island at a time, successive points occur at each of the three island locations in turn, until they appear to form three smooth curves.

If we choose a value of $x_0$ that places the starting point at the centre of the island that straddles the $\dot{x} = 0$ line, then the trajectory would appear as a succession of three points, one at the centre of each island in turn. This is because the centre of each island corresponds to a starting condition that places the test particle at the middle of the resonance. This is the type of resonant orbit illustrated in Fig. 8.15a for the 2:1 resonance. Such points are said to be periodic points of the Poincaré map because the system returns to the same point every third time the trajectory crosses the section. By moving the starting location further away from the centre the islands would get larger, corresponding to larger variations in *e* and *a*. Eventually some starting values would lead to trajectories that were not in resonant motion and there would no longer be distinct islands in the section plot.

### 9.3.3 Chaotic Orbits

Figure 9.6 shows the plots of *e* and *a* as a function of time for the test particle orbit with starting values $x_0 = 0.56$, $y_0 = 0$, and $\dot{x}_0 = 0$, and $\dot{y}$ determined from Eq. (9.5) with $C_J = 3.07$, the same value used above to illustrate the regular trajectory. The corresponding orbital elements are $a_0 = 0.6984$ and $e_0 = 0.1967$. Note that although these values are only slightly different from those used above, the nature of the variations in *e* and *a* are very different. The eccentricity undergoes irregular variations from 0.188 to 0.328 and the value of *a* does not remain close to the resonant value. This is a chaotic trajectory where there is no obvious pattern to the variations in the orbital elements; this alone does not necessarily mean that the orbit is chaotic. For example, a secular solution to the *N*-body problem can display complicated behaviour similar to that associated with chaotic motion and yet we know that it is derived from an analytical solution composed of a finite number of frequencies.

The identification of this particular orbit as chaotic becomes apparent from a study of its Poincaré surface of section (see Fig. 9.7). Note that the orbit covers a larger region of phase space than in the regular example (cf. Fig: 9.5). Furthermore, instead of the points lying on a smooth curve, they are beginning to fill an area of the phase space. Note also that there is a tendency for some points to "stick" to the edge of the 7:4 and other resonances; indeed the points

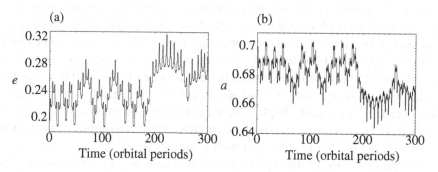

Fig. 9.6. The time variability of (a) eccentricity *e* and (b) semi-major axis *a* for initial values $a_0 = 0.6984$ and $e_0 = 0.1967$. The plots show a behaviour characteristic of chaotic orbits. (Adapted from Murray 1998.)

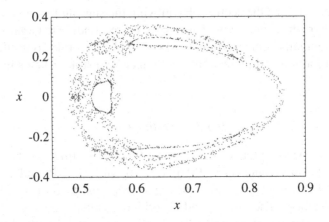

Fig. 9.7.  A surface of section for the chaotic trajectory shown in Fig. 9.6. The points are plotted whenever $y = 0$ with $\dot{y} > 0$, and the scale is the same as that in Fig. 9.5. (Adapted from Murray 1998.)

in the chaotic trajectory help to define several empty regions, each of which can be associated with a resonance. The stickiness phenomenon implies that if we examine the trajectory at particular times it may give the impression of regular motion. In fact, this is another characteristic of chaotic behaviour – it can give the impression of regular motion for long periods of time; this can make chaotic motion difficult to detect in some circumstances.

Although the chaotic orbit shown in Fig. 9.7 covers a large area of the $x$–$\dot{x}$ space, there is some evidence that there are still bounds to the motion. Thus, this orbit may illustrate the phenomenon of *bounded chaos* and this will be discussed in Sect. 9.4.

### 9.3.4 The Lyapounov Characteristic Exponent

Chaotic orbits have the characteristic that they are sensitively dependent on initial conditions. This is illustrated in Fig. 9.8 where we show part of the plot of $e$ as a function of time for two similar chaotic trajectories. The first corresponds to the trajectory shown in Figs. 9.6 and 9.7 where we used $x_0 = 0.56$; the second has $x_0 = 0.56001$ and we have adjusted the initial value of $\dot{y}$ to maintain the same value of $C_J$. It is clear that after 60 orbital periods the orbits have drifted apart. If we performed the same displacement experiment and followed the evolution of a regular orbit there would be some variation in the magnitude of the relative separation but we would not detect such a large drift.

We can make use of the type of divergence seen in Fig. 9.8 to measure the *maximum Lyapounov characteristic exponent* (LCE) of a dynamical system, giving us a quantitative measure of the rate of divergence of nearby trajectories (see, for example, Lichtenberg & Lieberman 1983). In a dynamical system such

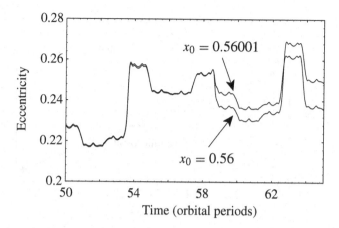

Fig. 9.8. Parts of two plots of the time evolution of the eccentricity for two nearby orbits in a chaotic region of the phase space. The initial values of $x$ differ by 0.00001 and a clear separation can be seen after $\sim 60$ orbital periods. (Adapted from Murray 1998.)

as the three-body problem there are a number of quantities called the Lyapounov characteristic exponents. For an arbitrary starting condition it can be shown that a measurement of the local divergence of nearby trajectories leads to an estimate of the largest of these exponents.

Consider two orbits separated in phase space by a distance $d_0$ at time $t_0$ (see Fig. 9.9a). Let $d$ be their separation at time $t$. The orbit is chaotic if $d$ is approximately related to $d_0$ by

$$d = d_0 \exp \gamma (t - t_0), \tag{9.6}$$

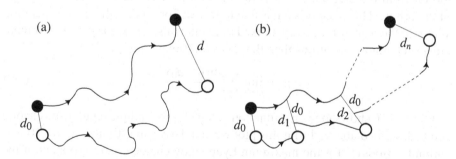

Fig. 9.9. Calculation of the maximum Lyapounov characteristic exponent by measuring the divergence of nearby trajectories. (a) The straightforward method of calculating the initial and final separations in phase space. (b) The renormalisation method whereby the displaced particle's trajectory is moved back along the separation vector to the original displacement distance.

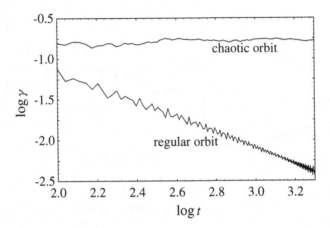

Fig. 9.10. Plots of $\log \gamma$ as a function of log time for the regular and chaotic trajectories discussed above. (Adapted from Murray 1998.)

where $\gamma$ is the maximum Lyapounov characteristic exponent. Note that we must have $\gamma > 0$ for otherwise the trajectories would approach one another as $t$ increased. We can estimate the value of $\gamma$ from the results of a numerical integration by means of the relation

$$\gamma = \lim_{t \to \infty} \frac{\ln (d/d_0)}{t - t_0}. \tag{9.7}$$

Monitoring the behaviour of $\gamma$ as a function of time on a log–log scale usually reveals a striking difference between regular and chaotic trajectories. For regular orbits the initial and final displacements are close to one another ($d \approx d_0$) and hence a log–log plot would have a slope of $-1$. However, if the orbit is chaotic, then $\gamma$ tends to a positive value. In cases where the displaced trajectory drifts too far from the original one $\gamma$ is no longer a measure of the local divergence of the orbits. Therefore it is advisable to rescale (or renormalise) the separation vector of the nearby trajectory at fixed time intervals, $\Delta t$ (see Fig. 9.9b). If there are $n$ such renormalisations then the revised estimate of $\gamma$ is

$$\gamma = \lim_{n \to \infty} \sum_{i=1}^{n} \frac{\ln (d_i/d_0)}{n \Delta t}. \tag{9.8}$$

Figure 9.10 shows $\log \gamma$ as a function of $\log t$ for the sample regular and chaotic orbits described above. From this plot we can derive an estimate of $\gamma = 10^{-0.77}$ (orbital periods)$^{-1}$ for the maximum Lyapounov characteristic exponent of the chaotic orbit. The corresponding *Lyapounov time*, the time for the displacement to increase by a factor e, is given by $1/\gamma$, which in this case is $\sim 6$ orbital periods. Therefore, at least for this starting condition, the chaotic nature of the orbit quickly becomes apparent. This is consistent with the demonstration shown in Fig. 9.8.

For simpler dynamical systems it can be possible to obtain all the Lyapounov characteristic exponents by analytical means. However, this is usually the exception in solar system dynamics. Furthermore, numerical experiments that follow the evolution of two nearby orbits may not always be successful, since our approximation to the maximum LCE, $\gamma$, is only defined in the limit as $t \to \infty$ or $n \to \infty$. For example, the phenomenon of some chaotic orbits sticking to the boundaries of resonances implies that there may be the appearance of a regular trajectory for long periods of time. Therefore, although the chaotic nature of a chaotic orbit may be readily apparent, it may never be possible to prove a regular orbit is truly regular.

## 9.4 Chaos in the Circular Restricted Problem

The planar, circular, restricted three-body problem is perhaps the simplest dynamical model that approximates the motion of real objects in the solar system and yet, as we have seen, it has orbital solutions that display a surprising degree of complexity. In the previous section we demonstrated the existence of regular and chaotic orbits for a small sample of orbits; yet the structure of the phase space in the circular restricted problem is far from being fully understood. In this section we consider some of the properties of the restricted problem for the specific mass ratio of $\mu_2 = 10^{-3}$, following the work of Winter & Murray (1994a,b). Note that for this value of $\mu_2$ the results given here for the motion of test particles approximate the motion of asteroids perturbed by a coplanar Jupiter moving on a circular orbit.

There have been several numerical studies of the circular restricted problem (see, for example, Hénon 1969 and Jefferys 1971). In the comprehensive survey by Winter & Murray (1994a,b) several hundred surfaces of section were produced for a variety of Jacobi constants and orbits both internal and external to the orbit of the perturber were examined. From these plots it is possible to measure properties such as the location and size of regular and chaotic regions and the maximum amplitude of resonant librations.

Recall that in the restricted problem neither the orbital energy nor the angular momentum are conserved because the potential experienced by the particle is explicitly time dependent. However, the dynamical system retains an integral of the motion, the Jacobi constant, given by Eq. (9.5). As we saw in Chapter 3, setting $\dot{x} = \dot{y} = 0$ in the definition of $C_J$ results in an equation that describes the zero-velocity curves associated with the value of $C_J$. These curves define the limits of prohibited regions in the physical space since particles with that value of $C_J$ cannot move inside such regions. Figure 9.11 shows zero-velocity curves and the corresponding regions from which the particle is excluded for a selection of values of $C_J$.

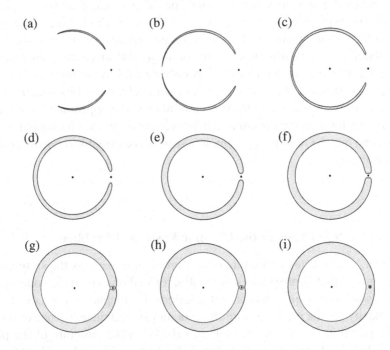

Fig. 9.11.   Plots of zero-velocity curves and the enclosed prohibited regions (shaded) values of the Jacobi constant equal to (a) 3, (b) 3.001, (c) 3.002, (d) 3.01, (e) 3.02, (f) 3.03, (g) 3.039, (h) 3.04, and (i) 3.05. The small points denote the locations of the two masses.

For low values of $C_J$ the prohibited regions are confined to a zone around the $L_4$ and $L_5$ Lagrangian points located 60° ahead of and behind the orbit of the perturbing mass. For larger values of $C_J$ these zones merge and encompass the $L_3$ point. However, there is still an open region close to the location of the perturbing mass. As $C_J$ increases the prohibited region broadens and encroaches upon the perturbing mass, eventually enclosing it. At this stage particle orbits that are initially interior to the perturbing mass must remain interior to it for all time; similarly those orbits that are initially exterior to it must remain exterior for all time. For some values of the Jacobi constant (e.g., $C_J = 3.050$) there is a small region of possible motion around the perturbing mass. In such cases a test particle orbiting the perturbing mass can never orbit the central mass or escape from its orbit around the perturbing mass. For larger values of $C_J$ this region decreases further and the radial extent of the forbidden region grows.

It is clear from Fig. 9.11 that for values of $C_J \geq 3.040$ a particle that is started in an orbit about the central mass cannot get closer to the perturber than a certain limit, which is determined by the extent of the prohibited region.

Figure 9.12 shows the minimum possible distance, $d_{\text{min}}$, between the particle and the perturbing mass as a function of the Jacobi constant, $C_{\text{J}}$.

In the computations of Winter & Murray the starting conditions were mostly chosen such that for each value of $C_{\text{J}}$ the values of $x$ were selected by taking $y = \dot{x} = 0$ with $\dot{y} > 0$ and Jupiter along the $x$ axis at conjunction with the particle. Hence the integration was usually started at either the pericentre or apocentre of the particle's orbit. In a few cases the starting conditions were chosen such that $\dot{x} \neq 0$. This was needed in order to obtain surfaces of section of regular areas related to the main resonances of even order.

There is always a choice of surfaces of section. In their study Winter & Murray chose a system where the values of $x$ and $\dot{x}$ were computed whenever the trajectory crossed the plane $y = 0$ with $\dot{y} > 0$. In the previous section we saw how the surface of section technique is good at illustrating the regular or chaotic nature of the trajectory. If there are smooth, well-defined islands, then the trajectory is likely to be regular and the islands correspond to libration around an exact resonance between the mean motions of the perturber and the particle; in general the number of islands corresponds to the order of the resonance. Exceptions to this rule include those orbits associated with near-zero eccentricity of the test particle and a class of eccentric orbits in the 1:1 resonance with the perturber. Any "fuzzy" distribution of points in the surface of section implies that the trajectory is chaotic. The surface of section is also useful in demonstrating the extent of the particle's motion throughout the phase space. Throughout this section we show a representative sample of orbits and the appearance of their Poincaré surfaces of section. In the surfaces of section periodic orbits give rise to fixed points that are the centre of islands of stability. The islands correspond to the quasi-periodic orbits librating around the stable positions.

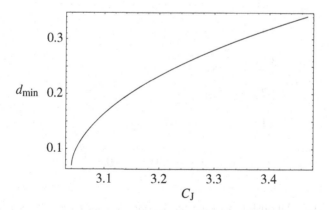

Fig. 9.12. The minimum possible distance $d_{\text{min}}$ between the test particle and the perturbing mass as a function of the Jacobi constant $C_{\text{J}}$. (Adapted from Winter & Murray 1994a.)

For a given point, $(x, \dot{x})$, on the surface of section one can use the value of the Jacobi constant, the fact that $y = 0$, and Eq. (9.5) to find the value of $\dot{y}$. Therefore, all four quantities are known and the semi-major axis and the eccentricity can be found using

$$a = \left[ \frac{2}{r_1} - \frac{V^2}{\mu_1} \right]^{-1},$$

(9.9)

$$e = \sqrt{1 - \frac{h^2}{\mu_1 a}},$$

(9.10)

where $V^2 = \dot{x}^2 + (\dot{y} + x + \mu_2)^2$ and $h = (x + \mu_2)(\dot{y} + x + \mu_2)$. Using these equations one can identify the initial mean motion frequencies associated with points on a surface of section and then identify the resonances associated with the islands of stability that appear in the plots, bearing in mind that the mean motion of the perturber is equal to unity and that generally the number of islands is equal to the order of the resonance.

The periodic orbits associated with the planar, circular, restricted three-body problem can be classified into two types. *Periodic orbits of the first kind* arise from the fact that the test particle is started in a circular orbit ($e_0 = 0$). Figure 9.13 shows the values of $C_J$ as a function of $x_0$ for starting conditions with $\dot{x}_0 = 0$ that give rise to periodic orbits of the first kind. This diagram can be used in conjunction with the surface of section plots to identify islands associated with such periodic orbits. The fact that this function has a minimum at $C_J = 3.027$ implies that for a Jacobi constant below this value there are no starting values $x_0$ that give rise to periodic orbits of the first kind. *Periodic orbits of the second*

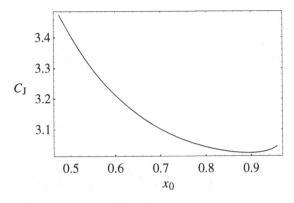

Fig. 9.13.   The relationship between the Jacobi constant $C_J$ and the starting value $x_0$ with $\dot{x}_0 = 0$, which give rise to periodic orbits of the first kind. (Adapted from Winter & Murray 1994a.)

*kind* arise from the fact that the particle is at the centre of a resonance; this is true for both primary and secondary resonances (see below).

The structure of the phase space of the planar, circular, restricted three-body problem is mainly determined by resonant phenomena (see Chapter 8). In these situations there is a commensurability between the mean motion of the particle and the mean motion of the perturbing mass, that is a relationship of the form $(p + q)n' \approx pn$, where $p$ and $q$ are integers and $n$ and $n'$ are the mean motions of the test particle and the perturber, respectively; $q$ is called the *order* of the resonance. The location of stable and unstable equilibrium points due to each resonance, as well as the strength of the resonances, define the location and size of regular and chaotic regions. The resonances can be classified into *primary* and *secondary resonances*. Primary resonances are due to commensurabilities between the mean motions of the perturbing mass and the test particle. In the restricted problem secondary resonances arise when the libration frequency of a primary resonance is commensurate with the circulation frequency of a higher order primary resonance.

A set of three islands are shown in Fig. 9.14a corresponding to a single trajectory with $C_J = 3.02$, $x_0 = 0.205$, and $\dot{x}_0 = 0$; this corresponds to $a_0 = 0.521$ and $e_0 = 0.606$. This trajectory is librating around the 5:2 primary resonance. Clusters of islands arising from a simpler island chain are features that appear in surfaces of section due to a trajectory librating around a secondary resonance. Figure 9.14b shows the surface of section of a trajectory that librates around the

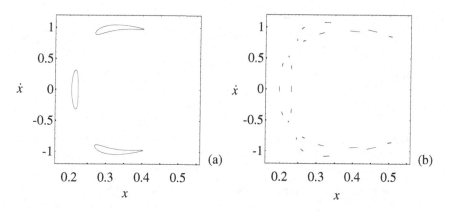

Fig. 9.14. Poincaré surfaces of section showing two quasi-periodic orbits for the same value of the Jacobi constant, $C_J = 3.02$. (a) Using $x_0 = 0.205$ the trajectory shows three islands associated with libration about the 5:2 primary resonance. (b) Using $x_0 = 0.23$ the trajectory shows the twenty-four islands associated with the 8:1 secondary resonance between the libration frequency of the 5:2 primary resonance and the circulation frequency of the 60:24 primary resonance. (Adapted from Winter & Murray 1994a.)

8:1 secondary resonance arising from libration in the 5:2 primary resonance and circulation in the 60:24 primary resonance.

The chaotic motion is associated with unstable equilibrium points and their accompanying separatrices. The existence of chaotic motion does not imply that the trajectory is unbounded. For example, Fig. 9.11 shows that orbits can remain confined for all time for certain values of the Jacobi constant. Chaotic trajectories can be localised or cover large regions of the $(x, \dot{x})$ plane. Two extreme examples are shown in Figure 9.15. The first shows the motion of a particle close to a narrow separatrix (Fig. 9.15a); the second case (Fig. 9.15b) shows a chaotic "sea" with the particle covering a large area. In fact chaotic trajectories can be used to highlight the locations of regular regions of the phase space.

An inspection of the surfaces of section on the following plots shows the clear increase in the extent of the chaos as $C_J$ decreases. This is connected with the change in the zero-velocity curves (see Fig. 9.11) that permits closer approaches between the particle and the perturber as the value of the Jacobi constant gets smaller. From Eq. (9.5) it is clear that if $(x, \dot{x})$ denotes a point on the surface of section, then the point $(x, -\dot{x})$ must also lie on the same surface. This explains the obvious symmetry of the surfaces of section about the $x$ axis.

In general, for the starting conditions considered by Winter & Murray (1994a,b), the orbits are initially prograde in the inertial and rotating reference frames. Most prograde orbits in the rotating frame only produce points on the surface of section with positive values of $x$. However, points with negative values of $x$ can also appear in a surface of section due to regular orbits with apocentric

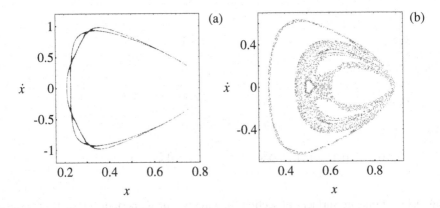

Fig. 9.15.  Poincaré surfaces of section for two chaotic orbits. (a) With $C_J = 3.111$ and $x_0 = 0.23$ the resulting trajectory is at the separatrix of the 8:3 resonance and remains confined to a well-defined area. (b) With $C_J = 3.051$ and $x_0 = 0.445$ the plot shows the large area covered in the $(x, \dot{x})$ space by a chaotic trajectory; notice the regular regions are being highlighted by the chaotic trajectory. (Adapted from Winter & Murray 1994a.)

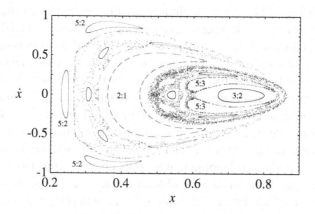

Fig. 9.16. Poincaré surface of section for a variety of trajectories with $C_J = 3.07$. The location of several resonant islands are identified. (Adapted from Murray 1998.)

distances larger than unity. Chaotic trajectories can achieve mean motions that are lower than that of the perturber or even orbits that are retrograde in the inertial frame. Both cases give rise to orbits that are retrograde in the rotating frame, producing points on the surface of section with negative values of $x$. These points have been deliberately excluded from the plots presented here.

The extent of the chaos in a system can depend on several factors. In the case of the circular restricted problem the key quantities are the values of $C_J$ and $\mu_2$. In Figs. 9.16 and 9.17 a number of trajectories are shown for two different values of the Jacobi constant but the same value of $\mu_2$. In the first case, Fig. 9.16, the value is $C_J = 3.07$ (the same as the value used in Figs. 9.5 and 9.7) while in Fig. 9.17 it is $C_J = 3.13$. It is clear that the extent of the chaos

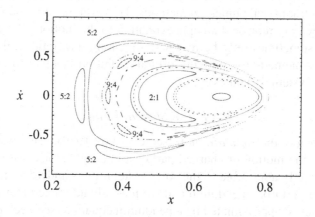

Fig. 9.17. Poincaré surface of section for a variety of trajectories with $C_J = 3.13$. The location of several resonant islands are identified. (Adapted from Murray 1998.)

is reduced in Fig. 9.17. We have already seen in Fig. 9.11 that the value of $C_J$ determines how close the test particle can get to the perturber. Indeed, for the case $\mu_2 = 0.001$ and $C_J > 3.04$ it is impossible for their orbits to intersect, although the perturbations can still be significant. We can interpret the behaviour seen in Figs. 9.16 and 9.17 in the following terms: As $C_J$ decreases, the extent of the chaotic separatrices associated with each resonance increases until the regular curves separating adjacent resonances are broken down and the neighbouring chaotic regions begin to merge. This can also be thought of as the overlap of adjacent resonances giving rise to chaotic motion and will be discussed further in Sect. 9.6. It is the overlapping that permits chaotic orbits to explore regions of the phase space that are inaccessible to the regular orbits. If we apply this observation to the Sun–Jupiter–asteroid problem, it implies that certain asteroids might undergo large changes in their orbital elements.

## 9.5 Algebraic Mappings

We have seen how trajectories of the test particle in systems such as the restricted three-body problem are derived by finding numerical solutions of the equations of motion. This allows us to calculate the position of the particle in phase space at some time $t$ from knowledge of its initial state at some time $t_0$. The numerical method to accomplish this (e.g., the standard fourth-order Runge–Kutta algorithm) consists of a sequence of algebraic instructions that take $x_0$, $y_0$, $\dot{x}_0$, $\dot{y}_0$ and return $x$, $y$, $\dot{x}$, and, $\dot{y}$. Any such numerical method is an example of an *algebraic mapping*, whereby the evolution of the system can be followed by means of a mapping from the initial state to the final state.

In this section we examine several algebraic mappings that have played a key role in helping us understand dynamical evolution in the solar system. Here we are not concerned with new numerical methods as such; these mappings represent tools that allow us to study orbital evolution over much longer timescales than permitted by traditional numerical methods. However, in common with all such methods they only represent an approximation to the motion. Because chaotic motion can sometimes only be revealed after long integrations, the advent of efficient new methods for studying orbital evolution has played an important role in solar system dynamics.

### 9.5.1 The Standard Map

First we consider the *standard map* (see Chirikov 1979). This was derived to approximate the motion of charged particles in accelerators. Although it may not seem directly applicable to the solar system, the standard map arises from a model of a perturbed pendulum and we have already noted that motion near resonance can be approximated by a pendulum equation (see Sect. 8.6). In fact, the derivation and dynamical properties of the standard map prove to be a simple yet useful guide to similar mappings that are even more applicable. Indeed, the complexity of orbits that result from such a simple system provides a foretaste of what can be expected from other mappings.

Using a suitable scaling of the coordinates we can write the Hamiltonian of the simple pendulum as

$$\mathcal{H} = \frac{I^2}{4\pi} + \frac{k_0}{2\pi} \cos\theta, \tag{9.11}$$

where $I$ and $\theta$ are action and angles variables derived from the original angular position and momentum (effectively the $\varphi$ and $\dot{\varphi}$ in our original pendulum model given in Eq. (8.48)). The details of the appropriate transformations need not concern us here since we are only interested in the properties of the system. Now suppose that the suspension point of the pendulum is subjected to an oscillation that can be represented by an infinite number of high-frequency terms such that the Hamiltonian is given by

$$\mathcal{H} = \frac{I^2}{4\pi} + \frac{k_0}{2\pi} \cos\theta + \sum_{n=1}^{\infty} k_n(I) \cos(\theta - nt). \tag{9.12}$$

This makes the problem difficult to handle since we have to deal with an infinite number of short-period terms. However, with suitable approximations we can make use of these infinite number of terms to introduce an impulse into the system. Effectively we will be considering a new system where the cosine term is only applied at discrete intervals. The resulting system can be approximated by an algebraic map. If we write

$$\frac{k_0}{2\pi} \cos\theta + \sum_{n=1}^{\infty} k_n(I) \cos(\theta - nt) \approx \frac{k_0}{2\pi} \sum_{n=0}^{\infty} \cos(\theta - nt) \tag{9.13}$$

then we can use the properties of the Dirac $\delta$-function to obtain

$$\mathcal{H} \approx \frac{I^2}{4\pi} + k_0 \cos\theta \, \delta_{2\pi}(t), \tag{9.14}$$

where $\delta_{2\pi}(t)$ is a $2\pi$-periodic $\delta$-function.

The equations of motion are then

$$\dot{I} = -\frac{\partial \mathcal{H}}{\partial \theta} = +k_0 \sin\theta \, \delta_{2\pi}(t), \tag{9.15}$$

$$\dot{\theta} = +\frac{\partial \mathcal{H}}{\partial I} = \frac{I}{2\pi}. \tag{9.16}$$

Integrating Eq. (9.15) gives

$$I' - I = k_0 \sin\theta \int_0^{2\pi} \delta_{2\pi}(t) \, dt = k_0 \sin\theta \tag{9.17}$$

while Eq. (9.16) gives

$$\theta' - \theta = \int_0^{2\pi} \frac{I}{2\pi} \, dt \approx I'. \tag{9.18}$$

Thus, with time steps of $2\pi$, we can write this approximate solution of the equations of motion as a mapping taking us from $(I, \theta)$ to $(I', \theta')$. This is the *standard map* given by

$$I' = I + k_0 \sin \theta,$$
$$\theta' = \theta + I'. \tag{9.19}$$

The standard map is a discretised approximation to the motion of the perturbed pendulum and defines a completely deterministic system; that is, after any number of iterations we can, in principle, state what the values of $I$ and $\theta$ will be.

If we always take $I$ and $\theta$ as modulo $2\pi$ (thanks to our original scaling) then we can consider the resulting motion of the perturbed pendulum as motion on the surface of a torus, with a different torus for each value of $k_0$; this is because any point on the torus can be defined by two angles. The parameter $k_0$ can be thought of as the magnitude of the perturbation on the pendulum.

Figure 9.18 shows thirty-six separate trajectories for two values of the constant $k_0$. The starting conditions were identical in each case. For both values of $k_0$ the island at $(0, \pi)$ is present. This is the stable equilibrium point in the original unperturbed pendulum that survives the short-period perturbations. However, even for $k_0 = 0.8$ we can see that a number of the islands have emerged. They seem to occur in chains and as with the circular restricted problem they represent different resonances between the fundamental frequency of the unperturbed pendulum and the infinite number of short-period perturbing frequencies. Again the number of islands in the chain is related to the order of the resonance. At $k_0 = 1.2$ the unusual behaviour of the orbits at the edges of the resonant islands becomes obvious; this is chaotic motion near the separatrix. As $k_0$ increases it seems as though the adjacent islands are starting to merge as the extent of the "fuzzy" separatrix motion increases. For $k_0 = 1.2$ there are few islands left and

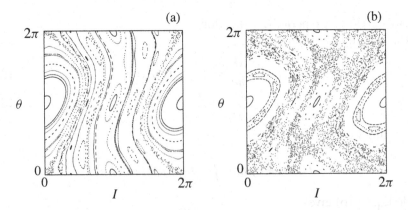

Fig. 9.18. The evolution of thirty-six different trajectories of the standard map with different values of $k_0$. (a) $k_0 = 0.8$. (b) $k_0 = 1.2$.

the irregular motion trajectories are starting to dominate, and they appear to have few bounds. Note how the maximum extent of the islands become well defined as the trajectory traces them out for short periods. Most of these phenomena have already been observed in our study of surfaces of section in the circular restricted problem, and yet here they arise from a much simpler system.

### 9.5.2 Resonance Maps

Resonance maps are derived from the averaged disturbing function in order to model the evolution of a test particle at a given resonance. They rely on the fact that the secular part of the disturbing function can be solved analytically (see Sect. 7.9) while the individual resonant terms can be approximated by a sequence of impulses applied in succession. The technique for converting the resonant terms to impulses follows that used by Chirikov (1979) to derive the standard map.

Wisdom (1982) pioneered this method when he devised an algebraic mapping to study the orbital evolution of test particles at the 3:1 jovian resonance in the asteroid belt. In particular he was interested in explaining the origin of the prominent gap in the asteroid distribution associated with this resonance. In its original form Wisdom's map included variations in the inclination and allowed for secular evolution of Jupiter's orbit using the theory developed by Brouwer & van Woerkom (1950). In a subsequent paper Wisdom (1983) presented a revised version of the map to study the planar case with fixed values of Jupiter's eccentricity and pericentre. Here we show how the planar version of the map was derived; examples of its use in understanding the structure of the 3:1 gap are given in Sect. 9.8.2.

Wisdom's starting point was a second-order expansion of the disturbing function containing the secular and resonant terms. He transformed the standard orbital elements to Poincaré elements given by

$$L = \sqrt{\mu_1 a}, \tag{9.20}$$

$$\rho = \sqrt{\mu_1 a} \left( 1 - \sqrt{1 - e^2} \right) \approx \sqrt{\mu_1 a} \frac{e^2}{2}, \tag{9.21}$$

where $\mu_1 = 1 - \mu$ and the conjugate elements are $\lambda$ and $\varpi$ respectively. He then transformed to the resonant variable $\varphi = \lambda - 3\lambda'$ and used $\omega = -\varpi$, $x = \sqrt{2\rho} \cos \omega$, and $y = \sqrt{2\rho} \sin \omega$ to produce the Hamiltonian

$$\mathcal{H} = -\frac{\mu_1^2}{2\Phi^2} - 3\Phi + \Lambda' + F_1(x^2 + y^2) + \bar{F}_1 e' x$$
$$- \left[ C_1(x^2 - y^2) + D_1 e' x + E_1 e'^2 \right] \cos \varphi$$
$$+ \left[ C_1 2xy + D_1 e' y \right] \sin \varphi, \tag{9.22}$$

where $\Lambda'$ is the variable conjugate to $\lambda'$, and $F_1, \bar{F}_1, C_1, D_1$, and $E_1$ are constants that are functions of Laplace coefficients. Here $e'$ is the eccentricity of Jupiter, which can be treated as constant or can be calculated from a known secular solution for the planets. In a procedure similar to the derivation of the standard map, Wisdom added high-frequency terms to the sine and cosine terms in $\mathcal{H}$. Since an infinite number of high-frequency terms are effectively ignored when producing an averaged disturbing function, the inclusion of a selected number of such high-frequency terms could be justified provided they did not change the nature of the problem. In this case they permitted the derivation of a map to study the motion.

The Hamiltonian becomes

$$
\begin{aligned}
\mathcal{H} = & -\frac{\mu_1^2}{2\Phi^2} - 3\Phi + \Lambda' + F_1(x^2 + y^2) + \bar{F}_1 e' x \\
& - \sum_{i=-\infty}^{\infty} C_1(x^2 - y^2) \cos[\varphi - i(t - \zeta_1)] + \sum_{i=-\infty}^{\infty} D_1 e' x \cos[\varphi - i(t - \zeta_2)] \\
& + \sum_{i=-\infty}^{\infty} E_1 e'^2 \cos[\varphi - i(t - \zeta_3)] + \sum_{i=-\infty}^{\infty} C_1 2xy \cos[\varphi - i(t - \gamma_1) - \pi/2] \\
& + \sum_{i=-\infty}^{\infty} D_1 e' y \cos[\varphi - i(t - \gamma_2) - \pi/2],
\end{aligned}
$$

(9.23)

where $\zeta_1$, $\zeta_2$, $\zeta_3$, $\gamma_1$, and $\gamma_2$ are all arbitrary constants. Each sum can now be expressed as a sum of Dirac $\delta$-functions. For example,

$$
\begin{aligned}
\sum_{i=-\infty}^{\infty} \cos[\varphi - i(t - \zeta_1)] &= \cos\varphi \sum_{i=-\infty}^{\infty} \cos[i(t - \zeta_1)] \\
&= \cos\varphi \sum_{i=-\infty}^{\infty} 2\pi\delta[(t - \zeta_1) - 2\pi i] \\
&= \cos\varphi \, 2\pi \, \delta_{2\pi}(t - \zeta_1),
\end{aligned}
$$

(9.24)

where, as with the standard map, $\delta_{2\pi}$ is a $2\pi$-periodic $\delta$-function.

The planar averaged Hamiltonian now becomes

$$
\begin{aligned}
\mathcal{H} = & -\frac{\mu_1^2}{2\Phi^2} - 3\Phi + \Lambda' + F_1(x^2 + y^2) + \bar{F}_1 e' x \\
& - C_1(x^2 - y^2) \cos\varphi \, 2\pi \, \delta_{2\pi}(t - \zeta_1) \\
& - D_1 e' x \cos\varphi \, 2\pi \, \delta_{2\pi}(t - \zeta_2) - E_1 e'^2 \cos\varphi \, 2\pi \, \delta_{2\pi}(t - \zeta_3) \\
& + C_1 2xy \sin\varphi \, 2\pi \, \delta_{2\pi}(t - \gamma_1) + D_1 e' y \sin\varphi \, 2\pi \, \delta_{2\pi}(t - \gamma_2).
\end{aligned}
$$

(9.25)

Wisdom chose $\zeta_1 = \zeta_2 = \zeta_3 = 0$ and $\gamma_1 = \gamma_2 = \pi/2$, which allowed the impulses to act at two different times in the course of one Jupiter period.

Now consider just the secular part of the Hamiltonian. We have

$$
\mathcal{H}^{(\text{sec})} = -\frac{\mu_1^2}{2\Phi^2} - 3\Phi + \Lambda' + F_1(x^2 + y^2) + \bar{F}_1 e' x,
$$

(9.26)

and the equations of motion governing the secular evolution are

$$\dot{x} = -\frac{\partial \mathcal{H}^{(\text{sec})}}{\partial y} = -2F_1 y, \qquad \dot{\Phi} = -\frac{\partial \mathcal{H}^{(\text{sec})}}{\partial \varphi} = 0,$$

$$\dot{y} = +\frac{\partial \mathcal{H}^{(\text{sec})}}{\partial x} = +2F_1 x + \bar{F}_1 e', \qquad \dot{\varphi} = +\frac{\partial \mathcal{H}^{(\text{sec})}}{\partial \Phi} = \frac{\mu_1^3}{\Phi^3} - 3. \tag{9.27}$$

The solutions for $\Phi$ and $\varphi$ are trivial:

$$\Phi = \Phi(t_0) = \Phi_0, \qquad \varphi = \left(\frac{\mu_1^2}{\Phi_0^3} - 3\right)(t - t_0) + \varphi_0, \tag{9.28}$$

while the solutions for $x$ and $y$ are

$$x = x(t_0) \cos[2F_1(t - t_0)] - y(t_0) \sin[2F_1(t - t_0)]$$
$$- \frac{\bar{F}_1 e'}{2F_1}[1 - \cos[2F_1(t - t_0)]],$$
$$y = x(t_0) \sin[2F_1(t - t_0)] + y(t_0) \cos[2F_1(t - t_0)]$$
$$+ \frac{\bar{F}_1 e'}{2F_1} \sin[2F_1(t - t_0)]. \tag{9.29}$$

Thus, as we saw in Chapter 7, there is an analytical solution for the motion under purely secular perturbations.

Now consider the resonant terms. Consider just the first term where the Hamiltonian is given by

$$\mathcal{H}_1^{(\text{res})} = -\frac{2\pi C_1}{\Delta}(x^2 - y^2) \cos \varphi, \tag{9.30}$$

where

$$\delta(t - t_i) \leftarrow \lim_{\Delta \to 0} \begin{cases} 1/\Delta, & \text{if } t_i \leq t < t_i + \Delta, \\ 0, & \text{otherwise.} \end{cases} \tag{9.31}$$

The equations of motion for this particular resonant term are

$$\dot{x} = -\frac{\partial \mathcal{H}_1^{(\text{sec})}}{\partial y} = -\frac{4\pi C_1}{\Delta} y \cos \varphi = -\frac{R_1}{\Delta} y_1, \tag{9.32}$$

$$\dot{y} = +\frac{\partial \mathcal{H}_1^{(\text{sec})}}{\partial x} = -\frac{4\pi C_1}{\Delta} x \cos \varphi = -\frac{R_1}{\Delta} x_1, \tag{9.33}$$

$$\dot{\Phi} = -\frac{\partial \mathcal{H}_1^{(\text{sec})}}{\partial \varphi} = -\frac{2\pi C_1}{\Delta}(x^2 - y^2) \sin \varphi, \tag{9.34}$$

$$\dot{\varphi} = +\frac{\partial \mathcal{H}_1^{(\text{sec})}}{\partial \Phi} = 0, \tag{9.35}$$

where $R_1 = 4\pi C_1 \cos \varphi$ and is a constant since $\dot{\varphi} = 0$ for this Hamiltonian. The solution for $x$ and $y$ is

$$x = x(t_1) \cosh\left[\frac{R_1(t - t_1)}{\Delta}\right] - y(t_1) \sinh\left[\frac{R_1(t - t_1)}{\Delta}\right], \tag{9.36}$$

$$y = y(t_1) \cosh \left[ \frac{R_1(t - t_1)}{\Delta} \right] - x(t_1) \sinh \left[ \frac{R_1(t - t_1)}{\Delta} \right]. \tag{9.37}$$

Now set $t = t_1 + \Delta$ and let $\Delta \to 0$. The corresponding changes in the elements are

$$\Delta x = x(t_1) \left( \cosh [4\pi C_1 \cos \varphi] - 1 \right) - y(t_1) \sinh[4\pi C_1 \cos \varphi], \tag{9.38}$$
$$\Delta y = y(t_1) \left( \cosh [4\pi C_1 \cos \varphi] - 1 \right) - x(t_1) \sinh[4\pi C_1 \cos \varphi]. \tag{9.39}$$

Similarly we have

$$\Delta \Phi = -2\pi C_1 \left( [x(t_1)]^2 - [y(t_1)]^2 \right) \sin \varphi(t_1). \tag{9.40}$$

This procedure can be repeated for each of the four other resonant terms given by the Hamiltonians

$$\mathcal{H}_2^{(\text{res})} = -\frac{2\pi D_1}{\Delta} e' x \cos \varphi, \tag{9.41}$$

$$\mathcal{H}_3^{(\text{res})} = -\frac{2\pi E_1}{\Delta} e'^2 \cos \varphi, \tag{9.42}$$

$$\mathcal{H}_4^{(\text{res})} = \frac{4\pi C_1}{\Delta} xy \sin \varphi, \tag{9.43}$$

$$\mathcal{H}_5^{(\text{res})} = \frac{2\pi D_1}{\Delta} e' y \sin \varphi. \tag{9.44}$$

For each of these Hamiltonians we can derive the instantaneous changes in the orbital elements just as we have done for the first resonant Hamiltonian, $\mathcal{H}_1^{(\text{res})}$, above.

Let the initial conditions be $x^{(0)}, y^{(0)}, \Phi^{(0)}, \varphi^{(0)}$ at time $t = 0$. The first mapping is applied at $t = 0$ (as determined by the constant $\zeta_1$) and from Eqs. (9.38)–(9.40) we obtain

$$x^{(1)} = x^{(0)} \cosh \left[ 4\pi C_1 \cos \varphi^{(0)} \right] - y^{(0)} \sinh \left[ 4\pi C_1 \cos \varphi^{(0)} \right], \tag{9.45}$$

$$y^{(1)} = y^{(0)} \cosh \left[ 4\pi C_1 \cos \varphi^{(0)} \right] - x^{(0)} \sinh \left[ 4\pi C_1 \cos \varphi^{(0)} \right], \tag{9.46}$$

$$\Phi^{(1)} = \Phi^{(0)} - 2\pi C_1 \left( \left[ x^{(0)} \right]^2 - \left[ y^{(0)} \right]^2 \right) \sin \varphi^{(0)}, \tag{9.47}$$

$$\varphi^{(1)} = \varphi^{(0)}. \tag{9.48}$$

Adopting similar procedures for $\mathcal{H}_2^{(\text{res})}$ and $\mathcal{H}_3^{(\text{res})}$, both of which are applied at time $t = 0$, we derive the second and third steps of the mapping:

$$x^{(2)} = x^{(1)}, \tag{9.49}$$
$$y^{(2)} = y^{(1)} - 2\pi D_1 e' \cos \varphi^{(1)}, \tag{9.50}$$
$$\Phi^{(2)} = \Phi^{(1)} - 2\pi D_1 e' x^{(1)} \sin \varphi^{(1)}, \tag{9.51}$$
$$\varphi^{(2)} = \varphi^{(1)} \tag{9.52}$$

and

$$x^{(3)} = x^{(2)}, \tag{9.53}$$

$$y^{(3)} = y^{(2)}, \tag{9.54}$$

$$\Phi^{(3)} = \Phi^{(2)} - 2\pi E_1 e'^2 \sin \varphi^{(2)}, \tag{9.55}$$

$$\varphi^{(3)} = \varphi^{(2)}. \tag{9.56}$$

The remaining two resonant impulses are applied at time $t = \pi/2$ and so we make use of the analytical solution of the secular part to move the solution ahead to that point. We have $t - t_0 = \pi/2$ and Eqs. (9.28) and (9.29) give us

$$x^{(4)} = x^{(3)} \cos \pi F_1 - y^{(3)} \sin \pi F_1 - \frac{\bar{F}_1 e'}{2F_1} (1 - \cos \pi F_1), \tag{9.57}$$

$$y^{(4)} = x^{(3)} \sin \pi F_1 + y^{(3)} \cos \pi F_1 + \frac{\bar{F}_1 e'}{2F_1} \sin \pi F_1, \tag{9.58}$$

$$\Phi^{(4)} = \Phi^{(3)}, \tag{9.59}$$

$$\varphi^{(4)} \varphi^{(3)} + \frac{\pi}{2} \left( \frac{\mu_1^2}{[\Phi^{(3)}]^3} - 3 \right) \tag{9.60}$$

for the fourth part of the mapping. We can now proceed to apply the impulses due to $\mathcal{H}_4^{(\text{res})}$ and $\mathcal{H}_5^{(\text{res})}$. This gives

$$x^{(5)} = x^{(4)} \exp[+4\pi C_1 \sin \varphi^{(4)}], \tag{9.61}$$

$$y^{(5)} = y^{(4)} \exp[-4\pi C_1 \sin \varphi^{(4)}], \tag{9.62}$$

$$\Phi^{(5)} = \Phi^{(4)} + 4\pi C_1 x^{(4)} y^{(4)} \cos \varphi^{(4)}, \tag{9.63}$$

$$\varphi^{(5)} = \varphi^{(4)} \tag{9.64}$$

and

$$x^{(6)} = x^{(5)} + 2\pi D_1 e' \sin \varphi^{(5)}, \tag{9.65}$$

$$y^{(6)} = y^{(5)}, \tag{9.66}$$

$$\Phi^{(6)} = \Phi^{(5)} + 2\pi D_1 e' y^{(5)} \cos \varphi^{(5)}, \tag{9.67}$$

$$\varphi^{(6)} = \varphi^{(5)}. \tag{9.68}$$

Finally, we apply the solution to the secular part to move the orbit from $t = \pi/2$ to $t = 2\pi$, to ready it for the next application of the $\mathcal{H}_1^{(\text{res})}$ and to allow the next cycle of the mapping to begin. Hence the final mapping is

$$x^{(7)} = x^{(6)} \cos 3\pi F_1 - y^{(6)} \sin 3\pi F_1 - \frac{\bar{F}_1 e'}{2F_1} (1 - \cos 3\pi F_1), \tag{9.69}$$

$$y^{(7)} = x^{(6)} \sin 3\pi F_1 + y^{(6)} \cos 3\pi F_1 + \frac{\bar{F}_1 e'}{2F_1} \sin 3\pi F_1, \tag{9.70}$$

$$\Phi^{(7)} = \Phi^{(6)}, \tag{9.71}$$

$$\varphi^{(7)} = \varphi^{(6)} + \frac{3\pi}{2} \left( \frac{\mu_1^2}{\left[ \Phi^{(6)} \right]^3} - 3 \right). \tag{9.72}$$

Wisdom's map relies on (i) the fact that the secular problem is integrable for low eccentricity and inclination (recall that secular terms greater than second order are ignored) and (ii) the replacement of the resonant terms in the disturbing function by impulses applied at specific times. This accounts for the speed of the map, which Wisdom (1982) estimated at $10^3$ times faster than solving the full equations of motion. Murray & Fox (1984) compared the behaviour of the map with that derived from an integration of the averaged equations from which the map is derived and a full integration of the equations of motion. They showed that the map gave excellent agreement with the averaged system and only differed from the full system in a quantitative manner. However, the map could be used to predict the regular or chaotic nature of the true orbit, and the map's speed made it a powerful tool for the study of chaotic orbits in the asteroid belt.

Since the 3:1 resonance is a second-order resonance we require terms up to at least order 2 in the disturbing function. Note that the secular terms that contribute to precession only appear at second order and higher. If we wish to study first-order resonances using a map then we should include resonant terms up to at least second order to be consistent with the secular contribution. Murray (1986) took this approach when he used Wisdom's method to derive maps for the 2:1 and 3:2 resonances. As well as the resonant arguments, which contained $2\lambda' - \lambda$ (and $3\lambda' - 2\lambda$), for consistency he also included the second-order arguments which contained $4\lambda' - 2\lambda$ (and $6\lambda' - 4\lambda$). Šidlichovsky & Melendo (1986) derived a similar map for the 5:2 resonance in the asteroid belt. Although they were dealing with a third-order resonance and used second-order secular terms in their derivation, this was still consistent since the higher-order secular terms only come in at order 4 and above. Note that a common problem with all these approaches is the convergence of the disturbing function expansion for large values of $\alpha = a/a'$ (see the discussion in Sect. 9.8.3).

### 9.5.3 Encounter Maps

Resonance maps use a disturbing function approach to calculate the averaged effect of the perturbation over time. In direct contrast an encounter map approximates the direct effect on the test particle as it encounters a perturber. The technique was employed by Duncan et al. (1989) in their derivation of a map to study the long-term evolution of orbits in the outer solar system. Since we are dealing with perturbed motion of the test particle in the vicinity of the perturber it is natural to choose a coordinate system centred on the perturber. Therefore we can make use of Hill's equations (see Sect. 3.13).

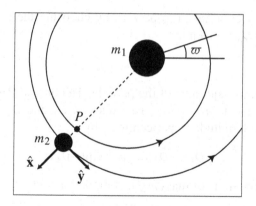

Fig. 9.19. Geometry of the conjunction between a test particle $P$ on a near-circular interior orbit and a perturber on a circular orbit.

Consider the motion of a test particle orbiting a central body of mass $m_1$. The particle is subject to perturbations from a planet in a circular, coplanar orbit with semi-major axis $a'$ and mass $m_2 \ll m_1$. In a frame centred on the mass $m_2$ with the $x$ axis pointing radially outwards and the $y$ axis at $90°$ to it in the direction of motion (see Fig. 9.19) the approximate equations of motion are

$$\ddot{x} - 2n_2\dot{y} - 3n_2^2x = -\frac{\mathcal{G}m_2x}{(x^2+y^2)^{3/2}}, \tag{9.73}$$

$$\ddot{y} + 2n_2\dot{x} = -\frac{\mathcal{G}m_2y}{(x^2+y^2)^{3/2}}, \tag{9.74}$$

where $n_2^2 = \mathcal{G}m_1/a_2^3$ and $n_2$ denotes the mean motion of the mass $m_2$; these equations should be compared with Eqs. (3.202) and (3.203). There is also an integral of the motion that is equivalent to the Jacobi integral in the restricted problem. It is given by (cf. Eq. (3.205))

$$C_{\mathrm{H}} = 3n_2^2x^2 - \dot{x}^2 - \dot{y}^2 + \frac{2\mathcal{G}m_2}{(x^2+y^2)^{3/2}}. \tag{9.75}$$

Hénon & Petit (1986) showed that in the limit $m_2 \to 0$ Hill's equations have the solution

$$x = D_1 \cos n_2t + D_2 \sin N_2t + D_3, \tag{9.76}$$

$$y = -2D_1 \sin n_2t + 2D_2 \cos n_2t - \frac{3}{2}D_3n_2t + D_4, \tag{9.77}$$

where $D_1$, $D_2$ $D_3$, and $D_4$ are constants of integration. Substituting these values in the equation for $C_{\mathrm{H}}$ with $m_2 \to 0$ gives the limiting value of the constant as

$$C_{\mathrm{H}} = n_2^2 \left( \frac{3}{4}D_3^2 - D_1^2 - D_2^2 \right). \tag{9.78}$$

The constant $D_4$ can be taken to be zero by choosing the time origin at each conjunction, and $D_3$ is given by

$$D_3 = a - a_2, \tag{9.79}$$

where $a$ is the semi-major axis of the particle. Hénon and Petit (1986) showed that to lowest order in $|a - a_2|/a_2$ the values of $D_1$ and $D_2$ are related to the eccentricity $e$ and longitude of pericentre $\varpi$ by

$$D_1 + iD_2 = -a_2 z \exp(-i\lambda_c), \tag{9.80}$$

where $\lambda_c$ is the longitude of mass $m_2$ at conjunction and

$$z = e \exp(i\varpi) \tag{9.81}$$

is a complex eccentricity.

Now we need to calculate the changes in $D_1$, $D_2$, and $D_3$ that occur due to the conjunction. For this step Duncan et al. (1989) followed the work of Julian & Toomre (1966) and Hénon & Petit (1986). For small initial eccentricity the relevant changes are

$$\Delta D_1 = 0, \tag{9.82}$$

$$\Delta D_2 = -g \, \text{sign}\,(D_3) \frac{m_2}{m_1} \frac{a_2^3}{D_3^2}, \tag{9.83}$$

$$\Delta D_3 = 0, \tag{9.84}$$

where the constant $g$ is given by

$$g = \frac{8}{9} [2K_0(2/3) + K_1(2/3)] = 2.239566674, \tag{9.85}$$

where $K_0$ and $K_1$ are modified Bessel functions of the second kind. Therefore the effect of the conjunction is to change the complex eccentricity by an amount

$$\Delta z = \frac{ig \exp(i\lambda_c)}{\varepsilon^2} \text{sign}\,(\varepsilon) \frac{m_2}{m_1}, \tag{9.86}$$

where $\varepsilon$ is the relative difference in semi-major axis given by

$$\varepsilon = \frac{a - a_2}{a_2}. \tag{9.87}$$

Duncan et al. pointed out that the same result can be derived without using Hill's equations, by means of Gauss's equations or Lagrange's equations.

The fact that $\Delta D_3$ is zero to lowest order implies that there is no change in semi-major axis. However, the modified form of the Jacobi constant given in Eq. (9.78) allows the derivation of a more accurate expression. Let $a'$ and $z' = z + \Delta z$ denote the values of the semi-major axis and complex eccentricity immediately after the conjunction with $\varepsilon' = (a' - a_2)/a_2$. Hence, the expression for $C_H$ gives

$$\varepsilon'^2 = \varepsilon^2 + \frac{4}{3}\left(|z'|^2 - |z|^2\right). \tag{9.88}$$

Using Eqs. (9.86) and (9.88) we can calculate the values of $a'$ and $z'$ from $a$ and $z$. We will assume that the new values remain constant until the next conjunction occurs. The time between successive conjunctions (the synodic period) is given by

$$T_s = \frac{2\pi}{|n_2 - n|} = \frac{2\pi}{n_2}\left|\left(\frac{a_2}{a'}\right)^{3/2} - 1\right|^{-1} \tag{9.89}$$

and hence the next longitude at conjunction, $\lambda'_c$, is given by

$$\lambda'_c = \lambda_c + 2\pi\, f(\varepsilon'), \tag{9.90}$$

where

$$f(\varepsilon) = \left|(1 + \varepsilon)^{-3/2} - 1\right|^{-1}. \tag{9.91}$$

Let $\lambda_n$ denote the longitude of the $n$th conjunction and let $z_n$, and $\varepsilon_n$ denote the values of the complex eccentricity and relative semi-major axis just before the $n$th conjunction. Putting all three changes together we have the complete map as given by Duncan et al. (1989):

$$z_{n+1} = z_n + \frac{i g \exp(i\lambda_n)}{\varepsilon_1}\, \mathrm{sign}(\varepsilon_1)\frac{m_2}{m_1}, \tag{9.92}$$

$$\varepsilon_{n+1} = \varepsilon_n\sqrt{1 + \frac{4\left(|z_{n+1}|^2 - |z_n|^2\right)}{3\varepsilon_n^2}}, \tag{9.93}$$

$$\lambda_{n+1} = \lambda_n + 2\pi\, f(\varepsilon_{n+1}). \tag{9.94}$$

Note the occurrence of $\varepsilon_1$ rather than $\varepsilon_n$ in Eq. (9.92). The change, which has little effect on the accuracy of the map, was made by Duncan et al. in order to make the map area preserving.

Duncan et al. (1989) give a number of alternative forms of the map, each of which illustrates a particular property of the system. By introducing $\Delta\varepsilon_n = \varepsilon_n - \varepsilon_1$, considering the case where $\varepsilon_1 > 0$, and expanding the right hand sides of Eqs. (9.93) and (9.94), we obtain

$$z_{n+1} = z_n + \frac{i g \exp(i\lambda_n)}{\varepsilon_1}\frac{m_2}{m_1}, \tag{9.95}$$

$$\Delta\varepsilon_{n+1} = \frac{2\left(|z_{n+1}|^2 - |z_n|^2\right)}{3\varepsilon_1}, \tag{9.96}$$

$$\lambda_{n+1} = \lambda_n + \frac{4\pi}{3\varepsilon_1} - \frac{4\pi\,\Delta\varepsilon_{n+1}}{3\varepsilon_1^2}. \tag{9.97}$$

The map derived by Duncan et al. was also modified to include the effect of a perturber moving in an elliptical orbit. They showed that all that was required was to replace the original definition of $z$ by

$$z = e \exp(i\varpi) - e_2 \exp(i\varpi_2), \tag{9.98}$$

where $e_2$ and $\varpi_2$ are the eccentricity and longitude of pericentre respectively of the perturbing body. Everything else remains the same. In a further modification Duncan et al. showed how the map could be adapted to include the effect of perturbations from two planets on circular orbits.

Namouni et al. (1996) included higher order terms in the eccentricity and inclination in the expansion of the interaction, leading to an improved mapping for encounters in the Hill's approximation. In all of these cases, as with the resonance maps, the power of the encounter map lies in its ability to "jump" instantly to the next perturbation by means of an analytical solution. We will discuss the uses of the map in Sect. 9.6.

### 9.5.4 N-Body Maps

The addition of an infinite number of high-frequency terms to the Hamiltonian was a necessary part of the derivation of the resonance maps. These maps also rely on truncated expansions of the planetary disturbing function and its inherent problems. Wisdom & Holman (1991) derived a map for the $N$-body problem that made use of the trick involving the Dirac $\delta$-function but avoided the need for series expansions in orbital elements. Wisdom & Holman's map is an important tool because of (i) its speed and (ii) its *symplectic* nature, which results in no secular changes in the energy of the system. The use of this map was to have a profound effect on studies of long-term stability in the solar system and its derivation is worth examining in some detail.

The starting point for the $N$-body map is the Hamiltonian of the gravitational $N$-body problem where we are required to find the solution for a system of $N$ mutually perturbed bodies. Let the $i$th body have a mass $m_i$, a linear momentum $\mathbf{p}_i = m_i \mathbf{v}_i$, and a separation from the $j$th body of $r_{ij} = |\mathbf{r}_j - \mathbf{r}_i|$. The Hamiltonian of the system is given by (cf. Eq. (2.170))

$$\mathcal{H} = \sum_{j=0}^{N-1} \frac{p_j^2}{2m_j} - \sum_{j=0}^{N-2} \sum_{k=j+1}^{N-1} \frac{\mathcal{G} m_j m_k}{r_{jk}}. \tag{9.99}$$

To derive an $N$-body map this Hamiltonian has to be separated into a keplerian part and an interaction part (Wisdom & Holman 1991). A Hamiltonian is defined to be keplerian (see Sect. 2.10) if it is possible to write it in the form

$$\mathcal{H} = \frac{p^2}{2m} - \frac{\mathcal{G} M m}{r} \tag{9.100}$$

or as the sum of such forms. In order to achieve the proper separation of the Hamiltonian, it is necessary to choose a new system of variables called the *Jacobian coordinates* (see, for example, Plummer 1918).

Consider $N - 1$ bodies of mass $m_1, m_2, \ldots, m_{N-1}$ orbiting a central body of mass $m_0$. With respect to a fixed origin at $O$ the $N$ bodies have position vectors $\mathbf{r}_0, \mathbf{r}_1, \ldots, \mathbf{r}_{N-1}$ (see Fig. 9.20a for the case where $N = 4$). Let $\mathbf{R}_0$ denote the position vector of the centre of mass of a system consisting of the central mass alone (i.e., $\mathbf{R}_0 = \mathbf{r}_0$) with $\mathbf{R}_1$ denoting the centre of mass of the system composed of the masses $m_0$ and $m_1$, and so on. Hence $\mathbf{R}_i$ denotes the position vector of the centre of mass of all the bodies with index up to and including $i$ (see Fig. 9.20b). We can write this definition as

$$\mathbf{R}_i = \frac{1}{\eta_i} \sum_{j=0}^{i} m_j \mathbf{r}_j, \tag{9.101}$$

where

$$\eta_i = \sum_{j=0}^{i} m_i. \tag{9.102}$$

The first Jacobian position vector, $\tilde{\mathbf{r}}_0$ (the tilde denotes a value in the Jacobian system), is the position vector of the centre of mass of the whole system. Hence

$$\tilde{\mathbf{r}}_0 = \mathbf{R}_{N-1} \tag{9.103}$$

for our $N$-body system of $N - 1$ bodies orbiting the central mass. The remaining Jacobian position vectors are given by

$$\tilde{\mathbf{r}}_i = \mathbf{r}_i - \mathbf{R}_{i-1}. \tag{9.104}$$

Hence $\tilde{\mathbf{r}}_1$ denotes the position vector of $m_1$ with respect to $m_0$, $\tilde{\mathbf{r}}_2$ denotes the position vector of $m_2$ with respect to the centre of mass of $m_0$ and $m_1$, etc. (see Fig. 9.20b).

In the Jacobian coordinate system the momentum vectors conjugate to the position vectors $\tilde{\mathbf{r}}_i$ are

$$\tilde{\mathbf{p}}_i = \tilde{m}_i \tilde{\mathbf{v}}_i, \tag{9.105}$$

where $\tilde{\mathbf{v}}_i = \dot{\tilde{\mathbf{r}}}_i$ and the $\tilde{m}_i$ are given by

$$\tilde{m}_i = \begin{cases} (\eta_{i-1}/\eta_i)\, m_i, & \text{if } 0 < i < N, \\ \eta_{N-1} = M_{\text{total}}, & \text{if } i = 0, \end{cases} \tag{9.106}$$

where $M_{\text{total}}$ is the total mass of the system. Expressions for the Jacobian momentum vectors in terms of their Cartesian counterparts can be found by differentiating Eqs. (9.103) and (9.104) and multiplying by $\tilde{m}_0$ and $\tilde{m}_i$ respectively.

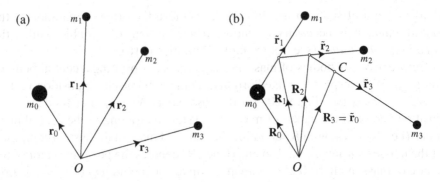

Fig. 9.20. The Jacobian coordinate system for a four-body problem where the position of the $i$th body is referred to the centre of mass of the system of the previous $i - 1$ bodies. (a) The position vectors $\mathbf{r}_i$ of four masses with respect to the origin at $O$. (b) The position vectors $\mathbf{R}_i$ of the centres of mass of the different subsystems (denoted by small open circles) and the position vectors $\tilde{\mathbf{r}}_i$ in the Jacobian coordinate system relative to the centres of mass for the same system. $C$ denotes the centre of mass of the system.

This gives

$$\tilde{\mathbf{p}}_0 = \sum_{j=0}^{N-1} \mathbf{p}_j, \qquad (9.107)$$

which is the total linear momentum of the system, and

$$\tilde{\mathbf{p}}_i = \left(\frac{\eta_{i-1}}{\eta_i}\right) \mathbf{p}_i - \left(\frac{m_i}{\eta_i}\right) \sum_{j=0}^{i-1} \mathbf{p}_j \qquad (0 < i < N) \qquad (9.108)$$

(see Saha & Tremaine 1994).

   To express $\mathcal{H}$ in terms of the Jacobian coordinates and momenta we need to find suitable expressions that relate the standard Cartesian system to the Jacobian system. By writing out expressions for each of the $\tilde{\mathbf{r}}_i$, multiplying each by $\eta_{i-1}$, and subtracting consecutive pairs of equations, it can be shown (Tom Kehoe, private communication) that the standard coordinates can be written as

$$\mathbf{r}_0 = \tilde{\mathbf{r}}_0 - \sum_{j=1}^{N-1} \left(\frac{m_j}{\eta_j}\right) \tilde{\mathbf{r}}_j, \qquad (9.109)$$

$$\mathbf{r}_i = \tilde{\mathbf{r}}_0 + \left(\frac{\eta_{i-1}}{\eta_i}\right) \tilde{\mathbf{r}}_i - \sum_{j=i+1}^{N-1} \left(\frac{m_j}{\eta_j}\right) \tilde{\mathbf{r}}_j \qquad (0 < i < N - 1), \qquad (9.110)$$

$$\mathbf{r}_{N-1} = \tilde{\mathbf{r}}_0 + \left(\frac{\eta_{N-2}}{\eta_{N-1}}\right) \tilde{\mathbf{r}}_{N-1}. \qquad (9.111)$$

These equations can be used to derive the following expressions for the relative position vectors:

$$\mathbf{r}_{i0} = \tilde{\mathbf{r}}_i + \sum_{j=1}^{i-1} \left( \frac{m_j}{\eta_j} \right) \tilde{\mathbf{r}}_j \qquad (0 < i < N), \tag{9.112}$$

$$\mathbf{r}_{ij} = \left( \frac{\eta_{i-1}}{\eta_i} \right) \tilde{\mathbf{r}}_i - \sum_{k=i+1}^{j-1} \left( \frac{m_k}{\eta_k} \right) \tilde{\mathbf{r}}_k - \tilde{\mathbf{r}}_j \qquad (0 < i < j < N). \tag{9.113}$$

By differentiating Eqs. (9.109)–(9.111) with respect to time and multiplying them by $m_0$, $m_i$, and $m_{N-1}$ respectively, we derive the following expressions for the corresponding momenta:

$$\mathbf{p}_0 = \left( \frac{m_0}{\eta_{N-1}} \right) \tilde{\mathbf{p}}_0 - \sum_{j=1}^{N-1} \left( \frac{m_0}{\eta_{j-1}} \right) \tilde{\mathbf{p}}_j, \tag{9.114}$$

$$\mathbf{p}_i = \left( \frac{m_i}{\eta_{N-1}} \right) \tilde{\mathbf{p}}_0 + \tilde{\mathbf{p}}_i - \sum_{j=i+1}^{N-1} \left( \frac{m_i}{\eta_{j-1}} \right) \tilde{\mathbf{p}}_j \qquad (0 < i < N - 1), \tag{9.115}$$

$$\mathbf{p}_{N-1} = \left( \frac{m_{N-1}}{\eta_{N-1}} \right) \tilde{\mathbf{p}}_0 + \tilde{\mathbf{p}}_{N-1}. \tag{9.116}$$

With these definitions the kinetic energy term of the Hamiltonian in Eq. (9.99) can now be written as

$$\sum_{j=0}^{N-1} \frac{p_j^2}{2m_j} = \frac{\tilde{p}_0^2}{2\tilde{m}_0} + \sum_{j=1}^{N-1} \frac{\tilde{p}_j^2}{2\tilde{m}_j}. \tag{9.117}$$

Plummer (1918) gives an alternative derivation of this expression without the need for the intermediate expressions for the momentum vectors. In terms of the Jacobian coordinates the full Hamiltonian is

$$\mathcal{H} = \frac{\tilde{p}_0^2}{2\tilde{m}_0} + \sum_{j=1}^{N-1} \frac{\tilde{p}_j^2}{2\tilde{m}_j} - \sum_{j=0}^{N-2} \sum_{k=j+1}^{N-1} \frac{\mathcal{G}m_j m_k}{r_{jk}}. \tag{9.118}$$

Because $r_{ij}$ does not depend on $\tilde{\mathbf{r}}_0$ (the position vector of the centre of mass), the total Jacobian momentum $\tilde{\mathbf{p}}_0$ has to be an integral of the motion. This is simply another way of stating the well-known result that in the gravitational $N$-body system with no external forces the centre of mass is either at rest or is moving with uniform velocity in a straight line. Consequently, we can ignore the first term in Eq. (9.118). Adding and subtracting the quantity $\sum_{j=1}^{N-1} \mathcal{G}m_0 m_j / \tilde{r}_j$ to our expression for $\mathcal{H}$ means that we can now write the Hamiltonian as

$$\mathcal{H} = \mathcal{H}_{\text{Kepler}} + \mathcal{H}_{\text{interaction}}, \tag{9.119}$$

where

$$\mathcal{H}_{\text{Kepler}} = \sum_{j=1}^{N-1} \left( \frac{\tilde{p}_j^2}{2\tilde{m}_j} - \frac{\mathcal{G}\tilde{M}_j \tilde{m}_j}{\tilde{r}_j} \right), \tag{9.120}$$

$$\mathcal{H}_{\text{interaction}} = \sum_{j=1}^{N-1} \frac{\mathcal{G}m_0 m_j}{\tilde{r}_j} - \sum_{j=0}^{N-2} \sum_{k=j+1}^{N-1} \frac{\mathcal{G}m_j m_k}{r_{jk}}, \tag{9.121}$$

where, in $\mathcal{H}_{\text{Kepler}}$, we have expressed the quantity $\sum_{i=j}^{N-1} \mathcal{G}m_0 m_j / \tilde{r}_j$ in terms of the Jacobian masses $\tilde{m}_i$ and $\tilde{M}_i = (\eta_i / \eta_{i-1}) m_0$ for $0 < i < N$. Now we can consider the Kepler and interaction parts of $\mathcal{H}$ separately.

Note that $\mathcal{H}_{\text{Kepler}}$ depends on the individual $\tilde{r}_i$ and $\tilde{p}_i$. Thus $\mathcal{H}_{\text{Kepler}}$ has now been expressed as the sum of $N - 1$ individual two-body problem Hamiltonians as desired. Generalising the expressions given in Eq. (2.170) we have

$$\left(\dot{\tilde{\mathbf{r}}}_i\right)_{\text{Kepler}} = +\nabla_{\tilde{\mathbf{p}}_i}\mathcal{H}_{\text{Kepler}}, \qquad \left(\dot{\tilde{\mathbf{p}}}_i\right)_{\text{Kepler}} = -\nabla_{\tilde{\mathbf{r}}_i}\mathcal{H}_{\text{Kepler}} \tag{9.122}$$

for the evolution of the Jacobian position and momentum vectors under $\mathcal{H}_{\text{Kepler}}$ alone. Hence

$$\left(\dot{\tilde{\mathbf{r}}}_i\right)_{\text{Kepler}} = \frac{\mathbf{p}_i}{\tilde{m}_i}, \qquad \left(\dot{\tilde{\mathbf{p}}}_i\right)_{\text{Kepler}} = -\frac{\mathcal{G}\tilde{M}_i \tilde{m}_i}{\tilde{r}_i^3}\tilde{\mathbf{r}}_i \tag{9.123}$$

for all $0 < i < N$ (cf. Eq. (2.174)). Equations (9.123) combine to give the equation of motion of the $i$th Jacobian body (of mass $\tilde{m}_i$) under the effect of $\mathcal{H}_{\text{Kepler}}$ as

$$\frac{\mathrm{d}^2\tilde{\mathbf{r}}_i}{\mathrm{d}t^2} + \frac{\mathcal{G}\tilde{M}_i}{\tilde{r}_i^3}\tilde{\mathbf{r}}_i = 0 \tag{9.124}$$

for all $0 < i < N$ (cf. Eq. (2.168)). This can be interpreted as the equation of motion of a body about a fixed mass $\tilde{M}_i$ located at the centre of mass, $\mathbf{R}_{i-1}$, of the previous $i - 1$ bodies. It is also the equation of relative motion of two bodies with total mass $\tilde{M}_i$ and vector separation $\tilde{\mathbf{r}}_i$.

The first point to note about $\mathcal{H}_{\text{interaction}}$ is that it depends on neither $\mathbf{p}_i$ nor $\tilde{\mathbf{p}}_i$ and hence, from Hamilton's equations, $\mathbf{r}_i$ and $\tilde{\mathbf{r}}_i$ are constants of the motion under the effect of $\mathcal{H}_{\text{interaction}}$. The next point to note is that $\mathcal{H}_{\text{interaction}}$ can itself be expressed as a sum. We can write

$$\mathcal{H}_{\text{interaction}} = \mathcal{H}_{\text{Jacobian}} + \mathcal{H}_{\text{Cartesian}}, \tag{9.125}$$

where

$$\mathcal{H}_{\text{Jacobian}} = \sum_{j=1}^{N-1} \frac{\mathcal{G}m_0 m_j}{\tilde{r}_j}, \qquad \mathcal{H}_{\text{Cartesian}} = -\sum_{j=0}^{N-2} \sum_{k=j+1}^{N-1} \frac{\mathcal{G}m_j m_k}{r_{jk}}. \tag{9.126}$$

Therefore

$$\dot{\tilde{\mathbf{p}}}_i = -\nabla_{\tilde{\mathbf{r}}_i}\mathcal{H}_{\text{Jacobian}} - \nabla_{\tilde{\mathbf{r}}_i}\mathcal{H}_{\text{Cartesian}} = \left(\dot{\tilde{\mathbf{p}}}_i\right)_{\text{Jacobian}} + \left(\dot{\tilde{\mathbf{p}}}_i\right)_{\text{Cartesian}}, \tag{9.127}$$

where it is easy to show that

$$\left(\dot{\tilde{\mathbf{p}}}_i\right)_{\text{Jacobian}} = \frac{\mathcal{G}m_0 m_i}{\tilde{r}_i^3}\tilde{\mathbf{r}}_i. \tag{9.128}$$

To calculate $\dot{\tilde{\mathbf{p}}}_i$ under the effect of the Cartesian part of the Hamiltonian, we first find $\dot{\mathbf{p}}_i$ under the same Hamiltonian and then substitute the resulting expression in the time derivative of Eq. (9.108). This procedure gives

$$(\dot{\mathbf{p}}_i)_{\text{Cartesian}} = -\nabla_{\mathbf{r}_i} \mathcal{H}_{\text{Cartesian}} = \sum_{\substack{k=0 \\ (k\neq i)}}^{N-1} \frac{\mathcal{G} m_i m_k}{r_{ik}^3} \mathbf{r}_{ik}, \tag{9.129}$$

$$\dot{\tilde{\mathbf{p}}}_i = \left(\frac{\eta_{i-1}}{\eta_i}\right) \dot{\mathbf{p}}_i - \left(\frac{m_i}{\eta_i}\right) \sum_{j=0}^{i-1} \dot{\mathbf{p}}_j \tag{9.130}$$

and hence

$$\left(\dot{\tilde{\mathbf{p}}}_i\right)_{\text{Cartesian}} = \left(\frac{\eta_{i-1}}{\eta_i}\right) \sum_{\substack{k=0 \\ (k\neq i)}}^{N-1} \frac{\mathcal{G} m_i m_k}{r_{ik}^3} \mathbf{r}_{ik} - \left(\frac{m_i}{\eta_i}\right) \sum_{j=0}^{i-1} \sum_{\substack{k=0 \\ (k\neq j)}}^{N-1} \frac{\mathcal{G} m_j m_k}{r_{jk}^3} \mathbf{r}_{jk} \tag{9.131}$$

for all $0 < i < N$. Combining Eqs. (9.128) and (9.131) and dividing through by $\tilde{m}_i$ and using $\tilde{M}_i = (\eta_i/\eta_{i-1}) m_0$ gives

$$\left(\dot{\tilde{\mathbf{v}}}_i\right)_{\text{interaction}} = \frac{\mathcal{G}\tilde{M}_i}{\tilde{r}_i^3} \tilde{\mathbf{r}}_i + \sum_{\substack{k=0 \\ (k\neq i)}}^{N-1} \frac{\mathcal{G} m_k}{r_{ik}^3} \mathbf{r}_{ik} - \frac{1}{\eta_{i-1}} \sum_{j=0}^{i-1} \sum_{\substack{k=0 \\ (k\neq j)}}^{N-1} \frac{\mathcal{G} m_j m_k}{r_{jk}^3} \mathbf{r}_{jk}. \tag{9.132}$$

It can be shown (Tom Kehoe, private communication) that many of the terms on the right-hand side of Eq. (9.132) cancel out and the equation simplifies to

$$\left(\dot{\tilde{\mathbf{v}}}_i\right)_{\text{interaction}} = \mathcal{G}\tilde{M}_i \left[\frac{\tilde{\mathbf{r}}_i}{\tilde{r}_i^3} - \frac{\mathbf{r}_{0i}}{r_{0i}^3}\right] - \left(\frac{\eta_i}{\eta_{i-1}}\right) \sum_{j=1}^{i-1} \frac{\mathcal{G} m_j}{r_{ji}^3} \mathbf{r}_{ji}$$

$$+ \sum_{j=i+1}^{N-1} \frac{\mathcal{G} m_j}{r_{ij}^3} \mathbf{r}_{ij} - \frac{1}{\eta_{i-1}} \sum_{j=0}^{i-1} \sum_{k=i+1}^{N-1} \frac{\mathcal{G} m_j m_k}{r_{jk}^3} \mathbf{r}_{jk}. \tag{9.133}$$

Care must be taken when evaluating the term in square brackets because of the numerical dangers of subtracting nearly equal quantities. The final part of taking account of the interaction Hamiltonian is the most trivial. In a time step $\Delta t$ the Jacobian velocity vector of the $i$th body changes by an amount $\Delta\tilde{\mathbf{v}}_i$ given by

$$\Delta\tilde{\mathbf{v}}_i = \Delta t \left(\dot{\tilde{\mathbf{v}}}_i\right)_{\text{interaction}} \qquad (0 < i < N) \tag{9.134}$$

while all the $\tilde{\mathbf{r}}_i$ and $\mathbf{r}_i$ remain constant.

The speed of a map comes from its ability to make use of an integrability (or an approximation to it) within the system. In the case of resonance maps it was the fact that the ever-present secular perturbations are integrable and the resonant perturbations are applied as impulses. For the encounter map it is assumed that there is unperturbed motion between the successive impulses at conjunction. In the case of the Wisdom & Holman $N$-body map we exploit the essentially Keplerian nature of the orbit. The Hamiltonian is now in a form suitable for the addition of a single $\delta$-function and the derivation of a map. This can be achieved in a similar way to that discussed for the resonance maps in Sect. 9.5.2. The basis of Wisdom & Holman's $N$-body map is to replace the Hamiltonian given in Eq. (9.119) by

$$\mathcal{H} = \mathcal{H}_{\text{Kepler}} + 2\pi \delta_{2\pi}(\Omega t)\mathcal{H}_{\text{interaction}}, \qquad (9.135)$$

where $\delta_{2\pi}(\Omega t)$ is a $2\pi$-periodic Dirac $\delta$-function with mapping frequency $\Omega$; this use of $\delta$-functions is identical to that discussed in Sect. 9.5.2 for resonance maps, although in that case $\Omega = 1$. Note again that because the $\mathcal{H}_{\text{interaction}}$ is independent of the momenta, the coordinates remain constant over the application of the impulse and it is only the momenta that change. Between applications of the $\delta$-functions (associated with evolution under $\mathcal{H}_{\text{interaction}}$ given in Eq. (9.134)) the system is stepped forward to the next jump by means of the $f$ and $g$ series (associated with the evolution under $\mathcal{H}_{\text{Kepler}}$ given by Eq. (9.124)). This is based on the principle that at any time $t_0$ the position vector $\mathbf{r}$ and the velocity vector $\mathbf{v}$ determine an orbital plane and so the values of $\mathbf{r}$ and $\mathbf{v}$ at time $t$ can be written as the linear combination

$$\mathbf{r}(t) = f(t)\,\mathbf{r}(t_0) + g(t)\,\mathbf{v}(t_0) \qquad (9.136)$$

and hence

$$\mathbf{v}(t) = \dot{f}(t)\,\mathbf{r}(t_0) + \dot{g}(t)\,\mathbf{v}(t_0). \qquad (9.137)$$

The functions $f$, $g$, $\dot{f}$, and $\dot{g}$ are defined in Eqs. (2.69) and (2.71), each of which involves the difference in eccentric anomalies, $\Delta E = E - E_0$. The difference form of Kepler's equation can be written

$$n\,\Delta t = \Delta E - e\cos E_0 \sin \Delta E + e\sin E_0(1 - \cos \Delta E) \qquad (9.138)$$

and this has to be solved for each mass to advance the system to the next impulse. Using $f$ and $g$ functions the Kepler step and the interaction step can both be evolved using Cartesian coordinates, which, as Wisdom & Holman point out, avoids the need for computationally expensive coordinate transformations between steps (Wisdom & Holman 1991). Methods for the solution of Kepler's equation are discussed in Sect. 2.4.

Wisdom & Holman point out that there are two choices regarding optimization of the map. Either one can choose to have the Hamiltonian of the map and the true Hamiltonian to agree as much as possible or, alternatively, one can have agreement in the Taylor series solution for the real system and the map over one time step. Let the Hamiltonian for the map have the form

$$\mathcal{H}_{\text{Map}} = \mathcal{H}_{\text{Kepler}} + \Phi(\Omega t)\mathcal{H}_{\text{interaction}}, \qquad (9.139)$$

where, because the central mass is usually much larger than the perturbing mass, $\mathcal{H}_{\text{Kepler}} \gg \mathcal{H}_{\text{interaction}}$, and

$$\Phi(t) = 2\pi \sum_{i=0}^{k-1} a_i\, \delta_{2\pi}(t - 2\pi d_i). \qquad (9.140)$$

This means that there are $k$ $\delta$-functions for each mapping period. The functions have amplitudes $a_i$ and phases $d_i$, where $0 < d_i < 1$. In terms of Fourier series,

$$\Phi(t) = \sum_{i=0}^{k-1} a_i \sum_{n=-\infty}^{\infty} \cos[n(t - 2\pi d_i)] \tag{9.141}$$

$$= \sum_{n=-\infty}^{\infty} A_n \cos(nt) + \sum_{n=-\infty}^{\infty} B_n \sin(nt), \tag{9.142}$$

where

$$A_n = \sum_{i=0}^{k-1} a_i \cos(2\pi n d_i), \qquad B_n = \sum_{i=0}^{k-1} a_i \cos(2\pi n d_i). \tag{9.143}$$

If we want to get the best agreement between the two Hamiltonians then, for a given value of $k$ (the order of the map), we have to find the values of the $2k$ quantities $a_i$ and $d_i$ that optimise the agreement with the true Hamiltonian. We can do this by placing conditions on the $A_i$ and $B_i$. Since the averaged value of the mapping Hamiltonian over one mapping period has to equal the true Hamiltonian, this implies

$$A_0 = \sum_{i=0}^{k-1} a_i = 1. \tag{9.144}$$

Similarly, for the two Hamiltonians to agree we must have

$$A_n = \sum_{i=0}^{k-1} a_i \cos(2\pi n d_i) = 0, \qquad B_n = \sum_{i=0}^{k-1} a_i \cos(2\pi n d_i) = 0 \tag{9.145}$$

for $n \neq 0$. Wisdom & Holman point out that suitable solutions to these equations for a given order $k$ are $a_i = 1/k$ with the $d_i$ having equal separations. Note that no useful constraint is placed on the absolute phase; all that is required is that the phases are equally spaced.

The alternative approach is to optimise agreement between the series solutions. However, Wisdom & Holman showed that this can be achieved for a second-order (i.e., error is $\mathcal{O}(\Delta t^3)$) map by taking $a_0 = 1$ with $d_0 = 1/2$ and $d_1 = 1$, which simultaneously satisfies the condition for agreement of the Hamiltonians. Let $\Delta t = 2\pi/\Omega$ denote the time interval over which one complete cycle of the map is to be applied. In the "generalised leapfrog" application proposed by Wisdom & Holman all the bodies in the system are evolved as independent two-body Kepler orbits for a time interval $\frac{1}{2}\Delta t$. The impulses from the perturbing planets are then applied, leading to the changes in the Jacobian velocity vectors, $\Delta \tilde{v}_i$, given by Eq. (9.134). This is followed by evolving all bodies on independent Kepler orbits again for the remaining time $\frac{1}{2}\Delta t$ to complete the time step.

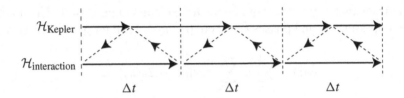

Fig. 9.21. Schematic illustration of the generalised leapfrog method applied to the $k = 1$ case for three equal time intervals, $\Delta t$. Solid lines represent evolution under the particular Hamiltonian; dashed lines show the sequence of application.

Figure 9.21 schematically illustrates the continued application of this process for the case $k = 1$. Note that the $\mathcal{H}_{\text{interaction}}$ steps (i.e., those involving the application of the $\delta$-functions) always involve equally spaced whole intervals whereas the $\mathcal{H}_{\text{Kepler}}$ steps always start and end with a half interval. The map proposed by Wisdom & Holman satisfies the best of both worlds. This is because it preserves the condition $\sum a_i = 1$ while having the $\delta$-functions evenly spaced; furthermore the use of an absolute phase allows it to be an accurate representation of the solution to second order. In fact, it can be shown that the map follows exactly the dynamics of a surrogate Hamiltonian that differs from $\mathcal{H}$ by terms of $\mathcal{O}(\Delta t^2)$. This ensures that there is no secular trend in the energy of the evolved system (Saha & Tremaine 1994).

The use of symplectic integrators of the type developed by Wisdom & Holman has now become widespread. Higher-order methods have been derived (see, for example, Kinoshita et al. 1991) and there have been other enhancements such as the reduction in long-term errors (Saha & Tremaine 1992), the introduction of separate time steps for each body (Saha & Tremaine 1994), a symplectic algorithm that allows for close approaches between bodies (Levison & Duncan 1994), and one that allows for dissipation (Malhotra 1994).

## 9.6 Separatrices and Resonance Overlap

We have already met the concept of a separatrix in Sect. 8.6 where a particular choice of energy in the pendulum model led to motion with infinite period. Lower values of the energy led to libration and higher values to circulation (see Fig. 8.6b). We have also seen how the pendulum model is a good approximation to motion near resonance in the averaged problem. In both cases the problem is integrable. However, the addition of short-period terms to the pendulum model, leading to the standard map, showed that chaotic motion can arise. A similar type of behaviour is seen in the surfaces of section of the planar restricted three-body problem. In both cases the chaos is associated with motion near the separatrix. Orbits well inside the separatrix and close to the island centres are regular. The chaotic orbits are essentially unpredictable, although they are derived from a deterministic system. Why does motion at the separatrix in the perturbed system produce such chaotic orbits? To answer this we will consider a perturbed version of the separatrix diagram originally shown for the unperturbed pendulum in Fig. 8.6b.

Consider Fig. 9.22 where we have plotted some of the curves of constant Hamiltonian for the perturbed case. The unstable equilibrium point can be thought of as being made up of two branches of the curve. The $\mathcal{H}^+$ branches indicate motion towards the unstable point, whereas the $\mathcal{H}^-$ curves indicate motion away from the point. It is the combination of these two branches that leads to near-hyperbolic motion very close to the unstable equilibrium point. However, since the left-hand unstable point is the same as the right-hand unstable point (the angular coordinate is $2\pi$-periodic), the $\mathcal{H}^-$ branch emanating from the left-hand point eventually has to meet the $\mathcal{H}^+$ branch entering the right-hand point. In the unperturbed case the meeting is smooth, but in the perturbed case things are different.

Consider a point $x$ at a crossing point of the two branches. If we consider the motion of the perturbed system in terms of a map $F$, for example the standard map discussed above, then after one application of the map the new value is $x' = F(x)$, after two applications it is $x'' = F(x') = F^2(x)$, etc. Hence if $x$ is on the $\mathcal{H}^+$ branch the unstable point is defined by $\lim_{n\to\infty} F^n(x)$, where $F^n(x)$ denotes the $n$th application of the map $F$. Because we are dealing with motion on the separatrix we can expect the points to get closer together as $n \to \infty$. However, the unstable point can also be thought of as the $\lim_{n\to\infty} T^{-n}x$ if we are considering motion on the $\mathcal{H}^-$ branch. Since we are dealing with a Hamiltonian system the motion is area preserving. Thus, when the curves from the different branches

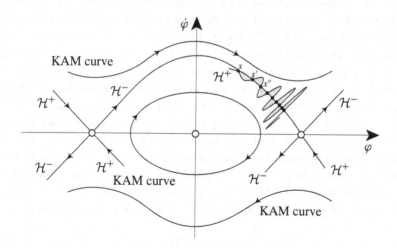

Fig. 9.22. The stable ($\mathcal{H}^+$) and unstable ($\mathcal{H}^-$) branches of the separatrix intersect to generate chaotic motion near the separatrix for a perturbed pendulum. The equilibrium points are denoted by unfilled circles. (Adapted from Lichtenberg & Lieberman 1983.)

meet, the area enclosed by their intersection (the shaded areas in Fig. 9.22) has to remain constant. But to reconcile this with the fact that the iterated points are getting closer together, we must have the $\mathcal{H}^-$ branch of the curve oscillating more and more wildly, as shown in Fig. 9.22. Eventually, as the unstable point is approached the oscillation will be nearly parallel to the final $\mathcal{H}^-$ curves coming from the unstable point.

Recall also that the period of the unperturbed motion tends to $\infty$ as the separatrix is approached. This gives rise to numerous resonances with the other frequencies. For example, in Fig. 9.14 we saw how large-amplitude librations at the 5:2 resonance have a frequency that matches the circulation frequency of the 60:24 resonance, producing another eight islands within each of the three main islands. Of course, each of these new islands will have its own chaotic separatrix. It is this mechanism combined with the complicated behaviour of the Hamiltonian curves that gives rise to the chaotic separatrix. The extent of the chaotic region for a given resonance can be determined using methods originally developed by Melnikov (1963) (see, for example, Lichtenberg & Lieberman (1983) and Guckenheimer & Holmes (1983)).

In Fig. 9.18 we noted that apart from the chaotic motion near the separatrix there were still regular orbits left, although they were becoming rare as $k_0$ increased. The same was also true in the surfaces of section plots for the circular restricted problem (see Figs. 9.16 and 9.17). These are represented by the curves labelled "KAM curves" in Fig. 9.22. (The initials refer to Kolmogorov, Arnold, and Moser who derived an important result on the continued existence of regular orbits as an integrable system became subject to a small perturbation.)

Chaotic motion can also be thought of as arising from the overlap of adjacent resonances. Consider the case of the circular restricted three-body problem. If we consider the averaged problem in the vicinity of a resonance then, according to the theory we have developed in Chapter 8, there is a well-defined separatrix and maximum width to the resonance. As a first approximation we could consider the phase space as being made up of a succession of such resonances, each independent of the others with no short-period terms to affect the separatrix. An obvious example is the sequence of first-order interior resonances of the form $p + 1 : p$. In Sect. 8.7 we derived expressions for the maximum libration width of such resonances, each treated in isolation. Since each resonance has a well-defined width in semi-major axis, and since the separation of adjacent resonances becomes smaller as the perturber is approached, there will come a point at which these overlap. Wisdom (1980) showed that in the circular restricted problem for small eccentricity ($e \leq 0.15$) of the particle this point is reached when

$$p_{\text{overlap}} \approx 0.51 \mu_2^{-2/7}, \tag{9.146}$$

where $\mu_2 = m_2/(m_1 + m_2)$ is the mass of the perturbing secondary in dimensionless units. Using Kepler's third law we find that this corresponds to a separation in semi-major axis of

$$\Delta a_{\text{overlap}} \approx 1.3\mu_2^{2/7}a',$$ (9.147)

where $a'$ is the semi-major axis of the perturber. Therefore we would expect particles in the region $a' \pm \Delta a_{\text{overlap}}$ to be on chaotic orbits, having close encounters with the perturber and being removed from the zone. Note that this result predicts a scaling of $\Delta a_{\text{overlap}}$ as the two-sevenths power of the mass ratio of the system.

Duncan et al. (1989) give a simple heuristic derivation of the same result, using the properties of the encounter map discussed in Sect. 9.5.3. They note that in their system chaos occurs when the perturbation in semi-major axis at a single encounter is sufficiently large to change the perturbation in orbital phase by $\pi$ before the next encounter. In such a case there is no correlation between the perturbations in longitudes at successive conjunctions. This implies that the system "forgets" the previous longitude and the motion is chaotic. In terms of the quantities used in Eqs. (9.95)–(9.97) this implies that chaos sets in when

$$\frac{4\pi\,\Delta\varepsilon_{n+1}}{3\varepsilon_1^2} \geq \pi$$ (9.148)

or

$$|\Delta\varepsilon| \geq \frac{3}{4}\varepsilon_1^2,$$ (9.149)

where we recall that $\varepsilon$ is the relative difference in semi-major axis between the particle and the perturber. If we take the initial eccentricity to be zero and apply the condition after the first encounter, assuming that increasing $|\Delta\varepsilon|$ leads to increasing $|z| = e$, we have $|z_1| = 0$ and, from Eq. (9.96),

$$|z_2| = \frac{g}{\varepsilon_1^2}\frac{m_2}{m_1}.$$ (9.150)

Since

$$\varepsilon\,\Delta\varepsilon = \frac{2}{3}|z_2|^2$$ (9.151)

the condition for chaos reduces to

$$\varepsilon_1 \leq \left(\frac{8}{9}g^2\right)^{1/7}\left(\frac{m_2}{m_1}\right)^{2/7}.$$ (9.152)

Putting in numerical values and making use of the fact that for the critical case $\varepsilon_1 = \Delta a_{\text{overlap}}/a'$, we obtain the result

$$\Delta a_{\text{overlap}} \approx 1.24\mu_2^{2/7}a'.$$ (9.153)

This result is in good agreement with that originally derived by Wisdom (1980) using a different method.

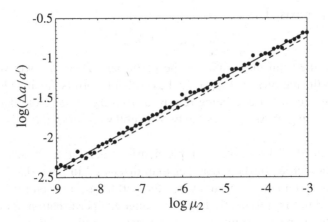

Fig. 9.23.  The results of a series of numerical experiments to measure $\Delta a_{\text{overlap}}/a'$ using the encounter map devised by Duncan et al. (1989). The solid line is the best fit to the data. The dashed line denotes the theoretical result of Wisdom (1980).

We have checked the validity of this result by carrying out the same numerical experiment as performed by Duncan et al. (1989). This involved starting test particles on circular orbits at different relative separations $\Delta a/a'$ from the circular orbit of the perturber. The last value of $\Delta a/a'$ for which the orbit did not achieve an eccentricity $e > \Delta a/a'$ within $10^3$ encounters was noted for a range of values of $\mu_2$. The results are plotted in Fig. 9.23. The solid straight line corresponds to the best-fit solution of the form $\Delta a/a' = 1.57\mu_2^{0.286}$ and the dashed line denotes Wisdom's relation. Although there is good agreement between the slopes of each line, the values of $\Delta a/a'$ from the numerical experiment are slightly larger. In any case the Wisdom result seems to provide a lower bound on the numerical results.

## 9.7 The Rotation of Hyperion

We first introduced the concept of resonance in the context of spin–orbit coupling between a satellite and planet. In Sect. 5.4 we discussed the equation of motion governing the orientation of a satellite and showed how in certain circumstances we can obtain approximate analytical solutions to the motion. In Fig. 5.17b we plotted the values of $\dot{\theta}/n$ and $\theta$ (where $\theta$ is the orientation angle, $\dot{\theta}$ is its time derivative, and $n$ is the mean motion) for a number of different starting conditions for a slightly nonspherical satellite moving in orbit with eccentricity $e = 0.1$. We noted the "fuzzy" appearance of some of the points near separatrices. These we can now explain as being due to a small chaotic zone. We were also able to derive approximate analytical solutions that mimicked the numerical results (see Figs. 5.17a), although not the chaos.

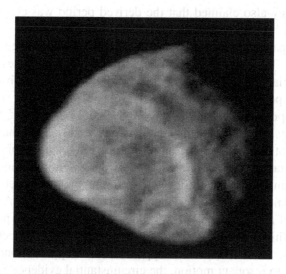

Fig. 9.24. A *Voyager 2* image of the saturnian satellite Hyperion showing its irregular shape. The longest dimension of the satellite is ∼ 350 km.

As pointed out in Chapter 5, most natural satellites in the solar system are in synchronous rotation, keeping one face pointed towards the planet. The timescale, $T_{\text{despin}}$, required for most satellites to settle into this configuration is considerably less than the age of the solar system (from Table 5.2 the value for the Moon is $3 \times 10^7$ y). However, it was shown in Sect. 5.5 that the value of $T_{\text{despin}}$ is inversely proportional to the magnitude of the tidal torque, $N_{\text{s}}$, which is itself a strong function of the size of the satellite and inversely proportional to the sixth power of its distance from the planet (see Eq. (5.2)). Clearly, small satellites orbiting far from the parent planet may not have had time to evolve into a spin–orbit resonance. The saturnian satellite Hyperion has an unusual shape (see Fig. 9.24) with approximate radial dimensions of 175 km × 120 km × 100 km. Hyperion's orbital eccentricity is 0.1 and it has an orbital radius of 24.5 Saturn radii making it one of the most distant satellites of Saturn. Therefore we would expect such a small object at this distance to have a large value of $T_{\text{despin}}$.

Prior to the *Voyager* encounters with the saturnian system in 1980 and 1981, calculations of the tidal despinning timescales for the satellites suggested that Hyperion was a satellite that may not have had sufficient time to despin to a synchronous state. Initial observations of the light curve of Hyperion by Thomas et al. (1984) suggested that the *Voyager 1* and *2* data were best fitted by a spin period of 13 d. Since Hyperion's orbital period is 21.3 d, this implied that the rotation was nonsynchronous. The data also suggested that the spin axis was parallel to the orbital plane. To derive the spin period Thomas et al. (1984) had assumed a fixed rotation rate over a 61 d interval during the *Voyager 2* encounter. They also claimed that the derived period was consistent with the *Voyager 1* observations some 220 d earlier.

However, a paper by Wisdom, Peale & Mignard (1984) suggested a more unusual scenario. They carried out numerical and analytical studies of the solution to the differential equation describing the variation of $\theta$ given by Eq. (5.56). In the case of Hyperion the *Voyager* observations of its unusual shape suggested that a more appropriate value was $\omega_0 = 0.89 \pm 0.22$ (see Eq. (5.85)), assuming uniform density. This, combined with the large eccentricity of the orbit ($e = 0.1$), implied that not only was Hyperion not trapped in a spin–orbit resonance, but also that its rotation was *chaotic* and underwent essentially random but nevertheless deterministic changes. Since an averaging method is not appropriate for Hyperion, it is necessary to carry out a numerical integration. A surface of section for $e = 0.1$ and $\omega_0 = 0.89$ using values appropriate for Hyperion, is shown in Fig. 9.25. This shows the results of integrating a single trajectory for 20,000 orbital periods. Unless Hyperion happened to be trapped in one of the islands corresponding to resonant motion, the circumstantial evidence for a chaotic rotation is convincing. Note that by following the evolution of a single chaotic trajectory we can help to define the locations of the remaining regular regions; the same technique was employed to produce Fig. 9.15b for a completely different system. The result should be compared with Fig. 5.8 where seven different trajectories are shown and regular motion dominates.

The surface of section shown in Fig. 9.25 is similar to that found by Wisdom et al. and suggests that a large chaotic zone surrounds the $p = 1/2$ and $p = 2$

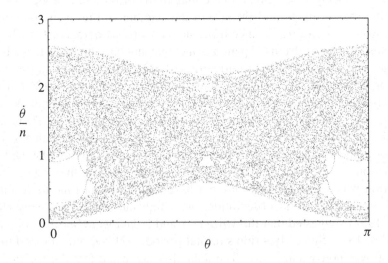

Fig. 9.25. A surface of section plot for a single trajectory in the $(\theta, \dot{\theta}/n)$ coordinate system using $\omega = 0.89$ and $e = 0.1$. The starting value is $\theta = 0.0$ and $\dot{\theta}/n = 0.28$ and the trajectory is followed for 20,000 orbital periods. (Adapted from Murray 1998.)

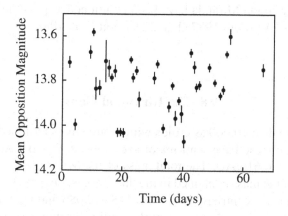

Fig. 9.26. Ground-based observations of Hyperion's light curve obtained by Klavetter (1989). Although the amplitude is large, there is no obvious periodicity and this is consistent with chaotic rotation.

spin states and also that libration in the $p = 3/2$ state is not possible. Wisdom et al. went on to show that the satellite was also likely to be *attitude unstable* such that it was tumbling in space. Ground-based observations by Klavetter (1989) appear to confirm these results (see Fig. 9.26).

It is interesting to note that the chaotic nature of Hyperion's rotation is a consequence both of its unusual shape and its large orbital eccentricity, although it is primarily due to the former. We have already seen that, left to itself, the tides raised on a satellite by a planet act to damp the eccentricity of the satellite's orbit. However, in the case of Hyperion the eccentricity cannot damp since it is a forced eccentricity due to its 4:3 orbit–orbit resonance with the nearby massive satellite Titan. Therefore, while the orbit of Hyperion appears to be stable because of the 4:3 resonance with Titan, its rotation is chaotic because of the competing effects of a variety of spin–orbit resonances.

However, the whole question of the rotation of Hyperion has still to be resolved. According to the work of Black et al. (1995), based on an improved shape model and rotation rate for Hyperion, although the spin state may be technically chaotic, the satellite can give the impression of regular rotation for long periods of time.

It is entirely likely that other irregularly shaped satellites have experienced chaotic rotation at some time in their history (Wisdom 1987a,b). In order for a satellite to get captured into spin–orbit resonance it has to cross chaotic separatrices and this is certain to have induced an episode of chaotic rotation. This could also have occurred if the satellite suffered a large impact that altered its rotational properties. Such episodes could also have produced significant internal heating and resurfacing events in some satellites. The Martian moon Phobos and the uranian moon Miranda have been mentioned as possible candidates for this process (see Wisdom 1987a,b and Dermott et al. 1988).

## 9.8 The Kirkwood Gaps

There are more than 10,000 asteroids with accurately determined orbits and the vast majority of these have semi-major axes between 2 and 4 AU, with a mean of $\langle a \rangle = 2.74 \pm 0.616$ AU. The total mass of the belt is $\sim 10^{-9}$ Earth masses, with $\sim 80\%$ of the mass contained in the three largest objects: Ceres, Pallas, and Vesta. The mean eccentricity is $\langle e \rangle = 0.148 \pm 0.086$ and the mean inclination is $\langle I \rangle = 8.58° \pm 6.62°$. A more useful quantity for comparison is the sine of the inclination; for the asteroid belt, $\langle \sin I \rangle = 0.148 \pm 0.111$. Therefore the mean and the standard deviation of the eccentricities and sine of the inclinations are comparable. Because the asteroids are small and numerous and undergo the significant gravitational effect of one major perturber (Jupiter) and several minor ones, they provide a natural laboratory for the study of dynamical processes in the solar system. We have already seen in Sect. 7.10 and Sect. 7.11 that there is good evidence for collisional evolution within the asteroid belt. This is not surprising since typical encounter velocities between asteroids are $\sim 5$ km s$^{-1}$, too large for accretion to occur.

The first asteroid was discovered by Guiseppe Piazzi on the first day of the nineteenth century. By the time Daniel Kirkwood carried out his study (Kirkwood 1867) of the distribution of the semi-major axes of asteroid orbits, he was able to make use of data on ninety-one objects. Kirkwood ordered the known asteroid orbits by semi-major axis and noted that the three widest separations occurred in the vicinity of the 9:4, 5:2, and 3:1 jovian resonances. He also calculated the location of the 7:2, 7:3, and 2:1 resonances but did not specifically associate them with gaps. However, Kirkwood was the first person to note that the distribution was nonrandom and was likely to have been influenced by jovian perturbations. He was of the opinion that *"As in the case of the perturbation of Saturn's ring by the interior satellites, the tendency of Jupiter's influence would be to form gaps or chasms in the primitive ring"* (Kirkwood 1867). It is clear that the inspiration for Kirkwood's study was the association of the Cassini division in Saturn's rings with the 2:1 resonance with Mimas. The mechanism that Kirkwood proposed involved the effect of repeated conjunctions. He believed that the likely buildup in the asteroid's eccentricity would lead to collision and coalescence with nearby objects causing depletion of material at the resonance.

As an historical footnote to the discovery of the gaps, we note that Kopal (1972) claimed that the existence of the gaps was first pointed out by a Prague astronomer, K. Hornstein, prior to the discovery by Kirkwood. We have found no documentary evidence to support Kopal's claim.

### 9.8.1 Resonant Structure of the Asteroid Belt

Since Kirkwood's time the available data on the asteroid belt has grown by more than an order of magnitude. Now there are well-determined orbits for more than 10,000 asteroids. As well as orbital data there is a wealth on information available on asteroid colours, shapes, rotational properties, and, in the case of (243) Ida, satellites (see Fig. 2.6). All this information can be used to understand better the dynamical evolution of the asteroid belt.

Figure 9.27a shows a plot of the projected location of the asteroids in $x$–$y$ space (the plane of the ecliptic) at 12h UTC on 18 December 1997. The limits on this plot exclude asteroids beyond $\sim 4$ AU (for example, objects at the 4:3 and 1:1 jovian resonances) but there is still evidence of a well-defined ring of material. This can be highlighted by plotting a similar diagram with the semi-major axis and longitude of perihelion denoting the radial and angular positions respectively. Figure 9.27b shows clearly the resonant structure of the main belt, with a number of cleared "lanes" corresponding to the major Kirkwood gaps. The inner and outer edges of the belt are sharp and associated with the 4:1 and 2:1 resonances respectively (see Fig. 1.6), but the clustering at the 3:2 resonance (the Hilda group of asteroids) is clearly visible. The plot illustrates another feature of the asteroid belt – a clear bias in the angular distribution. Since the angular component of each point denotes the perihelion position of an asteroid, it is clear that there is a marked preference for longitudes of perihelion close to Jupiter's value (indicated by the arrow). This is simply a consequence of the fact that the precession rate of an asteroid dominated by jovian perturbations is at a

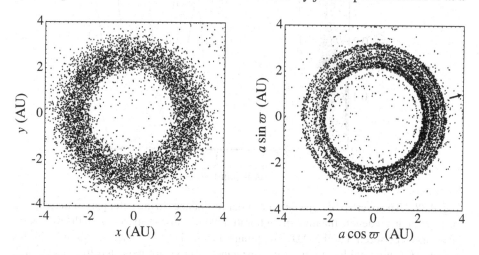

Fig. 9.27. (a) A plot of the $x$–$y$ distribution of the known asteroids on 18 December 1997. (b) A plot of the $a\cos\varpi$–$a\sin\varpi$ distribution of the known asteroids using the same orbital data as in (a). Note the clear suggestion of a band of material corresponding to the main belt.

minimum when the pericentres are aligned and a maximum when they differ by 180° (see Eq. (7.94)).

An important feature of the asteroid distribution is the decline in the number of asteroids beyond the 3:2 resonance (at 3.97 AU), apart from a small number of objects at the 4:3 resonance (at 4.29 AU). This absence of objects is also a resonant effect, caused by the overlapping of the sequence of first-order resonances as Jupiter is approached. Application of the resonance overlap criterion discussed in Sect. 9.6 (see Wisdom 1980) shows that in the Sun–Jupiter case $p_{overlap} \approx 4$, corresponding to a chaotic zone extending 0.9 AU inside Jupiter's orbit. Therefore we would expect a cleared zone in the asteroid belt beyond 4.3 AU; this is in good agreement with observations (see Fig. 9.28).

A number of hypotheses for the origin of the resonant gaps have emerged since Kirkwood's original discovery. It has to be pointed out that any viable theory must explain not only the gaps at certain resonances but also the groupings at others. Greenberg & Scholl (1979) classified the theories into four main groups:

- *The statistical hypothesis.* This is the claim that the gaps are simply an illusion. Since the dynamical equations governing the motion of objects in resonance are similar to those of a pendulum, if an asteroid is in resonance it will spend most of its time at the extremes of its motion. Consequently, at a given time, statistically the asteroid is more likely to be at the extremes of its librational motion (i.e., at the edges of the resonance rather than in the middle). With the

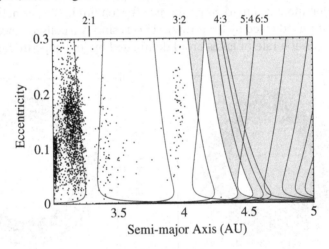

Fig. 9.28.  The overlap of adjacent first-order resonances in the outer part of the asteroid belt. The curves denote the maximum libration widths in *a–e* space for all the first-order resonances between 3 and 5 AU. The points denote the *a–e* values of asteroids in this region. The nominal locations of the resonance centres are indicated for the 2:1, 3:2, 4:3, 5:4, and 6:5 resonances. The shaded areas denote regions of resonance overlap. Note the prominent gap at the 2:1 resonance and the grouping at the 3:2 resonance.

advent of digital computers it became feasible to investigate this hypothesis by numerically integrating the equations of motion of known asteroids near the Kirkwood gaps. Apart from a few objects that were discovered to be librating, there was no indication that the gaps were illusory.

- *The collisional hypothesis.* This was originally proposed by Kirkwood. It suggests that the changes in orbital elements at a resonance cause the perturbed asteroids to collide with nearby objects. This could cause objects to accrete away from the gaps (unlikely since the the relative velocities would be too high) or a general breakdown of material due to collisional disruption. The latter can be checked by looking for evidence of a mass effect whereby the average mass of objects, as well as their number density, decreases as the resonance is approached.

- *The cosmogonic hypothesis.* Cosmogonic theories propose that the gaps are regions where asteroids failed to form or that they reflect processes that operated in the early stages of the formation of the solar system but have now ceased. For example, one idea is that *resonance sweeping* occurred when Jupiter's orbit started to expand, perhaps due to it orbiting a less massive solar nebula. Such a mechanism can be investigated using the resonance model discussed in Chapter 8. This may be able to explain the formation of some gaps, but not features such as the difference between the 2:1 and 3:2 resonances. Its effectiveness depends on unknown parameters such as the extent of orbital change in the early solar system.

- *The gravitational hypothesis.* According to this hypothesis the gaps can be understood in the context of the Sun–Jupiter–asteroid three-body problem. Research over the past two decades has shown that this is the most likely origin of the gaps and that chaotic motion is involved. However, the exact mechanism is not yet entirely understood.

Dermott & Murray (1983) showed that when the distribution of asteroid orbits is plotted in $a$–$e$ space and the maximum libration widths of the strong resonances are superimposed, there is a good correspondence between the width of the gaps and the width of the libration zone (see Fig. 9.29). Note that the presence of an asteroid within the libration zone for a given resonance does not necessarily imply that it is in libration. This is because the data points for the asteroids are all plotted for the same epoch. A more correct comparison for an individual resonance would involve an integration of the orbit until the resonant angle for maximum libration had the appropriate value. Although Dermott & Murray provided no mechanism for the removal of asteroids, they concluded that the gravitational hypothesis was the most likely to be correct.

If the gaps can be understood in the context of the Sun–Jupiter–asteroid restricted problem then a simple, long-term integration of the equations should be sufficient to search for a removal mechanism. We have already carried out numerical investigations of this problem when we examined the surfaces of

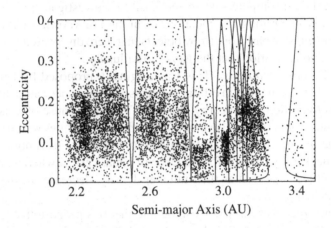

Fig. 9.29. The maximum libration widths in *a–e* space of the strong jovian resonances superimposed on the distribution of asteroids in the main belt. Note the correspondence between the widths of the gaps and the widths of the resonances.

section of the circular restricted problem for a mass ratio of $10^{-3}$ (see Sect. 9.4). In that case chaotic motion was detected. However, the central parts of each major resonance (i.e., each island in the surfaces of section) were regular in appearance. This leads one to believe that the Kirkwood gaps cannot be explained in the context of the planar, circular restricted problem alone.

Unfortunately long-term integrations of more complicated models (such as the elliptical and nonplanar problems) were almost impossible until the arrival of fast (and cheap) computers. The limited integrations that had been carried out previously did not reveal any dramatic behaviour in the motion of hypothetical asteroids at resonances. A major breakthrough occurred when Wisdom (1982) derived his algebraic mapping to study asteroid motion at the 3:1 resonance (see Sect. 9.5.2).

### 9.8.2 The 3:1 Resonance

In a series of papers Wisdom (1982, 1983, 1985a) provided convincing evidence of the role of chaos in the origin of the 3:1 Kirkwood gap. He used the map (Wisdom 1982) to show that for some starting conditions a test particle at the 3:1 resonance could give the appearance of regular behaviour for several thousand years and then undergo a large increase in its orbital eccentricity. At the location of the resonance ($a \approx 2.5$ AU) the asteroid would cross the orbit of Mars and eventually be removed by direct perturbations. Wisdom proposed that although the chaos in the motion is caused by Jupiter, it is Mars that actually removes the object. A sample plot of the chaotic evolution of a test particle at the 3:1 resonance is shown in Fig. 9.30. This was obtained using Wisdom's mapping technique.

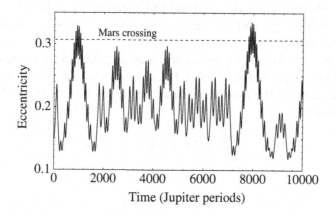

Fig. 9.30. The evolution of a test particle's eccentricity at the 3:1 resonance showing the chaotic nature of the orbit and how it can cross the orbit of Mars. The initial values were $a_0/a' = 0.481$, $e_0 = 0.15$, $\varpi_0 - \varpi' = 0$, and $3\lambda' - \lambda_0 = \pi$, corresponding to the chaotic, "representative" plane identified by Wisdom (1982). (Adapted from Murray 1998.)

In his initial paper Wisdom (1982) included the effect of secular perturbations on the orbit of Jupiter, as well as a full three-dimensional calculation. He went on to show (Wisdom 1983) that the observed chaos was not due to the secular evolution of Jupiter's orbit and that it occurred even in the planar case. However, Jupiter's eccentricity played a crucial role. Murray & Fox (1984) showed that the chaos was not an artifact of the map but was a genuine reflection of the type

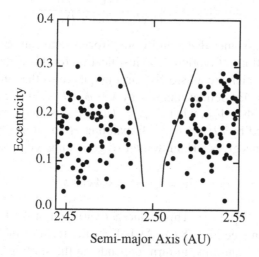

Fig. 9.31. The positions of asteroids at the 3:1 resonance compared with the extent of the chaotic region. (Adapted from Wisdom 1983.)

of motion inherent in the averaged and full equations of motion. The excellent agreement between the width of the chaotic region and the width of the observed is illustrated in Fig. 9.31.

In further work Wisdom (1985a) provided a convincing analytical basis for the observed chaotic motion at the 3:1 resonance. He took the standard form of the Hamiltonian for the averaged, planar, restricted three-body problem at the 3:1 resonance:

$$
\begin{aligned}
\mathcal{H} = & -\frac{\mu_1^2}{2\Phi^2} - 3\Phi + \mu F(x^2 + y^2) + e'\mu G x \\
& - \left[ C(x^2 - y^2) + e'Dx + e'^2 E \right] \cos\phi \\
& - \mu(C2xy + e'Dy)\sin\phi,
\end{aligned}
\tag{9.154}
$$

where $\phi = \lambda - 3\lambda'$ and $C$, $D$, $E$, and $F$ are constants with the other variables having their usual definitions. The Hamiltonian can also be written as

$$
\begin{aligned}
\mathcal{H} = & -\frac{\mu_1^2}{2\Phi^2} - 3\Phi + \mu F(x^2 + y^2) + e'\mu G x \\
& - \mu \left[ C(x^2 - y^2) + e'Dx + e'^2 E \right] \cos\phi \\
& - \mu A(x, y)\cos[\phi - P(x, y)],
\end{aligned}
\tag{9.155}
$$

where

$$
A(x, y) = \sqrt{\left[ C(x^2 - y^2) + e'Dx + e'^2 E \right]^2 + (2xyC + e'Dy)^2}
\tag{9.156}
$$

and

$$
\tan P(x, y) = \frac{2xyC + e'Dy}{C(x^2 - y^2) + e'Dx + e'^2 E}.
\tag{9.157}
$$

Wisdom then assumes that $x$ and $y$ are "frozen" and can be treated as parameters in the equations of motion. The justification for this is that $x$ and $y$ depend on $a$, $e$, and $\varpi$, which are all slowly varying quantities that only change on resonant timescales. He defines a new $\lambda$ by $\lambda = \phi - P(x, y) - \pi$ and its conjugate momentum $\Lambda = \Phi - \Phi_{res}$, where $\Phi_{res} = (\mu_1^2/3)^{1/3}$ is the nominal centre of the resonance defined by $\partial\mathcal{H}/\partial\Phi = 0$. Expanding $\mathcal{H}$ about $\Phi_{res}$ and keeping only the quadratic terms gives a Hamiltonian $\mathcal{H}'$ governing the variations in $\lambda$ and $\Lambda$:

$$
\mathcal{H}' = \frac{1}{2}\alpha\Lambda^2 + \mu A\cos\phi,
\tag{9.158}
$$

where $\alpha = -3\mu_1^2/\Phi_{res}^4 < 0$. This is another variation of the Hamiltonian of the simple pendulum (see Sect. 8.6). As before, the form of the solution depends on the value of $\mathcal{H}'$ and this, in turn, depends on the starting conditions. Three situations can arise: (i) If $\mathcal{H}' < -A$ the angle $\lambda$ circulates, (ii) if $-A < \mathcal{H}' < A$ then $\lambda$ librates about zero, and (iii) if $\mathcal{H}' = A$ then the motion is on the separatrix.

In the case of libration the solution for $\lambda$ is

$$\lambda = 4 \sum_{n=1}^{\infty} \frac{\sin(2n-1)\omega t}{(2n-1)\cosh[(n-1/2)\pi K'/K]}, \tag{9.159}$$

where $K(k)$ and $K'(k)$ are the complete elliptical integrals of the first kind defined by

$$K(k) = \int_0^{\pi/2} (1 - k \sin^2\theta)^{-1/2} d\theta \tag{9.160}$$

and

$$K'(k) = K(1-k) = \int_0^{\pi/2} \left[1 - (1-k)\sin^2\theta\right]^{-1/2} d\theta \tag{9.161}$$

with modulus $k_L = \sqrt{(\mu A - \mathcal{H}')/2\mu A}$ and $\omega = \pi\omega_0/2K$. The frequency $\omega_0$ is the frequency of the small-amplitude oscillations given by $\omega_0 = \sqrt{-\alpha\mu A}$.

In the case of circulation the solution for $\lambda$ is

$$\lambda = \omega t + 2 \sum_{n=1}^{\infty} \frac{\sin n\omega t}{n \cosh(n\pi K'/K)}, \tag{9.162}$$

with $\omega = \pi\omega_C/K$ and $\omega_C = [-\alpha(\mu A - \mathcal{H}')/2]^{1/2}$. In this case the elliptic modulus is $k_C = [2\mu A/(\mu A - \mathcal{H}')]^{1/2}$.

Using the Hamiltonian in Eq. (9.155) the equations of motion can be written

$$\frac{dx}{dt} = -2\mu F y - \mu \frac{\partial A}{\partial y}\cos\lambda - \mu A \frac{\partial P}{\partial y}\sin\lambda, \tag{9.163}$$

$$\frac{dy}{dt} = 2\mu F x + e'\mu G + \mu \frac{\partial A}{\partial x}\cos\lambda + \mu A \frac{\partial P}{\partial x}\sin\lambda, \tag{9.164}$$

$$\frac{d\lambda}{dt} = \alpha\Lambda - \frac{\partial P}{\partial x}\frac{dx}{dt} - \frac{\partial P}{\partial y}\frac{dy}{dt}, \tag{9.165}$$

$$\frac{d\Lambda}{dt} = \mu A \sin\lambda. \tag{9.166}$$

From Eqs. (9.163) and (9.164) the timescale for significant evolution of $x$ and $y$ is proportional to $\mu^{-1/2}$. In a sense only the averaged effect of the rapidly oscillating terms containing $\cos\lambda$ and $\sin\lambda$ will contribute to the large variations of $x$ and $y$. Wisdom suggested splitting $x$ and $y$ into their long- and short-period parts by writing $x = \bar{x} + \xi$ and $y = \bar{y} + \eta$ with the resulting equations of motion,

$$\frac{d\bar{x}}{dt} = -2\mu F\bar{y} - \mu \frac{\partial A(\bar{x}, \bar{y})}{\partial \bar{y}}\langle\cos\lambda\rangle - \mu A(\bar{x}, \bar{y})\frac{\partial P}{\partial \bar{y}}\langle\sin\lambda\rangle, \tag{9.167}$$

$$\frac{d\bar{y}}{dt} = 2\mu F\bar{x} + e'\mu G + \mu \frac{\partial A(\bar{x}, \bar{y})}{\partial \bar{x}}\langle\cos\lambda\rangle + \mu A(\bar{x}, \bar{y})\frac{\partial P}{\partial \bar{x}}\langle\sin\lambda\rangle, \tag{9.170}$$

where the angled brackets indicate an average over one oscillation period of $\lambda$. These averages are

$$\langle \cos \lambda \rangle = \frac{1}{T} \int_0^T \cos \lambda \, dt = \frac{2E(k_L)}{K(k_L)} - 1, \tag{9.171}$$

$$\langle \sin \lambda \rangle = \frac{1}{T} \int_0^T \sin \lambda \, dt = 0 \tag{9.172}$$

for libration and

$$\langle \cos \lambda \rangle = \frac{1}{T} \int_0^T \cos \lambda \, dt = \frac{2E(k_C)}{k_C^2 K(k_C)} + 1 - \frac{2}{k_C^2}, \tag{9.173}$$

$$\langle \sin \lambda \rangle = \frac{1}{T} \int_0^T \sin \lambda \, dt = 0 \tag{9.174}$$

for circulation. In each case $E(k)$ is the complete elliptical integral of the second kind defined by

$$E(k) = \int_0^{\pi/2} (1 - k \sin^2 \theta)^{1/2} d\theta. \tag{9.175}$$

The equations of motion for $x$ and $y$ are then

$$\frac{d\bar{x}}{dt} = -2\mu F \bar{y} - \mu \frac{\partial A(\bar{x}, \bar{y})}{\partial \bar{y}} \langle \cos \lambda \rangle, \tag{9.176}$$

$$\frac{d\bar{y}}{dt} = 2\mu F \bar{x} + e' \mu G + \mu \frac{\partial A(\bar{x}, \bar{y})}{\partial \bar{x}} \langle \cos \lambda \rangle. \tag{9.177}$$

These equations effectively describe the motion of the guiding centre of trajectories at the 3:1 resonance. Wisdom used plots in $\bar{x}$–$\bar{y}$ space to illustrate the origin of the chaotic motion at the resonance. In particular he showed how certain starting conditions close to separatrices can give the impression of regular motion for long periods of time, before displaying sudden increases in eccentricity.

Meteorites are believed to be the fragments of material broken up by collisions within the asteroid belt, and there is certainly a good correlation between the reflectance properties of meteorite samples and asteroid spectra. The length of time taken for the meteorite to reach the Earth can be estimated from the *cosmic ray exposure age* of a sample. This is the length of time that the meteorite has been exposed to the general cosmic ray background of the solar system. Prior to the work of Wisdom (1985b) there had been a problem reconciling known exposure ages of $\sim 20 \times 10^6$ y of the ordinary chondrite class of meteorites with possible delivery mechanisms based on planetary perturbations. Wisdom (1985b) showed that the 3:1 resonance was the most likely source of meteorites. When the perturbations of the other planets are included, he showed that it is possible for objects at the 3:1 resonance to get eccentricities as large as 0.6, in which case they can cross the orbit of the Earth. At the same time Wetherill

(1985) had shown that there had to be a source for the ordinary chondrites near 2.5 AU.

### 9.8.3 Other Resonances

The structure of the phase space at other jovian resonances has been investigated using maps, as well as numerical and analytical approaches. Murray (1986) derived a map for first-order resonances and used it to study the 2:1 and 3:2 jovian resonances. However, there is a severe problem with the convergence of the disturbing function at these resonances for any but the lowest values of $e$. The results of Murray's map were only valid for low $e$ at the 2:1 resonance and probably invalid at the 3:2 resonance. In any case Murray failed to demonstrate any fundamental difference between the two resonances that could account for a concentration of objects at the 3:2 and a lack of them at the 2:1. However, making use of a computer called the Digital Orrery, which was built specifically for this purpose, Wisdom (1987a) showed that chaos is again responsible. From the results of numerical integrations he deduced the presence of a large chaotic zone at the centre of the 2:1 resonance and the absence of such a zone at the 3:2 resonance.

At this stage it is important to point out that not all chaotic orbits result in gross instabilities that lead to planet-crossing trajectories. The study by Milani & Nobili (1992) of the motion of asteroid (522) Helga showed that it has a measurable Lyapounov exponent of $1/6600 \text{ y}^{-1}$; yet numerical integrations showed no sign of it having a close approach to Jupiter or any of the other planets. This is another example of the phenomenon of bounded chaos mentioned in Sect. 9.3.3. In the case of (522) Helga the chaos may be connected with its proximity to the 12:7 jovian resonance. Murray & Holman (1997) showed that Helga's evolution is governed by a random walk.

It seems likely that the dynamical structure of the asteroid belt has been shaped by planetary perturbations, most notably those from Jupiter. Therefore the current, evolved population of asteroids must represent those orbits from the original population that have managed to survive the cumulative effects of chaos and collisions over the age of the solar system.

A general study of the dynamical evolution of planet-crossing asteroids was undertaken by Milani et al. (1989). They carried out numerical integrations of 410 asteroids for $2 \times 10^5$ y, including the gravitational perturbations from all the planets except Mercury and Pluto. Their results showed that the long-term motion of some of these asteroids is remarkably complex. For example, a plot of the semi-major axis of the Earth-crossing asteroid (1620) Geographos shows that it undergoes temporary trapping in a number of high-order resonances with the Earth in the course of its chaotic evolution (see Fig. 9.32). A study of high-order, jovian mean motion resonances in the outer asteroid belt (Holman & Murray 1996) has shown that even weak resonances can give rise to significant chaotic zones.

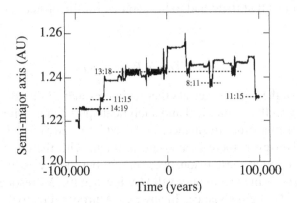

Fig. 9.32. The orbital evolution of the semi-major axis of (1620) Geographos. (Adapted from Milani et al. 1989.)

The advances in high-speed, cheap digital computers and the use of a variety of mapping techniques have led to many developments in the study of the long-term motion of asteroids and comets under planetary perturbations. The view that the evolution of such objects is dominated by a single planet (Jupiter in the case of the asteroids) or even a single resonance has now been found to be too narrow. Wisdom (1985b) had already shown that once perturbations from the other planets are included objects at the 3:1 jovian resonance could achieve eccentricities > 0.6 and so could become Earth-crossing. This strongly suggested that secular interactions between the planets also played a role in determining long-term stability. The results of many recent integrations have shown that secular resonances (see Sect. 7.11 and Fig. 7.19) rather than mean motion resonances can be the major perturbing influences (see, for example, Moons & Morbidelli 1995, Froeschlé et al. 1995). In fact, as shown by Gladman et al. (1997), the dynamical evolution of objects injected into mean motion resonances can be extremely complicated.

## 9.9 The Neptune–Pluto System

Pluto has the largest eccentricity ($e = 0.25$) of any planet and a plot of its orbital path appears to show an intersection with the orbit of Neptune (see Fig. 2.15b). However, plots that show the projection of Pluto's orbit in the plane of the ecliptic fail to bring out the effect of the inclination. Figure 9.33 shows the orbits and positions of Neptune and Pluto in 1989 when Pluto was at perihelion. The dashed curves in the plot indicate those portions of the orbit which lie below the plane of the ecliptic and the straight lines are the lines of nodes for each orbit. Neptune moves on a near-circular orbit with an inclination of < 2° (see Table

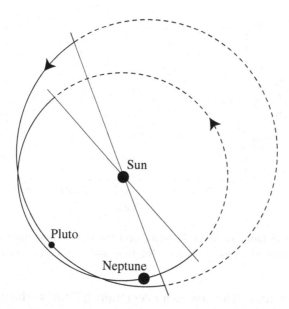

Fig. 9.33. The orbital configurations of Neptune and Pluto when Pluto was at perihelion in 1989. The straight lines denote the lines of nodes of each orbit. The dashed part of each orbit denotes the parts of the orbit below the plane of the ecliptic.

A1.2) and so it is always within 1 AU of the ecliptic. In contrast, Pluto has an eccentric orbit with a large inclination (17°), which implies significant motion out of the ecliptic. Furthermore, Neptune and Pluto are in a 3:2 resonance and the combination of effects means that when Neptune and Pluto are at conjunction, a close approach is avoided.

In Fig. 9.34 we plot the separation in AU between Neptune and Pluto over a timescale of some 1,000 years, centred on 1989. This shows that even though the orbits appear to intersect, the two planets never approach closer than $\sim 20$ AU, significantly larger than that suggested by Fig. 9.33. Note that this plot is based on two separate two-body problems and no mutual perturbations are included. However, it does illustrate that the geometry of the orbits is important in determining the closest approach distance.

Pluto was discovered by Clyde Tombaugh in 1930 and it has an orbital period of 248 y. The discovery of the Edgeworth–Kuiper belt (see, for example, Luu & Jewitt 1996) has shown that Pluto is just one member of a whole class of objects orbiting in this region of the solar system, some of which (the *Plutinos*; see Jewitt & Luu 1996) are also known to be in a 3:2 resonance with Neptune. However, none of the objects has yet to be observed over a complete orbital period. This implies that dynamical studies of possible resonances have to be accomplished by

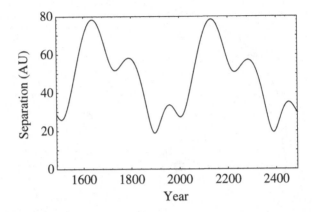

Fig. 9.34. The distance between Neptune and Pluto as a function of time based on individual two-body motions of the Sun–Neptune and Sun–Pluto systems.

numerical integration. The first such integration of Pluto's orbit was undertaken by Cohen & Hubbard (1965). On the basis of a backward integration covering 120,000 y they deduced that Pluto and Neptune are involved in a resonance with argument $\varphi = 3\lambda' - 2\lambda - \varpi'$, where $\lambda$ and $\lambda'$ denote the mean longitudes of Neptune and Pluto respectively and $\varpi'$ is the longitude of perihelion of Pluto. Cohen & Hubbard showed that in the Neptune–Pluto system $\varphi$ librates about 180° with an amplitude of $\sim$ 76° and a period of 19,670 y. Figure 9.35 shows the approximate path of Pluto in a frame rotating with the radius vector from the Sun to Neptune.

Sussman & Wisdom (1988) carried out a 845 million year integration of the outer planets using the Digital Orrery and showed that the motion of Pluto is chaotic. In a previous integration covering 200 million years, Applegate et al.

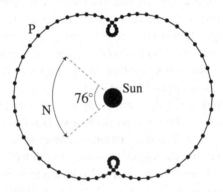

Fig. 9.35. The path of Pluto (P) in a frame rotating with the radius vector from the Sun to Neptune (N). The points along the path are placed at equal time intervals. The arc denotes the 76° amplitude of $\varphi$ about 180°.

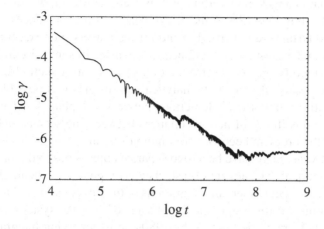

Fig. 9.36. Calculation of $\log \gamma$ as a function of $\log t$ for Pluto, where $t$ is the time in years and $\gamma$ is an estimate of Pluto's maximum Lyapounov exponent.

(1986) showed that Pluto's argument of perihelion is modulated with a period of 34 million years and $h' = e' \sin \varpi'$ has significant long-period variations with a period of 137 million years. The variation in $h'$ is due to a near-commensurability between the frequency of circulation of Pluto's longitude of ascending node and one of the eigenfrequencies of the system. The longer integration by Sussman & Wisdom (1988) confirmed these periods and also showed evidence for a component with a period of $\sim 600$ million years. Sussman & Wisdom also calculated a maximum Lyapounov exponent, $\gamma$, for the system and found that $\gamma \approx 10^{-7.3} \text{ y}^{-1}$, implying an e-folding timescale for exponential divergence of $\sim 20$ million years. Figure 9.36 (cf. Fig. 9.9) shows the results of a similar calculation based on the output of a numerical integration over $10^9$ y. Notice that the slope of the plot is approximately $-1$ for $\sim 10^{7.5}$ y before $\gamma$ tends to a fixed value (in this case a value larger than that found by Sussman & Wisdom). This demonstrates how a chaotic orbit can give the appearance of being regular for long time intervals before its chaotic nature becomes apparent. Despite the existence of chaos in Pluto's orbit, there is no evidence that it will ever leave the resonance with Neptune, even in integrations over timescales as long as the age of the solar system (Kinoshita & Nakai 1996). An understanding of the origin of this apparently stable, yet chaotic resonance may involve the orbital migration of Neptune in the early history of the solar system (Malhotra 1993).

## 9.10 The Stability of the Solar System

In our discussion of secular perturbation theory, we showed that the solar system is stable according to the Laplace–Lagrange theory. This is because our simplifying assumptions enabled us to derive an integrable solution to the problem of the secular changes in the orbits of $N$ planets. However, we also pointed out that these assumptions (low $e$ and $I$, no mean motion resonances, no effect of short-period terms) are not strictly valid for the planets. For example, we know that Jupiter and Saturn are in a 5:2 near-resonance. In some circumstances it is possible to modify the secular theory to take account of such effects, but it is important to realise that in such analytical approaches it is possible to neglect some potentially important interactions between the planets. If we go on to consider the possibility of near-resonances between the secular frequencies of the system the problem becomes even more difficult.

In recent years there have been two separate approaches to the problem, both of them numerical. The advent of fast, cheap computers has meant that it is now possible to carry out numerical integrations of the full equations of motion of the planetary orbits for times approaching the age of the solar system. A number of such projects were carried out in the 1980s, most involving integrations of the outer planets only (Jupiter, Saturn, Uranus, Neptune, and Pluto). These include the work by Kinoshita & Nakai (1984), Applegate et al. (1986), and Roy et al. (1988). The reason for the early focus on the orbits of the outer planets was the problem with timescales. Any numerical integration of the entire solar system would have to use a time step that was suitable for following the orbits of both Mercury and Pluto simultaneously. These time steps are determined by the orbital period of the innermost (i.e., fastest-moving) object. There are also significant problems with truncation error and the buildup of round-off error, although some of these can be alleviated with a suitable choice of time step (see Sussman & Wisdom 1988). As a test of their symplectic mapping, Wisdom & Holman (1991) re-did the calculation of Sussman & Wisdom (1988) and extended the integration of the outer solar system to one billion years.

An alternative approach to full integration (with or without a mapping technique) is to generate a system of averaged equations using appropriate terms from the disturbing function. Such a system of equations is an approximation to the real system and requires significant algebraic calculations (usually performed by computer) before the numerical integration can begin. However, the equations can be integrated using a much larger time step because the short-period effects have been averaged out. For example, 40 d is a typical time step in a full integration of the outer planets whereas a typical value for an averaged system is 500 y. In his study of the evolution of all the planets over 10 million years, Laskar (1988) integrated the averaged equations. He showed that the inner planets are also chaotic and have maximum Lyapounov exponents of $\sim 10^{-6.7}$ $y^{-1}$, giving e-folding timescales of 5 million years.

All these numerical integrations show a consistent picture of planets moving on chaotic orbits. However, none show any sign of gross instabilities in the system for integrations over timescales comparable to the age of the solar system. Therefore, although the planets move on chaotic orbits the chaos is subtle. It implies that there are fundamental limits to our ability to predict the positions of the planets for long time intervals. If we had the starting positions and velocities of the planets with infinite precision and then had a perfect integrator, we would be able to calculate the future or past orbits. However, since any physical measurement has finite precision, there will always be an in-built error in the calculations. In a chaotic system this error will propagate exponentially. In the case of planets such as the Earth this means that we cannot predict the future position of the Earth in its orbit. For example, a maximum Lyapounov exponent of $10^{-6.7}$ means that an error of 1 cm in the position of the Earth today propagates such that we cannot predict its location 200 million years in the future.

Laskar (1994) extended his integrations by investigating the planetary orbits for 10 billion years into the past and 15 billion years into the future. These showed a close coupling of the Earth and Venus in terms of their variations of eccentricity and inclination. The results also showed dramatic, chaotic variations in the orbits of Mercury ($0.1 < e < 0.5$, $8° < I < 21°$) and Mars ($0.1 < e < 0.2$, $4° < I < 10°$). Although an eccentricity of 0.5 is not sufficient for Mercury to cross the orbit of Venus, it prompted Laskar (1994) to carry out further integrations, shifting the initial position of the Earth by 150 m. Some of the integrations showed that Mercury's eccentricity could reach a value $e \approx 1$ (i.e., near-hyperbolic) at epochs 3.5 billion years in the future or 6.6 billion years in the past. However, when assessing the significance of these results it is important to remember that the physical model of point (constant) mass objects orbiting in a conservative system may no longer be applicable when such timescales are considered.

With the possible exception of Mercury (see above), all integrations show that the planets remain close to their current orbits for timescales approaching a billion years or more, even though the orbits are technically chaotic. In this sense the solar system is stable and it is likely that motions of the planets represent another case of bounded chaos. As yet there is no analytical proof of the solar system's stability or the origin of the chaos. Analyses of the stability properties of other planetary systems are just beginning (see, for example, Holman et al. 1997).

## Exercise Questions

**9.1** Write a computer program to solve numerically the Cartesian equations of motion of the planar, circular, restricted three-body problem in the rotating reference frame (Eqs. (3.16) and (3.17)). Your program should take initial values of the orbital elements, convert them to Cartesian position and velocity in the rotating frame (choosing units such that the mean motion of the two masses is unity), solve the equations to find the new position and velocity at fixed time intervals, and convert them to orbital elements for output. Consider the motion of a test particle in the Sun–Jupiter system with $\mu_2 = 10^{-3}$, where Jupiter is always started at zero initial longitude and the unit of distance is the radius of Jupiter's orbit. The particle has initial semi-major axis $a_0 = 0.8$, eccentricity $e_0 = 0.4$, and longitude of perihelion $\varpi_0 = 295°$. Numerically integrate the equations of motion for one Jupiter period using initial mean longitude $\lambda_0 = 293°$ and $\lambda_0 = 293.3°$ (see Fig. 9.1), giving the differences between the initial and final values of $a$, $e$, and $\varpi$ in each case. By varying the value of $\lambda_0$ in steps of $0.1°$, find the value of $\lambda_0$ that gives (i) the largest increase and (ii) the largest decrease in semi-major axis. Explain your results in terms of the geometry of the Jupiter encounter.

**9.2**     Write a computer program to solve numerically the Cartesian equations of motion of the planar, circular, restricted three-body problem in the rotating reference frame (Eqs. (3.16) and (3.17)) to produce a sequence of pairs of values of $x$ and $\dot{x}$ whenever $y = 0$ with $\dot{y} > 0$. Use your program with $\mu_2 = 10^{-3}$ and initial values $\dot{x} = 0$ with $0.2 \leq x \leq 0.8$ (at intervals of 0.01) to produce a Poincaré surface of section for prograde, interior orbits with a Jacobi constant $C_J = 3.03$. Each integration should be followed until 500 points have been produced on the surface of section *or* the trajectory gets within a distance $\alpha = (\mu_2/3)^{1/3} \approx 0.07$ of the secondary mass *or* the radial distance exceeds 3. Identify any resonant "islands" that appear in the resulting surface of section.

**9.3**     Use the planar form of Wisdom's resonance map given in Sect. 9.5.2 to write a computer program to study the motion of asteroidal test particles at the 3:1 resonance with Jupiter. The program should take initial values of the semi-major axis ($a$), eccentricity ($e$), longitude of perihelion ($\varpi$), and mean longitude ($\lambda$) of the asteroid and the equivalent values $a'$, $e'$, $\varpi'$, and $\lambda'$ for Jupiter; calculate the initial values of the quantities $x$, $y$, $\Phi$, and $\varphi$ defined in Sect. 9.5.2; calculate the final values of these quantities after the seven steps of the map; and return the new elements for the asteroid one Jupiter period later, taking Jupiter's orbit to be fixed. Use the program with initial values $a/a' = 0.481$, $e = 0.15$, $e' = 0.048$, $\varpi = \varpi' = 0$, and $3\lambda' - \lambda = 180°$ to plot the evolution of the asteroid's eccentricity over $10^4$ Jupiter periods (cf. Fig. 9.30). Repeat the run with the same initial values of $a$, $a'$, $e'$, $\varpi$, $\varpi'$, and $3\lambda' - \lambda = 180°$ but using $e = 0.1500001$ to produce a similar plot of $e$ as a function of time and comment on the difference between the two plots. Plot the difference in eccentricity between the two runs as a function of time and use this to estimate the maximum Lyapounov characteristic exponent.

**9.4**     The computer program written for Question 9.3 can be used to carry out a study of asteroid motion at the 3:1 jovian resonance. Take an array of starting conditions in the "representative" plane using $0.47 \leq a/a' \leq 0.49$ (at intervals of 0.001), $0 \leq e \leq 0.3$ (at intervals of 0.01), with $\varpi = \varpi' = 0$, $3\lambda' - \lambda = 180°$, and $e' = 0.048$, where $e'$ and $\varpi'$ are taken to be constant. Follow each trajectory and plot those starting values of $a$ and $e$ that give rise to orbits with $e > 0.3$ (i.e., approximate Mars-crossing orbits) within $10^4$ Jupiter periods.

**9.5** Use the encounter map of Duncan et al. given in Eqs. (9.92)–(9.94) to write a computer program to study the orbital evolution of test particles due to successive conjunctions with Jupiter, where the orbit of Jupiter (mass ratio $m_2/m_1 = 9.545 \times 10^{-4}$ and semi-major axis $a' = 1$) and the initial orbit of the test particle are taken to be circular, and where the objects are initially at conjunction. For each initial value of the particle's semi-major axis $a$, starting at $a = 0.7$ with increments of $0.001$, have the program follow the particle for $10^4$ encounters. What is the last initial value of $\Delta a = a' - a$ for which the given starting condition did not achieve $e > \Delta a/a'$ within $10^4$ encounters and how does this compare with the value predicted on the basis of the resonance overlap criterion?

**9.6** Write a computer program to implement the Wisdom & Holman $N$-body map discussed in Sect. 9.5.4. Use the program to integrate the orbits of the planets for 50 years starting at the epoch of J2000 (JD 2451545.0) using the data given in Table A.2 and the masses given in Table A.4. Use Eq. (A.14) and Table A.3 to calculate the mean longitude of each planet at 100 day intervals for 50 years using a mapping interval $\Delta t = 5$ days. Compare these values with those obtained using the map with a mapping interval of (a) 1 day and (b) 10 days, listing the maximum difference over the 50 year interval in each case.

# 10

## Planetary Rings

Hast any philosophy in thee, shepherd?

William Shakespeare, *As You Like It, III, ii*

### 10.1 Introduction

The first ring system to be observed in the solar system was discovered around
Saturn by Galileo in 1610. Unsure of the nature of the phenomenon he had observed, he originally interpreted the ring ansae as two moons, one on each side
of the planet. In a Latin anagram sent to fellow scientists he announced, *"I have
observed the most distant planet to have a triple form"*. Galileo was surprised
to find that the phenomenon had disappeared by 1612, only to reappear again
soon afterwards. Huygens (1659) correctly attributed the varying appearance
as being due to the different views of a thin disk of material surrounding Saturn. It was Maxwell (1859) who provided a mathematical proof that the rings
could not be solid; they had to be composed of individual particles orbiting the
planet.

The rings of Uranus were detected serendipitously in March 1977 by astronomers observing an occultation of a star by the planet. The *Voyager* spacecraft detected a faint ring around Jupiter (Smith et al. 1979a), and occultations of
stars by Neptune led to the discovery of the ring arcs of Neptune, subsequently
shown to be the optically thicker parts of a faint ring system. The flybys of the
outer planets by the *Voyager* spacecraft and the continuing ground- and space-
based observations of the ring systems have provided evidence of a wide variety
of dynamical phenomena, which provide an ideal testing ground for some of the
concepts covered in this book.

In this chapter we do not attempt to give a complete summary of knowledge
concerning planetary rings, nor even to provide a summary of the dynamics of

ring systems.  Instead, we concentrate on those aspects of ring dynamics that illustrate resonant phenomena and can be understood using simple extensions of the methods already discussed in previous chapters.  Therefore it is important to realise that many of the properties of planetary ring systems can only be understood by undertaking a fluid dynamical approach to the problem.  The chapters in Greenberg & Brahic (1984) provide comprehensive surveys of the field in 1984; Nicholson & Dones (1991) give a more recent summary of ring observations and research.

## 10.2  Planetary Ring Systems

Data on planetary rings are given in Appendix A. The relative sizes and locations of the different systems are shown in Fig. 10.1 on a uniform scale of planetary radii, following the example of Nicholson & Dones (1991).  Figure 10.1 also shows the radial locations of the satellites that orbit in the vicinity of the rings. The synchronous orbit for each planet (dashed line) is dynamically important since under tidal forces all prograde satellites orbiting beyond it would move away from the planet and those orbiting inside it would approach the planet and eventually break up.  Representative *Voyager* images of the ring systems of the outer planets are shown in Fig. 10.2.

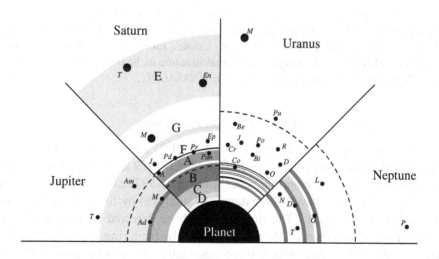

Fig. 10.1.  The ring systems of the outer planets and their accompanying satellites on a uniform scale of planetary radii. The location of the synchronous orbit for each planet is denoted by a dashed curve. (Adapted from Nicholson & Dones (1991).)

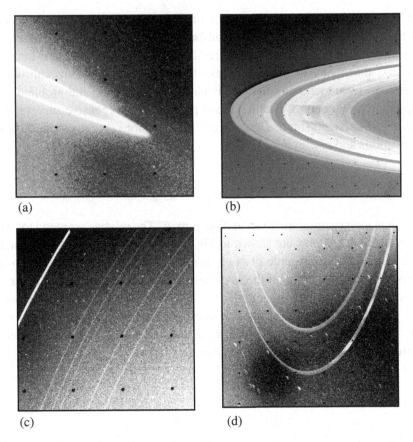

(a)                                    (b)

(c)                                    (d)

Fig. 10.2.  Representative *Voyager* images of the ring systems of the outer planets: (a) Jupiter, showing the halo extending above the inner part of the main ring, (b) Saturn, showing the main rings (with "spokes"), (c) Uranus, showing six of the known narrow rings, and (d) Neptune, showing the azimuthal structure in the outermost ring. The scale in each image is different. *(Images courtesy of NASA/JPL.)*

### 10.2.1  The Rings of Jupiter

The optically thin (optical depth, $\tau < 3 \times 10^{-6}$) ring system of Jupiter was discovered by the *Voyager 1* spacecraft in March 1979 and further images were obtained by *Voyager 2* later that year (Smith et al. 1979a,b). The set of twenty-five Voyager images have been extensively analysed by Showalter (1985) and Showalter et al. (1987). The ring has also been detected from ground-based observations (Nicholson & Matthews 1991). The main ring, centred at 129,130 km, is ~7,000 km wide with a sharp outer edge and a faint "gossamer" ring extending outwards beyond the orbit of Thebe. At the inner edge there is a toroidal-shaped halo of ring material, which is brighter close to the main ring and may extend more than halfway to the planet. The vertical structure extends symmetrically for perhaps 10,000 km above and below the equatorial plane. A schematic representation of the ring system is shown in Fig. 10.3.

The optical forward-scattering properties of the ring are indicative of the presence of a large population of micron-sized particles. Dust in the jovian environment is subjected to significant nongravitational perturbations resulting in lifetimes of $\sim 10^{3\pm1}$ y, implying a readily available source of material (Burns et al. 1984). Two small satellites, Metis (radius $\sim 20$ km) and Adrastea (radius $\sim 10$ km), orbit in the outer part of the ring and it is the impact of meteoroids on these and other smaller objects that is thought to be the source of most of the ring material. Observations of the gossamer ring by the *Galileo* spacecraft show two separate bands of material, starting at the orbits of Thebe and Amalthea. The structure of these bands is strikingly similar to that detected in the zodiacal dust bands (see Fig. 7.17) and there is clear evidence that Thebe and Amalthea are the source objects for this jovian dust (Ockert-Bell *et al.* 1999, Burns *et al.* 1999).

Consolmagno (1983) and Burns et al. (1985) suggested that the Lorentz force experienced by a charged dust grain could help explain some of the structure of the jovian ring system. In particular, commensurate relationships between the

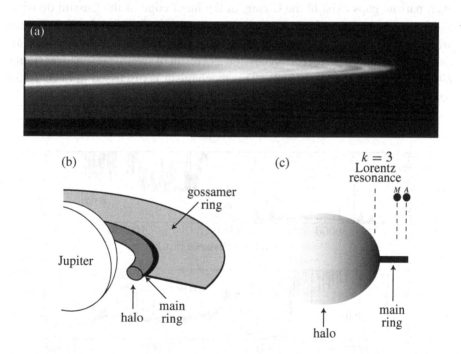

Fig. 10.3.   (a) A *Galileo* image of Jupiter's main ring *(image courtesy of NASA/JPL)*. A schematic illustration of the ring system showing (b) the main component with halo and gossamer ring and (c) the locations of Metis (M), Adrastea (A), and the suspected effect of the $k = 3$ Lorentz resonance at the inner edge of the main ring.

orbital period of the grain and the period of the electromagnetic force give rise to the so-called Lorentz resonances (Burns et al. 1985), which occur at semi-major axes given by

$$a_k = \left( \frac{k \mp 1}{k} \right)^{2/3} r_{\rm s}, \qquad k = 1, 2, \ldots, \infty, \tag{10.1}$$

where $r_{\rm s}$ is the synchronous orbital radius. The $k = 3$ interior Lorentz resonance occurs close to the transition point between the main ring and the halo (Schaffer & Burns 1987).

### 10.2.2 The Rings of Saturn

The main ring system of Saturn (see Fig. 10.1 and Fig. 10.2b) consists of the broad A and B rings separated by the Cassini division and the optically thinner C and D rings. Exterior to the main rings lie the narrow, "braided" F ring and the broader diffuse G and E rings. It is important to note that the main ring system contains very few actual gaps; most of the ring features that appear in the occultation profile (Fig. 10.4) are fluctuations in surface density. However, narrow gaps exist in the C ring, at the inner edge of the Cassini division, and in the outer part of the A ring. Gaps in the C ring and Cassini division are known to contain narrow, eccentric rings (Porco 1990), two of which are thought to be subject to forced precession as a result of satellite resonances (Porco 1983, Nicholson & Porco 1988). This mechanism is discussed in Sect. 10.5.4 below.

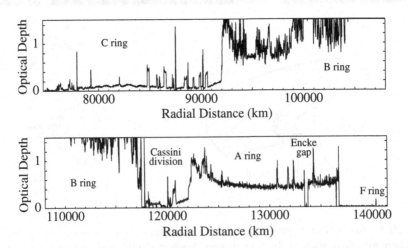

Fig. 10.4. Optical depth profiles of the Saturn ring system obtained from the *Voyager* photopolarimeter experiment (see Lane et al. 1982).

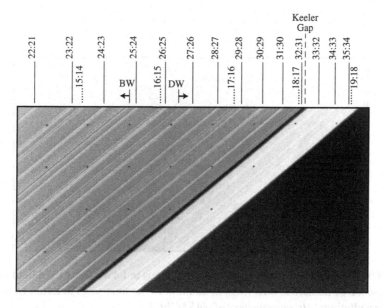

Fig. 10.5. The outer part of the A ring of Saturn showing the location of resonances with the F ring shepherding satellites Pandora (dotted lines) and Prometheus (solid lines). The locations and directions of propagation of a spiral bending wave (BW) and a spiral density wave (DW) associated with the Mimas 8:5 resonance are also indicated. *(Image courtesy of NASA/JPL.)*

Most of the structure in the A ring of Saturn seen in Fig. 10.4 can be explained by resonances with the smaller saturnian satellites. Figure 10.5 shows part of a *Voyager 2* image of the outer part of the A ring with annotation indicating the positions of all the resonances of the form $p + 1 : p$ with Pandora and Prometheus, which lie within the region. The ring features are caused by a succession of tightly wound, unresolved spiral density waves propagating away from the location of the resonances (see Sect. 10.4). There are also two features associated with the 8:5 resonance with Mimas. The outer one produces a spiral density wave propagating outwards and the inner one produces a spiral bending wave propagating inwards. Although the structure of the B ring is still poorly understood, there has been some progress in explaining some of the C ring structure in terms of resonances with planetary oscillations (Marley 1990, Marley & Porco 1993). Recently Hamilton & Burns (1994) have attempted to explain the properties of the extended E ring by including the effects of nongravitational forces on the micron-sized particles.

### 10.2.3 The Rings of Uranus

Prior to the *Voyager 2* encounter with Uranus in 1986, ground-based stellar occultation experiments had detected the presence of nine, narrow (typical widths

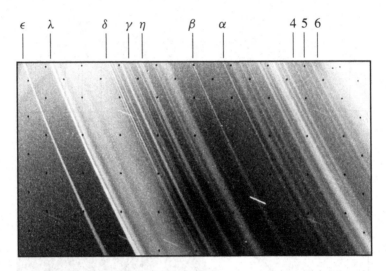

Fig. 10.6.   A *Voyager 2* image of the rings of Uranus in forward scattered light showing the presence of numerous dust rings in addition to the nine rings detected from Earth-based occultations. *(Image courtesy of NASA/JPL.)*

of < 10 km) rings around the planet. *Voyager* images in back-scattered light confirmed the existence of these rings as well as the $\lambda$ ring orbiting between the $\delta$ and $\epsilon$ rings and 1986U2R orbiting interior to ring 6. A 96 s exposure in forward-scattered light revealed the presence of a large number of dust rings (Fig. 10.6). French et al. (1991) provide a summary of the observations and dynamical models of the uranian ring system.

The shepherding satellite model (Goldreich & Tremaine 1979a; see also Sect. 10.5.3) for narrow rings has had considerable success in explaining the confinement of the $\epsilon$ ring by the satellites Cordelia and Ophelia, which orbit interior and exterior to the ring. Cordelia has a 24:25 outer eccentric resonance with the inner edge of the $\epsilon$ ring and Ophelia has a 14:13 inner eccentric resonance at the outer edge of the ring (Porco & Goldreich 1987). However, despite extensive searches, no other satellites have been found in the main ring system, although there is some evidence for their existence (see Murray & Thompson (1990)).

The $\epsilon$ ring is one of several in the uranian system to exhibit a variable width. The occultation data demonstrate that each of these rings can be modelled by two aligned ellipses of different semi-major axes and eccentricities, which precess uniformly under the zonal gravity harmonics at a rate determined by the central elliptical path. The two models proposed to explain this behaviour do not involve the presence of unseen satellites. Instead they rely on either the self-gravity of the ring (Goldreich & Tremaine 1979b) or the effectiveness of a "pinch" mechanism (see Sect. 10.5.4) within the ring (Dermott & Murray 1980). The prediction of a mean particle radius of 20–30 cm using the self-gravity model is in conflict with the *Voyager* observations of sizes > 70 cm.

### 10.2.4 The Rings of Neptune

Following the discovery of the rings of Uranus by ground-based, stellar occultation experiments in 1977, numerous attempts were made to discover a Neptunian ring system using similar methods. Hubbard et al. (1986) provided the first convincing evidence for the existence of ring material but only in ~ 10% of cases did such experiments yield any signatures, and in those a feature was detected on only one side of the planet. This led to the suggestion that Neptune had a system of "arcs" of ring material extending over ~ 10° in longitude. The *Voyager 2* images (see, for example, Fig. 10.2d) show that Neptune has at least three, distinct rings (the Galle, Le Verrier, and Adams rings) and that the arcs are just the optically thicker parts ($\tau \approx 0.04$) of the outermost Adams ring (Smith et al. 1989). Additional rings include the Lassell ring extending outwards from the Le Verrier ring and bounded by the Arago ring and an unnamed ring that appears to share the same orbit as the moon Galatea. The three main arcs in the Adams ring were named Liberté, Egalité, and Fraternité, although it is now known that there are at least two more arcs, Egalité 2 and Courage, in the same ring (see the review article by Porco et al. 1995).

### 10.2.5 Rings and Satellites

It is clear from Fig. 10.1 that there is a close association between the location of ring systems and the orbits of small satellites. We know from our study of tidal evolution of satellites that objects in prograde orbits inside the synchronous orbit will evolve inwards due to the effect of the tides they raise on the planet. We also know that the Roche zone for each planet lies in the range of 1.44 to 2.24 planetary radii. Therefore we would expect satellites brought in under tidal effects to break up as they approach the planet. In such cases the satellites act as sources of ring material. However, the rings are also subject to the gravitational perturbations from the satellite remnants. These can have the effect of confining some rings while causing other ring material to be disrupted. These processes for narrow rings are discussed in Sect. 10.5 below. In all such cases it is important to recognise the role of resonance in determining ring structure.

## 10.3 Resonances in Rings

The close proximity of rings and satellites in each system (see Fig. 10.1) suggests that there could be significant mutual gravitational perturbations. In the case of Saturn's rings, for example, virtually all the structure in the A ring can be understood in the context of resonances between ring particles and the satellites that orbit nearby (see Fig. 10.5). There are also resonant features in other ring systems. Therefore it is important to examine the nature of these resonant interactions in order to understand the observed features.

In Sect. 2.6 we showed that elliptical motion about a spherical central object can be considered in terms of motion around an epicentre which moves at a uniform rate equal to the mean motion $n$ around the central object, provided the eccentricity of the orbit is small. The secondary object traces out a closed, centred ellipse about the guiding centre. In Sect. 6.11 we considered the case of an object orbiting an oblate planet with an axisymmetric potential. As well as the standard $1/r$ potential, the particle experiences the effect of the zonal harmonic coefficients $J_2$, $J_4$, etc. We showed that these terms cause the orbit to rotate in space, giving rise to three separate frequencies, $n$, $\kappa$, and $\nu$ (see Eqs. (6.244)–(6.246)). These quantities are the modified mean motion, the epicyclic frequency, and the vertical frequency, respectively.

Now consider the gravitational effect of a perturbing satellite with semi-major axis $a'$. The satellite will have its own set of frequencies $n'$, $\kappa'$, and $\nu'$ given by Eqs. (6.244)–(6.246) with $a$ replaced by $a'$. The satellite's potential can be expanded in a Fourier series (see Chapter 6). For each argument in the expansion the *pattern speed*, $\Omega_p$, is defined as the angular frequency of a reference frame in which this argument is stationary. This will depend on the exact combination of frequencies under consideration and may be written as

$$m\Omega_p = mn' + k\kappa' + p\nu' \qquad (10.2)$$

or, because $\kappa' = n' - \dot{\varpi}'$ and $\nu' = n' - \dot{\Omega}'$,

$$m\Omega_p = (m + k + p)n' - k\dot{\varpi}' - p\dot{\Omega}', \qquad (10.3)$$

where $m$, $k$, and $p$ are integers and $m$ is nonnegative. The strongest resonances occur when an integer multiple of the difference between $n$ and $\Omega_p$ is equal to zero (for *corotation resonances*) or to the natural frequency of the radial or vertical oscillations of the ring particle (for *eccentric* (or *Lindblad*) *resonances* and *vertical resonances* respectively). The new terminology has its origins in galactic dynamics rather than solar system dynamics. We shall see, by means of specific examples, how these resonances are related to those described in Chapter 8. A summary of this work is given in Murray (1999).

### 10.3.1 Perturbations in Semi-major Axis and Corotation Resonances

A corotation resonance occurs where the pattern speed of the perturbing potential matches the orbital frequency of the particle. In this case

$$m(n - \Omega_p) = 0. \qquad (10.4)$$

From the definition of $\Omega_p$ this implies that the resonance condition is

$$(m + k + p)n' - mn - k\dot{\varpi}' - p\dot{\Omega}' = 0. \qquad (10.5)$$

If we ignore the variation of the mean longitude at epoch, this in turn can be considered as setting the condition $\dot{\varphi}_{cr} = 0$, where $\varphi_{cr}$ is the resonant angle given by

$$\varphi_{cr} = (m + k + p)\lambda' - m\lambda - k\varpi' - p\Omega'. \tag{10.6}$$

If we adopt our usual notation whereby the coefficient of $\lambda'$ in the resonant angle is always $j$ (see Chapter 6 and Appendix B) then

$$\varphi_{cr} = j\lambda' + (k + p - j)\lambda - k\varpi' - p\Omega', \tag{10.7}$$

where $j = m + k + p$. We can see that this argument satisfies the d'Alembert relation and we already know that for this to be a valid argument $|p|$ must be even; this is because the inclinations always occur in even powers and hence the sum of the coefficients of the nodes must always be even. Furthermore, this is the argument of a resonance of order $|k + p|$. The lowest order term associated with this resonant argument will be $\mathcal{O}(e'^{|k|} I'^{|p|})$ (see Sect. 6.7 and Appendix B). The 1:1 resonance (where $p = k = 0$) is clearly a special case of a corotation resonance.

Neither the longitude of pericentre nor the longitude of ascending node of the ring particle are involved in the resonant argument, $\varphi_{cr}$, of a corotation resonance. From Lagrange's equations this means that only the ring particle's semi-major axis is affected by a corotation resonance.

The relevant part of the disturbing function (cf. Eq. (8.26)) is

$$\mathcal{R} = \frac{\mathcal{G}m'}{a'} f_d(\alpha) e'^{|k|} s'^{|p|} \cos \varphi_{cr}, \tag{10.8}$$

where $\varphi_{cr}$ is given in Eq. (10.7). The exact form of $f_d(\alpha)$ depends on the resonance in question (see Table 8.1). From the definition of the corotation resonance the resonant angle can be written as

$$\varphi_{cr} = -m(\lambda - \Omega_p t) + \text{constant}, \tag{10.9}$$

and therefore there are $m$ equilibrium points in the frame rotating with the pattern speed $\Omega_p$ where libration is possible.

The geometry of a simple corotation resonance is shown in Fig. 10.7 for the case of a 3:2 resonance with $j = 3$, $m = 2$, $k = 1$, $p = 0$, and $e' = 0.25$. Because $m = 2$ there are two possible libration points in the frame rotating with the pattern speed of the perturbing satellite. In this frame the path of the satellite (the outer curve in Fig. 10.7) is closed (cf. Fig. 5.10 and Fig. 8.4c for the 2:3 resonance). Because the particle is close to but not at the exact resonance (i.e., it is not located at one of the two equilibrium points) it will librate about the equilibrium point. We have discussed this behaviour in Chapter 8.

We can calculate the maximum width of a corotation resonance by using the pendulum model developed in Chapter 8. For example, in the case of the 3:2 resonance discussed above we have $k = 1$ and $p = 0$, which corresponds to the

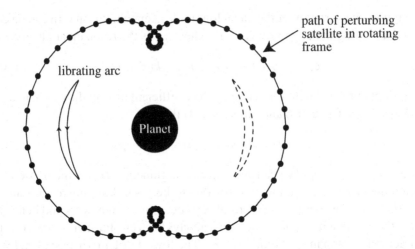

Fig. 10.7. The geometry of a 3:2 ($j = 3, m = 2, k = 1, p = 0$) corotation resonance for the case when $e' = 0.25$. In a frame corotating with the pattern speed of the perturbing satellite, $\Omega_p = n' + \kappa'/2$, there are two equilibrium points at this resonance. A ring particle close to one of these points will librate about a longitude that is stationary in the rotating frame (solid curve). Libration is also possible about the other equilibrium point (dashed curve).

argument 4D1.2 with $j = 3$. Taking the largest term in our previous expression for the maximum change in semi-major axis in a resonant libration (Eq. (8.58)) the width, $W_{cr}$, of a general corotation resonance can be written as

$$W_{cr} = 8 \left( \frac{a\,|\mathcal{R}|}{3\mathcal{G}m_p} \right)^{\frac{1}{2}} a, \qquad (10.10)$$

which agrees with the expressions given by Goldreich & Tremaine (1981) and Dermott (1984).

A feature associated with the Mimas 2:1 corotation resonance has been observed in Saturn's rings (Molnar & Dunn 1995). Corotation resonances have also been invoked to explain damping of eccentricities in narrow rings (Goldreich & Tremaine 1981) and the mutual repulsion of rings and satellites (Goldreich & Tremaine 1980).

Note that although we have formulated these expressions for the case where the satellite is an external perturber, the theory can still be applied to the case where the satellite orbits interior to the ring, provided we make minor changes in notation. As we shall see, an external corotation resonance is thought to be involved in the azimuthal confinement of the arcs in Neptune's Adams ring (see Sect. 10.8).

### 10.3.2 Perturbations in Eccentricity and Lindblad Resonances

A Lindblad resonance occurs when the pattern speed of the perturbing potential matches the radial frequency of the particle. In this case

$$m(n - \Omega_p) = \pm\kappa, \tag{10.11}$$

where the upper and lower signs correspond to the inner (ILR) and outer (OLR) Lindblad resonance respectively. The use of $\pm$ permits us to consider a ring particle that is orbiting inside or outside the orbit of the perturbing satellite. The resonance condition can also be written as

$$(m \mp 1)n \pm \dot{\varpi} - m\Omega_p = 0 \tag{10.12}$$

or

$$(m + k + p)n' - (m \mp 1)n - k\dot{\varpi}' \mp \dot{\varpi} - p\dot{\Omega}' = 0. \tag{10.13}$$

In terms of the resonant angle, $\varphi_{Lr}$, we have

$$\varphi_{Lr} = (m + k + p)\lambda' - (m \mp 1)\lambda - k\varpi' \mp \varpi - p\Omega'. \tag{10.14}$$

Using our standard notation we can write this as

$$\varphi_{Lr} = j\lambda' + (k + p \pm 1 - j)\lambda - k\varpi' \mp \varpi - p\Omega', \tag{10.15}$$

where, as before, $j = m + k + p$. Note that this is a resonant argument of order $|k + p \pm 1|$. Also, because the sum of the coefficients of $\Omega$ and $\Omega'$ must be even, $p$ has to be even for all Lindblad resonances. To lowest order the term associated with $\varphi_{Lr}$ is $\mathcal{O}(ee'^{|k|}I'^{|p|})$.

To help understand the mechanism of a Lindblad resonance we introduce the concept of a *streamline*, whereby we consider the motion of a collection of particles with the same semi-major axis and eccentricity. To lowest order in $e$ the equation of an ellipse can be written as

$$r = a[1 - e\cos(\theta - \varpi)], \tag{10.16}$$

where $\theta$ is the true longitude. However, we know that $\dot{\theta} = n$ and $\dot{\varpi} = n - \kappa$. Therefore we can write the condition given in Eq. (10.11) as

$$m\frac{d}{dt}(\theta - \Omega_p t) = \pm\frac{d}{dt}(\theta - \varpi). \tag{10.17}$$

Integration of this equation shows that for a given semi-major axis and eccentricity the longitude, $\theta_c = \theta - \Omega_p t$, of the particles in a frame corotating with the pattern speed is given by

$$m\theta_c = \theta - \varpi + \text{constant}. \tag{10.18}$$

Hence the paths, or streamlines, are described by the equation

$$r = a[1 - e\cos(m\theta_c - \text{constant})]. \tag{10.19}$$

The resulting streamlines for the cases $m = 0$ (a circle), $m = 1$ (a keplerian ellipse), $m = 2$ (a centred ellipse), and $m = 7$ (a seven-lobed curve) are shown in Fig. 10.8. In each case we have chosen the phase constant to be zero. Note that in the case of a keplerian ellipse, Fig. 10.8b, the path looks circular. This is because, to $\mathcal{O}(e)$, a keplerian ellipse is simply a circle displaced from its centre (see Eq. (2.101) and the discussion in Sect. 2.6). It is clear that the streamlines always produce an $m$-lobed pattern. Note the similarity with the paths in the rotating frame for interior first-order resonances shown in Fig. 8.3a. Because the Lindblad resonance involves only the first power of the eccentricity of the ring particle a streamline never intersects itself.

When discussing sharp edges near a discrete resonance in planetary rings it is important to define the physical width of a Lindblad resonance. The effect of the Lindblad resonance is to induce a forced eccentricity on the ring particles such that, at a given semi-major axis, the particles move in streamline motion. We have already noted that the resulting pattern in the rotating frame is a wavy, $m$-lobed ring. The size of the forced eccentricity will decrease as the distance from the exact resonance increases, with a phase change of $180°$ on opposite sides of the resonance. This phenomenon was illustrated in Fig. 8.11. The width of the resonance is determined by the separation from the exact resonance such that the value of the forced eccentricity is just sufficient for the outer streamline to intersect the inner one. The mechanism is discussed by Porco & Nicholson (1987). The resulting paths are illustrated in Fig. 10.9 for the case of the 7:6 Lindblad resonance with $k = p = 0$ and hence $\Omega_p = n'$.

In Sect. 8.8.1 we discussed the location of the equilibrium points for the case of large values of $\bar{\delta}$, corresponding to motion far from the resonance (see Fig. 8.11). There we noted that for the $e$ resonance the relevant part of the disturbing function is

$$\mathcal{R} = \frac{\mathcal{G}m'}{a'} f_{\rm d}(\alpha) e \cos\varphi_{\rm Lr}, \qquad (10.20)$$

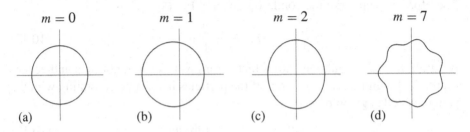

Fig. 10.8. Streamline paths obtained for (a) $m = 0$, $\Omega_p = k\kappa'$, (b) $m = 1$, $\Omega_p = n' + k\kappa'$, (c) $m = 2$, $\Omega_p = n' + (k/2)\kappa'$, and (d) $m = 7$, $\Omega_p = n' + (k/7)\kappa'$ using the same semi-major axis and $e = 0.08$ in (a), (c), and (d) and $e = 0.2$ in (b).

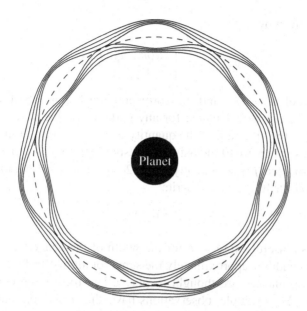

Fig. 10.9. A schematic diagram illustrating the streamlines for the 7:6 Lindblad resonance ($j = 7$, $m = 7$, $k = 0$, $p = 0$) with $\Omega_p = n'$. The curves denote the streamlines of particle orbits on either side of exact resonance, denoted by the dashed circle. The amplitude of each streamline is a linear function of the forced eccentricity, and the width of the resonance is determined by the distance from the resonance at which streamlines on opposite sides just touch.

where $\varphi_{\text{Lr}} = j\lambda' - (j-1)\lambda - \varpi$ and $f_{\text{d}} = (1/2)\left[-2j - \alpha D\right] b_{1/2}^{(j)}$ (see Table 8.1). This is the $k = p = 0$ Lindblad resonance in our new terminology. The location of the equilibrium point gives the value of the scaled forced eccentricity and is simply $2/\bar{\delta}$, where $\bar{\delta}$ is a measure of the proximity to resonance. Using the scaling relation given in Eq. (8.132) and the definitions of $\bar{\alpha}$, $\bar{\beta}$, and $\bar{\epsilon}$ given in Eqs. (8.103)–(8.105) for first-order resonances, we can write the resulting expression for the forced eccentricity as

$$e_{\text{f}} = \left| \frac{n\alpha(m'/m_{\text{p}})f_{\text{d}}}{[jn' - (j-1)n]} \right|. \tag{10.21}$$

If we write $a = a_{\text{res}} + \Delta a$ (where $\Delta a \ll a_{\text{res}}$) for the semi-major axis of the particle we can make use of Kepler's third law and a series expansion in $\Delta a/a_{\text{res}}$ to write the amplitude of the resulting forced wave as

$$ae_{\text{f}} = \frac{2\alpha a^2(m'/m_{\text{p}})|f_{\text{d}}|}{3(j-1)\,|a - a_{\text{res}}|} \tag{10.22}$$

for the case $k = p = 0$. For a critical value of the semi-major axis, $a_{\text{crit}}$, the amplitude of the ring wave equals the separation in semi-major axis from the exact resonance. For this value the full width of the $k = p = 0$ Lindblad resonance is given by

$$W_{\mathrm{Lr},0} = 4a \left[ \frac{2\alpha (m'/m_{\mathrm{p}})|f_{\mathrm{d}}|}{3(j-1)} \right]^{\frac{1}{2}}. \tag{10.23}$$

The values of $\alpha = a/a'$ and the corresponding Laplace coefficients in the definition of $f_{\mathrm{d}}$ can be calculated for any particular value of $j$. Figure 10.10 shows a plot of the variation of the quantity $2\alpha f_{\mathrm{d}}/(j-1)$ as a function of $j$ for the resonances 2:1 to 50:49 inclusive (corresponding to $j = 2$ to $j = 50$). This shows that provided $j$ is sufficiently large, $2\alpha f_{\mathrm{d}}/(j-1) \approx 1.6$ and hence, from our expression for $W_{\mathrm{Lr},0}$, we can write

$$W_{\mathrm{Lr},0} \approx 2.9 (m'/m_{\mathrm{p}})^{\frac{1}{2}} a. \tag{10.24}$$

This gives us a useful expression for the width of the general $k = p = 0$ ILR. Note that this width is approximately the same for all first-order resonances.

Lindblad resonances are thought to play a key role in the confinement of narrow rings. For example, observations have shown that the outer and inner edges of the $\epsilon$ ring of Uranus are located at the Ophelia 14:13 ILR and the Cordelia 24:25 OLR. It is also known that the existence of Lindblad resonances with satellites such as Mimas, Janus, Pandora, and Prometheus in Saturn's rings gives rise to density waves in the immediate vicinity of the resonance. This is discussed in Sect. 10.4 below. The theory given above for first-order resonances

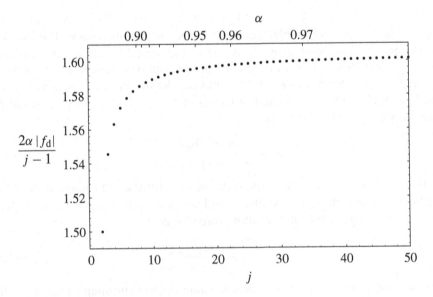

Fig. 10.10. A plot of $2\alpha |f_{\mathrm{d}}|/(j-1)$ as a function of $j$ (lower scale) and $\alpha$ (upper scale) for all the $j : j - 1$ ILRs with $k = p = 0$ from the 2:1 to the 50:49.

can easily be extended to handle higher order resonances using the results given in Sect. 8.8.

### 10.3.3 Perturbations in Inclination and Vertical Resonances

A vertical resonance occurs where the pattern speed of the perturbing potential matches the vertical frequency of the particle. In this case

$$m(n - \Omega_p) = \pm \nu, \tag{10.25}$$

where the upper and lower signs correspond to the inner (IVR) and outer (OVR) vertical resonances respectively. The resonance condition can also be written as

$$(m \mp 1)n \pm \dot{\Omega} - m\Omega_p = 0 \tag{10.26}$$

or

$$(m + k + p)n' - (m \mp 1)n - k\dot{\varpi}' \mp \dot{\Omega} - p\dot{\Omega}' = 0. \tag{10.27}$$

In terms of the resonant angle, $\varphi_{vr}$, we have

$$\varphi_{vr} = (m + k + p)\lambda' - (m \mp 1)\lambda - k\varpi' \mp \Omega - p\Omega'. \tag{10.28}$$

Using our standard notation we can write this as

$$\varphi_{vr} = j\lambda' + (k + p \pm 1 - j)\lambda - k\varpi' - p\Omega' \mp \Omega, \tag{10.29}$$

where, as before, $j = m + k + p$. Note that this is a resonant argument of order $|k + p \pm 1|$. Also, since the sum of the coefficients of $\Omega$ and $\Omega'$ must be even, $p$ has to be odd for all vertical resonances. The lowest order term associated with $\varphi_{vr}$ is $\mathcal{O}(e'^{|k|} I I'^{|p|})$.

Consider the case of the strong $k = 0$, $p = 1$ vertical resonance. This is equivalent to the $II'$ resonance discussed in Sect. 8.8.4. The relevant part of the disturbing function is

$$\mathcal{R} = \frac{\mathcal{G}m'}{a'} f_d(\alpha) s s' \cos \varphi_{vr}, \tag{10.30}$$

where $s = \sin \frac{1}{2} I$, $s' = \sin \frac{1}{2} I'$, $\varphi_{vr} = j\lambda' - (j - 2)\lambda - \Omega' - \Omega$, and $f_d = -\alpha b_{3/2}^{(j-1)}$ (see Table 8.1). Just as the Lindblad resonance is due to a forced eccentricity, so the vertical resonance is due to a forced inclination. There is also a phase shift of 180° in the longitude of ascending node on either side of the vertical resonance, as in the case of the longitude of pericentre in the Lindblad resonance. The location of the equilibrium point as a function of $\bar{\delta}$ (the distance from exact resonance), and hence the value of the scaled forced inclination, is given by Eq. (8.187). When $\bar{\delta}$ is large this gives a value of $2c/\bar{\delta}$. Using the standard scaling relationships gives a forced inclination of

$$I_f = \left| \frac{n\alpha(m'/m_p) \sin I' f_d}{4[jn' - (j - 1)n]} \right|. \tag{10.31}$$

The resulting amplitude of the forced vertical wave can be written as

$$a I_{\mathrm{f}} = \frac{\alpha a^2 (m'/m_{\mathrm{p}}) \sin I' |f_{\mathrm{d}}|}{6(j-2)|a - a_{\mathrm{res}}|}. \tag{10.32}$$

Although it is tempting to use the analogy with the Lindblad resonance to calculate an expression for the width of a vertical resonance, in reality the concept is meaningless partly because adjacent streamlines never intersect.

An OVR with Galatea is thought to play a role in the radial confinement of the Adams ring of Neptune. It is also known that the existence of IVRs with Mimas ($I' \approx 1.5°$) in Saturn's rings gives rise to bending waves in regions immediately interior to the resonance. This is discussed in Sect. 10.4.

### 10.3.4  Locations of Resonances

Using Kepler's third law, the approximate semi-major axis of an internal $p+q : p$ resonance is given by $a = [p/(p+q)]^{2/3} a'$, where $a'$ is the semi-major axis of the perturbing object. However, as we have seen in Chapter 8, the semi-major axis of the *exact* resonance depends on the particular resonant argument under consideration. Consider a general resonant argument of the form

$$\varphi = j_1 \lambda' + j_2 \lambda + j_3 \varpi' + j_4 \varpi + j_5 \Omega' + j_6 \Omega. \tag{10.33}$$

This argument is associated with the general $|j_1| : |j_2|$ resonance. If we ignore the variation of the mean longitude at epoch, the location of the exact resonance is given by the semi-major axis that satisfies the equation

$$j_1 n' + j_2 n + j_3 \dot{\varpi}' + j_4 \dot{\varpi} + j_5 \dot{\Omega}' + j_6 \dot{\Omega} = 0. \tag{10.34}$$

In the case of resonances with planets (e.g., jovian resonances in the asteroid belt) the dominant contribution to the $\dot{\varpi}$ and $\dot{\Omega}$ terms comes from the secular perturbations from the perturber itself and tends to be small. This means that the locations of the exact resonances corresponding to the various resonant arguments are all close to one another. However, in the case of resonances with planetary satellites the effect of the planet's oblateness invariably dominates the motion of the perturbed object's pericentre and node, especially in regions close to the planet. This leads to the phenomenon of resonance splitting discussed in Sect. 8.11, whereby the locations of the exact resonances at a given commensurability will be spread out. Furthermore, because $\dot{\varpi}$ and $\dot{\Omega}$ are approximately equal in magnitude but opposite in sign ($\varpi$ precesses and $\Omega$ regresses), those resonances involving only the pericentres or only the nodes will be equidistant (outside and inside respectively) from the location of the nominal resonance given by $a = |j_2/j_1|^{2/3} a'$, or, more precisely, $n = |j_1/j_2| n'$. Unfortunately $n$, $\dot{\varpi}$, and $\dot{\Omega}$ depend on $a$ in a nonlinear manner and so a numerical method is required to find the location of the exact resonance for a given argument.

Table 10.1. The type, classification, and parameters for the Mimas 3:1, 2:1, 4:2, 5:3, and 8:5 resonances in Saturn's rings.

| Resonant angle, $\varphi$ | Type | Class. | $j$ | $m$ | $k$ | $p$ | $a$ (km) | Feature |
|---|---|---|---|---|---|---|---|---|
| $3\lambda' - \lambda - 2\Omega$ | $I^2$ | — | 3 | — | — | — | 88,029 | |
| $3\lambda' - \lambda - \Omega' - \Omega$ | $II'$ | IVR | 3 | 2 | 0 | 1 | 88,705 | • |
| $3\lambda' - \lambda - 2\Omega'$ | $I'^2$ | CIR | 3 | 1 | 0 | 2 | 89,356 | |
| $3\lambda' - \lambda - 2\varpi'$ | $e'^2$ | CER | 3 | 1 | 2 | 0 | 89,562 | |
| $3\lambda' - \lambda - \varpi' - \varpi$ | $ee'$ | ILR | 3 | 2 | 1 | 0 | 90,193 | • |
| $3\lambda' - \lambda - 2\varpi$ | $e^2$ | — | 3 | — | — | — | 90,802 | |
| $4\lambda' - 2\lambda - 2\Omega$ | $I^2$ | — | 4 | — | — | — | 116,512 | |
| $4\lambda' - 2\lambda - \Omega' - \Omega$ | $II'$ | IVR | 4 | 3 | 0 | 1 | 116,725 | |
| $4\lambda' - 2\lambda - 2\Omega'$ | $I'^2$ | CIR | 4 | 2 | 0 | 2 | 116,935 | |
| $4\lambda' - 2\lambda - 2\varpi'$ | $e'^2$ | CER | 4 | 2 | 2 | 0 | 117,138 | |
| $2\lambda' - \lambda - \varpi'$ | $e'$ | CER | 2 | 1 | 1 | 0 | 117,138 | • |
| $4\lambda' - 2\lambda - \varpi' - \varpi$ | $ee'$ | ILR | 4 | 3 | 1 | 0 | 117,346 | • |
| $4\lambda' - 2\lambda - 2\varpi$ | $e^2$ | — | 4 | — | — | — | 117,552 | |
| $2\lambda' - \lambda - \varpi$ | $e$ | ILR | 2 | 2 | 0 | 0 | 117,552 | • |
| $5\lambda' - 3\lambda - 2\Omega$ | $I^2$ | — | 5 | — | — | — | 131,793 | |
| $5\lambda' - 3\lambda - \Omega' - \Omega$ | $II'$ | IVR | 5 | 4 | 0 | 1 | 131,900 | • |
| $5\lambda' - 3\lambda - 2\Omega'$ | $I'^2$ | CIR | 5 | 3 | 0 | 2 | 132,007 | |
| $5\lambda' - 3\lambda - 2\varpi'$ | $e'^2$ | CER | 5 | 3 | 2 | 0 | 132,191 | |
| $5\lambda' - 3\lambda - \varpi' - \varpi$ | $ee'$ | ILR | 5 | 4 | 1 | 0 | 132,298 | • |
| $5\lambda' - 3\lambda - 2\varpi$ | $e^2$ | — | 5 | — | — | — | 132,404 | |
| $8\lambda' - 5\lambda - \varpi' - 2\Omega$ | $e'I^2$ | — | 8 | — | — | — | 135,582 | |
| $8\lambda' - 5\lambda - \varpi' - \Omega' - \Omega$ | $e'II'$ | IVR | 8 | 6 | 1 | 1 | 135,642 | • |
| $8\lambda' - 5\lambda - \varpi - 2\Omega$ | $eI'^2$ | — | 8 | — | — | — | 135,642 | |
| $8\lambda' - 5\lambda - \varpi' - 2\Omega'$ | $e'I'^2$ | — | 8 | — | — | — | 135,701 | |
| $8\lambda' - 5\lambda - \varpi - \Omega' - \Omega$ | $eII'$ | — | 8 | — | — | — | 135,702 | |
| $8\lambda' - 5\lambda - \varpi - 2\Omega'$ | $eI'^2$ | ILR | 8 | 6 | 0 | 2 | 135,761 | |
| $8\lambda' - 5\lambda - 3\varpi'$ | $e'^3$ | CER | 8 | 5 | 3 | 0 | 135,820 | |
| $8\lambda' - 5\lambda - 2\varpi' - \varpi$ | $ee'^2$ | ILR | 8 | 6 | 2 | 0 | 135,879 | • |
| $8\lambda' - 5\lambda - \varpi' - 2\varpi$ | $e^2e'$ | — | 8 | — | — | — | 135,938 | |
| $8\lambda' - 5\lambda - 3\varpi$ | $e^3$ | — | 8 | — | — | — | 135,998 | |

Notes: $\varphi$ is the resonant angle. In the classification column ILR, IVR, CER, and CIR denote inner Lindblad resonance, inner vertical resonance, corotation eccentricity resonance, and corotation inclination resonance respectively. The "•" symbol in the final column denotes that there is an associated feature in Saturn's rings at this location.

As an example, Table 10.1 shows the results of such calculations for a selection of 3:1, 2:1, 4:2, 5:3, and 8:5 Mimas resonances that lie in Saturn's rings. In order to derive the locations of the resonances we have taken the mean motion, perichrone rate, and node rate of Mimas to be $381.9945°\text{d}^{-1}$, $1.0008°\text{d}^{-1}$, and $-0.9995°\text{d}^{-1}$ respectively (Harper & Taylor 1993). We have taken $\mathcal{G}m_\text{p}$ for Saturn to be $3.79312 \times 10^7$ km$^3$s$^{-2}$ and also included secular perturbations from Mimas in the calculation, although the latter are negligible. In the table we indicate the "type" of the resonance, using the terminology adopted in Chapter 8, as well as the "classification" of the resonance (denoted by "Class" in the table) using the terminology of ring dynamics. The effect of the oblateness causes the resonances at a given commensurability to be spread out in semi-major axis. Note that two pairs of resonances (the $e$ and $e^2$ resonances, and the $e'$ and $e'^2$ resonances) have the same location because the respective 4:2 arguments are simply twice the 2:1 arguments and therefore they have the same equation corresponding to exact resonance. Note that a number of resonances are not classified under the terminology of ring dynamics.

## 10.4 Density Waves and Bending Waves

The effect of a satellite resonance in a ring of sufficiently high surface density is to introduce an azimuthal variation in the gravitational potential. For example, we have already seen that a $j : j - 1$ inner Lindblad resonance gives rise to a $j$-lobed pattern in the streamlines of particle paths in the vicinity of the resonance (see Fig. 10.9). In general, it is the value of $m$ rather than $j$ that determines the number of lobes, although for first-order ILRs, $j = m$ (see Table 10.1). Beyond the resonance width, as the distance from exact resonance increases, each streamline shifts in azimuth but the whole pattern remains stationary in the rotating reference frame. This gives rise to a spiral pattern with $m$ spiral arms (see Fig. 10.11), even though each individual particle remains on a keplerian ellipse. The pattern is called a *spiral density wave*. Because the resulting gravitational potential varies with the same period as that due to the satellite, the spiral pattern is reinforced. In practice the spiral is very tightly wound, but a radial profile would always show the characteristic decrease in wavelength (the distance from peak to peak or trough to trough) with increasing distance from exact resonance.

There is an equivalent phenomenon due to vertical resonances. In this case the forced oscillations are in the vertical rather than the radial direction (i.e., there is a forced inclination) and the systematic shifts in azimuth lead to the formation of *spiral bending waves*. For an inner vertical resonance the result is trailing bending waves that propagate inwards from the exact resonance (see Fig. 10.12).

The basic theory of density waves was originally developed to explain the spiral structure of some galaxies, but Goldreich & Tremaine (1978b) adapted it to explain the formation of the Cassini division in Saturn's rings. Since then

(a)  (b)

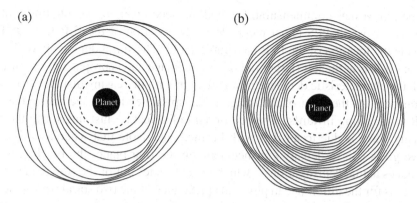

Fig. 10.11. Schematic diagrams of the coplanar particle paths that give rise to trailing spiral density waves near a resonance with an exterior satellite. (a) The two-armed spiral density wave associated with the 2:1 ($m = 2$) inner Lindblad resonance. (b) The seven-armed density wave associated with the 7:6 ($m = 7$) inner Lindblad resonances. The pattern rotates with the angular velocity of the satellite and propagates outwards from the exact resonance (denoted by the dashed circle).

significant advances in the theory have been made (see, for example, the review by Shu (1984) and the work of Rosen (1989).

Spiral density and bending waves have now been detected at many locations in Saturn's rings (Holberg et al. 1982, Lane et al. 1982, Shu et al. 1983). In Table 10.1 we list some examples of Mimas resonances in Saturn's rings. Because Mimas has an appreciable inclination (1.5°) its vertical resonances are comparable in strength to its Lindblad resonances. When strong resonances occur in rings of sufficient surface density it is possible to see the effects of resonance. Figure 10.13 shows part of a *Voyager 2* image of Saturn's A ring in the vicinity

(a)  (b)

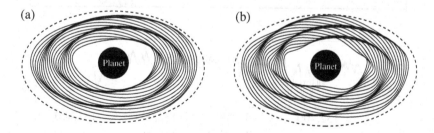

Fig. 10.12. Schematic diagrams showing an oblique view of the three-dimensional particle paths that give rise to trailing spiral bending waves near a resonance with an exterior satellite. (a) The two-armed spiral bending wave associated with the 3:1 ($m = 2$) inner vertical resonance. (b) The four-armed bending wave associated with the 5:3 ($m = 4$) inner vertical resonances. The pattern rotates with the angular velocity of the satellite and propagates inwards from the exact resonance (denoted by the dashed curve).

of the 5:3 Mimas commensurability and the accompanying occultation profile
from the *Voyager* PPS experiment. Here the location of the ILR lies 398 km
outside the location of the IVR (see Table 10.1). The spiral density wave due to
the 5:3 ILR is at the left (larger radius) of the image and is propagating outwards
with the characteristic decrease in wavelength with increasing distance from the
resonance. At the right of the image is the spiral bending wave due to the 5:3
IVR propagating inwards. The additional radial feature just interior to the 5:3
density wave is due to an ILR with Prometheus. Note that whereas a density
wave gives rise to density enhancements in the ring plane, the bending wave
produces actual "corrugations", which take particles out of the ring plane. This
accounts for the difference in physical appearance of the two spiral waves shown
in Fig. 10.13. The more extreme contrast visible of the bending wave was caused
by the effect of alternate brightness and shadow due to the vertically displaced
ring material being illuminated by the Sun, which was 8° above the ring plane
at the time the image was taken.

Further examples of density and bending waves are shown in Fig. 10.5. Although there is insufficient radial resolution in the image to detect individual
peaks and troughs, the effects of the density waves due to Prometheus and

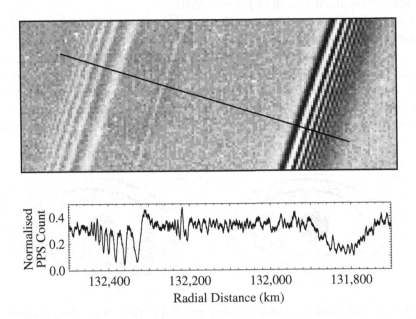

Fig. 10.13. *Voyager* image and associated photopolarimeter (PPS) occultation profile of
spiral density waves (left) and spiral bending waves (right) produced by the 5:3 Mimas
resonance in Saturn's A ring. The density and bending waves propagate outwards
and inwards respectively from the locations of exact resonance. *(Image courtesy of
NASA/JPL. PPS data courtesy of NASA PDS Rings node.)*

Pandora ILRs near the outer part of Saturn's A ring are clearly visible. Furthermore, the density and bending waves associated with the 8:5 Mimas commensurability are also evident.

## 10.5 Narrow Rings and Sharp Edges

Prior to the discovery of the rings of Uranus in 1977, the only known ring system belonged to Saturn. Ground-based observations showed that Saturn had a broad ring system, whereas the stellar occultation data for Uranus showed a system of narrow, sharp-edged rings. Some of the rings of Uranus were clearly eccentric and there was also evidence of nonzero inclination. These observations were to pose the first of many dynamical problems for ring theorists because narrow, eccentric, inclined rings should spread on timescales significantly shorter than the age of the solar system. The fact that these rings are observed suggests that either we are living at some special epoch when such rings exist or that there is some confinement mechanism that can maintain their properties over long timescales. In this section we examine the timescale problem and show how the dynamical effects of small satellites orbiting close to rings have been invoked to account for the unusual observations.

### 10.5.1 Spreading Timescales

Unconstrained narrow rings should spread due to a number of different effects. These include (i) collisions between ring particles, (ii) Poynting–Robertson light drag, (iii) plasma drag, and (iv) differential precession (Goldreich & Tremaine 1979a, 1982). Here we examine each of these processes in turn and obtain an expression for the timescale associated with each mechanism.

Consider a narrow ring of uniform surface density, mass $m$, orbital radius $r$, and radial width $W$ (with $W \ll r$). Brahic (1977) has shown that for a fixed angular momentum, the energy $E$ of the ring can be written as

$$E = -\frac{\mathcal{G} m_\mathrm{p} m W^2}{32 r^3} + \text{constant}, \tag{10.35}$$

where $m_\mathrm{p}$ is the mass of the planet. For a fixed $r$, $\dot{E} \propto -\dot{W}$ and hence $E$ is a maximum when $W$ is a minimum. Therefore any loss of energy due to collisions will result in spreading of the ring on a timescale $t_\mathrm{coll}$ given by

$$t_\mathrm{coll} = \frac{W}{\dot{W}} = -\frac{m n^2 W^2}{16 \dot{E}}, \tag{10.36}$$

where $n$ is the mean motion or mean angular velocity of the ring. Goldreich & Tremaine (1982) showed that $\dot{E}$ is related to the frequency of particle collisions $\omega$ by

$$\dot{E} = -3 v^2 m \omega (1 - \epsilon^2), \tag{10.37}$$

where $v$ is the one-dimensional random velocity of the particles and $\epsilon$ is their coefficient of restitution.

If the ring is considered as a differentially rotating fluid of density $\rho$ then the shear stress $S$ in the ring can be written as

$$S = \rho v r \frac{dn}{dr}, \tag{10.38}$$

where $v$ is the effective kinematic viscosity. This stress produces a torque that will transfer angular momentum in a direction and at a rate determined by the gradient of the angular velocity, $dn/dr$ (Safronov 1972, Lynden-Bell & Pringle 1974). In this case $\dot{E}$ is the work done by the torque and $v$ can be written as

$$v = \frac{v^2}{n} \frac{0.46\,\tau}{1 + \tau^2}, \tag{10.39}$$

where $\tau$ is a dimensionless quantity called the optical depth of the ring (Cook & Franklin 1964, Goldreich & Tremaine 1978a). If $d$ is the characteristic radius of the particles that make up the ring, then both the energy and shear stress approaches give the same result: $t_{coll}$ is approximately equal to the time it takes a particle to complete a random walk across the ring (Brahic 1977, Goldreich & Tremaine 1978a). This is given by

$$t_{coll} \approx \frac{1}{\tau n} \left( \frac{W}{d} \right)^2. \tag{10.40}$$

The second effect to consider is Poynting–Robertson (PR) light drag. This is caused by the nonisotropic reemission of incident radiation on a moving particle (see, for example, Burns et al. 1979). It is particularly important for particles with dimensions comparable to the wavelength of the incident radiation (i.e., several $\mu$m) and causes gradual orbital decay. For circumsolar motion PR drag causes small particles to spiral in towards the Sun, while for circumplanetary motion the effect causes particles to spiral towards the planet. The approximate timescale for decay under PR drag of a particle of radius $d$, density $\rho$, semi-major axis $a$, and inclination $I$, in orbit around a planet, is

$$t_{decay} \approx \frac{8\rho d c^2}{3 \left( L/4\pi a^2 \right) Q_{PR} (5 + \cos^2 I)}, \tag{10.41}$$

where $L$ is the solar luminosity, $c$ is the speed of light, and $Q_{PR}$ is a coefficient given by

$$Q_{PR} = Q_{abs} + Q_{sca} \left( 1 - \langle \cos \alpha \rangle \right), \tag{10.42}$$

where $Q_{abs}$, $Q_{sca}$, and $\alpha$ denote the absorption coefficient, the scattering coefficient (both dimensionless quantities), and the scattering angle respectively (see Burns et al. 1979). For a given orbit $t_{decay}$ is a linear function of $d$, the size of the particles. Therefore the orbits of small particles will decay faster than those of large particles. However, the effect of collisions between the particles should tend to average out these rates. If the surface density of the ring is nonuniform then the ring would be expected to spread on a timescale

$$t_{PR} \approx t_{decay} \left( \frac{W}{r} \right) \tag{10.43}$$

(see Dermott 1984).

A comparison of the formulae for the spreading timescales due to collisions (Eq. (10.40)) and PR drag (Eq. (10.43)) shows that for a ring of a given width and optical depth, $t_{coll} \propto d^{-2}$ whereas $t_{PR} \propto d$, where $d$ is the particle size. Therefore, decreasing the particle size will increase the timescale for spreading due to collisions but decrease the timescale for spreading due to PR drag. Increasing the particle size will have the opposite effect. For the parameters of the narrow rings of Uranus the longest lifetime is only $\sim 10^7$ y, or 0.2% of the age of the solar system. This is the essence of the problem with narrow rings and is the reason that several confinement mechanisms were proposed.

The planets Saturn and Uranus are known to have a corotating magnetosphere, which is maintained by the planet's magnetic field. In these circumstances the absorption of plasma by the ring particles results in spreading on a timescale

$$t_{plasma} \approx \frac{2dp_r}{3r\rho_p} \frac{n}{(n - \Omega_p)}, \tag{10.44}$$

where $\Omega_p$ is the angular velocity of the planet and $\rho_p$ is the plasma density (Burns et al. 1980). Plasma drag acts in a similar way to tidal drag by driving the ring particles away from synchronous orbit (see Grun et al. 1984); this is because the planet is the source of angular momentum. Therefore, inside synchronous orbit plasma drag enhances the effect of PR drag while outside synchronous orbit the two drag forces oppose one another.

Planetary oblateness causes an eccentric orbit to precess at an approximate rate given by

$$\dot{\varpi} = \frac{3}{2} J_2 \left( \frac{R_p}{a} \right)^2 n, \tag{10.45}$$

where $J_2$ is the first zonal harmonic, $R_p$ is the radius of the planet, $n$ is the mean motion of the object in orbit, and $m_p$ is the mass of the planet (cf. Eq. (6.256)). Therefore the difference in $\dot{\varpi}$ between the inner and outer edges of an eccentric ring with a width $\delta a$ in semi-major axis is given by

$$\delta \dot{\varpi} \approx -\frac{7}{2} \dot{\varpi} \frac{\delta a}{a}. \tag{10.46}$$

Hence differential precession should occur because the inner edge should be precessing more rapidly than the outer edge. The ring should be smeared out on a timescale of $2\pi/|\delta\dot{\varpi}|$, where $\varpi$ is measured in radians. For example, a 10 km-wide eccentric ring orbiting Uranus would be smeared out within ~2,500 years. Therefore, unless some other mechanism is operating, eccentric rings should have short lifetimes. There is a similar argument for the case of inclined rings where differential regression of the lines of nodes should again cause ring smearing.

### 10.5.2 *Localised Effects of Satellite Perturbations*

In Sect. 3.13 we carried out a numerical study of the paths of test particles on circular orbits being perturbed by a nearby satellite. The resulting diagram (Fig. 3.30) clearly shows that for particles at relatively large radial separations from the satellite, the encounter produces the appearance of a wave on the particle's trajectory. This process is relatively easy to understand. Because all particles encountering the satellite move on initially circular orbits, every particle at a given semi-major axis will receive exactly the same eccentricity from the satellite. Furthermore, because the perturbation is applied over such a short time interval, the orbits of the postencounter particles will be unperturbed keplerian ellipses of identical $a$ and $e$ but with systematic differences in phase. In a rotating reference frame the path of an individual particle moving on an elliptical orbit resembles a sinusoidal curve, as the particle moves from its pericentre to apocentre over the course of an orbital period. The following particle will follow the same path and so on for all subsequent particles. The resulting effect is that in the rotating frame the satellite appears to produce a standing wave on the ring's edge. This wave propagates upstream for interior particles and downstream for exterior particles (see Fig. 3.30). In the nonrotating frame the wave will be seen to move at the same rate as the mean motion of the satellite. Because the wave is a consequence of the particles' new-found eccentricity, the amplitude, $A$, of the wave is simply half the difference between the apocentric and pericentric distances. Hence $2A = a(1 + e) - a(1 - e) = ae$. We can make use of the Tisserand relation and simple perturbation theory to calculate the value of $e$.

Let $a$ and $a'$ denote the semi-major axes of the particle and satellite respectively, where we take $a' > a$ and $a'$ is assumed to be fixed. Let $m'$ and $m_p$ denote the masses of the satellite and planet respectively, with $m' \ll m_p$. We also assume that the satellite is moving on a circular orbit exterior to the particle (lower part of Fig. 3.30). By Kepler's third law the particle will have a larger angular velocity and so it will catch up and eventually pass the satellite. Since the mass of the satellite is so small, the particle is unaware of the satellite until encounter takes place. At conjunction the particle receives an impulse from the satellite and this leads to changes in the orbital elements of the particle. At this stage we can ignore the effect of the particle on the satellite.

In the context of the two-body problem the angular momentum per unit mass of the particle is given by

$$h = \sqrt{\mathcal{G}m_p a(1 - e^2)}, \tag{10.47}$$

and hence the changes in the orbital elements are related to the change in angular momentum by

$$2\frac{\delta h}{h} = \frac{\delta a}{a} - 2e^2,$$
(10.48)

where, because the orbit is initially circular, we have taken $\delta e = e$.

Although $ea \gg \delta a$, we can show (see Eq. (10.52)) that $\delta a/a \gg e^2$ and so $\delta h$ is determined by $\delta a$. However, $\delta a/a$ is $\mathcal{O}(m'/m_p)^2$ and can be calculated by deriving $\delta e$, a quantity of $\mathcal{O}(m'/m_p)$, from the Jacobi constant or the Tisserand relation (see Sect. 3.4) for the planet–satellite–particle three-body system. In the case of zero inclination and where the semi-major axis of the perturbing body is not taken to be unity, Eq. (3.46) can be written as

$$\frac{a'}{a} + 2\left(\frac{a}{a'}\right)^{1/2}\left(1 - e^2\right)^{1/2} = C,$$
(10.49)

where $C$ is a constant up to terms of $\mathcal{O}(m'/m_p)$. Note that so far we are not considering the effect of the encounter, just the relationship between the orbital elements of the perturber and the particle. Letting $\Delta a = a - a'$ denote the small separation of the semi-major axes in our model, we can expand Eq. (10.49) to give

$$\frac{3}{4}\left(\frac{\Delta a_n}{a'}\right)^2 - e_n^2 \approx C - 3,$$
(10.50)

where we have now generalised the equation such that $\Delta a_n$ and $e_n$ are the values of $\Delta a$ and $e$ after the $n$th encounter with the satellite (see Dermott & Murray 1981a). Note that there are no linear terms in $\Delta a_n$ in Eq. (10.50). A similar version of this equation is derived in Sect. 3.13 where we discuss Hill's equations (cf. Eq. (3.213)).

From Eq. (10.50) it is clear that an increase in $e$ (the inevitable outcome since initially the particle is on a circular orbit) always results in an increase in $|\Delta a|$, the separation in semi-major axis of the particle and the satellite. We can quantify the relationship by letting $\delta a = \Delta a_1 - \Delta a_0$ and substituting in Eq. (10.50). This gives

$$\frac{3}{4}\left(\frac{\Delta a_0}{a'}\right)^2 = \frac{3}{4}\left(\frac{\delta a + \Delta a_0}{a'}\right)^2 - e^2,$$
(10.51)

where we have used the relation $\Delta a_1 = \delta a + \Delta a_0$. Writing $a' = a - \Delta a_0$ and carrying out a binomial expansion with the assumption that $\delta a \ll \Delta a_0$ (i.e., the change in the particle's semi-major axis due to the encounter is negligible compared with the separation from the satellite) we obtain

$$\frac{\delta a}{a} = \frac{2}{3}\frac{a}{\Delta a}e^2.$$
(10.52)

Since $a/\Delta a_0$ is small, $\delta a/a \gg e^2$ and we can neglect the $e^2$ term in the equation for $\delta h/h$ above.

It only remains to calculate the eccentricity that the particle receives. This can be done using the perturbation equations derived in Sect. 2.9. For small eccentricities the variation of the $e$ is

$$\frac{de}{dt} \approx \frac{1}{na}\left(\bar{R}\sin f + 2\bar{T}\cos f\right), \qquad (10.53)$$

where $\bar{R}$ and $\bar{T}$ denote the radial and tangential components of the perturbing force (per unit mass) and $f$ is the true anomaly of the particle (cf. Eq. (2.150)). If the particle's interaction with the satellite is treated as an radial impulse lasting a time $\Delta t$ then, from Eq. (10.53), the resulting eccentricity is

$$e \approx \frac{1}{na}\bar{R}\,\Delta t\,\sin f. \qquad (10.54)$$

Following Dermott (1984) we consider an impulse that lasts for a time interval $\Delta t \approx 0.2P$, where $P = 2\pi/n$ is the orbital period of the particle. From Kepler's third law the relative angular velocity of the particle with respect to the satellite is

$$U = \frac{3}{2}\frac{n}{a}\Delta a_0 \qquad (10.55)$$

and hence

$$\Delta t = \frac{2\Delta a_0}{Ua}. \qquad (10.56)$$

If we assume that the particle has a true anomaly $f \approx \pi/2$ (i.e., the particle is close to quadrature) then $\sin f \approx 1$ and the resulting eccentricity is

$$e \approx \frac{4}{3}\frac{m'}{m_p}\left(\frac{a}{\Delta a_0}\right)^2 \qquad (10.57)$$

(cf. Lin & Papaloizou 1979). In fact the coefficient in this equation is 2.24 rather than 4/3 (Julian & Toomre 1966) since our calculation neglected the tangential force on the particle.

The particle's mean motion is $n$ and that of the satellite is $n+U$. Therefore the number of orbits between encounters is $n/U$. In that time the particle will have covered a distance of $2\pi a$ with respect to the satellite. Hence the wavelength is

$$\ell = \frac{2\pi a}{n/U} = 3\pi\,\Delta a_0. \qquad (10.58)$$

Note that the amplitude of the wave is a function of $\Delta a_0$ and $m'$, whereas the wavelength is a function of $\Delta a_0$ alone. Therefore observations of an edge wave allow determination of the mass and radial distance of the satellite that produced it, even though the satellite may not have been detected directly.

The variation of $A$ and $\ell$ with $\Delta a_0$ leads to an additional effect in nearby ring material – the formation of a wake. This is illustrated schematically in Fig. 10.14 for the case of an exterior satellite. Although each wave has a fixed wavelength the effect of successive waves with decreasing wavelengths is that of peaks and troughs in quasi-parallel lines that are not in line with the mean ring edge. The existence of a wake provides another means of indirect detection of a satellite and one that is particularly suited to the use of data obtained from occultations.

Another phenomenon that can occur in certain circumstances is the formation of shocks in rings. Consider two rings with a constant, presatellite separation of $W$ (see Fig. 10.15); these could be isolated rings or simply the inner and outer edges of a narrow ring. An orbiting satellite will produce a wave on the rings but each wave will have a slightly different wavelength due to the different separations.

Let the radial distances of each ring be $r_1$ and $r_2$. Before the encounter $r_1 = a_1$ and $r_2 = a_2$, where $a_1$ and $a_2$ are the constant semi-major axes of the rings. After an encounter with an interior satellite we have

$$r_1(D) = a_1 - A \sin(2\pi D / \ell_1), \qquad (10.59)$$

$$r_2(D) = a_2 - A \sin(2\pi D / \ell_2), \qquad (10.60)$$

where $D$ is the distance along the orbit from the satellite and the two wavelengths are $\ell_1 = 3\pi(\Delta a + W/2)$ and $\ell_2 = 3\pi(\Delta a - W/2)$. Here we are assuming that both waves have the same amplitude $A$. Solving for $r_1 = r_2$ gives

$$D \approx \frac{3\Delta a^2}{2A} \qquad (10.61)$$

for the distance at which the waves intersect and a shock forms. Therefore, shocks will survive until the next encounter provided $D < 2\pi a$, or, using the modified expression for $A$, provided

$$\Delta a < 1.75 \left( \frac{m'}{m_p} \right)^{1/4} a. \qquad (10.62)$$

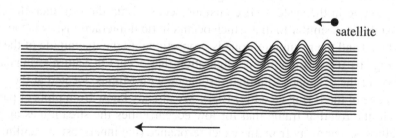

Fig. 10.14. A schematic diagram showing how the variation of the amplitude and wavelength with separation from a satellite leads to the formation of a wake in adjacent ring material upstream from an exterior satellite. The lower arrow indicates the direction of motion of the particles with respect to the satellite.

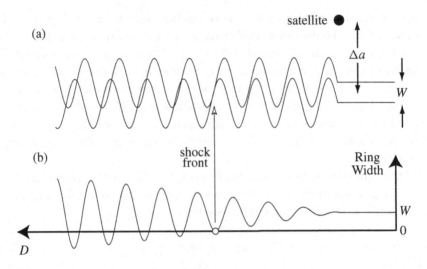

Fig. 10.15. A schematic diagram showing how a shock forms in a perturbed ring due to the intersection of streamlines (adapted from Dermott 1984). (a) Waves of approximately the same amplitude but differing wavelength generated by a satellite encounter; eccentricity damping is ignored. (b) The radial width of the ring as a function of distance $D$; zero width occurs and a shock forms where the rings intersect (denoted by the open circle on the axis).

We have already noted that the $\epsilon$ ring of Uranus has Lindblad resonances with small satellites at its inner and outer edges. These edges are extremely sharp. Indeed, occultation profiles even show diffraction effects caused by the sharpness (see, for example, Elliot & Nicholson 1984). Borderies et al. (1989) have shown that the sharp edges of some rings (broad as well as narrow) are a consequence of the presence of such resonances. We discussed the mechanism of Lindblad resonance in Sect. 10.3.2 and showed how the intersection of streamlines due to the forced eccentricity determines the width of the resonance in the undamped system. The symmetry of the configuration (see Fig. 10.9) ensures that there is no net torque in the system. If collisions occur within the ring then the effect of dissipation is similar to that which occurs in tidal interactions: A lag angle is introduced and the resultant torque is proportional to the magnitude of the lag.

In Fig. 10.16 we show the configuration that occurs in Saturn's rings at the location of the 2:1 inner Lindblad resonance with Mimas, located at the sharp, outer edge of the B ring (see Fig. 10.4). We already know from plots of particle paths in the rotating frame that for low eccentricities the streamlines are centred ellipses. Far away from the exact resonance (the innermost streamline) the forced eccentricity is negligible and the path is circular with uniform outward flow of angular momentum. As the resonance is approached (middle streamline) the path acquires an eccentricity giving it the appearance of a centred ellipse in the rotating frame. Note that the short axis of this and the other ellipses lags

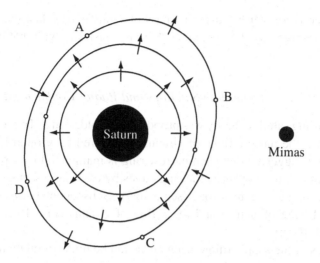

Fig. 10.16.   Three streamlines at the 2:1 inner Lindblad resonance with Mimas; the resonance is located at the outer edge of Saturn's B ring. The pattern is not symmetric with respect to the Saturn–Mimas line because of the lag in the minor axis of symmetry introduced by dissipation. The arrows show the direction of angular momentum flow, which is determined by the gradient of the local angular velocity. The behaviour of this flow is discussed in the text. (Figure adapted from Dermott 1984; originally due to P. Goldreich, personal communication.)

behind the Saturn–Mimas line, owing to the effect of particle collisions. For this and streamlines further out, the angular momentum flow is nonuniform. This particular streamline is critical since it contains two positions (marked by open circles) where the flow is zero. The outermost streamline has a larger forced eccentricity (since it is closer to the exact resonance) and the flow varies around the path and can have positive or negative signs. Along this streamline we mark the four points (labelled A, B, C, and D) where the flow is zero. If the outward angular momentum flow between A and B and between C and D is equal to the inward flow between A and D and between B and C, then the net flow across the streamline is zero. According to Borderies et al. (1989) the streamline for which this is true determines the edge of the ring and leads to a sharp boundary.

The outer edge of Saturn's B ring is located at 1.95 Saturn radii; the edge of the A ring at 2.27 Saturn radii is also sharp. In each case there is a satellite resonance in the vicinity. Porco et al. (1984a) have shown that the B ring edge has the shape of a "2-lobed pattern of radial oscillations" (a centred ellipse) consistent with the effect of the 2:1 resonance with Mimas. Similarly the A ring edge is located close to the 7:6 inner Lindblad resonance with Janus, one of the coorbital satellites, and so a 7-lobed pattern would be expected. Although the resonance is weaker because the mass of Janus is less than that of Mimas, the pattern observed by Porco et al. (1984a) is consistent with that of a 7-lobed figure.

### 10.5.3 Shepherding Satellites and Radial Confinement

The shepherding satellite theory was proposed by Goldreich & Tremaine (1979a) to account for the narrow rings of Uranus discovered by ground-based occultations. They suggested that each narrow ring is maintained by a pair of small satellites that act to "shepherd" the ring edges by exerting a secular torque that balances the ring's viscous torque (see below). Subsequent *Voyager* images of the narrow F ring of Saturn and the $\epsilon$ ring of Uranus (Fig. 10.17) appear to confirm their theory.

The basics of the shepherding mechanism can be understood by reexamining our model of an encounter between a single satellite and ring particle initially on a circular orbit. We have already calculated the eccentricity that the particle receives from the encounter but we also noted that there is a corresponding decrease in the particle's semi-major axis when it encounters an exterior satellite. Therefore the exterior satellite appears to repel the orbit of an interior ring. In the case of an interior satellite the ring particle's semi-major axis will increase due to the encounter. Therefore, the combined effect of a satellite on either side of a narrow ring is to confine the ring in semi-major axis (Fig. 10.18).

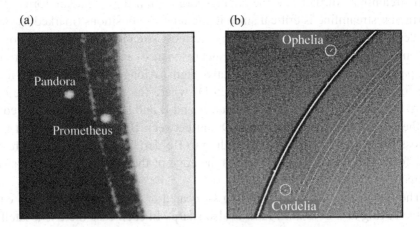

Fig. 10.17. *Voyager* images showing satellites shepherding narrow rings. (a) A low-resolution *Voyager 2* image of the F ring of Saturn with Pandora and Prometheus orbiting on either side. (b) A long-exposure *Voyager 2* image of the narrow rings of Uranus showing the outermost $\epsilon$ ring being shepherded by the satellites Ophelia and Cordelia (circled trails). *(Images courtesy of NASA/JPL.)*

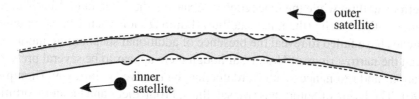

Fig. 10.18. A schematic diagram (adapted from Dermott 1984) illustrating the basics of the shepherding mechanism whereby a narrow ring (shaded area) is confined by two small satellites on either side of it.

In Fig. 10.18 we have assumed that the excited eccentricity produced by the encounter is eventually reduced to zero by the effect of collisions with other ring particles. This results in the wave being damped such that the particle is again in a circular orbit when it next encounters the satellite at a time $2\pi/U$ later. Whether or not the wave is damped obviously depends on the properties of the ring. If damping occurs we can estimate the torque, $\Gamma$, that the satellite exerts on the ring by calculating the rate of change of the orbital angular momentum, $L = mh$, where $m$ is the mass of the ring. We have

$$\Gamma = \dot{L} = \frac{\delta h}{2\pi/U} m. \tag{10.63}$$

We have shown above that

$$2\frac{\delta h}{h} \approx \frac{\delta a}{a} \approx \frac{2}{3}\frac{a}{\Delta a_0}e^2, \tag{10.64}$$

and hence the torque is given by

$$\Gamma = 0.399 \left(\frac{\mathcal{G}m'}{n\,\Delta a_0^2}\right)^2 m. \tag{10.65}$$

The effect of particle collisions is modelled by a *viscous torque*, which tends to spread the narrow ring in a radial direction. The original shepherding mechanism of Goldreich & Tremaine (1979a) was based on the balance between the viscous and satellite torque. However, the picture is more complicated because it can be shown that the shepherding satellites transfer sufficient energy to disrupt the ring (Borderies et al. 1984). A more complete model of shepherding was developed by Borderies et al. (1989) where both the satellite torque and the viscous torque in perturbed regions near the ring edges contribute to radial confinement.

The shepherding theory was originally proposed by Goldreich & Tremaine (1979a) in an attempt to solve the problem of the short lifetimes of the narrow uranian rings. When the *Voyager 2* spacecraft reached Uranus in January 1986 two small satellites, Cordelia and Ophelia, were detected on either side of the outermost $\epsilon$ ring of the planet. Analysis of their orbits showed that the outer edge of the $\epsilon$ ring is located at the 14:13 inner Lindblad resonance with Ophelia (the exterior satellite) while the inner edge is located at the 24:25 outer Lindblad resonance with Cordelia (the inner satellite) (Porco & Goldreich 1987). Extensive searches have failed to reveal the presence of additional shepherding moons between the narrow rings of Uranus, although there appear to be several preferred locations and resonances with Cordelia may be involved (Murray & Thompson 1990). The F ring of Saturn has two satellites, Prometheus and Pandora, orbiting on either side of it. This is thought to be another example of the shepherding mechanism although the situation is less straightforward (see Sect. 10.7).

### 10.5.4 Eccentric and Inclined Rings

Although in theory it is possible for a keplerian orbit in the solar system to be circular, in practice all bounded orbits are elliptical. Therefore at first glance the existence of eccentric narrow rings should not come as a surprise. However, the discovery of the narrow rings of Uranus in 1977 and the discoveries of the *Voyager* spacecraft at Saturn in 1980 and 1981 provided evidence that several narrow rings were eccentric and, more importantly, they were observed to be precessing uniformly.

Consider an eccentric narrow ring defined by an inner and outer ellipse. Let the midpoint of the ring have semi-major axis $a$ and eccentricity $e$ with the differences in these elements between the inner and outer edges being $\delta a$ and $\delta e$. If we take the equation of an ellipse to be $r \approx a(1 - e \cos f)$, where $f$ is the true anomaly, and assume that the ellipses have the same longitude of pericentre, the radial width of the ring as a function of $f$ is given by

$$W = \delta a \left[1 - (g + e) \cos f\right], \tag{10.66}$$

where we define

$$g = a \frac{\delta e}{\delta a} \tag{10.67}$$

to be the *eccentricity gradient* across the ring. The equation for $W$ has the same form (but with different constants) as the equation for $r$. Therefore the variations in $W$ and $r$ are in phase and $W$ is a linear function of $r$. This relationship has now been directly observed in a number of narrow rings using occultation data. For example, values of these quantities for the $\alpha$, $\beta$, and $\epsilon$ rings of Uranus are given in Table 10.2, together with the resulting minimum and maximum radial widths. All these rings can be accurately modelled by assuming aligned pericentres of their inner and outer edges.

Goldreich & Tremaine (1979a,b) have proposed that the observed pericentre alignment is maintained by the self-gravity of the ring. In this mechanism the mass within the ring has distributed itself such that the mass exterior to the inner edge acts to reduce its precession rate, while the mass interior to the outer edge acts to increase its precession rate. The net effect is that the ring precesses at

Table 10.2. *Orbital parameters of three eccentric rings of Uranus.*

| Ring | $\alpha$ | $\beta$ | $\epsilon$ |
|---|---|---|---|
| *a* (km) | 44758 | 45701 | 51188 |
| $\delta a$ (km) | $7.5 \pm 0.2$ | $7.8 \pm 0.3$ | $58.0 \pm 0.4$ |
| *e* | $7.8 \times 10^{-4}$ | $4.3 \times 10^{-4}$ | $7.94 \times 10^{-3}$ |
| $\delta e$ | $5.8 \times 10^{-5}$ | $6.0 \times 10^{-5}$ | $7.4 \times 10^{-4}$ |
| *g* | $0.35 \pm 0.03$ | $0.35 \pm 0.05$ | $0.65 \pm 0.01$ |
| Max. width (km) | 10.1 | 10.6 | 96.3 |
| Min. width (km) | 4.9 | 5.0 | 19.8 |

Note: The data are taken from Elliot & Nicholson (1984).

a uniform rate appropriate to its midpoint. It should be pointed out that the self-gravity model for eccentric rings is separate from the models proposed by Goldreich & Tremaine to account for the confinement of narrow rings. If the self-gravity mechanism works then the eccentricity gradient has to be related to quantities such as the mean width, $\langle W \rangle$, by the equation

$$g \approx +2.3e\langle W \rangle J_2 \left( \frac{R_p}{a} \right)^5 \frac{\rho_p}{\sigma}, \qquad (10.68)$$

where $\rho_p$ is the density of the planet and $\sigma$ is the surface density of the ring (Dermott 1984). This implies that all eccentric rings should have positive eccentricity gradients. In the case of the eccentric rings observed in the saturnian system the derived surface densities are compatible with values obtained independently from analysis of occultation data. However, in the case of the $\epsilon$ ring of Uranus the model predicts a typical particle radius of 20–30 cm whereas the observed value is at least 50 cm and may be as large as several metres (see Nicholson & Dones 1991). Therefore the $\epsilon$ ring may be too massive for the mechanism to work.

Table 10.3 shows that all the values of *g* for the eccentric uranian rings are close to unity. The same is true for most of the eccentric rings in the Saturn system. Yet *g* should be a function of the three presumably independent quantities *e*, $\langle W \rangle$, and $\sigma$. Dermott & Murray (1980) argued that such rings may be close-packed at their pericentres and that this would prevent differential precession. They proposed the *precessional pinch* model for eccentric rings (see Fig. 10.19 for a heuristic description of the mechanism). The model proposes that self-gravitation, particle collisions, and close-packing at pericentre combine to transform a narrow eccentric ring of uniform width into a ring with a large value of *g* and alignment of pericentres. If *a*, $\langle W \rangle$, and $\sigma$ are not independent quantities then it may not be necessary to invoke the precessional pinch mechanism. Work by Borderies et al. (1983) suggests that there is some coupling between these parameters with

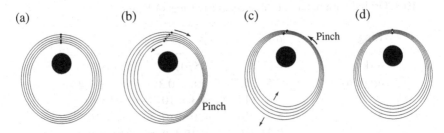

Fig. 10.19. The precessional pinch model for uniform precession of narrow eccentric rings (adapted from Dermott & Murray 1980). In each diagram the pericentres are denoted by small filled circles, with a small open circle for the pericentre of the central ellipse. (a) Initially the ring has aligned pericentres and uniform width. (b) Differential precession results in nonuniform width and the formation of a "pinch" at $f \approx -\pi/2$. (c) Self-gravity of the ring now acts to increase the eccentricity gradient $g$ (radial arrows) and movement of the pinch. (d) The stable equilibrium configuration is with the pinch located at the pericentre and the orbits aligned.

their evolution only ceasing when the close-packing of the particles at pericentre halts the growth of the mean eccentricity. However, the $\epsilon$ ring of Uranus still poses serious dynamical problems.

The narrowness and high surface density of the uranian rings give rise to the phenomenon of *normal modes*, whereby the self-gravity of the ring leads to the creation of density waves of arbitrary values of $m$ (Borderies et al. 1985). The excited waves become unstable but the existence of sharp ring edges can cause the waves to become self-sustaining and even detectable. The behaviour of these modes is identical to that described in Sect. 10.3.2 for streamlines at inner Lindblad resonances with the $m = 0$, $m = 1$, and $m = 2$ modes corresponding to streamlines with the shapes of a circle, a keplerian ellipse, and a centred ellipse respectively. There is also a pattern speed for the modes given by

$$m\Omega_p = (m - 1)n + \dot{\varpi}, \qquad (10.69)$$

which is identical to Eq. (10.12) for an ILR. However, in the case of normal modes there is no perturbing satellite even though there is a pattern speed. Indeed an eccentric, uniformly precessing ring maintained by self-gravity or the precessional pinch model exhibits an $m = 1$ mode; distortions with $m \neq 1$ should only be expected in the narrowest and densest of rings (French et al. 1991). There is now convincing evidence that the $\delta$ ring of Uranus exhibits an $m = 2$ distortion of amplitude $ae = 3.11$ km superimposed on the $m = 1$ mode. Furthermore, the $\gamma$ ring exhibits an $m = 0$ mode with an amplitude of 5.15 km (French et al. 1988, 1991).

Saturn's ring system contains at least seven examples of eccentric rings. Perhaps the most unusual is the Titan ringlet at $1.2908 R_S$ ($1 R_S = 60,330$ km) in the inner part of the C ring (Porco et al. 1984b). The ring lies in the outer half of a 184 km wide gap and has been observed to be a uniformly precessing keplerian ellipse with $\delta a = 25 \pm 3$ km and $\delta e = (1.4 \pm 0.4) \times 10^{-4}$. The measured precession rate is $\dot{\varpi} = (22.57 \pm 0.06)°\text{d}^{-1}$ and this is close to the mean motion of Titan ($22.57698°\text{d}^{-1}$). This suggests the unusual resonance argument

$$\varphi = \lambda' - \varpi \tag{10.70}$$

derived from 4D1.1 with $j = 1$ (see Appendix B), where $\lambda'$ is the mean longitude of Titan. Observations show that the ringlet's pericentre leads Titan's mean longitude by $13° \pm 5°$; this may be a result of energy dissipation (see discussion in Sect. 10.5.3 above). Because the ring is so close to Saturn it is very sensitive to the effects of the planet's zonal harmonics. This property was used by Nicholson & Porco (1988) to derive new values for $J_2$, $J_4$, and $J_6$ and place constraints on $J_8$, $J_{10}$, $J_{12}$, $J_{14}$, and $J_{16}$. The ring's eccentricity is consistent with it being forced by the Titan 1:0 resonance.

There are three other eccentric rings in Saturn's C ring: the Maxwell ringlet at $1.45 R_S$ (see Porco et al. 1984b) and rings at $1.470 R_S$ and $1.495 R_S$ (see Fig. 10.20). In each case the ring is located in a clear gap in the surrounding ring material. The Maxwell ringlet has a width of $\sim 64$ km and lies in the 268 km wide Maxwell gap. The ring appears to be freely precessing and there is no known resonance with a satellite. The ring at $1.470 R_S$ has a mean width of $\sim 16$ km and lies in a gap that is only $\sim 30$ km wide. Similarly the ring at $1.495 R_S$ has a mean width of $\sim 61$ km and the gap at its outer edge is only $\sim 15$ km (Porco & Nicholson 1987). The ring properties are summarised in Table 10.3.

The three remaining eccentric rings are all contained within the Cassini division. Of these the most interesting is the Huygens ringlet at $1.95 R_S$; although it is thought to be freely precessing, its behaviour still poses significant problems (Porco 1983). None of the seven eccentric rings at Saturn is known to have edges maintained by shepherding satellites.

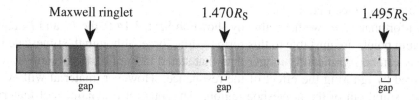

Fig. 10.20. A *Voyager 2* image of part of Saturn's C ring with the location of three gaps and eccentric rings indicated. *(Image courtesy of NASA/JPL.)*

Table 10.3. Orbital parameters of four eccentric rings in Saturn's C ring.

| Ring | Titan | Maxwell | $1.470R_S$ | $1.495R_S$ |
|---|---|---|---|---|
| $a$ (km) | 77871 | 87491 | 88716 | 90171 |
| $\delta a$ (km) | 25 | 64 | 16 | 61 |
| $e$ | $2.6 \times 10^{-4}$ | $3.4 \times 10^{-4}$ | $2.3 \times 10^{-5}$ | $3.1 \times 10^{-5}$ |
| $\delta e$ | $1.4 \times 10^{-4}$ | $3.4 \times 10^{-4}$ | $4.4 \times 10^{-5}$ | $0.9 \times 10^{-5}$ |
| $g$ | 0.44 | 0.46 | 0.24 | 0.01 |
| Max. width (km) | 37 | 88 | 20 | 62 |
| Min. width (km) | 13 | 40 | 18 | 60 |

Note: The data are taken from Porco et al. (1984b) and Porco & Nicholson (1987).

### 10.5.5 Embedded Satellites and Horseshoe Orbits

The discovery of the narrow rings of Uranus by Elliot et al. (1977) led to the development of several theories for the confinement of narrow rings against the effects of collisions and PR drag. The shepherding satellite mechanism discussed above has proved to be the most likely explanation for maintaining narrow rings, especially since the *Voyager 2* observations of Cordelia and Ophelia orbiting on either side of the $\epsilon$ ring of Uranus.

Dermott & Gold (1977) originally proposed that the rings of Uranus were located at the sites of three-body resonances with the known satellites. However, Goldreich & Nicholson (1977b) pointed out that Dermott & Gold had neglected three-body resonances with Miranda; although Miranda is smaller than the other four major satellites, it is closer to the rings and produces stronger resonances.

A new theory for the uranian rings was proposed by Dermott et al. (1979) and subsequently extended to other ring systems by Dermott et al. (1980). They suggested that each narrow ring was maintained by a small satellite $(m'/m_p) < 10^{-9}$ orbiting within the ring such that the ring particles were undergoing the type of horseshoe libration in the 1:1 resonance (see Sect. 3.8). Because very small mass ratios lead to orbits that are highly symmetric with respect to the satellite's orbit (see below), the orbital decay of particles under PR drag in the outer half of the orbit would be almost exactly cancelled by that experienced in the inner half (see Fig. 3.31).

Unfortunately, as we have already shown in Sect. 3.14.6, the $L_4$ and $L_5$ equilibrium points are unstable to this form of drag force, although the timescale for decay can be significantly longer than that stated in Eq. (10.41) because it was derived by ignoring the effect of the resonance. However, material will eventually spiral out of the horseshoe region. The long-term dynamics of material in horseshoe and tadpole orbits is not well understood, even in the absence of drag forces. The Jacobi constant and the existence of forbidden regions in the circular restricted problem can provide some limits to the motion of the particle, but usually there are insufficient constraints on long-term evolution.

We can understand some of these properties by returning to our discussion of the three-body problem. In Chapter 3 we demonstrated that as the mass ratio decreases the critical zero-velocity curve associated with horseshoe orbits passes through the $L_1$ and $L_2$ equilibrium points. Therefore, for motion in the horseshoe regime we can write the Jacobi constant as

$$C = 3 + \alpha \left( \frac{m'}{m_p} \right)^{2/3},$$ (10.71)

where $\alpha \leq 3^{4/3}$ is a constant (cf. Eqs. (3.94) and (3.95)). Dermott & Murray (1981a) have shown that $|\Delta a| = |a' - a|$ can be written as

$$|\Delta a| = 2 \left( \frac{\alpha}{3} \right)^{1/2} \left( \frac{m'}{m_p} \right)^{1/3} a'$$ (10.72)

and that the closest approach of the particle's guiding centre to the satellite is

$$d_{min} = \frac{2}{\alpha} \left( \frac{m'}{m_p} \right)^{1/3} a'.$$ (10.73)

In effect the quantity $\alpha$ is an impact parameter. If $\alpha$ is large ($\alpha > 1$) the particle will either hit the satellite or come close enough to be scattered. For smaller values ($\alpha < 1$) there is an apparent repulsion of the particle by the satellite at their closest approach. The smaller the value of $\alpha$ the more symmetric the path of the guiding centre about the semi-major axis of the satellite. This is well illustrated in Fig. 3.30 for motion in the vicinity of the satellite. The extent of the symmetry was quantified by Dermott et al. (1980) and Dermott & Murray (1981a). Writing

$$\left| \frac{\Delta a_0}{a'} \right| - \left| \frac{\Delta a_j}{a'} \right| = \pm \left( \frac{m'}{m_p} \right)^i,$$ (10.74)

where the subscript $j$ refers to the number of consecutive encounters with the satellite that the particle experiences, they found by numerical integration that for small values of $\alpha$ the orbits are nearly periodic after two encounters with $i \approx 1$. Therefore $\Delta a_0$ and $\Delta a_2$ differ by $\mathcal{O}(m'/m_p)$, implying a high degree of symmetry for small mass ratios. In that case drag forces will have little effect on the actual dynamics at each encounter and this leads to the type of behaviour shown in Fig. 3.31.

Dermott et al. (1980) suggested that the evolution of horseshoe orbits was governed by a random walk in $|\Delta a_0| - |\Delta a_2|$. If $i = 1$ in Eq. (10.74) then this would imply that ring particles (or small satellites) would escape from horseshoe orbits on a timescale

$$T_{escape} \approx \frac{T}{(m'/m_p)^{5/3}},$$ (10.75)

where $T$ is the orbital period of the perturbing satellite. Therefore it is the small satellites that are more likely to have associated coorbital material. Dermott (1984) notes that those satellites for which horseshoe companions are known or thought to exist (the saturnian satellites Janus and Mimas) all have $T_{escape} >$ $5 \times 10^9$ y.

A collisionally evolved system has a characteristic size distribution that ensures that there is a small number of large bodies and a large number of small bodies. The question arises as to how the larger bodies would affect the smaller ones in a disc of material such as an extended ring system? We have already shown how a satellite acts to "repel" ring particles approaching it on near-circular orbits. Therefore the likely result of placing a satellite in a uniform ring of small particles is that it will push away material from orbits that lie immediately interior and exterior to itself. This is the natural consequence of ring shepherding. However, there could still be material in the orbit of the satellite that could be maintained in horseshoe and tadpole orbits. The material need not have been formed at the same time as the satellite; for example, it could have arisen from collisional debris flung from the satellite, or from smaller moons already coorbital with it. In any case, according to the analysis shown above, we would expect such material to be associated with small rather than large satellites. There is now good evidence that at least one moon, the saturnian satellite Pan, is associated with ring material maintained in a 1:1 resonance.

### 10.6 The Encke Gap and Pan

*Voyager* observations of the Encke gap in Saturn's A ring showed the presence of a narrow, kinked ring or ring arc at its centre (see Fig. 10.21a). The 325 km wide gap has remarkably sharp edges and yet it is not associated with any strong satellite resonances. By reprojecting images of the gap edges it was possible to detect edge waves; wake phenomena were also detected in some images, as well as in the occultation data (Cuzzi & Scargle 1985, Showalter et al. 1986). Although *Voyager* did not provide complete longitudinal coverage of the gap the analysis by Cuzzi & Scargle (1985) showed that the amplitude of the wave, when detected, was not constant around the gap. All the evidence pointed to the presence of a small (radius $\sim$ 10 km) satellite orbiting in the gap. The satellite probably created the gap by a shepherding mechanism while encounters with ring particles at the inner and outer edges give rise to edge waves upstream and downstream from the satellite, respectively.

Figure 10.21 illustrates the types of phenomena detectable in the *Voyager* images. The general region of the gap is shown in Fig. 10.21a. The image was taken from above the ring plane and longitude increases from top right to bottom left. Other radial structure in this part of the A ring is caused by resonance with satellites orbiting exterior to the rings. The highlighted region

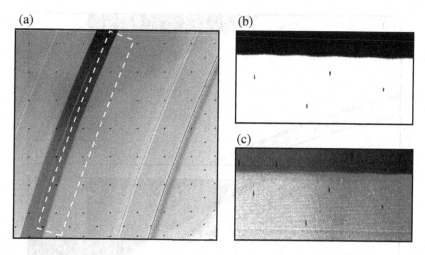

Fig. 10.21. Three views of the Encke gap in Saturn's rings. (a) A *Voyager 2* image taken in August 1981 showing the 325 km gap and a narrow ring close to its centre. *(Image courtesy of NASA/JPL.)* (b) Reprojection of a selected region covering 400 km in radius showing a wave at the inner edge of the gap. (c) The same region as shown in (b) but processed to highlight the wake detectable in the region interior to the edge wave. The image covers a longitude of 3.4° and the observed wavelength is ~ 0.7°. Observations of an edge wave and a wake are indicative of the presence of a small satellite orbiting in the Encke gap.

in Fig. 10.21a at the inner edge of the gap is reprojected with different image enhancements in Figs. 10.21b,c. The presence of an edge wave is clearly visible in Fig. 10.21b while a different enhancement of the same region shows a wake in the surrounding material (Fig. 10.21c). The observed wavelength is ~ 0.7°, corresponding to a satellite separation of ~ 170 km (using Eq. (10.58)). Note that the wake's troughs and peaks are parallel to neither the edge of the ring nor the resonance feature visible closer to the planet; this is consistent with what is to be expected from a wake caused by a satellite (cf. Fig. 10.14).

Direct evidence of the putative satellite, subsequently named Pan after the Greek god of shepherds, was finally detected in *Voyager* images (see Fig. 10.22) by Showalter (1991) who determined a semi-major axis of 133,582.8±0.8 km. The images used by Showalter had low resolution with a typical width for the Encke gap of two or three picture elements, or pixels. Yet the *Voyager* camera was capable of *detecting* although not *resolving* the satellite (see insets in Fig. 10.22). Therefore it was possible to derive an accurate mean motion for Pan by tracking its unresolved detections as the satellite moved in the gap; this in turn led to an accurate determination of the semi-major axis by means of Kepler's third law suitably modified to include the effects of Saturn's oblateness (cf. Eq. (6.244)).

As noted above, the Encke gap also contains an incomplete ring, similar in some respects to the arcs detected in the Adams ring of Neptune (see Sect. 10.8).

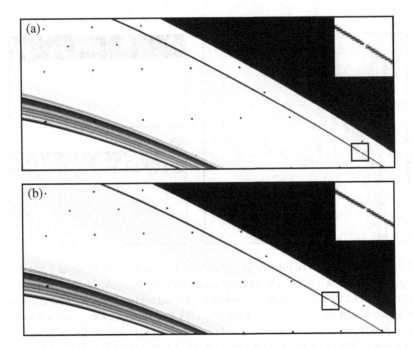

Fig. 10.22.   Two *Voyager 2* images that form part of the discovery sequence of the satellite Pan, showing its movement in the Encke gap in Saturn's A ring.  The time between each image is 5 minutes. *(Images courtesy of NASA/JPL.)*

The coincidence of this ring with the satellite's orbit (Showalter 1991) suggests that Pan is maintaining ring material in horseshoe and tadpole orbits along the lines proposed by Dermott et al. (1979) for the rings of Uranus. The maximum radial width of the ring ($\sim 20$ km) is compatible with the maximum extent of the horseshoe region (30 km) for this satellite (Dermott & Murray 1981a). If this is the case then the Encke gap system exhibits characteristics of both the shepherding and horseshoe orbit models of ring confinement. This is illustrated schematically in Fig. 10.23. Note that the neptunian satellite Galatea also appears to share its orbit with a faint ring (Showalter & Cuzzi 1992) and similar processes may be operating.

From an analysis of *Voyager* images Cooke (1991) has documented evidence for variable width in the 35 km wide Keeler gap at 136,488 km, close to the outer edge of Saturn's A ring. In many respects the Keeler gap is even more puzzling than the Encke gap. If the gap has been produced by an embedded satellite then it must be small, and yet the amplitude of wave features found by Cooke suggests the action of a satellite large enough to have been easily detected. The amplitude of the edge waves appears to be variable around the ring, perhaps implying the presence of more than one satellite. Cooke (1991) points out that there are a number of strong inner Lindblad resonances in the vicinity: the 18:17

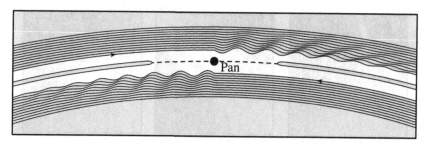

Fig. 10.23. A schematic diagram showing how Pan is responsible for edge waves as it maintains the Encke gap by a shepherding mechanism, while simultaneously keeping coorbital material in horseshoe orbits.

resonance with Pandora at 136,457 km and the 32:31 resonance with Prometheus at 136,481 km (see Fig. 10.5). However, the mechanism responsible for the phenomena at the Keeler gap remains unclear.

## 10.7 The F Ring of Saturn

Saturn's F ring was first detected in low resolution images by the *Pioneer 11* spacecraft (Gehrels et al. 1980). Subsequent observations by *Voyager* (Smith et al. 1981, 1982) provided more detailed images of this narrow ring, which lies 3,400 km beyond the edge of the A ring. Figure 10.24 shows three *Voyager* images of the F ring. Several *Voyager 1* images taken in November 1980 (see, for example, Fig. 10.24a) showed multiple strands with an apparent "braid" formed from two interacting strands. Various other structure in the ring, including apparent discontinuities, or "kinks", in the ring and "clumps" of material that moved with keplerian velocities, was also observed by *Voyager 1*. In contrast the *Voyager 2* images taken nine months later showed parallel strands (see, for example, Figs. 10.24b,c) with evidence for a braid in only one image.

During the passage of Saturn's ring plane through the Earth in 1995 several new features were detected in the region of the F ring using observations from the *Hubble Space Telescope*. Although it was originally thought that some of these features were newly discovered satellites it is now believed that all were clumps or transient features in the F ring and its vicinity (Bosh & Rivkin 1996, Nicholson et al. 1996). Therefore it appears that the F ring's structure is variable on timescales of months.

It is likely that the satellites Pandora and Prometheus play a significant role in confining the F ring and act to produce at least some of the observed structure. The separations in semi-major axis of Prometheus and Pandora from the F ring are $\sim 800$ km and $\sim 1,540$ km respectively. If the satellites have densities of $1.2$ g cm$^{-3}$ then the satellite/planet mass ratios are $1.15 \times 10^{-9}$ and $7.66 \times 10^{-10}$ respectively. Using the data in Table A1.8 and the formulae in Sect. 10.5.2 the expected amplitudes of the waves produced by Prometheus and Pandora are

Fig. 10.24.   Contrast enhanced *Voyager* images of Saturn's narrow F ring. (a) A *Voyager 1* image taken in November 1980 showing the braided appearance of at least two ring components and an inner, fainter component. (b) and (c) Two *Voyager 2* images taken in August 1981 showing multiple strands in the F ring that appear to extend over a longitudinal range of approximately 45°. Pandora is visible at the centre right of (c). The strands cover a radial distance of $\sim$ 300 km. *(Images courtesy of NASA/JPL.)*

$\sim$ 10 km and $\sim$ 2 km respectively. In fact, the amplitude of the wave is a critical function of the longitude of conjunction because both the ring and the satellites are in eccentric orbits. Application of Eq. (10.62) suggests that we would expect shocks formed in the F ring due to a conjunction with Prometheus would survive until the next encounter.

The structure visible in the *Voyager 2* images covers $\sim$ 300 km in radius, probably extends over $\sim$ 45° in longitude, and appears to consist of uniformly precessing, nearly aligned, nonintersecting elliptical strands (see Table 10.4). Murray et al. (1997) suggest that the small offsets in perichrones produce a "pinch" and strand intersections at longitudes examined by *Voyager 1* while at other longitudes the strands would appear parallel. This could account for the differences in the *Voyager 1* and 2 observations. Although a massive F ring core (the F-$\gamma$ strand in the notation of Murray et al. 1997) could maintain near alignment of the surrounding strands, the radial structure cannot be explained by resonances with the known satellites (Murray et al. 1997). Prometheus's resonances overlap in the vicinity of the F ring; Pandora's resonances are weaker and more widely separated but there is no obvious correlation between ring edges and resonances. The locations of the Lindblad and corotation resonances are almost identical for the shepherding satellites. The strongest resonances in the region are due to the coorbital satellites Janus and Epimetheus and they may play some role in determining the location of the innermost $\alpha$ strand. However, there is the added complication of the switch in the coorbital's orbits that occurs every four years.

One possible explanation for the F ring structure is that this region contains several small satellites (radii $\sim$ 5 km) that have acted to clear material from their orbits just as Pan has cleared the Encke gap. There is independent evidence

Table 10.4. *The orbital parameters of the F ring strands, their separations $\Delta a$ from the core (F-$\gamma$) strand, and their approximate radial widths, W. The formal errors in the eccentricity are all $< 0.0001$.*

| Strand | $a$ (km) | $e$ | $\varpi$ (°) | $\Delta a$ (km) | $W$ (km) |
|--------|----------|-----|--------------|-----------------|----------|
| F-$\alpha$ | 140,089±13 | 0.00268 | 242 ± 3 | −130 | 53 ± 5 |
| F-$\beta$ | 140,176±14 | 0.00282 | 238 ± 2 | −43 | 48 ± 5 |
| F-$\gamma$ | 140,219± 4 | 0.00279 | 235 ± 1 | — | 50 ± 5 |
| F-$\delta$ | 140,366±24 | 0.00266 | 240 ± 3 | 149 | 55 ± 5 |

Note: Data are taken from the paper by Murray et al. (1997).

from charged particle data for the existence of a moonlet belt in the vicinity of the F ring (Cuzzi & Burns 1988). Also, from a Fourier analysis of azimuthal structure, Kolvoord et al. (1990) deduced that the presence of a satellite orbiting at a radial distance of 1,180 km from the F ring.

Figure 10.25 shows a plot of the relative configurations of the F ring and the shepherding satellites in August 1981, the time of the *Voyager 2* encounter. It is clear from this plot that even if the eccentricities of their orbits remained constant, the fact that the apocentric distance of Prometheus is close to the pericentric distance of the innermost part of the F ring implies that differential precession could cause intersection of the orbits with the F ring experiencing significant perturbations. This was first noted by Synnott et al. (1983) and discussed further by Borderies et al. (1983). The individual precession rates of

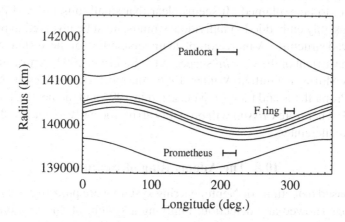

Fig. 10.25. A plot of orbital radius as a function of orbital longitude for Prometheus, Pandora, and the F ring strands at the time of the *Voyager 2* encounter in August 1981. The horizontal bars associated with each plot indicate the $1\sigma$ uncertainty in the longitude of pericentre of each orbit. (Adapted from Murray 1994a.)

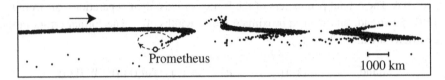

Fig. 10.26. The results of a numerical simulation of the effects of an encounter between Prometheus and the F ring. The path of Prometheus in the frame rotating with its mean motion is shown as a dashed ellipse. The arrow denotes the direction of motion of the particles in the rotating frame. (Adapted from Murray 1994a.)

orbits at this distance from Saturn are large ($\sim 3^\circ$ $d^{-1}$) but the relative precession rates are $\sim 50$ times smaller.

The 1995 ring plane crossing observations showed that Prometheus was lagging behind its expected position by $\sim 19^\circ$. This prompted Murray & Giuliatti Winter (1996) to carry out a more detailed analysis of the secular interactions of the system. They concluded that Prometheus can enter parts of the F ring every 19 years, the relative precession period, in good agreement with the earlier result of Borderies et al. (1983). Furthermore, intersection of the orbits can only be prevented if the F ring has a mass $> 25\%$ that of Prometheus. It is unlikely that a single collision can explain the observed lag in Prometheus's orbit but other factors may be involved.

Giuliatti Winter (1994) has investigated the effect of repeated encounters between F ring particles and Prometheus. A representative plot is shown in Fig. 10.26. As Prometheus approaches the ring it creates a gap as it scatters ring particles.

The F ring remains one of the most intriguing of the narrow rings, with many features yet to be explained. It seems clear that small, unseen satellites, some of them possibly embedded in individual strands, must be involved in producing the observed structure. A more complete understanding of the system may have to await the arrival of the *Cassini* spacecraft at Saturn in 2004. Unfortunately, as pointed by Murray & Giuliatti Winter (1996), the most recent series of encounters between Prometheus and the F ring occurred in 1994 and the next will not occur until 2013. Therefore *Cassini* will probably arrive too early to witness the effects of the next encounters.

## 10.8 The Adams Ring of Neptune

Ground-based detections of Neptune's ring system were puzzling: Only 10% of occultations showed any detection, implying a system of one or more narrow, short arcs of material. Therefore a mechanism was required that would lead to azimuthal as well as radial confinement. Lissauer (1985) proposed that the arcs were located at the $L_4$ or $L_5$ points of a small satellite and therefore had azimuthal confinement; suitable resonances from an additional satellite would act to confine the ring in radius. Goldreich et al. (1986) proposed that the arcs were being maintained by a pair of resonances: a corotation resonance supplying a number of equilibrium points and a Lindblad resonance exciting particle eccentricities and providing energy to the system to counteract the spreading effect of collisions.

In 1989 the images of Neptune's ring system sent back by *Voyager 2* showed that Neptune had an extensive, dusty ring system. The suspected ring arcs were just the optically thicker parts of a complete ring, now called the Adams ring (see Fig. 10.27a). Furthermore, the satellite Galatea was discovered orbiting at a distance of ～ 980 km interior to the ring. Porco (1991) proposed that resonances with Galatea were responsible for the arcs. The Adams ring lies close to a 42:43 commensurability with Galatea. Porco suggested that the arcs were located at some of the 84 equilibrium points of the 42:43 corotation inclination resonance. At such points the resonant argument $\varphi = 86\lambda' - 84\lambda - 2\Omega$ (where $\lambda'$ and $\lambda$ denote the mean longitude of the ring particle and Galatea, and where $\Omega$ is the longitude of ascending node of Galatea) would librate, each with a maximum extent of $360/84 = 4.3°$ and a radial width of 0.6 km. However, only some of these equilibrium sites are occupied, and only some of those are actually filled. The radial confinement is achieved by the existence of the 42:43 outer Lindblad resonance located 1.5 km interior to the Adams ring. Porco calculates that the effect of the forced eccentricity from this resonance produces a 29.6 km radial distortion in the arcs with a characteristic azimuthal wavelength of ～ 9°. There is good evidence for this model in the imaging data (Porco 1991, Porco

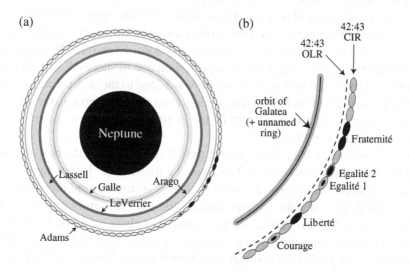

Fig. 10.27. A schematic diagram of the rings of Neptune as viewed from the south pole showing (a) the location and names of the main rings and (b) the theory proposed by Porco (1991) to explain the existence of arcs in the Adams ring as the filled centres of libration of the Galatea 42:43 resonance. (Adapted from Murray 1994a.)

et al. 1995). Work by Horanyi & Porco (1993) has highlighted the importance
of including the variation in the mean longitude at epoch in calculations of the
resonant locations.

## 10.9 The Evolution of Rings

In Sect. 10.5.1 we discussed the timescale problem associated with the spread
of narrow rings and how this led to various theories of ring confinement. We
have already pointed out that optically thin rings, such as the jovian ring system,
have lifetimes as short as $10^3$ y due to plasma drag and gas drag, implying a
replenishing source of material. Similarly the effect of the extended hydrogen
atmosphere at Uranus (Broadfoot et al. 1986) implies collapse of the ring system unless shepherding satellites can counteract the effects of the gas drag and
collisions between the ring particles.

One of the most important dynamical problems concerns the lifetime of Saturn's rings (Borderies et al. 1984, Nicholson & Dones 1991). We have already
seen how the small satellites orbiting at the edge of the main rings raise spiral
density and bending waves. As discussed in Sect. 10.5.2, an encounter between
a test particle and an external satellite in near-circular orbits results in an effective repulsion that decreases the particle's semi-major axis. In the context of the
restricted three-body problem, the (almost) massless ring particle experiences
the biggest effect, but in the full problem angular momentum must be conserved
and therefore the satellite's semi-major axis will increase, albeit by a minute
amount. However, there are a lot of ring particles; indeed the mass of Saturn's
rings is comparable to that of a satellite with a 200 km radius. Therefore, even
if other effects are ignored, the long-term consequence of ring–satellite encounters is to cause the collapse of the entire ring system. Even if the Prometheus–F
ring–Pandora system has reached an equilibrium due to successful shepherding,
the entire system will still be evolving outwards due to the density waves raised
by Prometheus and Pandora in the A ring; even the most optimistic timescale
for the spreading of the outer part of the A ring to its inner edge is $\sim 10^8$ y
(Nicholson & Dones 1991).

Borderies et al. (1984) point out that the entire collapse of the rings could be
avoided if Pandora was in the 3:2 resonance with Mimas. The prospect is tan-
talising since currently the 3:2 resonance lies $\sim 60$ km beyond the semi-major
axis of Pandora, a distance comparable to the longest radius of the satellite.
If Pandora and Mimas were in the resonance then there could be an exchange
of angular momentum with Mimas acting as a sink for the angular momentum
gained by Pandora and the resonance locking between Mimas and Tethys rein-
forcing the stability of Pandora's orbit. Borderies et al. (1984) suggest that if
Pandora was close enough to the chaotic separatrix of the resonance then that
might be sufficient to allow aperiodic transfers of angular momentum and prevent the collapse of the rings over the age of the solar system ($4.5 \times 10^9$ y). It is

intriguing to note that Pandora lies in the middle of six second-order resonances associated with the Mimas 6:4 commensurability (see Fig. 10.28). Therefore, since Pandora and Mimas are both receding from Saturn, it is likely that resonance passage has occurred or will occur at some time in their dynamical evolution.

In principle it is possible to measure the evolution of the orbits of the small satellites. For example, it has been estimated that Prometheus should have an acceleration of $-5.4 \times 10^{-20}\,\mathrm{s}^{-2}$ (Lissauer et al. 1985), implying that its semi-major axis should increase by $\sim 3$ m in the course of the four-year tour of Saturn by the *Cassini* spacecraft. While it may not be possible to detect such a small change in $a$, the increase will cause Prometheus to lag behind its expected position by $\sim 0.02°$ after four years. Provided all the perturbations on the satellite can be modelled, including those due to the major satellites and any mass in the F ring, then measurements of the lag will yield an expansion rate and hence the timescale for collapse of the rings. However, when Prometheus was detected during the ring plane crossing events of 1995, it was found to have a lag that was more than two orders of magnitude larger than that predicted by theory; furthermore the lag did not increase between the May and August and November crossings (Bosh & Rivkin 1996, Nicholson et al. 1996). Murray & Giuliatti Winter (1996) have proposed that collisions with the F ring or interactions with a coorbital satellite may be responsible, although the observations now place tight constraints on any mechanism.

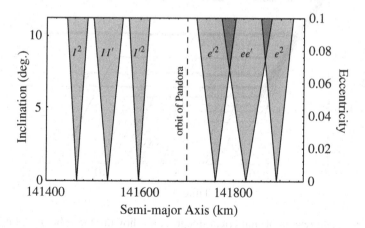

Fig. 10.28.  The locations and widths of the six second-order resonances associated with the Mimas 6:4 resonance and their proximity to the orbit of Pandora (dashed line). Note that the Mimas 3:2 $e$ and $e'$ exact resonances (not shown) are coincident with the location of the $e^2$ and $e'^2$ resonances

## 10.10  The Earth's Dust Ring

In many respects the asteroid belt behaves like a moonlet belt around the Sun. We
have already shown that it is likely that collisions within the belt have produced
the Hirayama families of asteroids, recognised by their clustering of proper
elements (see Sect. 7.10). Some of the dust from the same collisions is also
detectable as the bands detected by the *IRAS* spacecraft (see Sect. 7.11). Any
dust formed in the asteroid belt will spiral in towards the Sun due to the effects
of PR drag. Dermott et al. (1994) studied the orbital evolution of 12 $\mu$m dust
and showed that it can get temporarily trapped in a series of exterior first-order
resonances with the Earth (see Fig. 10.29).

  The dynamical trapping of dust in an exterior resonance is best viewed in a
frame rotating with the mean motion of the perturber. Figure 10.30a shows the
case of a dust particle trapped in the 5:6 Earth resonance without the presence
of PR drag. Note how the particle's path is symmetric about the Sun–Earth
line. When drag is introduced the resulting path is similar in many respects
but the equilibrium points have shifted and the path is no longer symmetrical
with respect to the Sun–Earth line (see Fig. 10.30b). Note that the presence of
"loops" in the region immediately trailing the Earth implies a slowing down of
the particle's motion in the rotating frame. Hence there would be a tendency for
a system of particles in this resonance to cluster at this location.

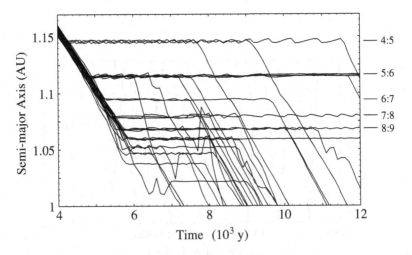

Fig. 10.29. The results of numerical integrations showing the orbital evolution of 12
$\mu$m dust particles spiralling in towards the Sun. As they approach the Earth's external
resonances (the locations of which are indicated at the right-hand side) some particles
get trapped in resonance thereby halting the orbital decay. Most of these particles then
undergo large changes in eccentricity causing them to escape from the resonance.

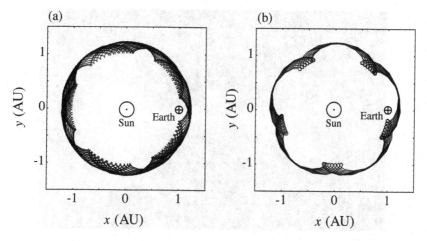

Fig. 10.30. Path of a test particle in a frame rotating with the mean motion of the Earth for two values of $\beta$, the ratio of radiation pressure to gravitational force. (a) The zero drag case ($\beta = 0$) where the path is symmetrical about the Sun–Earth line. (b) The drag case ($\beta = 0.037$) showing that the path is no longer symmetric about the Sun–Earth line.

Similar behaviour occurs with particles trapped in other external resonances with the Earth. The net result is that the spatial density of the dust will be enhanced in the region immediately trailing the Earth and diminished in the region immediately leading the Earth. Dermott et al. (1994) showed this effect by extensive numerical integrations (see Fig. 10.31). However, they also pointed out that there is independent evidence for the phenomenon from infrared measurements made by *IRAS*. These show that the infrared flux in the trailing direction was consistently 3 or 4% brighter than that in the leading direction, independent of the waveband and time of year. Further confirmation of the effect came with the publication of data from the *COBE* satellite (Reach et al. 1995). Therefore it has been confirmed that the Earth lies in the middle of a dust ring, probably composed of material of asteroidal origin.

In the frame rotating with the mean motion of the Earth, the path of the Earth is a small ellipse with axes of length $2ae$ and $4ae$, where $a$ is the semimajor axis of the orbit and $e$ is the eccentricity. Therefore it is likely that the Earth will preferentially enter the trailing dust cloud as it moves between aphelion and perihelion (i.e., from July to January). Furthermore, since the Earth's orbit undergoes perturbations due to the other planets, it is possible that the flux of material entering the Earth's atmosphere varies seasonally and secularly. Although the effects of asteroidal and cometary bombardment on the evolution of life on Earth are now appreciated, little is known so far about the effects of asteroidal dust.

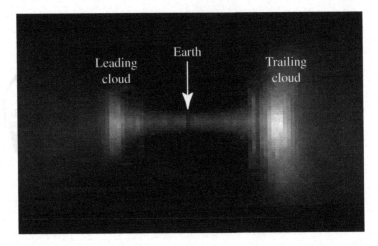

Fig. 10.31.  A representation of the leading–trailing asymmetry in the Earth's dust ring as viewed from the Sun–Earth line. The intensity of each part of the image is a function of the spatial density of the dust particles. The left and right regions denote the leading and trailing regions of the Earth's dust ring.

## Exercise Questions

**10.1**    Assuming that orbital precession is dominated by Saturn's $J_2$, calculate the semi-major axis and maximum radial width (both in km) of Janus's 7:6 inner Lindblad resonance.  If the outer edge of Saturn's A ring is located at a mean radius of 136,774.4 km, calculate the amplitude of the radial distortion of the A ring's edge produced by Janus's 7:6 inner Lindblad resonance.

**10.2**    *Voyager* images and occultation profiles of Saturn's A ring show the presence of a spiral wave feature that propagates close to 132,210 km from the planet's centre (see Fig. 10.13). Identify the satellite and the resonance responsible for this feature.

**10.3**    The saturnian satellite Pan has a semi-major axis of 133,583 km and lies in the middle of the 325 km-wide Encke gap. It is likely that Pan created the gap by the scattering and shepherding of nearby ring material.  If Pan has a radius of 10 km and the half-width of the gap corresponds to two and a half times the distance from the satellite to its inner Lagrangian point, $L_1$, what is the mass and mean density of Pan? For your value of the mass, use Eq. (9.147) to calculate the value of $p$ at which adjacent first-order resonances overlap.  What is the corresponding separation in semi-major axis from Pan? A narrow ring or arc has been observed in the same orbit as Pan. If the ring particles are being maintained in horseshoe orbits by Pan, what is the maximum radial width of the ring? Another ring or arc has been observed between Pan and the inner edge of the Encke gap. Comment on the possibility that this could be material being maintained at an interior corotation resonance.

**10.4** Consider the relative widths, $W_{cr}$ and $W_{Lr}$, of the $j : j - 1$ corotation ($e'$-) and inner Lindblad ($e$-) resonances of a satellite with orbital eccentricity $e'$ moving exterior to a ring system. Use the definitions of the maximum widths given in Sect. 10.3 together with the two appropriate expressions for the direct terms in the disturbing function, $f_d$, to derive an expression for $W_{cr}/W_{Lr}$. Hence, by evaluating the Laplace coefficients using $\alpha = [(j-1)/j]^{2/3}$ for each resonance pair, show that $W_{Lr} > W_{cr}$ except when $j \geq 0.5/e'$.

**10.5** The $\epsilon$ ring of Uranus is thought to be radially confined by the actions of the shepherding satellites Cordelia and Ophelia. By including the effects of precession, use the physical and orbital data given in Tables A.4 and A.11 to determine the locations and widths of all the first-order outer Lindblad resonances (OLRs) of Cordelia and inner Lindblad resonances (ILRs) of Ophelia that have semi-major axes between 51,000 and 51,300 km. If the inner and outer edges of the $\epsilon$ ring are located at 51,149±29 km, show that the nearest resonances to the edges are the 24:25 Cordelia OLR and the 14:13 Ophelia ILR. Use the differences between the locations of these resonances and the ring edges to estimate the amplitude (in km) of the radial distortions at each edge.

**10.6** The "arcs" of optically thicker material in Neptune's Adams ring are believed to be confined radially and azimuthally by resonances with the neptunian satellite Galatea. (a) Use the data in Table A.13 to calculate the semi-major axis of the exact 42:43 corotation inclination resonance (CIR) with resonant argument $86\lambda' - 84\lambda - 2\Omega$, where $\lambda'$, $\lambda$, and $\Omega$ are the mean longitude of a ring particle, the mean longitude, and the longitude of ascending node of Galatea respectively. (b) What is the maximum libration width (in km) of the CIR? (c) Use Lagrange's equations to derive an approximate expression for the shift in the location of the CIR due to the variation in the mean longitude of epoch. (d) Calculate the semi-major axis of Galatea's 42:43 outer Lindblad resonance (OLR). (e) If the Adams ring is located at the semi-major axis of the CIR calculated in (a), what is the amplitude of the wave on the arcs due to the OLR? (f) Sketch a schematic diagram of the arcs showing the effects of both resonances on a section of the ring.

# Appendix A
## Solar System Data

And he that calls on thee, let him bring forth
Eternal numbers to outlive long date.

William Shakespeare, *Sonnet XXXVIII*

## A.1 Introduction

This appendix contains lists of important astronomical constants, information about the use of the Julian date, orbital data, and physical properties of the known planets and satellites, as well as limited information about some of the minor bodies that make up the solar system.

The data are taken from a number of sources including *The Astronomical Almanac for the Year 1995* (HMSO, 1994), *The Explanatory Supplement to the Astronomical Almanac* (Seidelmann, 1992), and the article by Yoder (1995) and the references therein. The data from the first of these publications is reproduced with permission from HMSO. Other sources of data are indicated in the appropriate sections.

## A.2 Astronomical Constants

In 1976 the International Astronomical Union (IAU) defined a system of astronomical constants. The IAU system has units of length (the astronomical unit), mass (the mass of the Sun), and time (the day). If the units of length, mass, and time are the astronomical units of these quantities then the astronomical unit of length is the length for which the Gaussian gravitational constant $k$ has the value 0.01720209895. In effect, if the gravitational constant $\mathcal{G}$ is expressed in

526

Table A.1. Astronomical constants.

| Constant | Value |
|---|---|
| Gaussian gravitational constant | $k = 0.01720209895$ |
| Speed of light | $c = 299792458 \text{ m s}^{-1}$ |
| Constant of gravitation | $\mathcal{G} = 6.672 \times 10^{-11} \text{ m}^3 \text{ kg}^{-1} \text{ s}^{-2}$ |
| Astronomical unit of distance | $1 \text{ AU} = 1.495978707 \times 10^{11} \text{ m}$ |
| Mass of the Sun | $M_{Sun} = 1.98911 \times 10^{30} \text{ kg}$ |
| Radius of the Sun | $R_{Sun} = 6.960 \times 10^5 \text{ km}$ |

the astronomical units of length, mass, and time then $k^2 = \mathcal{G}$. Some of the 1976 IAU constants are given in Table A1.1.

The IAU has also adopted values for other quantities such as the mass ratios of the Sun to the planets, the masses of certain planetary satellites, radii of the planets, and the lower harmonics of the gravity fields of the planets (see HMSO, 1994). Unless otherwise stated, these are the values used in the following tables.

## A.3 Julian Date

To calculate the position of a solar system body at a particular time it is necessary to make use of the calendar system based on a fixed day, the *Julian day,* consisting of 86,400 s. Similarly, the *Julian year* consists of 365.25 Julian days and the *Julian century* has 36,525 Julian days.

The *Julian date* (JD) is the number of Julian days since $12^{\text{h}}$ Universal Time (noon at Greenwich) on 1 January 4713 B.C. The Julian date for any calendar date can be calculated by using the algorithm given by Montenbruck (1989). If $Y$, $M$, $D$, and $UT$ denote the year, month, day, and Universal Time, then the first step is to define the auxiliary quantities $y$ and $m$ using

$$y = Y - 1 \quad \text{and} \quad m = M + 12 \quad \text{if} \quad M \leq 2,$$
$$y = Y \quad \text{and} \quad m = M \quad \text{if} \quad M > 2 \quad \text{(A.1)}$$

and the quantity $B$ using

$$B = -2 \qquad \text{up to and including 4 Oct 1582,}$$
$$B = \text{Int}[y/400] - \text{Int}[y/100] \qquad \text{from and including 15 Oct 1582,} \quad \text{(A.2)}$$

where the function $\text{Int}[x]$ denotes the largest whole number that is smaller than or equal to $x$ (for example, $\text{Int}[3.4] = 3$ and $\text{Int}[-3.4] = -4$). The reasons for the peculiar definition of $B$ are to account for the "lost days" in October 1582, when the Gregorian calendar replaced the Julian calendar in Europe, and to deal with the introduction of a leap day in the Gregorian calendar. The Julian date is then given by

$$\text{JD} = \text{Int}[365.25 \, y] + \text{Int}[30.6001 \, (m + 1)]$$
$$+ B + 1720996.5 + D + UT/24. \quad \text{(A.3)}$$

Note that the integer value of the Julian date is incremented at noon. Care must be taken in dealing with years before A.D. 1 as noted by Montenbruck (1989). The astronomical year $-4712$ corresponds to 4713 B.C. and the year 0 to 1 B.C.

As an example, consider the calculation of the Julian date corresponding to $10^h24^m$ on 4 February 1946. In this case $Y = 1946$, $M = 2$, $D = 4$, and $UT = 10.4$. The auxiliary quantities are $y = 1945$, $m = 14$, and $B = -15$, giving JD 2431855.933.

The reverse procedure (the calculation of a calendar date from a given Julian date) can be accomplished using the algorithm given by Montenbruck (1989). If we define the following auxiliary quantities:

$$a = \text{Int}[\text{JD} + 0.5], \tag{A.4}$$

$$c = a + 1524 \qquad \text{if } a < 2299161 \tag{A.5}$$

$$b = \text{Int}[(a - 1867216.25)/36524.25] \qquad \text{if } a \geq 2299161, \tag{A.6}$$

$$c = a + b - \text{Int}[b/4] + 1525 \qquad \text{if } a \geq 2299161, \tag{A.7}$$

$$d = \text{Int}[(c - 122.1)/365.25], \tag{A.8}$$

$$e = \text{Int}[365.25\, d], \tag{A.9}$$

$$f = \text{Int}[(c - e)/30.6001], \tag{A.10}$$

then the calendar date is given by

$$D = c - e - \text{Int}[30.6001\, f] + \text{Frac}[\text{JD} + 0.5], \tag{A.11}$$

$$M = f - 1 - 12\,\text{Int}[f/14], \tag{A.12}$$

$$Y = d - 4715 - \text{Int}[(7 + M)/10], \tag{A.13}$$

where the function $\text{Frac}[x]$ denotes the fractional part of $x$. For example, $\text{Frac}[3.4] = 0.4$ and $\text{Frac}[-3.4] = -0.4$.

As an example, consider the calculation of the calendar date corresponding to JD 2434903.75. The auxiliary quantities are $a = 2434904$, $b = 15$, $c = 2436441$, $d = 6670$, $e = 2436217$, and $f = 7$, giving $D = 10.25$, $M = 6$, and $Y = 1954$. This corresponds to $6^h$ UT on 10 June 1954.

Because the equatorial and orbital planes of the Earth are not fixed, it is necessary to define the position of solar system bodies with reference to a coordinate system that specifies the mean equinox and ecliptic at a particular date or epoch. One of the standard epochs is referred to as J2000 and it denotes the reference frame at JD 2451545.0 ($12^h$ UT on 1 January 2000). Prior to the adoption of J2000 it was common to use the epoch of B1950, the reference frame at JD 2433282.423 ($22^h9^m$ UT on 31 December 1949).

## A.4 Orbital Elements of the Planets and Their Variation

The orbital elements of the planets change over time owing to their mutual perturbations (see Chapter 7). Tables A.2 and A.3 give the orbital elements of the planets and their variations at the epoch of J2000 (JD 2451545.0) with respect to the mean ecliptic and equinox of J2000. The data are taken from Standish et al. (1992).

To calculate the approximate elements at other times the following formulae are used:

$$a = a_0 + \dot{a}\, t \quad \text{AU}, \tag{A.14}$$

$$e = e_0 + \dot{e}\, t, \tag{A.15}$$

$$I = I_0 + (\dot{I}/3600)\, t \quad \text{degrees}, \tag{A.16}$$

$$\varpi = \varpi_0 + (\dot{\varpi}/3600)\, t \quad \text{degrees}, \tag{A.17}$$

$$\Omega = \Omega_0 + (\dot{\Omega}/3600)\, t \quad \text{degrees}, \tag{A.18}$$

$$\lambda = \lambda_0 + (\dot{\lambda}/3600 + 360 N_r)\, t \quad \text{degrees}, \tag{A.19}$$

where $t$ is the time in Julian centuries since JD 2451545.0. These are approximate formulae that have maximum errors of 600 arcseconds over the interval 1800–2050 (see Standish et al. 1992).

Table A.2. Planetary orbital elements at the epoch of J2000 (JD 2451545.0) with respect to the mean ecliptic and equinox of J2000. The quantities $a_0$, $e_0$, $I_0$, $\varpi_0$, $\Omega_0$, and $\lambda_0$ denote the semi-major axis (in AU), eccentricity, inclination, longitude of perihelion, longitude of ascending node, and mean longitude, respectively, where the angular quantities are in degrees. The data for the Earth are actually for the Earth–Moon barycentre.

| Planet | $a_0$ (AU) | $e_0$ | $I_0$ (°) | $\varpi_0$ (°) | $\Omega_0$ (°) | $\lambda_0$ (°) |
|---|---|---|---|---|---|---|
| Mercury | 0.38709893 | 0.20563069 | 7.00487 | 77.45645 | 48.33167 | 252.25084 |
| Venus | 0.72333199 | 0.00677323 | 3.39471 | 131.53298 | 76.68069 | 181.97973 |
| Earth | 1.00000011 | 0.01671022 | 0.00005 | 102.94719 | 348.73936 | 100.46435 |
| Mars | 1.52366231 | 0.09341233 | 1.85061 | 336.04084 | 49.57854 | 355.45332 |
| Jupiter | 5.20336301 | 0.04839266 | 1.30530 | 14.75385 | 100.55615 | 34.40438 |
| Saturn | 9.53707032 | 0.05415060 | 2.48446 | 92.43194 | 113.71504 | 49.94432 |
| Uranus | 19.19126393 | 0.04716771 | 0.76986 | 170.96424 | 74.22988 | 313.23218 |
| Neptune | 30.06896348 | 0.00858587 | 1.76917 | 44.97135 | 131.72169 | 304.88003 |
| Pluto | 39.48168677 | 0.24880766 | 17.14175 | 224.06676 | 110.30347 | 238.92881 |

*Appendix A*

Table A.3. Rates of change of planetary orbital elements at the epoch of J2000 (JD 2451545.0) with respect to the mean ecliptic and equinox of J2000. The quantities $\dot{a}$, $\dot{e}$, $\dot{I}$, $\dot{\varpi}$, $\dot{\Omega}$, and $\dot{\lambda}$ denote the change per century of the semi-major axis ($\times 10^8$), eccentricity ($\times 10^8$), inclination, longitude of perihelion, longitude of ascending node, and mean longitude, respectively, where the angular rates are measured in arcseconds per century ($1° = 3,600$ arcsecond). The quantity $N_r$ is used in determining the mean longitude (see Eq. (A.19)). The data for the Earth are for the Earth–Moon barycentre.

| Planet | $\dot{a}_0$ | $\dot{e}_0$ | $\dot{I}_0$ | $\dot{\varpi}_0$ | $\dot{\Omega}_0$ | $\dot{\lambda}_0$ | $N_r$ |
|---|---|---|---|---|---|---|---|
| Mercury | 66 | 2527 | −23.51 | 573.57 | −446.30 | 261628.29 | 415 |
| Venus | 92 | −4938 | −2.86 | −108.80 | −996.89 | 712136.06 | 162 |
| Earth | −5 | −3804 | −46.94 | 1198.28 | −18228.25 | 1293740.63 | 99 |
| Mars | −7221 | 11902 | −25.47 | 1560.78 | −1020.19 | 217103.78 | 53 |
| Jupiter | 60737 | −12880 | −4.15 | 839.93 | 1217.17 | 557078.35 | 8 |
| Saturn | −301530 | −36762 | 6.11 | −1948.89 | −1591.05 | 513052.95 | 3 |
| Uranus | 152025 | −19150 | −2.09 | 1312.56 | 1681.40 | 246547.79 | 1 |
| Neptune | −125196 | 2514 | −3.64 | −844.43 | −151.25 | 786449.21 | 0 |
| Pluto | −76912 | 6465 | 11.07 | −132.25 | −37.33 | 522747.90 | 0 |

As an example, consider the calculation of the orbital elements of Venus in the J2000 frame at $11^h$ British Summer Time (one hour ahead of Greenwich Mean Time) on 19 October 1996. The corresponding Julian date is JD 2450375.91667, giving $t = -0.03200775$. Using Tables A.4 and A.5 together with the formulae in Eqs. (A.14)–(A.19) gives $a = 0.723332$ AU, $e = 0.00677481$, $I = 3.39474°$, $\varpi = 131.534°$, $\Omega = 76.6896°$, and $\lambda = 108.956°$.

Note that Tables A.2 and A.3 have a limited range of validity and cannot be used to calculate the orbital elements of planets at remote epochs. The existence of chaotic motion (see Chapter 9), especially among the inner planets, places fundamental limits on the ability to predict the future state of the solar system.

## A.5 Planets, Satellites, and Rings

In Table A.4 we list the physical properties of each of the planets. Data are taken from Yoder (1995). The quoted value of $R$ in Table A.4 is the reference radius used in the determination of $J_2$ and $J_4$. This is the value of the planetary radius to be used in calculations of precession due to oblateness.

In Tables A.5–A.15 we list data relating to the satellites and, where appropriate, ring systems of each planet. In these tables $a$ is the semi-major axis, $T$ is the sidereal orbital period (with a minus sign denoting retrograde orbital motion), $e$ is the eccentricity, $I$ is the inclination of the satellite's orbit to the planet's equator (except where noted), $\langle R \rangle$ is the mean radius in km (with a preceding $*$ denoting that it is an irregularly shaped body), and $m$ is the mass of the satellite. Values of $e$ and $I$ preceded by the letter "f" indicate that the value is a forced quantity, usually due to a resonance. Values of $I$ preceded by the letter "e" indicate that the inclination is with respect to the ecliptic. The letter "c" after a satellite name denotes that it is a coorbital object. For rings $\tau$ denotes the optical depth of the ring.

Most of the data are taken from the tables contained in Yoder (1995) with minor corrections. Additional sources of data are noted in the table footnotes.

Table A.4. Physical properties of the planets.

| Planet | $m_p$ ($10^{24}$kg) | $R$ (km) | $T_{rot}$ (h) | $\langle\rho\rangle$ | $\epsilon$ (°) | $J_2$ | $J_4$ |
|---|---|---|---|---|---|---|---|
| Mercury | 0.3302 | 2,440 | 1047.51 | 5.427 | ~ 0.1 | 60 | |
| Venus | 4.8685 | 6,052 | −5832.444 | 5.204 | 177.3 | 4 | 2 |
| Earth | 5.9736 | 6,378 | 23.93419 | 5.515 | 23.45 | 1083 | −2 |
| Mars | 0.64185 | 3,394 | 24.622962 | 3.933 | 25.19 | 1960 | −19 |
| Jupiter | 1898.6 | 71,398 | 9.92425 | 1.326 | 3.12 | 14736 | −587 |
| Saturn | 568.46 | 60,330 | 10.65622 | 0.6873 | 26.73 | 16298 | −915 |
| Uranus | 86.832 | 26,200 | −17.24 | 1.318 | 97.86 | 3343 | −29 |
| Neptune | 102.43 | 25,225 | 16.11 | 1.638 | 29.56 | 3411 | −35 |
| Pluto | 0.0127 | 1,137 | −153.29 | 2.06 | 112.52 | | |

Notes: $m_p$ is the mass of the planet, $R$ is its reference radius, $T_{rot}$ is the sidereal rotational period (referred to a fixed frame of reference) with a minus sign denoting retrograde rotation, $\langle\rho\rangle$ is the mean density (in g cm$^{-3}$), $\epsilon$ is the obliquity, and $J_2$ ($\times10^{-6}$) and $J_4$ ($\times10^{-6}$) denote the first two even gravitational harmonics.

Table A.5. The satellite of Earth.

| Satellite | $a$ (km) | $T$ (d) | $e$ | $I$ (°) | $\langle R\rangle$ | $m$ ($10^{20}$kg) |
|---|---|---|---|---|---|---|
| Moon | 384,400 | 27.321661 | 0.054900 | 5.15 | 1737.53 | 734.9 |

Table A.6. The satellites of Mars.

| Satellite | $a$ (km) | $T$ (d) | $e$ | $I$ (°) | $\langle R\rangle$ | $m$ ($10^{15}$kg) |
|---|---|---|---|---|---|---|
| Phobos | 9,377.2 | 0.318910 | 0.0151 | 1.082 | *11 | 10.8 |
| Deimos | 23,463.2 | 1.262441 | 0.00033 | 1.791 | *6 | 1.8 |

Table A.7. The satellites of Jupiter.

| Satellite | $a$ (km) | $T$ (d) | $e$ | $I$ (°) | $\langle R \rangle$ | $m$ ($10^{20}$kg) |
|---|---|---|---|---|---|---|
| Metis | 127,979 | 0.294780 | < 0.004 | ~ 0 | 20 | |
| Adrastea | 128,980 | 0.29826 | ~ 0 | ~ 0 | 10 | |
| Amalthea | 181,300 | 0.498179 | 0.003 | 0.40 | *86 | |
| Thebe | 221,900 | 0.6745 | 0.015 | 0.8 | 50 | |
| Io | 421,600 | 1.769138 | [f]0.0041 | 0.040 | 1,821 | 893.3 |
| Europa | 670,900 | 3.551810 | [f]0.0101 | 0.470 | 1,565 | 479.7 |
| Ganymede | 1,070,000 | 7.154553 | [f]0.0015 | 0.195 | 2,634 | 1482 |
| Callisto | 1,883,000 | 16.689018 | 0.007 | 0.281 | 2,403 | 1076 |
| Leda | 11,094,000 | 238.72 | 0.148 | 27 | 5 | |
| Himalia | 11,480,000 | 250.5662 | 0.163 | 25 | 85 | |
| Lysithea | 11,720,000 | 259.22 | 0.107 | 29 | 12 | |
| Elara | 11,737,000 | 259.6528 | 0.207 | 28 | 40 | |
| Ananke | 21,200,000 | −631 | 0.169 | 147 | 10 | |
| Carme | 22,600,000 | −692 | 0.207 | 163 | 15 | |
| Pasiphae | 23,500,000 | −735 | 0.378 | 148 | 18 | |
| Sinope | 23,700,000 | −758 | 0.275 | 153 | 14 | |

Note: The semi-major axis of Metis is taken from Nicholson & Matthews (1991).

Table A.8. The ring system of Jupiter.

| Ring | Inner Edge (km) | Outer Edge (km) | $\tau$ |
|---|---|---|---|
| halo | 89,400 | 123,000 | $3 \times 10^{-6}$ |
| main | 123,000 | 128,940 | $5 \times 10^{-6}$ |
| gossamer | 128,940 | 242,000 | $1 \times 10^{-7}$ |

Table A.9. The satellites of Saturn.

| Satellite | $a$ (km) | $T$ (d) | $e$ | $I$ (°) | $\langle R \rangle$ | $m$ ($10^{20}$kg) |
|---|---|---|---|---|---|---|
| Pan | 133,583 | 0.5750 | ~0 | ~0 | 10 | |
| Atlas | 137,640 | 0.6019 | ~0 | ~0 | *16 | |
| Prometheus | 139,350 | 0.612986 | 0.0024 | 0.0 | *50 | 0.0014 |
| Pandora | 141,700 | 0.628804 | 0.0042 | 0.0 | *42 | 0.0013 |
| Epimetheus[c] | 151,422 | 0.694590 | 0.009 | 0.34 | *59 | 0.0055 |
| Janus[c] | 151,472 | 0.694590 | 0.007 | 0.14 | *89 | 0.0198 |
| Mimas | 185,520 | 0.9424218 | 0.0202 | [f]1.53 | 199 | 0.385 |
| Enceladus | 238,020 | 1.370218 | [f]0.0045 | 0.02 | 249 | 0.73 |
| Tethys | 294,660 | 1.887802 | 0.0000 | [f]1.09 | 530 | 6.22 |
| Telesto[c] | 294,660 | 1.887802 | ~0 | ~0 | *11 | |
| Calypso[c] | 294,660 | 1.887802 | ~0 | ~0 | *10 | |
| Dione | 377,400 | 2.736915 | [f]0.0022 | 0.02 | 560 | 10.52 |
| Helene[c] | 377,400 | 2.736915 | 0.005 | 0.2 | *16 | |
| Rhea | 527,040 | 4.517500 | [f]0.0010 | 0.35 | 764 | 23.1 |
| Titan | 1,221,850 | 15.945421 | 0.0292 | 0.33 | 2,575 | 1345.5 |
| Hyperion | 1,481,100 | 21.276609 | [f]0.1042 | 0.43 | *143 | |
| Iapetus | 3,561,300 | 79.330183 | 0.0283 | 7.52 | 718 | 15.9 |
| Phoebe | 12,952,000 | −550.48 | 0.163 | 175.3 | 110 | |

Table A.10. The ring system of Saturn.

| Ring | $a$ (km) | $W$ (km) | $e$ | $\tau$ |
|---|---|---|---|---|
| D (inner edge) | 66,900 | 7,758 | | |
| C (inner edge) | 74,658 | 17,317 | | 0.05–0.35 |
| Titan ringlet | 77,871 | 25 | 0.00026 | |
| Maxwell ringlet | 87,491 | 64 | 0.00034 | |
| 1.470 $R_S$ ringlet | 88,716 | 16 | 0.000023 | |
| 1.495 $R_S$ ringlet | 90,171 | 61 | 0.000031 | |
| B (inner edge) | 91,975 | 25,532 | | 0.4–2.5 |
| B (outer edge) | 117,507 | | | |
| Huygens ringlet | 117,825 | 360 | 0.00040 | |
| A (inner edge) | 122,340 | 14,440 | | 0.4–1.0 |
| Encke ringlet | 133,589 | 20 | | |
| A (outer edge) | 136,780 | | | |
| F (core) | 140,219 | 50 | 0.0028 | 0.1 |
| G (inner edge) | 166,000 | 7,200 | | $1 \times 10^{-6}$ |
| E (inner edge) | 180,000 | 300,000 | | $1.5 \times 10^{-5}$ |

Note: Additional data are from Porco & Nicholson (1987) and Murray et al. (1997).

Table A.11. The satellites of Uranus.

| Satellite | $a$ (km) | $T$ (d) | $e$ | $I$ (°) | $\langle R \rangle$ | $m$ ($10^{20}$kg) |
|---|---|---|---|---|---|---|
| Cordelia | 49,752 | 0.3350331 | 0.000 | 0.1 | 13 | |
| Ophelia | 53,764 | 0.3764089 | 0.010 | 0.1 | 16 | |
| Bianca | 59,165 | 0.4345772 | 0.001 | 0.2 | 22 | |
| Cressida | 61,767 | 0.4635700 | 0.000 | 0.0 | 33 | |
| Desdemona | 62,659 | 0.4736510 | 0.000 | 0.2 | 29 | |
| Juliet | 64,358 | 0.4930660 | 0.001 | 0.1 | 42 | |
| Portia | 66,097 | 0.5131958 | 0.000 | 0.1 | 55 | |
| Rosalind | 69,927 | 0.5584589 | 0.000 | 0.3 | 29 | |
| Belinda | 75,255 | 0.6235248 | 0.000 | 0.0 | 34 | |
| Puck | 86,004 | 0.7618321 | 0.000 | 0.3 | 77 | |
| Miranda | 129,800 | 1.413 | 0.0027 | 4.22 | *235 | 0.659 |
| Ariel | 191,200 | 2.520 | 0.0034 | 0.31 | *579 | 13.53 |
| Umbriel | 266,000 | 4.144 | 0.0050 | 0.36 | 585 | 11.72 |
| Titania | 435,800 | 8.706 | 0.0022 | 0.10 | 789 | 35.27 |
| Oberon | 583,600 | 13.463 | 0.0008 | 0.10 | 761 | 30.14 |
| Caliban | 7,169,000 | −580 | 0.0823 | e139.67 | 40 | |
| Sycorax | 12,213,000 | −1289 | 0.5091 | e152.67 | 80 | |

Note: Data on Cordelia, Ophelia, Bianca, Cressida, Desdemona, Juliet, Portia, Rosalind, Belinda, and Puck are from Owen & Synnott (1987). Data on Caliban and Sycorax are from Marsden & Williams (1997).

Table A.12. The ring system of Uranus.

| Ring | $a$ (km) | $W$ (km) | $e$ | $I$ (°) | $\tau$ |
|---|---|---|---|---|---|
| 6 | 41,837 | ~ 1.5 | 0.00101 | 0.062 | ~ 0.3 |
| 5 | 42,234 | ~ 2 | 0.00190 | 0.054 | ~ 0.5 |
| 4 | 42,571 | ~ 2.5 | 0.00106 | 0.032 | ~ 0.3 |
| $\alpha$ | 44,718 | 4 → 10 | 0.00076 | 0.015 | ~ 0.4 |
| $\beta$ | 45,661 | 5 → 11 | 0.00044 | 0.005 | ~ 0.3 |
| $\eta$ | 47,176 | 1.6 | | | ≤ 0.4 |
| $\gamma$ | 47,627 | 1 → 4 | 0.00109 | 0.000 | ≥ 1.5 |
| $\delta$ | 48,300 | 3 → 7 | 0.00004 | 0.001 | ~ 0.5 |
| $\lambda$ | 50,024 | ~ 2 | 0.0 | 0.0 | ~ 0.1 |
| $\epsilon$ | 51,149 | 20 → 96 | 0.00794 | 0.000 | 0.5–2.3 |

Note: Data are from French et al. (1991). Note that the $\delta$ and $\gamma$ rings exhibit additional normal modes that affect their shape (see Sect. 10.5.4).

Table A.13. The satellites of Neptune.

| Satellite | $a$ (km) | $T$ (d) | $e$ | $I$ (°) | $\langle R \rangle$ | $m$ ($10^{20}$kg) |
|-----------|----------|---------|-----|---------|---------------------|-------------------|
| Naiad | 48,227 | 0.294396 | 0.000 | 4.74 | 29 | |
| Thalassa | 50,075 | 0.311485 | 0.000 | 0.21 | 40 | |
| Despina | 52,526 | 0.334655 | 0.000 | 0.07 | 74 | |
| Galatea | 61,953 | 0.428745 | 0.000 | 0.05 | 79 | |
| Larissa | 73,548 | 0.554654 | 0.000 | 0.20 | *94 | |
| Proteus | 117,647 | 1.122315 | 0.000 | 0.55 | *209 | |
| Triton | 354,760 | −5.876854 | 0.0004 | 156.834 | 1353 | 215 |
| Nereid | 5,513,400 | 360.13619 | 0.7512 | 7.23 | 170 | |

Note: Data on Naiad, Thalassa, Despina, Galatea, Larissa, and Proteus are from Owen et al. (1991).

Table A.14. The ring system of Neptune.

| Ring | $a$ (km) | $W$ (km) | $\tau$ |
|------|----------|----------|--------|
| Galle | 42,000 | 2,000 | $\sim 1 \times 10^{-4}$ |
| Le Verrier | 53,200 | <100 | 0.01 |
| Lassell (inner edge) | 53,200 | 4,000 | $\sim 1 \times 10^{-4}$ |
| Arago | 57,200 | | |
| [unnamed] | 61,953 | | |
| Adams | 62,932.57 | 15 | 0.01–0.1 |

Note: Data are from Porco et al. (1995).

Table A.15. The satellite of Pluto.

| Satellite | $a$ (km) | $T$ (d) | $e$ | $I$ (°) | $\langle R \rangle$ | $m$ ($10^{20}$kg) |
|-----------|----------|---------|-----|---------|---------------------|-------------------|
| Charon | 19,636 | 6.387223 | 0.0076 | 96.163 | 593 | 15 |

Note: Orbital data are from Tholen & Buie (1997). Note that Charon's inclination is measured with respect to the mean equator and equinox of J2000.

## A.6 Asteroids, Centaurs, Trans-Neptunian Objects, and Comets

Orbital and physical data given Tables A.16–A.18 are from the files maintained by Ted Bowell at *ftp://ftp.lowell.edu/pub/elgb/astorb.html*. The orbital data on the comets listed in Table A.19 were compiled from data maintained by JPL at *http://ssd.jpl.nasa.gov.*

*Appendix A*

Table A.16. The osculating orbital elements and diameters for a selection of asteroids.

| Asteroid | $a$ (AU) | $e$ | $I$ (°) | $\omega$ (°) | $\Omega$ (°) | $M$ (°) | $D$ (km) |
|---|---|---|---|---|---|---|---|
| (1) Ceres | 2.7678 | 0.0774 | 10.581 | 73.460 | 80.530 | 207.082 | 913 |
| (2) Pallas | 2.7739 | 0.2323 | 34.823 | 309.824 | 173.237 | 194.832 | 523 |
| (4) Vesta | 2.3607 | 0.0904 | 7.135 | 149.854 | 103.962 | 138.500 | 501 |
| (8) Flora | 2.2013 | 0.1561 | 5.886 | 284.974 | 111.057 | 1.473 | 141 |
| (24) Themis | 3.1284 | 0.1343 | 0.762 | 109.433 | 36.084 | 45.687 | — |
| (153) Hilda | 3.9791 | 0.1435 | 7.834 | 42.908 | 228.436 | 281.451 | 175 |
| (158) Koronis | 2.8695 | 0.0538 | 1.004 | 145.872 | 278.735 | 71.881 | 39.8 |
| (221) Eos | 3.0140 | 0.0978 | 10.861 | 195.404 | 142.200 | 222.137 | 110 |
| (243) Ida | 2.8619 | 0.0450 | 1.137 | 112.419 | 324.419 | 92.511 | 32.5 |
| (253) Mathilde | 2.6455 | 0.2661 | 6.709 | 156.214 | 179.869 | 55.377 | 61 |
| (279) Thule | 4.2851 | 0.0104 | 2.338 | 82.036 | 73.687 | 346.609 | 135 |
| (433) Eros | 1.4583 | 0.2230 | 10.831 | 178.642 | 304.435 | 0.987 | — |
| (588) Achilles | 5.1840 | 0.1485 | 10.328 | 131.702 | 316.601 | 309.872 | 147 |
| (617) Patroclus | 5.2300 | 0.1390 | 22.048 | 306.843 | 44.410 | 279.573 | 149 |
| (887) Alinda | 2.4859 | 0.5627 | 9.303 | 350.067 | 110.721 | 27.106 | — |
| (944) Hidalgo | 5.7613 | 0.6587 | 42.552 | 56.348 | 21.672 | 175.102 | — |
| (951) Gaspra | 2.2092 | 0.1737 | 4.102 | 129.454 | 253.360 | 232.902 | 15.5 |
| (1362) Griqua | 3.2264 | 0.3681 | 24.181 | 262.339 | 121.452 | 193.286 | 31.1 |
| (1566) Icarus | 1.0778 | 0.8269 | 22.870 | 31.251 | 88.131 | 170.730 | — |
| (1620) Geographos | 1.2455 | 0.3354 | 13.341 | 276.756 | 337.352 | 180.595 | — |
| (1862) Apollo | 1.4712 | 0.5600 | 6.356 | 285.670 | 35.908 | 236.685 | — |
| (1915) Quetzalcoatl | 2.5373 | 0.5739 | 20.448 | 347.879 | 163.053 | 61.568 | — |
| (2703) Rodari | 2.1936 | 0.0573 | 6.036 | 170.372 | 49.593 | 258.351 | — |
| (3200) Phaethon | 1.2714 | 0.8900 | 22.101 | 321.823 | 265.585 | 350.681 | 5.2 |
| (3352) McAuliffe | 1.8794 | 0.3689 | 4.774 | 15.710 | 107.497 | 198.128 | — |
| (3647) Dermott | 2.8005 | 0.1030 | 8.043 | 218.687 | 126.470 | 322.421 | — |
| (3753) Cruithne | 0.9978 | 0.5149 | 19.812 | 43.651 | 126.367 | 336.711 | — |
| (3840) Mimistrobell | 2.2491 | 0.0831 | 3.922 | 167.440 | 42.424 | 193.648 | — |
| (4660) Nereus | 1.4897 | 0.3603 | 1.424 | 157.908 | 314.785 | 273.058 | — |
| (5261) Eureka | 1.5234 | 0.0648 | 20.280 | 95.467 | 245.141 | 286.094 | — |
| (5598) Carlmurray | 2.1913 | 0.1129 | 5.043 | 173.374 | 283.727 | 237.967 | — |

Notes: $a$ is the semi-major axis, $e$ the eccentricity, $I$ the inclination, $\omega$ the argument of perihelion, $\Omega$ the longitude of ascending node, $M$ the mean anomaly, and $D$ the *IRAS* estimated diameter in kilometres. The orbital elements are for the epoch of JD 2450800.5 and are referred to the ecliptic and equinox of J2000.

Table A.17. The osculating orbital elements for a selection of Centaurs. (The notation and epoch are the same as in Table A.16.)

| Centaur | a (AU) | e | I (°) | ω (°) | Ω (°) | M (°) |
|---|---|---|---|---|---|---|
| (2060) Chiron | 13.6481 | 0.3806 | 6.937 | 339.483 | 209.381 | 13.177 |
| (5145) Pholus | 20.2272 | 0.5714 | 24.701 | 354.537 | 119.415 | 24.718 |
| (7066) Nessus | 24.5932 | 0.5192 | 15.654 | 170.628 | 31.386 | 17.304 |
| 1994TA | 16.8428 | 0.3036 | 5.396 | 154.724 | 137.607 | 68.837 |
| 1995DW2 | 24.9156 | 0.2430 | 4.152 | 6.271 | 178.246 | 7.492 |
| 1995GO | 18.0690 | 0.6215 | 17.641 | 290.171 | 6.124 | 338.421 |
| 1997CU26 | 15.7234 | 0.1700 | 23.426 | 242.537 | 300.478 | 324.766 |

Table A.18. The osculating orbital elements for a selection of trans-Neptunian objects. (The notation and epoch are the same as in Table A.16.)

| Object | a (AU) | e | I (°) | ω (°) | Ω (°) | M (°) |
|---|---|---|---|---|---|---|
| 1992QB1 | 44.2978 | 0.0770 | 2.179 | 2.050 | 359.385 | 5.044 |
| 1994JQ1 | 43.9587 | 0.0494 | 3.755 | 254.732 | 25.671 | 298.729 |
| 1994TB | 39.8452 | 0.3211 | 12.111 | 98.009 | 317.298 | 330.327 |
| 1995DC2 | 43.8434 | 0.0704 | 2.344 | 122.376 | 154.182 | 246.905 |
| 1995GJ | 42.7874 | 0.0447 | 30.217 | 205.498 | 338.964 | 340.334 |
| 1995KJ1 | 39.1803 | 0.1389 | 3.004 | 339.900 | 47.839 | 210.605 |
| 1996KW1 | 45.2482 | 0.0333 | 5.633 | 350.858 | 38.370 | 187.775 |
| 1996RQ20 | 44.2941 | 0.1149 | 31.579 | 339.813 | 11.537 | 17.195 |
| 1996TK66 | 43.1408 | 0.0323 | 3.310 | 41.882 | 44.656 | 283.179 |
| 1996TL66 | 84.4705 | 0.5849 | 23.951 | 183.536 | 217.762 | 358.610 |
| 1996TQ66 | 39.6471 | 0.1297 | 14.634 | 29.523 | 10.688 | 350.959 |
| 1996TS66 | 44.1702 | 0.1354 | 7.352 | 142.922 | 285.759 | 336.117 |
| 1997CQ29 | 43.8550 | 0.0754 | 2.945 | 342.399 | 132.951 | 35.570 |
| 1997CR29 | 40.9387 | 0.0327 | 20.597 | 209.746 | 127.786 | 157.828 |
| 1997CS29 | 43.7232 | 0.0194 | 2.254 | 253.617 | 304.311 | 279.437 |
| 1997CT29 | 45.0745 | 0.1611 | 1.013 | 321.491 | 74.466 | 80.113 |
| 1997CU29 | 43.4318 | 0.0593 | 1.461 | 358.362 | 349.884 | 119.567 |
| 1997CW29 | 40.1200 | 0.0579 | 18.688 | 203.668 | 110.557 | 164.015 |
| 1997QH4 | 45.6913 | 0.1000 | 12.548 | 352.514 | 355.732 | 6.230 |

Table A.19. Orbital elements for a selection of comets.

| Comet | $\tau$ | $q$ (AU) | $e$ | $I$ (°) | $\omega$ (°) | $\Omega$ (°) | Epoch |
|---|---|---|---|---|---|---|---|
| 1P/Halley | 1986 Feb 9.5 | 0.5871 | 0.9673 | 162.24 | 111.87 | 58.86 | 46480 |
| 2P/Encke | 1997 May 23.6 | 0.3314 | 0.8500 | 11.93 | 186.27 | 334.72 | 50600 |
| 4P/Faye | 1999 May 6.1 | 1.6570 | 0.5683 | 9.05 | 204.97 | 199.34 | 51320 |
| 6P/d'Arrest | 1995 Jul 27.3 | 1.3458 | 0.6140 | 19.52 | 178.05 | 138.99 | 49920 |
| 7P/Pons-Winnecke | 1996 Jan 2.5 | 1.2559 | 0.6344 | 22.30 | 172.31 | 93.43 | 50080 |
| 8P/Tuttle | 1994 Jun 25.3 | 0.9977 | 0.8241 | 54.69 | 206.70 | 270.55 | 49520 |
| 9P/Tempel 1 | 2000 Jan 2.6 | 1.5000 | 0.5190 | 10.54 | 178.91 | 68.97 | 51560 |
| 10P/Tempel 2 | 1999 Sep 8.4 | 1.4817 | 0.5228 | 11.98 | 195.02 | 118.21 | 51440 |
| 14P/Wolf | 2000 Nov 21.1 | 2.4126 | 0.4071 | 27.52 | 162.36 | 204.12 | 51880 |
| 21P/Giacobini-Zinner | 1998 Nov 21.3 | 1.0337 | 0.7065 | 31.86 | 172.54 | 195.40 | 51120 |
| 22P/Kopff | 1996 Jul 2.2 | 1.5796 | 0.5441 | 4.72 | 162.84 | 120.91 | 50280 |
| 23P/Brorsen-Metcalf | 1989 Sep 11.9 | 0.4788 | 0.9720 | 19.33 | 129.61 | 311.59 | 47800 |
| 26P/Grigg-Skjellerup | 1997 Aug 30.3 | 0.9968 | 0.6638 | 21.09 | 359.33 | 213.31 | 50680 |
| 27P/Crommelin | 2011 Aug 3.8 | 0.7479 | 0.9187 | 28.96 | 195.98 | 250.64 | 55760 |
| 28P/Neujmin 1 | 2002 Dec 27.4 | 1.5521 | 0.7756 | 14.18 | 346.92 | 347.03 | 52640 |
| 30P/Reinmuth 1 | 1995 Sep 03.3 | 1.8736 | 0.5025 | 8.13 | 13.29 | 119.74 | 49960 |
| 36P/Whipple | 1994 Dec 22.4 | 3.0939 | 0.2587 | 9.93 | 201.87 | 182.50 | 49720 |
| 39P/Oterma | 2002 Dec 21.8 | 5.4707 | 0.2446 | 1.94 | 56.36 | 331.59 | 52640 |
| 43P/Wolf-Harrington | 1997 Sep 29.2 | 1.5818 | 0.5440 | 18.51 | 187.13 | 254.76 | 50720 |
| 44P/Reinmuth 2 | 2001 Feb 20.0 | 1.8897 | 0.4645 | 6.98 | 46.10 | 296.07 | 51960 |
| 46P/Wirtanen | 1997 Mar 14.2 | 1.0638 | 0.6568 | 11.72 | 356.34 | 82.21 | 50520 |
| 49P/Arend-Rigaux | 1998 Jul 12.6 | 1.3686 | 0.6115 | 18.29 | 330.56 | 121.73 | 51000 |
| 65P/Gunn | 1996 Jul 24.4 | 2.4619 | 0.3163 | 10.38 | 196.82 | 68.52 | 50280 |
| 68P/Klemola | 1998 May 1.7 | 1.7545 | 0.6413 | 11.09 | 154.54 | 175.54 | 50920 |
| 75P/Kohoutek | 2001 Feb 27.3 | 1.7873 | 0.4962 | 5.91 | 175.69 | 269.68 | 51960 |
| 95P/Chiron | 1996 Feb 12.9 | 8.4530 | 0.3806 | 6.94 | 339.48 | 209.38 | 50800 |
| 96P/Machholz 1 | 1996 Oct 15.1 | 0.1247 | 0.9586 | 60.07 | 14.59 | 94.53 | 50360 |
| 107P/Wilson-Harrington | 1996 Dec 6.4 | 1.0004 | 0.6217 | 2.78 | 90.92 | 270.95 | 50800 |
| 109P/Swift-Tuttle | 1992 Dec 12.3 | 0.9582 | 0.9636 | 113.43 | 153.00 | 139.44 | 48960 |

Notes: $\tau$ is the time of perihelion passage, $q = a(1 - e)$ is the perihelion distance in AU, $e$ is the eccentricity, $I$ is the inclination, $\omega$ is the argument of perihelion, $\Omega$ is the longitude of ascending node, and the Epoch is given as (JD − 2400000.5). The orbital information on these comets is not accurate enough for use in numerical integrations. Note that for the purposes of this table Chiron and Wilson-Harrington are classified as comets.

# Appendix B
## Expansion of the Disturbing Function

> Rightly to be great
> Is not to stir without great argument
>
> William Shakespeare, *Hamlet, IV, iv*

In Chapter 6 we described methods for obtaining an expansion of the disturbing function $\mathcal{R}$. Here we give a literal expansion of $\mathcal{R}_D$, $\mathcal{R}_E$, and $\mathcal{R}_I$ (the direct and indirect parts of $\mathcal{R}$) complete to fourth degree in the eccentricities and inclinations of the two bodies. The notation here is the same as that used in Chapter 6. In addition we make use of the convention

$$A_j = b_{1/2}^{(j)}(\alpha), \qquad B_j = b_{3/2}^{(j)}(\alpha), \qquad C_j = b_{5/2}^{(j)}(\alpha), \qquad \text{(B.1)}$$

where $b_s^{(j)}(\alpha)$ denotes a Laplace coefficient of the ratio $\alpha$ of the semi-major axes (see Sect. 6.4).

In the following expansion each cosine argument has a label indicating (i) the order of the expansion, (ii) whether the term is associated with the direct (prefix D) or indirect (prefix E or I) part of the disturbing function, (iii) the order of the argument (and hence the order of the resonance associated with that argument), and (iv) an identification number for the argument. For example, 4D2.11 denotes the eleventh possible argument of the second-order direct part of the fourth-order expansion. In this case Table B.8 gives a cosine argument of $j\lambda' + (2-j)\lambda - 2\varpi - \Omega' + \Omega$ and a corresponding term of $e^2 ss' f_{72}$. Inspection of Table B.11 gives

$$f_{72} = \frac{1}{8}\left[-2\alpha - j\alpha + 4j^2\alpha + 4j\alpha^2 D + \alpha^3 D^2\right] B_{j-1}, \qquad \text{(B.2)}$$

where $D$ denotes the differential operator $d/d\alpha$ and $B_{j-1}$ denotes the Laplace coefficient $b_{3/2}^{(j-1)}$. Therefore the appropriate term can be written as

$$\frac{1}{8}\left\{\left(-2\alpha - j\alpha + 4j^2\alpha\right)b_{3/2}^{(j-1)} + 4j\alpha^2\frac{db_{3/2}^{(j-1)}}{d\alpha} + \alpha^3\frac{d^2b_{3/2}^{(j-1)}}{d\alpha^2}\right\}e^2ss'$$

$$\times\cos\left[j\lambda' + (2-j)\lambda - 2\varpi - \Omega' + \Omega\right]. \tag{B.3}$$

Table B.1. Zeroth-order arguments: direct part.

| ID | Cosine Argument | Term |
|----|-----------------|------|
| 4D0.1 | $j\lambda' - j\lambda$ | $f_1 + (e^2 + e'^2)f_2 + (s^2 + s'^2)f_3$ $+ e^4 f_4 + e^2 e'^2 f_5 + e'^4 f_6$ $+ (e^2 s^2 + e'^2 s^2 + e^2 s'^2 + e'^2 s'^2)f_7$ $+ (s^4 + s'^4)f_8 + s^2 s'^2 f_9$ |
| 4D0.2 | $j\lambda' - j\lambda + \varpi' - \varpi$ | $ee' f_{10} + e^3 e' f_{11} + ee'^3 f_{12}$ $+ ee'(s^2 + s'^2)f_{13}$ |
| 4D0.3 | $j\lambda' - j\lambda + \Omega' - \Omega$ | $ss' f_{14} + ss'(e^2 + e'^2)f_{15}$ $+ ss'(s^2 + s'^2)f_{16}$ |
| 4D0.4 | $j\lambda' - j\lambda + 2\varpi' - 2\varpi$ | $e^2 e'^2 f_{17}$ |
| 4D0.5 | $j\lambda' - j\lambda + 2\varpi - 2\Omega$ | $e^2 s^2 f_{18}$ |
| 4D0.6 | $j\lambda' - j\lambda + \varpi' + \varpi - 2\Omega$ | $ee' s^2 f_{19}$ |
| 4D0.7 | $j\lambda' - j\lambda + 2\varpi' - 2\Omega$ | $e'^2 s^2 f_{20}$ |
| 4D0.8 | $j\lambda' - j\lambda + 2\varpi - \Omega' - \Omega$ | $e^2 ss' f_{21}$ |
| 4D0.9 | $j\lambda' - j\lambda + \varpi' - \varpi - \Omega' + \Omega$ | $ee' ss' f_{22}$ |
| 4D0.10 | $j\lambda' - j\lambda + \varpi' - \varpi + \Omega' - \Omega$ | $ee' ss' f_{23}$ |
| 4D0.11 | $j\lambda' - j\lambda + \varpi' + \varpi - \Omega' - \Omega$ | $ee' ss' f_{24}$ |
| 4D0.12 | $j\lambda' - j\lambda + 2\varpi' - \Omega' - \Omega$ | $e'^2 ss' f_{25}$ |
| 4D0.13 | $j\lambda' - j\lambda + 2\varpi - 2\Omega'$ | $e^2 s'^2 f_{18}$ |
| 4D0.14 | $j\lambda' - j\lambda + \varpi' + \varpi - 2\Omega'$ | $ee' s'^2 f_{19}$ |
| 4D0.15 | $j\lambda' - j\lambda + 2\varpi' - 2\Omega'$ | $e'^2 s'^2 f_{20}$ |
| 4D0.16 | $j\lambda' - j\lambda + 2\Omega' - 2\Omega$ | $s^2 s'^2 f_{26}$ |

Table B.2. Zeroth-order arguments: indirect part (external and internal perturbers).

| ID | Cosine Argument | Term |
|---|---|---|
| 4E0.1, 4I0.1 | $\lambda' - \lambda$ | $-1 + \frac{1}{2}(e^2 + e'^2) + \frac{1}{64}(e^4 + e'^4)$ $-\frac{1}{4}e^2 e'^2 + s^2$ $-\frac{1}{2}(e^2 + e'^2)(s^2 + s'^2)$ $+s'^2 - s^2 s'^2$ |
| 4E0.2, 4I0.2 | $2\lambda' - 2\lambda - \varpi' + \varpi$ | $-ee' + \frac{3}{4}e^3 e' + \frac{3}{4}ee'^3$ $+ee's^2 + ee's'^2$ |
| 4E0.3, 4I0.3 | $\lambda' - \lambda - \Omega' + \Omega$ | $-2ss' + e^2 ss' + e'^2 ss'$ $+s^3 s' + ss'^3$ |
| 4E0.4, 4I0.4 | $\lambda' - \lambda - 2\varpi' + 2\varpi$ | $-\frac{1}{64}e^2 e'^2$ |
| 4E0.5, 4I0.5 | $3\lambda' - 3\lambda - 2\varpi' + 2\varpi$ | $-\frac{81}{64}e^2 e'^2$ |
| 4E0.6, 4I0.6 | $\lambda' - \lambda + 2\varpi - 2\Omega$ | $-\frac{1}{8}e^2 s^2$ |
| 4E0.7, 4I0.7 | $\lambda' - \lambda - 2\varpi' + 2\Omega$ | $-\frac{1}{8}e'^2 s^2$ |
| 4E0.8, 4I0.8 | $\lambda' - \lambda + 2\varpi - \Omega' - \Omega$ | $\frac{1}{4}e^2 ss'$ |
| 4E0.9, 4I0.9 | $2\lambda' - 2\lambda - \varpi' + \varpi - \Omega' + \Omega$ | $-2ee'ss'$ |
| 4E0.10, 4I0.10 | $\lambda' - \lambda - 2\varpi' + \Omega' + \Omega$ | $\frac{1}{4}e'^2 ss'$ |
| 4E0.11, 4I0.11 | $\lambda' - \lambda + 2\varpi - 2\Omega'$ | $-\frac{1}{8}e^2 s'^2$ |
| 4E0.12, 4I0.12 | $\lambda' - \lambda - 2\varpi' + 2\Omega'$ | $-\frac{1}{8}e'^2 s'^2$ |
| 4E0.13, 4I0.13 | $\lambda' - \lambda - 2\Omega' + 2\Omega$ | $-s^2 s'^2$ |

Table B.3. Zeroth-order arguments: functions of semi-major axis.

| $i$ | $f_i$ |
|---|---|
| 1 | $\frac{1}{2} A_j$ |
| 2 | $\frac{1}{8} \left[ -4j^2 + 2\alpha D + \alpha^2 D^2 \right] A_j$ |
| 3 | $\frac{1}{4} \left[ -\alpha \right] B_{j-1} + \frac{1}{4} \left[ -\alpha \right] B_{j+1}$ |
| 4 | $\frac{1}{128} \left[ -9j^2 + 16j^4 - 8j^2\alpha D - 8j^2\alpha^2 D^2 + 4\alpha^3 D^3 + \alpha^4 D^4 \right] A_j$ |
| 5 | $\frac{1}{32} \left[ 16j^4 + 4\alpha D - 16j^2\alpha D + 14\alpha^2 D^2 - 8j^2\alpha^2 D^2 + 8\alpha^3 D^3 + \alpha^4 D^4 \right] A_j$ |
| 6 | $\frac{1}{128} \left[ -17j^2 + 16j^4 + 24\alpha D - 24j^2\alpha D + 36\alpha^2 D^2 - 8j^2\alpha^2 D^2 \right.$ $\left. + 12\alpha^3 D^3 + \alpha^4 D^4 \right] A_j$ |
| 7 | $\frac{1}{16} \left[ -2\alpha + 4j^2\alpha - 4\alpha^2 D - \alpha^3 D^2 \right] (B_{j-1} + B_{j+1})$ |
| 8 | $\frac{3}{16} \left[ \alpha^2 \right] C_{j-2} + \frac{3}{4} \left[ \alpha^2 \right] C_j + \frac{3}{16} \left[ \alpha^2 \right] C_{j+2}$ |
| 9 | $\frac{1}{4} \left[ \alpha \right] (B_{j-1} + B_{j+1}) + \frac{3}{8} \left[ \alpha^2 \right] C_{j-2} + \frac{15}{4} \left[ \alpha^2 \right] C_j + \frac{3}{8} \left[ \alpha^2 \right] C_{j+2}$ |
| 10 | $\frac{1}{4} \left[ 2 + 6j + 4j^2 - 2\alpha D - \alpha^2 D^2 \right] A_{j+1}$ |
| 11 | $\frac{1}{32} \left[ -6j - 26j^2 - 36j^3 - 16j^4 + 6j\alpha D + 12j^2\alpha D - 4\alpha^2 D^2 \right.$ $\left. + 7j\alpha^2 D^2 + 8j^2\alpha^2 D^2 - 6\alpha^3 D^3 - \alpha^4 D^4 \right] A_{j+1}$ |
| 12 | $\frac{1}{32} \left[ 4 + 2j - 22j^2 - 36j^3 - 16j^4 - 4\alpha D + 22j\alpha D + 20j^2\alpha D \right.$ $\left. - 22\alpha^2 D^2 + 7j\alpha^2 D^2 + 8j^2\alpha^2 D^2 - 10\alpha^3 D^3 - \alpha^4 D^4 \right] A_{j+1}$ |
| 13 | $\frac{1}{8} \left[ -6j\alpha - 4j^2\alpha + 4\alpha^2 D + \alpha^3 D^2 \right] (B_j + B_{j+2})$ |
| 14 | $[\alpha] B_{j+1}$ |
| 15 | $\frac{1}{4} \left[ 2\alpha - 4j^2\alpha + 4\alpha^2 D + \alpha^3 D^2 \right] B_{j+1}$ |
| 16 | $\frac{1}{2} \left[ -\alpha \right] B_{j+1} + 3 \left[ -\alpha^2 \right] C_j + \frac{3}{2} \left[ -\alpha^2 \right] C_{j+2}$ |
| 17 | $\frac{1}{64} \left[ 12 + 64j + 109j^2 + 72j^3 + 16j^4 - 12\alpha D - 28j\alpha D - 16j^2\alpha D \right.$ $\left. + 6\alpha^2 D^2 - 14j\alpha^2 D^2 - 8j^2\alpha^2 D^2 + 8\alpha^3 D^3 + \alpha^4 D^4 \right] A_{j+2}$ |
| 18 | $\frac{1}{16} \left[ 12\alpha - 15j\alpha + 4j^2\alpha + 8\alpha^2 D - 4j\alpha^2 D + \alpha^3 D^2 \right] B_{j-1}$ |
| 19 | $\frac{1}{8} \left[ 6j\alpha - 4j^2\alpha - 4\alpha^2 D + 4j\alpha^2 D - \alpha^3 D^2 \right] B_j$ |
| 20 | $\frac{1}{16} \left[ 3j\alpha + 4j^2\alpha - 4j\alpha^2 D + \alpha^3 D^2 \right] B_{j+1}$ |
| 21 | $\frac{1}{8} \left[ -12\alpha + 15j\alpha - 4j^2\alpha - 8\alpha^2 D + 4j\alpha^2 D - \alpha^3 D^2 \right] B_{j-1}$ |
| 22 | $\frac{1}{4} \left[ 6j\alpha + 4j^2\alpha - 4\alpha^2 D - \alpha^3 D^2 \right] B_j$ |
| 23 | $\frac{1}{4} \left[ 6j\alpha + 4j^2\alpha - 4\alpha^2 D - \alpha^3 D^2 \right] B_{j+2}$ |
| 24 | $\frac{1}{4} \left[ -6j\alpha + 4j^2\alpha + 4\alpha^2 D - 4j\alpha^2 D + \alpha^3 D^2 \right] B_j$ |
| 25 | $\frac{1}{8} \left[ -3j\alpha - 4j^2\alpha + 4j\alpha^2 D - \alpha^3 D^2 \right] B_{j+1}$ |
| 26 | $\frac{1}{2} \left[ \alpha \right] B_{j+1} + \frac{3}{4} \left[ \alpha^2 \right] C_j + \frac{3}{2} \left[ \alpha^2 \right] C_{j+2}$ |

Table B.4. First-order arguments: direct part.

| ID | Cosine Argument | Term |
|---|---|---|
| 4D1.1 | $j\lambda' + (1-j)\lambda - \varpi$ | $ef_{27} + e^3 f_{28} + ee'^2 f_{29}$ $+ e(s^2 + s'^2) f_{30}$ |
| 4D1.2 | $j\lambda' + (1-j)\lambda - \varpi'$ | $e' f_{31} + e^2 e' f_{32} + e'^3 f_{33}$ $+ e'(s^2 + s'^2) f_{34}$ |
| 4D1.3 | $j\lambda' + (1-j)\lambda + \varpi' - 2\varpi$ | $e^2 e' f_{35}$ |
| 4D1.4 | $j\lambda' + (1-j)\lambda - 2\varpi' + \varpi$ | $ee'^2 f_{36}$ |
| 4D1.5 | $j\lambda' + (1-j)\lambda + \varpi - 2\Omega$ | $es^2 f_{37}$ |
| 4D1.6 | $j\lambda' + (1-j)\lambda + \varpi' - 2\Omega$ | $e's^2 f_{38}$ |
| 4D1.7 | $j\lambda' + (1-j)\lambda - \varpi - \Omega' + \Omega$ | $ess' f_{39}$ |
| 4D1.8 | $j\lambda' + (1-j)\lambda - \varpi + \Omega' - \Omega$ | $ess' f_{40}$ |
| 4D1.9 | $j\lambda' + (1-j)\lambda + \varpi - \Omega' - \Omega$ | $ess' f_{41}$ |
| 4D1.10 | $j\lambda' + (1-j)\lambda - \varpi' - \Omega' + \Omega$ | $e'ss' f_{42}$ |
| 4D1.11 | $j\lambda' + (1-j)\lambda - \varpi' + \Omega' - \Omega$ | $e'ss' f_{43}$ |
| 4D1.12 | $j\lambda' + (1-j)\lambda + \varpi' - \Omega' - \Omega$ | $e'ss' f_{44}$ |
| 4D1.13 | $j\lambda' + (1-j)\lambda + \varpi - 2\Omega'$ | $es'^2 f_{37}$ |
| 4D1.14 | $j\lambda' + (1-j)\lambda + \varpi' - 2\Omega'$ | $e's'^2 f_{38}$ |

Table B.5. First-order arguments: indirect part (external perturber).

| ID | Cosine Argument | Term |
|---|---|---|
| 4E1.1 | $\lambda' - 2\lambda + \varpi$ | $-\frac{1}{2}e + \frac{3}{8}e^3 + \frac{1}{4}ee'^2 + \frac{1}{2}es^2 + \frac{1}{2}es'^2$ |
| 4E1.2 | $\lambda' - \varpi$ | $\frac{3}{2}e - \frac{3}{4}ee'^2 - \frac{3}{2}es^2 - \frac{3}{2}es'^2$ |
| 4E1.3 | $2\lambda' - \lambda - \varpi'$ | $-2e' + e^2e' + \frac{3}{2}e'^3 + 2e's^2 + 2e's'^2$ |
| 4E1.4 | $2\lambda' - 3\lambda - \varpi' + 2\varpi$ | $-\frac{3}{4}e^2e'$ |
| 4E1.5 | $\lambda' - 2\varpi' + \varpi$ | $\frac{3}{16}ee'^2$ |
| 4E1.6 | $3\lambda' - 2\lambda - 2\varpi' + \varpi$ | $-\frac{27}{16}ee'^2$ |
| 4E1.7 | $\lambda' + \varpi - 2\Omega$ | $\frac{3}{2}es^2$ |
| 4E1.8 | $\lambda' - 2\lambda + \varpi - \Omega' + \Omega$ | $-ess'$ |
| 4E1.9 | $\lambda' - \varpi - \Omega' + \Omega$ | $3ess'$ |
| 4E1.10 | $\lambda' + \varpi - \Omega' - \Omega$ | $-3ess'$ |
| 4E1.11 | $2\lambda' - \lambda - \varpi' - \Omega' + \Omega$ | $-4e'ss'$ |
| 4E1.12 | $\lambda' + \varpi - 2\Omega'$ | $\frac{3}{2}es'^2$ |

Table B.6. First-order arguments: indirect part (internal perturber).

| ID | Cosine Argument | Term |
|---|---|---|
| 4I1.1 | $\lambda' - 2\lambda + \varpi$ | $-2e + \frac{3}{2}e^3 + ee'^2 + 2es^2 + 2es'^2$ |
| 4I1.2 | $\lambda - \varpi'$ | $\frac{3}{2}e' - \frac{3}{4}e^2e' - \frac{3}{2}e's^2 - \frac{3}{2}e's'^2$ |
| 4I1.3 | $2\lambda' - \lambda - \varpi'$ | $-\frac{1}{2}e' + \frac{1}{4}e^2e' + \frac{3}{8}e'^3 + \frac{1}{2}e's^2 + \frac{1}{2}e's'^2$ |
| 4I1.4 | $\lambda + \varpi' - 2\varpi$ | $\frac{3}{16}e^2e'$ |
| 4I1.5 | $2\lambda' - 3\lambda - \varpi' + 2\varpi$ | $-\frac{27}{16}e^2e'$ |
| 4I1.6 | $3\lambda' - 2\lambda - 2\varpi' + \varpi$ | $-\frac{3}{4}ee'^2$ |
| 4I1.7 | $\lambda + \varpi' - 2\Omega$ | $\frac{3}{2}e's^2$ |
| 4I1.8 | $\lambda' - 2\lambda + \varpi - \Omega' + \Omega$ | $-4ess'$ |
| 4I1.9 | $\lambda - \varpi' + \Omega' - \Omega$ | $3e'ss'$ |
| 4I1.10 | $\lambda + \varpi' - \Omega' - \Omega$ | $-3e'ss'$ |
| 4I1.11 | $2\lambda' - \lambda - \varpi' - \Omega' + \Omega$ | $-e'ss'$ |
| 4I1.12 | $\lambda + \varpi' - 2\Omega'$ | $\frac{3}{2}e's'^2$ |

Table B.7. First-order arguments: functions of semi-major axis.

| $i$ | $f_i$ |
|---|---|
| 27 | $\frac{1}{2}\left[-2j - \alpha D\right] A_j$ |
| 28 | $\frac{1}{16}\left[2j - 10j^2 + 8j^3 + 3\alpha D - 7j\alpha D + 4j^2\alpha D - 2\alpha^2 D^2 - 2j\alpha^2 D^2\right.$ $\left.-\alpha^3 D^3\right] A_j$ |
| 29 | $\frac{1}{8}\left[8j^3 - 2\alpha D - 4j\alpha D + 4j^2\alpha D - 4\alpha^2 D^2 - 2j\alpha^2 D^2 - \alpha^3 D^3\right] A_j$ |
| 30 | $\frac{1}{4}\left[\alpha + 2j\alpha + \alpha^2 D\right](B_{j-1} + B_{j+1})$ |
| 31 | $\frac{1}{2}\left[-1 + 2j + \alpha D\right] A_{j-1}$ |
| 32 | $\frac{1}{8}\left[4 - 16j + 20j^2 - 8j^3 - 4\alpha D + 12j\alpha D - 4j^2\alpha D + 3\alpha^2 D^2\right.$ $\left.+2j\alpha^2 D^2 + \alpha^3 D^3\right] A_{j-1}$ |
| 33 | $\frac{1}{16}\left[-2 - j + 10j^2 - 8j^3 + 2\alpha D + 9j\alpha D - 4j^2\alpha D + 5\alpha^2 D^2\right.$ $\left.+2j\alpha^2 D^2 + \alpha^3 D^3\right] A_{j-1}$ |
| 34 | $\frac{1}{4}\left[-2j\alpha - \alpha^2 D\right](B_{j-2} + B_j)$ |
| 35 | $\frac{1}{16}\left[1 - j - 10j^2 - 8j^3 - \alpha D - j\alpha D - 4j^2\alpha D + 3\alpha^2 D^2\right.$ $\left.+2j\alpha^2 D^2 + \alpha^3 D^3\right] A_{j+1}$ |
| 36 | $\frac{1}{16}\left[-8 + 32j - 30j^2 + 8j^3 + 8\alpha D - 17j\alpha D + 4j^2\alpha D - 4\alpha^2 D^2\right.$ $\left.-2j\alpha^2 D^2 - \alpha^3 D^3\right] A_{j-2}$ |
| 37 | $\frac{1}{4}\left[-5\alpha + 2j\alpha - \alpha^2 D\right] B_{j-1}$ |
| 38 | $\frac{1}{4}\left[-2j\alpha + \alpha^2 D\right] B_j$ |
| 39 | $\frac{1}{2}\left[-\alpha - 2j\alpha - \alpha^2 D\right] B_{j-1}$ |
| 40 | $\frac{1}{2}\left[-\alpha - 2j\alpha - \alpha^2 D\right] B_{j+1}$ |
| 41 | $\frac{1}{2}\left[5\alpha - 2j\alpha + \alpha^2 D\right] B_{j-1}$ |
| 42 | $\frac{1}{2}\left[2j\alpha + \alpha^2 D\right] B_{j-2}$ |
| 43 | $\frac{1}{2}\left[2j\alpha + \alpha^2 D\right] B_j$ |
| 44 | $\frac{1}{2}\left[2j\alpha - \alpha^2 D\right] B_j$ |

Table B.8. Second-order arguments: direct part.

| ID | Cosine Argument | Term |
|---|---|---|
| 4D2.1 | $j\lambda' + (2-j)\lambda - 2\varpi$ | $e^2 f_{45} + e^4 f_{46} + e^2 e'^2 f_{47}$ $+ e^2(s^2 + s'^2) f_{48}$ |
| 4D2.2 | $j\lambda' + (2-j)\lambda - \varpi' - \varpi$ | $ee' f_{49} + e^3 e' f_{50} + ee'^3 f_{51}$ $+ ee'(s^2 + s'^2) f_{52}$ |
| 4D2.3 | $j\lambda' + (2-j)\lambda - 2\varpi'$ | $e'^2 f_{53} + e^2 e'^2 f_{54} + e'^4 f_{55}$ $+ e'^2(s^2 + s'^2) f_{56}$ |
| 4D2.4 | $j\lambda' + (2-j)\lambda - 2\Omega$ | $s^2 f_{57} + e^2 s^2 f_{58} + e'^2 s^2 f_{59}$ $+ s^4 f_{60} + s^2 s'^2 f_{61}$ |
| 4D2.5 | $j\lambda' + (2-j)\lambda - \Omega' - \Omega$ | $ss' f_{62} + e^2 ss' f_{63} + e'^2 ss' f_{64}$ $+ s^3 s' f_{65} + ss'^3 f_{66}$ |
| 4D2.6 | $j\lambda' + (2-j)\lambda - 2\Omega'$ | $s'^2 f_{57} + e^2 s'^2 f_{58} + e'^2 s'^2 f_{59}$ $+ s^2 s'^2 f_{67} + s'^4 f_{60}$ |
| 4D2.7 | $j\lambda' + (2-j)\lambda + \varpi' - 3\varpi$ | $e^3 e' f_{68}$ |
| 4D2.8 | $j\lambda' + (2-j)\lambda - 3\varpi' + \varpi$ | $ee'^3 f_{69}$ |
| 4D2.9 | $j\lambda' + (2-j)\lambda - \varpi' + \varpi - 2\Omega$ | $ee' s^2 f_{70}$ |
| 4D2.10 | $j\lambda' + (2-j)\lambda + \varpi' - \varpi - 2\Omega$ | $ee' s^2 f_{71}$ |
| 4D2.11 | $j\lambda' + (2-j)\lambda - 2\varpi - \Omega' + \Omega$ | $e^2 ss' f_{72}$ |
| 4D2.12 | $j\lambda' + (2-j)\lambda - 2\varpi + \Omega' - \Omega$ | $e^2 ss' f_{73}$ |
| 4D2.13 | $j\lambda' + (2-j)\lambda - \varpi' - \varpi - \Omega' + \Omega$ | $ee' ss' f_{74}$ |
| 4D2.14 | $j\lambda' + (2-j)\lambda - \varpi' - \varpi + \Omega' - \Omega$ | $ee' ss' f_{75}$ |
| 4D2.15 | $j\lambda' + (2-j)\lambda - \varpi' + \varpi - \Omega' - \Omega$ | $ee' ss' f_{76}$ |
| 4D2.16 | $j\lambda' + (2-j)\lambda + \varpi' - \varpi - \Omega' - \Omega$ | $ee' ss' f_{77}$ |
| 4D2.17 | $j\lambda' + (2-j)\lambda - 2\varpi' - \Omega' + \Omega$ | $e'^2 ss' f_{78}$ |
| 4D2.18 | $j\lambda' + (2-j)\lambda - 2\varpi' + \Omega' - \Omega$ | $e'^2 ss' f_{79}$ |
| 4D2.19 | $j\lambda' + (2-j)\lambda + \Omega' - 3\Omega$ | $s^3 s' f_{80}$ |
| 4D2.20 | $j\lambda' + (2-j)\lambda - \varpi' + \varpi - 2\Omega'$ | $ee' s'^2 f_{70}$ |
| 4D2.21 | $j\lambda' + (2-j)\lambda + \varpi' - \varpi - 2\Omega'$ | $ee' s'^2 f_{71}$ |
| 4D2.22 | $j\lambda' + (2-j)\lambda - 3\Omega' + \Omega$ | $ss'^3 f_{81}$ |

Table B.9. Second-order arguments: indirect part (external perturber).

| ID | Cosine Argument | Term |
|---|---|---|
| 4E2.1 | $\lambda' - 3\lambda + 2\varpi$ | $-\frac{3}{8}e^2 + \frac{3}{8}e^4 + \frac{3}{16}e^2e'^2$ $+\frac{3}{8}e^2s^2 + \frac{3}{8}e^2s'^2$ |
| 4E2.2 | $\lambda' + \lambda - 2\varpi$ | $-\frac{1}{8}e^2 - \frac{1}{24}e^4 + \frac{1}{16}e^2e'^2$ $+\frac{1}{8}e^2s^2 + \frac{1}{8}e^2s'^2$ |
| 4E2.3 | $2\lambda' - \varpi' - \varpi$ | $3ee' - \frac{9}{4}ee'^3 - 3ee's^2 - 3ee's'^2$ |
| 4E2.4 | $\lambda' + \lambda - 2\varpi'$ | $-\frac{1}{8}e'^2 + \frac{1}{16}e^2e'^2 - \frac{1}{24}e'^4$ $+\frac{1}{8}e'^2s^2 + \frac{1}{8}e'^2s'^2$ |
| 4E2.5 | $3\lambda' - \lambda - 2\varpi'$ | $-\frac{27}{8}e'^2 + \frac{27}{16}e^2e'^2 + \frac{27}{8}e'^4$ $+\frac{27}{8}e'^2s^2 + \frac{27}{8}e'^2s'^2$ |
| 4E2.6 | $\lambda' + \lambda - 2\Omega$ | $-s^2 + \frac{1}{2}e^2s^2 + \frac{1}{2}e'^2s^2 + s^2s'^2$ |
| 4E2.7 | $\lambda' + \lambda - \Omega' - \Omega$ | $2ss' - e^2ss' - e'^2ss' - s^3s' - ss'^3$ |
| 4E2.8 | $\lambda' + \lambda - 2\Omega'$ | $-s'^2 + \frac{1}{2}e^2s'^2 + \frac{1}{2}e'^2s'^2 + s^2s'^2$ |
| 4E2.9 | $2\lambda' - 4\lambda - \varpi' + 3\varpi$ | $-\frac{2}{3}e^3e'$ |
| 4E2.10 | $2\lambda' - 3\varpi' + \varpi$ | $\frac{1}{4}ee'^3$ |
| 4E2.11 | $4\lambda' - 2\lambda - 3\varpi' + \varpi$ | $-\frac{8}{3}ee'^3$ |
| 4E2.12 | $2\lambda' - \varpi' + \varpi - 2\Omega$ | $3ee's^2$ |
| 4E2.13 | $\lambda' - 3\lambda + 2\varpi - \Omega' + \Omega$ | $-\frac{3}{4}e^2ss'$ |
| 4E2.14 | $\lambda' + \lambda - 2\varpi - \Omega' + \Omega$ | $-\frac{1}{4}e^2ss'$ |
| 4E2.15 | $2\lambda' - \varpi' - \varpi - \Omega' + \Omega$ | $6ee'ss'$ |
| 4E2.16 | $2\lambda' - \varpi' + \varpi - \Omega' - \Omega$ | $-6ee'ss'$ |
| 4E2.17 | $\lambda' + \lambda - 2\varpi' + \Omega' - \Omega$ | $-\frac{1}{4}e'^2ss'$ |
| 4E2.18 | $3\lambda' - \lambda - 2\varpi' - \Omega' + \Omega$ | $-\frac{27}{4}e'^2ss'$ |
| 4E2.19 | $2\lambda' - \varpi' + \varpi - 2\Omega'$ | $3ee's'^2$ |

Table B.10. Second-order arguments: indirect part (internal perturber).

| ID | Cosine Argument | Term |
|---|---|---|
| 4I2.1 | $\lambda' - 3\lambda + 2\varpi$ | $-\frac{27}{8}e^2 + \frac{27}{8}e^4 + \frac{27}{16}e^2e'^2$ $+\frac{27}{8}e^2s^2 + \frac{27}{8}e^2s'^2$ |
| 4I2.2 | $\lambda' + \lambda - 2\varpi$ | $-\frac{1}{8}e^2 - \frac{1}{24}e^4 + \frac{1}{16}e^2e'^2$ $+\frac{1}{8}e^2s^2 + \frac{1}{8}e^2s'^2$ |
| 4I2.3 | $2\lambda - \varpi' - \varpi$ | $3ee' - \frac{9}{4}e^3e' - 3ee's^2 - 3ee's'^2$ |
| 4I2.4 | $\lambda' + \lambda - 2\varpi'$ | $-\frac{1}{8}e'^2 + \frac{1}{16}e^2e'^2 - \frac{1}{24}e'^4$ $+\frac{1}{8}e'^2s^2 + \frac{1}{8}e'^2s'^2$ |
| 4I2.5 | $3\lambda' - \lambda - 2\varpi'$ | $-\frac{3}{8}e'^2 + \frac{3}{16}e^2e'^2 + \frac{3}{8}e'^4$ $+\frac{3}{8}e'^2s^2 + \frac{3}{8}e'^2s'^2$ |
| 4I2.6 | $\lambda' + \lambda - 2\Omega$ | $-s^2 + \frac{1}{2}e^2s^2 + \frac{1}{2}e'^2s^2 + s^2s'^2$ |
| 4I2.7 | $\lambda' + \lambda - \Omega' - \Omega$ | $2ss' - e^2ss' - e'^2ss' - s^3s' - ss'^3$ |
| 4I2.8 | $\lambda' + \lambda - 2\Omega'$ | $-s'^2 + \frac{1}{2}e^2s'^2 + \frac{1}{2}e'^2s'^2 + s^2s'^2$ |
| 4I2.9 | $2\lambda + \varpi' - 3\varpi$ | $\frac{1}{4}e^3e'$ |
| 4I2.10 | $2\lambda' - 4\lambda - \varpi' + 3\varpi$ | $-\frac{8}{3}e^3e'$ |
| 4I2.11 | $4\lambda' - 2\lambda - 3\varpi' + \varpi$ | $-\frac{2}{3}ee'^3$ |
| 4I2.12 | $2\lambda + \varpi' - \varpi - 2\Omega$ | $3ee's^2$ |
| 4I2.13 | $\lambda' - 3\lambda + 2\varpi - \Omega' + \Omega$ | $-\frac{27}{4}e^2ss'$ |
| 4I2.14 | $\lambda' + \lambda - 2\varpi - \Omega' + \Omega$ | $-\frac{1}{4}e^2ss'$ |
| 4I2.15 | $2\lambda - \varpi' - \varpi + \Omega' - \Omega$ | $6ee'ss'$ |
| 4I2.16 | $2\lambda + \varpi' - \varpi - \Omega' - \Omega$ | $-6ee'ss'$ |
| 4I2.17 | $\lambda' + \lambda - 2\varpi' + \Omega' - \Omega$ | $-\frac{1}{4}e'^2ss'$ |
| 4I2.18 | $3\lambda' - \lambda - 2\varpi' - \Omega' + \Omega$ | $-\frac{3}{4}e'^2ss'$ |
| 4I2.19 | $2\lambda + \varpi' - \varpi - 2\Omega'$ | $3ee's'^2$ |

Table B.11. Second-order arguments: functions of semi-major axis.

| $i$ | $f_i$ |
|---|---|

45 $\quad \frac{1}{8}\left[-5j + 4j^2 - 2\alpha D + 4j\alpha D + \alpha^2 D^2\right] A_j$

46 $\quad \frac{1}{96}\left[22j - 64j^2 + 60j^3 - 16j^4 + 16\alpha D\right.$
$\qquad -46j\alpha D + 48j^2\alpha D - 16j^3\alpha D - 12\alpha^2 D^2$
$\qquad \left. +9j\alpha^2 D^2 + 4j\alpha^3 D^3 + \alpha^4 D^4\right] A_j$

47 $\quad \frac{1}{32}\left[20j^3 - 16j^4 - 4\alpha D - 2j\alpha D + 16j^2\alpha D\right.$
$\qquad -16j^3\alpha D - 2\alpha^2 D^2 + 11j\alpha^2 D^2 + 4\alpha^3 D^3$
$\qquad \left. +4j\alpha^3 D^3 + \alpha^4 D^4\right] A_j$

48 $\quad \frac{1}{16}\left[2\alpha + j\alpha - 4j^2\alpha - 4j\alpha^2 D - \alpha^3 D^2\right](B_{j-1} + B_{j+1})$

49 $\quad \frac{1}{4}\left[-2 + 6j - 4j^2 + 2\alpha D - 4j\alpha D - \alpha^2 D^2\right] A_{j-1}$

50 $\quad \frac{1}{32}\left[20 - 86j + 126j^2 - 76j^3 + 16j^4 - 20\alpha D\right.$
$\qquad +74j\alpha D - 64j^2\alpha D + 16j^3\alpha D$
$\qquad \left. +14\alpha^2 D^2 - 17j\alpha^2 D^2 - 2\alpha^3 D^3 - 4j\alpha^3 D^3 - \alpha^4 D^4\right] A_{j-}$

51 $\quad \frac{1}{32}\left[-4 + 2j + 22j^2 - 36j^3 + 16j^4 + 4\alpha D\right.$
$\qquad +6j\alpha D - 32j^2\alpha D + 16j^3\alpha D$
$\qquad \left. -2\alpha^2 D^2 - 19j\alpha^2 D^2 - 6\alpha^3 D^3 - 4j\alpha^3 D^3 - \alpha^4 D^4\right] A_{j-1}$

52 $\quad \frac{1}{8}\left[-2j\alpha + 4j^2\alpha + 4j\alpha^2 D + \alpha^3 D^2\right](B_{j-2} + B_j)$

53 $\quad \frac{1}{8}\left[2 - 7j + 4j^2 - 2\alpha D + 4j\alpha D + \alpha^2 D^2\right] A_{j-2}$

54 $\quad \frac{1}{32}\left[-32 + 144j - 184j^2 + 92j^3 - 16j^4 + 32\alpha D\right.$
$\qquad -102j\alpha D + 80j^2\alpha D - 16j^3\alpha D - 16\alpha^2 D^2$
$\qquad \left. +25j\alpha^2 D^2 + 4\alpha^3 D^3 + 4j\alpha^3 D^3 + \alpha^4 D^4\right] A_{j-2}$

55 $\quad \frac{1}{96}\left[12 - 14j - 40j^2 + 52j^3 - 16j^4 - 12\alpha D\right.$
$\qquad -10j\alpha D + 48j^2\alpha D - 16j^3\alpha D + 6\alpha^2 D^2$
$\qquad \left. +27j\alpha^2 D^2 + 8\alpha^3 D^3 + 4j\alpha^3 D^3 + \alpha^4 D^4\right] A_{j-2}$

56 $\quad \frac{1}{16}\left[3j\alpha - 4j^2\alpha - 4j\alpha^2 D - \alpha^3 D^2\right](B_{j-3} + B_{j-1})$

57 $\quad \frac{1}{2}\left[\alpha\right] B_{j-1}$

58 $\quad \frac{1}{8}\left[-14\alpha + 16j\alpha - 4j^2\alpha + 4\alpha^2 D + \alpha^3 D^2\right] B_{j-1}$

59 $\quad \frac{1}{8}\left[2\alpha - 4j^2\alpha + 4\alpha^2 D + \alpha^3 D^2\right] B_{j-1}$

60 $\quad \frac{3}{4}\left[-\alpha^2\right] C_{j-2} + \frac{3}{4}\left[-\alpha^2\right] C_j$

61 $\quad \frac{1}{2}\left[-\alpha\right] B_{j-1} + \frac{3}{4}\left[-\alpha^2\right] C_{j-2} + \frac{15}{4}\left[-\alpha^2\right] C_j$

Table B.11. *(continued)*

| $i$ | $f_i$ |
|---|---|
| 62 | $[-\alpha] B_{j-1}$ |
| 63 | $\frac{1}{4} \left[ 14\alpha - 16j\alpha + 4j^2\alpha - 4\alpha^2 D - \alpha^3 D^2 \right] B_{j-1}$ |
| 64 | $\frac{1}{4} \left[ -2\alpha + 4j^2\alpha - 4\alpha^2 D - \alpha^3 D^2 \right] B_{j-1}$ |
| 65 | $\frac{1}{2} [\alpha] B_{j-1} + 3 [\alpha^2] C_{j-2} + \frac{3}{2} [\alpha^2] C_j$ |
| 66 | $\frac{1}{2} [\alpha] B_{j-1} + \frac{3}{2} [\alpha^2] C_{j-2} + 3 [\alpha^2] C_j$ |
| 67 | $\frac{1}{2} [-\alpha] B_{j-1} + \frac{15}{4} [-\alpha^2] C_{j-2} + \frac{3}{4} [-\alpha^2] C_j$ |
| 68 | $\frac{1}{96} \left[ 4 - 2j - 26j^2 - 4j^3 + 16j^4 - 4\alpha D \right.$ $-2j\alpha D + 16j^3\alpha D + 6\alpha^2 D^2 - 3j\alpha^2 D^2$ $\left. -2\alpha^3 D^3 - 4j\alpha^3 D^3 - \alpha^4 D^4 \right] A_{j+1}$ |
| 69 | $\frac{1}{96} \left[ 36 - 186j + 238j^2 - 108j^3 + 16j^4 - 36\alpha D \right.$ $+130j\alpha D - 96j^2\alpha D + 16j^3\alpha D + 18\alpha^2 D^2$ $\left. -33j\alpha^2 D^2 - 6\alpha^3 D^3 - 4j\alpha^3 D^3 - \alpha^4 D^4 \right] A_{j-3}$ |
| 70 | $\frac{1}{8} \left[ -14j\alpha + 4j^2\alpha - 8\alpha^2 D - \alpha^3 D^2 \right] B_{j-2}$ |
| 71 | $\frac{1}{8} \left[ -2j\alpha + 4j^2\alpha - \alpha^3 D^2 \right] B_j$ |
| 72 | $\frac{1}{8} \left[ -2\alpha - j\alpha + 4j^2\alpha + 4j\alpha^2 D + \alpha^3 D^2 \right] B_{j-1}$ |
| 73 | $\frac{1}{8} \left[ -2\alpha - j\alpha + 4j^2\alpha + 4j\alpha^2 D + \alpha^3 D^2 \right] B_{j+1}$ |
| 74 | $\frac{1}{4} \left[ 2j\alpha - 4j^2\alpha - 4j\alpha^2 D - \alpha^3 D^2 \right] B_{j-2}$ |
| 75 | $\frac{1}{4} \left[ 2j\alpha - 4j^2\alpha - 4j\alpha^2 D - \alpha^3 D^2 \right] B_j$ |
| 76 | $\frac{1}{4} \left[ 14j\alpha - 4j^2\alpha + 8\alpha^2 D + \alpha^3 D^2 \right] B_{j-2}$ |
| 77 | $\frac{1}{4} \left[ 2j\alpha - 4j^2\alpha + \alpha^3 D^2 \right] B_j$ |
| 78 | $\frac{1}{8} \left[ -3j\alpha + 4j^2\alpha + 4j\alpha^2 D + \alpha^3 D^2 \right] B_{j-3}$ |
| 79 | $\frac{1}{8} \left[ -3j\alpha + 4j^2\alpha + 4j\alpha^2 D + \alpha^3 D^2 \right] B_{j-1}$ |
| 80 | $\frac{3}{2} [\alpha^2] C_j$ |
| 81 | $\frac{3}{2} [\alpha^2] C_{j-2}$ |

Table B.12. Third-order arguments: direct part.

| ID | Cosine Argument | Term |
|----|----------------|------|
| 4D3.1 | $j\lambda' + (3 - j)\lambda - 3\varpi$ | $e^3 f_{82}$ |
| 4D3.2 | $j\lambda' + (3 - j)\lambda - \varpi' - 2\varpi$ | $e^2 e' f_{83}$ |
| 4D3.3 | $j\lambda' + (3 - j)\lambda - 2\varpi' - \varpi$ | $ee'^2 f_{84}$ |
| 4D3.4 | $j\lambda' + (3 - j)\lambda - 3\varpi'$ | $e'^3 f_{85}$ |
| 4D3.5 | $j\lambda' + (3 - j)\lambda - \varpi - 2\Omega$ | $es^2 f_{86}$ |
| 4D3.6 | $j\lambda' + (3 - j)\lambda - \varpi' - 2\Omega$ | $e's^2 f_{87}$ |
| 4D3.7 | $j\lambda' + (3 - j)\lambda - \varpi - \Omega' - \Omega$ | $ess' f_{88}$ |
| 4D3.8 | $j\lambda' + (3 - j)\lambda - \varpi' - \Omega' - \Omega$ | $e'ss' f_{89}$ |
| 4D3.9 | $j\lambda' + (3 - j)\lambda - \varpi - 2\Omega'$ | $es'^2 f_{86}$ |
| 4D3.10 | $j\lambda' + (3 - j)\lambda - \varpi' - 2\Omega'$ | $e's'^2 f_{87}$ |

Table B.13. Third-order arguments: indirect part (external perturber).

| ID | Cosine Argument | Term |
|----|----------------|------|
| 4E3.1 | $\lambda' - 4\lambda + 3\varpi$ | $-\frac{1}{3}e^3$ |
| 4E3.2 | $\lambda' + 2\lambda - 3\varpi$ | $-\frac{1}{24}e^3$ |
| 4E3.3 | $2\lambda' + \lambda - \varpi' - 2\varpi$ | $-\frac{1}{4}e^2 e'$ |
| 4E3.4 | $\lambda' + 2\lambda - 2\varpi' - \varpi$ | $-\frac{1}{16}ee'^2$ |
| 4E3.5 | $3\lambda' - 2\varpi' - \varpi$ | $\frac{81}{16}ee'^2$ |
| 4E3.6 | $2\lambda' + \lambda - 3\varpi'$ | $-\frac{1}{6}e'^3$ |
| 4E3.7 | $4\lambda' - \lambda - 3\varpi'$ | $-\frac{16}{3}e'^3$ |
| 4E3.8 | $\lambda' + 2\lambda - \varpi - 2\Omega$ | $-\frac{1}{2}es^2$ |
| 4E3.9 | $2\lambda' + \lambda - \varpi' - 2\Omega$ | $-2e's^2$ |
| 4E3.10 | $\lambda' + 2\lambda - \varpi - \Omega' - \Omega$ | $ess'$ |
| 4E3.11 | $2\lambda' + \lambda - \varpi' - \Omega' - \Omega$ | $4e'ss'$ |
| 4E3.12 | $\lambda' + 2\lambda - \varpi - 2\Omega'$ | $-\frac{1}{2}es'^2$ |
| 4E3.13 | $2\lambda' + \lambda - \varpi' - 2\Omega'$ | $-2e's'^2$ |

Table B.14. Third-order arguments: indirect part (internal perturber).

| ID | Cosine Argument | Term |
|---|---|---|
| 4I3.1 | $\lambda' - 4\lambda + 3\varpi$ | $-\frac{16}{3}e^3$ |
| 4I3.2 | $\lambda' + 2\lambda - 3\varpi$ | $-\frac{1}{6}e^3$ |
| 4I3.3 | $3\lambda - \varpi' - 2\varpi$ | $\frac{81}{16}e^2e'$ |
| 4I3.4 | $2\lambda' + \lambda - \varpi' - 2\varpi$ | $-\frac{1}{16}e^2e'$ |
| 4I3.5 | $\lambda' + 2\lambda - 2\varpi' - \varpi$ | $-\frac{1}{4}ee'^2$ |
| 4I3.6 | $2\lambda' + \lambda - 3\varpi'$ | $-\frac{1}{24}e'^3$ |
| 4I3.7 | $4\lambda' - \lambda - 3\varpi'$ | $-\frac{1}{3}e'^3$ |
| 4I3.8 | $\lambda' + 2\lambda - \varpi - 2\Omega$ | $-2es^2$ |
| 4I3.9 | $2\lambda' + \lambda - \varpi' - 2\Omega$ | $-\frac{1}{2}e's^2$ |
| 4I3.10 | $\lambda' + 2\lambda - \varpi - \Omega' - \Omega$ | $4ess'$ |
| 4I3.11 | $2\lambda' + \lambda - \varpi' - \Omega' - \Omega$ | $e'ss'$ |
| 4I3.12 | $\lambda' + 2\lambda - \varpi - 2\Omega'$ | $-2es'^2$ |
| 4I3.13 | $2\lambda' + \lambda - \varpi' - 2\Omega'$ | $-\frac{1}{2}e's'^2$ |

Table B.15. Third-order arguments: functions of semi-major axis.

| $i$ | $f_i$ |
|---|---|
| 82 | $\frac{1}{48}\left[-26j + 30j^2 - 8j^3 - 9\alpha D + 27j\alpha D - 12j^2\alpha D \right.$ $\left. +6\alpha^2 D^2 - 6j\alpha^2 D^2 - \alpha^3 D^3\right]A_j$ |
| 83 | $\frac{1}{16}\left[-9 + 31j - 30j^2 + 8j^3 + 9\alpha D - 25j\alpha D + 12j^2\alpha D \right.$ $\left. -5\alpha^2 D^2 + 6j\alpha^2 D^2 + \alpha^3 D^3\right]A_{j-1}$ |
| 84 | $\frac{1}{16}\left[8 - 32j + 30j^2 - 8j^3 - 8\alpha D + 23j\alpha D - 12j^2\alpha D \right.$ $\left. +4\alpha^2 D^2 - 6j\alpha^2 D^2 - \alpha^3 D^3\right]A_{j-2}$ |
| 85 | $\frac{1}{48}\left[-6 + 29j - 30j^2 + 8j^3 + 6\alpha D - 21j\alpha D + 12j^2\alpha D \right.$ $\left. -3\alpha^2 D^2 + 6j\alpha^2 D^2 + \alpha^3 D^3\right]A_{j-3}$ |
| 86 | $\frac{1}{4}\left[3\alpha - 2j\alpha - \alpha^2 D\right]B_{j-1}$ |
| 87 | $\frac{1}{4}\left[2j\alpha + \alpha^2 D\right]B_{j-2}$ |
| 88 | $\frac{1}{2}\left[-3\alpha + 2j\alpha + \alpha^2 D\right]B_{j-1}$ |
| 89 | $\frac{1}{2}\left[-2j\alpha - \alpha^2 D\right]B_{j-2}$ |

Table B.16. Fourth-order arguments: direct part.

| ID | Cosine Argument | Term |
|---|---|---|
| 4D4.1 | $j\lambda' + (4 - j)\lambda - 4\varpi$ | $e^4 f_{90}$ |
| 4D4.2 | $j\lambda' + (4 - j)\lambda - \varpi' - 3\varpi$ | $e^3 e' f_{91}$ |
| 4D4.3 | $j\lambda' + (4 - j)\lambda - 2\varpi' - 2\varpi$ | $e^2 e'^2 f_{92}$ |
| 4D4.4 | $j\lambda' + (4 - j)\lambda - 3\varpi' - \varpi$ | $ee'^3 f_{93}$ |
| 4D4.5 | $j\lambda' + (4 - j)\lambda - 4\varpi'$ | $e'^4 f_{94}$ |
| 4D4.6 | $j\lambda' + (4 - j)\lambda - 2\varpi - 2\Omega$ | $e^2 s^2 f_{95}$ |
| 4D4.7 | $j\lambda' + (4 - j)\lambda - \varpi' - \varpi - 2\Omega$ | $ee' s^2 f_{96}$ |
| 4D4.8 | $j\lambda' + (4 - j)\lambda - 2\varpi' - 2\Omega$ | $e'^2 s^2 f_{97}$ |
| 4D4.9 | $j\lambda' + (4 - j)\lambda - 4\Omega$ | $s^4 f_{98}$ |
| 4D4.10 | $j\lambda' + (4 - j)\lambda - 2\varpi - \Omega' - \Omega$ | $e^2 ss' f_{99}$ |
| 4D4.11 | $j\lambda' + (4 - j)\lambda - \varpi' - \varpi - \Omega' - \Omega$ | $ee' ss' f_{100}$ |
| 4D4.12 | $j\lambda' + (4 - j)\lambda - 2\varpi' - \Omega' - \Omega$ | $e'^2 ss' f_{101}$ |
| 4D4.13 | $j\lambda' + (4 - j)\lambda - \Omega' - 3\Omega$ | $s^3 s' f_{102}$ |
| 4D4.14 | $j\lambda' + (4 - j)\lambda - 2\varpi - 2\Omega'$ | $e^2 s'^2 f_{95}$ |
| 4D4.15 | $j\lambda' + (4 - j)\lambda - \varpi' - \varpi - 2\Omega'$ | $ee' s'^2 f_{96}$ |
| 4D4.16 | $j\lambda' + (4 - j)\lambda - 2\varpi' - 2\Omega'$ | $e'^2 s'^2 f_{97}$ |
| 4D4.17 | $j\lambda' + (4 - j)\lambda - 2\Omega' - 2\Omega$ | $s^2 s'^2 f_{103}$ |
| 4D4.18 | $j\lambda' + (4 - j)\lambda - 3\Omega' - \Omega$ | $ss'^3 f_{102}$ |
| 4D4.19 | $j\lambda' + (4 - j)\lambda - 4\Omega'$ | $s'^4 f_{98}$ |

Table B.17. Fourth-order arguments: indirect part (external perturber).

| ID | Cosine Argument | Term |
|---|---|---|
| 4E4.1 | $\lambda' - 5\lambda + 4\varpi$ | $-\frac{125}{384}e^4$ |
| 4E4.2 | $\lambda' + 3\lambda - 4\varpi$ | $-\frac{3}{128}e^4$ |
| 4E4.3 | $2\lambda' + 2\lambda - \varpi' - 3\varpi$ | $-\frac{1}{12}e^3 e'$ |
| 4E4.4 | $\lambda' + 3\lambda - 2\varpi' - 2\varpi$ | $-\frac{3}{64}e^2 e'^2$ |
| 4E4.5 | $3\lambda' + \lambda - 2\varpi' - 2\varpi$ | $-\frac{27}{64}e^2 e'^2$ |
| 4E4.6 | $2\lambda' + 2\lambda - 3\varpi' - \varpi$ | $-\frac{1}{12}ee'^3$ |
| 4E4.7 | $4\lambda' - 3\varpi' - \varpi$ | $8ee'^3$ |
| 4E4.8 | $3\lambda' + \lambda - 4\varpi'$ | $-\frac{27}{128}e'^4$ |
| 4E4.9 | $5\lambda' - \lambda - 4\varpi'$ | $-\frac{3125}{384}e'^4$ |
| 4E4.10 | $\lambda' + 3\lambda - 2\varpi - 2\Omega$ | $-\frac{3}{8}e^2 s^2$ |
| 4E4.11 | $2\lambda' + 2\lambda - \varpi' - \varpi - 2\Omega$ | $-ee' s^2$ |
| 4E4.12 | $3\lambda' + \lambda - 2\varpi' - 2\Omega$ | $-\frac{27}{8}e'^2 s^2$ |
| 4E4.13 | $\lambda' + 3\lambda - 2\varpi - \Omega' - \Omega$ | $\frac{3}{4}e^2 s s'$ |
| 4E4.14 | $2\lambda' + 2\lambda - \varpi' - \varpi - \Omega' - \Omega$ | $2ee' s s'$ |
| 4E4.15 | $3\lambda' + \lambda - 2\varpi' - \Omega' - \Omega$ | $\frac{27}{4}e'^2 s s'$ |
| 4E4.16 | $\lambda' + 3\lambda - 2\varpi - 2\Omega'$ | $-\frac{3}{8}e^2 s'^2$ |
| 4E4.17 | $2\lambda' + 2\lambda - \varpi' - \varpi - 2\Omega'$ | $-ee' s'^2$ |
| 4E4.18 | $3\lambda' + \lambda - 2\varpi' - 2\Omega'$ | $-\frac{27}{8}e'^2 s'^2$ |

Table B.18. Fourth-order arguments: indirect part (internal perturber).

| ID | Cosine Argument | Term |
|----|----------------|------|
| 4I4.1 | $\lambda' - 5\lambda + 4\varpi$ | $-\frac{3125}{384}e^4$ |
| 4I4.2 | $\lambda' + 3\lambda - 4\varpi$ | $-\frac{27}{128}e^4$ |
| 4I4.3 | $4\lambda - \varpi' - 3\varpi$ | $8e^3e'$ |
| 4I4.4 | $2\lambda' + 2\lambda - \varpi' - 3\varpi$ | $-\frac{1}{12}e^3e'$ |
| 4I4.5 | $\lambda' + 3\lambda - 2\varpi' - 2\varpi$ | $-\frac{27}{64}e^2e'^2$ |
| 4I4.6 | $3\lambda' + \lambda - 2\varpi' - 2\varpi$ | $-\frac{3}{64}e^2e'^2$ |
| 4I4.7 | $2\lambda' + 2\lambda - 3\varpi' - \varpi$ | $-\frac{1}{12}ee'^3$ |
| 4I4.8 | $3\lambda' + \lambda - 4\varpi'$ | $-\frac{3}{128}e'^4$ |
| 4I4.9 | $5\lambda' - \lambda - 4\varpi'$ | $-\frac{125}{384}e'^4$ |
| 4I4.10 | $\lambda' + 3\lambda - 2\varpi - 2\Omega$ | $-\frac{27}{8}e^2s^2$ |
| 4I4.11 | $2\lambda' + 2\lambda - \varpi' - \varpi - 2\Omega$ | $-ee's^2$ |
| 4I4.12 | $3\lambda' + \lambda - 2\varpi' - 2\Omega$ | $-\frac{3}{8}e'^2s^2$ |
| 4I4.13 | $\lambda' + 3\lambda - 2\varpi - \Omega' - \Omega$ | $\frac{27}{4}e^2ss'$ |
| 4I4.14 | $2\lambda' + 2\lambda - \varpi' - \varpi - \Omega' - \Omega$ | $2ee'ss'$ |
| 4I4.15 | $3\lambda' + \lambda - 2\varpi' - \Omega' - \Omega$ | $\frac{3}{4}e'^2ss'$ |
| 4I4.16 | $\lambda' + 3\lambda - 2\varpi - 2\Omega'$ | $-\frac{27}{8}e^2s'^2$ |
| 4I4.17 | $2\lambda' + 2\lambda - \varpi' - \varpi - 2\Omega'$ | $-ee's'^2$ |
| 4I4.18 | $3\lambda' + \lambda - 2\varpi' - 2\Omega'$ | $-\frac{3}{8}e'^2s'^2$ |

Table B.19. Fourth-order arguments: functions of semi-major axis.

| $i$ | $f_i$ |
|---|---|
| 90 | $\frac{1}{384}\left[-206j + 283j^2 - 120j^3 + 16j^4 - 64\alpha D\right.$ $+236j\alpha D - 168j^2\alpha D + 32j^3\alpha D + 48\alpha^2 D^2 - 78j\alpha^2 D^2$ $\left. +24j^2\alpha^2 D^2 - 12\alpha^3 D^3 + 8j\alpha^3 D^3 + \alpha^4 D^4\right]A_j$ |
| 91 | $\frac{1}{96}\left[-64 + 238j - 274j^2 + 116j^3 - 16j^4 + 64\alpha D\right.$ $-206j\alpha D + 156j^2\alpha D - 32j^3\alpha D - 36\alpha^2 D^2 + 69j\alpha^2 D^2$ $\left. -24j^2\alpha^2 D^2 + 10\alpha^3 D^3 - 8j\alpha^3 D^3 - \alpha^4 D^4\right]A_{j-1}$ |
| 92 | $\frac{1}{64}\left[52 - 224j + 259j^2 - 112j^3 + 16j^4 - 52\alpha D\right.$ $+176j\alpha D - 144j^2\alpha D + 32j^3\alpha D + 26\alpha^2 D^2 - 60j\alpha^2 D^2$ $\left. +24j^2\alpha^2 D^2 - 8\alpha^3 D^3 + 8j\alpha^3 D^3 + \alpha^4 D^4\right]A_{j-2}$ |
| 93 | $\frac{1}{96}\left[-36 + 186j - 238j^2 + 108j^3 - 16j^4 + 36\alpha D\right.$ $-146j\alpha D + 132j^2\alpha D - 32j^3\alpha D - 18\alpha^2 D^2 + 51j\alpha^2 D^2$ $\left. -24j^2\alpha^2 D^2 + 6\alpha^3 D^3 - 8j\alpha^3 D^3 - \alpha^4 D^4\right]A_{j-3}$ |
| 94 | $\frac{1}{384}\left[24 - 146j + 211j^2 - 104j^3 + 16j^4 - 24\alpha D\right.$ $+116j\alpha D - 120j^2\alpha D + 32j^3\alpha D + 12\alpha^2 D^2 - 42j\alpha^2 D^2$ $\left. +24j^2\alpha^2 D^2 - 4\alpha^3 D^3 + 8j\alpha^3 D^3 + \alpha^4 D^4\right]A_{j-4}$ |
| 95 | $\frac{1}{16}\left[16\alpha - 17j\alpha + 4j^2\alpha - 8\alpha^2 D + 4j\alpha^2 D + \alpha^3 D^2\right]B_{j-1}$ |
| 96 | $\frac{1}{8}\left[10j\alpha - 4j^2\alpha + 4\alpha^2 D - 4j\alpha^2 D - \alpha^3 D^2\right]B_{j-2}$ |
| 97 | $\frac{1}{16}\left[-3j\alpha + 4j^2\alpha + 4j\alpha^2 D + \alpha^3 D^2\right]B_{j-3}$ |
| 98 | $\frac{3}{8}\left[\alpha^2\right]C_{j-2}$ |
| 99 | $\frac{1}{8}\left[-16\alpha + 17j\alpha - 4j^2\alpha + 8\alpha^2 D - 4j\alpha^2 D - \alpha^3 D^2\right]B_{j-1}$ |
| 100 | $\frac{1}{4}\left[-10j\alpha + 4j^2\alpha - 4\alpha^2 D + 4j\alpha^2 D + \alpha^3 D^2\right]B_{j-2}$ |
| 101 | $\frac{1}{8}\left[3j\alpha - 4j^2\alpha - 4j\alpha^2 D - \alpha^3 D^2\right]B_{j-3}$ |
| 102 | $\frac{3}{2}\left[-\alpha^2\right]C_{j-2}$ |
| 103 | $\frac{9}{4}\left[\alpha^2\right]C_{j-2}$ |

# References

Adams, J. C. (1847). An explanation of the observed irregularities in the motion of Uranus, on the hypothesis of disturbances caused by a more distant planet; with a determination of the mass, orbit, and position of the disturbing body, *Mon. Not. R. Astron. Soc.* **7**, 149–152.

Aggarwal, H. R. and Oberbeck, V. R. (1974). Roche limit of a solid body, *Astrophy. J.* **191**, 577–588.

Aksnes, K. (1988). General formulas for three-body resonances, in *Long-Term Dynamical Behaviour of Natural and Artificial N-Body Systems*, ed. A. E. Roy (Kluwer, Dordrecht).

Alfvén, H. (1964). On the origin of asteroids, *Icarus* **3**, 52–56.

Allan, R. R. (1969). Evolution of the Mimas–Tethys commensurability, *Astron. J.* **74**, 497–506.

Anderson, J. D., Lau, E. L., Sjogren, W. L., Schubert, G., and Moore, W. B. (1996a). Gravitational constraints on the internal structure of Ganymede, *Nature* **384**, 541–543.

Anderson, J. D., Lau, E. L., Sjogren, W. L., Schubert, G., and Moore, W. B. (1997a). Europa's differentiated internal structure: Inferences from two Galileo experiments, *Science* **276**, 1236–1239.

Anderson, J. D., Lau, E. L., Sjogren, W. L., Schubert, G., and Moore, W. B. (1997b). Gravitational evidence for an undifferentiated Callisto, *Nature* **387**, 264–266.

Anderson, J. D., Schubert, G., Jacobson, R. A., Lau, E. L., Moore, W. B., and Sjogren, W. L. (1998). Distribution of rock, metals, and ices in Callisto, *Science* **280**, 1573–1576.

Anderson, J. D., Sjogren, W. L., and Schubert, G. (1996b). Galileo gravity results and the internal structure of Io, *Science* **272**, 709–712.

Applegate, J., Douglas, M. R., Gürsel, Y., Sussman, G. J., and Wisdom, J. (1986). The outer solar system for 200 million years, *Astron. J.* **92**, 176–194.

Baierlein, R. (1983). *Newtonian Dynamics* (McGraw-Hill, New York).

Bailey, M. E., Chamber, J. E., and Hahn, G. (1992). Origin of sungrazers: a frequent cometary end-state, *Astron. Astrophys.* **257**, 315–322.

557

Belton, M. J. S., Chapman, C. R., Thomas, P. C., Davies, M. E., Greenberg, R., Klaasen, K., Byrnes, D., D'Amario, L., Synnott, S., Johnson, T. V., McEwen, A., Merline, W., Davis, D. R., Petit, J.-M., Storrs, A., Veverka, J., and Zellner, B. (1995). Bulk density of asteroid 243-Ida from the orbit of its satellite Dactyl, *Nature* **374**, 785–788.

Bernal, J. D. (1969). *Science in History. Vol. 1. The Emergence of Science* (Pelican, Harmondsworth).

Black, G. J., Nicholson, P. D., and Thomas, P. C. (1995). Hyperion: Rotational dynamics, *Icarus* **117**, 149–171.

Blakely, R. J. (1995). *Potential Theory in Gravity and Magnetic Applications* (Cambridge University Press, Cambridge).

Blitzer, L. (1982). Dynamical stability and potential energy, *Am. J. Phys.* **50**, 431–434.

Boquet, F. (1889). Développement de la fonction perturbatrice, calcul des terms du huitième ordre, *Ann. Obs. Paris, Mém.* **19**, B1–B75.

Borderies, N. and Goldreich, P. (1984). A simple derivation of capture probabilities for the $j + 1 : j$ and $j + 2 : j$ orbit–orbit resonance problems, *Celest. Mech.* **32**, 127–136.

Borderies, N., Goldreich, P., and Tremaine, S. (1983). Perturbed particle disks, *Icarus* **55**, 124–132.

Borderies, N., Goldreich, P., and Tremaine, S. (1984). Unsolved problems in planetary ring dynamics, in *Planetary Rings*, ed. R. Greenberg and A. Brahic (University of Arizona Press, Tucson).

Borderies, N., Goldreich, P., and Tremaine, S. (1985). A granular flow model for dense planetary rings, *Icarus* **63**, 406–420.

Borderies, N., Goldreich, P., and Tremaine, S. (1989). The formation of sharp edges in planetary rings by nearby satellites, *Icarus* **80**, 344–360.

Bosh, A. and Rivkin, A. (1996). Observations of Saturn's inner satellites during the May 1995 ring-plane crossing, *Science* **272**, 518–521.

Bowman, F. (1958). *Introduction to Bessel Functions* (Dover, New York).

Brahic, A. (1977). Systems of colliding bodies in a gravitational field. I. Numerical simulation of the standard model, *Astron. Astrophys.* **54**, 895–907.

Bretagnon, P. (1974). Termes à longues périodes dans le système solaire, *Astron. Astrophys.* **30**, 141–154.

Bretagnon, P. (1982). Théorie du mouvement de l'ensemble des planètes. Solution VSOP-82, *Astron. Astrophys.* **114**, 278–288.

Broadfoot, A. L. et al. (1986). Ultraviolet spectrometer observations of Uranus, *Science* **233**, 74–79.

Brouwer, D. and Clemence, G. M. (1961). *Methods of Celestial Mechanics* (Academic Press, New York).

Brouwer, D. and van Woerkom, A. J. J. (1950). The secular variations of the orbital elements of the principal planets, *Astron. Papers Amer. Ephem.* **13**, 81–107.

Brown, E. W. and Shook, C. A. (1933). *Planetary Theory* (Cambridge University Press, Cambridge).

Bullen, K. E. (1975). *The Earth's Density* (Chapman and Hall, London).

Burkardt, T. M. and Danby, J. M. A. (1983). The solution of Kepler's equation, II, *Celest. Mech.* **31**, 317–328.

Burns, J. A. (1973). Where are the satellites of the inner planets? *Nature Physical Science* **242**, 23–25.

Burns, J. A. (1976). Elementary derivation of the perturbation equations of celestial mechanics, *Am. J. Phys.* **44**, 944–949.

Burns, J. A. (1986). The evolution of satellite orbits, in *Satellites*, ed. J. A. Burns and M. S. Matthews (University of Arizona Press, Tucson).

Burns, J. A., Lamy, P., and Soter, S. (1979). Radiation forces on small particles in the solar system, *Icarus* **40**, 1–48.

Burns, J. A., Schaffer, L. E., Greenberg, R. J., and Showalter, M. (1985). Lorentz resonances and the structure of the jovian ring, *Nature* **316**, 115–119.

Burns, J. A., Showalter, M. R., Cuzzi, J. N., and Pollack, J. B. (1980). Physical processes in Jupiter's ring: Clues to its origin by Jove! *Icarus* **44**, 339–360.

Burns, J. A., Showalter, M. R., Hamilton, D. P., Nicholson, P. D., de Pater, I., Ocketr-Bell, M. E., and Thomas, P. C. (1999). The formation of Jupiter's faint rings, *Science* **284**, 1146–1150.

Burns, J. A., Showalter, M. R., and Morfill, G. E. (1984). The ethereal rings of Jupiter and Saturn, in *Planetary Rings*, ed. R. Greenberg and A. Brahic (University of Arizona Press, Tucson).

Carpino, M., Milani, A., Nobili, A. M. (1987). Long-term numerical integrations and synthetic theories for the motion of the outer planets, *Astron. Astrophys.* **181**, 182–194.

Cauchy, A. L. (1827). Sur les moments d'inértie, *Exercises de Mathématiques* **2**, 93–103.

Champenois, S. and Vienne, A. (1999a). Chaos and secondary resonances in the Mimas–Tethys system, *Celest. Mech. Dyn. Astron.* in Press.

Champenois, S. and Vienne, A. (1999b). The role of secondary resonances in the evolution of the Mimas–Tethys system, *Icarus* (in press).

Chandrasekhar, S. (1987). *Ellipsoidal Figures of Equilibrium* (Dover, New York).

Chirikov, B. V. (1979). A universal instability of many-dimensional oscillator systems, *Phys. Rep.* **52**, 263–379.

Chodas, P. W. and Yeomans, D. K. (1996). The orbital motion and impact circumstances of Comet Shoemaker-Levy 9, in *The Collision of Comet Shoemaker-Levy 9 and Jupiter*, ed. K. S. Knoll, H. A. Weaver, and P. D. Feldman (Cambridge University Press, Cambridge).

Chree, C. (1896a). Forced vibrations in isotropic elastic solid spheres and spherical shells, *Cambridge Phil. Trans.* **16**, 14–57.

Chree, C. (1896b). Tides, on the "equilibrium theory", *Cambridge Phil. Trans.* **16**, 133–151.

Christou, A. A. (1999). A numerical survey of transient co-orbitals of the terrestrial planets, *Icarus* (submitted).

Christou, A. A. and Murray, C. D. (1997). A second order Laplace–Lagrange theory applied to the uranian satellite system, *Astron. Astrophys.* **326**, 416–427.

Clairaut, A. C. (1743). *Théorie de la Figure de la Terre, Tirée des Principes de l'Hydrastatique* (Durand, Paris).

Cochran, A. L., Levison, H. F., Stern, S. A., and Duncan, M. J. (1995). The discovery of Halley-sized Kuiper belt objects using the Hubble Space Telescope, *Astrophys. J.* **455**, 342–346.

Cohen, C. J. and Hubbard, E. C. (1965). Libration of the close approaches of Pluto to Neptune, *Astron. J.* **70**, 10–13.

Colombo, G., Franklin, F. A., and Munford, C. M. (1968). On a family of periodic orbits of the restricted three-body problem and the question of the gaps in the asteroid belt and in Saturn's rings, *Astron. J.* **73**, 111–123.

Colombo, G., Lautman, D. A., and Shapiro, I. I. (1966). The Earth's dust belt: Fact or fiction? 2. Gravitational focussing and Jacobi capture, *J. Geophys. Res.* **71**, 5705–5717.

Colwell, P. (1993). *Solving Kepler's Equation Over Three Centuries* (Willmann-Bell, Richmond).

Consolmagno, G. J. (1983). Lorentz forces on the dust in Jupiter's ring, *J. Geophys. Res.* **88**, 5607–5612.

Cook, A. F. and Franklin, F. A. (1964). Rediscussion of Maxwell's Adams prize essay on the stability of Saturn's rings, *Astron. J.* **69**, 173–200.

Cook, A. H. (1973). *Physics of the Earth and Planets* (Macmillan, London).

Cook, A. H. (1980). *Interiors of the Planets* (Cambridge University Press, Cambridge).

Cooke, M. L. (1991). Saturn's rings: Radial variation in the Keeler Gap and C ring photometry, Cornell University PhD thesis.

Cuzzi, J. N. and Burns, J. A. (1988). Charged particle depletion surrounding Saturn's F ring: Evidence for a moonlet belt? *Icarus* **74**, 284–324.

Cuzzi, J. N. and Scargle, J. D. (1985). Wavy edges suggest moonlet in Encke's Gap, *Astrophys. J.* **292**, 276–290.

Danby, J. M. A. (1987). The solution of Kepler's equation, III, *Celest. Mech.* **40**, 303–312.

Danby, J. M. A. (1988). *Fundamentals of Celestial Mechanics,* 2nd ed. (Willmann-Bell, Richmond).

Danby, J. M. A. and Burkardt, T. M. (1983). The solution of Kepler's equation, I, *Celest. Mech.* **31**, 95–107.

Darwin, G. H. (1899). The theory of the figure of the Earth carried to the second order in small quantities, *Mon. Not. R. Astron. Soc.* **60**, 82–124.

Darwin, G. H. (1908). Tidal friction and cosmogony, in *Scientific Papers* **2** (Cambridge University Press, Cambridge).

Deprit, A. and Deprit-Bartholomé, A. (1967). Stability of the triangular Lagrangian points, *Astron. J.* **72**, 173–179.

Deprit, A., Henrard, J., and Rom, A. (1971). Analytical lunar ephemeris: Delaunay's theory, *Astron. J.* **76**, 269–272.

Dermott, S. F. (1972). Bode's law and the preference for near-commensurability among pairs of orbital periods in the Solar System, in *The Origin of the Solar System,* ed. H. Reeves (CNRS, Paris).

Dermott, S. F. (1973). Bode's law and the resonant structure of the Solar System, *Nature Physical Science* **244**, 18–21.

Dermott, S. F. (1979a). Tidal dissipation in the solid cores of the major planets, *Icarus* **37**, 310–321.

Dermott, S. F. (1979b). Shapes and gravitational moments of satellites and asteroids, *Icarus* **37**, 576–586.

Dermott, S. F. (1984). Dynamics of narrow rings, in *Planetary Rings*, ed. R. Greenberg and A. Brahic (University of Arizona Press, Tucson).

Dermott, S. F. and Gold, T. (1977). The rings of Uranus: Theory, *Nature* **267**, 590–593.

Dermott, S. F., Gold, T., and Sinclair, A. T. (1979). The rings of Uranus: Nature and origin, *Astron. J.* **84**, 1225–1234.

Dermott, S. F., Gomes, R. S., Durda, D. D., Gustafson, B. A. S., Jayaraman, S., Xu, Y. L., and Nicholson, P. D. (1992). Dynamics of the zodiacal cloud, in *Chaos, Resonance and Collective Dynamical Phenomena in the Solar System*, ed. S. Ferraz-Mello (Kluwer, Dordrecht).

Dermott, S. F., Jayaraman, S., Xu, Y. L., Gustafson, B. Å. S., and Liou, J. C. (1994). A circumsolar ring of asteroidal dust in resonant lock with the Earth, *Nature* **369**, 719–723.

Dermott, S. F., Malhotra, R., and Murray, C. D. (1988). Dynamics of the uranian and saturnian satellite systems: A chaotic route to melting Miranda? *Icarus* **76**, 295–334.

Dermott, S. F. and Murray, C. D. (1980). Origin of the eccentricity gradient and apse alignment of the $\epsilon$ ring of Uranus, *Icarus* **43**, 338–349.

Dermott, S. F. and Murray, C. D. (1981a). The dynamics of tadpole and horseshoe orbits. I. Theory, *Icarus* **48**, 1–11.

Dermott, S. F. and Murray, C. D. (1981b). The dynamics of tadpole and horseshoe orbits. II. The coorbital satellites of Saturn, *Icarus* **48**, 12–22.

Dermott, S. F. and Murray, C. D. (1983). Nature of the Kirkwood gaps in the asteroid belt, *Nature* **301**, 201–205.

Dermott, S. F., Murray, C. D., and Sinclair, A. T. (1980). The narrow rings of Jupiter, Saturn and Uranus, *Nature* **284**, 309–313.

Dermott, S. F. and Nicholson, P. D. (1986). Masses of the satellites of Uranus, *Nature* **319**, 115–120.

Dermott, S. F., Nicholson, P. D., Burns, J. A., and Houck, J. R. (1984). Origin of the Solar System dust bands discovered by *IRAS*, *Nature* **312**, 505–509.

Dermott, S. F., Nicholson, P. D., Burns, J. A., and Houck, J. R. (1985). An analysis of *IRAS*' Solar System dust bands, in *Properties and Interactions of Interplanetary Dust*, ed. R. H. Giese and P. Lamy (D. Reidel, Dordrecht).

Dermott, S. F. and Sagan, C. (1995). Tidal effects of disconnected hydrocarbon seas on Titan, *Nature* **374**, 238–240.

Dermott, S. F. and Thomas, P. C. (1988). The shape and internal structure of Mimas, *Icarus* **73**, 25–65.

Domingo, V., Fleck, B., and Poland, A. I. (1995). The *SOHO* mission – An overview, *Solar Physics* **162**, 1–37.

Duncan, M., Quinn, T., and Tremaine, S. (1987). The origin of short-period comets, *Astrophys. J.* **328**, L69–73.

Duncan, M., Quinn, T., and Tremaine, S. (1989). The long-term evolution of orbits in the solar system: A mapping approach, *Icarus* **82**, 402–418.

Duriez, L. (1979). Approche d'une théorie générale planétaire en variables elliptiques héliocentriques, University of Lille PhD thesis.

Duxbury, T. C. and Callahan, J. (1982). Phobos and Deimos cartography, *Lunar Planet. Sci.* **XIII**, 190.

Edgeworth, K. E. (1943). The evolution of our planetary system, *J. Brit. Astron. Assoc.* **53**, 181–188.

Elliot, J. L., Dunham, E. W., and Mink, D. J. (1977). The rings of Uranus, *Nature* **267**, 328–330.

Elliot, J. L. and Nicholson, P. D. (1984). The rings of Uranus, in *Planetary Rings*, ed. R. Greenberg and A. Brahic (University of Arizona Press, Tucson).

Ellis, K. M. and Murray, C. D. (1999). The disturbing function in solar system dynamics, *Icarus* (submitted for publication).

Fernandez, J. A. (1997). The formation of the Oort cloud and the primitive galactic environment, *Icarus* **129**, 106–119.

Feynman, R. P., Leighton, R. B., and Sands, M. (1963). *The Feynman Lectures on Physics, Vol. 1* (Addison-Wesley, Reading, MA).

Field, J. V. (1988). *Kepler's Geometric Cosmology* (Athlone Press, London).

French, R. G., Elliot, J. L., French, L. M., Kangas, J. A., Meech, K. J., Ressler, M. E., Buie, M. W., Frogel, J. A., Holberg, J. B., Fuensalida, J. J., and Joy, M. (1988). Uranian ring orbits from Earth-based and *Voyager* occultation experiments, *Icarus* **73**, 349–378.

French, R. G., Nicholson, P. D., Porco, C. C., and Marouf, E. A. (1991). Dynamics and structure of the uranian rings, in *Uranus*, ed. J. Bergstrahl et al. (University of Arizona Press, Tucson).

Froeschlé, C., Hahn, G., Gonczi, R., Morbidelli, A., and Farinella, P. (1995). Secular resonances and the dynamics of Mars-crossing and near-Earth asteroids, *Icarus* **117**, 45–61.

Froeschlé, Ch. and Morbidelli, A. (1994). The secular resonances in the solar system, in *Asteroids, Comets, Meteors 1993*, ed. A. Milani, M. di Martino, and A. Cellino (Kluwer, Dordrecht).

Gehrels, T. et al. (1980). Imaging photopolarimeter on Pioneer Saturn, *Science* **207**, 434–439.

Giuliatti Winter, S. M. (1994). The dynamics of Saturn's F ring, University of London PhD thesis.

Gladman, B. J., Migliorini, F., Morbidelli, A., Zappala, V., Michel, P., Cellino, A., Froeschlé, C., Levison, H. F., Bailey, M., and Duncan, M. (1997). Dynamical lifetimes of objects injected into asteroid belt resonances, *Science* **277**, 197–201.

Goldreich, P. (1965). An explanation of the frequent occurrence of commensurable mean motions in the solar system, *Mon. Not. R. Astron. Soc.* **130**, 159–181.

Goldreich, P. (1966). Final spin states of planets and satellites, *Astron. J.* **71**, 1–7.

Goldreich, P. and Nicholson, P. D. (1977a). Turbulent viscosity and Jupiter's tidal $Q$, *Icarus* **30**, 301–304.

Goldreich, P. and Nicholson, P. D. (1977b). The revenge of tiny Miranda, *Nature* **269**, 783–785.

Goldreich, P. and Peale, S. J. (1966). Spin-orbit coupling in the solar system, *Astron. J.* **71**, 425–438.

Goldreich, P. and Peale, S. J. (1968). Dynamics of planetary rotations, *Annu. Rev. Astron. Astrophys.* **6**, 287–320.

Goldreich, P. and Soter, S. (1966). *Q* in the solar system, *Icarus* **5**, 375–389.

Goldreich, P. and Tremaine, S. (1978a). The velocity dispersion in Saturn's rings, *Icarus* **34**, 227–239.

Goldreich, P. and Tremaine, S. (1978b). The formation of the Cassini Division in Saturn's rings, *Icarus* **34**, 240–253.

Goldreich, P. and Tremaine, S. (1979a). Towards a theory for the uranian rings, *Nature* **277**, 97–99.

Goldreich, P. and Tremaine, S. (1979b). Precession of the $\epsilon$ ring of Uranus, *Astron. J.* **84**, 1638–1641.

Goldreich, P. and Tremaine, S. (1980). Disk–satellite interactions, *Astrophys. J.* **241**, 425–441.

Goldreich, P. and Tremaine, S. (1981). The origin of the eccentricities of the rings of Uranus, *Astrophys. J.* **243**, 1062–1075.

Goldreich, P. and Tremaine, S. (1982). The dynamics of planetary rings, *Annu. Rev. Astron. Astrophys.* **20**, 249–283.

Goldreich, P., Tremaine, S., and Borderies, N. (1986). Towards a theory for Neptune's arc rings, *Astron. J.* **92**, 490–494.

Greenberg, R. (1973). The inclination-type resonance of Mimas and Tethys, *Mon. Not. R. Astron. Soc.* **165**, 305–311.

Greenberg, R. (1975). On the Laplace relation among the satellites of Uranus, *Mon. Not. R. Astron. Soc.* **173**, 121–129.

Greenberg, R. (1976). The Laplace relation and the masses of Uranus' satellites, *Icarus* **29**, 427–433.

Greenberg, R. (1977). Orbit–orbit resonances among natural satellites, in *Planetary Satellites*, ed. J. A. Burns (University of Arizona Press, Tucson).

Greenberg, R. (1981). Apsidal precession of orbits about an oblate planet, *Astron. J.* **86**, 912–914.

Greenberg, R. and Brahic, A. (1984). *Planetary Rings* (University of Arizona Press, Tucson).

Greenberg, R. and Scholl, H. (1979). Resonances in the asteroid belt, in *Asteroids*, ed. T. Gehrels (University of Arizona Press, Tucson).

Grosser, M. (1979). *The Discovery of Neptune* (Dover, New York).

Grün, E., Morfill, G. E., and Mendis, D. A. (1984). Dust–magnetosphere interactions, in *Planetary Rings*, ed. R. Greenberg and A. Brahic (University of Arizona Press, Tucson).

Guckenheimer, J. and Holmes, P. (1983). *Nonlinear Oscillations, Dynamical Systems, and Bifurcations of Vector Fields* (Springer-Verlag, New York).

Hagihara, Y. (1970). *Celestial Mechanics I. Dynamical Principles and Transformation Theory* (MIT Press, Cambridge).

Hagihara, Y. (1972a). *Celestial Mechanics II, i. Perturbation Theory* (MIT Press, Cambridge).

Hagihara, Y. (1972b). *Celestial Mechanics II, ii. Perturbation Theory* (MIT Press, Cambridge).

Hagihara, Y. (1974a). *Celestial Mechanics III, i. Differential Equations in Celestial Mechanics* (Japan Society for the Promotion of Science, Tokyo).

Hagihara, Y. (1974b). *Celestial Mechanics III, ii. Differential Equations in Celestial Mechanics* (Japan Society for the Promotion of Science, Tokyo).

Hagihara, Y. (1975a). *Celestial Mechanics IV, i. Periodic and Quasi-Periodic Solutions* (Japan Society for the Promotion of Science, Tokyo).

Hagihara, Y. (1975b). *Celestial Mechanics IV, ii. Periodic and Quasi-Periodic Solutions* (Japan Society for the Promotion of Science, Tokyo).

Hagihara, Y. (1976a). *Celestial Mechanics V, i. Topology of the Three-Body Problem* (Japan Society for the Promotion of Science, Tokyo).

Hagihara, Y. (1976b). *Celestial Mechanics V, ii. Topology of the Three-Body Problem* (Japan Society for the Promotion of Science, Tokyo).

Hamilton, D. P. (1994). A comparison of Lorentz, planetary gravitational, and satellite gravitational resonances, *Icarus* **109**, 221–240.

Hamilton, D. P. and Burns, J. A. (1994). Origin of Saturn's E ring: Self-sustained, naturally, *Science* **264**, 550–553.

Hanel, R. et al. (1981). Infrared observations of the saturnian system from *Voyager 1*, *Science* **212**, 192–200.

Harper, D. and Taylor, D. B. (1993). The orbits of the major satellites of Saturn, *Astron. Astrophys.* **268**, 326–349.

Haughton, S. (1855). On the rotation of a solid body round a fixed point; being an account of the late Professor MacCullagh's lectures on that subject, *Trans. Royal Irish Academy* **22**, 139–154.

Hénon, M. (1969). Numerical exploration of the restricted problem. V. Hill's case: Periodic orbits and their stability, *Astron. Astrophys.* **1**, 223–238.

Hénon, M. and Petit, J.-M. (1986). Series expansions for encounter-type solutions of Hill's problem, *Celest. Mech.* **38**, 67–100.

Henrard, J. (1982). Capture into resonance: An extension of the use of adiabatic invariants, *Celest. Mech.* **27**, 3–22.

Henrard, J. and Lemaître, A. (1983). A second fundamental model of resonance, *Celest. Mech.* **30**, 197–218.

Her Majesty's Stationery Office (1994). *Astronomical Almanac for the Year 1995* (HMSO, London).

Hill, G. W. (1878). Researches in the lunar theory, *Am. J. Math.* **1**, 5–26, 129–147, 245–261.

Hill, G. W. (1897). On the values of the eccentricities and longitudes of the perihelia of Jupiter and Saturn for distant epochs, *Astron. J.* **17**, 81–87.

Hirayama, K. (1918). Groups of asteroid probably of common origin, *Astron. J.* **31**, 185–188.

Holberg, J. B., Forrester, W. T., and Lissauer, J. J. (1982). Identification of resonance features within the rings of Saturn, *Nature* **297**, 115–120.

Holman, M. J. and Murray, N. W. (1996). Chaos in high-order mean motion resonances in the outer asteroid belt, *Astron. J.* **112**, 1278–1293.

Holman, M. J., Touma, J., and Tremaine, S. (1997). Chaotic variations in the eccentricity of the planet orbiting 16 Cygni B, *Nature* **386**, 254–256.

Horanyi, M. and Porco, C. C. (1993). Where exactly are the arcs of Neptune? *Icarus* **106**, 525–535.

Hoyle, F. (1974). The work of Nicolaus Copernicus, *Proc. R. Soc. London. A* **336**, 105–114.

Hubbard, W. B., Brahic, A., Sicardy, B., Elicer, L.-R., Roques, F., and Vilas, F. (1986). Occultation detection of a neptunian ring-like arc, *Nature* **319**, 636–640.

Hughes, S. (1981). The computation of tables of Hansen coefficients, *Celest. Mech.* **25**, 101–107.

Huygens, C. (1659). *Systema Saturnium* (The Hague).

Jefferys, W. H. (1971). *An Atlas of Surfaces of Section for the Restricted Problem of Three Bodies* (University of Texas, Austin).

Jeffreys, H. (1929). *The Earth*, 2nd ed. (Cambridge University Press, Cambridge).

Jeffreys, H. (1961). The effect of tidal friction on eccentricity and inclination, *Mon. Not. R. Astron. Soc.* **122**, 339–343.

Jeffreys, H. (1970). *The Earth*, 5th ed. (Cambridge University Press, Cambridge).

Jewitt, D. and Luu, J. (1996). The Plutinos, in *Completing the Inventory of the Solar System, ASP Conference Series, Vol. 107*, ed. T. W. Rettig and J. M. Hahn (Astronomical Society of the Pacific, San Francisco).

Julian, W. H. and Toomre, A. (1966). Non-axisymmetric responses of differentially rotating disks of stars, *Astrophys. J.* **146**, 810–830.

Kaula, W. M. (1961). Analysis of gravitational and geometric aspects of geodetic utilization of satellites, *Geophys. J.* **5**, 104–133.

Kaula, W. M. (1962). Development of the lunar and solar disturbing functions for a close satellite, *Astron. J.* **67**, 300–303.

Kaula, W. M. (1966). *Theory of Satellite Geodesy* (Blaisdell, Waltham, MA.).

Kepler, J. (1596). *Mysterium Cosmographicum*, 1st ed. (Tübingen).

Kepler, J. (1609). *Astronomia Nova* (Heidelberg).

Kepler, J. (1610). *Dissertatio cum Nuncio Sidereo* (Prague).

Kepler, J. (1619). *Harmonices Mundi Libri V* (Linz).

Kepler, J. (1621). *Mysterium Cosmographicum*, 2nd ed. (Frankfurt).

Kinoshita, H. and Nakai, H. (1984). The motions of the perihelions of Neptune and Pluto, *Celest. Mech.* **34**, 203–217.

Kinoshita, H. and Nakai, H. (1996). Long-term behaviour of the motion of Pluto over 5.5 billion years, *Earth, Moon, and Planets* **72**, 165–173.

Kinoshita, H., Yoshida, H., and Nakai, H. (1991). Symplectic integrators and their application to dynamical astronomy, *Celest. Mech. Dyn. Astron.* **50**, 59–71.

Kirkwood, D. (1867). *Meteoric Astronomy* (Lippincott, Philadelphia).

Klavetter, J. J. (1989). Rotation of Hyperion. I. Observations, *Astron. J.* **97**, 570–579.

Kneević, Z. and Milani, A. (1994). Asteroid proper elements: The big picture, in *Asteroids, Comets, Meteors 1993*, ed. A. Milani, M. di Martino, and A. Cellino (Kluwer, Dordrecht).

Kolvoord, R. A., Burns, J. A., and Showalter, M. R. (1990). Periodic features in Saturn's F ring, *Nature* **345**, 695–697.

Kopal, Z. (1972). *The Solar System* (Oxford University Press, London).

Kozai, Y. (1957). On the astronomical constants of saturnian satellite system, *Ann. Tokyo Astron. Obs. 2nd Ser.* **5**, 73–106.

Kozai, Y. (1962). Secular perturbations of asteroids with high inclinations and eccentricities, *Astron. J.* **67**, 591–598.

Kuiper, G. (1944). Titan: A satellite with an atmosphere, *Astrophys. J.* **100**, 378–383.

Kuiper, G. (1951). The origin of the solar system, in *Astrophysics: A Topical Symposium*, ed. J. A. Hyner (McGraw-Hill, New York), pp. 357–406.

Lane, A. L., Hord, C. W., West, R. A., Esposito, L. W., Coffeen, D. L., Sato, M., Simmons, K. E., Pomphrey, R. B., and Morris, R. B. (1982). Photopolarimetry from *Voyager 2:* Preliminary results on Saturn, Titan, and the rings, *Science* **215**, 537–543.

Laskar, J. (1985). Accurate methods in general planetary theory, *Astron. Astrophys.* **144**, 133–146.

Laskar, J. (1986a). Secular terms of classical planetary theories using the results of general theory, *Astron. Astrophys.* **157**, 59–70.

Laskar, J. (1986b). A general theory for the uranian satellites, *Astron. Astrophys.* **166**, 349–358.

Laskar, J. (1988). Secular evolution of the Solar System over 10 million years, *Astron. Astrophys.* **198**, 341–362.

Laskar, J. (1989). A numerical experiment on the chaotic behaviour of the solar system, *Nature* **338**, 237–238.

Laskar, J. (1994). Large scale chaos in the solar system, *Astron. Astrophys.* **287**, L9–12.

Laskar, J., Quinn, T., and Tremaine, S. (1992). Confirmation of resonant structure in the solar system, *Icarus* **95**, 148–152.

Lemaître, A. and Morbidelli, A. (1994). Proper elements for highly inclined asteroidal orbits, *Celest. Mech. Dyn. Astron.* **60**, 29–56.

Le Verrier, U. J.-J. (1847). Sur la planète qui produit les anomalies observées dans le mouvement d'Uranus. Détermination de sa masse, de son orbite et de sa position actuelle, *Comptes Rendus* **23**, 428–438, 657–659.

Le Verrier, U. J.-J. (1855). Développement de la fonction qui sert de base au calcul des perturbations des mouvements des planètes, *Ann. Obs. Paris, Mém.* **1**, 258–331.

Levison, H. F. and Duncan, M. J. (1994). The long-term dynamical behavior of short-period comets, *Icarus* **108**, 18–36.

Lichtenberg, A. J. and Lieberman, M. A. (1983). *Regular and Stochastic Motion* (Springer-Verlag, New York).

Lieske, J. H. (1998). Galilean satellite ephemerides E5, *Astron. Astrophys. Suppl.* **129**, 205–217.

Lin, D. N. C. and Papaloizou, J. (1979). Tidal torques on accretion discs in binary systems with extreme mass ratios, *Mon. Not. R. Astron. Soc.* **186**, 799–812.

Lissauer, J. J. (1985). Shepherding model for Neptune's arc ring, *Nature* **318**, 544–545.

Lissauer, J. J., Goldreich, P., and Tremaine, S. (1985). Evolution of the Janus–Epimetheus coorbital resonance due to torques from Saturn's rings, *Icarus* **64**, 425–434.

Longuet-Higgins, M. S. (1950). A theory of the origin of microseisms, *Phil. Trans. R. Soc. London A* **243**, 1–35.

Love, A. E. H. (1911). *Some Problems of Geodynamics* (Cambridge University Press, Cambridge).

Love, A. E. H. (1944). *A Treatise on the Mathematical Theory of Elasticity*, 4th ed. (Dover, New York).

Low, F. J. et al. (1984). Infrared cirrus: New components of the extended infrared emission, *Astrophys. J.* **278**, L19–22.

Lunine, J. I. (1993). Does Titan have an ocean? A review of current understanding of Titan's surface, *Rev. Geophys.* **31**, 133–149.

Luu, J. (1994). The Kuiper belt, in *Asteroids, Comets, Meteors 1993*, ed. A. Milani, M. Di Martino, and M. Cellino (Kluwer, Dordrecht).

Luu, J., and Jewitt, D. (1996). Enlarging the solar system: The Kuiper belt, in *Completing the Inventory of the Solar System, ASP Conference Series, Vol. 107*, ed. T. W. Rettig and J. M. Hahn (Astronomical Society of the Pacific, San Francisco).

Lynden-Bell, D. and Pringle, J. E. (1974). The evolution of viscous discs and the origin of the nebular variables, *Mon. Not. R. Astron. Soc.* **168**, 603–637.

MacCullagh, J. (1844a). *Proc. Royal Irish Academy* **2**, 520–526.

MacCullagh, J. (1844b). *Proc. Royal Irish Academy* **2**, 542–545.

MacDonald, G. J. F. (1964). Tidal friction, *Rev. Geophys.* **2**, 467–541.

MacMillan, W. D. (1936). *Dynamics of Rigid Bodies* (McGraw-Hill, New York).

MacRobert, T. M. (1967). *Spherical Harmonics* (Pergamon Press, Oxford).

Malhotra, R. (1988). Some aspects of the dynamics of orbit–orbit resonances in the uranian satellite system, Cornell University PhD thesis.

Malhotra, R. (1993). The origin of Pluto's peculiar orbit, *Nature* **365**, 819–820.

Malhotra, R. (1994). A mapping method for the gravitational few-body problem with dissipation, *Celest. Mech. Dyn. Astron.* **60**, 373–385.

Malhotra, R. (1995). The origin of Pluto's orbit – Implications for the solar system beyond Neptune, *Astron. J.* **110**, 420–429.

Malhotra, R., Black, D., Eck, A., and Jackson, A. (1992). Resonant orbital evolution in the putative planetary system of PSR1257+12, *Nature* **356**, 583–585.

Malhotra, R., Fox, K., Murray, C. D., and Nicholson, P. D. (1989). Secular perturbations of the uranian satellites: Theory and practice, *Astron. Astrophys.* **221**, 348–358.

Marchal, C. (1990). *The Three-Body Problem* (Elsevier, Amsterdam).

Marley, M. S. (1990). Nonradial oscillations of Saturn: Implications for ring system structure, University of Arizona PhD thesis.

Marley, M. S. and Porco, C. C. (1993). Planetary acoustic mode seismology: Saturn's rings, *Icarus* **106**, 508–524.

Marsden, B. G. and Williams, G. V. (1997). *IAU Circular 6780*.

Maxwell, J. C. (1859). *On the Stability of the Motions of Saturn's Rings* (MacMillan, London).

Meech, K. and Belton, M. (1989), *IAU Circular 4770*.

Melnikov, V. K. (1963). On the stability of the center for time periodic perturbations, *Trans. Moscow Math. Soc.* **12**, 1–57.

Message, P. J. (1966). On nearly-commensurable periods in the restricted problem of three bodies, with calculations of the long-period variations of the interior 2:1 case, in *The Theory of Orbits in the Solar System and in Stellar Systems*, IAU Symposium No. 25, ed. G. Contopoulos (Academic Press, London).

Message, P. J. (1989). The use of computer algorithms in the construction of a theory of the long-period perturbations of Saturn's satellite Hyperion, *Celest. Mech.* **45**, 45–53.

Message, P. J. (1993). On the second order long-period motion of Hyperion, *Celest. Mech. Dyn. Astron.* **56**, 277–284.

Michel, P. and Thomas, F. (1995). The Kozai resonance for near-Earth asteroids with semimajor axes smaller than 2 AU, *Astron. Astrophys.* **307**, 310–318.

Mikkola, S., Innanen, K. A., Muinonen, K., and Bowell, E. (1994). A preliminary analysis of the orbit of the Mars trojan asteroid (5261) Eureka, *Celest. Mech. Dyn. Astron.* **58**, 53–64.

Milani, A. (1993). The Trojan asteroid belt: Proper elements, stability, chaos and families, *Celest. Mech.* **57**, 59–94.

Milani, A., Carpino, M., Hahn, G., and Nobili, A. M. (1989). Dynamics of planet-crossing asteroids: Classes of orbital behaviour, *Icarus* **78**, 212–269.

Milani, A. and Knežević, Z. (1990). Secular perturbation theory and computation of asteroid proper elements, *Celest. Mech.* **49**, 347–411.

Milani, A. and Nobili, A. M. (1992). An example of stable chaos in the solar system, *Nature* **357**, 569–571.

Molnar, L. A. and Dunn, D. E. (1995). The Mimas 2:1 eccentric corotational resonance in Saturn's outer B ring, *Icarus* **116**, 397–408.

Montenbruck, O. (1989). *Practical Ephemeris Calculations* (Springer-Verlag, Heidelberg).

Moons, M. and Morbidelli, A. (1995). Secular resonances in mean motion commensurabilities – the 4/1, 3/1, 5/2, and 7/3 cases, *Icarus* **114**, 33–50.

Morabito, L., Synnott, S. P., Kupferman, P. N., and Collins, S. A. (1979). Discovery of current active extraterrestrial volcanism, *Science* **204**, 972.

Murray, C. D. (1985). A note on Le Verrier's expansion of the disturbing function, *Celest. Mech.* **36**, 163–164.

Murray, C. D. (1986). The structure of the 2:1 and 3:2 jovian resonances, *Icarus* **65**, 70–82.

Murray, C. D. (1994a). Planetary ring dynamics, *Phil. Trans. R. Soc. London* A **349**, 335–344.

Murray, C. D. (1994b). The dynamical effects of drag in the circular restricted three-body problem: I. The location and stability of the Lagrangian equilibrium points, *Icarus* **112**, 465–484.

Murray, C. D. (1996). Real and imaginary Kirkwood gaps, *Mon. Not. R. Astron. Soc.* **279**, 978–986.

Murray, C. D. (1998). Chaotic motion in the solar system, in *Encyclopedia of the Solar System*, ed. T. Johnson, P. Weissman, and L. McFadden (Academic Press, Orlando).

Murray, C. D. (1999). The dynamics of planetary rings and small satellites, in *Dynamics of Small Bodies in the Solar System: A Major Key to Solar System Studies*, ed. B. Steves and A. E. Roy (Kluwer, Dordrecht).

Murray, C. D. and Fox, K. (1984). Structure of the 3:1 jovian resonance: A comparison of numerical methods, *Icarus* **59**, 221–233.

Murray, C. D. and Giuliatti Winter, S. M. (1996). Periodic collisions between the moon Prometheus and Saturn's F ring, *Nature* **380**, 139–141.

Murray, C. D., Gordon, M. K., and Giuliatti Winter, S. M. (1997). Unraveling the strands of Saturn's F ring, *Icarus* **129**, 304–316.

Murray, C. D. and Thompson, R. P. (1990). Orbits of shepherd satellites deduced from structure of the rings of Uranus, *Nature* **348**, 499–502.

Murray, N. and Holman, M. (1997). Diffusive chaos in the outer asteroid belt, *Astron. J.* **114**, 1246–1259.

Nakamura, Y., Latham, G. V., and Dorman, H. J. (1976). Seismic structure of the Moon, *Proc. Lunar Sci. Conf.* **7**, 602–603.

Namouni, F. (1999). Secular interactions of coorbiting objects, *Icarus* **137**, 293–314.

Namouni, F., Christou, A. A., and Murray, C. D. (1999). New coorbital dynamics in the solar system, *Astro-ph/9904016*.

Namouni, F., Luciani, J. F., Tabachnik, S., and Pellat, R. (1996). A mapping approach to Hill's distant encounters, *Astron. Astrophys.* **313**, 979–992.

Neugebauer, G., Beichmann, C. A., Soifer, B. T., Aumann, H. H., Chester, T. J., Gautier, T. N., Gillet, F. C., Hauser, M. G., Houck, J. R., Lonsdale, C. J., Low, F. J., and Young, E. (1984). Early results from the *Infrared Astronomical Satellite*, *Science* **224**, 14–21.

Newcomb, S. (1895). A development of the perturbative function in cosines of multiples of the mean anomalies and of angles between the perihelia and common node and in powers of the eccentricities and mutual inclination, *Astron. Papers Am. Ephem.* **5**, 5–48.

Newton, I. (1687). *Philosophiae Naturalis Principia Mathematica* (Royal Society, London).

Nicholson, P. D. and Dones, L. (1991). Planetary rings, *Rev. Geophys., Suppl.* **29**, 313–327.

Nicholson, P. D., Hamilton, D. P., Matthews, K., and Yoder, C. F. (1992). New observations of Saturn's coorbital satellites, *Icarus* **100**, 464–484.

Nicholson, P. D. and Matthews, K. (1991). Near-infrared observations of the jovian ring and small satellites, *Icarus* **93**, 331–346.

Nicholson, P. D. and Porco, C. C. (1988). A new constraint on Saturn's zonal gravity harmonics supplied by *Voyager* observations of an eccentric ringlet, *J. Geophys. Res.* **93**, 10209–10224.

Nicholson, P. D., Showalter, M. R., Dones, L., French, R. G., Larson, S. M., Lissauer, J. J., McGhee, C. A., Seitzer, P., Sicardy, B., and Danielson, G. E. (1996). Observations of Saturn's ring-plane crossings in August and November 1995, *Science* **272**, 509–515.

Nieto, M. M. (1972). *The Titius–Bode Law of Planetary Distances: Its History and Theory* (Pergamon Press, Oxford).

Null, G. W., Owen, W. M., and Synnott, S. P. (1993). Masses and densities of Pluto and Charon, *Astron. J.* **105**, 2319–2335.

Ockert-Bell, M. E., Burns, J. A., Daubar, I. J., Thomas, P. C., Veverka, J., Belton, M. J. S., and Klaasen, K. P. (1999). The structure of Jupiter's ring system as revealed by the *Galileo* imaging experiment, *Icarus* **138**, 188-213.

Oikawa, S. and Everhart, E. (1979). Past and future orbit of 1977UB, Object Chiron, *Astron. J.* **84**, 134–139.

Owen, W. M. and Synnott, S. P. (1987). Orbits of the ten small satellites of Uranus, *Astron. J.* **93**, 1268–1271.

Owen, W. M., Vaughan, R. M., and Synnott, S. P. (1991). Orbits of the six new satellites of Neptune, *Astron. J.* **101**, 1511–1515.

Peale, S. J. (1976). Orbital resonances in the solar system, *Annu. Rev. Astron. Astrophys.* **14**, 215–245.

Peale, S. J. (1986). Orbital resonances, unusual configurations and exotic rotation states among planetary satellites, in *Satellites*, ed. J. A. Burns and M. S. Matthews (University of Arizona Press, Tucson).

Peale, S. J. (1988). Speculative histories of the uranian satellite system, *Icarus* **74**, 153–171.

Peale, S. J. (1993a). On the verification of the planetary system around PSR1257+12, *Astron. J.* **105**, 1562–1570.

Peale, S. J. (1993b). The effect of the nebula on the Trojan precursors, *Icarus* **106**, 308–322.

Peale, S. J., Cassen, P. and Reynolds, R. T. (1979). Melting of Io by tidal dissipation, *Science* **203**, 892–894.

Peirce, B. (1849). Development of the perturbative function of planetary motion, *Astron. J.* **1**, 1–8, 31–36.

Pettengill, G. H. and Dyce, R. B. (1965). A radar determination of the rotation of the planet Mercury, *Nature* **206**, 1240.

Plummer, H. C. (1918). *An Introductory Treatise on Dynamical Astronomy* (Cambridge University Press, Cambridge).

Poincaré, H. (1892). *Les Méthodes Nouvelles de la Mécanique Céleste I* (Gauthier-Villars, Paris).

Poincaré, H. (1893). *Les Méthodes Nouvelles de la Mécanique Céleste II* (Gauthier-Villars, Paris).

Poincaré, H. (1899). *Les Méthodes Nouvelles de la Mécanique Céleste III* (Gauthier-Villars, Paris).

Poincaré, H. (1902). Sur les planètes du type d'Hécube, *Bulletin Astronomique* **19**, 289–312.

Poincaré, H. (1905). *Leçons de Mécanique Céleste I* (Gauthier-Villars, Paris).

Porco, C. C. (1983). Voyager observations of Saturn's rings, California Institute of Technology PhD thesis.

Porco, C. C. (1990). Narrow rings: Observation and theory, *Adv. Space Res.* **10**, 221–229.

Porco, C. C. (1991). An explanation for Neptune's ring arcs, *Science* **253**, 995–1001.

Porco, C. C., Danielson, G. E., Goldreich, P., Holberg, J. B., and Lane, A. L. (1984a). Saturn's non-axisymmetric ring edges at $1.95R_S$ and $2.27R_S$, *Icarus* **60**, 17–28.

Porco, C. C. and Goldreich, P. (1987). Shepherding of the uranian rings. I. Kinematics, *Astron. J.* **93**, 724–729.

Porco, C. C. and Nicholson, P. D. (1987). Eccentric features in Saturn's outer C ring, *Icarus* **72**, 437–467.

Porco, C. C., Nicholson, P. D., Borderies, N., Danielson, G. E., Goldreich, P., Holberg, J. B., and Lane, A. L. (1984b). The eccentric saturnian ringlets at $1.29R_S$ and $1.45R_S$, *Icarus* **60**, 1–16.

Porco, C. C., Nicholson, P. D., Cuzzi, J. N., Lissauer, J. J., and Esposito, L. W. (1995). Neptune's ring system, in *Neptune and Triton*, ed. D. P. Cruikshank (University of Arizona Press, Tucson).

Proudman, J. (1953). *Dynamical Oceanography* (Methuen, London).

Quinn, T. R., Tremaine, S., and Duncan, M. (1991). A three million year integration of the Earth's orbit, *Astron. J.* **101**, 2287–2305.

Radau, R. (1885). Sur la loi des densités à l'intérieur de la Terre, *C.R. Acad. Sci. Paris* **100**, 972–974.

Ramsey, A. S. (1937). *Dynamics. Part 2* (Cambridge University Press, Cambridge).

Ramsey, A. S. (1940). *An Introduction to the Theory of Newtonian Attraction* (Cambridge University Press, Cambridge).

Rappaport, N., Bertotti, B., Giampieri, G., and Anderson, J. D. (1997). Doppler measurements of the quadrupole moments of Titan, *Icarus* **126**, 313–323.

Rasio, F. A., Nicholson, P.D., Shapiro, S.L., and Teukolsky, S.A. (1992). An observational test for the existence of a planetary system orbiting PSR1257+12, *Nature* **355**, 325–326.

Reach, W. T., Franz, B. A., Weiland, J. L., Hauser, M. G., Kelsall, T. N., Wright, E. L., Rawley, G., Stemwedel, S. W., and Spiesman, W. J. (1995). Observational confirmation of a circumsolar dust ring by the *COBE* satellite, *Nature* **374**, 521–523.

Reid, M. (1973). The tidal loss of satellite-orbiting objects and its implication for the lunar surface, *Icarus* **20**, 240–248.

Rosen, P. A. (1989). Waves in Saturn's rings probed by radio occultation, Stanford University PhD thesis.

Roseveare, N. T. (1982). *Mercury's Perihelion from Le Verrier to Einstein* (Clarendon Press, Oxford).

Roy, A. E. (1988). *Orbital Motion* (Adam Hilger, Bristol).

Roy, A. E. and Ovenden, M. W. (1954). On the occurrence of commensurable mean motions in the solar system, *Mon. Not. R. Astron. Soc.* **114**, 232–241.

Roy, A. E., Walker, I. W., Macdonald, A. J., Williams, I. P., Fox, K., Murray, C. D., Milani, A., Nobili, A. M., Message, P. J., Sinclair, A. T., and Carpino, M. (1988). Project LONGSTOP, *Vistas in Astronomy* **32**, 95–116.

Safronov, V. S. (1972). *Evolution of the Protoplanetary Cloud and Formation of the Earth and Planets* (Israel Program for Scientific Translation, Jerusalem).

Sagan, C. (1994). *Pale Blue Dot* (Random House, New York).

Sagan, C. and Dermott, S. F. (1982). The tide in the seas of Titan, *Nature* **300**, 731–733.

Saha, P. and Tremaine, S. (1992). Symplectic integrators for solar system dynamics, *Astron. J.* **104**, 1633–1640.

Saha, P. and Tremaine, S. (1994). Long-term planetary integration with individual time steps, *Astron. J.* **108**, 1962–1969.

Schaffer, L. E. and Burns, J. A. (1987). The dynamics of weakly charged dust: Motion through Jupiter's gravitational and magnetic fields, *J. Geophys. Res.* **92**, 2264–2280.

Schuerman, D. (1980). The restricted three-body problem including radiation pressure, *Astrophys. J.* **238**, 337–342.

Sears, W. D. (1995). Tidal dissipation in oceans on Titan, *Icarus* **113**, 39–56.

Seidelmann, P. K. (1992) (ed.) *Explanatory Supplement to the Astronomical Almanac* (University Science Books, Mill Valley, CA).

Shoemaker, E. M., Shoemaker, E. S., and Wolfe, R. F. (1989). Trojan asteroids: Populations, dynamical structure and origin of the $L_4$ and $L_5$ swarms, in *Asteroids II*, ed. R. P. Binzel, T. Gehrels, and M. S. Matthews (University of Arizona Press, Tucson).

Showalter, M. R. (1985). Jupiter's ring system resolved: Physical properties inferred from the *Voyager* images, Cornell University PhD thesis.

Showalter, M. R. (1991). Visual detection of 1981S13, Saturn's eighteenth satellite, and its role in the Encke gap, *Nature* **351**, 709–713.

Showalter, M. R., Burns, J. A., Cuzzi, J. N., and Pollack, J. B. (1987). Jupiter's ring system: New results on structure and particle properties, *Icarus* **69**, 458–498.

Showalter, M. R. and Cuzzi, J. N. (1992). Physical properties of Neptune's ring system, *Bull. Am. Astron. Soc.* **24**, 1029.

Showalter, M. R., Cuzzi, J. N., Marouf, E. A., and Esposito, L. W. (1986). Satellite wakes and the orbit of the Encke gap moonlet, *Icarus* **66**, 297–323.

Shu, F. (1984). Waves in planetary rings, in *Planetary Rings*, ed. R. Greenberg and A. Brahic (University of Arizona Press, Tucson).

Shu, F., Cuzzi, J. N., and Lissauer, J. J. (1983). Bending waves in Saturn's rings, *Icarus* **53**, 185–206.

Šidlichovsky, M. and Melendo, B. (1986). Mapping for the 5/2 asteroidal commensurability, *Bull. Astron. Inst. Czech.* **37**, 65–80.

Simmons, J. F. L., McDonald, A. J. C., and Brown, J. C. (1985). The restricted 3-body problem with radiation pressure, *Celest. Mech.* **35**, 145–187.

Sinclair, A. T. (1972). On the origin of the commensurabilities amongst the satellites of Saturn, *Mon. Not. R. Astron. Soc.* **160**, 169–187.

Smith, B. A., et al. (1979a). The Jupiter system seen through the eyes of *Voyager 1*, *Science* **204**, 951–972.

Smith, B. A., et al. (1979b). The Galilean satellites and Jupiter: *Voyager 2* imaging science results, *Science* **206**, 927–950.

Smith, B. A., et al. (1981). Encounter with Saturn: *Voyager 1* imaging science results, *Science* **212**, 163–191.

Smith, B. A., et al. (1982). A new look at the Saturn system: The *Voyager 2* images, *Science* **215**, 504–537.

Smith, B. A., et al. (1989). *Voyager 2* at Neptune: Imaging science results, *Science* **246**, 1422–1449.

Spohn, T. (1997). Tides of Io, in *Tidal Phenomena*, ed. H. Wilhelm, W. Zurn, and H.-G. Wenzel (Springer, Berlin).

Standish, E. M. and Hellings, R. W. (1989). A determination of the masses of Ceres, Pallas, and Vesta from their perturbations upon the orbit of Mars, *Icarus* **80**, 326–333.

Standish, E. M., Newhall, X. X., Williams, J. G., and Yeomans, D. K. (1992). Orbital ephemerides of the Sun, Moon, and planets, in *Explanatory Supplement to the Astronomical Almanac*, ed. P. K. Seidelmann (University Science Books, Mill Valley, CA).

Stevenson, D. J. (1983). Anomalous bulk viscosity of two-phase fluids and implications for planetary interiors, *J. Geophys. Res.* **88**, 2445–2455.

Street, R. O. (1925). Oceanic tides as modified by a yielding Earth, *Mon. Not. R. Astron. Soc. Geophys. Suppl.* **1**, 292–306.

Sussman, G. J. and Wisdom, J. (1988). Numerical evidence that Pluto is chaotic, *Science* **241**, 433–437.

Synnott, S. P., Terrile, R. J., Jacobson, R. A., and Smith, B. A. (1983). Orbits of Saturn's F-ring and its shepherding satellites, *Icarus* **53**, 156–158.

Szebehely, V. (1967). *The Theory of Orbits* (Academic Press, New York).

Taylor, D. B. (1981). Horseshoe periodic orbits in the restricted problem of three bodies for a Sun–Jupiter mass ratio, *Astron. Astrophys.* **103**, 288–294.

Tholen, D. J. and Buie, M. W. (1997). The orbit of Charon. 1. New Hubble Space Telescope observations, *Icarus* **125**, 245–260.

Tholen, D. J., Hartmann, W. K., and Cruikshank, D. P. (1988). *IAU Circular 4554*.

Thomas, F. and Morbidelli, A. (1996). The Kozai resonance in the outer solar system and the dynamics of long-period comets, *Celest. Mech. Dyn. Astron.* **64**, 209–229.

Thomas, P., Veverka, J., and Dermott, S. F. (1986). Small satellites, in *Satellites*, ed. J. A. Burns and M. S. Matthews (University of Arizona Press, Tucson).

Thomas, P., Veverka, J., Wenkert, D., Danielson, G. E., and Davies, M. (1984). Hyperion: 13-day rotation from *Voyager* data, *Nature* **307**, 716–717.

Thomson, W. (Lord Kelvin) (1863). Dynamical problems regarding elastic spheroidal shells; and On the rigidity of the Earth, *Phil. Trans. R. Soc. London* **153**, 573–616.

Tisserand, F.-F. (1896). *Traité de Mécanique Céleste IV* (Gauthier-Villars, Paris).

Tittemore, W. C. and Wisdom, J. (1988). Tidal evolution of the uranian satellites. I. Passage of Ariel and Umbriel through the 5:3 mean-motion commensurability, *Icarus* **74**, 172–230.

Tittemore, W. C. and Wisdom, J. (1989). Tidal evolution of the uranian satellites. II. An explanation of the anomalously high orbital inclination of Miranda, *Icarus* **78**, 63–89.

Tittemore, W. C. and Wisdom, J. (1990). Tidal evolution of the uranian satellites. III. Evolution through the Miranda–Umbriel 3:1, Miranda–Ariel 5:3 and Ariel–Umbriel 2:1 mean-motion resonances, *Icarus* **85**, 394–443.

Veeder, G. J., Matson, D. L., Johnson, T. V., Blaney, D. L., and Goguen, J. D. (1994). Io's heat flow from infrared radiometry: 1983–1993, *J. Geophys. Res.* **99**, 17095–17162.

Ward, W. R. and Reid, M. J. (1973). Solar tidal friction and satellite loss, *Mon. Not. R. Astron. Soc.* **164**, 21–32.

Webb, D. J. (1982). Tides and the evolution of the Earth–Moon system, *Geophys. J. R. Astron. Soc.* **70**, 261–271.

Westfall, R. S. (1980). *Never at Rest. A Biography of Isaac Newton* (Cambridge University Press, Cambridge).

Wetherill, G. W. (1985). Asteroidal source of ordinary chondrites, *Meteoritics* **18**, 1–22.

Wiegert, P. A., Innanen, K. A., and Mikkola, S. (1997). An asteroidal companion to the Earth, *Nature* **387**, 685–686.

Williams, J. G. (1969). Secular perturbations in the solar system, University of California Los Angeles PhD thesis.

Williams, J. G. (1979). Proper elements and family membership of the asteroids, in *Asteroids*, ed. T. Gehrels (University of Arizona Press, Tucson).

Williams, J. G. and Faulkner, J. (1981). The positions of secular resonance surfaces, *Icarus* **46**, 390–399.

Winter, O. C. and Murray, C. D. (1994a). Atlas of the planar, circular, restricted three-body problem. I. Internal orbits, *QMW Maths Notes*, No.16, Queen Mary and Westfield College, London.

Winter, O. C. and Murray, C. D. (1994b). Atlas of the planar, circular, restricted three-body problem. II. External orbits, *QMW Maths Notes*, No.17, Queen Mary and Westfield College, London.

Winter, O. C. and Murray, C. D. (1997). Resonance and chaos. I. First-order interior resonances, *Astron. Astrophys.* **319**, 290–304.

Wisdom, J. (1980). The resonance overlap criterion and the onset of stochastic behavior in the restricted three-body problem, *Astron. J.* **85**, 1122–1133.

Wisdom, J. (1982). The origin of the Kirkwood gaps: A mapping technique for asteroidal motion near the 3/1 commensurability, *Astron. J.* **87**, 577–593.

Wisdom, J. (1983). Chaotic behavior and the origin of the 3/1 Kirkwood gap, *Icarus* **56**, 51–74.

Wisdom, J. (1985a). A perturbative treatment of motion near the 3/1 commensurability, *Icarus* **63**, 272–289.

Wisdom, J. (1985b). Meteorites may follow a chaotic route to Earth, *Nature* **315**, 731–733.

Wisdom, J. (1987a). Urey Prize Lecture: Chaotic dynamics in the solar system, *Icarus* **72**, 241–275.

Wisdom, J. (1987b). Chaotic behaviour in the solar system, *Proc. R. Soc. London.* A **413**, 109–129.

Wisdom, J. and Holman, M. (1991). Symplectic maps for the *N*-body problem, *Astron. J.* **102**, 1528–1538.

Wisdom, J., Peale, S. J., and Mignard, F. (1984). The chaotic rotation of Hyperion, *Icarus* **58**, 137–152.

Wolszczan, A. (1994). Confirmation of Earth-mass planets orbiting the millisecond pulsar PSR B1257+12, *Science* **264**, 538–542.

Wolszczan, A. and Frail, D. A. (1992). A planetary system around the millisecond pulsar PSR1257+12, *Nature* **355**, 145–147.

Woltjer, J. (1928). The motion of Hyperion, *Annalen van de Sterrewacht te Leiden* **XVI**, Pt. 3.

Yoder, C. F. (1973). On the establishment and evolution of orbit–orbit resonances, University of California Santa Barbara PhD thesis.

Yoder, C. F. (1995). Astrometric and geodetic properties of Earth and the solar system, in *Global Earth Physics. A Handbook of Physical Constants*, ed. T. Ahrens (American Geophysical Union, Washington).

Yoder, C. F., Colombo, G., Synnott, S. P., and Yoder, K. A. (1983). Theory of motion of Saturn's coorbiting satellites, *Icarus* **53**, 431–443.

Yoder, C. F. and Peale, S. J. (1981). The tides of Io, *Icarus* **47**, 1–35.

Yoshikawa, M. (1989). A survey of the motions of asteroids in the commensurabilities with Jupiter, *Astron. Astrophys.* **213**, 436–458.

Yuasa, M. (1973). Theory of secular perturbations of asteroids including terms of higher order and higher degree, *Publ. Astron. Soc. Japan* **25**, 399–445.

# Index

Printed in the United States
By Bookmasters